# Ullmann's Polymers and Plastics

# Excellence since 1914
**Built from generations of expertise, for generations to come.**

**ULLMANN'S ENCYCLOPEDIA OF INDUSTRIAL CHEMISTRY**

If you want to learn more about novel technologies in biochemistry or nanotechnology or discover unexpected new aspects of seemingly totally familiar processes in industrial chemistry – ULLMANN'S Encyclopedia of Industrial Chemistry is your first choice.

For over 100 years now, this reference provides top-notch information on the most diverse fields of industrial chemistry and chemical engineering.

When it comes to definite works on industrial chemistry, it has always been ULLMANN'S.

### Welcome to the ULLMANN'S ACADEMY

**Kick-start your career!**

Key topics in industrial chemistry explained by the ULLMANN'S Encyclopedia experts – for teaching and learning, or for simply refreshing your knowledge.

### What's new?

The Smart Article introduces new and enhanced article tools for chemistry content. It is now available within ULLMANN'S.

For further information on the features and functions available, go to wileyonlinelibrary.com/thesmartarticle

### Only interested in a specific topic?

ULLMANN'S Energy
3 Volume Set • ISBN: 978-3-527-33370-7

ULLMANN'S Fine Chemicals
3 Volume Set • ISBN: 978-3-527-33477-3

ULLMANN'S Fibers
2 Volumes • ISBN: 978-3-527-31772-1

ULLMANN'S Modeling and Simulation
ISBN: 978-3-527-31605-2

ULLMANN'S Renewable Resources
ISBN: 978-3-527-33369-1

ULLMANN'S Agrochemicals
2 Volume Set • ISBN: 978-3-527-31604-5

ULLMANN'S Biotechnology and Biochemical Engineering
2 Volume Set • ISBN: 978-3-527-31603-8

ULLMANN'S Reaction Engineering
2 Volume Set • ISBN: 978-3-527-33371-4

ULLMANN'S Industrial Toxicology
2 Volume Set • ISBN: 978-3-527-31247-4

ULLMANN'S Chemical Engineering and Plant Design
2 Volume Set • ISBN: 978-3-527-31111-8

## The information you need in the format you want.

### DVD

- Released once a year.
- Fully networkable for up to 200 users.
- Time-limited access for up to 14 months (expires March 2016).
- 2015 Edition
  ISBN: 978-3-527-33754-5

### Online

- Over 1,150 articles available online.
- Over 3,000 authors from over 30 countries have contributed.
- Offers flexible access 24/7 from your library, home, or on the road.
- Updated 4 times per year.
  ISBN: 978-3-527-30673-2

### Print

- Available as a comprehensive 40 Volume-Set.
- The 7th Edition published in Aug 2011.
- ISBN: 978-3-527-32943-4

## Visit our website to

Find sample chapters

Learn more about the history, the Editor-in-Chief and the Editorial Advisory Board

Discover the ULLMANN'S ACADEMY and more...

**wileyonlinelibrary.com/ref/ullmanns**

---

Chemistry that delivers... Continuous product innovation
# Create Innovate Inspire

**WILEY-VCH**          **WILEY**

# Ullmann's Polymers and Plastics

Products and Processes

Volume 4

Verlag GmbH & Co. KGaA

**Editor in Chief:**

Dr. Barbara Elvers, Hamburg, Germany

All books published by **Wiley-VCH** are carefully produced. Nevertheless, authors, editors, and publisher do not warrant the information contained in these books, including this book, to be free of errors. Readers are advised to keep in mind that statements, data, illustrations, procedural details or other items may inadvertently be inaccurate.

**Library of Congress Card No.:**
applied for

**British Library Cataloguing-in-Publication Data**
A catalogue record for this book is available from the British Library.

**Bibliographic information published by the Deutsche Nationalbibliothek**
The Deutsche Nationalbibliothek lists this publication in the Deutsche Nationalbibliografie; detailed bibliographic data are available on the Internet at <http://dnb.d-nb.de>.

© 2016 Wiley-VCH Verlag GmbH & Co. KGaA, Boschstr. 12, 69469 Weinheim, Germany

All rights reserved (including those of translation into other languages). No part of this book may be reproduced in any form – by photoprinting, microfilm, or any other means – nor transmitted or translated into a machine language without written permission from the publishers. Registered names, trademarks, etc. used in this book, even when not specifically marked as such, are not to be considered unprotected by law.

**Print ISBN:** 978-3-527-33823-8
**ePDF ISBN:** 978-3-527-68595-0
**ePub ISBN:** 978-3-527-68596-7
**Mobi ISBN:** 978-3-527-68597-4

**Cover Design** Grafik-Design Schulz, Fußgönheim, Germany
**Typesetting** Thomson Digital, Noida, India
**Printing and Binding** Markono Print Media Pte Ltd, Singapore

Printed on acid-free paper

# Preface

This handbook features selected articles from the 7$^{th}$ edition of *ULLMANN'S Encyclopedia of Industrial Chemistry*, including newly written articles that have not been published in a printed edition before. True to the tradition of the ULLMANN'S Encyclopedia, polymers and plastics are addressed from an industrial perspective, including production figures, quality standards and patent protection issues where appropriate. Safety and environmental aspects which are a key concern for modern process industries are likewise considered.

More content on related topics can be found in the complete edition of the ULLMANN'S Encyclopedia.

# About ULLMANN'S

ULLMANN'S Encyclopedia is the world's largest reference in applied chemistry, industrial chemistry, and chemical engineering. In its current edition, the Encyclopedia contains more than 30,000 pages, 15,000 tables, 25,000 figures, and innumerable literature sources and cross-references, offering a wealth of comprehensive and well-structured information on all facets of industrial chemistry.

1,100 major articles cover the following main areas:

- Agrochemicals
- Analytical Techniques
- Biochemistry and Biotechnology
- Chemical Reactions
- Dyes and Pigments
- Energy
- Environmental Protection and Industrial Safety
- Fat, Oil, Food and Feed, Cosmetics
- Inorganic Chemicals
- Materials
- Metals and Alloys
- Organic Chemicals
- Pharmaceuticals
- Polymers and Plastics
- Processes and Process Engineering
- Renewable Resources
- Special Topics

First published in 1914 by Professor Fritz Ullmann in Berlin, the *Enzyklopädie der Technischen Chemie* (as the German title read) quickly became the standard reference work in industrial chemistry. Generations of chemists have since relied on ULLMANN'S as their prime reference source. Three further German editions followed in 1928–1932, 1951–1970, and in 1972–1984. From 1985 to 1996, the 5$^{th}$ edition of ULLMANN'S Encyclopedia of Industrial Chemistry was the first edition to be published in English rather than German language. So far, two more complete English editions have been published in print; the 6$^{th}$ edition of 40 volumes in 2002, and the 7$^{th}$ edition in 2011, again comprising 40 volumes. In addition, a number of smaller topic-oriented editions have been published.

Since 1997, *ULLMANN'S Encyclopedia of Industrial Chemistry* has also been available in electronic format, first in a CD-ROM edition and, since 2000, in an enhanced online edition. Both electronic editions feature powerful search and navigation functions as well as regular content updates.

# Contents

**Volume 1**
Symbols and Units ... IX
Conversion Factors ... XI
Abbreviations ... XIII
Country Codes ... XVIII
Periodic Table of Elements ... XIX

**Part 1: Fundamentals** ... 1
Plastics, General Survey, 1. Definition, Molecular Structure and Properties ... 3
Plastics, General Survey, 2. Production of Polymers and Plastics ... 149
Plastics, General Survey, 3. Supermolecular Structures ... 187
Plastics, General Survey, 4. Polymer Composites ... 205
Plastics, General Survey, 5. Plastics and Sustainability ... 223
Plastics, Analysis ... 231
Polymerization Processes, 1. Fundamentals ... 265
Polymerization Processes, 2. Modeling of Processes and Reactors ... 315
Plastics, Processing, 1. Processing of Thermoplastics ... 367
Plastics, Processing, 2. Processing of Thermosets ... 407
Plastics Processing, 3. Machining, Bonding, Surface Treatment ... 439
Plastics, Properties and Testing ... 471
Plastics, Additives ... 527
Plasticizers ... 581

**Volume 2**
**Part 2: Organic Polymers** ... 601
Fluoropolymers, Organic ... 603
Polyacrylamides and Poly(Acrylic Acids) ... 659
Polyacrylates ... 675
Polyamides ... 697
Polyaspartates and Polysuccinimide ... 733
Polybutenes ... 747
Polycarbonates ... 763
Polyester Resins, Unsaturated ... 781
Polyesters ... 791
Polyethylene ... 817
Polyimides ... 859
Polymethacrylates ... 885

Polyoxyalkylenes ... 899
Polyoxymethylenes ... 911
Poly(Phenylene Oxides) ... 927
Polypropylene ... 937

**Volume 3**
Polystyrene and Styrene Copolymers ... 981
Polyureas ... 1029
Polyurethanes ... 1051
Poly(Vinyl Chloride) ... 1111
Polyvinyl Compounds, Others ... 1141
Poly(Vinyl Esters) ... 1165
Poly(Vinyl Ethers) ... 1175
Poly(Vinylidene Chloride) ... 1181
Polymer Blends ... 1197
Polymers, Biodegradable ... 1231
Polymers, Electrically Conducting ... 1261
Polymers, High-Temperature ... 1281
Reinforced Plastics ... 1325
Specialty Plastics ... 1343
Thermoplastic Elastomers ... 1365

**Volume 4**
**Part 3: Films, Fibers, Foams** ... 1405
Films ... 1407
Fibers, 4. Polyamide Fibers ... 1435
Fibers, 5. Polyester Fibers ... 1453
Fibers, 6. Polyurethane Fibers ... 1487
Fibers, 7. Polyolefin Fibers ... 1495
Fibers, 8. Polyacrylonitrile Fibers ... 1513
Fibers, 9. Polyvinyl Fibers ... 1529
Fibers, 10. Polytetrafluoroethylene Fibers ... 1539
High-Performance Fibers ... 1541
Foamed Plastics ... 1563

**Part 4: Resins** ... 1595
Alkyd Resins ... 1597
Amino Resins ... 1615
Epoxy Resins ... 1643
Phenolic Resins ... 1733
Resins, Synthetic ... 1751

**Part 5: Inorganic Polymers** ... 1775
Inorganic Polymers ... 1777

**Author Index** ... 1817
**Subject Index** ... 1823

# Symbols and Units

Symbols and units agree with SI standards (for conversion factors see page XI). The following list gives the most important symbols used in the encyclopedia. Articles with many specific units and symbols have a similar list as front matter.

| Symbol | Unit | Physical Quantity |
|---|---|---|
| $a_B$ | | activity of substance B |
| $A_r$ | | relative atomic mass (atomic weight) |
| $A$ | $m^2$ | area |
| $c_B$ | $mol/m^3$, mol/L (M) | concentration of substance B |
| $C$ | C/V | electric capacity |
| $c_p$, $c_v$ | $J\,kg^{-1}\,K^{-1}$ | specific heat capacity |
| $d$ | cm, m | diameter |
| $d$ | | relative density ($\varrho/\varrho_{water}$) |
| $D$ | $m^2/s$ | diffusion coefficient |
| $D$ | Gy (=J/kg) | absorbed dose |
| $e$ | C | elementary charge |
| $E$ | J | energy |
| $E$ | V/m | electric field strength |
| $E$ | V | electromotive force |
| $E_A$ | J | activation energy |
| $f$ | | activity coefficient |
| $F$ | C/mol | Faraday constant |
| $F$ | N | force |
| $g$ | $m/s^2$ | acceleration due to gravity |
| $G$ | J | Gibbs free energy |
| $h$ | m | height |
| $\hbar$ | $W \cdot s^2$ | Planck constant |
| $H$ | J | enthalpy |
| $I$ | A | electric current |
| $I$ | cd | luminous intensity |
| $k$ | (variable) | rate constant of a chemical reaction |
| $k$ | J/K | Boltzmann constant |
| $K$ | (variable) | equilibrium constant |
| $l$ | m | length |
| $m$ | g, kg, t | mass |
| $M_r$ | | relative molecular mass (molecular weight) |
| $n_D^{20}$ | | refractive index (sodium D-line, 20 °C) |
| $n$ | mol | amount of substance |
| $N_A$ | $mol^{-1}$ | Avogadro constant ($6.023 \times 10^{23}\,mol^{-1}$) |
| $P$ | Pa, bar* | pressure |
| $Q$ | J | quantity of heat |
| $r$ | m | radius |
| $R$ | $J\,K^{-1}\,mol^{-1}$ | gas constant |
| $R$ | Ω | electric resistance |
| $S$ | J/K | entropy |
| $t$ | s, min, h, d, month, a | time |
| $t$ | °C | temperature |
| $T$ | K | absolute temperature |
| $u$ | m/s | velocity |
| $U$ | V | electric potential |

Symbols and Units (Continued from p. IX)

| Symbol | Unit | Physical Quantity |
| --- | --- | --- |
| $U$ | J | internal energy |
| $V$ | m$^3$, L, mL, µL | volume |
| $w$ | | mass fraction |
| $W$ | J | work |
| $x_B$ | | mole fraction of substance B |
| $Z$ | | proton number, atomic number |
| $\alpha$ | | cubic expansion coefficient |
| $\alpha$ | Wm$^{-2}$K$^{-1}$ | heat-transfer coefficient (heat-transfer number) |
| $\alpha$ | | degree of dissociation of electrolyte |
| $[\alpha]$ | $10^{-2}$deg cm$^2$g$^{-1}$ | specific rotation |
| $\eta$ | Pa·s | dynamic viscosity |
| $\theta$ | °C | temperature |
| $\varkappa$ | | $c_p/c_v$ |
| $\lambda$ | Wm$^{-1}$K$^{-1}$ | thermal conductivity |
| $\lambda$ | nm, m | wavelength |
| $\mu$ | | chemical potential |
| $\nu$ | Hz, s$^{-1}$ | frequency |
| $\nu$ | m$^2$/s | kinematic viscosity ($\eta/\varrho$) |
| $\pi$ | Pa | osmotic pressure |
| $\varrho$ | g/cm$^3$ | density |
| $\sigma$ | N/m | surface tension |
| $\tau$ | Pa (N/m$^2$) | shear stress |
| $\varphi$ | | volume fraction |
| $\chi$ | Pa$^{-1}$ (m$^2$/N) | compressibility |

*The official unit of pressure is the pascal (Pa).

# Conversion Factors

| SI unit | Non-SI unit | From SI to non-SI multiply by |
|---|---|---|
| *Mass* | | |
| kg | pound (avoirdupois) | 2.205 |
| kg | ton (long) | $9.842 \times 10^{-4}$ |
| kg | ton (short) | $1.102 \times 10^{-3}$ |
| *Volume* | | |
| $m^3$ | cubic inch | $6.102 \times 10^4$ |
| $m^3$ | cubic foot | 35.315 |
| $m^3$ | gallon (U.S., liquid) | $2.642 \times 10^2$ |
| $m^3$ | gallon (Imperial) | $2.200 \times 10^2$ |
| *Temperature* | | |
| °C | °F | °C × 1.8 + 32 |
| *Force* | | |
| N | dyne | $1.0 \times 10^5$ |
| *Energy, Work* | | |
| J | Btu (int.) | $9.480 \times 10^{-4}$ |
| J | cal (int.) | $2.389 \times 10^{-1}$ |
| J | eV | $6.242 \times 10^{18}$ |
| J | erg | $1.0 \times 10^7$ |
| J | kW·h | $2.778 \times 10^{-7}$ |
| J | kp·m | $1.020 \times 10^{-1}$ |
| *Pressure* | | |
| MPa | at | 10.20 |
| MPa | atm | 9.869 |
| MPa | bar | 10 |
| kPa | mbar | 10 |
| kPa | mm Hg | 7.502 |
| kPa | psi | 0.145 |
| kPa | torr | 7.502 |

# Powers of Ten

| | | | |
|---|---|---|---|
| E (exa) | $10^{18}$ | d (deci) | $10^{-1}$ |
| P (peta) | $10^{15}$ | c (centi) | $10^{-2}$ |
| T (tera) | $10^{12}$ | m (milli) | $10^{-3}$ |
| G (giga) | $10^9$ | μ (micro) | $10^{-6}$ |
| M (mega) | $10^6$ | n (nano) | $10^{-9}$ |
| k (kilo) | $10^3$ | p (pico) | $10^{-12}$ |
| h (hecto) | $10^2$ | f (femto) | $10^{-15}$ |
| da (deca) | 10 | a (atto) | $10^{-18}$ |

# Abbreviations

The following is a list of the abbreviations used in the text. Common terms, the names of publications and institutions, and legal agreements are included along with their full identities. Other abbreviations will be defined wherever they first occur in an article. For further abbreviations, see page IX, Symbols and Units; page XVII, Frequently Cited Companies (Abbreviations), and page XVIII, Country Codes in patent references. The names of periodical publications are abbreviated exactly as done by Chemical Abstracts Service.

| | | | |
|---|---|---|---|
| abs. | absolute | BGA | Bundesgesundheitsamt (Federal Republic of Germany) |
| a.c. | alternating current | | |
| ACGIH | American Conference of Governmental Industrial Hygienists | BGBl. | Bundesgesetzblatt (Federal Republic of Germany) |
| ACS | American Chemical Society | BIOS | British Intelligence Objectives Subcommittee Report (see also FIAT) |
| ADI | acceptable daily intake | | |
| ADN | accord européen relatif au transport international des marchandises dangereuses par voie de navigation interieure (European agreement concerning the international transportation of dangerous goods by inland waterways) | BOD | biological oxygen demand |
| | | $bp$ | boiling point |
| | | B.P. | British Pharmacopeia |
| | | BS | British Standard |
| | | ca. | circa |
| | | calcd. | calculated |
| ADNR | ADN par le Rhin (regulation concerning the transportation of dangerous goods on the Rhine and all national waterways of the countries concerned) | CAS | Chemical Abstracts Service |
| | | cat. | catalyst, catalyzed |
| | | CEN | Comité Européen de Normalisation |
| | | cf. | compare |
| ADP | adenosine 5'-diphosphate | CFR | Code of Federal Regulations (United States) |
| ADR | accord européen relatif au transport international des marchandises dangereuses par route (European agreement concerning the international transportation of dangerous goods by road) | cfu | colony forming units |
| | | Chap. | chapter |
| | | ChemG | Chemikaliengesetz (Federal Republic of Germany) |
| AEC | Atomic Energy Commission (United States) | C.I. | Colour Index |
| | | CIOS | Combined Intelligence Objectives Subcommitee Report (see also FIAT) |
| a.i. | active ingredient | | |
| AIChE | American Institute of Chemical Engineers | CLP | Classification, Labelling and Packaging |
| | | CNS | central nervous system |
| AIME | American Institute of Mining, Metallurgical, and Petroleum Engineers | Co. | Company |
| | | COD | chemical oxygen demand |
| ANSI | American National Standards Institute | conc. | concentrated |
| AMP | adenosine 5'-monophosphate | const. | constant |
| APhA | American Pharmaceutical Association | Corp. | Corporation |
| API | American Petroleum Institute | crit. | critical |
| ASTM | American Society for Testing and Materials | CSA | Chemical Safety Assessment according to REACH |
| ATP | adenosine 5'-triphosphate | CSR | Chemical Safety Report according to REACH |
| BAM | Bundesanstalt für Materialprüfung (Federal Republic of Germany) | | |
| | | CTFA | The Cosmetic, Toiletry and Fragrance Association (United States) |
| BAT | Biologischer Arbeitsstofftoleranzwert (biological tolerance value for a working material, established by MAK Commission, see MAK) | DAB | Deutsches Arzneibuch, Deutscher Apotheker-Verlag, Stuttgart |
| | | d.c. | direct current |
| Beilstein | Beilstein's Handbook of Organic Chemistry, Springer, Berlin – Heidelberg – New York | decomp. | decompose, decomposition |
| | | DFG | Deutsche Forschungsgemeinschaft (German Science Foundation) |
| BET | Brunauer – Emmett – Teller | dil. | dilute, diluted |

| | | | |
|---|---|---|---|
| DIN | Deutsche Industrienorm (Federal Republic of Germany) | | (regulation in the Federal Republic of Germany concerning the transportation of dangerous goods by rail) |
| DMF | dimethylformamide | | |
| DNA | deoxyribonucleic acid | GGVS | Verordnung in der Bundesrepublik Deutschland über die Beförderung gefährlicher Güter auf der Straße (regulation in the Federal Republic of Germany concerning the transportation of dangerous goods by road) |
| DOE | Department of Energy (United States) | | |
| DOT | Department of Transportation – Materials Transportation Bureau (United States) | | |
| DTA | differential thermal analysis | | |
| EC | effective concentration | GGVSee | Verordnung in der Bundesrepublik Deutschland über die Beförderung gefährlicher Güter mit Seeschiffen (regulation in the Federal Republic of Germany concerning the transportation of dangerous goods by sea-going vessels) |
| EC | European Community | | |
| ed. | editor, edition, edited | | |
| e.g. | for example | | |
| emf | electromotive force | | |
| EmS | Emergency Schedule | | |
| EN | European Standard (European Community) | GHS | Globally Harmonised System of Chemicals (internationally agreed-upon system, created by the UN, designed to replace the various classification and labeling standards used in different countries by using consistent criteria for classification and labeling on a global level) |
| EPA | Environmental Protection Agency (United States) | | |
| EPR | electron paramagnetic resonance | | |
| Eq. | equation | | |
| ESCA | electron spectroscopy for chemical analysis | | |
| esp. | especially | GLC | gas-liquid chromatography |
| ESR | electron spin resonance | Gmelin | Gmelin's Handbook of Inorganic Chemistry, 8th ed., Springer, Berlin – Heidelberg – New York |
| Et | ethyl substituent ($-C_2H_5$) | | |
| et al. | and others | | |
| etc. | et cetera | GRAS | generally recognized as safe |
| EVO | Eisenbahnverkehrsordnung (Federal Republic of Germany) | Hal | halogen substituent ($-F, -Cl, -Br, -I$) |
| | | Houben-Weyl | Methoden der organischen Chemie, 4th ed., Georg Thieme Verlag, Stuttgart |
| exp (...) | $e^{(...)}$, mathematical exponent | | |
| FAO | Food and Agriculture Organization (United Nations) | | |
| | | HPLC | high performance liquid chromatography |
| FDA | Food and Drug Administration (United States) | | |
| | | H statement | hazard statement in GHS |
| FD&C | Food, Drug and Cosmetic Act (United States) | IAEA | International Atomic Energy Agency |
| | | IARC | International Agency for Research on Cancer, Lyon, France |
| FHSA | Federal Hazardous Substances Act (United States) | | |
| | | IATA-DGR | International Air Transport Association, Dangerous Goods Regulations |
| FIAT | Field Information Agency, Technical (United States reports on the chemical industry in Germany, 1945) | | |
| | | ICAO | International Civil Aviation Organization |
| Fig. | figure | | |
| fp | freezing point | i.e. | that is |
| Friedländer | P. Friedländer, Fortschritte der Teerfarbenfabrikation und verwandter Industriezweige Vol. 1–25, Springer, Berlin 1888–1942 | i.m. | intramuscular |
| | | IMDG | International Maritime Dangerous Goods Code |
| | | IMO | Inter-Governmental Maritime Consultive Organization (in the past: IMCO) |
| FT | Fourier transform | | |
| (g) | gas, gaseous | Inst. | Institute |
| GC | gas chromatography | i.p. | intraperitoneal |
| GefStoffV | Gefahrstoffverordnung (regulations in the Federal Republic of Germany concerning hazardous substances) | IR | infrared |
| | | ISO | International Organization for Standardization |
| GGVE | Verordnung in der Bundesrepublik Deutschland über die Beförderung gefährlicher Güter mit der Eisenbahn | IUPAC | International Union of Pure and Applied Chemistry |
| | | i.v. | intravenous |

| | | | |
|---|---|---|---|
| Kirk-Othmer | Encyclopedia of Chemical Technology, 3rd ed., 1991–1998, 5th ed., 2004–2007, John Wiley & Sons, Hoboken | no. | number |
| | | NOEL | no observed effect level |
| | | NRC | Nuclear Regulatory Commission (United States) |
| (l) | liquid | | |
| Landolt-Börnstein | Zahlenwerte u. Funktionen aus Physik, Chemie, Astronomie, Geophysik u. Technik, Springer, Heidelberg 1950–1980; Zahlenwerte und Funktionen aus Naturwissenschaften und Technik, Neue Serie, Springer, Heidelberg, since 1961 | NRDC | National Research Development Corporation (United States) |
| | | NSC | National Service Center (United States) |
| | | NSF | National Science Foundation (United States) |
| | | NTSB | National Transportation Safety Board (United States) |
| $LC_{50}$ | lethal concentration for 50 % of the test animals | OECD | Organization for Economic Cooperation and Development |
| LCLo | lowest published lethal concentration | OSHA | Occupational Safety and Health Administration (United States) |
| $LD_{50}$ | lethal dose for 50 % of the test animals | | |
| LDLo | lowest published lethal dose | p., pp. | page, pages |
| ln | logarithm (base e) | Patty | G.D. Clayton, F.E. Clayton (eds.): Patty's Industrial Hygiene and Toxicology, 3rd ed., Wiley Interscience, New York |
| LNG | liquefied natural gas | | |
| log | logarithm (base 10) | | |
| LPG | liquefied petroleum gas | | |
| M | mol/L | PB report | Publication Board Report (U.S. Department of Commerce, Scientific and Industrial Reports) |
| M | metal (in chemical formulas) | | |
| MAK | Maximale Arbeitsplatzkonzentration (maximum concentration at the workplace in the Federal Republic of Germany); cf. Deutsche Forschungsgemeinschaft (ed.): Maximale Arbeitsplatzkonzentrationen (MAK) und Biologische Arbeitsstofftoleranzwerte (BAT), WILEY-VCH Verlag, Weinheim (published annually) | | |
| | | PEL | permitted exposure limit |
| | | Ph | phenyl substituent (—$C_6H_5$) |
| | | Ph. Eur. | European Pharmacopoeia, Council of Europe, Strasbourg |
| | | phr | part per hundred rubber (resin) |
| | | PNS | peripheral nervous system |
| | | ppm | parts per million |
| | | P statement | precautionary statement in GHS |
| max. | maximum | q.v. | which see (quod vide) |
| MCA | Manufacturing Chemists Association (United States) | REACH | Registration, Evaluation, Authorisation and Restriction of Chemicals (EU regulation addressing the production and use of chemical substances, and their potential impacts on both human health and the environment) |
| Me | methyl substituent (–$CH_3$) | | |
| Methodicum Chimicum | Methodicum Chimicum, Georg Thieme Verlag, Stuttgart | | |
| MFAG | Medical First Aid Guide for Use in Accidents Involving Dangerous Goods | | |
| | | ref. | refer, reference |
| MIK | maximale Immissionskonzentration (maximum immission concentration) | resp. | respectively |
| | | $R_f$ | retention factor (TLC) |
| min. | minimum | R.H. | relative humidity |
| mp | melting point | RID | réglement international concernant le transport des marchandises dangereuses par chemin de fer (international convention concerning the transportation of dangerous goods by rail) |
| MS | mass spectrum, mass spectrometry | | |
| NAS | National Academy of Sciences (United States) | | |
| NASA | National Aeronautics and Space Administration (United States) | | |
| | | RNA | ribonucleic acid |
| NBS | National Bureau of Standards (United States) | R phrase (R-Satz) | risk phrase according to ChemG and GefStoffV (Federal Republic of Germany) |
| NCTC | National Collection of Type Cultures (United States) | | |
| | | rpm | revolutions per minute |
| NIH | National Institutes of Health (United States) | RTECS | Registry of Toxic Effects of Chemical Substances, edited by the National Institute of Occupational Safety and Health (United States) |
| NIOSH | National Institute for Occupational Safety and Health (United States) | | |
| NMR | nuclear magnetic resonance | (s) | solid |

| | | | |
|---|---|---|---|
| SAE | Society of Automotive Engineers (United States) | | der Technischen Chemie, 4th ed., Verlag Chemie, Weinheim 1972–1984; 3rd ed., Urban und Schwarzenberg, München 1951–1970 |
| SAICM | Strategic Approach on International Chemicals Management (international framework to foster the sound management of chemicals) | USAEC | United States Atomic Energy Commission |
| s.c. | subcutaneous | USAN | United States Adopted Names |
| SI | International System of Units | USD | United States Dispensatory |
| SIMS | secondary ion mass spectrometry | USDA | United States Department of Agriculture |
| S phrase (S-Satz) | safety phrase according to ChemG and GefStoffV (Federal Republic of Germany) | U.S.P. | United States Pharmacopeia |
| | | UV | ultraviolet |
| STEL | Short Term Exposure Limit (see TLV) | UVV | Unfallverhütungsvorschriften der Berufsgenossenschaft (workplace safety regulations in the Federal Republic of Germany) |
| STP | standard temperature and pressure (0°C, 101.325 kPa) | | |
| $T_g$ | glass transition temperature | VbF | Verordnung in der Bundesrepublik Deutschland über die Errichtung und den Betrieb von Anlagen zur Lagerung, Abfüllung und Beförderung brennbarer Flüssigkeiten (regulation in the Federal Republic of Germany concerning the construction and operation of plants for storage, filling, and transportation of flammable liquids; classification according to the flash point of liquids, in accordance with the classification in the United States) |
| TA Luft | Technische Anleitung zur Reinhaltung der Luft (clean air regulation in Federal Republic of Germany) | | |
| TA Lärm | Technische Anleitung zum Schutz gegen Lärm (low noise regulation in Federal Republic of Germany) | | |
| TDLo | lowest published toxic dose | | |
| THF | tetrahydrofuran | | |
| TLC | thin layer chromatography | | |
| TLV | Threshold Limit Value (TWA and STEL); published annually by the American Conference of Governmental Industrial Hygienists (ACGIH), Cincinnati, Ohio | | |
| | | VDE | Verband Deutscher Elektroingenieure (Federal Republic of Germany) |
| | | VDI | Verein Deutscher Ingenieure (Federal Republic of Germany) |
| TOD | total oxygen demand | vol | volume |
| TRK | Technische Richtkonzentration (lowest technically feasible level) | vol. | volume (of a series of books) |
| | | vs. | versus |
| TSCA | Toxic Substances Control Act (United States) | WGK | Wassergefährdungsklasse (water hazard class) |
| TÜV | Technischer Überwachungsverein (Technical Control Board of the Federal Republic of Germany) | WHO | World Health Organization (United Nations) |
| TWA | Time Weighted Average | Winnacker-Küchler | Chemische Technologie, 4th ed., Carl Hanser Verlag, München, 1982-1986; Winnacker-Küchler, Chemische Technik: Prozesse und Produkte, Wiley-VCH, Weinheim, 2003–2006 |
| UBA | Umweltbundesamt (Federal Environmental Agency) | | |
| Ullmann | Ullmann's Encyclopedia of Industrial Chemistry, 6th ed., Wiley-VCH, Weinheim 2002; Ullmann's Encyclopedia of Industrial Chemistry, 5th ed., VCH Verlagsgesellschaft, Weinheim 1985–1996; Ullmanns Encyklopädie | | |
| | | wt | weight |
| | | $ | U.S. dollar, unless otherwise stated |

# Frequently Cited Companies (Abbreviations)

| | | | |
|---|---|---|---|
| Air Products | Air Products and Chemicals | IFP | Institut Français du Pétrole |
| Akzo | Algemene Koninklijke Zout Organon | INCO | International Nickel Company |
| | | 3M | Minnesota Mining and Manufacturing Company |
| Alcoa | Aluminum Company of America | Mitsubishi Chemical | Mitsubishi Chemical Industries |
| Allied | Allied Corporation | | |
| Amer. Cyanamid | American Cyanamid Company | Monsanto | Monsanto Company |
| | | Nippon Shokubai | Nippon Shokubai Kagaku Kogyo |
| BASF | BASF Aktiengesellschaft | | |
| Bayer | Bayer AG | PCUK | Pechiney Ugine Kuhlmann |
| BP | British Petroleum Company | PPG | Pittsburg Plate Glass Industries |
| Celanese | Celanese Corporation | Searle | G.D. Searle & Company |
| Daicel | Daicel Chemical Industries | SKF | Smith Kline & French Laboratories |
| Dainippon | Dainippon Ink and Chemicals Inc. | SNAM | Societá Nazionale Metandotti |
| Dow Chemical | The Dow Chemical Company | Sohio | Standard Oil of Ohio |
| | | Stauffer | Stauffer Chemical Company |
| DSM | Dutch Staats Mijnen | Sumitomo | Sumitomo Chemical Company |
| Du Pont | E.I. du Pont de Nemours & Company | Toray | Toray Industries Inc. |
| Exxon | Exxon Corporation | UCB | Union Chimique Belge |
| FMC | Food Machinery & Chemical Corporation | Union Carbide | Union Carbide Corporation |
| GAF | General Aniline & Film Corporation | UOP | Universal Oil Products Company |
| W.R. Grace | W.R. Grace & Company | VEBA | Vereinigte Elektrizitäts- und Bergwerks-AG |
| Hoechst | Hoechst Aktiengesellschaft | Wacker | Wacker Chemie GmbH |
| IBM | International Business Machines Corporation | | |
| ICI | Imperial Chemical Industries | | |

# Country Codes

The following list contains a selection of standard country codes used in the patent references.

| | | | |
|---|---|---|---|
| AT | Austria | IL | Israel |
| AU | Australia | IT | Italy |
| BE | Belgium | JP | Japan* |
| BG | Bulgaria | LU | Luxembourg |
| BR | Brazil | MA | Morocco |
| CA | Canada | NL | Netherlands* |
| CH | Switzerland | NO | Norway |
| CS | Czechoslovakia | NZ | New Zealand |
| DD | German Democratic Republic | PL | Poland |
| DE | Federal Republic of Germany | PT | Portugal |
|    | (and Germany before 1949)* | SE | Sweden |
| DK | Denmark | SU | Soviet Union |
| ES | Spain | US | United States of America |
| FI | Finland | YU | Yugoslavia |
| FR | France | ZA | South Africa |
| GB | United Kingdom | EP | European Patent Office* |
| GR | Greece | WO | World Intellectual Property |
| HU | Hungary | | Organization |
| ID | Indonesia | | |

*For Europe, Federal Republic of Germany, Japan, and the Netherlands, the type of patent is specified: EP (patent), EP-A (application), DE (patent), DE-OS (Offenlegungsschrift), DE-AS (Auslegeschrift), JP (patent), JP-Kokai (Kokai tokkyo koho), NL (patent), and NL-A (application).

# Periodic Table of Elements

element symbol, atomic number, and relative atomic mass (atomic weight)

- 1A    "European" group designation and old IUPAC recommendation
- 1    group designation to 1986 IUPAC proposal
- IA    "American" group designation, also used by the Chemical Abstracts Service until the end of 1986

| 1A<br>1<br>IA | 2A<br>2<br>IIA | 3A<br>3<br>IIIB | 4A<br>4<br>IVB | 5A<br>5<br>VB | 6A<br>6<br>VIB | 7A<br>7<br>VIIB | 8<br>8<br>VIII | 8<br>9<br>VIII | 8<br>10<br>VIII | 1B<br>11<br>IB | 2B<br>12<br>IIB | 3B<br>13<br>IIIA | 4B<br>14<br>IVA | 5B<br>15<br>VA | 6B<br>16<br>VIA | 7B<br>17<br>VIA | 0<br>18<br>VIIIA |
|---|---|---|---|---|---|---|---|---|---|---|---|---|---|---|---|---|---|
| 1<br>**H**<br>1.0079 | | | | | | | | | | | | | | | | | 2<br>**He**<br>4.0026 |
| 3<br>**Li**<br>6.941 | 4<br>**Be**<br>9.0122 | | | | | | | | | | | 5<br>**B**<br>10.811 | 6<br>**C**<br>12.011 | 7<br>**N**<br>14.007 | 8<br>**O**<br>15.999 | 9<br>**F**<br>18.998 | 10<br>**Ne**<br>20.180 |
| 11<br>**Na**<br>22.990 | 12<br>**Mg**<br>24.305 | | | | | | | | | | | 13<br>**Al**<br>26.982 | 14<br>**Si**<br>28.086 | 15<br>**P**<br>30.974 | 16<br>**S**<br>32.066 | 17<br>**Cl**<br>35.453 | 18<br>**Ar**<br>39.948 |
| 19<br>**K**<br>39.098 | 20<br>**Ca**<br>40.078 | 21<br>**Sc**<br>44.956 | 22<br>**Ti**<br>47.867 | 23<br>**V**<br>50.942 | 24<br>**Cr**<br>51.996 | 25<br>**Mn**<br>54.938 | 26<br>**Fe**<br>55.845 | 27<br>**Co**<br>58.933 | 28<br>**Ni**<br>58.693 | 29<br>**Cu**<br>63.546 | 30<br>**Zn**<br>65.409 | 31<br>**Ga**<br>69.723 | 32<br>**Ge**<br>72.61 | 33<br>**As**<br>74.922 | 34<br>**Se**<br>78.96 | 35<br>**Br**<br>79.904 | 36<br>**Kr**<br>83.80 |
| 37<br>**Rb**<br>85.468 | 38<br>**Sr**<br>87.62 | 39<br>**Y**<br>88.906 | 40<br>**Zr**<br>91.224 | 41<br>**Nb**<br>92.906 | 42<br>**Mo**<br>95.94 | 43<br>**Tc***<br>98.906 | 44<br>**Ru**<br>101.07 | 45<br>**Rh**<br>102.91 | 46<br>**Pd**<br>106.42 | 47<br>**Ag**<br>107.87 | 48<br>**Cd**<br>112.41 | 49<br>**In**<br>114.82 | 50<br>**Sn**<br>118.71 | 51<br>**Sb**<br>121.76 | 52<br>**Te**<br>127.60 | 53<br>**I**<br>126.90 | 54<br>**Xe**<br>131.29 |
| 55<br>**Cs**<br>132.91 | 56<br>**Ba**<br>137.33 | | 72<br>**Hf**<br>178.49 | 73<br>**Ta**<br>180.95 | 74<br>**W**<br>183.84 | 75<br>**Re**<br>186.21 | 76<br>**Os**<br>190.23 | 77<br>**Ir**<br>192.22 | 78<br>**Pt**<br>195.08 | 79<br>**Au**<br>196.97 | 80<br>**Hg**<br>200.59 | 81<br>**Tl**<br>204.38 | 82<br>**Pb**<br>207.2 | 83<br>**Bi**<br>208.98 | 84<br>**Po***<br>208.98 | 85<br>**At***<br>209.99 | 86<br>**Rn***<br>222.02 |
| 87<br>**Fr***<br>223.02 | 88<br>**Ra***<br>226.03 | | 104<br>**Rf***<br>261.11 | 105<br>**Db***<br>262.11 | 106<br>**Sg**<br>| 107<br>**Bh** | 108<br>**Hs** | 109<br>**Mt** | 110<br>**Ds** | 111<br>**Rg** | 112<br>**Cn** | 113<br>**Uut**$^a$ | 114<br>**Fl** | 115<br>**Uup**$^a$ | 116<br>**Lv** | | 118<br>**Uuo**$^a$ |

$^a$ provisional IUPAC symbol

| 57<br>**La**<br>138.91 | 58<br>**Ce**<br>140.12 | 59<br>**Pr**<br>140.91 | 60<br>**Nd**<br>144.24 | 61<br>**Pm***<br>146.92 | 62<br>**Sm**<br>150.36 | 63<br>**Eu**<br>151.97 | 64<br>**Gd**<br>157.25 | 65<br>**Tb**<br>158.93 | 66<br>**Dy**<br>162.50 | 67<br>**Ho**<br>164.93 | 68<br>**Er**<br>167.26 | 69<br>**Tm**<br>168.93 | 70<br>**Yb**<br>173.04 | 71<br>**Lu**<br>174.97 |
|---|---|---|---|---|---|---|---|---|---|---|---|---|---|---|
| 89<br>**Ac***<br>227.03 | 90<br>**Th***<br>232.04 | 91<br>**Pa***<br>231.04 | 92<br>**U***<br>238.03 | 93<br>**Np***<br>237.05 | 94<br>**Pu***<br>244.06 | 95<br>**Am***<br>243.06 | 96<br>**Cm***<br>247.07 | 97<br>**Bk***<br>247.07 | 98<br>**Cf***<br>251.08 | 99<br>**Es***<br>252.08 | 100<br>**Fm***<br>257.10 | 101<br>**Md***<br>258.10 | 102<br>**No***<br>259.10 | 103<br>**Lr***<br>260.11 |

\* radioactive element; mass of most important isotope given.

# Part 3

# Films, Fibers, Foams

# Films

PETER SCHMITZ, Hoechst AG, Werk Kalle, Wiesbaden, Federal Republic of Germany

SIEGFRIED JANOCHA, Hoechst AG, Werk Kalle, Wiesbaden, Federal Republic of Germany

| | | |
|---|---|---|
| 1. | Introduction | 1407 |
| 2. | Production Technology | 1408 |
| 2.1. | Solution Casting | 1409 |
| 2.2. | Thermoplastic Film Forming | 1410 |
| 2.2.1. | Calendering | 1410 |
| 2.2.2. | Extrusion | 1411 |
| 2.3. | Stretching | 1412 |
| 2.3.1. | Monoaxial Stretching | 1412 |
| 2.3.2. | Biaxial Stretching | 1413 |
| 2.4. | Film Modification and Converting | 1414 |
| 2.4.1. | Corona Treatment | 1414 |
| 2.4.2. | Coating | 1415 |
| 2.4.3. | Lamination | 1415 |
| 3. | Additives | 1415 |
| 4. | Delivery Forms | 1416 |
| 5. | Test Methods | 1416 |
| 6. | Film Products | 1418 |
| 6.1. | Cellulose Derivatives | 1418 |
| 6.1.1. | Cellulose Hydrate (→ Cellulose) | 1418 |
| 6.1.2. | Cellulose Esters (→ Cellulose Esters) | 1418 |
| 6.2. | Hydrocarbon Polymers | 1423 |
| 6.2.1. | Low-Density Polyethylene (LDPE) (→ Polyethylene) | 1423 |
| 6.2.2. | High-Density Polyethylene (HDPE) (→ Polyethylene) | 1423 |
| 6.2.3. | Linear Low-Density Polyethylene (LLDPE) (→ Polyethylene) | 1423 |
| 6.2.4. | Polypropylene (PP) (→ Polypropylene) | 1423 |
| 6.2.5. | Poly(4-methyl-1-pentene) (PMP) (→ Poly(4-Methyl-1-Pentene)) | 1424 |
| 6.2.6. | Ionomers (→ Specialty Plastics) | 1424 |
| 6.2.7. | Polystyrene (PS) (→ Polystyrene and Styrene Copolymers) | 1424 |
| 6.3. | Halogenated Hydrocarbon Polymers | 1424 |
| 6.3.1. | Poly(Vinyl Chloride) (PVC) [→ Poly(Vinyl Chloride)] | 1424 |
| 6.3.1.1. | Rigid PVC | 1425 |
| 6.3.1.2. | Plasticized PVC | 1425 |
| 6.3.2. | Poly(Vinylidene Chloride) (PVDC) [→ Poly(Vinylidene Chloride)] | 1425 |
| 6.3.3. | Fluoropolymers (→Fluoropolymers,Organic) | 1425 |
| 6.3.3.1. | Poly(Vinyl Fluoride) (PVF) | 1426 |
| 6.3.3.2. | Poly(Vinylidene Fluoride) (PVDF) | 1426 |
| 6.3.3.3. | Polytetrafluoroethylene (PTFE) | 1426 |
| 6.3.3.4. | Polychlorotrifluoroethylene (PCTFE) | 1426 |
| 6.4. | Poly(Methyl Methacrylate) (PMMA) (→ Polyacrylates) | 1426 |
| 6.5. | Poly(Vinyl Alcohol) (PVA) (→ Polyvinyl Compounds, Others) | 1427 |
| 6.6. | Ethylene–Vinyl Alcohol Copolymer (EVOH) | 1427 |
| 6.7. | Polyamide (PA) (→ Polyamides) | 1427 |
| 6.8. | Polyester (→ Polyesters) | 1427 |
| 6.9. | Polycarbonate (PC) (→ Polycarbonates) | 1428 |
| 6.10. | Polysulfone (PSU) (→ Specialty Plastics) | 1428 |
| 6.11. | Polyimide (PI) (→ Polyimides) | 1429 |
| 6.12. | Other Films | 1429 |
| 6.12.1. | Polyetherimide (PEI) | 1429 |
| 6.12.2. | Polyethersulfone (PES) (→ Specialty Plastics) | 1429 |
| 6.12.3. | Poly(ether ketone) (PEK) | 1430 |
| 6.12.4. | Poly(Phenylene Sulfide) (PPS) (→ Specialty Plastics) | 1430 |
| 6.13. | Composite Films | 1430 |
| 7. | Summary of Uses | 1430 |
| 8. | Environmental Aspects | 1432 |
| | References | 1433 |

## 1. Introduction

The term *film* is applied to thin materials present in a continuous planar form, which are at the same time self-supporting and flexible. A film may consist of one or more layers, usually of differing composition. The term generally encompasses both metallic and polymeric substances, which

**Table 1.** Worldwide consumption of some of the most important types of film in 2005[*]

| Material | Consumption, $10^6$ t |
|---|---|
| Low-density polyethylene (LDPE) | 11.00 |
| Polypropylene, biaxially oriented (BOPP) | 3.70 |
| Poly(vinyl chloride) (PVC), rigid and plasticized | 0.8 |
| Polyesters | 2.20 |
| Cellophane | 0.07 |

[*] http://www.pardos-marketing.com

may be either homogeneous or made heterogeneous by the presence of incompatible additives or blending components. In this article, a film is assumed to be a polymeric material which can be wound into a roll. Film thicknesses range from 0.5 µm to ca. 1 mm.

The characteristics of a film are determined primarily by the raw materials employed, but they are additionally affected by processing methods, modification, and converting (e.g., stretching, coating and lamination).

Film materials are of considerable economic importance. In 2005, the annual worldwide consumption of film for packaging and technical applications was ca. $50 \times 10^6$ t. Figures for the worldwide consumption of the more important film materials are listed in Table 1.

Films are used as flexible packaging materials, as insulation for electrical conductors, as protective layers on sensitive surfaces, and as windable base materials for subsequent application of magnetic coatings, electrically conducting, and light-sensitive layers as well as printing inks. Multitudes of film and sheet types have been developed to meet the increasing demands of these applications. These products provide a broad spectrum of mechanical, optical, and electrical properties, as well as various degrees of permeability and stability to temperature and aggressive media.

**Historical Aspects** [1]. Polymeric films were first developed at the end of the 19th century, when cellulose nitrate was introduced as a base material for photographic emulsions. Other cellulose-based films followed, the most important of which were derived from cellulose acetate and cellulose hydrate. Cellulose triacetate films quickly replaced the highly flammable cellulose nitrate films in the photographic industry; cellulose hydrate films (cellophanes) became important packaging materials, maintaining their dominance until polyolefin films were introduced in the 1950s. Cellophane coated with cellulose nitrate or poly(vinylidene chloride) also played a key role in this respect due to its low permeability to water vapor and oxygen, coupled with desirable sealing properties.

All cellulose-based films are prepared from polymer solutions, however the subsequent introduction of totally synthetic polymers has led to new film production methods, as well as new types of films with unique properties. These developments have been encouraged by the close working relationship between fiber and film producers, particularly in terms of polymer compositions and processing technology.

Important milestones in the development of films included the introduction of

1. poly(vinyl chloride) films in the late 1930s
2. low-density polyethylene films in the late 1940s
3. biaxially oriented polyester and polypropylene films in the 1950s

Parallel and subsequent to these developments, a series of specialty films has emerged. These films are distinguished either by high mechanical or thermal stability, or by special properties such as impermeability to oxygen, advantageous electrical characteristics, or resistance to chemical attack.

## 2. Production Technology [2]

Generally speaking, the manufacture of films from both natural and synthetic polymers relies on two basic techniques:

1. solution casting
2. thermoplastic film forming

The method chosen is generally determined by the behavior of the starting materials with respect to solvents and temperature. However, in certain cases the desired product characteristics are decisive.

For *solution casting*, the raw materials must first be dissolved in an appropriate solvent. The solution is then pressed through a slot die either onto a moving belt or into a precipitation bath.

*Thermoplastic film forming* can be divided into two categories: (1) calendering (film formation in a rolling process) and (2) film formation from a melt (an extrusion process).

## 2.1. Solution Casting

Solution casting is applicable to soluble polymers that are difficult or impossible to thermoform. It also has advantages in the preparation of films for optical applications. Cellulose acetate, polycarbonate, and polyimide films are prepared in this way.

Solution casting consists of three stages:

1. preparation of the casting solution
2. film formation on a moving surface (drum or endless belt) or in a precipitation bath
3. removal of residual solvent

Figure 1 is a schematic illustration of a belt casting machine that is suitable for the production of cellulose acetate film.

The casting solution is prepared in a large agitator vessel, if possible using a low-boiling solvent. The solution is normally very viscous (ca. 20 Pa · s) and is passed through a filter press equipped with a cotton filter. After the viscosity has been checked, the solution finally enters the casting reservoir. Entrapped air is removed either by evacuation or by heating.

The solution is fed from the reservoir under pressure through the die and is cast onto a rotating drum or a moving metal belt. The die gap is adjusted according to the desired thickness of the film.

The metal belt may be made of copper, nickel, or steel, and its surface finish is of critical importance in determining the characteristics of the resulting film. For optical-grade films, the surface must be polished and filtration must be highly effective. Belt dimensions range from 25 to 55 m in length and 1 to 2 m in width; belt speed varies from 1 to 60 m/min, depending on the desired thickness of the film.

The film is dried on the belt by a countercurrent of warm air and is then removed, further dried, and wound. Typical thicknesses of films prepared in this way are 15 – 250 µm.

In some cases, the polymer solution emerges from a slot die directly into a liquid precipitation bath. The polymer solvent is miscible with the bath fluid, whereas the polymer is insoluble in this fluid. Consequently, the polymer coagulates in the bath to form a gel film with a microporous structure. Thus, this process is frequently employed for the preparation of membranes (→ Membranes and Membrane Separation Processes, 3. Membrane Preparation and Membrane Module Constructions).

Cellulose hydrate (cellophane) films [3] are produced by a modification of this process in which the gel film undergoes a chemical reaction. The casting solution is prepared by reacting cellulose with caustic soda solution and carbon disulfide to form viscose. The resulting xanthate solution is then pressed into a spinning bath containing sulfuric acid, causing decomposition of the xanthate and forming a gel web composed of regenerated cellulose (see also → Cellulose).

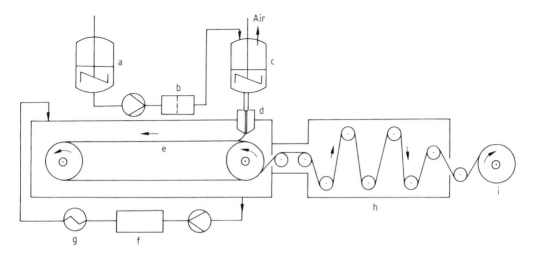

**Figure 1.** Film production by solution casting on an endless belt
a) Agitator vessel; b) Filter; c) Degasser/casting reservoir; d) Die/caster; e) Metal belt; f) Solvent recovery; g) Air heater; h) Final dryer; i) Winding

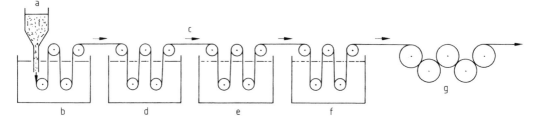

**Figure 2.** Production of cellulose hydrate film by casting into a precipitation bath
a) Casting solution, slot die; b) Precipitation bath; c) Regenerated film web; d) Regeneration bath; e) Wash and bleaching bath; f) Plasticizer bath; g) Drying rolls

A slot die is immersed in the precipitation bath (see Fig. 2). Shortly after its formation, the gel film possesses sufficient stability to permit its transport over rollers. The film is subsequently washed and bleached, passed through plasticizer baths, dried, and wound.

Film thickness ranges from 12 to 45 µm with a typical width of ca. 2 m. Production rates of 80 m/min are common for films in the mid-thickness range. The normal slot die may be replaced by an annular orifice, in which case the product is a tube of regenerated cellulose [4]. Further processing is as described above for flat cellulose hydrate films.

## 2.2. Thermoplastic Film Forming

### 2.2.1. Calendering [5–7]

Only polymers with a broad range of plasticity and high viscosity can be processed by the calendering method; these criteria are met by homo- and copolymers of poly(vinyl chloride) (PVC).

The production of calendered (e.g., PVC) film is shown in Figure 3; three steps are involved:

1. preparation of a blend of the components in a mixer
2. homogenization and plastification
3. film formation in the calender

The powdered mixture is usually converted into a fused mass by continuous compounding in a kneader or planetary-gear extruder using heat, shearing forces, and pressure. The mass is then rolled into a film web, ideally with a four- or five-roll calender, in which the nip rolls are mounted in an "L"-form. These rolls are made of steel or cast iron and they generally have a smooth surface. The nip rolls are heated to 180 – 220 °C, depending on the process and the polymer mixture; heating must be uniform. The calendered web then passes over cooling rolls followed by an edge-trimmer, in some cases a stretching facility, and finally a winding unit. Calendering equipment operates at high

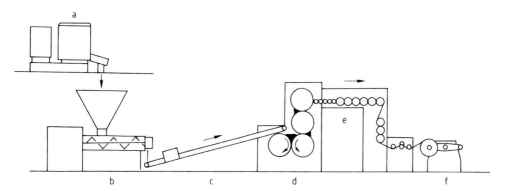

**Figure 3.** Film production by calendering
a) Mixer; b) Kneader; c) Feed band; d) Calender; e) Cooling and tempering rolls; f) Winder

pressure and must, therefore, be solidly constructed, making it very expensive. The largest production facilities for processing plasticized PVC have nip rolls 85 cm in diameter and are 280 cm wide. Calenders for rigid PVC utilize rolls up to 70 cm in diameter and are 250 cm wide. Production rates of 50 – 100 m/min are achieved depending on the desired film thickness; with stretching, line speeds approach 300 m/min. These figures correspond to an output of 500 – 2000 kg/h.

### 2.2.2. Extrusion [8]

Advantages of extrusion include a wide range of attainable film thickness, large-scale processing, and the fact that recovery systems for expensive solvents are not needed.

The technique utilizes an extruder to melt the polymer, a die to confer shape to the melt, and a cooling system to solidify it. The process is applicable only to raw materials for which there is a sufficiently large temperature difference between melting and the onset of decomposition (e.g., polyolefins, polyamides, and polyesters).

*Chill Roll Extrusion.* Figure 4 is a schematic illustration of a chill roll extrusion line. Moisture-containing raw materials (moisture content usually ≪1 %) are subjected to continuous or discontinuous drying, preferably utilizing heated air or a vacuum, prior to melting in the extruder.

Various extruder designs can be used including single-screw and twin-screw extruders, the latter with either corotating or contrarotating screws. Extremely high outputs of 4 – 5 t/h can be achieved with two extruders combined in sequence (tandem extruder). Single-screw extruders are the most common. Screw geometry must be adapted to the raw materials being processed and has a major effect on the output and quality of the melt. The largest single-screw extruders currently in use have a screw diameter ($D$) of 30 cm and a screw length of 30 $D$.

The melt is filtered in one of several ways depending on the desired purity: wire mesh, sintered metal, and wire–felt disks are all used. The filtered raw materials are forced through a slot die. The required pressure can normally be achieved with the extruder, however in special cases a gear pump is installed before the filter to ensure constant metering of the melt. In coextrusion (see Section 2.4.2), the die combines several different melt streams. The width of the die and the die gap determine the width and thickness of the film, respectively. In large production units, the dies may be as wide as 3.5 m. Adjustment of the die gap permits the film thickness profile to be corrected during production. Initial adjustment and readjustment can be performed either manually or automatically, e.g., by thermal expansion of adjusting screws that are heated as a function of the relative film thickness.

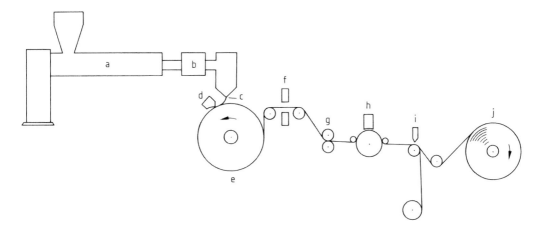

**Figure 4.** Film production by chill roll extrusion
a) Extruder; b) Filter; c) Slot die; d) Air knife; e) Cooling roll; f) Device for measuring film thickness; g) Tension rolls; h) Corona treatment; i) Edge trimming; j) Winding

When the melt film emerges from the die, it is cast onto a highly polished, chrome-plated chill roll. The takeoff speed depends on both film thickness and output and may be as high as 100 m/min. Entrapment of air between the melt and the roller surface is prevented by using an air jet (air knife), electrostatic "pinning", or a vacuum device.

The thickness of the film is measured with the aid of a traversing radiation source and detector. The film is then edge-trimmed and wound to a roll. If the film is to be printed, corona pretreatment can also be carried out (see Section 2.4.1).

The film is usually cut to the designated width in a separate process—the actual width depending on the application. If monoaxial or biaxial orientation is desired, the film is fed directly into a stretching unit (see Section 2.3) without intermediate winding.

***Film Blowing (Blown-Bubble Extrusion).*** In the film blowing process, the filtered melt is pressed through an annular die instead of a slot die; this results in the production of a tubular film (Fig. 5). The tube (or "bubble") is inflated by means of air blown through the center of the die. The film is thereby expanded at a temperature close to its melting point and is then cooled by an airstream located above the die. A number of techniques are employed for air cooling; the most common involves passing the tube through a cooling ring. Supplementary internal cooling is also usually used if a high throughput is required. If biaxial stretching is to be performed, the melt must be solidified immediately after it emerges from the die (see Section 2.3.2).

The cooled tube is collapsed by transporting it through a convergent collapsing frame and a set of nip rolls. The film can then either be wound directly as a flattened tube or slit open on both sides and directed to two separate winding stations. The largest film blowing systems utilize annular dies with a diameter of 2 m and an output of 1.5 t/h.

The film blowing process is employed mainly for polyolefin and polyamide films. Multilayer films can be produced by coextrusion of tubular films (see Section 2.4.2).

## 2.3. Stretching [8]

Many films acquire their desired properties only after stretching. Stretching below the softening temperature orients the polymer molecules and crystallites in the direction of stretch, leading to changes in the stress–strain characteristics of the material. Both the modulus of elasticity and the ultimate strength are increased by this process, but elongation at break and dimensional stability with respect to temperature decrease. A number of other characteristics may be altered as well, depending on the type of film involved, e.g. gas permeability, electrical properties, heat sealability, abrasion resistance, coefficient of friction, puncture resistance, density, and optical properties. Stretching can be conducted either solely in the machine (longitudinal) direction (monoaxial stretching) or in both the machine and the transverse directions (biaxial stretching).

*Monoaxial stretching* (see Section 2.3.1), which is also commonly used in the production of high-strength fibers, produces films of enhanced strength in the direction of manufacture. The resulting films are thus anisotropic.

*Biaxial stretching* (see Section 2.3.2) can be conducted in one or two steps (simultaneous and sequential stretching, respectively). Sequential stretching may result in either anisotropic or isotropic films depending on the ratio of the final length to the initial length (the draw ratio). Simultaneous stretching tends to produce isotropic films.

The majority of films prepared from orientable polymers are amenable to either sequential or simultaneous biaxial stretching. However, films derived from polymers with a high rate of crystallization during orientation (stress crystallization) [e.g., some polyamides, poly(butylene terephthalate), and ethylene–tetrafluoroethylene copolymer] can only be oriented by simultaneous stretching.

### 2.3.1. Monoaxial Stretching

Monoaxial stretching is accomplished by subjecting a film to an increasing velocity gradient as it passes through one or more gaps between temperature-controlled rolls. Numerous stretching machines are available; they differ with respect to the number and arrangement of the rolls. Monoaxial stretching is of particular importance in the manufacture of high-strength tape from polyethylene and polypropylene, as well as for the orientation of poly(vinyl

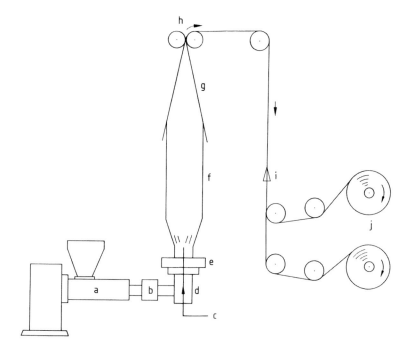

**Figure 5.** Film production by the blowing method
a) Extruder; b) Filter; c) Blowing air; d) Annular die; e) Air cooling ring; f) Film tube (bubble); g) Flattening; h) Nip rolls; i) Tube separation (slitting); j) Winding

chloride), polycarbonate, and poly(vinylidene fluoride) films. Longitudinal stretching units are also incorporated in film production lines that employ sequential stretching technology for the manufacture of biaxially oriented films (see Section 2.3.2).

### 2.3.2. Biaxial Stretching

*Sequential Stretching.* Sequential biaxial stretching is currently the preferred orienting method for example with polyester and polypropylene films. Figure 6 illustrates a production line that uses longitudinal–lateral stretching (the so-called machine direction—transverse direction, MD–TD sequence) for the manufacture of biaxially stretched film. The film line consists essentially of a chill roll facility coupled with forward and transverse stretching units.

The polymer is first dried, melted in an extruder, passed through a slot die, and vitrified on a chill roll. Longitudinal stretching follows using a system of rollers, whereas drawing in the transverse direction occurs in an air-heated oven (stenter) comprising several temperature-controlled zones. When the longitudinally stretched film enters the stenter, it is grasped at both edges by two series of continuously circulating clips which are linked together in a chain. In the zone where the film reaches the stretching temperature, the two series of parallel-running clips are gradually drawn apart and hence stretch the film in the transverse direction (Fig. 6). The film is then heat-set (crystallized), cooled, released from the clips, and wound to a roll. Draw ratios for polyester lie typically between 3:1 and 4:1 [9, 10].

Facilities for edge-trimming, thickness determination, and, where appropriate, corona treatment (see Section 2.4.1) are located between the stenter and the winding unit. Winding speeds range from 80 to 250 m/min. Film thickness varies from 1 to 350 µm.

Polyester films with a high modulus of elasticity in the machine direction are best prepared by a TD–MD stretching sequence, or in a three-step process with MD–TD–MD stretching.

*Simultaneous Stretching.* Simultaneous biaxial orientation is most easily achieved by inflation of a tubular film. In the continuous

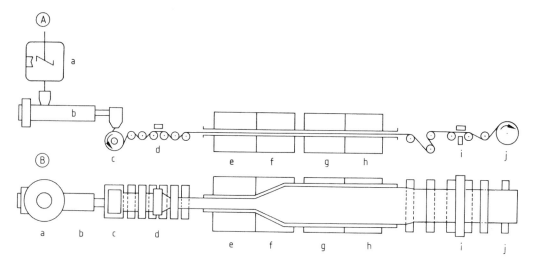

**Figure 6.** Production of biaxially oriented film by sequential stretching in a machine direction–transverse direction (MD–TD) sequence [production unit viewed from the side (A) and from above (B)]
a) Dryer; b) Extruder; c) Slot die and chill roll; d) Stretching in the machine direction (MD); e) Preheating zone; f) Zone for stretching in the transverse direction (TD); g) Heat-setting (crystallizing) zone; h) Cooling zone; i) Thickness measurement; j) Winding

process, a tube is first created by cooling the melt in a water bath or with a cooling mandrel. This tube is subsequently heated to the stretching temperature with a ring of radiant heaters. It is then inflated, collapsed, and transported as described for blown films (see Section 2.2.2). This technique is widely used for preparing biaxially oriented poly(vinylidene chloride), polyamide, polyester, and especially polypropylene tubular films; all of these films have a high shrinking potential. For this reason, oriented polypropylene tubular films are normally slit, separated, and then heat-set, either with the aid of heated rollers or in a hot-air tunnel.

Simultaneous biaxial stretching of a flat film is technically more complicated than that of tubular materials. Normally, a special stretching (stenter) frame is used in which the clips are not linked in a chain as described for sequential stretching. The clips are individually accelerated as they pass through the transverse-stretching zone by means of a progressive transport system.

## 2.4. Film Modification and Converting

Films are normally intermediates from which end products are manufactured in subsequent processing steps. The terms *film modification* and *converting* refer to operations such as surface treatment, coating, or lamination, which enhance the value of a film. Commonly encountered end products derived from films include packaging materials (e.g., printed bags and adhesive tapes), capacitors, magnetic tapes, and photographic films.

Untreated films often lack the specific characteristics that are required in the final product, e.g., stability during handling, heat sealability, wettability, adhesion of printing ink or other coatings, and barrier properties. Further processing is then needed to remedy these deficiencies. Sometimes the desired results can be obtained with additives (cf. Chap. 3), which include antistatic agents and lubricants. However, surface treatment or coating is often necessary, and in some cases two or more films must be combined to form a laminate.

### 2.4.1. Corona Treatment

Corona treatment is the principal method of preparing a film surface for printing or coating. A continuously moving film is exposed to an ionized atmosphere, which is produced by an alternating electrical field [11]. The changes induced in the film surface enhance both its wettability and adhesion properties. Corona treaters can be integrated into the film production process (see Sections 2.2.2 and 2.3.2) and are widely utilized in the pretreatment of polyolefin films.

### 2.4.2. Coating

Films may be coated with dissolved, dispersed, emulsified, or molten materials [8, 12]. The techniques employed differ considerably, depending on whether the material is applied as a liquid or as a melt.

***Liquid materials*** are normally coated on one or both sides of a moving film web either by immersion or with the aid of applicator rolls. The coated film is then dried in a heated-air oven or by means of rollers.

Coating is normally performed separately from film production (*off-line coating*), although in some special cases, the two processes can be integrated (*in-line coating*) [13]. In in-line coating (e.g., of polyester films), the coating assemblies are preferentially placed between the MD and TD stretching units.

Coating materials such as nitrocellulose, polyacrylates, and poly(vinylidene chloride) are used to improve heat sealability and gas barrier properties. Adhesion is improved by coating films with polyurethanes, cross-linkable polyacrylate, copolyesters, or epoxides. Other coatings include antistatic or slip agents, scratchresistant and antiadhesive materials.

Films with a different functional coating on each surface possess a combination of properties (e.g., sealing and barrier properties) and can be produced on a large scale.

***Molten materials*** are applied to films with the aid of a die by a process analogous to chill extrusion (see Section 2.2.2). The film itself is supported by a roller. Extrusion coating can be conducted either off- or in-line; the latter approach is employed with biaxially stretched films.

*Coextrusion* represents a special case of melt coating [14, 15] in which two or more polymer melt streams are combined either in the melt pipes or in a multimanifold die just before the material emerges from the die lips. In the former case, special adapters of varying design ensure that the melt streams do not mix in the die and hence leave the die gap in the form of a multilayer film analogous to that produced by a multimanifold die. This process is also adaptable to annular die systems.

Coextrusion has gained considerable importance. Sealable, biaxially oriented polypropylene film with a triple-layer structure is a particularly important product in the packaging industry. The thickness of the sealable layer is typically 0.5 µm.

Coextruded, double-layer films are commonly employed for magnetic tape and capacitors; they have a smooth side which can be coated, and a rough side, which ensures a low coefficient of friction for adequate processability.

### 2.4.3. Lamination [8, 12]

Two or more different films can be bonded to form a composite or multilayer film. Their individual properties are thus combined, resulting in a variety of high-quality materials for both the packaging industry and the technical sector. For example, a heat-sealable film may be laminated to a film that acts as a barrier to gases and/or to a high-strength film. The potential number of property combinations is very large.

Laminated composite films are commonly prepared by the *dry lamination process*, in which a dissolved or dispersed adhesive is coated on the surface of the film. After evaporation of the solvent in a dryer, the coated film is bonded to a second film web in a laminating unit consisting of a pair of nip rolls, one made of rubber and the other of steel.

The simplest laminating system is composed of two unwinding stations, an adhesive applicator, a dryer, a laminating unit, and a windup station. However, many different systems are in use, depending on the desired number of laminate layers and the choice of adhesive. In some cases, adhesion can be achieved without using solvents, e.g., by using hot melts, reactive adhesives, or systems that can be hardened by irradiation.

## 3. Additives

Nearly all films contain additives which perform two main functions: (1) stabilization of the polymers and (2) improvement of processability and the potential uses of the film [16, 17] (see also → Plastics, Additives, → Plasticizers).

Additives can be introduced when the polymer is being prepared or processed, or during film production. An additive must withstand all stages of processing without any loss of activity. Additives used in films that come into contact with food must also comply with legal health and safety requirements.

***Processing Stabilizers.*** The thermal decomposition of some film raw materials must be suppressed by addition of processing stabilizers so that the melt can be thermoformed. The choice of stabilizer depends on the decomposition mechanism involved. Thus, sterically hindered phenols effectively stabilize polypropylene, while poly(vinyl chloride) responds to treatment with metal salts (in particular organic tin salts) or *N,N'*-diphenylthiourea [102-08-9]. Additive concentrations range from <1 to 3 wt %.

***Plasticizers.*** Plasticizers are particularly useful for adjusting the flexibility of poly(vinyl chloride), cellulose acetate, and cellulose hydrate films. The plasticizer concentration may be as high as 40 % (plasticized PVC); typical examples include esters of phthalic and phosphoric acids.

***Slip or Antiblocking Agents.*** Films with smooth surfaces possess a high coefficient of friction and therefore tend to block during winding and other processing operations. To improve the frictional properties of these films, their surface is finely structured by adding small amounts of finely dispersed pigments. The surface of biaxially oriented polyester films can be improved by adding inert inorganic pigments (e.g., calcium carbonate, silicon dioxide, or calcium phosphate) or particles composed of cross-linked polyacrylates. The particles are < 1 µm in diameter, and are used at a concentration of < 0.5 wt %. Lubricants such as fatty amides or polysiloxanes may also be introduced.

***Fillers and Dyes.*** The optical properties of films can be altered by the incorporation of fillers and dyes. A wide range of concentrations is employed. Many fillers are inert inorganic substances, e.g., titanium dioxide, barium sulfate, kaolin, and carbon black. Colored films are normally produced from raw materials that have been dyed in bulk. However, in special cases the films themselves may be dyed either by a diffusion process or by printing the entire surface.

***Antistatic Agents.*** Films are rendered antistatic by mixing the virgin polymer with a substance bearing hydrophilic groups. The additive migrates to the surface layer. Quaternary ammonium compounds, ethyl sulfonates and, especially with polyolefin films, ethoxylated alkyl amines are used as antistatic agents.

***Weathering Stabilizers.*** Most films for outdoor applications are prone to photooxidative degradation and are stabilized by adding UV-absorbing substances. Piperidyl derivatives, for example, are effective stabilizers of polypropylene.

## 4. Delivery Forms

Commercial quantities of films are available in various thicknesses (ca. 1 – 1000 µm) and are supplied in the form of rolls with cores of a standard diameter (e.g., 6″).

Standardized thicknesses are demanded for many applications. Attempts are being made to establish a standardized system, since this also simplifies production planning and stock control. Films are offered in widths ranging from a few millimeters to several meters. Standard widths have only been established in a few areas of application.

The maximum length of a film depends on the intended application as well as on the film thickness. For example, with poly(ethylene terephthalate) (PET), films with a thickness of 5 µm are supplied in rolls with a typical length of 30 km.

Films that are precut to size are also marketed, although only to a very limited extent.

## 5. Test Methods [18–24]

Many of the test methods employed for process control in film manufacture have been standardized. However, the results obtained are not always compatible because the test conditions specified in different standards often vary.

The *chemical composition* of a film is normally apparent from its IR spectrum. Trace elements are determined by atomic absorption spectroscopy, mass spectrometry, or neutron activation analysis. Relative molecular masses can be measured from the viscosity of a film solution. The identity of a film can often be determined more simply with the aid of systematic analytical schemes (e.g., [20], Appendix B).

The *average thickness* of a film is established according to DIN 53 370 or ASTM E 252. No standard method for measuring thickness in both the machine and transverse directions exists as yet. During manufacture, film thickness can be determined by measuring the absorption of either β- or IR radiation or by a dielectric method. The uniformity of the thickness of a film sample can be determined mechanically, electrically, or optically. The accuracy of measurement in all cases is 0.1 µm.

Methods for determining the *mechanical properties* of films are described in ASTM D 882, D 1822, D 1894, D 1922; ISO 1184 (1983). Individual methods are as follows: tear strength, DIN 53 455; elongation at break, DIN 53 457; modulus of elasticity, DIN 53 448; impact strength, DIN 53 374; tear resistance, DIN 53 375; and tear propagation resistance, DIN 40 634 and DIN 53 363.

The *coefficient of friction* is of considerable importance in the converting of films, and conventional methods for its determination are described in DIN 53 375 and ASTM D 1894. Special techniques have been developed for some applications (e.g., for magnetic tapes).

*Film density* is determined according to ASTM 1505 IC.

The *electrical properties* of films are tested according to VDE 0345/1.65 and 0303/7/1.65; ASTM D 149, D 150, D 229, and D 252; and DIN/sec 15 C. Parameters that are measured include the dielectric strength, dielectric constant, dissipation factor, resistivity, surface resistance, electrolytic corrosion, and glow discharge ignition voltage.

The *melting point, glass transition temperature, crystallization* and *softening behavior*, and *decomposition temperature* are easily determined by differential thermal analysis. The *long-term thermal stability* of films designed for use as electrical insulators is normally determined according to VDE 0530 or IEC 216; the films can then be classified according to the temperature at which they may be used, i.e., Class T (90 °C), A (105 °C), E (120 °C), B (130 °C), F (155 °C), H (180 °C), and C (> 180 °C).

The *coefficient of expansion, dimensional stability*, and *thermal conductivity* can be determined by reference to the following standards, which are not, however, exclusively designed for films: DIN 1341, 40 634, 52 328; ASTM D 229, D 1204; VDE 0530/1/1.66.

The following standards apply to the measurement of *optical properties* (transparency, haze, gloss, refractive index, and color): DIN 5033, 53 491; ASTM 1003, D 307, D 1925, D 2457. The measurements are easy to perform, but interpretation of the results can be difficult.

Determination of *permeability* with respect to liquids, vapors, and gases is relatively complicated, but this parameter is especially valuable, since films are widely used as packaging materials. Appropriate procedures are described in DIN 53 122 and 53 386, and ASTM D 1434 and E 96 (E).

*Chemical resistance* is most easily determined by examining the changes induced in ultimate strength and elongation at break as a result of exposure to the substances concerned. However, the following standards are also applicable: DIN 53 476; ASTM D 543, D 756, D 1435, E 96; ISO/R 175 and R 511.

*Water absorption* is determined according to DIN 53 472 and ASTM 0570.

The most useful test of *surface structure* is carried out with a profilometer. Since the 1980s this technique has proved itself to be particularly useful for examining film substrates used in the manufacture of magnetic tapes. The best known procedures are those described in DIN 4760 – 4774, the Japanese Industry Standard JIS 0601–70, and ISO/DIS 4267/1. In special circumstances, additional information can be obtained by using microscopy (interference methods and scanning electron microscopy).

The *roll hardness* of film rolls can be evaluated by using the Shore hardness as a criterion (DIN 53 505, ISO 12 868). A better procedure makes use of a falling sphere [24].

The *flammability* of films is evaluated according to DIN 4102, ASTM D 1433, ISO R 1715, and Underwriters Laboratory Standard UL 94 VTM.

Measurement of the *extraction of low molecular mass components* of films with solvents has not been standardized. Most studies are based on special procedures developed by individual manufacturers and end users. Interest centers on films designed for the packaging industry. Information regarding the extractable components of a film often serves as the basic for its legal approval (e.g., FDA approval) [25].

# 6. Film Products [20, 26, 27]

A remarkable variety of film materials has been developed as a result of several factors: Film manufacturers have a wide choice of potential raw materials and can also use special processing steps such as stretching to modify certain properties of a film during its manufacture. Further variants are made possible by the introduction of additives, by the application of functional coatings, or by lamination of two or more types of film. The following discussion is limited to representative examples from characteristic film groups. Important physical properties of these materials are compiled in Tables 2–5. (The lists of trade names supplied are by no means exhaustive.)

## 6.1. Cellulose Derivatives

### 6.1.1. Cellulose Hydrate (→ Cellulose)

Cellophane [9005-81-6] is a trade name applied to films based on cellulose hydrate (regenerated cellulose).

Cellophane is prepared from cellulose via the sodium xanthate derivative in a solution casting process (see Section 2.1) and is available in thicknesses ranging from 12 to 45 μm. Untreated cellophane film is crystal clear, glossy, and highly permeable to water vapor; it is unstable with respect to water, but stable to fats and oils. Cellophane is not heat-sealable, but can be bonded; it is odorless, tasteless, and authorized for use as a food packaging material by the FDA and similar institutions. It possesses desirable mechanical characteristics at the normal water content of 7 – 8 %. Properties are listed in Table 2.

Cellophane is increasingly being replaced by oriented polypropylene films; only a few niche applications are left.

**Manufacturers.** Wolff Walsrode, Germany; Chemiefaser Lenzing, Austria; British Cellophane, United Kingdom; Flexel, Atlanta, United States; Futamura, Japan.

### 6.1.2. Cellulose Esters (→ Cellulose Esters)

Films derived from cellulose acetate [9004-35-7] (CA) (2-acetate, 2 1/2-acetate, and 3-acetate), cellulose acetate propionate [9004-39-1] (CAP), and cellulose acetate butyrate [9004-36-8] (CAB) are available; they all contain plasticizers—phthalate esters, aliphatic alcohols, triphenyl phosphate. Solution casting is the usual production method. Methylene chloride [75-09-2] is used as a solvent [9]. The casting solution has a cellulose acetate concentration of 18 – 26 % and a viscosity of 15 – 30 Pa · s.

Cellulose acetate films are the most important of the cellulose ester films. The properties of cellulose acetate vary as a function of the number of acetate groups per cellulose molecule. The properties of cellulose acetate film (2 1/2 acetate) are listed in Table 2. Films with the highest acetate concentration, corresponding to the minimum number of hydroxyl groups, show reduced sensitivity to moisture as well as minimal absorption of water and low water-vapor permeability. These films are crystal clear in their untreated state; they are dimensionally stable at ambient temperature and humidity and possess good electrical properties. Special surface characteristics (e.g., gloss, roughness) can be conferred during manufacture by casting on an appropriate type of metal belt.

Cellulose acetate films are used as supports for photographic coatings, as transparent windows, release films, decorative materials, and as electrical insulators.

**Table 2.** Properties of selected films[a]

| | Cellophane (uncoated) | Cellulose acetate (CA)[b] | Polyethylene (PE) low-density (LDPE) | Polyethylene (PE) high-density (HDPE) | Poly(4-methyl-1-pentene) (PMP) | Poly-propylene (BOPP) |
|---|---|---|---|---|---|---|
| *Mechanical properties* | | | | | | |
| Thicknesses, µm | 12–45 | 12–350 | 25–200 | 50–1000 | 75–300 | 4–80 |
| Density, g/cm³ | 1.45 | 1.3 | 0.92 | 0.95 | 0.83 | 0.91 |
| Ultimate strength | | | | | | |
|   longitudinal, N/mm² | 100 | 100 | 25[g] | 23[g] | 24 | 140 |
|   lateral, N/mm² | 60 | 100 | 20[g] | 23[g] | 24 | 280 |
| Elongation at break | | | | | | |
|   longitudinal, % | 20 | 30 | 400[g] | 400 | 10 | 140 |
|   lateral, % | 60 | 30 | 700[g] | 400 | 10 | 50 |
| Modulus of elasticity | | | | | | |
|   longitudinal, N/mm² | 5 300 | 1500 | 170 | 900 | 1100 | 2 000 |
|   lateral, N/mm² | 2 800 | 1500 | 170 | 900 | 1100 | 4 000 |
| *Electrical properties* | | | | | | |
| Dielectric strength[c], kV/mm | 100 | 150 | 200 | 200 | 65 | 340 |
| Dielectric constant at 1 kHz | 3.2 | 4.5 | 2.3 | 2.2 | 2.1 | 2.2 |
| Dissipation factor | 0.015 | 0.02 | 0.0002 | 0.0004 | 0.003 | 0.0002 |
| Volume resistivity, Ω cm | $10^{11}$ | $10^{14}$ | $10^{17}$ | $10^{17}$ | $10^{16}$ | $5\times 10^{17}$ |
| *Thermal properties* | | | | | | |
| mp, °C | | | 110 | 130 | 235 | 165 |
| Application temperature range[d], °C | −15 to +150 | −15 to +120 | −60 to +95 | −50 to +100 | −20 to +130 | −20 to +120 |
| Optical properties $n_D^{20}$ | 1.6 | 1.4–1.5 | 1.50 | 1.50 | 1.46 | 1.50 |
| *Dimensional stability* | | | | | | |
| Coefficient of thermal expansion, K⁻¹ | $6\times 10^{-5}$ | $6\times 10^{-5}$ | | | | |
| Shrinkage at 100 °C (30 min) | | | | | | |
|   longitudinal, % | 0.7–3.0 | 0.4 | 1.0 (80 °C) | 0.5 | 1.6 (160 °C) | 3 |
|   lateral, % | 0.7–3.0 | 0.2 | 0 | 0 | 0 | 1 |
| *Physicochemical properties* | | | | | | |
| Water absorption in 24 h, wt % | 45–115 | 5 | <0.1 | <0.1 | <0.1 | <0.1 |
| Stability toward | | | | | | |
|   daylight | good | good | limited | limited | poor | limited |
|   organic solvents | moderate | unstable | good | good | good | good |
|   acid | unstable | unstable | good | good | very good | good |
|   alkali | unstable | unstable | good | good | very good | good |
|   salt solutions | unstable | stable | good | good | very good | good |
| *Permeability* | | | | | | |
| Water vapor, g m⁻² d⁻¹ | very high | 350 | 2.5 | 1.0 | 0.6 | 1.5 |
| Oxygen[e], cm³ m⁻² d⁻¹ bar⁻¹ | 10 | 1 500 | 4 000 | 1 600 | 38 000 | 600 |
| Carbon dioxide[e], cm³ m⁻² d⁻¹ bar⁻¹ | 100 | 10 000 | 16 000 | 7 000 | 110 000 | 1 800 |
| Nitrogen[f], cm³ m⁻² d⁻¹ bar⁻¹ | 12 | 300 | 1 300 | 400 | 9 400 | 140 |

[a] Standard values determined as described in Chapter 5.
[b] Cellulose acetate film (2 1/2-acetate) containing plasticizer.
[c] At 50 Hz, film thickness = 40 µm, plate/plate.
[d] Temperature range over which no significant loss in strength or stiffness occurs.
[e] Film thickness = 40 µm, 23 °C.
[f] Film thickness = 200 µm.
[g] Film thickness = 50 µm.
[h] Decomposition temperature.
[i] For biaxially stretched (balanced) film.
[j] 20 °C, 0 % relative humidity.
[k] Film thickness = 25 µm.
[l] 40 °C, 90 % relative humidity.
[m] Film thickness = 100 µm.

**Table 3.** Properties of selected films[a]

| | Polystyrene (PS) | Poly(vinyl chloride) (PVC) | | Poly(vinylidene chloride) (PVDC) | Polytetrafluoroethylene (PTFE) | Poly(vinyl fluoride) (PVF) |
|---|---|---|---|---|---|---|
| | | rigid | plasticized | | | |
| *Mechanical properties* | | | | | | |
| Thicknesses, μm | 4 – 500 | 30 – 700 | 10 – 1 000 | 25 | 12 – 1 000 | 12 – 50 |
| Density, g/cm$^3$ | 1.06 | 1.38 | 1.24 – 1.3 | 1.66 – 1.75 | 2.1 – 2.2 | 1.38 – 1.57 |
| Ultimate strength | | | | | | |
|   longitudinal, N/mm$^2$ | 60 | 50 | 20 – 35 | 30 – 50 | 10 – 30 | 60 – 150 |
|   lateral, N/mm$^2$ | 60 | 50 | 20 – 35 | 30 – 50 | 10 – 30 | 60 – 150 |
| Elongation at break | | | | | | |
|   longitudinal, % | 10 | 70 | 250 | 150 – 250 | 100 – 350 | 100 – 200 |
|   lateral, % | 10 | 40 | 250 | 150 – 250 | 100 – 350 | 100 – 200 |
| Modulus of elasticity | | | | | | |
|   longitudinal, N/mm$^2$ | 3 000 | 2 500 | 300 – 1 000 | 230 | 400 | 1 400 |
|   lateral, N/mm$^2$ | 3 000 | 2 500 | 300 – 1 000 | 230 | 400 | 1 400 |
| *Electrical properties* | | | | | | |
| Dielectric strength[c], kV/mm | 200 | 60[f] | 50 – 100[f] | | 28[m] | 140 |
| Dielectric constant at 1 kHz | 2.5 | 3.5 | 3.3 – 5.3 | 4.5 – 6.0 | 2.0 – 2.1 | 8.5 |
| Dissipation factor | 0.0005 | 0.02 | 0.03 – 0.1 | 0.06 – 0.07 | 0.0002 | 1.6 |
| Volume resistivity, Ω cm | 10$^{16}$ | 10$^{15}$ | 10$^{11}$ – 10$^{14}$ | 10$^{13}$ – 10$^{14}$ | 10$^{18}$ | 10$^{13}$ |
| *Thermal properties* | | | | | | |
| mp, °C | 240 | 160[h] | | 150 | 330 | 200 |
| Application temperature range[d], °C | – 60 to + 100 | – 15 to + 75 | – 30 to + 50 | 0 to + 140 | – 190 to +250 | – 70 to +110 |
| Optical properties $n_\mathrm{D}^{20}$ | 1.59 | 1.54 | | 1.62 | | |
| *Dimensional stability* | | | | | | |
| Shrinkage at 100 °C (30 min) | | | | | | |
|   longitudinal, % | 25 | 3 | 0.5 | 10 – 35 | | 1 |
|   lateral, % | 25 | 3 | 0.5 | 10 – 35 | | 1 |
| *Physicochemical properties* | | | | | | |
| Water absorption in 24 h, wt % | < 0.1 | < 0.1 | < 0.1 | < 0.1 | < 0.1 | < 0.5 |
| Stability toward | | | | | | |
|   daylight | limited | good | good | limited | very good | very good |
|   organic solvents | limited | good | good | limited | very good | very good |
|   acid | good | good | good | good | very good | very good |
|   alkali | good | good | good | good | very good | very good |
|   salt solutions | good | good | good | good | very good | very good |
| *Permeability* | | | | | | |
| Water vapor, g m$^{-2}$ d$^{-1}$ | 30 | 7.5 | 10 – 100 | 1.5 | thickness < 400 μm contains micropores | 32 |
| Oxygen[e], cm$^3$ m$^{-2}$ d$^{-1}$ bar$^{-1}$ | 3 000 | 80 | 700 – 1 000 | 30 | | 30 |
| Carbon dioxide[e], cm$^3$ m$^{-2}$ d$^{-1}$ bar$^{-1}$ | 17 000 | 160 | 3 000 – 5 000 | 200 | | 110 |
| Nitrogen[f], cm$^3$ m$^{-2}$ d$^{-1}$ bar$^{-1}$ | 500 | 12 | 150 – 1 000 | 5 | | 25 |

[a] Standard values determined as described in Chapter 5.
[b] Cellulose acetate film (2 1/2-acetate) containing plasticizer.
[c] At 50 Hz, film thickness = 40 μm, plate/plate.
[d] Temperature range over which no significant loss in strength or stiffness occurs.
[e] Film thickness = 40 μm, 23 °C.
[f] Film thickness = 200 μm.
[g] Film thickness = 50 μm.
[h] Decomposition temperature.
[i] For biaxially stretched (balanced) film.
[j] 20 °C, 0 % relative humidity.
[k] Film thickness = 25 μm.
[l] 40 °C, 90 % relative humidity.
[m] Film thickness = 100 μm.

**Table 4.** Properties of selected films[a]

| | Poly(ethylene terephthalate) (PET) | Polycarbonate (PC), monoaxially stretched | Nylon 6 (PA), biaxially stretched | Poly(methyl methacrylate) (PMMA) | Ethylene–vinyl alcohol copolymer (EVOH), biaxially stretched |
|---|---|---|---|---|---|
| *Mechanical properties* | | | | | |
| Thicknesses, µm | 1.0 – 350 | 2 – 500 | 12 – 25 | 50 – 500 | 12 – 25 |
| Density, g/cm$^3$ | 1.4 | 1.21 | 1.16 | 1.14 | 1.19 |
| Ultimate strength | | | | | |
| longitudinal, N/mm$^2$ | 200[i] | 250 | 260 | 40 | 210 |
| lateral, N/mm$^2$ | 200[i] | 8 | 260 | 40 | 210 |
| Elongation at break | | | | | |
| longitudinal, % | 100[i] | 30 | 100 – 120 | 75 | 50 |
| lateral, % | 100[i] | 100 | 100 – 120 | 75 | 50 |
| Modulus of elasticity | | | | | |
| longitudinal, N/mm$^2$ | 4 000[i] | 2 100 | 1 400 | 1 900 | 3 600 |
| lateral, N/mm$^2$ | 4 000[i] | 800 | 1 400 | 1 900 | 3 600 |
| *Electrical properties* | | | | | |
| Dielectric strength[c], kV/mm | 320 | 300 | 50 | 90 | |
| Dielectric constant at 1 kHz | 3.3 | 2.8 | 3.4 – 3.7 | 4.4 | |
| Dissipation factor | 0.005 | 0.0012 | 0.016 | 0.046 | |
| Volume resistivity, Ω cm | $10^{18}$ | $2 \times 10^{17}$ | $5 \times 10^{17}$ | $10^{16}$ | |
| *Thermal properties* | | | | | |
| mp, °C | 260 | 260 | 200[h] | 160 | 181 |
| Application temperature range[d], °C | –269 to +180 | –140 to +140 | –200 to +180 | –20 to +120 | |
| Optical properties $n_D^{20}$ | 1.60 | 1.60 | | | |
| *Dimensional stability* | | | | | |
| Coefficient of thermal expansion, K$^{-1}$ | $2 \times 10^{-5}$ | $7 \times 10^{-5}$ | | | |
| Shrinkage at 100 °C (30 min) | | | | | |
| longitudinal, % | < 1 | < 1 | < 1 | –11 | –4 (140 °C) |
| lateral, % | < 1 | < 1 | < 1 | 0 | –0.5 |
| *Physicochemical properties* | | | | | |
| Water absorption in 24 h, wt % | 0.5 | 0.35 | 5 | 0.5 | 5.9 |
| Stability toward | | | | | |
| daylight | limited | limited | good | very good | |
| organic solvents | moderate | moderate | good | good (aliph.) poor (arom.) | good |
| acid | moderate | moderate | poor | very good | |
| alkali | stable only at low concentration | stable only at low concentration | good | very good | |
| salt solutions | good | good | good | very good | |
| *Permeability* | | | | | |
| Water vapor, g m$^{-2}$ d$^{-1}$ | 8 | 35 | 100 | 75 | 40 |
| Oxygen[e], cm$^3$ m$^{-2}$ d$^{-1}$ bar$^{-1}$ | 50 | 8 000 | 26 | | 0.5[j] |
| Carbon dioxide[e], cm$^3$ m$^{-2}$ d$^{-1}$ bar$^{-1}$ | 250 | 25 000 | 100 | | 12 |
| Nitrogen[f], cm$^3$ m$^{-2}$ d$^{-1}$ bar$^{-1}$ | 12 | 250 | 9 | | 0.08 |

[a] Standard values determined as described in Chapter 5.
[b] Cellulose acetate film (2 1/2-acetate) containing plasticizer.
[c] At 50 Hz, film thickness = 40 µm, plate/plate.
[d] Temperature range over which no significant loss in strength or stiffness occurs.
[e] Film thickness = 40 µm, 23 °C.
[f] Film thickness = 200 µm.
[g] Film thickness = 50 µm.
[h] Decomposition temperature.
[i] For biaxially stretched (balanced) film.
[j] 20 °C, 0 % relative humidity.
[k] Film thickness = 25 µm.
[l] 40 °C, 90 % relative humidity.
[m] Film thickness = 100 µm.

**Table 5.** Properties of selected films[a]

|  | Polysulfone (PSU) | Polyimide (PI) | Polyetherimide (PEI) | Polyethersulfone (PES) | Polyetherketone (PEK) | Poly(phenylenesulfide) (PPS) |
|---|---|---|---|---|---|---|
| *Mechanical properties* | | | | | | |
| Thicknesses, μm | 25 – 1 000 | 7.5 – 125 | 75 – 500 | 25 – 300 | 75 – 250 | 2.0 – 125 |
| Density, g/cm$^3$ | 1.24 | 1.42 | 1.27 | 1.37 | 1.24 – 1.30 | 1.35 |
| Ultimate strength | | | | | | |
| longitudinal, N/mm$^2$ | 80 | 180 | 100 | 85 | 120 | 220 |
| lateral, N/mm$^2$ | 80 | 180 | 100 | 85 | 120 | 220 |
| Elongation at break | | | | | | |
| longitudinal, % | 50 | 70 | 70 | 15 | 200 | 55 |
| lateral, % | 50 | 70 | 70 | 15 | 200 | 55 |
| Modulus of elasticity | | | | | | |
| longitudinal, N/mm$^2$ | 2 500 | 2 700 | 3 000 | 2 400 | 330 | 4 000 |
| lateral, N/mm$^2$ | 2 500 | 2 700 | 3 000 | 2 400 | 330 | 4 000 |
| *Electrical properties* | | | | | | |
| Dielectric strength[c], kV/mm | 200 | 280[k] | 145 | 125 | 170 | 240 |
| Dielectric constant at 1 kHz | 3.14 | 3.5 | 3.56 | 3.90 | 3.5 | 3.0 |
| Dissipation factor | 0.001 | 0.003 | 0.0013 | 0.003 | 0.002 | 0.0006 |
| Volume resistivity, Ω cm | 5×10$^{16}$ | 10$^{18}$ | 3.4×10$^{16}$ | 1.7×10$^{16}$ | 10$^{17}$ | 10$^{17}$ |
| *Thermal properties* | | | | | | |
| mp, °C | 370 | 406[h] | 216[h] | 223[h] | 334 | 285 |
| Application temperature range[d], °C | – 80 to + 170 | – 260 to + 400 | ≤ 170 | ≤ 180 | ≤ 240 | ≤ 180 |
| Optical properties $n_D^{20}$ | 1.63 | 1.78 | | | | |
| *Dimensional stability* | | | | | | |
| Coefficient of thermal expansion, K$^{-1}$ | 5.6×10$^{-5}$ | 2×10$^{-5}$ | 4.5×10$^{-5}$ | 4.4×10$^{-5}$ | 4×10$^{-5}$ | 3×10$^{-5}$ |
| Shrinkage at 100 °C (30 min) | | | | | | |
| longitudinal, % | < 0.1 | < 0.1 | < 0.1 | < 0.1 | < 0.1 | < 0.1 |
| lateral, % | < 0.1 | < 0.1 | < 0.1 | < 0.1 | < 0.1 | < 0.1 |
| *Physicochemical properties* | | | | | | |
| Water absorption in 24 h, wt % | 0.4 | 1.3 – 2.9 | 1.0[l] | 1.4[l] | 0.4[j] | 0.05 |
| Stability toward | | | | | | |
| daylight | good | good | good | | | good |
| organic solvents | moderate | moderate | satisfactory | satisfactory | very good | very good |
| acid | good | good | good | good | very good | very good |
| alkali | good | stable only at low concentration | good | good | very good | very good |
| salt solutions | good | good | good | good | very good | very good |
| *Permeability* | | | | | | |
| Water vapor, g m$^{-2}$ d$^{-1}$ | 98 | 84 | 43.5 | 100 | 28.5 | 5 |
| Oxygen[e], cm$^3$ m$^{-2}$ d$^{-1}$ bar$^{-1}$ | | 3.8 | | 200 | 55 | 100 |
| Carbon dioxide[e], cm$^3$ m$^{-2}$ d$^{-1}$ bar$^{-1}$ | | 6.9 | | | | 500 |
| Nitrogen[f], cm$^3$ m$^{-2}$ d$^{-1}$ bar$^{-1}$ | | 0.9 | | | | 50 |

[a] Standard values determined as described in Chapter 5.
[b] Cellulose acetate film (2 1/2-acetate) containing plasticizer.
[c] At 50 Hz, film thickness = 40 μm, plate/plate.
[d] Temperature range over which no significant loss in strength or stiffness occurs.
[e] Film thickness = 40 μm, 23 °C.
[f] Film thickness = 200 μm.
[g] Film thickness = 50 μm.
[h] Decomposition temperature.
[i] For biaxially stretched (balanced) film.
[j] 20 °C, 0 % relative humidity.
[k] Film thickness = 25 μm.
[l] 40 °C, 90 % relative humidity.
[m] Film thickness = 100 μm.

## 6.2. Hydrocarbon Polymers

### 6.2.1. Low-Density Polyethylene (LDPE) (→ Polyethylene)

Low-density polyethylene [9002-88-4] (LDPE) has a molecular mass of > 50 000 and is prepared by the high-pressure process. It is converted into film mainly by the film blowing process at a melt temperature of 150 – 210 °C. For properties of the film, see Table 2.

Film thickness of LDPE ranges between 25 and 200 µm. The films are transparent to crystal clear, depending on the choice of relative molecular mass and process technology. They are soft, flexible, and resistant to creases and show high tear and tear propagation resistance. Mechanical properties are stable at low temperature and chemical resistance is very high. Films made from LDPE display low permeability to water vapor but high permeability to gases. They are nontoxic and contain neither plasticizers nor odorous materials. Low-density polyethylene films are easily welded and thermoformed. Specific shrinking characteristics can be imparted to the films by stretching at low temperature.

Films made from LDPE are used for packaging and as protective coverings. They are also employed in the production of composite films. Low-density polyethylene has become commercially important. Worldwide annual production is estimated at about $11\times10^6$ t.

**Trade Names.** Petrothene (Lyondell Basell, United States); Cryovac L (Cryovac, United States).

Film derived from a copolymer of ethylene with vinyl acetate (EVM film) is used similarly as a packaging material. Its exceptional adhesive properties make it particularly suitable as a cling and stretch-wrap film.

**Trade Name.** Montothene (American Can, United States).

### 6.2.2. High-Density Polyethylene (HDPE) (→ Polyethylene)

High-density polyethylene [9002-88-4] (HDPE) prepared by the low-pressure method is also gaining importance as a raw material for films. Films made from HDPE are produced both by chill roll extrusion and by film blowing. Characteristics of the films are given in Table 2.

Films derived from HDPE have a higher chemical resistance than LDPE films and can be employed over a somewhat wider temperature range; HDPE films are used in construction and civil engineering as insulators. High-density polyethylene film can be welded and bonded, and is readily thermoformable.

The market for HDPE films in 2005 was estimated to be $9\times10^6$ t.

**Trade Names.** Petrothene (Lyondell Basell United States); Fortiflex (Ineos, United Kingdom).

### 6.2.3. Linear Low-Density Polyethylene (LLDPE) (→ Polyethylene)

The 1980s witnessed the development of films derived from linear low-density polyethylene (LLDPE), which are produced in the same way as LDPE films. Linear low-density polyethylene films have better mechanical properties (greater strength and higher ultimate strength) than ordinary LDPE films. As a result, LLDPE films have found applications as sacks for transporting heavy materials. Since LLDPE also has a higher melting point (118 °C) than LDPE (110 °C), these sacks can be filled with hot materials (e.g., freshly produced cement). Stretch-wrap packaging represents another expanding market. Like other polyethylene films, LLDPE films can be combined with additional materials to form composite films.

### 6.2.4. Polypropylene (PP) (→ Polypropylene)

Polypropylene [9003-07-0] (PP) film is available in an unoriented form as well as in both mono- and biaxially oriented versions.

*Unstretched PP film* is prepared either by chill roll casting or by blowing. Its properties are comparable to those of LDPE film, but with the advantages of greater transparency and a wider application temperature range. Unstretched PP film is used exclusively for packaging purposes. West European consumption in 1986 was 80 000 t.

*Monoaxially stretched PP film* exhibits increased strength in the direction of orientation and is used for three major applications: (1) cable insulation, (2) tapes and industrial fabrics, and (3) yarns and ropes.

*Biaxially oriented polypropylene (BOPP) film* is the most important type of polypropylene film; its properties are listed in Table 2. Worldwide consumption in 2010 is estimated reach $6\times10^6$ t of which $1\times10^6$ t be attributed to can China.

Biaxial orientation can be achieved by any of the standard methods (see Section 2.3.2), but sequential stretching is preferred because this method permits a wide range of thicknesses. Most BOPP packaging films are provided with a separate sealing layer, formed either by coextrusion or in an off-line coating process. Coextruded sealing coatings normally consist of polyolefin copolymers, whereas polyacrylate or poly(vinylidene chloride) is used for off-line coating. Poly(vinylidene chloride) has the added advantage that it simultaneously improves the oxygen barrier properties of the film.

Biaxially oriented polypropylene films range in thickness from 4 to 80 μm. They are used industrially in laminates and as release and capacitor films.

**Trade Names.** Trespaphan (Treophan, Germany); Propafilm (Innovia, United Kingdom); Walothen (Walothen Germany); Bicor (Exxon Mobil, Belgium); Moplefan (Treofan, Germany).

### 6.2.5. Poly(4-methyl-1-pentene) (PMP) [29] (→ Poly(4-Methyl-1-Pentene))

Poly(4-methyl-1-pentene) [9016-80-2] (PMP) films are produced by extrusion and possess the lowest density of all available films (0.83 g/cm$^3$). They have a higher softening point (*mp* 240 °C) than PP films (*mp* 165 °C). Film thickness ranges from 50 to 300 μm. For properties, see Table 2.

PMP films have been used only in combination with other materials, mainly with cardboard for the packaging of oven-ready meals.

### 6.2.6. Ionomers (→ Specialty Plastics)

The term *ionomer* describes a family of polymers incorporating both covalent and ionic bonds. Ionomers [25608-26-8] are based on polyethylene and have carboxyl groups located along their polymer chains. These groups provide ionic cross-links between chains [20]. The corresponding films are prepared by extrusion.

Many of the properties of ionomer films are comparable to those of LDPE films. However, ionomer films are more resistant at room temperature to fats and oils than LDPE films, and they possess exceptionally good sealing properties. They are also extremely tough and are ideally suited for packaging.

**Trade Name.** Surlyn A (Du Pont, United States).

### 6.2.7. Polystyrene (PS) (→ Polystyrene and Styrene Copolymers)

Polystyrene [9003-53-6] (PS) films are available in unoriented as well as in mono- and biaxially stretched form. They are produced by chill roll extrusion or by blowing. For properties, see Table 3. For foamed PS films see → Foamed Plastics.

Oriented films are crystal clear with a high gloss but are extremely brittle. Their most important application is for envelope windows. The polymer must be modified with synthetic rubber to increase its strength. However, the resulting increase in impact strength is obtained at the expense of reduced transparency. Film thicknesses are between 4 and 500 μm.

The favorable dielectric characteristics of PS film result in its use in high-voltage capacitors and as an insulator for cables.

**Trade Names.** Styroflex (Norddeutsche Seekabelwerke, Germany); Trycite (Dow Chemical, United States).

## 6.3. Halogenated Hydrocarbon Polymers

### 6.3.1. Poly(Vinyl Chloride) (PVC) [6] [→ Poly(Vinyl Chloride)]

Films formed from poly(vinyl chloride) [9002-86-2] (PVC) may be either rigid or flexible. Rigid films consist solely of PVC; flexible films are produced by addition of plasticizers or by copolymerization with acrylate esters or vinyl acetate. Both rigid and plasticized PVC films are produced by calendering processes, although chill roll casting and film blowing are also possible. Some PVC films are also subjected to stretching.

Rigid and plasticized PVC films are of great economic importance. Worldwide consumption is estimated at $0.8 \times 10^6$ t/a.

#### 6.3.1.1. Rigid PVC

Rigid PVC film is available in a wide range of thicknesses (30 – 700 µm); important properties are listed in Table 3. The material varies from crystal clear to transparent and may be colored by the addition of pigments. Rigid PVC is unaffected by water and is virtually impermeable to odors. It possesses outstanding chemical resistance and is also highly flame resistant. The film can be welded, heat-sealed, embossed, thermoformed, stamped, bonded, imprinted, and (under high vacuum) metallized. These remarkable properties combine to produce a broad spectrum of applications for rigid PVC film both in packaging and the industrial sector. Adhesive tapes account for a particularly large share of the market.

#### 6.3.1.2. Plasticized PVC

Plasticized PVC films have an even wider range of applications than rigid PVC because the properties of plasticized PVC can be modified by varying the type and amount of plasticizer added. Plasticized PVC films are available with a thickness of 10 – 1000 µm; for important properties, see Table 3. The material can be strained and shows little tendency to wrinkle; its consistency varies from soft to hard. Plasticized PVC film is resistant to both tear and tear propagation, and it can be embossed, welded, bonded, and stitched. It is resistant to most inorganic and organic chemicals, as well as to water and light. It also shows good aging and weathering properties. Permeability to water vapor and gases is minimal.

Plasticized PVC film is used mainly in packaging and the building industry, as a protective covering, in laminated films, and for the modification of wood surfaces. It is also processed into commercial adhesive tapes.

**Trade Names.** Genotherm, (Klockker Pentaplast, Germany); Mipolam (Gerflor, Germany).

### 6.3.2. Poly(Vinylidene Chloride) (PVDC) [27, 30] [→ Poly(Vinylidene Chloride)]

Poly(vinylidene chloride) [9002-85-1] (PVDC) films are made from a vinylidene chloride–vinyl chloride–vinyl acetate copolymer [9003-20-7]. The comonomers render the product sufficiently stable for extrusion. Plasticizers are also added to improve processing properties. The films are manufactured by the tubular drawing process. Since the stability of the melt is low, the film tube must be cooled as it emerges from the die prior to orientation by blowing (see Section 2.3.2). For properties, see Table 3.

Poly(vinylidene chloride) film has been on the market under the trade name Saran since the late 1940s and is still widely used because of its outstanding water vapor and oxygen barrier properties. Its principal application is in the packaging of perishable foodstuffs.

The material is supplied as a shrinkable monofilm or coextruded film; PVDC laminates are also available. Coextruded products are prepared in both tubular and flat form. The PVDC generally is sandwiched between other polymeric layers which impart strength and toughness. Adhesion problems are overcome by simultaneously extruding an adhesive tie layer.

**Trade Names.** Saran, Saranex (Dow Chemical, United States); Cryovac (Cryovac, United States); Krehalon (Kureha, Japan); Supralon (Supralon, Liechtenstein).

### 6.3.3. Fluoropolymers [26] (→ Fluoropolymers, Organic)

Incorporation of fluorine into hydrocarbon chains results in polymers that display antiadhesive characteristics, thermal stability, flame resistance, insulating properties, and, most important of all, extreme stability with respect to weathering and aggressive media. As a result of these properties, which depend on the degree of substitution, these films are used mainly under extreme conditions.

From a chemical standpoint, the fluoropolymers used for film production fall into three categories:

1. perfluorinated polymers,
2. partially fluorinated polymers,
3. polymers containing chlorine as well as fluorine

Polytetrafluoroethylene (PTFE) is a perfluorinated polymer that cannot be processed from a melt by conventional techniques. Polytetrafluoroethylene films are prepared by isostatically sintering the powdered material, followed by "skiving" (paring) of the sintered block. All other fluoropolymers—including copolymers of PTFE with perfluorinated vinyl components—are extrudable and can, therefore, be made into films either by the chill roll process or by film blowing.

The market for fluoropolymer films in the United States is dominated by films made from poly(vinyl fluoride) (PVF), PTFE and polychlorotrifluoroethylene (PCTFE). Extrudable copolymer films made from tetrafluoroethylene–hexafluoropropylene copolymer (FEP), perfluoroalkoxy polymer (PFA), and ethylene–tetrafluoroethylene copolymer (ETFE) have also become somewhat important, or at least have a considerable potential for development. Film made of poly(vinylidene fluoride) (PVDF) is also worth mentioning because of its piezoelectric properties.

### 6.3.3.1. Poly(Vinyl Fluoride) (PVF)

Poly(vinyl fluoride) [24981-14-4] (PVF) film is commercially available in both transparent and pigmented form in thicknesses of 12 – 50 μm. For properties, see Table 3. Its antiadhesive characteristics, together with its resistance to mechanical, chemical, and weathering effects, make PVF film suitable for use as a protective lamination on both interior and exterior paneling. It also serves as a release film in the pressing of synthetic resin.

**Trade Name.** Tedlar (Du Pont, United States).

### 6.3.3.2. Poly(Vinylidene Fluoride) (PVDF) [31]

Poly(vinylidene fluoride) [24937-79-9] (PVDF) films are solvent cast or extruded with a thickness of 50 –500 μm. They possess excellent resistance to weathering and aggressive media.

Poly(vinylidene fluoride) film has an exceptionally high dielectric constant and a relatively high dissipation factor. Because of its piezoelectric properties, the monoaxially stretched film is often employed as a mechanical–electrical transducer in membranes. The consumption of film for piezoelectric applications is small (ca. 1 t); its price is extremely high.

**Trade Name.** Fluorex (Soliant, United States); Kynar (Arkema, France).

### 6.3.3.3. Polytetrafluoroethylene (PTFE)

Polytetrafluoroethylene [9002-84-0] (PTFE) films have the highest resistance to aggressive agents of any polymer film and are thermally stable up to 250 °C. Since PTFE films are produced via sintering and skiving, they are somewhat porous; some properties are listed in Table 3. Most of the PTFE film that is produced is used for sealing pipe joints and flanges.

Two related copolymers are extrudable: FEP [25036-53-7], derived from tetrafluoroethylene and hexafluoropropylene, and PFA [2645-79-6], a copolymer of tetrafluoroethylene and a perfluorovinyl ether. Films produced from these materials are not porous, but otherwise display essentially the same characteristics as PTFE itself. The mechanical and optical properties of the extruded films are superior to those of skived PTFE films. Commercial products are available in thicknesses from 12 –1000 μm. Their principal uses are as release films, in the electrical industry, and especially as container linings.

**Trade Name.** Hostaflon (Dyneon, Germany); Teflon (Du Pont, United States); Fluon (Asahi glass, Japan).

### 6.3.3.4. Polychlorotrifluoroethylene (PCTFE)

Films made from polychlorotrifluoroethylene [9002-83-9] (PCTFE) are used widely in the packaging of pharmaceuticals and medical supplies; this is because the films possess excellent moisture barrier properties and chemical resistance, coupled with good processability, sterilizability, and UV stability. These films are available in thicknesses ranging from 12 to 250 μm.

**Trade Name.** Aclar (Honey well, United States).

## 6.4. Poly(Methyl Methacrylate) (PMMA) [32] (→ Polyacrylates)

Poly(methyl methacrylate) [9011-14-7] (PMMA) films are prepared either by chill roll extrusion or by film blowing; thicknesses range from 50 to 500 μm. They show high resistance to heat and weathering, but their ultimate strength and elongation at break are relatively low. For properties, see Table 4.

Poly(methyl methacrylate) film is frequently used as a protective layer on metal and plastic, thereby competing with PVF film.

**Trade Name.** Korad (Polymer Extruded Products, United States); Plexiglas-Folie (Evonik, Germany); Shinkolite Film (Mitsubishi Rayon, Japan).

### 6.5. Poly(Vinyl Alcohol) (PVA) [33] (→ Polyvinyl Compounds, Others)

Poly(vinyl alcohol) [9002-89-5] (PVA) is produced by hydrolyzing poly(vinyl acetate). Both the fully and partially hydrolyzed polymers are utilized for film manufacture; the fully hydrolyzed polymer is water-soluble.

Poly(vinyl alcohol) film can be prepared by casting an aqueous PVA solution or by extrusion. Prior to extrusion, the poly(vinyl alcohol) is compounded into a paste with water and a plasticizer. Film thicknesses are in the range of 20 – 75 µm.

The characteristic features of PVA film include its high transparency and gloss, toughness, absence of static charge, and, in the absence of moisture, desirable oxygen barrier properties. Water solubility varies, depending on the vinyl acetate content.

Poly(vinyl alcohol) is employed as a monofilm, but can also be coated with PVDC or laminated with polyolefins. Biaxially oriented films can be produced by simultaneous stretching [34].

Poly(vinyl alcohol) films are used as packaging materials for food, textiles, and as release films in the manufacture of synthetic resin products. They are also employed in the manufacture of water-soluble sachets for disposable packaging of hygienic products and as a basis for polarized films.

**Trade Names.** Vinylon (Kuraray, Japan); Emblar-OV (Unitika, Japan); Bovlon (Nippon Gohsei, Japan).

### 6.6. Ethylene–Vinyl Alcohol Copolymer (EVOH)

A partially crystalline copolymer of ethylene and vinyl alcohol, EVOH [25067-34-9] is prepared by hydrolysis of an ethylene–vinyl acetate copolymer. This material first appeared on the market in the 1970s [35]; it can be processed into a film by extrusion.

Films derived from EVOH are excellent gas barriers. Biaxial orientation leads to increased strength and further reduction of gas permeability. Since the gas barrier properties of EVOH films are adversely affected by moisture, they are primarily used in the form of laminates or coextrudates. Some properties of oriented EVOH film are listed in Table 4.

The principal application of these materials is in food packaging.

**Trade Names.** Eval (Kuraray, Japan); Exceed (Okura, Japan).

### 6.7. Polyamide (PA) [20] (→ Polyamides)

Polyamide (PA) films (thickness 12 – 25 µm) are made mainly from nylon 6 [25038-54-4] and nylon 66 [32131-17-2] by chill roll extrusion or film blowing. The films display high tensile and tear strengths, as well as high resistance to puncture, abrasion, oils, and fats. Nylon films have a low permeability to gases, depending on the atmospheric humidity. Nylon 11 [25035-04-5] and nylon 12 [24937-16-4] are used for films with lower water-vapor permeability.

Orientation improves the properties of nylon films. Biaxially oriented films are produced from nylon 6 by simultaneous stretching. Properties of these films are listed in Table 4.

Polyamide films are used as monofilms but are employed mainly in composite structures produced by lamination, extrusion coating, or coextrusion with sealing or barrier resins. The world consumption of polyamide films is estimated at $320 \times 10^3$ t.

**Trade Names.** Capran (Honey well, United States); Dartek (Du Pont, United States); Bonyl (Kohjin, Japan); Emblem (Unitika, Japan).

### 6.8. Polyester [10] (→ Polyesters)

The term *polyester film* generally refers to films prepared from poly(ethylene terephthalate) [25038-59-9] (PET). However, other polyesters are also processed into film, e.g., poly(butylene terephthalate) [24968-12-5] (PBT) and poly (ethylene-2,6-naphthalene dicarboxylate)

[24968-11-4] (PEN) (trade name: Q-Film, Teijin/Japan) [36].

Polyester films are manufactured by extrusion; the majority are biaxially oriented and produced in flat form; only a small amount of tubular film is made.

Biaxially oriented PET films are prepared in thicknesses ranging from 1.0 to 350 μm and vary in appearance from clear to transparent, depending on their end use. Pigments can also be added for coloration.

Polyester films have favorable mechanical and electrical properties, are resistant to solvents, and display long-term thermal stability up to 130 °C (VDE 0530). They can be welded and bonded by special techniques. Furthermore, these films can be printed by flexographic, gravure, and screen printing techniques; they are also suitable for high-vacuum metallization without the need for special pretreatment. Some polyester films can be subjected to further deforming subsequent to their manufacture, for example, by thermoforming or thermally induced shrinking. Properties of PET film are listed in Table 4; its permeability to gases and water vapor can be significantly decreased by coating with PVDC.

Polyester films are used for more applications than any other type of film. They are employed as packaging materials, dielectrics, and electrical insulators, as base films for photographic and magnetic coatings, as release films in synthetic resin moldings, and as color and montage film in the graphics industry. These films also serve as a base for flexible printed circuits, as drafting films, and as decorative film. Polyester films have gained considerable importance in the manufacture of videotapes. The demand for polyester film as a base for thermal transfer tapes is also increasing.

Polyester films can be combined with other materials (PE film, paper webs, and aluminum foil) to produce excellent composites.

The economic importance of polyester film is formidable: The total consumption of PET in 2005 was estimated at $2.2 \times 10^6$ t because of sustained growth in packaging application.

**Trade Names.** Hostaphan (Mitsubishi Polyester film, Germany); Mylar (Du Pont, United States); Melinex (Du Pont Tejin films, United States); Lumirror (Toray, Japan); Tetoron (Du Pont Tejin films, Japan); Diafoil (Diafoil, Japan); Terphane (Terphane, United States); Espet (Toyobo, Japan).

## 6.9. Polycarbonate (PC) [20] (→ Polycarbonates)

Polycarbonate [25037-45-0] (PC) film first appeared on the market in 1957 and is prepared from bisphenol A and phosgene. The principal manufacturers of both polycarbonate and polycarbonate film are General Electric and Bayer.

Polycarbonate films are produced either by extrusion (chill roll) or by solution casting and are available with a thickness of 2 – 500 μm. The thinner grades are subjected to monoaxial stretching. The properties of monoaxially stretched PC film are listed in Table 4.

The special characteristics of these films include their sterilizability, dimensional stability, thermoformability, and long-term thermal stability up to 130 °C (VDE 0530). They are transparent to crystal clear in appearance, and possess good electrical properties. Applications have been found in packaging, the electrical industry, and above all in the field of graphics. Consumption of PC film in the United States is ca. 3000 t/a.

Increased usage is anticipated, partly as a result of new developments in the raw material sector.

**Trade Names.** Lexan (Sabic Innovative Plastics, United States); Makrofol (Bayer, Material Science Germany); Pokalon (Lofo, Switzerland).

## 6.10. Polysulfone (PSU) (→ Specialty Plastics)

The first commercially available polysulfone (PSU) was Udel, which was introduced by Union Carbide in 1965 (see also Section 6.12.2). Udel polysulfone [25135-51-7] is produced by reacting bisphenol A and 4,4′-dichlorodiphenyl sulfone. The material is amorphous, with a glass transition temperature of 190 °C, and can be converted into film either by extrusion or by solution casting.

Polysulfone films are available in thicknesses ranging from 25 to 1000 µm. They are highly transparent, possess high heat and dimensional stabilities, and can be thermoformed. Their electrical properties are comparable to those of biaxially oriented polyester films. Polysulfone films are suitable for long-term use at temperatures up to 150 °C, making them superior in this respect to polyester films. Polysulfone films are also stable to alkali, acid, oils, and detergents, although they are attacked by ketones and chlorinated or aromatic hydrocarbons. The properties of Udel PSU film are listed in Table 5.

Principal applications are in packaging, the electrical industry, adhesive tape manufacture, and display technology [37].

**Trade Names.** Sumilite FS 1200 (Sumitomo Bakelite, Japan).

## 6.11. Polyimide (PI) (→ Polyimides)

The most important polyimide [25036-53-7] (PI) for film production is derived from pyromellitic acid and 4,4′-diaminodiphenyl ether [38]. Films made from this material do not melt, and they have the highest thermal stability of any synthetic film. They are suitable for long-term applications at temperatures up to 240 °C and short exposure to peak temperatures of up to 400 °C. For properties, see Table 5.

Polyimide films are prepared by dissolving the appropriate monomers in a suitable solvent (e.g., $N,N$-dimethylacetamide), in which they react to form a soluble intermediate (polyamic acid). The film is cast from this solution and is subsequently converted into the insoluble polyimide by intramolecular condensation at 300 °C.

The combination of high thermal stability and flame resistance, along with stability to radiation and favorable electrical properties, makes this film attractive for applications in the aerospace industry (e.g., as an electrical insulator), and the manufacture of electronic components [39]. Commercial grades range in thickness from 7.5 to 125 µm.

Worldwide capacity in 1986 was 1400 t/a with an expected annual growth rate of 10 %.

**Trade Names.** Kapton (Du Pont, United States; Toray/Du Pont, Japan); Upilex (Ube Industries, Japan); Novax (Mitsubishi Chemical, Japan).

## 6.12. Other Films

### 6.12.1. Polyetherimide (PEI)

The backbone of polyetherimides (PEI) contains both ether and amide linkages. The PEI used for film production was first marketed by General Electric in 1982 under the trade name Ultem and is an amorphous thermoplastic polymer with the structure:

[61128-24-3]

Poly(etherimide) film is prepared by the chill roll process at an extrusion temperature of 350 °C and is available with a thickness of 75–500 µm.

This film is especially suited for electrical applications due to its favorable mechanical and electrical properties. The maximum temperature for long-term use is 170 °C; thus, thermal stability is substantially less than that of conventional polyimides. Nonetheless, stability is adequate for many purposes and significantly higher than that of polyester film [37]. The film is used mainly for electrical insulation. For properties, see Table 5.

**Trade Names.** Ultem Films (Sabic Innovative films, United States); Sumilite FS-1400 (Sumitomo Bakelite, Japan); Europlex PEI (Evonik, Germany).

### 6.12.2. Polyethersulfone (PES) (→ Specialty Plastics)

The poly(ethersulfone) (PES) used in film manufacture is an amorphous, thermoplastic material manufactured by ICI under the trade name Victrex. It consists of aromatic rings linked alternately by ether and sulfone groups:

[25667-42-9]

Poly(ethersulfone) film is normally manufactured by the chill roll process, although it can also be cast from solution. Thicknesses vary from 25 to 300 µm. Its range of applications is determined by its electrical and optical properties combined with its thermostability and flame resistant behavior [37, 40, 41].

Poly(ethersulfone) films are used in electrical insulation, interior aircraft fittings, and liquid crystal displays.

**Trade Names.** Victrex PES-5200 G (Victrex, United Kingdom); Sumilite FS 1300 (Sumitomo Bakelite, Japan); Europlex-Folie (Evonik, Germany).

### 6.12.3. Poly(ether ketone) (PEK)

Poly(ether ketones) [2738-27-4] can be extruded into either fibers or films in the temperature range 350–420 °C.

Films made from PEK display high thermal stability (up to 240 °C), solvent and chemical resistance, good dimensional stability, flame resistance, and stability to both radiation and hydrolysis. Properties are listed in Table 5. Since PEK film also possesses favorable electrical properties and is a low-density material, it is used in the aerospace industry.

Both crystalline and amorphous unoriented films are commercially available in thicknesses from 75 to 250 µm [37, 41, 42]. Oriented films are still being developed.

**Trade Names.** Sumilite FS 1100 (Sumitomo Bakelite, Japan).

### 6.12.4. Poly(Phenylene Sulfide) (PPS) (→ Specialty Plastics)

Poly(phenylene sulfide) [9016-75-5] (PPS) was marketed as early as 1973 by the Phillips Petroleum Company under the trade name Ryton. However, it was not until 1985 that Phillips, in collaboration with Toray, succeeded in producing the material in film quality.

Poly(phenylene sulfide) is a partially crystalline polymer prepared from sodium sulfide and 1,4-dichlorobenzene. Poly(phenylene sulfide) film is manufactured by extrusion at 300 °C and is similar to PET film in that it can be biaxially oriented and heat-set. Properties of PPS film are given in Table 5. Thicknesses are in the range 2–125 µm.

Poly(phenylene sulfide) film combines favorable electrical characteristics with high thermal stability [43]. Thus, it is mainly used in the electrical industry, although it also has potential as a magnetic storage medium.

**Trade Names.** Torelina (Toray, Japan).

## 6.13. Composite Films [44]

The most important production methods for composite films are adhesive lamination, extrusion coating (see Section 2.4.3), and, more recently, the simultaneous extrusion (coextrusion) of several raw materials (see Section 2.4.2). The latter approach is unique in that it can produce very thin multilayer films (ca. 12 µm) that still display combined properties.

The incorporation of PET, BOPP, PA, and PVC into composite films permits optimization of ultimate strength, elongation at break, and puncture resistance.

Composite films are especially important as packaging materials, particularly for bulk materials regardless of their state, i.e., gases, liquids, and solids.

Composite films that contain polyethylene can be welded by either heat impulse or thermal sealing methods. Thin polyethylene layers (ca. 30 µm) may suffice for packaging lightweight goods, provided that no need exists for evacuation or a controlled atmosphere. Moderately thick layers ($\geq$ ca. 40 µm) are required for vacuum and inert atmosphere packings. Thicker PE film ($\geq$ ca. 60 µm) is employed if the material to be packaged is a liquid or a paste.

The combination of plastic films with aluminum foil produces semirigid composites with exceptionally low permeability to gases, water vapor, and odors. Special heat-sealable, polyolefin-based layers result in composite films that can be separated by peeling. Composite films incorporating ionomers are especially suitable for packaging oily materials.

## 7. Summary of Uses [2, 10, 26, 27]

The most important uses of individual films have been described in Chapter 6 and are summarized in Table 6. They can be divided into three main categories: packaging, electrical insulation, and use as bases for other active materials.

Table 6. Uses of films

| Application | Cellophane | CA | PE | BOPP | PMP | Ionomer | PS | PVC | PVDC | PVF | PTFE | PCTFE | PMMA | PVA | EVOH | PA | PET | PC | PSU | PI | PEI | PES | PEK | PPS |
|---|---|---|---|---|---|---|---|---|---|---|---|---|---|---|---|---|---|---|---|---|---|---|---|---|
| *Packaging* | | | | | | | | | | | | | | | | | | | | | | | | |
| Food (incl. snack and convenience foods) | x | | x | x | x | x | x | x | x | | | | | x | x | x | x | x | x | | | | | |
| Nonfood items | x | x | x | x | | x | x | x | | | | x | | | x | x | x | x | | | | | x | |
| *Electrical Insulation* | | | | | | | | | | | | | | | | | | | | | | | | |
| Capacitors | | | | x | | | x | | | | | | | | | | x | x | x | | x | x | x | x |
| Insulation | | x | | x | x | | x | | | | x | | | | | | x | x | x | x | x | x | x | x |
| Printed circuits | | | | | | | | | | | | | | | | | x | | | x | | | | x |
| *Base Film for Functional Coatings* | | | | | | | | | | | | | | | | | | | | | | | | |
| Adhesive tapes | x | x | | x | | | | x | | | | | | | | | x | x | x | | x | | | |
| Magnetic tapes | | | | | | | | | | | | | | | | | x | | | | | | | |
| Floppy disks | | | x | | | | | | | | | | | | | | x | | | | | | | |
| Photographic film | | x | | | | | | | | | | | | | | | x | | | x | | | | |
| Reprography | | x | | | | | | | | | | | | | | | x | x | | | | | | |
| Decoration | x | | x | | | | | x | | | | | | | | | x | | | | | | | |
| Transfer metallization | | | x | | | | | | | | | | | | | | x | | | | | | | |
| Graphics | x | | | | | | | | | | | | | | | | x | x | | | | | | |
| *Miscellaneous* | | | | | | | | | | | | | | | | | | | | | | | | |
| Office supplies | | | x | x | | | x | x | | | | | | | | | x | | | | | | | |
| Heat mirrors | | | | | | | | | | | | | | | | | x | | | | | | | |
| Protective lamination | | | x | x | x | | | x | | x | x | | x | | | | x | | | | x | x | | |
| Solar collectors | | | | | | | | | | x | | | | | | | x | | | | | | | |
| Release film | x | | x | x | x | | | | | | x | | | x | | | x | | | x | | | | |
| Protective covers | | | x | | | | | x | | | | | | | | | | | | | | | | |
| Display items | | | | | | | | | | | | | x | | | | | x | | | | | | |

*For explanation of abbreviations, see abbreviations list.

***Packaging.*** Packaging is the most important application of films; virtually every type of film finds some application in the packaging field. Plain and printed films are commonly used in packaging a wide range of foods and nonfood articles. Legal requirements for food packaging are described in [25] (see also → Foods, 4. Food Packaging).

Tubular film prepared either directly by blowing or by the subsequent sealing of flat film is used for the manufacture of a wide variety of bags, sacks, and blister or cushion packs. Thicker films are thermoformed into bottles, containers, cups, trays, and bowls. Shrinkable film is employed in palette packaging. Film is also the basis of skin packings, packing tapes, and even tear-off bands. Decorative iridescent packaging films are available which consist of layers of two or more different polymers with different refractive indices. The iridescent color is produced by light interference [45]. Special characteristics such as impermeability to gases and water vapor can be obtained by applying coatings or using laminated structures.

***Electrical Insulation.*** Films with a high dielectric constant and a low dielectric loss factor are particularly useful in a.c. and d.c. capacitors. Such films may be as thin as 1.0 µm. The films themselves are dielectrics, but they can be converted to electrodes either by vacuum metallizing or by combining them with aluminum foil. Polyester and PI films, either alone or in combination with paper or synthetic nonwoven materials serve as insulating material in electrical equipment. Polymer films are also employed in cable sheathing; thin strips of film are often used as electrical insulation or labels; they also create plasticizer barriers, provide thermal and mechanical insulation, and, in combination with metallic film, confer protection against static charging. Polyester (PET), PC, and PI films are used in the production of flexible printed circuits because of their exceptional dimensional and thermal stability. Film switches represent a new area of application, especially for PET film.

***Base Films for Functional Coatings.*** Polyester film in particular has acquired major significance as a base material for magnetic coatings. Special polyester films combining extra strength with defined surface roughness have been developed for this purpose. Magnetic tapes based on these films are marketed internationally for audio cassette players, video recorders, computers, and other commercial and industrial recording equipment.

Films with a high dimensional stability (e.g. PET) are used as substrates for drafting and photocopying. Films are also used in the printing industry as carriers for transferring color to paper or plastics (stamping foil). Film coated with deformable materials are used as embossing films. Iridescent films and films coated with firmly bonded layers of metal and/or lacquers are frequently used for decorative or display purposes. Films are also the basis for most adhesive tapes. The introduction of a wide range of film-based typewriter ribbons are another application.

***Miscellaneous.*** The furniture industry employs films as veneers for the surface treatment of sheet materials. Thicker films (sheets) are thermoformed into ceiling and wall panels, as well as decorative fixtures and reproductions.

The use of films in heat mirrors and solar collectors (→ Solar Technology) is a more recent development. Heat mirrors employ coated films (mostly PET) that transmit visible light and reflect infrared light. In solar collectors, solar energy is absorbed and passed to a flowing heat-transfer medium (e.g., water). Solar collectors may be constructed from UV-stable film.

Film is frequently used as a protective covering for equipment and materials. Many modern greenhouses are constructed almost entirely from transparent film material. The building industry generally makes frequent use of films on account of their waterproofing properties. Small amounts of films are used for rainwear and clothing.

# 8. Environmental Aspects

Most of the scrap rolls and waste generated in film production as a result of edge trimming and manufacturing problems is collected and recycled. Special facilities are required for shredding the film and compressing it into granules suitable for reprocessing.

Film constitutes 1 – 2 % of normal household waste. Separate collection and recycling of waste film is in theory possible, but is expensive and requires considerable organization. All films can be regarded as environmentally harmless, provided that they are disposed of properly. Combustion of film waste presents no particular technological problems. Even PVC film can be safely incinerated, although public controversy suggests otherwise; modern combustion facilities are capable of trapping any hydrogen chloride vapor that is formed.

Films can create litter problems; a number of film additives and other chemical modifications have been proposed for accelerating the rate at which films undergo photooxidative decomposition. However, apart from one exception these suggestions have not been implemented, primarily because the required precision in formulation for a given application is difficult to achieve. In addition they conflict with the present ideas of litter prevention. The sole exception is the *mulch film* employed in agriculture for stimulating seed germination and growth; this product is designed to have a lifetime of only one growing season.

**Abbreviations used in this article:**

| | |
|---|---|
| BOPP | biaxially oriented polypropylene |
| CA | cellulose acetate |
| CAB | cellulose acetate butyrate |
| CAP | cellulose acetate propionate |
| ETFE | ethylene–tetrafluoroethylene copolymer |
| EVM | ethylene–vinyl acetate copolymer |
| EVOH | ethylene–vinyl alcohol copolymer |
| FEP | tetrafluoroethylene–hexafluoropropylene copolymer |
| HDPE | high-density polyethylene |
| LDPE | low-density polyethylene |
| LLDPE | linear low-density polyethylene |
| MD | machine direction |
| PA | polyamide |
| PBT | poly(butylene terephthalate) |
| PC | polycarbonate |
| PCTFE | polychlorotrifluoroethylene |
| PEI | polyetherimide |
| PEK | polyetherketone |
| PEN | poly(ethylene-2,6-naphthalene dicarboxylate) |
| PES | polyethersulfone |
| PET | poly(ethylene terephthalate) |
| PFA | perfluoroalkoxy copolymer |
| PI | polyimide |
| PMMA | poly(methyl methacrylate) |
| PMP | poly(4-methyl-1-pentene) |
| PP | polypropylene |
| PPS | poly(phenylene sulfide) |
| PS | polystyrene |
| PSU | polysulfone |
| PTFE | polytetrafluoroethylene |
| PVA | poly(vinyl alcohol) |
| PVC | poly(vinyl chloride) |
| PVDC | poly(vinylidene chloride) |
| PVDF | poly(vinylidene fluoride) |
| PVF | poly(vinyl fluoride) |
| TD | transverse direction |

## References

1. J. H. Stickelmeyer: History of Plastic Films, in W. R. R. Park (ed.): *Plastics Film Technology*, Van Nostrand Reinhold Company, New York 1969, pp. 3 – 9.
2. O. J. Sweeting: *The Science and Technology of Polymer Films*, vol. I, II, Interscience, New York 1968, 1970.
3. J. Voss: Aus Lösungen hergestellte Hydratcellulose-Kunststoffe in R. Vieweg, E. Becker (eds.): *Kunststoffhandbuch*, vol. III, Hanser Verlag, München 1965, pp. 84 – 165.
4. E. Karmas: *Sausage Casing Technology*, Noyes Data Corp., New Jersey 1974, pp. 146 – 177.
5. L. I. Nass: *Encyclopedia of PVC*, vol. 3, Marcel Dekker, New York – Basel 1977.
6. G. W. Becker, D. Braun (eds.): *Kunststoffhandbuch, Polyvinylchlorid 2/2*, 2nd ed., Hanser Verlag, München – Wien 1985; M. Kreutzer, "Extrudieren von Folien und Platten," pp. 957 – 977; M. Kreutzer, C. F.-J. Holthausen, F. Pilz, "Kalandrieren von Folien," pp. 1001 – 1029.
7. R. A. Elden: Calandering of Film and Sheeting in A. Whelan, J. L. Craft (eds.): *Developments in PVC Production and Processing – 1*, Applied Science Publ., London 1977, pp. 205 – 221.
8. F. Hensen, W. Knappe, H. Potente (eds.): *Handbuch der Kunststoff-Extrusionstechnik*, "Extrusionsanlagen," vol. II, Hanser Verlag, München – Wien 1986; G. Winkler: "Extrusion von Blasfolien," pp. 76 – 106; R. Hessenbruch: "Coextrusion von Blasfolien," pp. 107 – 124; H. Bongaerts: "Extrusion von Flachfolien nach dem Kühlwalzenverfahren," pp. 125 – 186; F. Hensen: "Extrusion von gereckten Folien," pp. 243 – 270; E. Schöllkopf, K. Roesch: "Beschichten und Kaschieren durch Extrusion," pp. 320 – 353.
9. S. Janocha: "Herstellung von Polyethylenterephthalat Folien," in *Winnacker-Küchler*, **7**, 520 – 523.
10. E. Werner, S. Janocha, M. J. Hopper, K. J. Mackenzie: Polyesters, Films in *Encyclopedia of Polymer Science and Engineering*, 2nd ed., vol. 12, J. Wiley & Sons, New York 1988, pp. 193 – 216.
11. R. J. Ashley, D. Briggs, K. S. Ford, R. S. A. Kelley: Adhesion Problems in the Packaging Industry in D. M. Brewis and D. Briggs (eds.): *Industrial Adhesion Problems*, Orbital Press, Oxford 1985, pp. 213 – 218.
12. H. L. Weiss: *Coating and Laminating Machines*, Converting Technology Company, Milwaukee 1983.

13. ICI, GB 1 264 338, 1969 (G. A. E. Pears).
14. S. Levy: *Plastics Extrusion Technology Handbook*, Industrial Press Inc., New York 1981.
15. R. H. Cramm, W. R. Sibbach: *Coextrusion Coating and Film Fabrication*, Tappi Press 1983.
16. R. Gächter, H. Müller: *Plastics Additives Handbook*, Hanser Verlag, Munich–Vienna–New York 1985.
17. J. Siepek, H. Daoust: *Additives for Plastics*, Springer Verlag, New York–Heidelberg–Berlin 1983.
18. B. Carlowitz: *übersicht über die Prüfung von Kunststoffen*, Kunststoff-Verlag, München 1983.
19. R. Hummel: *Atlas der Polymer- und Kunststoffanalyse*, Verlag Chemie, Weinheim, Germany 1978–1983.
20. J. H. Briston, L. L. Katan: *Plastics Films*, 2nd ed. Longman Scientific & Technical, Harlow, United Kingdom, 1983.
21. ASTM Standards, Parts 26, 27, 1967.
22. DIN-Taschenbuch 18, 21, Beuth-Verlag, Berlin 1986.
23. J. V. Schmitz, W. E. Brown: *Testing of Polymers*, vol. **I – IV**, Interscience, New York–London–Sydney 1965 – 1969.
24. Beloit, US 3 425 267, 1969 (J. D. Pfeiffer).
25. Office of Federal Register, *Code of Federal Regulations, 21 CFR*, National Archives and Records Administration, Washington 1986.
26. C. R. Oswin: *Plastic Films and Packaging*, Applied Science Publ., London 1975.
27. C. J. Benning: *Plastic Films for Packaging*, Technomic Publishing Co., New York 1983.
28. Hoechst, EP 004 633, 1979 (D. Gebhardt, G. Crass, S. Janocha).
29. Data Sheet 1982 A, Mitsui Petrochemical Ind., Tokyo.
30. R. A. Wessling: *Polyvinylidene Chloride*, Gordon and Breach Science Publishers, New York–London–Paris 1977, pp. 176 – 180.
31. Data Sheet, Kynar Film, Westlake Plastics Company.
32. Data Sheet, Korad A, Georgia Pacific Corp. Data Sheet, Shinkolite-Film, Mitsubishi Rayon, Tokyo.
33. C. A. Finch: *Polyvinyl Alcohol*, J. Wiley & Sons, London–New York–Sydney–Toronto 1973, pp. 378 –387.
34. Data Sheet, Emblar-OV, Biaxially oriented PVA-Film, Unitika Ltd., Osaka. Data Sheet, Bovlon, PVA, Biaxially Oriented Film, Nippon Gohsei, Osaka.
35. H. Iwasaki, K. Sato, K. Akao, K. Watanabe: "Development and Commercial Production of EVAL-Resin," *CEER Chem. Econ. & Eng. Rev.* **9** (1977) no. 10, 32 – 37. Data Sheet, EVAL EF, Biaxial Oriented EVAL-Film, Kuraray, Osaka.
36. Teijin, DE-AS 2 337 815, 1973 (S. Shimotsuma, T. Asai, M. Hosoi, K. Sagamihara *et al.*).
37. Data Sheet, *Highly Functional Engineering Plastics Films*, Sumilite FS-1000 Series, Sumitomo Bakelite Co., Tokyo.
38. C. E. Sroog, *Macromol. Rev.* **11** (1976) 161 – 208.
39. Data Sheet, E 36 484–2, Kapton, Du Pont, Genf, Switzerland 1983.
40. Data Sheet SR TD 10, Stabar S, Polyethersulfone Film, ICI, Welwyn Garden City, England 1984.
41. Data Sheet, *High Performance Film Talpa-1000/2000*, Mitsui Toatsu Chemicals, Tokyo.
42. Data Sheets SR TD 11, SR TD 12, *Stabar K/XK, Polyetherketone Films*, ICI, Welwyn Garden City, England 1985. A. S. Wood, "Unmatched Performance, a Build-up in Ketone Resins," *Modern Plastics*, April 1987, pp. 46 – 48.
43. Toray, DE 2 916 841, 1979 (T. Asakura, Y. Noguchi, O. Hiroaki). "PPS Film Torelina," Technical Information, Toray Industries, Tokyo 1987.
44. Data Sheets Hoechst, Werk Kalle, Wiesbaden 1980 – 1987.
45. "Mearl Iridescent Films," Technical Information Sheet TIB-G, The Mearl Corporation, 1050 Lower South Street, Peekskill, New York 10566, April 1980.

# Polyamide Fibers

LELAND L. ESTES, E. I. Du Pont de Nemours & Co., Nashville, Tennessee, USA

MICHAEL SCHWEIZER, ITCF Denkendorf, Denkendorf, Germany

| | | | | |
|---|---|---|---|---|
| 1. | Properties | 1435 | 4. | Modifying Additions ........... 1443 |
| 2. | Polyamides for Synthetic Fibers ... 1436 | | 5. | Technology of Melt Spinning ..... 1445 |
| 2.1. | Aminocarboxylic Acid Type Polyamides ................. 1437 | | 5.1. | Melt Spinning of Nylon ......... 1445 |
| | | | 5.2. | Special Techniques and Products .. 1447 |
| 2.2. | Diamine–Dicarboxylic Acid Type Polyamides ................. 1440 | | 6. | Uses ....................... 1448 |
| | | | 7. | Economic Aspects ............. 1448 |
| 2.3. | Copolyamides ............... 1442 | | | References................... 1450 |
| 3. | Special Requirements for Polyamides ................. 1442 | | | |

This article deals predominantly with linear polyamide fibers derived from aliphatic monomers, all of the generic class called *nylon*. Linear, aliphatic polyamides are conveniently divided into two groups: those made from aminocarboxylic acids and those made from diamine–dicarboxylic acids. These are characterized by the number of carbon atoms in the monomeric starting material. For example, nylon 11 is the polycondensate of 11-aminoundecanoic acid [$H_2N(CH_2)_{10}COOH$]. Nylon 610 is the polycondensate of 1,6-diaminohexane [$H_2N(CH_2)_6NH_2$] and sebacic acid [$HOOC(CH_2)_8COOH$]. Fibers of aromatic polyamides in which the amide linkages attach directly to two aromatic rings are called aramids, and are covered under → High-Performance Fibers. The history of the beginning and growth of nylon has been reviewed [1].

## 1. Properties

Nylon fibers are superior to all natural and regenerated fibers with respect to high tenacity and elongation, abrasion resistance, and insensitivity to rotting and moth attack. By heat treatment, fabrics of nylon fiber can be set into stable shapes through formation of additional hydrogen bonds. The highly polar nature of carbonamide groups also results in swelling of nylon fibers in water or in polar solvents, and permits simple dyeing processes with disperse, acid, and metallized dyes. Nylon fibers absorb moisture and swell longitudinally. This limits "outdoor" use because of poor dimensional stability.

Other general properties of nylon are as follows:

- Shrinkage: dependent on conditions; 1–15% when setting in water or hot air
- Abrasion resistance: outstanding; especially suitable for textile floor coverings
- Electrostatic charge: high at low relative humidity; reduced significantly in antistatic fibers (addition of carbon black or polyglycol ether compounds)
- Solubility: at room temperature in concentrated inorganic acids, *m*-cresol, *o*-chlorophenol, and phenol–alcohol mixtures; at the boil in benzyl alcohol and phenylethyl alcohol
- Dyeability: with acid, metal-complex, disperse, and reactive dyes; special types with basic dyes
- Stability:
  - *Light stability:* highly dependent on pigmentation and presence or absence of stabilizers

**Table 1.** Physical properties of some Nylon fibers

| Property | Nylon 6 | Nylon 66 | Nylon 11 | PACM-12[*] |
|---|---|---|---|---|
| mp, °C | 214–220 | 255–260 | 190 | 275 |
| $\varrho$, g/cm$^3$ | 1.14 | 1.14 | 1.04 | 1.03 |
| Water absorption,% | | | | |
|   65% R.H., 21 °C | 3.5–4.5 | 3.5–4.5 | 1.3 | 1.5–2 |
|   Water retention | 9–15 | 9–15 | 2.9 | ca. 3 |
| Tenacity, cN/dtex | 3–9 | 3–9 | 4–7 | 2.5–3 |
| Elongation at break,% | 80–15 | 80–15 | 40–15 | 30–25 |
| Wet–dry strength,% | 80–90 | 80–90 | 100 | NA[**] |
| Loop–straight strength,% | 70–90 | 70–90 | NA[**] | NA[**] |
| Elastic modulus, cN/dtex | | | | |
|   Textile fibers | 5–30 | 5–30 | 45 | 20–30 |
|   Industrial fibers | 60–90 | 60–90 | NA[**] | NA[**] |
| Setting temperature, °C | 185–195 | 200–230 | NA[**] | 190 |

[*] Nylon from bis(4-aminocyclohexyl)methane and dodecanedicarboxylic acid.
[**] Not applicable.

- *Thermal oxidation:* yellowing and molecular degradation reduced by stabilizers (Cu$^+$ or antioxidants)
- *Weathering:* relatively poor outdoor life, especially with fine deniers
- *Dilute acid:* adequate resistance
- *Dilute alkali:* good resistance
- *Biological stability:* good to very good resistance against decay-producing bacteria, mold fungi, and moth larvae
- *Shape stability:* good crease resistance with dry materials. On absorbing water, nylon fibers swell in the direction of the fiber axis. The lengthening must be considered in the manufacture of clothing, especially with nylon 6 and nylon 66. With felts for floor coverings, dimensional stability is improved by blending with nonswelling fibers such as polypropylene.

Physical properties of four nylon fibers are given in Table 1.

## 2. Polyamides for Synthetic Fibers

The high molecular mass, linear polyamides that are useful for synthetic fibers contain mainly methylene groups between the carbonamide groups. They comprise those compositions that can be melted and spun into filaments without decomposing.

The introduction of cyclic segments into an aliphatic polyamide chain generally stiffens the chain and raises the melting point (Table 2). The effect is most pronounced when the ring structure is aromatic and symmetrical, and when the amide group is connected directly to an aromatic ring. Connection of two aromatic rings to the amide group defines the aramid class of polyamides, which usually degrade without melting and cannot be melt-spun. Melt temperature is also strongly affected by ring symmetry, as seen in the 90 °C difference in melting points of the *meta* and *para* isomers of the two xylene diamines.

Introduction of C-alkyl side chains in nylons disrupts molecular order and leads to both a reduction in melt temperature and increased

**Table 2.** Comparison of melting points for various polyamides

| Diamine | Diacid | CAS registry no.[*] | mp, °C |
|---|---|---|---|
| *m*-Xylylene-$\alpha,\alpha'$-diamine | adipic | [25805-74-7] | 243 |
| 1,6-Diaminohexane | adipic | [9011-55-6] | 265 |
| *p*-Xylylene-$\alpha,\alpha'$-diamine | adipic | [24938-72-5] | 333 |
| 1,6-Diaminohexane | terephthalic | [24938-03-2] | 370 |
| *p*-Xylylene-$\alpha,\alpha'$-diamine | terephthalic | [24938-64-5] | > 400 |

[*] For the polyamide from the diamine and diacid.

solubility in organic solvents. The influence of N-alkyl substituents is even greater, because they eliminate the possibility of hydrogen bonding between polyamide molecules. The polyamide of piperazine and adipic acid, which melts at 185 °C, is of scientific interest because the effect of the missing hydrogen bonds is largely compensated for by chain stiffening through the piperazine radical.

The effect of C-methyl and N-methyl groups on the melting temperature of nylons made from adipic acid and a diamine is as follows:

| | |
|---|---|
| $H_2N-(CH_2)_4-NH_2$ | 251 °C |
| $H_2N-(CH_2)_2-CH(CH_3)-(CH_2)_3-NH_2$ | 180 °C |
| $HN(CH_3)-(CH_2)_6-NH_2$ | 145 °C |
| $HN(CH_3)-(CH_2)_6-NH(CH_3)$ | −75 °C |

Substitution of oxygen or sulfur for a methylene group in aliphatic polyamides also results in a lower melting point. In the low molecular mass region, nylon melt temperatures increase significantly with increasing molecular mass. This effect, however, can be neglected at the high molecular masses necessary for fiber production. The strength of nylon fibers also increases with increased molecular mass, but melt spinning has technological limits because of the difficulty of handling very high melt viscosities.

The melting temperature of nylon decreases with increased ratio of methylene groups to carbonamide groups, and approaches the melting temperature of polyethylene (mp 130–150 °C, depending on molecular mass). Thus, nylon 22 melts at 145 °C. However, melting temperatures of the two types of nylon do not decrease uniformly, because nylons in the homologous series with an even number of methylene groups between carbonamide groups melt at higher temperatures than the adjacent nylons with an odd number of methylene groups (Tables 3 and 5). This is because the geometrical symmetry of the even-numbered configuration permits tighter chain packing and better hydrogen bonding. Water absorption and nylon density also decrease with increased number of methylene groups, i.e., as the molecule becomes more alkane-like (Table 3).

## 2.1. Aminocarboxylic Acid Type Polyamides

Nylons of the aminocarboxylic type can be prepared by the following methods:

1. Polycondensation of $\omega$-aminocarboxylic acids having more than five carbon atoms
2. Hydrolytic polymerization of lactams having more than six ring members
3. Anionic polymerization of lactams with exclusion of water

Tables 1 and 3 list the most important *physical properties* of this type. Fibers are characterized by a high tenacity of 6–7 g/dtex and a break elongation of 20–30%. An exception is nylon 4, with a tenacity of ca. 4 g/dtex.

***Nylon 1*** *[32010-01-8]*. Nylon 1 falls into this class by virtue of the linear connection of carbonamide groups; it is a polyamide based on its chemical structure [2]:

−CO−NH−CO−NH−CO−NH−.

However, derivatives prepared from monoisocyanates have no polyamide properties.

***Nylon 2*** *[25734-27-4]*. The polyamide from aminoacetic acid is really the first member of the series defined by the general formula $H[HN-(CH_2)_n-CO]_xOH$ ($n=1$). So far,

**Table 3.** Physical properties of some aminocarboxylic acid type polyamides

| Polyamide | mp, °C | Water absorption, % | | Wet strength as a | $\varrho$, g/cm³ |
|---|---|---|---|---|---|
| | | 65% R. H. | 100% R. H. | % of dry strength | |
| 4 | 265 | 9.1 | 28.0 | 70–80 | |
| 6 | 223 | 4.3 | 9.5 | 85–90 | 1.14 |
| 7 | 233 | 2.8 | 5.0 | | 1.10 |
| 8 | 200 | 1.7 | 4.0 | 93–97 | 1.08 |
| 9 | 209 | 1.5 | 3.3 | | 1.06 |
| 10 | 188 | 1.4 | 3.0 | | 1.04 |
| 11 | 190 | 1.3 | 2.8 | | 1.04 |
| 12 | 179 | 1.3 | 2.7 | 100 | |

preparing nylon 2 with sufficiently high molecular mass for the production of fibers has not been possible. A high order of crystallization would be expected, and spinning would only be possible by a solvent spinning process. Filaments of nylon 2 could resemble natural silk very closely in physical properties.

*High molecular mass nylon 2 derivatives* alkylated on the α-carbon atom may be also viewed as polypeptides of α-aminocarboxylic acids and can be prepared by ionic polymerization of N-carboxylic acid anhydrides:

$$H_2N-CHR-COOH \xrightarrow{COCl_2} \text{(N-carboxyanhydride)} \xrightarrow{-CO_2} \text{\textthreesuperior}(HN-CHR-CO)_x$$

Polymerization can be initiated with a variety of catalysts such as alkali hydroxides and carbonates, phenols, and amines. Spinnable polymers have been obtained only from pure optical isomers, i.e., L- or D-aminocarboxylic acid derivatives. The Japanese have produced semiworks quantities of polypeptide fibers based on L-glutamic ω-ester, which are reported to have unusual properties.

**Nylon 3** *[25513-34-2]*. Nylon 3 would be expected to have good fiber properties, but no economically attractive route to its preparation has been found. It melts at 330 °C and should be spinnable to fibers from superheated aqueous solutions [3].

*C-Alkyl derivatives of nylon 3* are of interest for fibers because of new syntheses of corresponding β-lactams [4].

Chlorocyanogen and sulfur trioxide react to form the easily accessible N-carbonylsulfamidic acid chloride. Under mild conditions, this reacts with olefins to form the β-lactam-N-sulfonyl chloride, which can be converted to the β-lactam through hydrolysis in a weakly acidic medium:

$$\text{C=C} + \text{N=C=O} \longrightarrow \text{(β-lactam-N-SO}_2\text{Cl)} \xrightarrow{H^+} \text{(β-lactam)}$$

Because both the β-lactams and the corresponding polyamides generally decompose above 250 °C, these nylons are prepared preferably through anionic polymerization in solution, at temperatures of −15 to 30 °C. For copolymerization, the β-lactams must have reasonably similar polymerization rates. Polymerizations are anhydrous with alkaline catalysts, e. g., alkali lactams or N-alkyl lactams. In contrast to melt-spun polyamides in which molecular masses of 15 000 to 25 000 are adequate for good fiber properties, molecular masses of > 200 000 are required for β-lactams, where fibers are made by solution spinning. Fibers with good textile properties and remarkably good oxidative stability can be prepared from (β,β-dimethyl-β-propriolactam), spun from a ca. 15% solution. Typical solvents are trifluoroethanol, phenol, and dimethyl sulfoxide.

**Nylon 4** *[24938-56-5]*. The polycondensate of 4-aminobutyric acid is prepared by anionic polymerization of γ-butyrolactam in the presence of alkaline catalysts at < 60 °C.

The reaction is carried out in an inert hydrocarbon solvent, and conversion is ca. 80–85%. Precipitated polymer must be washed thoroughly to remove residual monomer, traces of which accelerate decomposition of the polymer at the melting point. Nylon 4 can be melt-spun at 279–280 °C, but residence time in the melt must be short to avoid decomposition or reversion to γ-butyrolactam. Nylon 4 can also be spun from a 45% formic acid solution, but lower tenacity is obtained. Because of its relatively high water absorption (9% at 65% R.H.), nylon 4 can be dyed easily, but it has relatively poor crease resistance. The fibers have not been significant economically, in part because of their poor temperature sensitivity in spinning [5].

**Nylon 6** *[25038-54-4]*. While not ideal for optimum fiber properties, nylon 6 represents about half of the commercial production of nylon fibers and occupies a transitional position in this series. Whereas 4-aminobutyric and 5-aminovaleric acid proceed smoothly to the corresponding lactams on heating above melt temperature, 6-aminocaproic acid forms a temperature-dependent equilibrium condensate. At 250 °C, this consists of ca. 89% linear polyamide, 8.5% caprolactam, and 2.5% larger ring amides. This equilibrium condensate is identical with the equilibrium polymer that results from hydrolytic polymerization of ε-caprolactam.

Problem-free continuous polymerization of caprolactam (→ Caprolactam) can be carried out in perpendicular flow-through tubes, with small additions of water, at 240–270 °C and atmospheric pressure. Batch polymerization in pressure autoclaves can also be carried out, which is a preferred technique for small production units or for polymer with special additives. Anhydrous caprolactam can also be polymerized with alkaline catalysts or cocatalysts, but this "fast polymerization" is not used in the manufacture of fibers. Excess caprolactam and higher ring amides must be removed from the equilibrium mixture before further processing to fibers. The equilibrium melt is extruded, solidified, cut to granules or flake, washed repeatedly with boiling water, and then vacuum dried. This can be done batchwise or continuously. Careful exclusion of oxygen is essential in all processing steps. Nylon 6 can also be polymerized and spun directly by passing the equilibrium polymer through a continuous, thin-film evaporator to remove impurities and then pumping it directly to the spinning machine [5, pp. 121–134].

Because of its relatively low melting point and high water absorption, and the lack of a technically simple way to convert monomer to a continuously spinnable melt, nylon 6 is not the best aminocarboxylic acid type for producing synthetic fibers. The reason for its great importance and large production volume is economic. The high-purity caprolactam required for fiber production is made from favorably priced raw materials (benzene, cyclohexane, or phenol).

***Nylon 7 [25035-01-2].*** Nylon 7 is generally considered to be the optimum aminocarboxylic acid type polyamide for fibers. In both the condensation of 7-aminoheptanoic acid and the hydrolytic polymerization of enantholactam, yields of linear polymer exceed 99%. This means that nylon 7 can be produced continuously in a technically simple polymerization-spinning process. Textile properties of nylon 7 fibers are superior to those of nylon 6 and, to a degree, those of nylon 66. Of particular interest, compared to nylon 6, are the 10 °C higher melting temperature and reduced water absorption. These two properties lead to important advantages, i.e., a considerably higher wet modulus and improved crease recovery (Table 4).

To date, no economically competitive route for production of the monomer has been reported. Small quantities of nylon 7 are produced in Eastern Europe. For this, 7-aminoheptanoic acid is reportedly prepared by telomerization of carbon tetrachloride and ethylene, via the intermediate 7-chloroheptenoic acid.

***Nylon 8 [25035-02-3].*** Nylon 8, which melts at 200 °C, is prepared by hydrolytic polymerization of caprolactam. The equilibrium polymerizate contains only traces of ring amide, so that fibers can be produced by using a continuous polymerization-spinning process. Because of the reduced number of carbonamide groups, water absorption is low, which provides relatively good wet strength and wet elasticity. However, raw materials for the synthesis of caprolactam are too expensive for this polymer to be attractive economically.

Table 4. Physical and textile properties of high-strength filaments from nylons 6, 7, and 66 [6]

| Property | Nylon 6 | Nylon 7 | Nylon 66 |
|---|---|---|---|
| Tenacity, cN/dtex | 8.0–8.5 | 8.1–8.5 | 8.2–8.7 |
| Wet strength,% of dry | 81 | 92 | 85 |
| Elongation at break,% | 24 | 14–16 | 18 |
| Elastic modulus, cN/dtex[*] | 41 | 46–49 | 54 |
| Wet modulus,% of dry[**] | 27 | 62 | 31 |
| Boil-off shrinkage,% | 12 | 8–10 | 10 |
| $\varrho$, g/cm$^3$ | 1.14 | 1.10 | 1.14 |
| mp, °C | 214–220 | 225–233 | 255–260 |
| Water absorption,% | | | |
| 65% R.H. | 4.3 | 2.8 | 4–4.5 |
| 100% R.H. | 9–10 | 5 | 9–10 |

[*] At 1% elongation.
[**] In 25 °C water.

***Nylon 11 [25035-04-5].*** The synthesis of nylon 11 is based on a five-step process from castor oil to 11-aminoundecanoic acid. A 30–40% aqueous dispersion of the acid can be polycondensed in three steps, and the nylon 11 formed can be isolated and melt-spun. Addition of phosphoric or phosphorous acid facilitates spinning by raising the melt viscosity and limiting side-chain growth. As expected from the longer hydrocarbon chain, nylon 11 shows hardly any sensitivity or dimensional change in the presence of moisture. The fibers are too expensive for general use and are restricted to a few special applications in luggage, ropes, transmission belts, and bristles.

***Nylon 12 [24937-16-4].*** Nylon 12 is prepared by hydrolytic polymerization of lauryl lactam, made by a four-step process from butadiene. The reaction to give polymer is essentially complete, so spun products can be prepared by using a continuous polymerization-spinning process. Because nylon 12 has good thermal stability, polymerization temperatures can be raised to 310–340 °C [7]. Except for a 10 °C lower melting point, fiber properties of nylon 12 closely correspond to those of nylon 11. These fibers are not made in large volume because of cost, and their use is restricted mainly to industrial specialty areas.

## 2.2. Diamine–Dicarboxylic Acid Type Polyamides

Nylons of this group are prepared for fiber use primarily by thermal polycondensation of equimolar amounts of diamines and dicarboxylic acids or their salts. Monomers that can undergo intramolecular condensation, like glutaric and succinic acids, are not suitable for thermal polycondensation. In commercial practice, aqueous solutions or suspensions of the salt are processed in batchwise or continuous operation. Polycondensation is induced initially by treatment of the concentrated salt under pressure to avoid loss of diamine, and then under vacuum to increase molecular mass to the degree required for fiber production.

This type of nylon can also be prepared from diamines and dicarboxylic acid dichlorides through interfacial polymerization. Some nylons, with melting temperatures in a range suitable for melt spinning, 180–280 °C (Table 5), are obtained by combination of aliphatic diamines and dicarboxylic acids. However, only nylon 66 has any economic significance.

***Nylon 46 [24936-71-8].*** The fiber from 1,4-diaminobutane and adipic acid is claimed to have better high-temperature properties than nylon 66 [8]. Good strength retention and low shrinkage suggest industrial applications. The advantages of nylon 46 fibers concern the heat capacity, dimensional stability, hydrolytic resistance, processing, fabric softness and dyeability [9]. Further promising end-use opportunities are seen to open up including protective clothing, V-belts, conveyor belts, cord fabrics, and sewing threads.

***Nylon 66 [32131-17-2].*** Because raw materials are favorably priced for producing both adipic acid (cyclohexane) and 1,6-diaminohexane (acrylonitrile, butadiene, and adipic acid), nylon

**Table 5.** Melting temperature of nylons

| Diamine | Dicarboxylic acid | Nylon | mp, °C |
|---|---|---|---|
| 1,4-Diaminobutane | adipic | 46 | 278/295 |
| | pimelic | 47 | 233 |
| | suberic | 48 | 250 |
| | azelaic | 49 | 233 |
| | sebacic | 410 | 239 |
| 1,5-Diaminopentane | glutaric | 55 | 198 |
| | adipic | 56 | 223 |
| | pimelic | 57 | 183 |
| | suberic | 58 | 202 |
| | azelaic | 59 | 179 |
| | sebacic | 510 | 185–195 |
| 1,6-Diaminohexane | adipic | 66 | 250–265 |
| | pimelic | 67 | 202–228 |
| | suberic | 68 | 220–232 |
| | azelaic | 69 | 185–226 |
| | sebacic | 610 | 209–221 |
| 1,7-Diaminoheptane | adipic | 76 | 226–250 |
| | pimelic | 77 | 196–214 |
| | sebacic | 710 | 187–208 |
| 1,8-Diaminooctane | adipic | 86 | 235–250 |
| | suberic | 88 | 205–225 |
| | sebacic | 810 | 197–210 |
| 1,9-Diaminononane | adipic | 96 | 205 |
| | azelaic | 99 | 165 |
| | sebacic | 910 | 179 |
| 1,10-Diaminodecane | adipic | 106 | 230–236 |
| | suberic | 108 | 208–217 |
| | sebacic | 1010 | 194–203 |

66 has outstanding economic importance in the production of synthetic fibers (→ Adipic Acid; → Cyclohexane).

For the production of fiber-grade polymer, a 50–60% aqueous nylon salt solution is heated in a pressure autoclave, with careful exclusion of oxygen. The pressure is released slowly, and the resultant low molecular mass polymer is heated under vacuum in the same vessel to increase molecular mass and melt viscosity to the level required for fiber spinning. Polycondensation can also be carried out continuously [10] and the resultant melt pumped directly to the spinning machine.

***Nylon 610 [9008-66-6].*** Nylon 610 is prepared under conditions similar to those for nylon 66. Because of low water absorption and high elasticity, nylon 610 is especially suitable for production of bristles. The cost of sebacic acid, produced from castor oil, is fairly high and limits the use of this polymer.

***Nylons Containing Aliphatic or Aromatic Rings.*** There is currently no significant commercial production of linear polyamides containing aliphatic ring structures. Many have been investigated for use in tires for improved flat spotting, for better carpet aesthetics, and for improved wrinkle resistance and tactile aesthetics in nylon apparel fabrics.

One ring-containing nylon fiber, T-472 [26403-85-0], was marketed commercially in the early 1970s by Du Pont. It was based on bis (4-aminocyclohexyl)methane and dodecanoic acid.

The diamine is obtained by catalytic hydrogenation of 4,4′-diaminodiphenylmethane (made from aniline and formaldehyde), resulting in a mixture of *cis* and *trans* isomers. The mixture must be fractionally crystallized to yield the preferred 70% *trans,trans* product [11]. This isomeric mixture is polycondensed with dodecanoic acid to yield a melt-spinnable polymer ($mp$ 275 °C) and fibers with high tenacity and resilience, and low water sensitivity.

The luxury, silklike fabric made from T-472 fiber was marketed under the registered trade name Qiana. The product was withdrawn from the market in the early 1980s for economic reasons.

Nylons containing aromatic rings are called aramids. Aramids are produced in principle by the polycondensation process [12]. Melt condensation as in the case of the aliphatic polyamides is not possible here because the melting point lies above the decomposition temperature. Aramids can be produced [13] by reaction of diamines with acid dichlorides by means of solution polycondensation, interfacial polycondensation, or emulsion polycondensation. Solution polycondensation has proved successful for the production of aramids. Strong polar organic liquids such as dimethylacetamide, dimethyl sulfoxide, *N*-methylpyrrolidone, etc. are used as solvents. To increase the solubility of the polyamide being produced, lithium chloride, calcium chloride, or other inorganic salts can be added. The reaction is strongly exothermic so the reactive solution has to be cooled and stirred intensively. Reaction times are between 2 and 20 h. After the condensation reaction the byproducts (hydrochloric acid) must be neutralized. This is done, for example, using calcium hydroxide or calcium carbonate. The neutral solution is degassed and as a rule can be used directly in the spinning process.

The base polymers of the aramids which are currently the most important are poly-*m*-phenyleneisophthalamide (MPIA) used to produce the meta-type fibers, poly-*p*-phenylene-terephthalamide (PPTA) used to produce the para-type fibers and poly-*p*-phenylene/3,4-diphenyleneterephthalamide used to produce the copolymer fibers of the para-type [14].

MPIA is produced by polycondensation of *m*-phenylenediamine and the dichloride of isophthalic acid (isophthaloyl dichloride) (Du Pont method). The freshly distilled *m*-phenylenediamine is dissolved in dimethylacetamide (DMAC) and a small quantity of aniline is added as a chain-breaking reagent. After cooling, the solution is placed in a reaction vessel where the polycondensation reaction takes place at a temperature of 70°C. Quicklime is used to neutralize the hydrochloric acid.

PPTA is formed from the components *p*-phenylenediamine and terephthaloyl dichloride (methods used by Du Pont and Akzo/Enka AG). Here too solution polymerization is used with the aid of a mixture of organic solvents such as *N*-methyl pyrrolidone and hexamethyl phosphoramide. For the process to be successful it is important that pure diamine (> 99.5%) and a diacyl chloride (> 92.5%) are used and that the solvent contains only a small amount of water (< 0.05%). Since the reaction product is a solid, the polymer synthesized must be isolated from the solvent and dried. After that the polymer is dissolved again in concentrated sulfuric acid at a concentration which allows the spinning of liquid crystals as spinning solution [15].

For copolymerization, as in the production of PPTA, terephthalic acid chloride, *p*-phenylenediamine and additionally 3,4-diaminodiphenyl ether (DAPE), which is an aromatic diamine and ether compound, are used. Compared with PPTA this polymer is soluble in organic solvents. Polymerization and spinning are linked directly together and no separation of the polymer is necessary [15]. Because of the low concentration of the spinning solution, hot drawing is necessary here to obtain good fiber properties.

## 2.3. Copolyamides

Nylon copolymers are formed easily by polycondensation of monomer mixtures. Compared to the corresponding pure polyamides, they have lower melt temperatures and higher solubilities. In the case of copolyamides of 6-aminocaproic acid and 1,5-hexanediammonium adipate, the melt temperature minimum is ca. 160 °C at 60 wt% of the amino acid. The melting temperature of the copolyamides can be reduced still further through ternary monomer mixtures. Low-melting copolyamides are suitable, for example, for thermal bonding of needle felts and nonwoven fabrics. Copolyamides of $\epsilon$-caprolactam and 11-aminoundecanoic acid are especially suitable for fishing lines because of good flexibility and increased transparency.

New low-emission copolyamides were developed for the automotive industry [16].

## 3. Special Requirements for Polyamides

The *purity* of all nylon starting materials is especially important because reactive impurities result in uncontrolled chain growth and can reduce the quality of the spun product, e.g., by increased yellowing tendency. Mechanical impurities must be removed carefully by filtration before polymerization. The polymer melt is also filtered to remove mechanical impurities and pigment agglomerates, by passing it through a bed of sand or powdered metal and fine screens, just before it is extruded in filament form. This is especially important in the production of fine-denier filaments in which coarse impurities can reduce quality and yield through filament breaks.

Because nylons are particularly sensitive to oxidation at elevated temperature, *oxygen* must be scrupulously excluded in both the polymerization and the spinning steps. The oxygen content of the inert gases used, primarily carbon dioxide and nitrogen, should be maintained below 0.001%. In both the hydrolytic polymerization of lactams and the polycondensation of aminocarboxylic acids, an equilibrium exists whose position is influenced substantially by the water content of the polyamide melt:

$$H[HN-(CH_2)_n-CO]_a OH + H[HN-(CH_2)_n-CO]_b OH \rightleftarrows$$
$$H[HN-(CH_2)_n-CO]_{a+b} OH + H_2O$$

*Water removal* leads to an increase in average molecular mass; water addition reduces average molecular mass. The same is true for nylons of the diamine–dicarboxylic acid type. *Viscosity stabilizers* are added to the monomeric starting materials to control the degree of polymerization and prevent an undue increase in viscosity in remelt. Both monofunctional and bifunctional compounds are used to limit chain growth. Especially suitable for this purpose are primary and secondary aliphatic amines or diamines, aliphatic and aromatic carboxylic acids, and dicarboxylic acids. Examples are cyclohexylamine, 1,6-diaminohexane, acetic acid, benzoic acid, and adipic acid. Trifunctional compounds are undesirable because they lead to molecular branching.

Good *analytical control* of polymer quality is essential to good spinning continuity and high fiber quality for all synthetic polymers. A narrow molecular mass distribution is desirable. Broad distributions can result from improper design or improper operation of a continuous polymerizer for nylon, and this frequently leads to inferior fiber properties.

*Molecular mass distribution* can be determined by using the ultracentrifuge [17], by turbidity titration, by fractional precipitation or partition between two liquids, or by chromatography. From the molecular mass distribution, molecular mass can be calculated as the number average ($M_n$) or the mass average ($M_m$). The ratio $M_m/M_n$ is always greater than 1 and is a numerical measure of molecular nonuniformity. The absolute value of $M_n$ can be determined cryoscopically, osmotically, or through end-group analysis; $M_m$ can be determined by light scattering [18] or by the hypsometric distribution of macromolecules in ultracentrifugal sedimentation.

The simplest and most commonly used method for obtaining relative values of mass-average molecular mass ($M_m$) is measurement of solution viscosity [17, pp. 49–67] in a capillary viscometer at constant temperature, where

$$\eta_{\text{relative}} = \text{flow time of the solution divided by flow time of the solvent} \quad (1)$$

and

$$\eta_{\text{specific}} = 1/\eta_{\text{relative}} \quad (2)$$

Suitable solvents are concentrated sulfuric acid, formic acid, and phenol. The specific solution viscosity, $\eta_{\text{specific}}/c$, referred to the polyamide concentration in grams per 100 mL, is concentration dependent. Plotting viscosity as a function of decreasing concentration and extrapolating to zero yield the intrinsic viscosity, which is directly proportional to $M_m$. In the production of nylon fibers, sufficient control of the process can be achieved by routinely measuring and controlling the viscosity of a standard solution, e.g., 1 g/100 mL. The melt viscosity, of practical importance for melt spinning, can be determined in rotary, torsional viscometers and capillary viscometers, or by using the falling ball method. In measuring melt viscosity, oxygen and moisture must be rigorously excluded and temperature held constant. On melting, polyamides can change by amide ring formation, molecular mass increase, and degradation, so melt viscosity is likely to vary according to method and measuring conditions. In general, the melt viscosity of polyamides for melt spinning is ca. 50–500 Pa · s.

The *free amino and carboxyl end groups* of polyamides are determined by potentiometric or conductometric titration with dilute perchloric acid and alkali hydroxide solution, respectively. A solvent mixture of phenol, ethanol, and water is suitable for amino-group titration. A supersaturated polyamide solution, in benzyl alcohol or α-phenylethyl alcohol, is used for carboxyl end-group determination. This analytical procedure cannot be used for all polyamides.

*Water analysis* is carried out by heating the flake or fiber, and released water vapor is carried with dry nitrogen into Karl Fischer reagent. In this method, both the physically bound water and that formed by chemical reaction are determined.

*Extractable materials* in nylon 6, caprolactam, and higher ring amides can be determined by extracting the finely divided polymer for several hours in water or lower alcohols. However, in this relatively simple method, the high ring amides are not extracted completely. For quantitative determination and separation of ring amides, the polymer is dissolved in formic acid and the linear polyamide molecules are then precipitated by addition of methanol. Ring amides can be determined quantitatively in a methanol solution by using gel chromatography.

## 4. Modifying Additions

*Dyeing* of nylons that absorb water and swell, such as nylons 6 and 66, can be accomplished relatively easily with disperse, acid, or metallized dyes. The dye molecules migrate readily into the fiber structure. The dyeing of nylon fibers with acid dyes results from reaction with the amino end groups. The number of such reactive sites is influenced by $M_n$ but can be modified by additives such as viscosity stabilizers. For example, carboxylic stabilizers will block amino groups and reduce affinity for acid dyes. The use of primary aliphatic amines or diamines provides additional basic sites and increases acid dyeability. Introduction of tertiary amine groups into the chain strongly increases absorption of acid dyes.

Nylons can be modified to accept basic dyes by introduction of strong acidic groups such as sulfonic acids, e.g., 5-sulfoisophthalic acid, in equimolar mixture with 1,6-diaminohexane [19]. Polycolor effects (graduated color depth, variable color and shades) can be achieved by dyeing fabrics of mixed nylon fibers with different dyeabilities in a single dye bath containing acid, basic, and disperse dyes.

*Delustering* is the practice of altering the optical properties of a fiber, usually by incorporation of a white pigment. This is most often done for textile uses. The most common delusterant is titanium dioxide [13463-67-7], added in amounts of 0.03–2%, depending on the effect desired. Of the two forms of titanium dioxide available, the anatase variety is preferred for continuous-filament yarns because the harder rutile modification causes excessive abrasion and wear on guide surfaces and machine parts over which the yarn runs. The rutile form can be used with staple fibers in which guide wear potential is not so serious. Other pigments, such as clay, can also be used. Delustering can also be accomplished by surface roughening or by introducing voids into the filament.

Size distribution [20] of titanium dioxide pigments and pigment aggregates is generally 0.1–1.0 μm, with a frequency maximum in the region of optimum turbidity at 0.2–0.4 μm. However, commercial pigments also contain larger agglomerates, which must be broken down through wet grinding or separated from the aqueous dispersion by fractional sedimentation. Pigment agglomeration must be avoided in pigment addition to the monomer or to the polymer melt, because the particles result in filter plugging as well as weak sections in the filament. For continuous production of nylon flake or fiber, adding a pigment–polymer concentrate of 20–40% titanium dioxide to the melt just before spinning is advantageous. This permits quick conversion to different degrees of delustering, but also imposes strict requirements on control and mixing.

The magnitude of pigment addition can be checked by ash determination or photometric titanium analysis. The uniformity of pigment distribution, as well as the number and size distribution of pigment agglomerates, can be analyzed by microscopic examination of thin fiber cross sections.

*Aging and light protection* must be provided for some nylons by addition of special additives [21]. Manganese(II) compounds, in amounts as low as $5 \times 10^{-4}\%$, are especially effective for nylons 6 and 66. Titanium dioxide pigments, especially of the anatase type, reduce light resistance, and one variety of anatase $TiO_2$ is coated with manganese to offset this effect. Many manganese compounds, such as the acetate, sulfate, phosphate, polyphosphate, and glycerophosphate salts have been proposed. In selecting the type and amount of additive, undesirable side effects, such as yellowing of the filaments and extraction of the manganese (II) during dyeing or wet finishing, must be considered in addition to the light-protective ability desired. Use of a manganese salt with a reducing anion, such as manganese(II) hypophosphite, improves the whiteness of fibers. Addition of sodium phenylphosphinate, in combination with manganese(II) hypophosphite, substantially improves both whiteness and quality [22].

The *heat-aging resistance* of nylons must be improved for uses such as tire cord. The thermooxidative degradation of nylon can be reduced considerably by adding antioxidants (0.1–1.0%), such as substituted aromatic diamines, and their action can be increased by combining them with trialkyl phenyl phosphites. Textile nylon, for which yellowing cannot be tolerated, requires special phenolic antioxidants substituted with alkyl radicals in the positions *ortho* to the — OH group. Examples are the amine salts of 3,5-dialkyl-4-hydroxyphenylcarboxylic acids. Copper(I) salts, in amounts as low as $5 \times 10^{-4}\%$ copper, are also effective. Many synergistically acting stabilizer mixtures are described in the literature, e.g., complex compounds of copper(I) iodide and hydrocarbon phosphines or phosphites are said to be especially effective.

Inorganic and organic pigments for *spin dyeing* of nylons must be stable under melt processing conditions and must give high colorfastness [23]. Pigment particle size should be less than 1–2 μm to avoid impairment of fiber strength, and agglomerates must be carefully milled or separated by sedimentation. Colored pigments are generally added to the starting monomer as aqueous dispersions. They may, however, be added to the melt as pigment–polymer concentrates [24]. This technique is desirable for short commercial runs but requires precise metering and good mixing.

Suitable additives for spun-dyed nylon are the yellow to red cadmium pigments, anhydrous iron oxide pigments, ultramarine, copper phthalocyanine, perylene, and quinacridone pigments. Because nylon melts generally have a strong reducing action, perylene pigments can be reduced on the surface with a significant change in color. However, reoxidation often occurs in air shortly after extrusion of the filament. Some azo and anthraquinone dyes are also soluble in nylon melts and are sufficiently stable for spin dyeing.

*Dyeing of nylon flake* in aqueous baths, before spinning, is an economical way to produce a large range of colors from one flake type. However, the resulting colors do not have the outstanding fastness of pigment-spun fibers.

*Black nylon fibers* are obtained by using carbon black pigments with an average particle size of 10–30 µm. Carbon black pigment in too high a concentration can block carboxyl groups, reduce the degree of polymerization, and impair flow behavior of the melt.

*Optical brighteners* are frequently used to improve aesthetics of textile fibers, especially to hide the effects of soiling. The more stable brighteners are added to the monomeric starting material, the less stable ones, such as the coumarin type, are injected into the melt.

## 5. Technology of Melt Spinning

### 5.1. Melt Spinning of Nylon

All commercially significant nylons are melt-spun at temperatures up to ca. 300 °C [25]. Processes may be continuous from monomer to spun fiber, or they can be staged. When a variety of small-volume products is required, producing nylon flake and remelting it for the spinning process are economically advantageous. This is especially true for nylon 6, which must be washed to extract polycondensation byproducts.

In modern plants, flake is usually stored in large hoppers and then transported pneumatically or in smaller hoppers to the spinning area. Prior to spinning, the flake must be conditioned to proper moisture content because this can have a strong effect on melt viscosity. Great care is taken to store and transport flake with exclusion of moisture. Also, before being delivered to the melt unit, fines and excessively large cuts, as well as contaminating metal particles, are removed.

Nylon flake is commonly melted by using screw extruders or, in the older process, heated grids or grates. In either case, oxygen must be rigorously excluded. Spinning temperature depends mainly on the thermal stability and melt viscosity of the nylon. With nylon 6 and nylon 66, temperatures are usually kept at 280–290 °C. After leaving the melt unit, polymer is delivered to a metering pump and then to a spin pack. The pack contains filter media, passages for distributing polymer within the pack, and a spinneret. The filter section serves to remove mechanical impurities and to provide high shear for the melt. In a typical filter section, the molten polymer passes through layers of either powdered metal or fine sand, then through layers of very fine mesh screen, and finally through a photo-etched plate before reaching the spinneret.

The spinneret is a metal plate, usually 6–10 mm thick, filled with holes, each hole forming one filament. Filaments from round spinneret holes have circular cross sections. The cross section can be altered to almost any desired filament shape to achieve unusual luster effects, as well as altered physical and tactile properties. Because the surface tension of the melt leads to a rounded filament, melt temperature, melt viscosity, and cooling rate must all be adjusted to achieve the desired filament cross section.

Spinneret holes of 200–400 µm can be used to produce nylon filament and staple yarns. The molten polymer extruded from the hole is pulled away by downstream rolls, reducing the diameter 5- to 50-fold. The diameter, or the mass per unit length, of the spun filament is determined by the rate at which polymer is metered through the spinneret hole and the rate at which it is pulled away. Small particles that pass through the filter leave weak spots in the filament, which may break in subsequent processing.

Nylons up to ca. 30 dtex/filament can be spun into air, but higher diameters require spinning into water or other liquids. Except for a few specialty uses, such large diameter filaments are generally made for nontextile applications. Turbid monofilaments are usually obtained when spinning into water, because of crystallization and surface corrosion. Clear filaments can be prepared by spinning nylon melts into cooled organic liquids such as carbon tetrachloride.

Unusual and useful filament structures can be prepared with specially constructed spinnerets [25]. Two melts are fed simultaneously to a spinneret hole, which results in a bicomponent filament that crimps spontaneously after drawing and finishing. Suitable melt pairs for this are nylon 66 and a 66–copolyamide [27] or a copolyamide 612 [28]. Spinnerets with two concentric holes yield sheath-core filaments from two melts. Injection of an inert gas into the center hole causes a void to develop and results in a hollow filament cross section.

*Grid spinning* was the original process for melt spinning of nylon. Flake is gravity-fed from an overhead hopper and melted on a spiral coil or grid heated with diphenyl. An atmosphere of superheated steam is generally provided to control melt equilibration. In starting up a grid-spinning head, melted polymer dropping from the melting grid collects in a conical sump, flows into a pressure pump, and is fed to a metering pump where it is metered through the filter pack and spinneret. Because pump demand must always be smaller than the maximum melt capacity, the sump fills until the melt grid is immersed in the melt, thus automatically regulating melting via the free-grid surface. When the process is interrupted, the grid heat must be shut off, otherwise molten polymer will rise into the cooler zone and solidify to form a bridge that blocks the flake feed. The melting capacity of a grid-spinning unit with a grid diameter of 160–200 mm is ca. 30–60 g/min. Higher capacity can be achieved with larger grids. One melting grid can feed four or more spinnerets.

*Extrusion Spinning.* More modern spinning equipment uses melt extruders instead of grid units.

Extruders provide a shorter dwell time in the melt, better mixing, and better viscosity control. Their main advantage, however, is the large melt capacity, up to ca. 1500 kg/d for a 150 mm-diameter screw.

Screws with length/diameter ratios of 20–30 are preferred to assure a more uniform melt temperature through the longer compression zone. An additional mixing zone connected to the compression zone improves homogeneity of the melt and permits increased throughput [29].

The melt is fed at constant pressure from the extruder through a jacket-heated transfer line to the spinning machine. One melt extruder can supply 30–40 individual spinning positions, each of which might spin eight or more individual thread lines. Use of a central filter for the melt, located between the extruder and the spinning block, is considered advantageous for some processes. This type of filtration is practical for nylon 6 but not as effective for nylon 66 because of the tendency of the latter to degrade and form gels in, and after, the filter.

Melted polymer must be fed to the individual spinnerets with the same average dwell time. Distribution piping is designed in cascade form to provide the same melt travel distance to each spinneret. In addition, dwell time differences arising from parabolic velocity distribution of melt inside the pipe must be balanced with static mixers ahead of each branch. Local or chronological nonuniformity of the melt can usually be detected first as dyeing differences in filament yarns.

*Direct Spinning.* Economic and quality advantages result by feeding continuously polymerized melt directly to the spinning machine. This is most easily accomplished with nylon 66, which produces only small amounts of low molecular mass cyclic amides. Direct spinning becomes most attractive when a few products can be made in high volume and such products can be made with little change in overall throughput. Large transitional changes are to be avoided. In such systems, introducing additives into the melt downstream of the polymerization unit is also desirable. This requires additional expenditures for the preparation of melt concentrates, as well as for the melting, metering, and homogenization equipment required for injection.

As noted earlier, nylon 6 fibers can also be made by direct spinning, but the equilibrium mixture, containing substantial amounts of caprolactam and higher cyclic amides, must be dealt with. This is accomplished through *thin-film evacuation*. The melt, prepared continuously in a polymerization tube, is passed to an evacuating assembly at 270 °C [30]. It is distributed through an annular slit onto a cylindrical wall, as a coherent film of melt. As the melt flows down the wall, volatile components (caprolactam and water) evolve at a pressure of ca. 10 Pa. At the bottom of the assembly, the evacuated nylon 6 melt is homogenized through a mixing stirrer and conveyed by a compression screw through the discharge and to the spinning machine.

## 5.2. Special Techniques and Products

***Fine-Denier Nylon Filament Yarns.*** Most fine-denier filament yarns are used in wearing apparel and household upholstery. Products are in the region of 10–400 dtex total, with filament counts from 1 to > 100. Single filaments run from 1.5 to ca. 6 dtex (9–36 µm).

Spinnerets for fine-denier textile filaments are usually made of stainless steel, 40–70 mm in diameter and 10–15 mm thick. Because of the high viscosity of the melt, the spinneret hole initially has a large diameter (2–3 mm), called the counterbore, which tapers in a conical section to a smaller exit hole (200–400 µm) from which the melt is extruded.

With a windup speed of 600–1200 m/min, solidification in the quench (cooling) chimney is completed within ca. 50 cm of the spinneret. Cool air is blown at right angles across the filament path, and careful attention must be given to avoiding pulsation or turbulence in air flow.

Filaments leaving the quench chimney are moistened with a spinning finish (oils emulsified in water) to provide static protection and lubrication for further processing. Many filaments are combined after finish application to form the final, multifilament bundle. If the yarn is to be drawn in a separate step, it is normally passed through a steam tube to promote development of the proper crystalline structure. For coupled spinning and drawing, yarn can be forwarded from the feed roll, which pulls the filaments away from the spinneret, directly to a set of faster moving draw rolls and then to a package-winding unit. Textile yarns are commonly drawn three to four times the original spun length to develop the required tensile–elongation properties. Depending on the process and the end-use, packages of ca. 1–20 kg are produced.

Economics of the spinning operation have been improved over the years by winding multiple packages on one winding unit, by winding larger packages for each thread line, and by operating the entire process at higher speed. Filament yarns are now prepared at spinning speeds of 5000 m/min, and useful fiber properties are now achieved in some cases without a separate drawing step [31]. Spinning, drawing, and texturing may also be combined in one continuous operation.

***Coarse-Denier Nylon Filament Yarns.*** Coarse nylon yarns range from 400 to > 35 000 dtex total and from 6 to ca. 20 dtex per filament. Rectangular spinnerets are commonly used, spinning several hundred holes per spinneret. In the production of high-strength nylon yarns (e.g., tire cord), draw ratios of 5–6 are used and drawing is commonly done in two stages, with heating, to develop maximum tenacity and low elongation. Tenacities of 8.0 cN/dtex and 15–17% elongation are common for nylon 66 tire yarns. Package weights of 40 kg are common, and ends from several hundreds of these packages are combined and wound on a larger spool (beam) for shipment to the customer.

Nylon filament yarns for carpets and other floor coverings are usually sold in the range of 1000–3000 dtex total and 15 dtex per filament. Nylon for netting and ropes reach 35 000 dtex total, with 20 dtex per filament. These yarns are drawn four to six times their original length with heat assists and are deposited in rope form in a can. Some of these are then stretched further in a hot water bath to develop maximum tenacity.

***Nylon Wires and Bristles*** Coarser nylon filaments (0.05–5 mm) are generally made with a higher degree of polymerization. Copolyamides are often more suitable for fishing lines.

Several processing steps are frequently coupled in one continuous operation. Flake is melted in a single-screw extruder, and the melt is extruded through spinneret holes (1–10 mm in diameter). The coarse filaments are solidified in a water bath and drawn between rollers in air or hot water at a speed of 200–400 m/min. Thicker wires may be drawn through dies to set the draw point and round the cross section. To improve dimensional stability and stiffness, coarse filaments for bristles are heat-set in boiling water, either continuously or on reels or bobbins.

## 6. Uses

Nylon fibers are used in a wealth of final products in appropriate make-up form [33]. They are used in applications that need high heat capacitance, high strength, and resilience [34]. Fine count nylon filament yarns with a total count up to approximately 200 dtex are processed largely into stockings, tights, ladies underwear and corset fabrics. Important products from textured nylon yarns are swimwear and sportwear, women's dresses, blouses, shirtings, elastic outerwear, jacket and overcoat fabrics, umbrella, décor and upholstery cover fabrics. In the technical sector, woven fabrics for airbags [32] and parachutes are produced from fine, high-tenacity nylon filament yarns. Coarse count nylon filament yarns are used mainly in carpets and other floor coverings. Other end-uses are high-tenacity tire cord, nets, fishing equipment, as reinforced fibers in driving belts and conveyor belts and also surgical sewing threads, luggage and footwear (so-called Oxford yarns) belts of all types (e.g., car seat belts), tapes and ropes. Nylon staple fiber yarns are processed particularly in blends with cotton, wool and other man-made fibers. By blending with nylon, the strength and wearing properties of a fabric are improved without changing the general character. Typical products are men's socks, warp knitted fabrics, clothing fabrics and also industrial clothing and children's wear, lining fabrics and knitting yarns. Nylon staple fibers are used in carpets and other floor coverings, as an admixture for needle felts, particularly in the running layer, for producing nonwovens and as cut fibers for electrostatic flocking. Nylon staple fiber yarns are processed into technical woven fabrics, filter fabrics, cord belts and inlay fabrics for the rubber industry. Nylon monofils and bristles are used in numerous application fields, such as tennis racket strings, fishing nets, fishing lines, abseiling ropes, safety lines, jump fabrics, screen cloths and other technical fabrics, bristles, tooth brushes (frequently of nylon 610) and paint brushes.

## 7. Economic Aspects

In recent years, nylon filament yarn and staple has lost share in the global fibers market. Although in 1980 nylon accounted for more than 10% of global fiber production, by 2000 the figure had fallen to 7.4% with a further decline to 5.1% by 2009. In 2020, Tecnon OrbiChem forecasts that the share will have dropped below 4%. Although share has declined, production has been increasing, from around $3 \times 10^6$ t in 1980 to almost $3.7 \times 10^6$ t in 2009. By 2020, total worldwide production of nylon filament yarns and staple fibers is forecast to reach more than $4.4 \times 10^6$ t [35].

Filament yarn is used in a wide range of applications including carpets and industrial end-uses such as high-performance tires and airbags where demand is strengthening. Unlike staple, filament yarn production is more of a global business, with manufacturers in all regions of the world.

In North America, production of nylon filament yarn has fallen by an average of 5.2% per annum since ca. 2005; but the rate is expected to ease to around 3.5% per annum during the forecast period. The change reflects the fact that more than 80% of nylon filament consumption in the USA is in carpets, a product that is less susceptible to substitution by imports because of the highly capital intensive nature of the manufacturing process and the considerable transport costs associated with the bulky product.

Production of nylon filament yarn in Western Europe has been declining more slowly than in North America, falling by an average 2.5% per annum since ca. 2005.

Although now the largest producer of PA filament yarn in the world, China is also the number one importer, accounting for 24% of the world total in 2008, although volumes are in decline as domestic production increases. With nylon 6 being the dominant polymer produced in Asia, the majority of growth in global nylon production this decade has been in fiber produced using this technology. Production of nylon 6 fiber is expected to grow by around 8% per annum during the forecast period with output of nylon 66 in the region remaining stable. As far as exports are concerned, Taiwan continues to be the major exporting nation, accounting for around 20% of the global total, of which more than 80% goes to China. Italy, and Germany also export significant quantities — each accounting for around 10% of the global

**Table 6.** Trade names and producers of nylon fibers [36]

| Trade name | Nylon | Manufacturer |
|---|---|---|
| Acelan | 6 | Tae Kwang Industrial Co./Korea |
| Aliaf | 6 | Sherkat Sahami Aliaf/Iran |
| Alto | 6 | Aquafil SpA/Italy |
| Amilan | 66/66 | Toray Industries Inc./Japan Toray-Monofilament Co./Japan |
| Amni | 66 | Rhodia Poliamida/Brazil |
| Anso | 6 | Shaw Industries/USA |
| Antron | 66 | Invista Canada Inc./Canada Invista Deutschland GmbH/Germany Invista Inc./USA Invista (Australia) Pty/Australia |
| Aqualon | 6 | Aquafil SpA/Italy |
| Arafelle | 66 | Nilit Ltd./Israel |
| Asahikasei Nylon | 66 | Asahi Chemical Industry Co./Japan |
| Asota | 6 | Asota GmbH/Austria |
| Perlon-Draht | 6 | Perlon-Monofil GmbH/Germany |
| Borgolon | 6 | Borgolon SpA/Italy |
| Bri-Nylon | 66 | SANS Fibres (Pty.) Ltd/South Africa |
| Carbyl | 6 | Inquitex SA/Spain |
| Chainlon | 6 | Chain Yarn Co./Taiwan |
| Coolan | 6 | Hualon Corp./Taiwan |
| Cordura | 66 | Nilit Germany GmbH/Germany |
| Danamid | 6 | Zoltek Magyar Viscosa RT/Hungary |
| Dilon | 6 | Dilon Ltd./Pakistan |
| Dimlon | 6 | Tekstiplik Sanayii A.S./Turkey |
| Dorix | 6 | Reinhold GmbH/Germany |
| Enka NylonEnka Nylon | 666 | Polyamide High Performance GmbH/Germany Polyamide High Performance, Inc./USA |
| Enkalon | 6 | Polyamide High Performance, Inc./USA Enka de Colombia SA/Colombia Polyamide High Performance GmbH/Germany Century Enka Ltd./Italy |
| Fabelon | 6/66 | Domo Fibers NV/Belgium |
| Garware | 6 | Garware Nylons Ltd./India |
| Grilon Grilon | 6, 12, 6.106 | Ems-Chemie AG/Switzerland Nicieza y Taverna SA/Argentina |
| Islon | 6 | Insa AS/Turkey |
| Jayanka | 6 | J.K. Synthetics Ltd./India |
| Jacord | 6 | J.K. Synthetics Ltd./India |
| Jaykalon | 6 | J.K. Synthetics Ltd./India |
| Jinfan | 6 | Guandong Xinhui Meida Nylon/China |
| Krasil | 6 | Dilon Ltd./Pakistan |
| Kugafil | 6 | Fugafil - saran GmbH/Germany |
| Laufaron | 6 | Laufaron GmbH/Germany |
| Leona | 66 | Asahi Chemical Industry Co./Japan |
| Lilion | 6/666 | Nylstar SpA/ItalySNIACE S.A./Spain |
| Long Life | 6 | Manufacturas del Sur SA/Peru |
| Meryl | 666/666/6666 | Nylstar SA /France Nylstar Inc./USA Nylstar SpA/Italy Nylstar SA/Spain |
| Misrnylon | 6 | Misr Rayonne Co./Egypt |
| Mistilon | 6 | Manufacturas del Sur SA/Peru |
| Modipon | 6 | Modipon Ltd./India |
| Nakron | 6 | Thai Toray Synthetics Co./Thailand |
| Nexilon | 6/66 | Nexis Fibers GmbH/Germany |
| Nicord | 6 | Nirlon Ltd./India |
| Nirlon | 6 | Nirlon Ltd./India |
| Noval | 666 | La Fibra Srl/Italy Sarl Lamato/France |
| Nurel | 6 | Nurel SA/Spain |
| Nylhair | 6 | Inquitex SA/Spain |
| Nylpak | 6 | Bengal Fibre Industries/Pakistan |
| N.R.C. | 6 | National Rayon Corp./India |
| Palmylon | 6 | Kohap Chemtec Corp./Korea |
| Pao | 6 | Jang Dah Nylon Corp./Taiwan |
| Passorea | 66 | Rhodia Performance Fibres /France |
| Perlon | 6 | Perlon-Monofil GmbH/Germany |
| Prakarn | 6 | Asia Fiber Ltd./Thailand |
| Promilan | 66 | Toray Industries, Inc./Japan |
| Pylon | 6 | Pylon Industries Ltd./Bangladesh |

| | | |
|---|---|---|
| Queen | 6 | Cheng Chi Fibre Co./Taiwan |
| Radilon | 6/66 | RadiciFibres SpA/Italy |
| Reilon | 6 | Reinhold GmbH/Germany |
| Sakorn | 6 | Hantex Corp./Thailand |
| Schwarzafil | 6 | TFG Thüringer Filamente GmbH/Germany |
| Schwarzalon | 6 | TFG Thüringer Filamente GmbH/Germany |
| Sensil | 66 | Nilit Ltd./Israel |
| Settestelle | 6 | Golden Lady SpA/Italy |
| Shreelon | 6 | Shree Synthetics Ltd/India |
| Sifas | 6 | SIFAS AS/Turkey |
| Snialon | 6 | Baroda Rayon Corp./India |
| Sniater | 6 | Baroda Rayon Corp./India |
| Stanyl | 4.6 | DSM High Performance Fibers BV/Netherlands |
| Stanylenka | 4.6 | Polyamide High Performance GmbH/Germany |
| Sunylon | 6 | Formosa Chemicals & Fibre Corp./Taiwan |
| Supplex | 66 | Invista Brasil/Brazil |
| | 6.66 | Invista Srl/Argentina |
| Sylkharesse | 66 | Sarl Lamato/France |
| Tactel | 66 | Nilit Ltd./Israel, Nilit Germany GmbH/Germany Invista Brasil/Brazil Du Pont (UK) Ltd./UK |
| Tactel | 6, 66 | Invista Srl, Argentina |
| Taftilon | 6, 66 | INSA AS/Turkey |
| Tasinlon | 6 | Chung Shing Textile Co./Taiwan |
| Taslon | 6 | Acelon Chemicals & Fiber Corp./Taiwan |
| Texfiber | 6 | Hilon Corp./Philippines |
| Three Gun | 6 | Chang Shing Textile Co./Taiwan |
| Timbrelle | 66 | TWD GmbH/Germany |
| TNT | 6 | Toray Nylon Thai Co./Thailand |
| Toplon | 6, 66 | Hyosung Corp./Korea |
| Toray Nylon | 6 | Toray-Monofilament Co./Japan |
| Toyobo Nylon | 6 | Toyobo Co./Japan |
| Tufcord | 6 | SRF Ltd./India |
| Tynex | 6, 12 | DuPont Filaments/USA |
| Ultralon | 6 | Aquafil SpA/Italy |
| Ultron | 66 | Solutia Inc./USA |
| Unitika Nylon | 6 | Unitika Ltd./Japan |
| Vanylon | 6 | Vanylon SA/Colombia |
| Wellon | 6, 66 | Wellman International Ltd./Ireland Wellman, Inc./USA |
| Wellstrand | 6, 66 | Wellman, Inc./USA |
| Yuan Pao | 6 | Jang Dah Nylon Industrial Co./Taiwan |

total — though volumes from another former major player, Canada, are in decline.

Trade names and producers of nylon fibers are listed in Table 6. World production of synthetic nylon fibers (yarn and staple) are listed in Table 7.

**Table 7.** World production of synthetic nylon fibers (yarn and stable)

| Year | Production, t |
|---|---|
| 1980 | 3 185 000 |
| 1990 | 3 776 000 |
| 1995 | 3 911 000 |
| 2000 | 4 118 000 |
| 2005 | 3 924 000 |
| 2009 | 3 467 000 |

## References

1. O.L. Shealy, *Fiber Prod. Int.* **12** (1984) no. 2, 32–41.
2. P. Schlack, *Chemiefasern* **13** (1963) 560–562.
3. Kurashiki Rayon Co., JP 406628, 1965 (K. Masubayashi, T. Kawaguchi).
4. R. Graf et al., *Angew. Chem.* **74** (1962) 523–530.
5. J.S. Robinson (ed.): *Fiber Forming Polymers-Recent Advances*, Noyes Data Corp., Park Ridge, N. J. 1980, pp. 135–165.12.
6. C.F. Horn et al., *Angew. Chem.* **74** (1962) 531–540.
7. Monsanto, FR 1419132, 1964 (J.E. Tate). Monsanto, US 3402152, 1968 (E.T. Erignac, R.T. Wright).
8. D. O'Sullivan, *Chem. Eng. News* **62** 1984, no. 21, 21–22.
9. A. Konopik et al., *Chem. Fib. Int.* **48** (1998) 207–209.
10. BASF, US 4060517, 1977 (F. Mertes et al.).
11. Du Pont, US 3502624, 1970 (C.F. Black).
12. B. Wulfhorst et al., Aramid Fibers, Fiber tables according to P.-A. Koch, First Issue1989, 11–16.
13. T. Skwarski, *Lenzinger Ber.* **45** (1978) 28–36.
14. V. Kacvinsky, *Techn. Text.* **31** (1988) 131–136.
15. H. Imuro et al., *Chemiefasern/Text. Ind.* **37/39** (1987), T4–T9.

16. *Tech. Text.* **5** (2009) E206.
17. F. Happy (ed.): *Applied Fiber Science*, vol. 2, Academic Press, New York 1981, pp. 1–44.
18. F. Happy (ed.): *Applied Fiber Science*, vol. 1, Academic Press, New York 1978, pp. 381–416.
19. Du Pont, US 3184436, 1965 (E.E. Magat).
20. I.M. Kolthoff, P.J. Elving, F.H. Stross (eds.): *Treatise on Analytical Chemistry*, vol. 4, J. Wiley & Sons, New York 1977, part 3,pp. 273–313.
21. N.S. Allen, J.F. McKellar, *J. Polym. Sci., Macromol. Rev.* **13** (1978) 241–281.
22. W. Schnabel: *Polymer Degradation–Principles and Practical Applications*, Macmillan Publ., New York 1981, pp. 40–44.
23. BASF, DE-OS 1570261, 1965 (C. Dorfner et al.).
24. Glanzstoff, DE-OS 1494765, 1965 (E. Neyer, H. Linhartt).
25. M. Sittig: *Polyamide Fiber Manufacture*, Chemical Process Review no. 61, Noyes Data Corp., Park Ridge, N. J. 1972, pp. 183–254.
26. C. Placek: *Multicomponent Fibers*, Noyes Data Corp., Park Ridge, N. J. 1971, pp. 40–52.
27. Du Pont, US 3779853, 1973 (E.H. Olsen).
28. ICI, US 3675408, 1972 (D.M. Pacini et al.).
29. K. Riggert et al., *Chemiefasern Text. Anwendungstech./Text.-Ind.* **23** (1973) 1183–1186.
30. BASF, DE-OS 1495177, 1964 (P. Maahs et al.).
31. R.A. Vaidya, *Indian Text. J.* **94** (1984) no. 10, 53–58.
32. M. Bongartz, *Chem. Fib. Int.* **58** (2008) 98–101.
33. S. Zaremba et al., *Chem. Fib. Int.* **47** (1997) 442–464.
34. I. Julian, *Int. Fib. J.* **20** (2005) 24–36.
35. N. Bywater, *Chem. Fib. Int.* **60** (2010) 60–62.
36. H.-J. Koslowski, *Chemiefaserlexikon*, 12.Auflage, Deutscher Fachverlag GmbH, Frankfurt am Main 2008, pp. 281–294.

# Polyester Fibers

HELMUT SATTLER, Königstein-Schneidhain, Germany

MICHAEL SCHWEIZER, ITCF Denkendorf, Denkendorf, Germany

| | | |
|---|---|---|
| 1. | Production | 1454 |
| 1.1. | Starting Materials and Polymer | 1454 |
| 1.2. | Spinning | 1455 |
| 1.3. | Drawing | 1457 |
| 1.4. | Aftertreatment | 1458 |
| 1.5. | Dyeing of Poly(Ethylene Terephthalate) Fiber Products | 1459 |
| 2. | Structure and Properties | 1461 |
| 2.1. | Polymer | 1461 |
| 2.2. | As-Spun Yarns | 1463 |
| 2.3. | Drawn Materials | 1464 |
| 2.4. | Dyed Poly(Ethylene Terephthalate) Material | 1464 |
| 3. | Poly(Ethylene Terephthalate) Fibers | 1465 |
| 3.1. | Staple Fibers | 1465 |
| 3.2. | Continuous-Filament Yarns for Apparel and Home Furnishings | 1468 |
| 3.3. | Continuous-Filament Yarns for Industrial Applications | 1471 |
| 3.4. | Spunbonds | 1474 |
| 3.5. | Monofilaments | 1475 |
| 4. | Other Polyester Fibers | 1476 |
| 4.1. | Poly(Tetramethylene Terephthalate) | 1476 |
| 4.2. | Poly(Trimethylene Terephthalate) | 1477 |
| 4.3. | Poly(1,4-Dimethylenecyclohexane Terephthalate) | 1479 |
| 5. | Uses | 1479 |
| 6. | Economic Aspects | 1480 |
| | References | 1484 |

*History.* The first spinnable high molecular mass polyesters were prepared by W. H. CAROTHERS and co-workers in the laboratories of E. I. du Pont de Nemours & Co. around 1930 [1]. Dicarboxylic acids in use then were sebacic acid, hexadecanedicarboxylic acid, and others; some diols used were ethylene glycol and propylene glycol. The discovery that hardened spun strands could be drawn to several times their original length and, thereby strengthened, was made with these polyesters [2]. However, their melting points were ca. 75 °C, too low for practical application. For this reason, CAROTHERS turned his attention to the more promising polyamides [3].

In the late 1930s, investigations showed that terephthalic acid and other aromatic acids could yield useful polyamides. Therefore, polyesters containing terephthalic acid as the dicarboxylic acid were studied at several locations [4]. Poly(ethylene terephthalate) was discovered in the laboratories of Calico Printers Association. Beginning in 1940, WHINFIELD and DICKSON devoted themselves mainly to poly(ethylene terephthalate) [5, 6], while SCHLACK was working on poly(tetramethylene terephthalate) [7].

In 1947, ICI and Du Pont acquired patent rights for Great Britain and the United States, respectively, and initiated the commercial development of polyester fibers [4]. In the early 1960s, ICI licensed its technology worldwide to some large fiber producers whose own developments led to today's mass production. Thus, poly(ethylene terephthalate) has gradually become the most important fiber polymer [8].

According to DIN definition 60 001, fibers are designated polyester if the polymer chain contains at least 85 % of an ester of a diol and terephthalic acid. Poly(ethylene terephthalate) [25038-59-9] (PET), poly(oxy-1,2-ethanediyloxycarbonyl-1,4-phenylene-carbonyl), is a polyester with terephthalic acid [100-21-0] as the dicarboxylic acid and ethylene glycol as the diol ($n = 80 – 150$):

$$H{\left[OCH_2CH_2O-\underset{\underset{O}{\|}}{C}-\!\!\!\left\langle\!\!\!\bigcirc\!\!\!\right\rangle\!\!\!-\underset{\underset{O}{\|}}{C}\right]}_n OCH_2CH_2OH$$

For a detailed description of the chemistry, raw materials, production, and analysis of polyesters, see → Polyesters. The following discussion contains information pertinent to an understanding of the production and properties of these fibers. Section 2.1 covers aspects of production common to all types of PET fibers; specific information can be found under the individual fibers (Chap. 4).

# 1. Production

## 1.1. Starting Materials and Polymer

*Ethylene Glycol [107-21-1].* Ethylene glycol is produced commercially by the reaction of ethylene oxide with water, which in turn is obtained by the oxidation of ethylene (→ Ethylene Glycol; → Ethylene). Prerequisites for smooth polycondensation to poly(ethylene terephthalate) and for the desired polymer properties are extreme purity of ethylene glycol and other raw materials and the absence of water.

*Dimethyl Terephthalate [120-61-6] (DMT).* Until the end of the 1960s, economical production of terephthalic acid in the purity required for fibers was difficult. For that reason, the more easily available dimethyl ester of terephthalic acid was generally chosen as the starting component for PET (→ Terephthalic Acid, Dimethyl Terephthalate, and Isophthalic Acid).

This material is produced commercially primarily by the *Witten process* (identical to the Imhausen or Katzschmann process [9]). *p*-Xylene [106-42-3] is subjected to a two-step, liquid-phase oxidation in air. The *p*-toluic acid thus produced is esterified with methanol to the methyl ester of *p*-toluic acid, which is then esterified with methanol to DMT.

*Terephthalic Acid [100-21-0] (TPA)* (→ Terephthalic Acid, Dimethyl Terephthalate, and Isophthalic Acid).

Since the early 1970s, extra pure terephthalic acid of fiber quality has been produced commercially primarily by the *Amoco process* [10].

Inhibition of the oxidation of the second methyl group of *p*-xylene is suppressed with the aid of added bromine-containing promoters as cocatalysts. As with DMT, the oxidation takes place in air and produces raw terephthalic acid, which is dissolved at high temperature under pressure in water, hydrated, and thus purified.

*Polymer.* The polymer is prepared in two steps:

1. Transesterification of dimethyl terephthalate (**1**) with ethylene glycol to bis-glycol terephthalate (**2**) (continuous or batchwise) or direct esterification of terephthalic acid.
2. Polycondensation (continuous or batchwise) to give poly(ethylene terephthalate) (**3**).

$$CH_3OOC-\langle\bigcirc\rangle-COOCH_3 + 2HOCH_2CH_2OH \xrightarrow[150-200\,°C]{Cat.}$$
$$\mathbf{1}$$

$$HOCH_2CH_2OOC-\langle\bigcirc\rangle-COOCH_2CH_2OH + 2CH_3OH$$
$$\mathbf{2}$$

$$n\,(\mathbf{2}) \xrightarrow[\substack{Vacuum\\280-290\,°C}]{Cat.}$$

$$H{-}\!\!\left[OCH_2CH_2OOC-\langle\bigcirc\rangle-CO\right]_{\!\!n}\!\!{-}OCH_2CH_2OH$$
$$\mathbf{3}$$

$$+ (n\text{-}1)\,HOCH_2CH_2OH$$

The batch process is preferred for special polymer types, particularly for small amounts of polymer. The continuous process is used for standard polymer types in large volume, especially if spinning immediately follows polycondensation.

*Poly(Ethylene Terephthalate) from Dimethyl Terephthalate.* Commercial transesterification of dimethyl terephthalate with subsequent polycondensation was initially carried out batchwise.

*Batchwise Process.* First, DMT is esterified as quantitatively as possible with an excess of ethylene glycol (molar ratio range of DMT to ethylene glycol 1:2.5 to 1:4.5) between 150 and 180 °C. Catalysts are primarily zinc, calcium, and manganese acetates. Bis-glycol terephthalate, as well as higher homologues, are formed. Excess glycol is distilled at temperatures up to 250 °C.

Polycondensation is carried out at higher temperatures, up to ca. 285 °C, under vacuum.

In this step, the catalysts are primarily antimony trioxide or oxides of similar amphoteric metals. Phosphorus compounds are added as stabilizers to inactivate the transesterification catalyst.

Other additives, e.g., titanium dioxide for delustering or carbon black for black coloration, can be added to the melt prior to polycondensation. Once polycondensation is complete, the condensate is forced out of the vessel with nitrogen and quenched in water. The melt, which is thus solidified largely in the amorphous state, can then be formed into chips.

*Continuous Process.* Continuous transesterification and polycondensation occur in several, mostly horizontal, reactors in a manner similar to the batchwise process. As the polyester melt is formed during condensation, care is taken to keep its layer thickness small and its surface correspondingly large. This is necessary because diffusion of excess glycol determines the rate of polycondensation. Discharge from the last reactor occurs by means of an extruder or a pump. The polyester melt is then either passed to the spinning units and spun immediately or chilled in water and cut into chips. Continuous processing units are generally quite large, with throughputs of up to 100 t/d and average holdup times of the material in the unit of 5 – 10 h.

**Poly(Ethylene Terephthalate) from Terephthalic Acid.** Since highly purified, fiber-grade terephthalic acid has become available at low cost, it is increasingly being esterified directly with ethylene glycol (molar ratio of terephthalic acid to ethylene glycol 1:1.1 to 1:1.5). The reaction occurs at about 260 °C under pressure and generally requires no catalyst. Lower linear oligomers are formed. After the glycol has been distilled, polycondensation follows. Its course is similar to that with DMT as the starting product. A polycondensation catalyst is also required. The economic advantages of this route are that circulation of methanol is eliminated, substantially less excess glycol is necessary, and above all, the amount (in grams) of starting material required is less by ca. 14 % because of the lower molecular mass of TPA vs. DMT. However, direct esterification results in somewhat more diethylene glycol byproduct. This process is used in continuously operating modern plants.

**Poly(Ethylene Terephthalate) with Very High Molecular Mass.** Industrial end uses require very high molecular masses (corresponding to very high average lengths of molecules) to attain the necessary tenacity and modulus. Because removal of excess glycol in polycondensation is rate-determining, reducing the thickness of the melt layer as the reaction progresses would be appropriate. However, due to the melt viscosity increase during polycondensation, this becomes more and more difficult. Holdup time at high melt temperature is thus increased, which produces thermal degradation in parallel with polymerization. Polymer whiteness thus deteriorates, and COOH end groups are formed in undesirable concentrations in addition to the OH end groups. To get around these problems, PET of very high molecular mass is now generally produced by *postcondensation in the solid phase* (solid condensation).

Granulate PET, produced by melt condensation, is dried and exposed to heat either batchwise in vacuum or continuously under an anhydrous gaseous protective blanket free of oxygen (generally nitrogen) at temperatures up to ca. 250 °C (i.e., somewhat below the melting point). A lamellar structure with noncrystalline regions and voids forms. Polycondensation occurs in parallel with recrystallization and additional crystallization. The glycol – water mixture produced diffuses through the noncrystalline regions and mainly through the interlamellar voids that are being formed. The result, in addition to acceptable polymer color, is primarily the desired low concentration of COOH end groups. Once the desired average molecular mass has been attained, the granulate is chilled and packaged. Alternatively, it can be melted directly with the exclusion of water and sent to the spinning units.

## 1.2. Spinning

The melting point of poly(ethylene terephthalate) (265 °C) is clearly below the decomposition temperature of the melt; therefore, PET can be melt-spun. During melt spinning, a molecular orientation is created in the filaments by the drawdown tension; this orientation is frozen during cooling. Here, PET exhibits a special characteristic which distinguishes it from other commercial melt-spinnable polymers: because of its relatively high glass transition temperature (70 °C), the spun material solidifies completely in the spinning shaft to form a stable, supercooled melt with molecular orientation more or less in the yarn direction [4]. This orientation, measured by birefringence, is proportional to the tensile stress along the path of the yarn at the time and place of solidification, [11, 12].

As filaments solidify, they also attain their final speed (winding speed) and thus their final linear density (as-spun fineness). Where the solidification point lies along the path of the as-spun yarn depends on spinning conditions; distances between 20 and 150 cm from the spinneret have been reported [12].

Because of the relatively low tendency of PET to crystallize at room temperature, the as-spun yarn is amorphous as long as drawdown tensions are not too great. Stress-induced crystallization occurs only at higher tensile stress as a result of spinning speeds above ca. 3000 – 4000 m/min or of very intense air quench; this crystallinity also is frozen into the yarn. Under these processing conditions, the as-spun yarn is partially crystalline [13, 14].

At very high spinning speeds (at about 8000 m/min), the crystallinity of the as-spun yarn can be as high as 50 % [15, 16], and the distribution of directions of the crystallites becomes even more narrow. At the same time, the layered structure of crystalline order, which resembles that of PET spun slowly and then drawn, is slowly transformed into a more fibrillar structure of larger ordered regions. Despite the high orientation of the crystalline fractions, the total orientation again decreases somewhat; therefore, orientation of the amorphous material between the fibrils is assumed to decrease [17].

At spinning speeds > 4000 m/min, an additional sheath-core orientation occurs in the filaments [15]. This effect can cause the orientation in the sheath, measured by birefringence, to be almost twice as great as in the filament core [18, 19]. Correspondingly, the sheath of the filaments is also more highly crystalline. These differences remain during further processing of the as-spun yarn and can lead to nonuniformities in dyeing if they are not controlled in the process [20].

As mentioned, the orientation and even the crystallinity of filaments depend on spinning conditions and increase with increasing spinning speed, stronger air quench, and higher molecular mass. Both characteristics decrease with increasing spinning temperature, with higher mass transport, and thus with greater as-spun linear density. If the spinning speed is increased and the fineness is held constant by a proportional increase in mass transport, the effect of spinning speed predominates. The orientation and even the crystallinity increase [12–14]. The diameter of the spinneret holes, hence the spin-to-draw ratio, has relatively little effect on the final spin orientation and crystallinity except when spinning speeds are ≥5000 m/min [17].

Orientation and crystallinity of the filaments are characteristics of the as-spun yarn that must be considered in further processing because of their effect on the properties of the end product. Both factors are held as constant as possible with respect to time and between the spinning units.

As described in Section 1.1, PET is produced either continuously or batchwise. In the continuous process, the spinning and polymerization units can be connected. Thus, freshly condensed melt flows directly to the spinning units. Because the economic advantages of direct spinning are obvious, its use for high-volume end products, e.g., for staple fibers and standard filament yarns, which are always made by the same process, is increasing constantly [21].

The PET granulate that is delivered to a spinning plant is first stored in silos. The granulate must be dried immediately before melting because even the smallest amount of water in the polymer causes hydrolytic degradation of the melt. Residual water content < 0.01 % is desirable. Drying is carried out mostly in continuous shaft dryers. The granulate moves slowly downward, while a hot, often predried, countercurrent air stream blows over it. Since amorphous PET granulate tends to become sticky during heating, a crystallizer in which the chips are slightly crystallized, at least on the surface, often precedes the dryer [22]. Many dryer designs are known, with throughputs amounting to several tons per hour. For some products, drying the PET granulate in a vacuum is better, for example, in batches in large tumble dryers.

From the dryer, the granulate reaches the melt extruder, which transports the melt to one or more spinning machines. Because of thermal degradation of the melt, maintaining equal holdup times in the hot lines between the extruder and the attached spinning units is important; the temperature of these lines must also be the same. Only then can the uniformity of the as-spun yarn and the end products be assured.

Spinning units differ considerably in construction and processing data, depending on the end product. The same is true for the method of cooling, finishing, drawing down, and collecting the as-spun yarn [4]. Common operational winding speed ranges from ca. 500 m/min for industrial yarns to 6500 m/min for filament yarns spun at the highest speed. Thick filaments, e.g., monofilaments, are spun at a much lower speed. Common to all is the spinning temperature, which is 270 – 310 °C.

## 1.3. Drawing

The glass transition temperature of PET is ca. 70 °C, which is somewhat above room temperature. Therefore, when PET yarn spun at relatively low speed is elongated slowly, cold drawing occurs with a draw point [23]. From the linear densities on both sides of the draw point or from the length of the plateau of the stress – strain diagram (see Section 2.2), a "natural draw ratio" can be specified. This ratio depends on the orientation of the as-spun yarn, which already represents a part of the total orientation [24]. The product of the orienting spin-to-draw and the natural draw ratio is constant to a first approximation and is equal to ca. 6 – 7 [25].

In commercial practice, as-spun PET yarn is always drawn at a temperature ≥ 80 °C. Because of the added work of drawing, the real yarn temperature in the draw zone is frequently higher than the external temperature, depending on the drawing arrangement.

During drawing [11], the molecular network of the as-spun yarn is oriented further. The increasing parallelism of the molecular chains initiates PET crystallization. With further drawing, the crystalline regions and the amorphous areas between them become more and more oriented [26]. Orientation, crystallinity, and strength of the drawn yarn increase with increasing draw ratio, whereas break elongation decreases [30]. This is illustrated by the stress – strain diagrams of yarns drawn differently (Fig. 1).

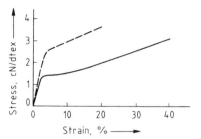

**Figure 1.** Stress – strain curves of two drawn PET yarns
——— Draw ratio 1 : 1.52; – – – – Draw ratio 1 : 1.83

In addition, the product of orienting spin-to-draw and draw ratio is approximately constant for the maximum possible draw ratio and for drawing to a given break elongation [25]. Thus, PET yarn spun at higher speeds needs to be drawn less to attain a given break elongation. However, this affects the structure of the drawn material [22, 25, 28]. Because less energy is added during drawing, the crystallinity is somewhat lower, and both the total orientation, as measured by birefringence, and the tenacity of the yarn are reduced [29]. These relations begin to change at spinning speeds of ca. 5000 m/min. Then the as-spun yarn is so highly crystalline that the ordered regions are no longer loosened during drawing but orient themselves as far as possible as a whole along the fiber axis. The crystallinity again increases [15].

The more the drawing temperature exceeds the glass transition temperature the more easily can PET be drawn [11]. Drawing tension drops if drawing temperature is increased and also if the glass transition temperature is decreased, either by water absorption by the as-spun yarn [4] or by modifications of the polymer. However, drawing temperature must not be set too high, for then the purely plastic and non-orienting elongation becomes too large a percentage of the total draw.

The draw ratio would have to be set very high to achieve high strength. The yarn in the draw zone would then no longer be able to absorb stresses and would break. As with other polymers, two-stage drawing is necessary; the predrawn yarn enters a second draw zone where it is then capable of withstanding the drawing tension.

In industrial practice, PET as-spun yarn is almost always drawn continuously between two godets or sets of rolls, whose speeds determine the draw ratio [23]. In so doing, the required draw temperature is imparted to the yarn by heaters or heated rolls [4]. Keeping the draw point stationary is important; otherwise the drawn product will contain differentially drawn areas, which produce defects, particularly nonuniform dye uptake. Details of each drawing arrangement are dictated by the product to be drawn (see also → Fibers, 3. General Production Technology, Section 2.2).

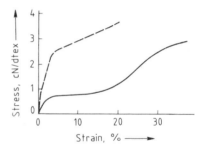

**Figure 2.** Stress – strain curves of drawn PET yarn, subsequently allowed to shrink freely

## 1.4. Aftertreatment

***Thermal Aftertreatment.*** At room temperature, oriented and noncrystallized segments of the PET molecular chains in a drawn threadlike structure are frozen and immobile; the entire structure is not in equilibrium. When the yarn is heated beyond the glass transition temperature, the amorphous chain segments begin to move so as to achieve an entropically more favorable, less oriented position, to the extent permitted by their restraints in the crystalline regions. These regions also lose orientation and, as an overall result, the yarn shrinks. The extent of shrinkage depends on the orientation and crystallinity of the drawn structure; it increases with increasing temperature and, initially, also with the duration of heat treatment [11, 27, 31]. Any plasticizers, e.g., water, that have diffused into the structure favor shrinkage.

The thermal instability of drawn PET yarn structures would be quite troublesome during further processing and in textile end products. Therefore, it is often reduced during fiber and yarn manufacture by a thermal aftertreatment called *heat-setting* [4, 32].

Structures that are allowed to shrink without restraint between 110 and 180 °C are thermally stable below the setting temperature. In this process, orientation decreases markedly, which manifests itself on the stress – strain curve as a saddle with a yield stress of ca. 1 cN/dtex (Fig. 2). If the product, which has been heat-set without restraint, is loaded to this yield stress saddle, it is permanently deformed. In general, normal textile applications rarely reach this amount of loading.

During unrestrained heat-setting, crystallinity increases considerably, primarily favoring the ordered regions already present. Degrees of crystallinity up to 50 % have been reported. The amorphous fraction is correspondingly reduced, but the chain segments are disoriented in response to shrinkage. With increasing crystallinity, maximum dye uptake (i.e., the saturation concentration) decreases but dye diffusion becomes much easier as the amorphous chain segments disorient. In this way, dyeability reaches a minimum at heat-setting temperatures of 150 – 180 °C [27, 31, 33].

If heat-setting takes place at constant length without any shrinkage, the effects are intensified with temperature and duration of heat-setting. Crystallinity also increases but less so than during unrestrained heat-setting. The PET products are not completely stable thermally but will retain some residual shrinkage [27, 32]. Therefore, the temperature at which this type of heat-setting is carried out is frequently very high, e. g., 220 °C. Cooling the material under tension after it leaves the heat-setting zone is important to prevent any shrinkage. Heat-setting at constant length decreases extensibility and increases strength.

Hot postdrawing of a previously drawn yarn represents heat-setting with negative shrinkage. Crystallinity and orientation increase markedly, and the initial slope of the stress – strain curve of the yarn and the tenacity increase. The extent of changes in shrinkability depends on processing conditions.

In practice, a middle course is often adopted, i.e., heat-setting while constant shrinkage is allowed [32]. This increases crystallinity and somewhat reduces orientation. The increase in extensibility of the material approximately corresponds to the shrinkage permitted; at the same time, a slight saddle is formed on the stress –

strain curve. The residual shrinkage of material heat-set in this way is smaller by about the amount of shrinkage permitted during setting. The many processing parameters, such as draw ratio, heat-setting temperature, and shrinkage allowed during setting, permit matching the properties of the PET product quite closely to the requirements of each end use.

Heat-setting does more than prevent thermal instability. The structural rearrangements it induces make prior changes in shape impressed on the yarn permanent [4]. This feature is used widely in producing and processing fiber products, e.g., in ironing or pleating when creases are to be set hot. However, in producing the fibrous structure and in further heat-setting, care must be taken to chose conditions which will assure that any subsequent heat-setting will still be effective. "Deadening" a fiber (or a textile structure formed from it), so that it can no longer be heat-set, should not be done.

During setting, primarily the orientation of the amorphous parts changes; however, growing crystalline regions also disorient if any shrinkage has been permitted. This leads to a corresponding change in the superstructure of the PET fiber. In general, the layered structure becomes more pronounced. The orientation angle of the crystalline layers with respect to the fiber axis increases [27].

Structural rearrangements during heat-setting and the resultant changes in textile properties depend in a complicated way on the structure and prior history of the PET yarn and on the conditions of the heat-setting process. Therefore, formulating general, quantitative relationships is impossible. These depend on the given heat-setting process, which differs from product to product. All setting processes are included in the manufacturing process immediately after drawing or deformation.

***Deformation.*** Melt-spun fibers including PET are generally flat and, in this condition, unsuitable for many applications [4]. Hence, all staple fibers and tows must be matched to natural fibers by *crimping*. PET filament yarns for apparel are *textured* → Fibers, 3. General Production Technology, Section 3.3 on a large scale. The processes and installations used in both cases are quite similar for all man-made fibers. The deformations produced are frequently heat-set, either without tension (staple fibers heat-set without restraint) or with tension (texturing), depending on the product.

## 1.5. Dyeing of Poly(Ethylene Terephthalate) Fiber Products

Dyeing of textiles and dyeing technology in general are discussed under → Textile Dyeing.

As a result of polycondensation and subsequent degradation, PET contains OH and COOH end groups. Neither is suitable for bonding ionic dyes. Therefore, PET is dyed with disperse dyes unless it has been modified to contain ionic end groups (→ Disperse Dyes). These dyes are available in powder or liquid form. They are only slightly soluble in water and are therefore used as aqueous dispersions in the dye bath or even as foam. Dye preparations necessarily contain dispersants.

Migration of dyes into the fiber is controlled by the laws of diffusion [34]. The parameter determining the rate of migration is thus the diffusion constant $D$, which depends on the size and structure of the dye molecule. The saturation concentration $c_\infty$, reached at very long dyeing times, is the second parameter required to describe the kinetics of dyeing.

Investigations have shown that $D$ is affected substantially by the orientation of the molecular segments in the noncrystalline region of the fiber. With increasing orientation, molecular mobility decreases, and the glass transition temperature $T_g$, a measure of mobility, increases. Beginning at ca. 68 °C in an isotropic, amorphous PET, $T_g$ can reach 100 °C in a highly oriented, partially crystalline fiber [35, 36].

The diffusion constant is also affected by the ordering of the molecules, though to a much lesser degree. This becomes evident if the temperature of heat treatment prior to dyeing exceeds 180 °C. Above this temperature, the noncrystalline regions (layers between crystalline layers) are loosened, which restores the mobility of the molecular segments in these regions; parts of molecules that up to then had been extended can coil up or even refold, which leads to a growth in crystalline regions transverse to the molecular axis and to an "emptying" of the noncrystalline regions. Material that has been spun at high speed and then drawn behaves similarly [20]. However, such ordered states in the noncrystalline regions of the fiber are first of all keys to controlling the saturation concentration $c_\infty$.

The fiber can take up dye to any significant degree only at temperatures above $T_g$. At these temperatures, the polymer is present as a highly viscous liquid in which the mobility of molecular segments in the noncrystalline regions is restricted. The saturation concentration evidently depends on the accessible free volume in the noncrystalline regions and in voids. It is also certainly affected by the structure of the interfaces of these volumes. A quantitative description is not yet possible.

Thus, with reference to fiber structure, the diffusion constant $D$ is the most important orientation-dependent kinetic parameter of dye migration through the "liquid" polymer regions. The saturation concentration is determined by the states of order of the supercrystalline structure (small-angle X-ray structure). However, practical dyeing behavior is essentially determined by the diffusion constant.

Poly(ethylene terephthalate) fibers can be dyed to deep shades without using auxiliary chemicals only at temperatures ca. 50 °C above the glass transition temperature. Such dyeing is carried out in aqueous media under pressure in a closed vessel and is called high-temperature (HT) dyeing. High-temperature dyeing of textiles made of 100 % PET or blends of PET with cotton or other cellulosic fibers presents no problems. Any wool in the blend, however, is damaged.

Carriers are used for open dyeing in boiling water, especially of blends of PET and wool. These chemicals are swelling agents, i.e., poor solvents, for the polyester fiber. They temporarily loosen intermolecular bonds in the fiber and thus increase the mobility of molecular segments. In other words, they lower the glass transition temperature. The problems connected with cleanup of wastewater are a disadvantage.

Therefore, PET fibers have been developed that can be dyed to deep shades in boiling water without carrier (see Section 2.1) [37]. These fibers contain block comonomers, generally polyglycols, and meet all requirements of the apparel and home furnishings trades. However, minor losses in light- and washfastness of the dyeings compared to unmodified PET are unavoidable.

Ionic dyeing for differential dyeing requires appropriate ionic end groups in the molecule. Sodium salts of 5-sulfoisophthalic acid dimethyl ester [138-25-0] and 5-sulfoisophthalic acid [22326-31-4] have been used; they can be added during transesterification of DMT or during direct esterification of TPA [38]. Fibers can be dyed with cationic dyes without difficulty, but the dyeings are much less lightfast.

Dyeing with acid dyes requires basic end groups. Thus far, building such end groups into PET, e.g., via amines, has not been possible without yellowing or high aminolytic degradation, resulting from the high polycondensation and spinning temperatures.

The most important *dyeing processes* are the exhaust process and the continuous process; aftertreatment of PET dyeings is very important for the quality of the dyeing.

The dyeing process most widely used with PET fiber products is the *exhaust process*. Depending on the situation in each dye plant (wastewater) and the chemical composition of the dyed goods (presence of wool), dyeing can be carried out under pressure at HT conditions (60 – 90 min at 125 – 140 °C), or it may be carried out open or under low pressure, respectively, with carrier (60 – 90 min at 95 – 110 °C). Combining both processes is rare and used only if highly oriented PET products are to be dyed to deep shades.

*Continuous processes* are also used. With the aid of saturated steam, superheated steam (pad steam process, 2 – 8 min holdup time at 160 – 180 °C), or hot air (pad thermosol process, 15 – 90 s holdup time at 180 – 220 °C), high throughputs, which improve the economics of dyeing, can be achieved. Blends of PET with cotton and other cellulosic fibers are preferably dyed by these processes.

*Aftertreatment* of dyed goods is essential for the quality of the dyeing. This is especially true with regard to their fastness, but also holds for other properties of dyed goods (fibers, yarns, and textile fabrics) such as processibility, appearance, hand, and mechanical properties. In deep dyeing, particularly, residual dye remains on the fiber surface after drying and heat-setting with deleterious effects, particularly on fastness. A reducing, alkaline, postdyeing, cleaning treatment removes these dyes, as well as any oligomers that are diffused from the interior to the surface of the fibers. Temperature and pH of the postcleaning bath must be carefully controlled because of the danger of hydrolytic degradation.

## 2. Structure and Properties

### 2.1. Polymer

*Chemical Structure.* *Homopolymer.* In addition to the OH end groups of the macromolecules, PET fibers contain 10 – 40 milliequivalents of carboxyl (COOH) end groups per kilogram of fiber. These are produced in the melt at elevated temperature (especially beginning at 280 °C) by breaking of the ester bonds. Because COOH end groups catalyze hydrolytic degradation of the polyester, they are occasionally protected (e.g., with epoxide compounds or carbodiimides), especially in products destined for industrial use [39].

Diethylene glycol, which is present in the PET chain in concentrations up to ca. 1 mol %, affects softening behavior of the polyester, dyeability of the fibers, and to a certain extent, lightfastness. For this reason, diglycol formation is suppressed as much as possible by adjusting the conditions of the reaction.

In addition to macromolecules, the polymer contains ca. 2 % oligomers, primarily cyclic trimers, but also linear oligomers. The former can adversely affect washing and dyeing.

Additional impurities in the polyester are acetaldehyde and degradation products. At higher thermal loading of the melt during further processing, these can lead to formation of gel particles and discoloration.

In the solid state, PET absorbs ca. 0.4 % of water from the atmosphere at 20 °C and 65 % relative humidity. Because even minute amounts of water can cause, or at least substantially favor, hydrolytic degradation, the polyester must be carefully dried to a moisture content below 0.01 % prior to spinning.

*Copolymer.* The principle of copolymerization of PET with suitable monomers is used to improve certain fiber properties.

The relatively high glass transition temperature of the fibers limits dyeability in boiling water to light and medium shades. Deep shades can be obtained by using higher dyeing temperature (HT dyeing at ca. 135 °C under pressure) or carriers. When these procedures are impractical, e.g., because of the presence of wool or the contamination of wastewater, fibers of block copolymers (PET with polyglycols) can be used [37, 38]. By loosening the noncrystalline regions, the glass transition temperature of these fibers is lowered far enough to attain both the required diffusion and the desired saturation concentration of disperse dyes in the fiber.

In differential dyeing (single bath dyeing of textile fabrics to various shades), fibers are needed that can be dyed with both cationic and disperse dyes. The sodium salt of 5-sulfoisophthalic acid [51876-65-4] is almost exclusively built into the PET molecule for this purpose [38].

The low pilling property of textile fabrics, necessary for practical use, requires low-pilling PET staple fibers. Low molecular mass PET is used for these fibers. In addition, cross-links sensitive to hydrolysis are also built into the polyester with the aid of, for example, organosilicon compounds [40].

A particular requirement for home furnishings is low flammability. For years PET fibers have been used in which flame retardants in the form of suitable phosphorus compounds are co-condensed with PET ($\rightarrow$ Flame Retardants) [41, 42]. In this way, flame protection becomes permanent; it cannot be lost in washing or dry cleaning.

*Additives.* In many cases, the polymer includes additives intended to change the normally transparent and brilliant white appearance of fiber products. The most frequently used additive is titanium dioxide [13463-67-7], mostly anatase but occasionally the harder rutile, which serve to dull the fiber. Particles with average grain sizes between 0.3 and 0.4 µm are added to the transesterification product in concentrations of 0.1 – 3 %.

Dyed fiber products are made by adding soluble dyes or pigments, if extreme lightfastness of dyed goods is required, to the melt. To produce black fibers, carbon black is usually added.

Many attempts have been made to achieve permanently antistatic polyester fibers by admixture of suitable compounds. Surface resistances of $10^{10}$ Ω have been attained [43]. However, not much practical use has as yet been made of these possibilities. In general, the desired electrical surface conductivity is obtained by finishes applied to the fiber.

***Physical - Chemical Structure.*** The most important structural variable of fiber-forming polymers is the *average molecular mass*. This describes the mass of the average macromolecule and is the product of the molecular mass of the monomer and the average degree of polymerization. A measure of the molecular mass and hence the length of molecules is the specific viscosity, $\eta_{\text{spec}}$, measured in dilute solution:

$$\eta_{\text{spec}} = (\eta - \eta_0)/\eta_0$$

where $\eta$ is the viscosity of the solution and $\eta_0$ the viscosity of solvent. For poly(ethylene terephthalate), $\eta_{\text{spe}}^c$ is between 0.5 and 1.5, depending on the application of the fiber products.

Generally, molecular mass is described by the intrinsic viscosity $[\eta]$, often designated as *IV*

$$[\eta] = \lim_{c \to 0} \eta_{\text{spec}}/c$$

where $c$ is the concentration of polymer in the solution being measured. Again, depending on the area of application, $[\eta]$ for PET ranges from 0.45 dL/g for low-pilling fibers to 1.05 dL/g for tire cord, which corresponds to a span of 50 to over 200 for the average degree of polymerization.

The *distribution of molecular mass* corresponds to that given by FLORY for condensation polymers [44]. Changes in this distribution function by degradation can be reversed by melting. The distribution function of molecular mass is rarely measured and is not used for product testing.

*Melt viscosity* is measured with capillary or rotary viscometers. In polymers intended for standard fibers, it is ca. 250 Pa·s at 285 °C. Melt viscosity increases rapidly with increasing $[\eta]$ and decreasing temperature. It is one of the basic variables of spinnability. The orientation of the macromolecules during filament formation, is particularly dependent on melt viscosity.

The *transition temperatures,* generally measured by differential thermal analysis, are the final determinants of the orientation, crystallization, and melting behavior of PET. The glass transition temperature is measured as ca. 68 °C on the isotropic and amorphous polymer. It depends on the orientation and crystallinity of the polymer and thus can be as high as 100 °C in highly oriented, semicrystalline, industrial, continuous-filament yarns. In the isotropic and amorphous polymer, the crystallization temperature is ca. 100 °C; the maximum rate of crystallization is observed between 140 and 180 °C. The melting point of PET, which depends on crystallinity and crystal size, lies between 255 and 265 °C.

***Physical Structure.*** Poly(ethylene terephthalate) quenched from the melt is isotropic and almost amorphous. The density of amorphous PET is 1.335 g/cm$^3$. The polymer can be crystallized primarily by the action of heat but also by the presence of agents such as dichloromethane or dioxane. The crystal structure is triclinic; the density of the crystalline regions is 1.515 g/cm$^3$ [45].

The configuration and conformation of monomeric terephthalic acid and ethylene glycol in the macromolecule are also known. Infrared spectroscopy shows that the terephthalic acid component is present in the trans configuration in the crystalline part, but in the *cis* configuration in the noncrystalline part. The IR method shows that the glycol component exists in the trans conformation in the undisturbed crystalline region and in both the gauche and the trans conformation in the noncrystalline region. The mass ratio of the glycol components in the gauche and trans conformations changes not only during crystallization but also during orientation and disorientation. This ratio correlates to the mechanical properties of shaped PET articles [46].

When PET granulate is melted and the melt is forced through spinneret holes, as-spun filaments are formed. The orientation of the macromolecules in these filaments, characterized by the birefringence $\Delta n$ of the filaments, depends to a large extent on windup speed. Thus, $\Delta n$ at a windup speed of 1500 m/min is only ca. 0.010, but at 3500 m/min it is as much as 0.055. Density, too, is subject to this dependence. At 1500 m/min it is ca. 1.34 g/cm$^3$, and at 3500 m/min it is 1.35 g/cm$^3$. The degrees of crystallinity behave similarly [47].

During drawing of as-spun filaments, the macromolecules are oriented further, and the birefringence of filament yarns for industrial uses increases to ca. 0.240. The ability of PET to crystallize is exploited in many ways during drawing and also in subsequent aftertreatment. Thus, the fiber density ranges from 1.36 g/cm$^3$ for drawn high-shrinkage staple fibers up to 1.41 g/cm$^3$ for highly drawn low-shrinkage filament yarns.

## 2.2. As-Spun Yarns

Nonuniformities of as-spun yarn along a filament or filament bundle, but also from filament to filament, from bundle to bundle, or even from plant to plant, are reflected in the structure of the final product. There, they appear as differences in textile properties, primarily in dye uptake. For this reason, yarn is carefully monitored during spinning. The glass transition temperature of PET (70 °C) causes the as-spun yarn to be quite stable at room temperature and makes measurement of orientation and even of crystallinity relatively easy. Intervening in the manufacturing process in case of deviation is then possible.

Many physical measurement techniques are available to investigate the structure of PET as-spun yarn. The total orientation of a filament is measured by *optical birefringence*. Except for highly crystalline as-spun yarn, the birefringence is proportional to the orientation factor [11]. If suitably measured, it provides indications of any possible sheath–core structure. Another measure of orientation is the velocity of sound and with it the *sonic modulus* [48, 49]. They change little at lower orientation and increase noticeably only at spinning speeds ≥ 2500 m/min [14].

The degree of crystallinity of as-spun yarn can be calculated easily from its *density*; care is required, however, because the densities of completely amorphous and completely crystalline PET, required in these calculations, are not entirely independent of orientation [47]. In addition, true separation of the amorphous and crystalline phases is presumed, which is sometimes questionable.

Much information can be gained from *X-ray* investigations. Wide-angle patterns or goniometry provide data on the size, quantity, and orientation of the crystallites, and the approximate degree of crystallization. With their aid, total orientation can be divided into orientation of the amorphous and of the crystalline fractions of the filament. Small angle X-ray scattering permits conclusions to be drawn concerning the superstructure, i.e., the mutual arrangement of crystalline and amorphous regions and also the state of order of the latter.

*Thermal Measurements.* If as-spun yarn or tow is heated at constant length beyond the softening point to ca. 100 °C, a *shrinkage stress* that corresponds to the orienting stress during spinning is created [25]. This stress is therefore a measure of orientation and is largely proportional to the optical birefringence. These two quantities diverge only at high spinning speeds.

If as-spun yarn is heated without restraint, it shrinks, sometimes very much. In so doing, the orienting elongation $D$, frozen in during spinning, is recovered; hence, $D$ can be calculated from the *spun yarn shrinkage* [50]. As expected from theory, the quantity $D^2 - 1/D$ is approximately proportional to birefringence and shrinkage stress at constant length, which are measures of orientation [50, 51]. However, this is true only for spinning speeds up to 1500 m/min with normal quench. Beyond that, shrinkage is impeded by crystalline regions developed in the sample during spinning or by the regions created during measurement. Thus, spun yarn shrinkage goes through a maximum of 60 – 70 % at spinning speeds between 2000 and 4000 m/min and then drops rapidly with increasing speed [13, 20, 25]. In this spinning range, spun yarn shrinkage is again recommended as a measure of crystallinity, even of dyeability of the final product [20, 29].

*Differential Thermal Measurements (DTA).* The glass transition temperature, the temperature of the maximum rate of crystallization, the melting point of crystallites, and the final melting point can all be determined from DTA curves [14].

*Mechanical Tests.* If a PET as-spun yarn is elongated at room temperature and its tensile stress is measured at the same time, the stress – strain curves shown in Figure 3 are obtained. For as-spun yarn of lower orientation, a saddle appears after the initial rise in stress, while a draw point crawls along the clamped yarn [23].

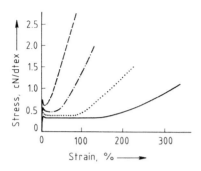

**Figure 3.** Stress – strain curves of PET as-spun yarns spun at four spinning speeds
Spinning speeds:
——— 1200 m/min
······ 2000 m/min
—·—·3000 m/min
– – – – 4000 m/min

The percentage of elongation reached at the end of this saddle is called the natural draw ratio $R_n$ of the as-spun yarn. With increasing spinning orientation, shown in the figure by higher spinning speeds, $R_n$ decreases; this is because a greater amount of extensibility has been anticipated during spinning [25]. If spinning orientation becomes too large, $R_n$ can no longer be specified [20, 28]. Then either the elongation of the as-spun yarn at a given load or the force necessary to produce a given elongation is used [25]. In practice, determining the operational draw ratio that imparts a definite property, such as break elongation or dye uptake to the final product might be useful. This, however, assumes that no changes occur in further processing.

To monitor the spinning process and possibly pinpoint defects in commercial practice, tests should be limited to those that can be most readily and accurately carried out and that provide averages over a sufficient quantity of as-spun yarn. Each kind of test, taken separately, is useful only for monitoring, not for control. If a test quantity exceeds the tolerance range, use of another kind of test is recommended to readjust the process conditions.

## 2.3. Drawn Materials

During drawing, PET crystallizes in the triclinic configuration mentioned in Section 2.1. The supermolecular structure is assumed to consist of somewhat parallel layers, inclined to the fiber axis. With increasing draw, the distance between the layers, measured in a direction normal to the layer surface, decreases, and the layers rotate more in the direction of the fiber axis [27, 52].

In practice, material that has been drawn but not processed further is almost never available as an intermediate product. A posttreatment almost always follows immediately, as described in Section 1.4.

## 2.4. Dyed Poly(Ethylene Terephthalate) Material

***Fastness Properties.*** The lightfastness of dyed textiles for apparel and home furnishings is important in preserving their coloration. A lightfastness rating of 6 can be achieved readily in articles of unmodified PET fibers with most commercially available disperse dyes. The lightfastness of PET fibers, modified to be dyeable to deep shades at the boil without carrier, may be half a rating lower. Dyeings of PET fibers, modified to be cationic-dyeable, generally do not rate higher than 4 – 5 in lightfastness.

Heat-setting fastness, rubbing fastness, and washfastness of dyeings of PET fibers meet all practical requirements. The only exception is the slightly lower washfastness of modified PET fiber, dyeable without carrier. A wash temperature of 40 °C should not be exceeded.

***Degradation and Mechanical Properties.*** In discussing alkaline postdyeing cleaning treatment, the danger of hydrolytic damage to PET fibers is mentioned. This damage consists of molecular degradation, which can extend to different degrees from the fiber skin to the core. The consequence, in general, is a reduction in tensile strength and breaking elongation of the fibers caused by the embrittlement connected with degradation. At the same time, the hand of the textile fabric can occasionally be changed. However, sometimes such changes of the hand are desired (e.g., by alkalization).

Another change in the properties of PET fibers or textiles is sometimes produced by shrinkage during dyeing, which may be accompanied by crystallization processes. Carefully controlling the dimensions of, and stresses in, textile fabrics as well as the drying and setting temperatures is required.

## 3. Poly(Ethylene Terephthalate) Fibers

### 3.1. Staple Fibers

*Special Features of the Manufacturing Process.* In the manufacture of PET staple fibers, polymers with intrinsic viscosities [$\eta$] of 0.65 – 0.75 are generally used; for special types, polymers of much lower viscosity with [$\eta$] ca. 0.5 are occasionally used. The equipment on which PET staple fibers are produced is usually large, with daily production up to 50 t and more. Therefore, feeding the spinning units directly from a condensation unit makes economic sense [53].

A spinning unit for PET staple fibers consists of 8 – 32 spinning positions including their accompanying spinning shafts, aligned side by side. Each position is supplied with melt, either simultaneously or in groups, by an extruder or directly from a condensation unit [4].

Depending on the type of air quench, spinnerets may be circular with uniformly arranged spinneret holes, they may be plates in which the holes are arranged rectangularly, and finally they may be circular with an annular arrangement of holes [53]. The number of holes is based on the as-spun fineness and the feed of the spinning position. The spinneret hole shape is not always round. For profiled fibers with non-round cross sections, such as triangles, five-pointed stars, or even hollow fibers, the holes must have profiles that exaggerate those of the final product because the profile impressed by the hole tends to blur somewhat during spinning.

The output of a spinning position has been increased from 500 g/min to 1500 g/min and more by increasing the number of spinneret holes, by forceful quenching, and by increasing the spinning speed.

For example, a filament bundle with a linear density of 8 dtex per filament is spun from an 1125-hole spinneret at a speed of 1500 m/min. The output per spinning position is 1350 g/min. A 16-position spinning machine then provides about 31 t/d of as-spun yarn in the form of a tow with total length of 2160 km and a linear density of 144 000 dtex, consisting of 18 000 individual filaments.

The length of the spinning shafts is determined by the need not only to ensure that the filaments are cooled below 70 °C but also to prevent the moving individual filaments from sticking together. The shafts are generally 5 m long but, with a suitable quench, they may be shorter [53].

Finish is applied to the filament bundle at the lower end of the shaft to ensure trouble-free packaging of the as-spun material. The spin finish is also used to impart surface characteristics to the fibers needed downstream during manufacturing and textile processing [54–56]. This is more economical and protects the environment better, but some compromises with respect to processability during tow drawing and textile processing are unavoidable.

After the finish application, strands coming down from the spinning units are turned horizontally and combined into a tow, which is drawn off and deposited in a spinning can. Depending on the installation, these cans hold 0.5 – 2 t of tow, all of which should have the same length.

The fibrous as-spun product is preoriented; in newer spinning units with very effective quench, for example, it is characterized by a birefringence of $(8 - 14) \times 10^{-3}$ or by a natural draw ratio of ca. 1:2.5.

High-speed spinning at ca. 3000 m/min, which is advantageous for textile filament yarns, has been attempted for production of PET staple fibers [57]. Problems still exist in developing a reliable process for depositing the tow in the spinning can [58]. A sheath – core effect in the filaments also presents difficulties.

As is customary in the production of staple fibers and tows from melt- or dry-spun synthetic polymers, the as-spun tows are combined and further processed continuously as a tow band. The tow band (up to ca. 1.20 m wide) is first immersed in a warm bath where the stresses in the individual strands are equalized; the tow is thoroughly wetted, and if necessary, additional finish is added. From the immersion bath, the tow enters the draw zone between two sets of rolls [54].

During drawing, the draw points or draw zones of all filaments in the band must be as close together as possible and must not wander [54]. A number of drawing processes are available for this [59]. If the tow is cold when it enters the draw zone, the yield stress must be lowered by sudden heating, e.g., via a steam jet, so that drawing can occur. In this process, the draw ratio is limited, and if necessary, a second drawing step must be provided.

If the tow is heated on hot feed rolls, drawing begins on the last or next-to-last feed roll and can even be facilitated, if necessary, by an immersion bath under the third roll from the end. It is also usual to pass the tow through an additional hot zone in the draw stand. With this process, reaching higher draws with one drawing is possible.

Drawing in water is a compromise in which only the last feed roll is heated and dipped into the hot water draw bath. Drawing begins at that point.

If postdrawing is to be accomplished in a separate draw zone, the draw rolls of the first stage are heated because they represent the feed rolls of the second draw, which is then carried out in a heating device.

The stress – strain behavior of the end product can be preselected by the draw ratio but is more critically affected by the type and place of heat-setting. Thus, fibers that are to exhibit a steep stress – strain curve with high initial modulus, similar to cotton, are relatively highly drawn, e.g., 1:4.5 to 1:6, depending on the as-spun yarn. After drawing, the tow is heat-set under tension, usually at constant length. The heat-setting temperature is between 160 and 230 °C, depending on the shrinkage desired. After setting, the tow traverses a cooling zone under tension and then enters the crimper. To facilitate crimping, it is first slightly moistened or steamed. The tow must not become too hot in the crimper; otherwise it relaxes and textile properties deteriorate. After crimping, finish can again be applied if necessary, in which case the tow must again be carefully dried [59].

Fibers that are to have a flat stress – strain curve similar to wool do not need to be so highly drawn; a draw ratio range of ca. 1:2.5 to 1:3 will suffice. After drawing, the tow is sent immediately to the crimper via a short steam zone. The crimp is then heat-set without tension, often on traveling grate machines. The heat-setting temperature, ca. 120 – 180 °C, is selected according to the amount of shrinkage desired in the end product. The flat stress – strain curve is formed by the tensionless setting, and the crimp becomes highly stable.

Finally, the crimped and, if necessary, shrinkage-stabilized tow is packaged. Either it is compressed into a bale and sold as endless tow, or it passes through a cutter and is cut into staple fibers, which are then baled. The usual mass of a bale can range up to 500 kg.

Staple fiber draw lines are up to 80 m long and produce from 20 to 250 t/d, the latter, of course, only for high-volume products. Delivery speeds range from 150 to 300 m/min, with tow sizes up to 600 ktex and more [60]. Investment and manufacturing costs per tonne per day are reduced considerably with increasing productivity of these draw lines [58].

Completely continuous installations with capacities of 12 – 15 t/d produce smaller lots. They are supposed to be superior to the usual tow band installations in terms of investment and manufacturing costs, but they produce fibers with three-dimensional crimp. Whether such fibers can be spun in blends with natural fibers and whether fabrics produced from such yarns will be acceptable in the marketplace remain open questions [58].

***Types of Fibers, Properties, and Uses.*** Polyester staple fibers encompass the broadest product spectrum of any fiber. They can be found in all areas of application [4, 61–63]. They are used in apparel for overcoats, jackets, leisure and sportswear, ladies' outerwear, men's suiting, work and protective clothing, children's wear, underwear, and sleepwear. In home furnishings, they range from drapery and curtain fabrics, through furniture coverings, pillows and pillow stuffing, table and bed linen, to wall and floor coverings. They have a whole host of industrial end uses along with filament yarns, spunbonded products, and monofilaments. A few examples of applications are in filtration, automobile interiors, geotextiles, and material for the sewing trade.

Yarns made from PET fibers can be processed without reservation in weaving and knitting as well as in tufting. The fibers are completely suitable for processing as unblended

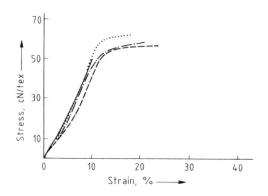

**Figure 4.** Stress – strain curves of PET cotton type staple fibers
——— Cotton
— — — — PET cotton type (normal shrinkage)
—·—·— PET cotton type (low shrinkage)
········ PET sewing thread type

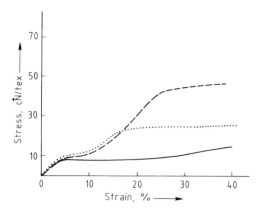

**Figure 5.** Stress – strain curves of PET wool types
——— Wool
— — — PET wool type (normal)
····· PET wool type (low pilling)

yarns (100 % PET), as well as in blends with cotton, wool, silk, cellulosic man-made fibers, acrylic fibers, and many more.

Of all polyester fibers, *cotton types* have the largest market share. These are processed into blended yarns, and these in turn into blended articles. These contain cotton, viscose, or modal fibers in what are today considered the "classical" blend ratios of 65:35 or 1:1 (PET to cotton) and in other ratios.

The stress – strain curve of cotton types is shown in Figure 4. Its initial slope corresponds approximately to that of cotton. Their finenesses of 0.4, 0.9, 1.3, and 2.4 dtex, and the major one of 1.7 dtex, at cut lengths of 30 and 38 mm are matched to cotton.

Fibers with normal high shrinkage and others with reduced shrinkage are available and used to vary the hand and appearance of textile fabrics over a wide range. With the additional possibilities offered by the choice of fineness and weave construction, the textile processor can exert a major influence on the permeability of, for example, tent, overcoat, shirt, and blouse fabrics. In comparison to wool types of polyester fiber, dyeing of highly oriented cotton types is somewhat slower but, like the former, presents no problems.

Special cotton-type fibers, which are also offered as continuous tow for processing on stretch – breaking converters, are intended for sewing. These fibers are especially highly drawn and, therefore, have the highest moduli (resistance to elongation) and tenacities.

*Wool types* are intended primarily for processing in blends with wool. The major blend ratio is 55:45 (PET to wool). The stress – strain curves of these fibers are matched to that of wool (see Fig. 5) by tensionless heat-setting during fiber manufacture, as described in Section 1.4. Shrinkage is therefore low, and dyeability is optimum. The spectrum of fineness and cut length ranges from 1.7 dtex (38 mm) to 6.7 dtex (120 mm). Wool types are also available as continuous tow for processing on cutter – converters.

The so-called pilling phenomenon, which occurs in blends with wool, has been responsible for considerable developmental cost. Small balls or pills are formed on the surface of textile fabrics during use. Suitable fibers have been developed to solve this problem [27, 62]. Their special structure eventually causes the fibers anchoring the pill to break. In this manner, the pill disappears soon after it is formed.

A third group consists of the *fiberfill types* commercially available as staple fibers but also as continuous tow. These fibers are used both in apparel (primarily for heat insulation) and in home furnishings (primarily in pillows, cushions, and bedding). Their crimp is two- or three-dimensional. To improve the hand of fiberfill, such fibers are often provided with a special finish, based primarily on silicone. The range of linear densities of these fibers is between 1.5 and 14 dtex at a cut length of 30 – 80 mm (most common length, 60 mm). This group also contains high-shrinkage fibers with a shrinkage at the boil of 50 %.

In home furnishings and automobile interiors, *polyester carpet types* are used. These also include poly(tetramethylene terephthalate) fibers, see Section 4.1. Carpet fibers based on PET are used as both normal and high-shrinkage types in home furnishings and in needle-punched felt. Such fibers are also made with various cross sections; their finenesses and cut lengths range from 6.7 dtex (135 mm) to 20 dtex (150 mm). An essential characteristic of polyester floor coverings is that their mechanical properties are unaffected by moisture.

*Special types* of PET staple fibers have secured a place in the market. Some can be dyed in an open vessel without carrier to deep shades [37]. Such staple fibers and tows meet special requirements of environmental protection where wool blends have to be used (carriers) and HT dyeing is therefore excluded. These fibers exist as standard types and as pill-resistant tows. Cationic dyeable staple fibers are used in relatively small amounts in the wool area and in polyester carpets.

Since the mid 1970s, *flame-retardant PET fibers* have rounded out the product spectrum [42]. These fibers are used in apparel and, especially, in home furnishings. They meet legal requirements and even exceed them in some cases. For these, PET copolymers with phosphorus compounds have proved particularly useful. Fibers of this type combine the comfort expected of unmodified PET fibers with modern safety requirements. They are distinguished by deep dyeability with disperse dyes. Because of their diverse applications, these fiber types are available in finenesses and cut lengths from 1.7 dtex (38 mm) to 13 dtex (150 mm).

## 3.2. Continuous-Filament Yarns for Apparel and Home Furnishings

***Special Features of the Manufacturing Process.*** Poly(ethylene terephthalate) that is to be spun into continuous-filament yarn for apparel generally exhibits an intrinsic viscosity $[\eta]$ of 0.70 – 0.75 [4]. It must be highly purified, especially when used in high-speed spinning, which is often the case. The polymer must also be dried carefully to < 0.01 % $H_2O$.

As in staple fiber units, the spinning positions of a spinning machine are aligned in a row. Quenching is mostly done by a transverse air flow with several spinnerets in a single shaft so that all filament bundles are affected uniformly. At the lower end of the spinning shafts, the filament bundles are individually combined into yarns while finish is applied to them. Finishing with a perforated yarn guide has been generally adopted. Below the spinning shafts, the yarns are usually drawn off by godets, the circumference and revolutions of which determine the spinning speed; the yarns are then wound individually on bobbins [4].

Originally, spinning speeds were 500 – 1500 m/min. The as-spun yarn was amorphous and characterized by an orientation birefringence $\Delta n$ of $(3 - 10) \times 10^{-3}$, a so-called *low orientation yarn (LOY)* [20, 62]. The as-spun yarn was drawn by passing it, with friction, over heated pins in the draw zone and then additionally, with surface contact, over a hot plate. This process was replaced by drawing on heated rolls with several wraps [4]. For heat-setting, the delivery roll of the set of draw rolls was also heated but at considerably higher temperature. Final drawoff was by a third pair of rolls. The drawn yarn was wound on a twister (draw twisting) or on a winder (draw winding); frequently it was then twisted, heat-set, and rewound. Such units, which involve slow spinning and separate drawing, are still operating but are probably no longer being built.

Improvements in winding frames for drawn yarn led to the next development, spin drawing. In this process, a set of draw rolls is mounted below each spinning shaft. The as-spun yarn is drawn down at ca. 700 m/min and finally wound at ca. 3000 m/min. Winders developed for high-speed spinning in 1983 permit drawing down at speeds up to ca. 4000 m/min with final windup speeds up to 6000 m/min [56]. This process is highly cost-effective, and product quality is good.

In a variant of the spin-drawing process, the draw rolls are mounted near the spinneret at the point where the filaments have just attained their final fineness and are still warm [65]. This process, too, is rated highly for cost and quality of product.

Spinning speeds have gradually been raised to 2800 – 4500 m/min to increase productivity of the spinning units. The result has been production of *partially oriented yarn (POY)* [20, 64]. These yarns are less sensitive because of their inherent slight crystallinity and require less drawing. Drawing can now be combined with texturing into a single process, with considerable savings in investment and operating costs [66].

Drawoff rolls or godets are not required in high-speed spinning, and drawing-off can be left to the winder; however, special measures must be taken to match the spinning tension to the optimum (mostly lower) windup tension [56].

Yarn spun at higher speed and then drawn takes up dye more readily; because it is more highly oriented, postdrawing is lower and entails less additional crystallinity. Material spun at very high speed is, moreover, said to have an especially loose structure, with fewer large crystalline regions, even after drawing [20].

Commercial high-speed spinning has become possible only through the development of suitable winders, which can also be used for as-spun yarn. They can go as high as 6000 m/min and, for experimental purposes, up to 10 000 m/min [67].

In the speed range of 3000 – 4000 m/min, productivity no longer increases substantially with spinning speed. Hence, yarn spun at 2800 – 4000 m/min has been adopted as the starting material for draw texturing [64]. Therefore, additional investment in increased spinning speed is probably not worthwhile [68]. Efforts have been directed at reducing spin orientation and thus increasing draw ratio and productivity. For example, a heating tube below the spinneret reduces spin orientation [69], as do minor additions to the polymer of low-viscosity polymers [70] or components with slight cross-linking action [71].

The drawing process is combined with texturing in draw texturing to produce textured yarns from material spun at high speed. In this process, the yarn is drawn in the texturing zone on the first heater of the texturing unit, i.e., between the delivery rolls and the twister. This places stringent requirements on the as-spun yarn. Its structural properties, such as orientation and crystallinity, must be uniform to ensure uniform dyeability of the end product. In addition, the as-spun yarn must not have any defects such as broken filaments. In most cases, the yarn is additionally heat-set on a second heater in the texturing unit with a 7 – 14 % contraction in length. This setting process reduces the crimp elasticity of the yarn, which is then dimensionally stable.

The major problem in the manufacture of textured PET yarns is uniform dyeability. The many process steps and, hence, variables require special attention during manufacture. Total testing costs are high.

A smaller fraction of PET filament yarns for apparel is processed without texturing. Flat yarns are obtained by the old process of separate spinning and drawing mentioned previously, as well as by spin drawing. More recently, flat yarns have also been spun at high speed [15, 20]. Several processes, such as high-speed spin drawing and warp drawing [72, 73], can accomplish this economically and with high product quality.

Because PET yarns spun at 6000 m/min still have flat stress – strain curves, they cannot be used as is [64]. Thermal treatment in the spinning shaft, e.g., a heating zone in the shaft, helps increase their orientation and crystallinity [74, 75]. The stress – strain behavior of this *highly oriented yarn (HOY)* then suffices for several special applications. At still higher spinning speed, the sheath – core effect of orientation and crystallinity can exert a negative influence [20, 76].

***Types of Continuous-Filament Yarn, Properties, and Uses.*** Continuous-filament yarns are products of the man-made fiber industry that can be converted directly to textile fabrics, i.e., without intermediate processing. Whereas staple fibers are produced by first cutting crimped filaments which then must be spun into a yarn, possibly in mixtures with other fibers, in a relatively expensive process, filament yarn is available immediately.

Filament yarn is offered flat, i.e., with silk-like, or variously crimped textures; thus, it provides opportunities to change the hand, bulk, and optical appearance of downstream goods and to accommodate styling requirements in apparel and home furnishings, which far exceed the range attainable with natural fibers [4, 61–63].

Partially oriented yarns (POY) have, since the introduction of high-speed spinning technology (see Section 1.2), led to successful intermediates for internal fiber production as well as to salable products. These filament yarns are spun and wound at speeds of ca. 3000 – 4500 m/min. They are precursors for pirn drawing (cops drawing), warp drawing, as well as warp drawing – sizing or draw texturing. Most POY products with a fineness of 300 dtex and 32, 34, or 36 filaments per yarn (dtex 300 f 32, etc.) are produced for draw texturing, which is the largest end use.

*Flat Yarns.* For direct textile processing in weaving and knitting, suitable flat *fully oriented yarns (FOY)* are available on pirns, bobbins, or as warp beams. They are sold with round, trilobal, triangular, or even octalobal cross sections with an individual filament fineness of 2 – 5 dtex and a yarn fineness of 22 – 167 dtex.

Pirn (cops) material intended for weaving and warp knitting is produced "classically" with twist from LOY or from POY. Material in bobbins, on the other hand, is used exclusively in weaving and comes primarily from one-step processes such as spin drawing or high-speed spinning (4000 – 6000 m/min) (see Section 1.2). The fineness of individual filaments made by this latter process ranges from 0.6 to 3 dtex. These HOY yarns are distinguished by uniform dyeability. A stress – strain curve, which, in contrast to standard yarn, exhibits higher tensile elongation, is shown in Figure 6. Warp beams, finally, are produced by the warp-drawing process from POY exclusively for warp knitting. For weaving, the warp draw – sizing process is used.

*Textured Yarns.* The major application for textured yarn, primarily false-twist textured, is in knitting and weaving for apparel. Air-textured yarn, on the other hand, is used predominantly in home furnishings and is processed by weaving.

False-twist textured yarns are draw-textured, starting with POY, mostly by the friction process, but in some cases by the spindle process. Their linear density is 50 – 200 dtex (singles yarn), with an individual filament fineness of 1.25 – 5 dtex. The round cross section of POY filaments is distorted by draw texturing. However, special yarn types are produced with a triangular profile, which is preserved during draw texturing. This type of individual filament cross section imparts a special luster, almost a glitter, to textile fabrics. The linear density of most false-twist textured yarns now used in knitting and weaving is ca. 167 dtex. Fine yarn down to 50 dtex/40 filaments is used in ladies' blouses, for example.

Air-textured yarn can be made from either POY or FOY. With POY, a drawing step precedes air texturing on the texturing machine, whereas FOY is textured directly. In this process the filament cross section, usually round, is preserved. The linear density of such yarn is 200 – 3000 dtex for single yarn, which is appropriate for its primary use in home furnishings. The fineness of individual filaments is 2 – 3 dtex.

*Special Types.* The extensive product line of continuous-filament yarns includes many special types. Examples are filament yarns with staple yarn character, i.e., filament yarns with the less regular and bulkier structure of staple yarns, which is achieved by the type and arrangement of individual filaments. In some cases, free fiber ends even protrude from the yarn.

A large variety of properties is possible today. Among the most important are yarns with high bulk character, which is achieved by differential shrinkage of individual filaments. The properties of such yarns, however, are closer to flat yarns. Other yarns are formed by cotexturing different types of individual

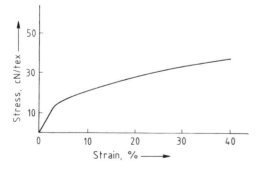

**Figure 6.** Stress – strain curve of highly oriented filament yarns (HOY)

filaments. These approach staple yarns, even with regard to the comfort of clothing made from them.

Extremely low fineness filaments with individual counts down to 0.1 dtex (produced by splitting thicker filaments) or 0.6 dtex (spun directly) are used in textiles with the appearance and feel of buckskin.

In differential dyeing cationic dyeable yarns are used in addition to normal disperse dyeable filament yarns. When yarns are made from blends of disperse dyeable individual filaments and cationic dyeable ones, they may be dyed in different colors.

Flame retardancy in home furnishings, especially in drapes and curtain fabrics, is achieved with filament yarns that consist of modified PET. By inserting a phosphorus compound into the fiber molecule, the flame retardancy of such yarns is permanent and cannot be removed in washing or dry cleaning.

Finally, filament yarns with permanent antistatic properties are available. In general, however, the surface conductivity required for processing and for some end uses is achieved by applying finishes.

## 3.3. Continuous-Filament Yarns for Industrial Applications

***Special Features of the Manufacturing Process.*** Filament yarns for industrial use are generally spun from higher molecular mass PET, which is the only way to achieve high strength. For low-fineness, high-tenacity yarn (final fineness < 1000 dtex) PET with an intrinsic viscosity $[\eta]$ of 0.75 – 0.80 is used; for coarse, high-tenacity yarns (final fineness ≥ 1000 dtex), the intrinsic viscosity $[\eta]$ of PET is 0.95 – 1.05. Such molecular masses are achieved either by continuous polycondensation with direct spinning or, preferably, by postcondensation in the solid state [4].

During condensation in the melt, COOH end groups, which catalyze hydrolytic degradation of PET in the melt, are formed in the polymer and can disarrange the structure in the final product. These end groups can be protected with carbodiimides [39] or similarly acting substances, which are added to the polymer in small amounts prior to spinning. No COOH groups are formed if solid-state condensation is carried out in a vacuum or under a protective gas and at temperatures below the melting point. Care must be taken, however, that the polymer does not absorb water during storage or during transportation from the condensation to the melting equipment.

Polymer with a high viscosity requires a rather high spinning temperature, e.g., ca. 290 °C for fine yarn and ca. 310 °C for coarse yarn [4]. To minimize thermal degradation of the melt, the lines carrying melt from extruder to spinneret must be as short as possible. Also, the melt in the distribution lines after the extruder is kept relatively "cold" and warmed to the higher spinning temperature only in the spinning pack [77].

Fine, high-tenacity yarn has a linear density similar to that of textile yarn. For this reason, it is spun on a similar machine. The yarn is either spun and drawn in separate process steps or spin-drawn continuously. The spinning speed for fine high-tenacity yarn in the two-step process is approximately 1000 m/min and 500 m/min and higher during spin drawing. The draw ratio is ca. 5; the yarn is usually allowed to shrink ca. 10 % during heat-setting at a hot roll temperature of ca. 240 °C. The delivery speed of the drawing machines is ca. 700 m/min, but during spin drawing, it is substantially higher (ca. 2500 – 3000 m/min).

A special, high-volume variant of fine, high-tenacity yarn is sewing yarn. By plying, twisting, and renewed heat-setting, throwsters turn this yarn into sewing thread, whose most important characteristic is the sewing length achievable. The goal is sewing yarn that needs no further plying and twisting to make it comparable in processability to sewing thread of staple fiber.

During spinning of particularly viscous polymers for coarse industrial yarn, very high drawing forces develop in the filaments. These forces can be relieved by delaying filament cooling with the aid of a heating tube directly below the spinneret, which reduces spin orientation and equalizes the structure of the filaments [69, 78]. In this way, noticeably higher final strength can be attained [79].

Coarse high-tenacity PET yarn is produced either in one step by spin drawing or in two steps by spinning and further treatment as a warp.

During spin drawing, the spinning speed is ca. 700 m/min; when spun on bobbins for warp processing, it is ca. 700 – 1000 m/min. Yarn has been spun at much higher speed, in which case the disturbing influences mentioned previously are compensated in other ways [80].

The spin-draw machines for coarse high-tenacity yarn resemble those for fine high-tenacity yarn except that some parts are even more massive [63]. Because of the high operating speed (up to 4000 m/min) the residence time on heat setting rolls and beyond is short. Therefore, preparing low-shrinkage material by spin drawing is difficult; however, the strength that can be achieved is considerable.

In the draw-line process, a warp of yarn is drawn in two steps and then heat-set. After the last set of rolls, which determines the setting shrinkage, the yarn is wound on individual bobbins. The delivery speed of such draw lines is 120 – 300 m/min, with a productivity of 4 – 10 t/d.

The ability of the end product to shrink and its shrinkage force can be reduced, both in spin drawing and in warp processing, by permitting shrinkage in the heat-setting step. This increases elongation of the yarn at loads important in practical applications. The much higher residence time of the yarn in postdrawing and setting in the warp process leads to lower residual shrinkage at equal relative elongation of the end product (relative elongation being the elongation at a given load). Inversely, at equal residual shrinkage, the relative elongation of the product from the warp process is lower. These relationships are influenced further by the polymer used, the spinning process, and the drawing conditions.

In certain end uses, coarse industrial PET yarn must exhibit good adhesion to rubber. For this purpose, an appropriate adhesive finish is applied during spinning or in warp processing prior to heat-setting. This finish may contain epoxy compounds (Epikote) or similarly acting material.

**Industrial Yarns, Properties, and Uses.** Of the three product forms used industrially, i.e., high-tenacity filament yarns, spunbonded products, and monofilaments, the filament yarns have by far the broadest application [4, 61–63]. This ranges from the building industry through machine and motor vehicle construction, chemistry, mining, agriculture and forestry, transportation, sporting goods, protection of the workplace and the environment, to apparel and home furnishings. Such yarns are offered on the market in three major categories: sewing yarn, fine yarn, and coarse yarn. Depending on end use, these yarns are employed either as normal-shrinkage or low-shrinkage types. In addition to fine and coarse white yarns, spun-dyed black types are also available. The extraordinarily broad product line is rounded out by plied fineness of coarse filament yarns. For an overview see Table 1.

*Sewing Yarns.* The value of sewn textiles, particularly their appearance and durability, depends to a large extent on seam performance. To best complement textile design and construction, the seam must meet many requirements, the most important of which are neat appearance, high tenacity and abrasion resistance, high lightfastness (especially to ultraviolet radiation), resistance to chemicals, and washfastness. In addition, good sewability is significant for the sewing yarns; this is defined as achieving a high sewing length at high sewing speeds without yarn breakage. All these requirements are met or exceeded by fine denier, high-tenacity PET filament yarns. They have performed well alongside PET staple yarns and PET stretch-broken yarns in 100 % PET sewing yarns as well as in core-spun yarns.

*Sewing Yarns with Normal Shrinkage.* Filament yarns with normal shrinkage are suitable for 100 % PET sewing yarns, which must be heatset (Table 1). Sewing threads prepared from them are used especially in high-performance areas, e.g., safety belts, cushions, leather shoes, luggage, and handbags.

*Sewing Yarns with Low Shrinkage.* Low-shrinkage filament yarns are offered in a broad range of finenesses (Table 1). The largest end use is in core-spun yarns. By spinning a sheath of cotton, viscose, or other polymeric staple fiber around a PET filament core, the properties of both fiber types are combined. The sheath assures sewability; the core, mechanical efficiency. Such yarns meet high requirements of aesthetics, dyeability, and temperature resistance.

**Table 1.** Filament yarns for industrial end uses

| Types | Shrinkage in hot air at 200°C, % | Fineness, dtex | Number of filaments | Filament fineness, dtex | Tenacity, cN/tex | Elongation, % |
|---|---|---|---|---|---|---|
| Yarns for sewing threads, normal shrinkage | 17 – 18 | 74 – 455 | 24 – 96 | 3.1 – 4.7 | 68 – 70 | 12 – 13 |
| Yarns for sewing threads, low shrinkage | 7 | 49 – 940 | 16 – 200 | 2.5 – 4.7 | 68 – 70 | 16 – 17 |
| Fine yarns, normal shrinkage | 18 – 19 | 280 – 940 | 48 – 200 | 3.3 – 5.8 | 65 – 70 | 12 – 14.5 |
| Fine yarns, normal shrinkage (spun-dyed black) | 19 – 22 | 550 – 940 | 50 – 100 | 9.4 – 11 | 66 – 69 | 19 – 22 |
| Fine yarns, low shrinkage | 4 – 14 | 140 – 550 | 24 – 96 | 5.8 | 60 | 21 |
| Coarse yarns, normal shrinkage | 15.5 – 22 | 1100 – 1670 | 100 – 200 | 5.5 – 11 | 72 – 73 | 13 – 15.5 |
| Coarse yarns, normal shrinkage (spun-dyed black) | 15.5 – 22 | 1100 – 1670 | 100 – 200 | 8.4 – 11 | 66 – 72 | 12.5 – 13.5 |
| Coarse yarns, low shrinkage | 3.5 – 5.0 (7.5 – 8.5)* | 1100 – 1670 | 200 | 5.5 – 8.4 | 72 (73 – 75)* | 16.5 – 17.5 (11.5 – 12.5)* |
| Coarse yarns for tire cord | 10 – 18 | 1100 – 1400 | 200 | 5.5 – 7 | 75 | 12 |
| Coarse yarns (plied fineness) | 22 | 2200 – 11 000 | 400 – 2000 | 5.5 | other values depend on single yarn | |

* With adhesive finish.

*Fine Filament Yarns.* Both normal- and low-shrinkage fine filament yarns are available for general use. Filament yarns with normal shrinkage are offered as white or spun-dyed black types. Both types are preferred in lashing straps, tapes, belts, ropes, nets, and fire hoses.

Low-shrinkage fine filament yarns are offered only in white (Table 1). They have a wide range of applications, extending from base fabrics for PVC or rubber coating for tarpaulins, membranes, containers, awnings, and tent roofs, through conveyor belts, V-belts, hoses, filter fabrics, and drying screens for paper making to sailcloth. Its use in sailcloth demonstrates the ability of PET to meet the requirements of low mass, high tenacity, and uniformity, as well as resistance to UV light, chemicals, water, and mildew.

*Coarse Filament Yarns.* Coarse yarns, like fine yarns, are offered with either normal or low shrinkage (Table 1). Yarns with normal shrinkage are available in white or spun-dyed black versions. White yarns, analogous to the corresponding fine yarns, are intended for lashing straps, belts, ropes, nets, hoses (especially pressure hoses), and safety belts. The situation is quite similar with respect to the black types, which are used primarily in safety belts and lashing straps.

Low-shrinkage coarse filament yarns, which are offered exclusively in white, have equally wide-ranging applications, such as base fabrics for PVC or rubber coating for tarpaulins, membranes, containers, awnings and tent roofs, hoses, filter fabrics, belts, conveyor belts, V-belts, and drying.

An especially interesting end use for low-shrinkage, high-tenacity, adhesively finished filament yarns is in automobile tires. The properties of such yarns make them highly suitable for the carcass of radial tires. Tires with these yarns no longer exhibit the disturbing phenomenon of flat spotting, observed in nylon-reinforced tires. (Flat spotting is the formation of an elliptical footprint in the tire during standing, resulting from molecular deformation of nylon because of its low glass transition temperature. It causes a bumpy ride until the tire warms up.) Since the early 1960s, much effort has been expended to solve the problems of PET tire-cord material. In addition to rubber adhesion, these included the hydrolytic stability of PET. High-tenacity filament yarns for tires are rather low in shrinkage. They have high moduli, exhibit high flexural strength, and thus produce little heat during mechanical loading. They, therefore, have all the prerequisites to utilize fully the high chemical stability (resistance to hydrolysis) provided by extremely low concentrations of COOH groups. Solid-phase polycondensation combined with chemical capping of the majority of any still-existing COOH groups (see Section 2.1) makes possible the high stability of the tire cord.

The high-tenacity, coarse-filament single yarns discussed so far are complemented for many end uses by a series of plied yarns. These include some that are adhesively finished for use with rubber, others that have low-shrinkage like the single yarns, and still others that have normal shrinkage. Their finenesses extend up to 11 000 dtex. These yarns are intended for use in the coating sector as well as in straps, belts, ropes, nets, and hoses.

## 3.4. Spunbonds

*Special Features of the Manufacturing Process.* Spunbonded products are consolidated random layers of 2- to 10-dtex continuous filaments [81]. They are produced as continuous sheets up to 5 m wide in a continuous process from polymers of average intrinsic viscosity ($[\eta]$ ca. 0.7). Their unit weights range from 20 to 500 g/cm$^2$ ($\rightarrow$ Fibers, 3. General Production Technology).

A spunbond production line usually consists of several rows of spinnerets with mutually displaced spinning packs. Each filament bundle is accelerated by a draw tube under the spinneret to a final speed of $\geq$ 5000 m/min and thus is drawn [82]. Spreading devices open the filament bundle and lay the filaments down on a moving belt in overlapping random patterns. Various designs are known for these spreading devices; i.e., mechanical, pneumatic, and electrostatic [81].

Properties of the filaments are the same as those of filament yarns spun at high speed: tenacity ca. 3.5 cN/dtex, break elongation ca. 70 %, and shrinkage at 200 °C ca. 3 %. The orientation is characterized by a birefringence $\Delta n$ of ca. 100×10$^{-3}$, and the degree of crystallinity is ca. 40 %.

The loose filament web, laid down as uniformly as possible, is then consolidated by using various processes singly or in combination, such as needle punching, calendering, and bonding. For certain applications, the web must be made thermally stable by additional heat-setting.

*Spunbonded Products, Properties, and Uses.* Polyester spunbonds are available in unit weights of 20 – 500 g/m$^2$ and in widths of 1 – 5 m. The consolidation discussed in the preceding section via needle punching, bonding, or calendering requires, in addition to PET and suitable binders, the use of polyesters of lower melting point or even of PET fibers with a sheath of a different polymer.

Lightweight webs (20 – 50 g/m$^2$) are available with individual filament counts of 2 – 3 dtex. They are used as cover webs in sanitary products and in horticulture.

Nonwoven products (50 – 130 g/m$^2$) have filaments with individual finenesses of 4 – 5 dtex. Their end uses are in construction and in shoe caps.

Materials with unit weights of 130 – 500 g/m$^2$ and individual filament finenesses up to 10 dtex are intended especially for use in construction (e.g., carriers for asphalt sheets for roofing), needle felts, embossed PVC coatings, underground construction, hydraulic engineering, and road building.

In addition to their large use in construction, polyester spunbonds are already being used successfully in automobiles (e.g., trunk linings) as well as in protective clothing and for environmental protection (covering of storage basins).

Web properties depend predominantly on the often different technologies for web production (laydown and bonding). No systematic data exist up to now that relate web structure to various end uses. However, in general, needle-punched webs are bulky, whereas thermally bonded webs are thin. Web strengths are ca. 2 – 4 N g$^{-1}$ m$^{-2}$ throughout (measured on strips 5 cm wide). Break elongations are 60 – 100 % for needle-punched webs, 20 – 60 % for bonded and needled webs, and generally < 30 % for thermally bonded webs.

## 3.5. Monofilaments

***Special Features of the Manufacturing Process.*** In the production of PET monofilaments, polymers with intrinsic viscosities [$\eta$] of 0.7 – 0.9 are used. Polymers of higher viscosity are often brought to their required value by solid-phase condensation. Because these polymers then have a lower content of COOH groups, they have advantages for use in hydrolysis-resistant monofilaments with increased hot – wet durability. Occasionally, the COOH end groups of the polymers are also protected chemically.

Monofilaments with final thicknesses of 0.03 – 2 mm are spun in air, cooled immediately in water, then drawn, heat-set, and packaged on a directly connected draw line [83]. Each extruder usually has its own spinning head and, accordingly, its own draw line. Various designs are recommended for spinning heads and spinnerets, such as circular spinnerets and spinnerets with straight rows of holes [84]. Feed for each spinneret hole should be the same. A few centimeters below the spinneret, the filaments enter a water bath, whose temperature is above the glass transition temperature of PET (70 °C). The filaments are cooled but remain soft. In the bath, they change direction, are brought out of the bath as a warp sheet, and are drawn off by a set of rolls at a rate of less than 100 m/min. The filaments have no spin orientation.

The first draw almost always takes place in water at 90 – 95 °C; the second draw, also takes place in a water bath, a higher boiling liquid, or a heater tunnel. Die plates are sometimes used for more accurate adjustment of the thickness. Heat-setting occurs in a heater tunnel, usually at 130 – 180 °C. Depending on the type desired, more or less shrinkage is permitted and the total draw ratio (ca. 1 : 4 to 1 : 6) is selected. At the end of the roll train, the monofilaments are taken up on individual bobbins on a creel.

In almost all applications of PET monofilaments a uniform diameter is the most important mark of quality. Tolerance of ± 3 % is required; if the process is carefully controlled, it will often be less. Additional quality requirements are smooth surface and, if necessary, uniform cross section. Textile properties of a given type must always remain constant, especially shrinkage tension and shrinkage. In monofilaments with improved resistance to hydrolysis, the residual strength after prolonged use should be preserved uniformly.

***Types of Monofilaments, Properties, and Uses.*** Monofilaments are produced with diameters of 0.034 – 2.00 mm These are thick filaments (the finenesses corresponding to these diameters are 12.5 – 43 000 dtex) and are, therefore, often called wires. A broad spectrum of types of PET monofilaments is found commercially in woven fabrics and zippers. These end uses place strict requirements on uniformity of diameter and properties, and especially on stress – strain behavior and shrinkage. The mechanical properties of typical PET monofilaments are evident from the stress – strain curves in Figure 7.

*Monofilaments for weaving* are used in the building industry in interior construction and interior installation. Other, often large, areas of application are paper machine sieves, safety belts, filter and screening fabrics, as well as conveyor belts. Fish nets, cordage, fishing lines, and medical applications all require the special properties of monofilaments for weaving.

The finest monofilaments, with diameters of 0.034 – 0.055 mm, have high elongation (see Fig. 7) and, at the same time, high shrinkage. Coarser wires range from 0.080 – 2.00. In addition, some wires have a rectangular cross section with diameters from 0.25 and 0.50 mm to 0.57 and 0.88 mm.

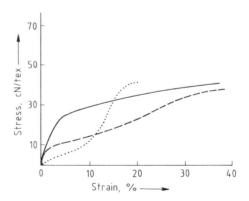

**Figure 7.** Stress – strain curves of typical PET monofilaments
───── Finest monofilament (0.055 mm diameter)
– – – Monofilament for weaving (weft type)
······ Monofilament for zippers

Monofilaments for weaving, especially for paper machine sieves are available as normal, hydrolysis-resistant, and chemically stabilized types. The most important representatives have high elongation (wires for the weft) or low elongation (wires for the warp) but low shrinkage.

Another important group comprises *zipper monofilaments*, with a diameter of 0.40 – 1.16 mm. This competes with polyamide wire. A major advantage with PET monofilaments is that their stress – strain properties and shrinkage behavior can be varied to a great extent. Thus, zipper wires are available over a whole gamut of combinations from low, elastic extensibility with high shrinkage to high, plastic extensibility with low shrinkage. The majority of these types exhibit low, elastic extensibility and medium shrinkage at the boil.

In general, the combination of modulus and shrinkage properties of zipper wire and wires for weaving is determined by the special manufacturing process. It is chosen according to the load – time program and the thermal exposure of monofilaments during use. The ability to vary their properties so widely makes possible the adaptation of PET monofilaments to any kind of processing and to any end use.

## 4. Other Polyester Fibers

### 4.1. Poly(Tetramethylene Terephthalate)

The polyester of terephthalic acid and 1,4-butanediol [110-63-4] is a comparatively old fiber polymer [7].

H─[O(CH$_2$)$_4$OC─⟨⟩─C─O(CH$_2$)$_4$]$_n$─OH

Because 1,4-butanediol has always been more expensive than ethylene glycol, poly(tetramethylene terephthalate) [30965-26-5], also called poly(butylene terephthalate) or PBT, remained more or less on the sidelines. Not until the middle of the 1960s were special PBT products developed by several producers; the advantages of these products compared to PET permitted use of the more expensive 1,4-butanediol [85]. Today, these PBT products have become well-entrenched in the marketplace.

Production of PBT is similar to that of PET. Dimethyl terephthalate is transesterified and subsequently polycondensed, either batchwise or continuously. However, the catalysts used with PET promote ring closure of 1,4-butanediol, leading to the potentially explosive tetrahydrofuran. Therefore, tetraisopropyl titanate [546-68-9], for example, is used as catalyst in both process steps.

Because the aliphatic sections of the molecular chains are longer and thus more mobile, the transition temperatures of PBT, *mp* ca. 230 °C and glass transition temperature ca. 25 °C, are lower than those of PET. In addition, PBT crystallizes more readily than PET. The PBT unit cell exists in two modifications. Of these, one is more extended, ca. 11 % longer, and reverts to the shorter one when unloaded. This is responsible for the better recovery of PBT products from stretching and bending.

Like PET, PBT is spun on standard machines from the melt. Because of the lower glass transition temperature, formation of spin orientation not only occurs along the spinning path but continues into the drawoff and the windup. In addition, clearly recognizable crystallization of PBT occurs during spinning, which in turn also affects orientation. Properties of the as-spun yarn thus depend in a complicated manner on spinning conditions.

After the usual drawing and heat-setting, PBT has a flatter stress – strain behavior than PET (Fig. 8). The permanent extension of a PBT product after straining, however, is much less (Fig. 9) [85]. For example, after a strain of 15 %, the permanent extension of PET is 8.5 %; that of PBT, only 2.5 %. For this reason, PBT is particularly suited for carpet fibers, which must exhibit good recovery after loading. Unlike PET, PBT can be dyed easily to deep shades at the boil without carrier because of its lower glass transition temperature.

In addition to its elastic behavior and easy dyeability, PBT has other favorable properties. For example, it tends to swell less than PET. All these factors, taken together, make PBT an interesting raw material for textile zippers. Finally, because of their resistance to alkali,

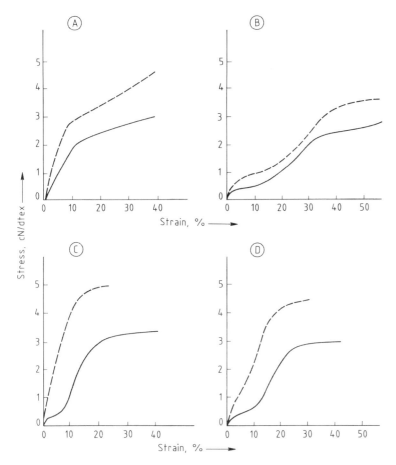

**Figure 8.** Stress – strain curves of drawn and heat-set PET and PBT yarns [83]
A) Drawn; B) Drawn and heat-set without clamping; C) Drawn and heat-set with clamping; D) Drawn, heat-set with clamping, plus heat-set without clamping
---- PET; ——— PBT

good abrasion resistance, and elastic behavior, PBT monofilaments are advantageously used in screen fabrics, which are constantly exposed to hot water or slightly alkaline media. In this way, fibers for carpets and monofilaments for zippers and screens have become products in which the more expensive polymer (PBT) pays for itself.

## 4.2. Poly(Trimethylene Terephthalate)

Poly (trimethylene terephthalate) (PTT) is a newly commercialized aromatic polyester fiber. Although PTT was available in commercial quantities only as recently as 1998 [88], it was one of the three high-melting-point aromatic polyesters first synthesized nearly 60 years ago [89, 90]. For a long time, the fiber industry had been aware of PTT having desirable properties for fiber applications. In a 1971 patent [91], Fiber Industries, Inc. found PTT fiber to have a lower modulus, better bending and work recoveries than PET, and was therefore more suitable than PET for making fiberfill and carpets. It was found that PTT had a better tensile elastic recovery and a lower modulus than PET and PBT [92]. These two properties are very desirable and are valued for making soft, stretch-fabrics with good hand and touch [93], and for resilient carpets [94].

In the early 1970s, the Shell Chemical Company, then a producer of 1,3-propanediol (PDO)

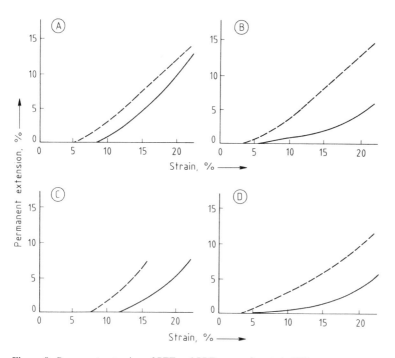

**Figure 9.** Permanent extension of PET and PBT yarns after strain [83]
A) Drawn; B) Drawn and heat-set without clamping; C) Drawn and heat-set with clamping; D) Drawn, heat-set with clamping, plus heat-set without clamping
---- PET; ——— PBT

via the acrolein route, explored the commercial potential of PDO and PTT by sampling PDO with several fiber companies. This led to a period of active research in PTT polymerization and applications [95–98].

Interest in PTT revived in the late 1980s when both Shell and Degussa made breakthroughs in two different PDO manufacturing technologies. Degussa was able to lower the cost of manufacturing PDO via the acrolein route and improve its purity to levels suitable for polymerization [99]. Shell developed an alternate synthesis route by hydroformylating ethylene oxide (EO) with a combination of CO and $H_2$ synthesis gas [100], leveraging their core competencies in hydroformylation technology and EO feedstock. In 1995, Shell announced the commercialization of PTT, and built a $73 \times 10^6$ t PDO plant in Geismar, Louisiana. This was followed by Du Pont announcing the retrofitting of an existing polyester plant in Kinston, North Carolina, to produce PTT using PDO obtained from Degussa. Du Pont and Genecor international collaborated to develop a potentially cheaper biological route for making PDO through glycerol fermentation. More than half a century after its synthesis, PTT finally joined PET and PBT, and became a commercial reality.

PTT is melt polymerized by either the transesterification of PDO with dimethyl terephthalate (DM) or by the direct esterification of PDO with purified terephthalic acid (PDA). The process is similar to that for PET. Because of PDO's lower reactivity, more active catalysts based on titanium and tin, which would discolor PET, are used to polymerize PTT. PTT is polymerized at much lower temperature between 250 and 275 °C. because of its higher melt degradation rate and faster crystallization rate, it requires special consideration in polymerization, pelletizing and solid-state treatment [89].

Like PET and PBT, PTT crystallizes into a triclinic structure. PTT chains appear zigzag, while PET chains are fully extended, and PBT chains look buckled. The highly contracted crystalline chain gives PTT a good tensile elastic recovery.

Most of the PTT application developments to date have focused on textile and carpet fibers. PTT fibers and yarns have bulk, resiliency, stretch-recovery, softness, hand and drape, properties which are similar to those of nylons and much better than those of PET. Such materials are inherently resistant to most stains which are acidic in nature because they not have dye sites. They also have a lower static propensity than nylons. PTT fibers are dyed with disperse dyes but at a lower temperature than PET because of the polymer's lower $T_g$. The combinations of these properties are attractive to carpet and textile manufacturers in some applications where PTT could replace nylon or PET. PTT also offers the potential of creating new fiber products by using the unique combinations of these properties not found in either nylon or PET alone.

### 4.3. Poly(1,4-Dimethylenecyclohexane Terephthalate)

Fibers of poly(1,4-dimethylenecyclohexane terephthalate) [25135-20-0], PDCT, have been on the market since the late 1950s under the trade name Kodel II (Tennessee Eastman Company, United States).

$$HOCH_2-\bigcirc-CH_2O-[C(=O)-\bigcirc-C(=O)-OCH_2-\bigcirc-CH_2O]_n-H$$

The intermediates are DMT (Section 1.1) and 1,4-bis(hydroxymethyl)cyclohexane [105-08-8] commonly known as 1,4-cyclohexanedimethanol. This diol is produced by the two-step hydrogenation of DMT at 160 – 180 °C and high pressure in the presence of catalysts (→ Alcohols, Polyhydric, Section 2.9). The diol can exist as either the *trans* or the *cis* isomer. By appropriate choice of reaction conditions, the *trans*: *cis* ratio can be adjusted to 7:3. To prepare the polymer, DMT is transesterified with the diol and after excess diol is distilled, polycondensation takes place in a manner analogous to PET.

The polymer is transparent and amorphous. It crystallizes in the triclinic system. The crystalline density is 1.265 g/cm³ for polymer with trans isomer and 1.303 g/cm³ for polymer with *cis* isomer [86, 87]. Wide-angle X-ray diagrams of crystalline samples show a continuous transition of the trans crystal lattice to the *cis* crystal lattice with increasing *cis*: *trans* ratio. The fiber density is 1.22 – 1.23 g/cm³. Small-angle X-ray investigations show a fibrillar structure with lattice layers at an angle to the fiber axis.

The glass transition temperature of the polymer is 76 °C; in drawn and crystalline fibers, it can be as high as 100 °C. The melting point is 285 – 295 °C [61].

Compared to PET, the tenacity of PDCT staple fibers is ca. 30 % less, with the same elongation at break. The abrasion resistance behaves similarly. The tensile or bending recovery of the fibers from equal deformation at room temperature, on the other hand, is greater for PDCT because of the higher glass transition temperature of the polymer [87]. The result is greater crimp stability of PDCT fibers and, hence, greater bulk of nonwovens and textile fabrics. Because of the higher glass transition temperature, however, PDCT fibers are more difficult to dye than PET fibers. Therefore, the dyeing temperature for PDCT fibers must be increased or suitable carriers and dyes chosen.

Applications of PDCT fibers are in areas such as fiberfill and carpets that require especially high crimp stability and generally high recovery forces following deformation at room temperature. These fibers are also used in blankets and synthetic furs [87].

## 5. Uses

The following properties are responsible for polyester fibers being used in all areas where fibers are applied:

1. The extremely broad range of mechanical properties of staple fibers and filament yarns
2. Their variable and easily adjustable shrinkage properties
3. Their ease of dyeing combined with high fastness properties
4. Their high chemical, solvent, and UV resistance

5. The low flammability of even the standard polymer, which can be further improved by modification without significantly impairing other valuable fiber properties
6. The deformability of the fibers and the fabrics made from them
7. The ability to make these deformations (e.g., crimp, creases, and pleats) permanent
8. The low moisture uptake and, hence, rapid drying
9. The fact that most soil remains on the surface where it is easily removed with water or solvents

These properties are largely insensitive to the presence of water at moderate temperature. Together, they are responsible for the dimensional stability and easy care of textiles made from polyester fibers.

Staple fibers, continuous-filament yarns, monofilaments, and spunbonds of polyester are found in all major textile markets. Among the polyesters, poly(ethylene terephthalate) is the most important polymer. Apparel, such as outerwear or sportswear, as well as work clothing and protective clothing, could not be imagined without PET fibers, either 100 % or in blends with other man-made or natural fibers. Similarly, PET fibers are widely used in an abundance of finished articles in adapted make-ups [102], e.g. home furnishings, such as drapes and curtains, furniture covers, table and bed linens, cushions, pillows, and carpets.

Polyester products have been particularly successful in industrial markets. Here, they are encountered everywhere fibers can be used. In many cases, because of their properties, especially their relatively low density (ca. 1.38 g/cm$^3$), they have provided elegant and, at the same time, economical solutions to problems. Examples are sewing thread, tarpaulins and awnings, safety belts, V-belts, tire cord, zippers, woven screens, geotextiles, roofing sheets, or book covers.

The extremely high performance of PET is reflected in the breadth of end uses [4, 27, 38, 61, 62]. This performance is based primarily on two properties of the fiber products: the dimensional stability of the textiles made from PET under mechanical and thermal stress, and their ease of care. Both properties are primarily determined by the glass transition temperature which, depending on the chemical structure of the polymer and the degree of orientation and order, lies between ca. 70 and 100 °C (see Section 2.1, → Fibers, 2. Structure). The fibers can be crystallized at any stage of their manufacture and during further processing. This permits production of any desired shape.

Polyester fibers are available in all possible forms in which natural and man-made fibers exist and can be processed on equipment used to manufacture yarns, woven fabrics, knit goods, and even nonwovens. Grouped broadly, the forms of PET are the following:

1. Staple fibers with a fineness of 0.4 – 20 dtex; short cut fibers, 6 mm length, fibers of 30 – 60 mm up to 150 mm; round or modified cross sections; crimped (except for flat cut fibers).
2. Textile continuous-filament yarns with a fineness of 20 – 200 dtex; individual filament fineness of 0.5 dtex (special types 0.1 dtex) to 5 dtex; flat, textured, or with staple yarn character; round or modified cross sections. The fineness of air-textured, continuous-filament yarns is 200 – 3000 dtex.
3. Fine high-tenacity continuous-filament yarns and yarns for sewing threads with a fineness of 50 – 940 dtex, with individual filaments of 3 – 10 dtex; flat with round cross sections.
4. Coarse high-tenacity continuous-filament yarns with a fineness of 1100 – 1670 dtex (plied fineness up to 11 000 dtex); individual filaments of 5 – 11 dtex; flat with round cross sections.
5. Spunbonds in widths of 1 – 5 m; unit weights of 20 – 500 g/m$^2$, with an individual filament fineness of 2 – 10 dtex.
6. Monofilaments with a diameter of 0.034 – ca. 2 mm; flat with round or modified cross sections.

# 6. Economic Aspects

Production of polyester fibers continues to grow at a faster rate than for all other fibers, both man-made and natural [101]. Thus, whereas during the 1990s, their share of world fiber production grew from 21.4 to 35.6%, it had increased to 45.8% by the end of 2009. Tecnon OrbiChem expects the upward trend to continue during the forecast period such that by the mid-2010s, driven by developments in China and other Asian countries, particularly India, polyester is expanded to account for more than 50% of worldwide fiber production, rising to more than 60% by 2020.

During the 1st decade of the 21st century, global production of polyester filament yarn has increased by an average rate of 7.9%/a, though the rate of growth did vary from a low of 1.4% in 2005 to a high of 14.9% the following year. Based on project growth in global GDP, it is expected that production of polyester filament yarn will increase by around 7.7%/a until 2010.

In comparison, global production of polyester staple fibers during the 1st decade of the 21st century has grown at an average rate of 5.5%/a, lower than filament yarns due mainly to the effects of low cotton prices during the early part of the decade. Tecnon OrbiChem believes that until 2010 production of polyester staple will grow slightly faster than previously at around 5.8%/a, which although slower than for filament yarn is nevertheless still well above the global figure for all fibers of 4.2%.

The world situation continues to be dominated by activity in China, where production of polyester filament represented less than a third of the global total. However, by 2005 the figure had increased to more than 50%, climbing to 71% by the end of 2009. Similarly, for staple, China accounted for slightly more than 25% of total world production, rising almost 50% in 2005 and 63% in 2009. Production of both filament and staple is expected to continue growing such that by 2020, China is forecast to account for more than 85% of the total world production of polyester fiber products. Overall, production of polyester filament yarn and staple fibers in Asia during 2009 represented 92% of global total (77% in 2000), and by 2020 production of polyester fiber in Asia is expected to account for 97% of the world total.

Because domestic consumption of both polyester filament and staple has failed to keep pace with increases in production, Chinese export trade has grown rapidly in recent years. Consequently, producers in other countries have come under pressure in both domestic and export markets. Thus whereas in 2004 China was ranked 8th and 6th in the league of exporting nations for filament and staple respectively, by 2008 it was number 1 for filament exports — with more than double the volume of the second placed nation Taiwan — and number 2 behind South Korea for staple. Exports of filament and staple amounted to almost $1.2 \times 10^6$ t in 2008, around a quarter of the world total. Meanwhile, imports of polyester filament yarn have fallen by 69% in four years to 184 000 t in 2008, whilst for staple fiber the fall has been 75% to 154 000 t.

Despite the rapid increase in output, the average plant utilization rate in China was only 72% in 2008, and resulted in some planned expansions being put on hold that year. However, with recovery evident in 2009, more than $1 \times 10^6$ t/a of new polyester filament capacity came on stream, with a further $1.5 \times 10^6$ t/a scheduled for 2010, and $3 \times 10^6$ t/a in 2011. With demand expected to grow less quickly, overcapacity looks set to remain a problem for filament manufacturers in China throughout the forecast period. In contrast, because polyester staple capacity has grown at a slower rate in recent years, the average utilization rate in 2008 was around 80%, a figure that is expected to rise during the forecast period even though new plants are likely to come on stream. However, recycled material is becoming an increasingly significant proportion of staple manufacture, thought to account for as much as 50% of total fiber production in 2009.

With Asia occupying such a dominant global position, the situation elsewhere in the world is less positive. In West Europe output of polyester filament yarn has been declining at a rate of almost 9%/a since 2005, with the expectation that the downward trend will accelerate to around 12%/a through 2010. The situation concerning staple fiber production has been better in the recent past, with output falling by around 1.5%/a since 2005. However, with competition from Asia growing as production for nonwoven applications increases, output of staple in Western Europe is expected to fall at around 15%/a in the forecast years. More than 90% of polyester staple consumption in West Europe is in nonwoven applications. However, the nonwovens industry in Asia is growing as multinational companies shift their manufacturing bases to lower cost regions.

Production of polyester filament yarns in North America has been falling by around 5.3%/a for since 2005 and is expected to decline at around the same rate during the forecast period. Because the carpet market is more resistant to import penetration than others, production of polyester BCF yarns will be more secure than perhaps the forecast suggests. There has been a definite movement towards the use of polyester BCF in the production of carpets in the USA, substituting staple polyester and nylon as well as polypropylene BCF and nylon BCF as well. In 2009, consumption of polyester BCF yarn grew by 26% in a weak carpet sector.

With raw material costs rising, many polyester fiber manufacturers in Europe, North America, and Asia are turning to recycled feedstock. Demand in China for polyester bottles has been increasing, with shortages reported in Asia early in 2008, which drove up prices.

Trade names and producers of polyester fibers are listed in Table 2. World production of synthetic polyester fibers (yarn and staple) are listed in Table 3.

Table 2. Trade names and producers of polyester fibers [103]

| Trade name | Type* | Manufacturer |
|---|---|---|
| A.C.E. | PET | Performance Fibers Inc./USA |
| Acelan | PET | Daehan Synthetic Fiber Co./Korea |
| Acelon | PET | Acelon Chemicals & Fiber Corp./Taiwan |
| Aliaf | PET | Terene Fibres India Ltd./India |
| Arachra | PET | Tongkook Corp./Korea |
| Asahikasei Ester | PET | Asahi Kasai Fibers Co./Japan |
| Bai Lu | PET | Xinxiang Bailu Chemical Fiber/China |
| Belira | PET | Incel/Bosnia-Herzegovina |
| Best | PET | Yizheng Chemical Fiber/China |
| Bonopoly | PET | Bongaigaon Ref. & Petrochemicals Ltd./India |
| Camellia | PET | Chanzhou Chemical Fiber/China |
| Centuron | PET | Tainan Spinning Co./Taiwan |
| Chiemlon | PET | Chiem Patana Synthetic Fibers Co./Thailand |
| CoolMax | PET | Advansa A.S./Turkey |
| Corterra | PTT | Setila SA/France |
| Dacron | PET | DAK Americas, LLC./USA |
| | | Suzhou (Far Eastern Textiles) Polyester/China |
| | | Advansa GmbH/Germany |
| | | Advansa A.S./Turkey |
| Di Guang | PET | Human Indi Chemical Fiber/China |
| Diolen | PET | Diolen Industrial Fibers GmbH/Germany |
| | | Polyenka SA/Brazil |
| | | Enka de Colombia SA/Colombia |
| Dragen | PET | Jiangxi Polyester Fiber/China |
| DSP | PET | Performance Fibers (Kaiping) Co./China |
| DSP Polyester | PET | Performance Fibers SA/France |
| | | Performance Fibers Inc./USA |
| Eastlene | PET | Far Eastern Textile Ltd./Taiwan |
| Eastlon | PET | Far Eastern Textile Ltd./Taiwan |
| EcoSpun | PET | Wellman, Inc./USA |
| Elana | PET | Elana SA/Poland |
| Eslon | PET | Saehan Industries Inc./Korea |
| Estrell | PET | Aquafil SpA/Italy |
| Eural | PET | Tergal Fibres SA/France |
| Fei Ma | PET | Foshan Polyester/China |
| Filwell | PET | Wellman International Ltd./Ireland |
| Flacron | PET | Ahmedabad & Calico Printing Co./India |
| Fortrel | PET | Wellman, Inc./USA |
| | | Mantex CA/Venezuela |
| Futura | PET | Futura Polyesters Ltd./India |
| Gatron | PET | Gatron Industries Ltd./Pakistan |
| Grisuten | PET | Märkische Faser GmbH/Germany |
| Hanslon | PET | Hankook Synthetic, Inc./Korea |
| Heiping | PBT | Yixing Huaya Chemical Fibre/China |

| | | |
|---|---|---|
| Hualon | PET | Hualon Corp./Taiwan |
| Huvis | PET | Huvis Corp./Korea |
| Hunvira | PET | Hung Chou Chemical Industry Ltd./Taiwan |
| Jailene | PET | Swadeshi Polytex Ltd./India |
| Jaykaylene | PET | J.K. Synthetics Ltd./India |
| Jekester | PET | Orissa Polyfibres Ltd./India |
| Jespan | PET | Hualon Corp (Malaysia) Sdn. /Malaysia |
| Jewelon | PET | Taiwan Spinning Co./Taiwan |
| Kintrel | PET | Kimex SA/Mexico |
| Kolon | PET | Kolon Industries Inc./Korea |
| Kreo | PET | Freudenberg Politex SpA/Italy |
| Kuraray Polyester | PET | Kuraray Co./Japan |
| Lalelen | PET | Sancak Tül A.S./Turkey |
| LSB | PET | La Seda de Barcelona SA/Spain |
| Luxlen | PET | Yu-Ho Fiber Industrial Corp./Taiwan |
| Mandal | PET | Sanghi Polyesters Ltd./India |
| Micrell | PET | Noyfil SA/Switzerland |
| | | Val Lesina SpA/Italy |
| Mirhon | PET | Miroglio SpA/Italy |
| Multisoft | PET | Ledervin/Brazil |
| Nakron | PET | Thai Toray Synthetics Co./Thailand |
| Nanlon | PET | Tainan Spinning Co./Taiwan |
| Navilon | PET | Kum Kang Co./Korea |
| Neopol | PTT | Hyosung Corp./Korea |
| Nergis | PET | Nergis A.S./Turkey |
| Nirester | PET | Nirlon Ltd./India |
| Oplene | PET | Jeil Synth. Fibers Co./Korea |
| Palmylon | PET | KP Chemtec Corp./Korea |
| Pathom | PET | Thai Toray Synthetics Co./Thailand |
| Pearly Lene | PET | Tung Ho Spinning Co./Taiwan |
| Pentex | PEN | Performance Fibers, Inc./USA |
| Polymisr | PET | Misr Rayonne Co./Egypt |
| Pontella | PET | Setila SA/France |
| Rebar | PET | Chia Hsin Food & Synthetic Fiber Co./Taiwan |
| Recron | PET | Reliance Industries Ltd./Indien |
| Sadrifill | PET | DS Fibres NV/Belgium |
| Sadriloft | PET | DS Fibres NV/Belgium |
| Sadrilux | PET | DS Fibres NV/Belgium |
| Savar | PET | Beximco Synthetics/Bangladesh |
| SENS | PTT | Miroglio SpA/Spain |
| Serill | PET | Aquafil SpA/Italy |
| Setila | PET | Setila SA/France |
| Shinlon | PET | Shinkong Synth. Fibers Corp./Taiwan |
| Shreester | PET | Shree Synthetics Ltd./India |
| Silkiss | PET | Miroglio SpA/Italy |
| Silver Ring | PET | Hubei Chemical Fiber/China |
| Skylon | PET | Sunkyong Industries Ltd./Korea |
| Slotera | PET | Slovensk Hodvab a.s./SR |
| Softfit | PTT | Nan Ya Plastics Corp./Taiwan |
| Solotex | PTT | Solotex Corp./Japan |
| Soluna | PET | Mitsubishi Rayon Co./Japan |
| Sorona | PTT | Huvis Corp./Korea |
| | | Saehan Industries, Inc./Korea |
| | | Far Eastern TEXTILE Ltd:/Taiwan |
| Startex | PET | Seong-An Synthetics Co./Korea |
| Tairilin | PET | Nan Ya Plastics Corp./Taiwan |
| | | Nan Ya Plastics Corp./USA |
| Tasinlon | PET | Chung Shing Textile Co./Taiwan |
| Tejin- Tetoron | PET | Teijin Ltd./Japan |
| | | Teijin Polyester (Thailand) Ltd./Thailand |
| Terene | PET | Terene Fibres India Ltd./India |
| Tergal | PET | Tergal Fibres SA/France |
| Terinda | PET | Advansa A.S./Turkey |
| Terital | PET | Fidion Srl/Italy |
| Terital Eco | PET | Fidion Srl/Italy |

| | | |
|---|---|---|
| Terlenka | PET | Enka del Colombie SA/Colombia |
| Tertex | PET | Inquitex SA/Spain |
| Terylene | PET | SANS Fibres (Pty.), Ltd./South Africa |
| Texlon | PET | Tongkook Corp./Korea |
| Thermolite | PET | Advansa A.S./Turkey |
| Tifico Polyester | PET | TIFICO/Indonesia |
| Toplon | PET | Hyosung Corp./Korea |
| Toray | PET | Toray Snthetic Fiber (Nantong) Co./China |
| Toray Tetoron | PET | Toray Industries, Inc./Japan |
| Torlen | PET | Elana/Poland |
| Toyobo Ester | PET | Toyobo Co./Japan |
| Trevira | PET | Hosaf Fibres (Pty) Ltd./South Africa Trevira GmbH Germany |
| Trevira CS | PET | Trevira GmbH/Germany |
| Tuntex | PET | Tuntex Fiber Corp./Taiwan |
| Ultrafil | PET | Pennine Fibre Industries Ltd/ United Kingdom |
| Unitika Ester | PET | Nippon Ester Co./Japan |
| Vadodora | PET | Yogi-Polyesters Ltd./India |
| Valena | PET | Val Lesina SpA/Italy |
| Vanlon | PET | Full Point Synthetic Industrial Co./Taiwan |
| Verde | PET | M & G Fibras e Resinas/Brazil |
| Wellene | PET | Wellman International Ltd./Ireland |
| Wolkiss | PET | Miroglio SpA/Italy |
| Yambolen | PET | Yambolen/Bulgaria |
| Zispan | PTT | Hankook Synthetic, Inc./Korea |

*PET = poly(ethylene terephthalate), PTT = poly(trimethylene terephthalate), PBT = poly(butylene terephthalate), PEN = poly(ethylene naphthalate).

Table 3. World Production of synthetic polyester fibers (yarn and staple)

| Year | Production, t |
|---|---|
| 1980 | 5 085 000 |
| 1990 | 8 445 000 |
| 1995 | 11 958 000 |
| 2000 | 18 912 000 |
| 2005 | 24 701 000 |
| 2009 | 32 412 000 |

# References

1. H. Mark, G.S. Whitby (eds.): *Collected Papers of W. H. Carothers on High Polymeric Substances*, Wiley-Interscience, New York 1940.
2. W.H. Carothers, J.W. Hill, *J. Am. Chem. Soc.* **54** (1932) 1579 – 1587.
3. H. Klare: *Geschichte der Chemiefaserforschung*, Akademie-Verlag, Berlin 1985.
4. H. Ludewig: *Polyesterfasern*, 2nd ed., Akademie-Verlag, Berlin 1975.
5. J.R. Whinfield, *Text. Res. J.* **23** (1953) 289 – 293.
6. Calico Printers' Assn., DE 973 553, 1940 – 1941 (J.R. Whinfield, J.T. Dickson).
7. P. Schlack, *Textil-Prax.* **8** (1953) 1055 – 1062.
8. International Rayon and Synthetic Fibres Committee (ed.): *Information on Man-Made Fibres*, vol. 22, Paris 1985.
9. E. Katzschmann, *Chem.-Ing.-Techn.* **38** (1966) 1 – 10.
10. Midcentury Corp., DE-AS 1 081 445, 1960 (S. Barker, A. Saffer).
11. A. Ziabicki: *Fundamentals of Fibre Formation*, Wiley-Interscience, London 1976.
12. I. Hamana, M. Matsui, S. Kato, *Melliand Textilber.* **4** (1969) 382 – 388, 499 – 503.
13. E. Liska, *Kolloid-Z. Z. Polym.* **251** (1973) 1028 – 1037.
14. H.M. Heuvel, R. Huisman, *J. Appl. Polym. Sci.* **22** (1978) 2229 – 2243.
15. A. Ziabicki, H. Kawai (ed.): *High-Speed Fiber Spinning*, Wiley-Intersience, New York 1985.
16. H.H. George, H. Matsui, H. Yasuda, Literature and Patent Survey, in A. Ziabicki, H. Kawai (ed.): *High-Speed Fiber Spinning*, Wiley-Intersience, New York 1985, pp. 553 – 565.
17. W. Dietrich, G. Reichelt, H. Renkert, *Chemiefasern/Text.-Ind.* **32/84** (1982) 612 – 625.
18. J. Shimizu, N. Okui, T. Kikutani, *Sen'i Gakkaishi* **37** (1981) T135 – T142.
19. M. Jambrich, A Diačiková, O. Ďurčová, A. Chvála, *Acta Polym.* **37** (1986) 118 – 121.
20. G.W. Davis, A.E. Everage, J.R. Talbot, *Fiber Prod.* **12** (1984) no. 2, 22 – 28.
21. H. Lückert, W. Stibal, *Chemiefasern/Text.-Ind.* **36/88** (1986) 24 – 29.
22. I. Marshall, A.B. Thompson, *J. Appl. Chem.* **4** (1954) 145 – 153.
23. K. Scharley, *Bühler Nachr.* **222** Bühler-Miag, Braunschweig 1983.
24. H.H. George, A. Holt, A. Buckley, *Polym. Eng. Sci.* **23** (1983) 95 – 99.

25. I. Jacob, H.-R. Schröder, *Chemiefasern/Text.-Ind.* **30/82** (1980) 114 – 119, 228 – 232.
26. J.H. Dumbleton, *J. Polym. Sci., Part A* **2** (1968) 795 – 800.
27. Rhovyl, FR 2 524 498, 1982 (G. Achard, P. Chion, J. Menault).
28. H. Brody, *J. Macromol. Sci., [B] Phys.* **22** (1983)no. 1, 19 – 41.
29. R.W. Miller, J.H. Southern, R.L. Ballman, *Text. Res. J.* **53** (1983) 670 – 677.
30. H. Berg, *Chemiefasern/Text. Ind.* **22/74** (1972) 215 – 222.
31. R. Huisman, H.M. Heuvel, *J. Appl. Polym. Sci.* **22** (1978) 943 – 965.
32. K. Riggert, *Chemiefasern* **19** (1969) 816 – 823.
33. J.H. Dumbleton, J.P. Bell, T. Murayama, *J. Appl. Polym. Sci.* **12** (1968) 2491 – 2508.
34. H. Güllemann, *Melliand Textilber.* **53** (1972) 910 – 914.
35. J.H. Dumbleton, T. Murayama, F.P. Bell, *Kolloid-Z. Z. Polym.* **228** (1968) 54 – 58.
36. V.B. Gupta, C. Ramesh, A.K. Gupta, *J. Appl. Polym. Sci.* **29** (1984) 3727 – 3739.
37. H. Zimmermann, *Melliand Textilber.* **57** (1976) 828 – 832.
38. F. Jakob, *Chemiefasern/Text.-Ind.* **74** (1972) 388 – 396.
39. A. Williams, I.T. Ibrahim, *Chem. Rev.* **81** (1981) 589 – 636.
40. S. Hoffrichter et al.,*Dtsch. Textiltech.* **20** (1970) 774 – 782.
41. V. Freudenberger, F. Jakob, *Angew. Makromol. Chem.* **105** (1982) 203 – 215.
42. H. Zimmermann, *Chemiefasern/Text.-Ind.* **80** (1978) 1054 – 1060.
43. Teijin, EPS 0 038 429, 1984 (K. Ozaki, M. Matsui).
44. P. Flory: *J. Amer. Chem. Soc.* **58** (1936) 1877 – 1885.
45. S. Fakirov, E.W. Fischer, G.F. Schmidt, *Makromol. Chem.* **176** (1975) 2459 – 2465.
46. H.M. Heuvel, R. Huisman, *J. Appl. Polym. Sci.* **30** (1985) 3069 – 3093.
47. B. von Falkai, W. Giessler, F. Schultze-Gebhardt, G. Spilgies, *Angew. Makromol. Chem.* **108** (1982) 9 – 39.
48. W.H. Charch, W.W. Moseley, *Text. Res. J.* **29** (1959) 525 – 535.
49. W.W. Moseley, *J. Appl. Polym. Sci.* **3** (1960) 266 – 276.
50. H. Sattler, *Kolloid-Z. Z. Polym.* **187** (1963) 12 – 18.
51. E. Liska, *Angew. Makromol. Chem.* **65** (1977) 147 – 168.
52. R. Bonart, *Kolloid-Z. Z. Polym.* **199** (1964) 136 – 144.
53. H. Lückert, W. Stibal, *Chemiefasern/Text.-Ind.* **36/88** (1986) 24 – 29.
54. L. Riehl, *Chemiefasern/Text.-Ind.* **28/80** (1978) 1039 – 1046; **29/81** (1979) 24 – 30.
55. N.H. Bezaz, *Chemiefasern/Text.-Ind.* **32/84** (1982) 262 – 265.
56. H. Lückert, M. Busch, *Chemiefasern/Text.-Ind.* **33/85** (1983) 29 – 38.
57. P. Dammann, *Chemiefasern/Text.-Ind.* **26/78** (1976) 521 – 522.
58. H. Lückert, *Chemiefasern/Text.-Ind.* **29/81** (1979) 1019 – 1022.
59. K. Riggert, *Chemiefasern/Text.-Ind.* **31/83** (1981) 638 – 648.
60. F. Schmitt, *Melliand Textilber.* **63** (1982) 830 – 835.
61. B. von Falkai (ed.): *Synthesefasern*, Verlag Chemie, Weinheim, Germany 1981.
62. J. Thimm, *Chemiefasern/Text.-Ind.* **75** (1973) 811 – 817.
63. G.W. Davis, A.E. Everage, J.R. Talbot, *Fiber Producer* **12** (1984)no. 6, 45 – 62.
64. H. Treptow, *Chemiefasern/Text.-Ind.* **35/87** (1985) 411 – 412.
65. Rhône-Poulenc-Textile, FR 1 551 878, 1967 (G. Barbe, P. Curtillat, C. Lequay).
66. Du Pont, DE-AS 2 241 718, 1971 (D.G. Petrille, M.J. Piazza, C. E. Reese).
67. *Fiber World* 1985, 22 – 51.
68. J. Lünenschloss, *Textiltechnik* **32** (1982) 232 – 235.
69. Du Pont, GB 900 009, 1960 (L. Cenzato).
70. Teijin Ltd., EP 159 875, 1984 (H. Yamada, N. Sayama, M. Kuno).
71. Monsanto, US 4 092 299, 1976 (D.L.G.R. MacLean, R.T.C. Estes).
72. F. Maag, *Chemiefasern/Text.-Ind.* **34/86** (1984) 173 – 178.
73. R.C. Mears, *Chemiefasern/Text.-Ind.* **35/87** (1985) 413 – 414.
74. Imperial Chem. Ind., DE 2 836 514, 1977 (P.L.I. Carr).
75. Imperial Chem. Ind., EP 42 664, 1980 (P.L.I. Carr).
76. Du Pont, US 4 134 882, 1977 (H.R.E. Frankfort, B.W. Knox).
77. K. Riggert, *Chemiefasern* **21** (1971) 379 – 384.
78. Hoechst AG, DE 2 115 312, 1971 (R. Johne, M. Bechter).
79. DuPont, US 3 216 187, 1962 (W.A. Chantry, A.F. Molini).
80. Hoechst AG, DE-OS 3 431 831, 1984 (H. Thaler).
81. L. Hartmann, *Chemiefasern/Text.-Ind.* **22/74** (1972) 231 – 236, 324 – 328.
82. L. Gerking, *Verfahrenstechnik (Mainz)* **10** (1976) 779 – 784.
83. F. Hensen, S. Braun, *Chemiefasern/Text.-Ind.* **29/81** (1979) 844 – 848.
84. H. Reinbold, *Chemiefasern/Text.-Ind.* **35/87** (1985) 420 – 422.
85. H. Sattler, *Textil-Prax.* **33** (1978) 1175 – 1178.
86. J.H. Semas, *Int. Fiber J.*, **12** (1997)no.1, 12.
87. H.H. Chuah: Synthesis, Properties and Applications of Poly (Trimethylene Terephthalate) inJ. Scheirs,T.E. Long (ed.): *Modern Polyesters: Chemistry and Technology of Polyesters and Copolyesters*, John Wiley & Sons, Wiley Series in Polymer Science, Chichester 2003, pp 361 – 397.
88. J.R. Whinfield, J.T. Dickson,GB 578079, 1946.
89. Fiber Industries, Inc., GB 1254826, 1971.
90. I.M. Ward et al., *J. Polym. Sci. Polym. Phys. Ed.*, **14** (1976) 263.
91. C. Heschmeyer, *Int. Fiber J.*, **15** (2000)no. 4, 66.
92. H.H. Chuah et al,*Int. Fiber J.*, **10** (1995)no. 5, 50.
93. E.I. du Pont de Nemours and Company, US 3584103, 1971 (M. E. Harris).
94. Teijin, US 39846001977, (S. Kawase, T. Kuratsuji).
95. Teijin, JP 51140992, 1976 (K. Takatoshi).
96. Teijin, JP 525320, 1977 (K. Takatoshi).
97. P. Nager: "Applications of 1,3-propanediol",*CHEMSPEC 99. Asia'91 Conference*,Tokyo, Japan,June 24–25, 1991, Specialty Chemicals Production, Marketing and Application.
98. Shell Oil Company, EP 478850 A1, 1992 (E. Drent).
99. E.V. Martin, H. Busch, *Angew. Chem.* **74** (1962) 624 – 628.
100. B. von Falkai, G. Spilgies, F. Schultze-Gebhardt, H. Wilsing, *Acta Polym.* **34** (1983) 86 – 98.
101. G. Tetzlaff, M. Dahmen, B. Wulfhurst: "Fiber tables according to P.-A. Koch", *Chemiefasern/Text.-Ind.* **43** (1993) 508 – 522.
102. N. Bywater, *Chem. Fib. Int.* **2** (2010) 60 – 62.
103. H.-J. Koslowski: *Chemiefaserlexikon*, 12. ed., Deutscher Fachverlag GmbH, Frankfurt am Main 2008, pp. 281 – 294.

# Polyurethane Fibers

KARL-HEINZ WOLF, Bayer AG, Dormagen, Germany

MICHAEL KAUSCH, Bayer AG, Dormagen, Germany

HANS SCHRÖER, Bayer AG, Dormagen, Germany

MICHAEL SCHWEIZER, ITCF Denkendorf, Denkendorf, Germany

| | | | | | |
|---|---|---|---|---|---|
| 1. | Physical Properties . . . . . . . . . . . . . | 1488 | 5. | Spinning Processes. . . . . . . . . . . . . . | 1491 |
| 2. | Chemical Properties. . . . . . . . . . . . . | 1489 | 6. | Uses . . . . . . . . . . . . . . . . . . . . . . . | 1492 |
| 3. | Raw Materials. . . . . . . . . . . . . . . . . | 1489 | | References. . . . . . . . . . . . . . . . . . . | 1493 |
| 4. | Production of Polymer. . . . . . . . . . . | 1491 | | | |

Following the discovery in 1937 of the diisocyanate addition polymerization process by O. BAYER and co-workers [1], this principle was applied to the production of polyurethane (PUR) fibers. Hard PUR fibers, first developed to compete with nylon 66, are no longer important. On the other hand, the same principle of synthesis was successfully applied to the manufacture of highly elastic yarns [2, 3].

Synthetic high-polymer fibers containing at least 85 wt% of segmented polyurethane are called *elasthane* [4]; the term *spandex* is used in the USA and in this article.

Spandex fibers are characterized by high elongation and elasticity. Like rubber yarns (elastodiene fibers), they belong to the class of elastomeric fibers. Spandex, produced as continuous-filament yarn, is used almost exclusively to manufacture elastic textiles and is always employed in combination with other fibers. Although spandex has gained considerable technical and economic importance, worldwide production, which amounted to 290 000 t in 2005, is relatively small compared to other textile fibers. Commercial products have distinctive raw materials bases, chemical structures, and manufacturing and spinning processes, which, in the final analysis, are responsible for the specific balance of properties [5–9].

Due to the rise in the significance of active wear in the world as compared to formal wear, spandex or PU filament consumption has been increasing [10]. Huge Chinese expansion has also led to the worldwide increase in the consumption of this filament type. It has become the first real mainstream significant usage filament from the non-melt spinning process. Dry spinning speeds of 1200 m/min have been reported, resulting in relatively mature low-cost manufacturing of this material for the applications.

The result has been widespread usage of this material for all kinds of applications – signifying how important economics are in popularizing an application development for a filament yarn product. As the product is used only as a small percentage in fabrics, popular thickness are more in the finer category. The bulk of the market consumption is in the dtex 33 – 78 range in terms of weight/volume being sold. It is significant that in recent years, the dtex 22 and dtex 17 PU products have also become important as the trend in fabrics is towards lighter close-to-the-skin products. Some recent products have been seen combined with very light dtex 33 polypropylene, resulting in extremely lightweight fabrics which are very popular in innerwear applications. A small but significant market has developed in the heavier deniers between dtex 330 and 470 for shoes, diapers, and medical applications. The growth scope in this area has leveled off recently. The major growth area for this material over the last few years has been in dtex 13 – 22 sector. Some other melt-spun stretch filaments have made an appearance but are still, in marketing terms, not significant. Here, XLA must be mentioned, which was marketed as a melt-spun stretch yarn which can be used as 100% alternative

Ullmann's Polymers and Plastics: Products and Processes
© 2016 Wiley-VCH Verlag GmbH & Co. KGaA, Weinheim
ISBN: 978-3-527-33823-8 / DOI: 10.1002/14356007.o10_o02

Table 1. Trade names and producers of polyurethane fibers

| Trade name | Type | Manufacturer |
|---|---|---|
| Acelan | EL | Tae Kwang Industrial Co./Korea |
| Creora | EL | Hyosung/China |
| Creora | EL | Hyosung Corp./Korea |
| Creora | EL | Hyosung/Turkey |
| Dorlastan | EL | Asahi Kasei Spandex Europe GmbH/Germany |
|  |  | Asahi Kasei Spandex America Inc./USA |
| Elaston | EL | Chemitex/Poland |
| Espa | EL | Toyobo Co./Japan |
| Fujibo Spandex | EL | Fujibo Kozak, Inc./Japan |
| Glospan | EL | Radici Spandex Corp/USA |
| Huastan | EL | Hualon Corp./Taiwan |
| Kopandex | EL | Kohap Ltd./Korea |
| Linel | EL | Fillatice SpA/Italy |
| Lycra | EL | Invista Inc./USA |
| Lycra | EL | Invista Brazil/Brazil |
| Lycra | EL | Invista (U.K.) Ltd./United Kingdom |
| Lycra | EL | Invista Canada Inc./Canada |
| Lycra | EL | Invista (China) Co./China |
| Lycra | EL | Fielmex SA de CV/Mexico |
| Lycra | EL | Invista/Mexico |
| Lycra | EL | Invista Singapore Fibres/Singapore |
| Mobilon | EL | Nisshinbo Industries, Inc./Japan |
| Opelon | EL | Opelontex Co./Japan |
| Roica | EL | Asahi Kasei Fibers Corp./Japan |
| Roica | EL | Thai Asahi Kasei Spandex Co./Thailand |
| Sheiflex | EL | Shei Heng Hsin Sheiflex Industrial Co./Taiwan |
| Texlon | EL | Tongkook Corp./Korea |
| Toplon | EL | Hyosung Corp./Korea |

*EL = elastane fibers.

# 1. Physical Properties

Spandex is highly extensible by application of an external force. When the load is removed, the material rapidly and completely reverts to its original state. This is a result of the two-phase structure of the hard and soft segments. The soft segments, which are highly mobile at the use temperature, are coiled in the unloaded state and stretched in the loaded state. The restoring force results from the increase in entropy during reversion to the coiled state (entropy elasticity). The crystalline hard segments of symmetrical aromatic urea groups form a network structure via hydrogen bonds and thus provide form stability [11–24].

Characteristic stress – strain curves of a commercial spandex are shown in Figure 1 in a cyclical loading – unloading experiment (stress – strain cycle). The mechanical properties of spandex filament yarns for the usual range of types and finenesses are listed in Table 3.

Below the glass transition temperature of the soft segments, spandex loses its elasticity. In commercial products, this temperature of 210 – 250 K is far below normal environmental temperature range. Critically important for textile processors is the behavior at high temperature (160 – 170 °C, which is the temperature during textile treatment and shaping processes).

to the material and need not to be used as a blend percentage as in case of PU yarns. Marketing and acceptance of this filament-making material held a lot of initial potential, but the cost reductions in PU filament enabled by the increasing of dry spinning speeds and the rising sophistication of the technology and production processes in PU have stalled the progress of XLA.

Trade names and producers of polyurethane fibers are listed in Table 1. World production of polyurethane fibers is listed in Table 2.

Table 2. World Production of polyurethane fibers

| Year | Production, t |
|---|---|
| 1995 | 70 000 |
| 2005 | 290 000 |

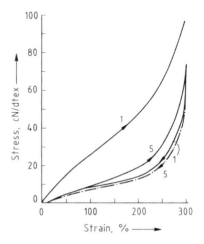

Figure 1. Stress – strain cycle of a 480-dtex spandex yarn (Dorlastan type V 400) *
* 1 is first curve; 5 is fifth curve; —— is elongation; —— is recovery.

**Table 3.** Mechanical properties of continuous filament spandex yarns

| Property | Range of values |
| --- | --- |
| Fineness | 20 – 5000 dtex |
| Elongation at maximum tensile load | |
|    Standard-elongation types | 420 – 570 % |
|    High-elongation types | 600 – 800 % |
| Tenacity | |
|    Referred to original fineness | 0.45 – 1.2 cN/dtex |
|    Referred to fineness at break [a] | 3 – 7 cN/dtex |
| Cyclic loading to 300 % elongation | |
|    Modulus, 150 % [b] | 0.04[c] – 0.1 [d] cN/dtex |
|    Modulus, 300 % [b] | 0.1[c] – 0.25 [d] cN/dtex |
| Residual elongation | 10 – 30 % |

[a] Fineness at maximum textile elongation.
[b] Referred to original fineness.
[c] High-elongation types.
[d] Standard-elongation types.

## 2. Chemical Properties

Compared to rubber yarns, spandex is substantially more stable to chemical influences during textile processing and subsequent use. Any possible damage manifests itself primarily in a loss of strength and an onset of yellowing. As expected, polyether spandex is more affected by oxidative processes and polyester spandex by hydrolysis. The stability of spandex subjected to various chemical treatments is given in Table 4.

Spandex-containing textiles can be cleaned chemically without problem. Sensitivity to photo-oxidative damage is increased by unsaturated fatty acids and cosmetic oils [25]. Nitrogen oxides lead to yellowing and, hence, are detrimental to white textiles that contain spandex. Compared to rubber yarns, spandex has a greater affinity for textile dyes.

## 3. Raw Materials

Segmented polyurethane, the raw material for spandex fibers, consists of long-chain dihydroxy compounds (macrodiols), aromatic diisocyanate compounds, and short-chain diamino or dihydroxy compounds (chain extenders). Because of their chain mobility, the long-chain diols (soft segments) are responsible for the extensibility. The urethane and urea groups, formed by aromatic diisocyanates and short-chain diols (hard segments), function as physical cross-links and are the source of yarn strength.

*Macrodiols.* Macrodiols suitable for the manufacture of spandex have molar masses of 1 – 3 kg/mol and melting points < 50 °C. Only polyether and polyester diol are important. The most frequently used polyether diol is poly (tetramethylene ether glycol), polymerized from tetrahydrofuran via ring opening (→ Polyoxyalkylenes, → Polyoxymethylenes). Spandex fibers with polyether soft segments are highly resistant to hydrolysis and alkali, and have excellent low-temperature properties. They are, however, sensitive to light, atmospheric oxygen, and the chlorinated water of swimming pools and must be specially protected from them.

Among the polyester diols used, adipic acid polyesters have gained great importance,

**Table 4.** Chemical resistance of spandex

| Chemical treatment | Polyether spandex[a] | Polyester spandex[a] | Remarks |
| --- | --- | --- | --- |
| HCl 10 %, 20 °C, 24 h | low | high | |
| $H_2SO_4$ 10 %, 20 °C, 24 h | low | high | |
| $CH_3COOH$ 10 %, 20 °C, 24 h | none | none | |
| $Cl_2$– $H_2O$ 20 mg/L, 20 °C, 24 h | high [b] | low | some yellowing |
| NaOH, followed by rinsing, suctioning, rinsing | none | none | |
| Cleaning in gasoline | low | low | |
| Xeno-test [c], 50 h | low | low | some yellowing |
| Xeno-test [c], 100 h | medium | low | little to moderate yellowing |

[a] In the classification used, the following strength loss is to be expected: none, 0 – 5 %; low, 5 – 20 %; medium, 20 – 40 %; high, 40 – 90 %; total, 90 – 100 %.
[b] Can largely be prevented by use of ZnO as spinning additive [24].
[c] Light source: Hanau 450 (UV-emitting xenon gas lamp).

whereas polycaprolactone diols have not. For polyadipates, mixtures of $C_2$ – $C_6$ diols, e.g., 1,2-ethanediol – 2,4-butanediol and 1,6-hexanediol – 2,2-dimethyl-1,3-propanediol are used most frequently. Polyester spandex is more resistant to chlorinated water than polyether types and is insensitive to oxidative damage. Its sensitivity to hydrolysis can be reduced by the use of diols with a greater number of carbon atoms. Polyesters of mixed diols give spandex with a low tendency to crystallize during extension and with improved elastic properties.

**Diisocyanates.** Commercial spandex is almost exclusively produced from 4,4′-methylenebis (phenyl isocyanate) [101-68-8] → Isocyanates, Organic. Toluene diisocyanate [26471-62-5] (TDI), which was used in the early years of spandex production, is rarely employed today. Although aliphatic diisocyanates lead to polyurethanes with increased light durability, they are of little use because of increased difficulties in processing the spinning solutions. However, 1,6-diisocyanatohexane can be employed as a modifier [26].

**Chain Extenders.** Most spandex producers use short-chain aliphatic diamines, such as 1,2-diaminoethane or 1,2-diaminopropane, as chain extenders because they give fibers with the best thermal and hydrothermal properties. Hydrazine hydrate, carbonic acid, dicarboxylic acid dihydrazides, aminocarboxylic acid hydrazides, and semicarbazidocarboxylic acid hydrazides [27, 28] were used earlier as chain extenders, but are no longer of any significance. Occasionally, small amounts of codiamines, such as 1,3-diaminocyclohexane [29] or N-methylbis(3-aminopropyl)amine [30], are used in combination with ethylenediamine. This favorably affects certain properties, e.g., solubility and storage behavior. 1,4-Butanediol, frequently used in thermoplastic polyurethanes, results in spandex with unsatisfactory properties. 1,4-Bis(2-hydroxyethoxy)benzene, on the other hand, produces a melt-spun spandex with improved properties [31].

**Solvents.** Spandex fibers based on polyurethane – polyurea decompose during melting; therefore, they are spun from solution. Only dipolar aprotic solvents, usually dimethylacetamide and dimethylformamide, can be used.

**Stabilizers and Adjuvants.** Because of the use of aromatic diisocyanates, light tends to discolor spandex [32]. Spandex in which polyethers are used as soft segment is, in addition, attacked photooxidatively. The following are used as stabilizers for these processes: phenolic antioxidants, sterically hindered amines, light-protective agents (e.g., benzotriazoles), and phosphites. They must meet special requirements such as low thermal volatility and good laundering resistance.

Spandex yellows under the influence of nitrogen oxides, which are contained primarily in combustion gases. Tertiary amine compounds have proved to be effective protectors against these; in addition, the patent literature describes a whole series of protective agents, e.g., carboxylic acid hydrazides [33], esters [34, 35], and phosphites [36, 37].

Polyether spandex, which is sensitive to chlorinated water, can be protected by tertiary amine compounds; polymers or polyurethanes with basic groups are used, such as poly($N,N$-dialkylamino)ethyl acrylates and methacrylates [29, 38]. Protective action is also obtained by inclusion of zinc oxide during spinning [23].

The dyeability of mixed articles (fabrics containing a mixture of yarns of polyamide and spandex) with acid dyes can be improved by adding basic compounds to the spandex. Alternatively, basic substituents, e.g., N-alkyldialkanolamines or N-alkyldiaminoalkylamines, can be built into the polymer molecule [39]. Previously, spandex was produced mainly as delustered yarns. Titanium dioxide was used as a delusterant, usually in concentrations of 2 – 5 %, to obtain as good a degree of whiteness as possible. Additional improvement in whiteness was obtained by adding nuancing dyes or pigments. However, nondelustered spandex types have gained greater prominence because they make possible more brilliant colors in dyed polyamide – spandex fabrics [40].

Freshly spun spandex yarn is slightly tacky. After being wound on the bobbin, cohesion between yarn layers develops with storage time. To counteract this and to improve the unwinding and processing behavior, separating and lubricating agents are used, e.g., polydimethylsiloxane and metal soaps [41].

## 4. Production of Polymer

The fiber raw material is produced by the two-stage prepolymer – chain extender process, frequently used in polyurethane chemistry.

***Production of Isocyanate Prepolymer.*** In this process step, an $\alpha,\omega$-OH functional polymer (macrodiol) is reacted with an aromatic diisocyanate (called a polymer-analogous reaction) to yield an $\alpha,\omega$-NCO functional polymer (macrodiisocyanate). The reaction is carried out in the melt or in a highly polar solvent (e.g., dimethylacetamide) at 50 – 80 °C in a batchwise or a continuous process. The molar mass of the prepolymer, which subsequently plays an important role in determining the elastic properties of the spandex, can be adjusted via the molar mass of the macrodiol and the molar ratio of NCO:OH [42]. The prepolymers have number-average molar masses of 4 – 8 kg/mol, as well as NCO contents of 2 – 4 wt%, and are generally processed immediately without intermediate storage.

***Chain Extension (Polymerization).*** In chain extension, the macrodiisocyanate is treated with a diamine in a highly polar solvent (e.g., dimethylacetamide) to give a high molar mass polyurethane – polyurea. The most frequently used short-chain aliphatic diamines react rapidly. To obtain a good polymer structure, rapid and intense mixing is, therefore, required during chain extension. Reactivity can be reduced by formation of a carbamate from the diamine and carbon dioxide prior to chain extension [43].

Both batchwise and continuous processes are used during chain extension. Spandex solutions with solids contents of ca. 20 – 35 % and viscosities of ca. 50 – 300 Pa · s (25 °C) are obtained. Molar mass is controlled during chain extension by using monoamines, e.g., diethylamine, as chain terminators. The polyurethane – ureas produced from diamines are linear polymers and, hence, soluble in the reaction media used.

With trifunctional starting components, covalently cross-linked polyurethanes are obtained, which are simultaneously chain extended and formed into a yarn [28] in the reaction-spinning process.

## 5. Spinning Processes

In contrast to elastic yarns based on rubber, spandex can be spun by using technologies generally known in the synthetic fiber industry. Because spandex based on polyurethane cannot be processed thermoplastically, solution-spinning processes are used to produce yarns.

***Solution Spinning.*** Of the solution-spinning processes (dry, wet, and reaction spinning), dry spinning has become the most important because it lends itself more readily than the others to production of the fine spandex yarns that are increasingly in demand today.

***Dry-Spinning Process*** (see Fig. 2). The high-viscosity elastomer solution is spun through multihole spinnerets (d) into heated spinning cells (e). During the process, the solvent evaporates, is absorbed by the entrained hot spin gases (air or inert gas), and is carried away. The solvent is separated by condensation and possibly scrubbing of the off-gas, distilled, and recycled. The individual filaments are cemented with the aid of a false-twisting device (j) to form a multifilament [44, 45]. Finish is applied to the

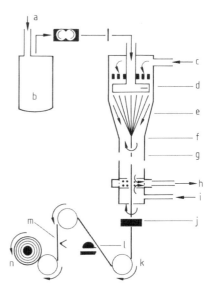

**Figure 2.** Dry-spinning process for spandex filament yarns [28]
a) Metering pump; b) Spinning solution; c) Hot spinning air; d) Spinneret; e) Heated spinning cell wall; f) Twisting point; g) Cell length, ca. 4 – 8 m; h) Spinning gas exhaust; i) Fresh gas; j) False-twisting device; k) Godet roll; l) Finish; m) Traverse; n) Windup device

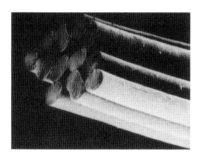

**Figure 3.** Dorlastan fiber cross sections, 160 dtex (12 individual capillaries) *
* ca. 50 µm.

yarn, which is wound up at a rate of 200 – 800 m/min. The drawoff speed is greater than the spinning speed (spin draw). The orientation thus produced affects yarn properties. To improve productivity, several multifilaments are spun from a single spinning cell at the same time. Figure 3 shows a cross section of a dry-spun spandex yarn.

***Wet-Spinning Process*** (see Fig. 4). In the wet-spinning process, the elastomer solution is spun into aqueous precipitation baths (d). In the process, the water-insoluble polyurethane coagulates, and the spinning solvent is carried off by the water. As in dry spinning, the individual filaments are cemented to a multifilament. After the filament has passed through wash baths and drying cabinets, finish is applied and the filament is wound. Windup speeds are significantly lower than in dry spinning, but equipment and energy costs are lower.

***Reaction-Spinning Process.*** In this process, chain extension of the isocyanate prepolymer by the amine and formation of the yarn occur simultaneously in the precipitation bath [46].

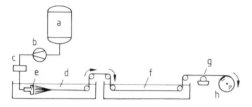

**Figure 4.** Schematic of a spandex wet- or reaction-spinning machine [45]
a) Spinning solution or prepolymer; b) Spinning pump; c) Filter; d) Precipitating or reaction spin bath; e) Spinneret; f) Wash or subsequent reaction bath; g) Finish; h) Windup

During spinning of the prepolymer into the reaction spin bath, a solid polymer skin forms immediately on the yarn surface and provides the stability necessary to draw off the yarn. Complete solidification is achieved in a second step. With this process, producing chemically cross-linked spandex by using trifunctional starting materials is also possible.

***Melt-Spinning Process.*** Only elastic yarns that are exclusively based on pure polyurethane can be prepared by the high-productivity melt-spinning process. These materials do not have the properties of commercial spandex based on polyurea – polyurethane and are, therefore, of little technical significance.

## 6. Uses

***Production of Elastic Yarns and Fabrics.*** Spandex fibers are processed and used exclusively in combination with hard fibers to impart high elasticity to textiles. In most cases, the spandex is used in bare form (i.e., unwrapped). Because of its rubberlike hand and poor frictional properties, spandex intended for certain end uses, such as woven elastic clothing fabrics [47], support stockings, and socks, is wrapped or core-spun with hard fibers. Of these elastic combination yarns [47], wrapped yarns are the most important. Wrapping can be carried out with one or two textile yarns. Core-spun yarns are produced on special secondary spinning machines on which staple fibers are twisted around the stretched spandex yarn. Core-spun yarns are distinguished by a remarkably soft hand. To improve the processability of the combination yarns, they can be changed by a combined stretch – steam process so that they (temporarily) lose their elastic extensibility. After being processed into fabric, the good elastic properties are restored by dyeing at the boil in the relaxed state.

The most important types of processing used to produce articles containing spandex are listed in Table 5. As can be seen, both knit (processing on circular or warp knitting machines) and woven fabrics can be produced. Articles with the highest extensibility can be prepared by the circular knitting process [48]. The elastic yarn is processed directly from the bobbin, the yarn being taken off over the end of the spool.

Table 5. End uses of spandex continuous-filament yarns

| End use | Manufacturing technology | Amount of spandex, wt % | Yarn type | Fineness, density, dtex |
|---|---|---|---|---|
| Lingerie and lingerie bands | circular knit | 5–12 | bare, wrapped, corespun | 22–400 |
| | warp knit | | bare, wrapped | 22–80 |
| Girdles | warp knit | 10–40 | bare, wrapped | 22–800 |
| Swim wear | warp knit | 10–32 | bare, | 22–80 |
| | circular knit | 6–12 | bare, wrapped | 44–160 |
| Hosiery bands, sock borders | circular knit | 10–20 | bare, wrapped | 135–400 |
| Support stockings | circular knit | 10–60 | bare, wrapped, corespun | 22–480 |
| Tapes | tape weaving | 10–50 | bare, wrapped | 320–5000 |
| Outerwear | weaving | 2–10 | rubber threads, wrapped yarns | 44–160 |

To prepare warp knitwear, the elastic yarns must first be assembled [49]. To do so, up to 1500 yarns are wound parallel, side by side, under constant extension onto a warp beam. Here, processing is exclusively in the bare form. By varying the fineness, the hard fiber yarn, and the pattern notation technique, the properties and appearance of the articles can be influenced in many ways with this process and can be adapted to both functional requirements and the demands of fashion [50, 51].

Processing of elastic yarns in woven fabrics has increased substantially. End uses are suit coat and pants fabrics for articles with increased wear comfort.

The major end uses for coarser spandex yarns and elastic wrapping yarns are tapes for various textile and engineering applications. In Western Europe, 45 % of spandex production is used in girdle and underwear manufacture, 20 % each in swim wear and hosiery, and ca. 15 % in other areas.

*Textile Properties of Elastic Fabrics.* The most important property of elastic fabrics is the combination of extensibility and elastic recovery. Spandex can elongate until the hard fiber construction obstructs further elongation. Extensibilities vary from ca. 25 % (woven fabrics) to > 200 % (knitwear). The elastic behavior of a warp-knitted fabric is shown in Figure 5.

By using spandex, the textile industry can fabricate elastic textiles in great variety, having given shaping and support forces, and also meeting all requirements with respect to end-use behavior, comfort, and fashion.

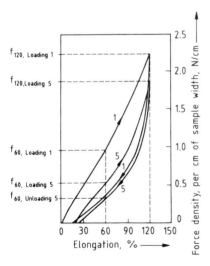

Figure 5. Curves of cyclic elongation of a tricot fabric (longitudinal direction) containing 18.5 % 45-dtex Dorlastan, unit mass 220 g/m², longitudinal extensibility 210 % [45]
* 1 is first curve; 5 is fifth curve; ——— is elongation; ——— is recovery.

## References

1. Bayer, DE 728981, 1937 (O. Bayer, H. Rinke, W. Siefken, L. Orthner *et al.*).
2. Bayer, DE 888766, 1951 (W. Brenschede).
3. Du Pont, US 2957852, 1955 (P.E. Frankenburg, A.H. Frazer).
4. DIN 60001, 1970.
5. H. Gall, K.H. Wolf in E. Becker, D. Braun (eds.): *Kunststoff-Handbuch*, vol. **7**, Carl Hanser Verlag München-Wien 1983, pp. 611 – 627.
6. M. Couper in M. Lewin, J. Preston (eds.): *Handbook of Fiber Science and Technology*, High Technology Fibers, vol. **3**, part A, Marcel Dekker, New York 1985, pp. 51 – 85.
7. H. Oertel in B. von Falkai (ed.): *Synthesefasern*, Verlag Chemie, Weinheim, Germany 1981, pp. 179 – 190.
8. P.A. Koch, *Chemiefasern/Text.-Ind.* **29/81** (1979) 95.
9. K.H. Wolf, *Text.-Prax. Int.* **36** (1981) 839.
10. D.J. Hourston, R. Meredith, *J. Appl. Polym. Sci.* **17** (1973) 3259.
11. R. Bonart, L. Morbitzer, *Kolloid-Z. Z. Polym.* **241** (1970) 909.

12. H. Hespe, E. Meisert, U. Eisele, L. Morbitzer et al., *Kolloid-Z. Z. Polym.* **250** (1972) 250.
13. R. Bonart, E.H. Müller, *J. Macromol. Sci. [B] Phys.* **10** (1974) 177.
14. R. Bonart, E.H. Müller, *J. Macromol. Sci. [B] Phys.* **10** (1974) 345.
15. R. Bonart, L. Morbitzer, E.H. Müller, *J. Macromol. Sci. [B] Phys.* **9** (1974) 447.
16. P. Hirt, H. Herlinger, *Angew. Makromol. Chem.* **40/41** (1974) 71.
17. H. Ishihara, J. Kimura, K. Saito, H. Ono, *J. Macromol. Sci. [B] Phys.* **10** (1974) 491.
18. J. Blackwell, K.H. Gardner, *Polymer* **20** (1979) 13.
19. L.L. Harrell, *Macromolecules* **2** (1969) 607.
20. H. Herlinger, P. Hirt, W. Aldinger, *Melliand Textilber.* 1979, 525.
21. L. Born, H. Hespe, J. Crone, K.H. Wolf, *Colloid Polym. Sci.* **260** (1982) 819.
22. L. Born, J. Crone, H. Hespe, E.H. Müller et al., *J. Polym. Sci., Polym. Phys. Ed.* **22** (1984) 163.
23. L. Born, H. Hespe, *Colloid Polym. Sci.* **263** (1985) 335.
24. Du Pont, DE 3124197, 1981 (E.M. Kenneth).
25. B. Küster, H. Herlinger, *Textil.-Prax. Int.* **36** (1981) 15.
26. Bayer, US 3377308, 1968 (H. Oertel, H. Rinke, W. Thoma).
27. H. Oertel, *Lenzinger Ber.* **45** (1978) 38 – 52.
28. H. Oertel, *Chemiefasern/Text.-Ind.* **27** (1979)no. 79, 1090 – 1096 **28** (1978)no. 80, 44 – 49.
29. Du Pont, DE 1918066, 1969 (Ch.G. Bottomley, O.L. Larry).
30. Amer. Cyanamid, US 3577388, 1969 (J.S. Megna).
31. Kanebo, JP 56110712, 1981.
32. T. Timm, *Kautsch. Gummi, Kunstst.* **37** (1984) 1021.
33. DE-AS 1184947, 1963 (F. Rosendahl, H. Rinke, H. Oertel).
34. Du Pont, EP 137407, 1985 (C.C. Ketterer).
35. Teijin, JP 4917449, 1974 (H. Moriga).
36. Du Pont, EP 137408, 1985 (W. Lewis).
37. Amer. Cyanamid, US 3573251, 1968 (J S. Megna, F.A.V. Sullivan).
38. Du Pont, DE 1669511, 1967 (O.L. Hunt).
39. Bayer, DE 1495830, 1964 (H. Oertel, H. Rinke, F. Moosmüller, F.-K. Rosendahl).
40. H. Gall, *Chemiefasern/Text.-Ind.* **32** (1982)no. 84, 510.
41. Du Pont, EP 0046073, 1980 (R.S. Hanzel, P. J. Sauer).
42. J.H. Saunders, K.C. Frisch: *Polyurethanes Chemistry and Technology in High Polymers*, vol. **16 – 1**, Interscience Publishers, New York 1962, p. 129.
43. Bayer, DE 1223154, 1963 (W. Thoma, H. Rinke, H. Oertel).
44. Bayer, DE OS 1288235, 1963 (H. Marzolph, H. Lenz, K. Bernklau, H. Scherzberg).
45. Du Pont, US 3094374, 1961 (P.M. Smith).
46. H.A. Pohl, *Text. Res. J.* **28** (1958) 473.
47. Bayer AG, Geschäftsbereich Fasern, Vertrieb: Produktinformation Dorlastan "Empfehlung zur Herstellung gewebter elastischer Bekleidungsstoffe mit Bayer Textilfaser Dorlastan," (available in English, "Suggestions on the Production of Woven Elastic Apparel Fabrics with Bayer Textile Fiber Dorlastan"), Bayer AG, Leverkusen, 1982.
48. Bayer AG, Geschäftsbereich Fasern, Vertrieb: Produktinformation Dorlastan "Bayer Textilfaser Dorlastan in der Rundstrickerei," (available in English, "Bayer Textile Fiber Dorlastan in Circular Knitting"), Bayer AG, Leverkusen, 1983.
49. K. Murenbeeld, *Textil.-Praxis Int.* **24** (1969) 335.
50. W. Müller, *Chemiefasern/Text.-Ind.* **20** (1970) 45.
51. Bayer AG, Geschäftsbereich Fasern, Vertrieb: Schrift "Bayer Textilfaser Dorlastan in der Kettenwirkerei," (available in English, "Suggestions on the Processing of Bayer Textile Fiber Dorlastan in the Warp-Knitting Industry"), Bayer AG, Leverkusen, 1977.

# Polyolefin Fibers

ANTONIO PELLEGRINI, Moplefan, Milano, Italy

PAOLO OLIVIERI, Moplefan, Milano, Italy

WERNER SCHOENE, BASF AG, Ludwigshafen, Germany

MICHAEL SCHWEIZER, ITCF Denkendorf, Denkendorf, Germany

| | | | | | |
|---|---|---|---|---|---|
| 1. | Polypropylene Fiber | 1495 | 2.1. | Raw Materials | 1504 |
| 1.1. | Chemical and Physical Properties | 1495 | 2.2. | Production of Tapes and Split Yarns | 1505 |
| 1.2. | Production | 1498 | 2.3. | Uses | 1507 |
| 1.2.1. | Raw Materials | 1498 | 3. | Monofilaments | 1508 |
| 1.2.2. | Staple Fiber Production Process | 1498 | 3.1. | Production | 1508 |
| 1.2.3. | Continuous Filament Yarn Production Process | 1500 | 3.2. | Uses | 1509 |
| 1.2.4. | Stabilization | 1500 | 4. | High-Modulus Polyolefin Yarns | 1509 |
| 1.2.5. | Pigmentation | 1502 | 4.1. | Production | 1509 |
| 1.3. | Uses | 1502 | 4.2. | Properties and Uses | 1510 |
| 1.4. | Economic Aspects | 1504 | | References | 1510 |
| 2. | Polyolefin Film Yarns and Split Yarns | 1504 | | | |

Polyolefin fibers, yarns, and monofilaments are usually produced by melt spinning. Film yarns and split yarns, obtained from films by uniaxial stretching, are also produced in large quantities. Polypropylene (PP) is the predominant raw material for fibers and yarns. High-modulus yarns of ultrahigh molecular mass high-density polyethylene (HDPE), commercialized in the mid 1980s, are an exception.

Both PP and HDPE are used to make film yarns and monofilaments; however, HDPE is used to a much lesser extent. Areas of application for PP fibers include carpet yarns, home furnishing fabrics, industrial fabrics, and nonwovens. Polyolefin tapes and monofilaments are used in bags, tarpaulin, industrial fabrics, carpet backing, twine, rope, and bristles. Trade names and producers of polyolefin fibers are listed in Table 1. World production of polyolefin fibers is listed in Table 2.

## 1. Polypropylene Fiber

Polypropylene fiber was developed in Italy by Montecatini during the second half of the 1950s [1]. Several other companies initially played an active role in its development, including Hercules, Reeves Brothers, Vectra, and Phillips in the United States; ICI in Europe; and Toray, Mitsubishi Rayon, and Toyobo in Japan. Some of them, like ICI, disappeared from the scene after the first few years; others improved their line of products and became extremely important in the market. There are many producers; in Western Europe they increased in number during the late 1970s, when the Montecatini patents expired.

### 1.1. Chemical and Physical Properties

*Chemical Structure.* Polypropylene is a hydrocarbon made up of $-CH_2CH(CH_3)-$ units. These propylene units form a stereoregular helix, with the methyl groups at an angular distance of 120° (isotactic polypropylene). A small percentage (ca. 2.5–5 %) is sterically random (atactic polypropylene).

Because its structure contains no polar groups, PP is resistant to most chemical agents, in particular to acids and alkalies; insensitive to water; and not dyeable with common types of dyes. Moreover, it has a low resistance to thermooxidative

**Table 1.** Trade names and Producers of polyolefin (PO) fibers

| Trade name | Type | Manufacturer |
|---|---|---|
| Arlene | PP | Aquafil SpA/Italy Aquafil USA Inc./ USA |
| Asota | PP/ PE | Asota GmbH/ Austria |
| Astra | PP | Drake Extrusion Ltd./ United Kingdom |
| Atlas | PP | Rheem Australia/ Australia |
| Bonafil | PP | Bonar Yarns & Fabrics Ltd./ United Kingdom |
| Charisma | PP | Drake Extrusion Ltd./ United Kingdom |
| Chisso Polypro | PP | Chisso Polypro Fiber Co./ Japan |
| Daiwabo Polypro | PP | Daiwabo Co./ Japan |
| Depas | PP | Depas A.S./ Turkey |
| Downspun | PP | PFE Ltd./ United Kingdom |
| Dryarn | PP | Aquafil SpA/ Italy |
| Duron | PP | Drake Extrusion Ltd./ United Kingdom Drake Extrusion/ USA |
| Dyneema | PE | DSM HPF B.V./ Netherlands DSM Dyneema LLC/ USA Nippon Dyneema Co./ Japan |
| Elsan | PP | Elsan A.S./ Turkey |
| Elustra | PP | Fiber Visions Inc./ USA |
| Emerlen | PP | MR 81 Srl/ Italy |
| Emu | PP | Boral Kinnears Pty/ Australia |
| ES Fiber | PP/PE | Fiber Visions A/S/ Denmark |
| ES Fiber | PP/PE | Chisso Corp./ Japan |
| Essera | PP | American Fibers and Yarns Co./ USA |
| Euralon | PP | Euroma SpA/ Italy |
| Gülsan | PP | Gülsan A.S./Turkey |
| Gymlene | PP | Drake Extrusion Ltd./ United Kingdom |
| Herculon | PP | Fiber Visions/ USA |
| Horizonte | PP | Fitesa/ Brazil |
| Hy | PP | Fiber Visions A/S/ Denmark Fiber Visions/ USA |
| Inhova | PP | American Fibers and Yarns Co./ USA |
| Istrona | PP | Istrochem/ SR |
| Krenit | PP | Chemfiber A/S Denmark |
| Kugafil | PP, PE | Fugafil – saran GmbH/ Germany |
| Leolene | PP | Drake Extrusion Ltd./ United Kingdom |
| Liplon | PP | Tai-Ray Co./ Japan |
| Mandal | PP | Sanghi Filaments Ltd./ India |
| Marquesa L | PP | American Fibers & Co./ USA |
| Meraklon | PP | Meraklon SpA/ Italy |
| Olefinesse | PP | Billermann KG/ Germany |
| Plasticel | PP | Plasticel SA de CV/ Mexico |
| Pliana | PP | Industrias Polifil, SA/ Mexico |
| Polyklon | PP | Meraklon SpA/ Italy |
| Polysilk | PP | Polysilk SA/ Spain |
| Polystar | PP | Nexcel Synthetics/ USA |
| Polysteen | PP, PE | Steen & Co. GmbH/ Germany |
| Prolen | PP | Chemosvit/ SR |
| Propilan | PP | Propilan SA/ Spain |
| Pylen | PP | Mitsubishi Rayon Co./ Japan |
| Radilene | PP | Deufil GmbH/ Germany |
| Reilen | PP | Reinhold GmbH/ Germany |
| Sasrifill | PP | DS Fibres NV/ Belgium |
| Salus | PP | Filament Fiber Technology Inc./ USA |
| Sarlon | PP | Sarlon Pty./Australia |
| Silver Ring | PP | Hubei Chemical Fiber/ China |
| Spectra | PE | Honeywell Advanced Fibers/ USA |
| Tecnisilk | PP | Polisilk SA/ Spain |
| Telar | PP | Filament Fiber Technology, Inc./ USA |
| Terclon | PP | Polyfil NV/ Belgium |
| Tiptolene | PE | Lankhorst Touwfabrieken BV /Netherlands |
| Toabo Polypro | PP | Toabo Material Co./ Japan |
| Unal Sentetik | PP | Unal Sentetik Sanayi A.S./ Turkey |
| Xanthi | PP | Thrace Plastics SA/ Greece |

PE = polyethylene
PP = polypropylene

**Table 2.** World production of polyolefin fibers

| Year | Production, t |
|---|---|
| 1990 | 1 000 000 |
| 1995 | 3 960 000 |
| 2000 | 5 984 000[a] |
| 2005 | 6 463 000[b] |
| 2009 | 2 600 000[b] |

[a] Including film fibers (tapes).
[b] Including film fibers and spunbonds.
[c] Excluding tapes.

and photooxidative degradation. The problem of nondyeability has been solved by mass dyeing techniques and the problem of degradation by the use of suitable additives. Because of the absence of polar groups, only van der Waals forces exist between macromolecules; these forces are much weaker than the hydrogen bonds found, for example, in polyamide. Because mechanical characteristics depend on these forces and on molecular entanglement, very high molecular masses ($M_r$ from ca. 170 000 to 300 000, compared to ca. 20 000 for polyamide) must be used to obtain satisfactory strength (fiber tenacity).

***Rheological Aspects [2].*** Polypropylene fiber is produced by melt spinning, i.e., by extruding the molten polymer through suitable spinnerets. Therefore, the rheological behavior of the polymer melt must be considered in selecting processing conditions or even in choosing a certain polymer for specific spinning conditions. The first aspect to be considered is that isotactic PP in the molten state is highly non-Newtonian and exhibits a strong pseudoplastic character; the apparent viscosity of the melt is not constant with varying shear stress but it decreases when shear stress increases. The deviation from Newtonian behavior is markedly dependent on the relative molecular mass ($M_r$) and polydispersity (molecular mass distribution $M_rD$); for polymers of comparable polydispersity, the melt viscosity increases with increasing viscosity–average molecular mass. An extremely useful, practical indication of $M_r$ is the melt flow index (MFI), which is a measure of flow under standard conditions (ASTM D 1238: grams extruded in 10 min through a standard orifice under a pressure of 2.16 kg/cm$^2$ at 230 °C); MFI is inversely proportional to melt viscosity. The MFI is measured at a shear rate close to those commonly met in usual spinning processes.

The initial MFI of PP polymers used in fiber production ranges from 2 to 35, depending on the process involved and the type of fiber produced. In addition to pseudoplasticity, PP melt exhibits appreciable signs of viscoelasticity. In a viscoelastic material the deformations undergone by the melt have not only a viscous but also an elastic component, which can therefore be recovered when the applied stress ceases, i.e., when the extruded polymer emerges from the spinneret capillary. The residual elastic energy, whose magnitude depends on the residence time in the capillary, can cause, among other phenomena, enlargement of the extrudate to a diameter larger than that of the capillary (die swell). The polymer swelling increases as the shear stress increases and as the ratio of length to capillary radius decreases.

Die swell also depends on the molecular mass distribution ($M_rD$): at comparable melt viscosities, die swell is larger for a polymer with broad $M_rD$ than for a polymer with a narrow $M_rD$. Swelling increases as temperature decreases. All these rheological characteristics are taken into account when spinneret hole geometry, polymer type (MFI, $M_rD$), and extrusion conditions are being selected.

***Structural Characteristics [2 pp. 327–337], [3].*** The stereoregular structure of isotactic PP macromolecules makes an ordered three-dimensional configuration (i.e., a crystalline structure) possible when the molten polymer is cooled to the solid state. Under normal conditions, isotactic PP crystallizes in the monoclinic system; if cooled rapidly to room temperature, it exhibits a paracrystalline form, called smectic. In practice, PP fiber is a partially crystalline material. The crystalline structure of the fiber depends on the conditions of extrusion and quenching, the take-up speed in the spinning step, and the draw ratio and annealing conditions in the finishing step. Normally, the crystallinity of the finished fiber does not exceed 70 %; values of 40–50 % are usual. Many physical properties are influenced by crystallinity. Mechanical properties, however, are affected less by the percentage of monoclinic material than by the molecular orientation imparted during drawing; orientation is evaluated by measurement of birefringence.

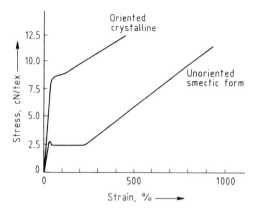

**Figure 1.** Stress – strain curves of as-spun polypropylene fibers

***Mechanical Properties.*** Depending on spinning conditions (extrusion temperature, quenching, spinning speed, and rate of deformation in the molten state), the mechanical properties of as-spun PP fibers can vary widely, as shown schematically in Figure 1. The mechanical properties of drawn fibers are influenced substantially by the applied draw ratio and, consequently, also vary widely (Fig. 2).

The relationship between tenacity and draw ratio depends on the polymer used and, in particular, its polydispersity. As polydispersity decreases, the tenacity obtained at a given draw ratio increases, but the drawability of the spun fiber decreases at the same time. Therefore, polymers with narrow $M_rD$ are not suitable for production of high-tenacity fibers; they do, however, have advantage in terms of spinning continuity at high spinning speeds.

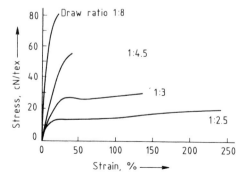

**Figure 2.** Stress – strain curves of drawn polypropylene fibers

## 1.2. Production

### 1.2.1. Raw Materials

Polypropylene fibers are obtained from PP homopolymer. They may also be based on PP and a propylene–ethylene random copolymer or a blend of polypropylene and polyethylene homopolymers. In the latter case, the definition "polyolefin fiber" or "olefin fiber" is more correct. However, polyolefin is also frequently used for PP homopolymer fiber.

Polypropylene is used in the production of side-by-side bicomponent polypropylene–polyethylene fibers. The Customs Cooperation Council of the European Economic Community (EEC) adopted the following definition of textile polypropylene:

Fibers composed of aliphatic saturated hydrocarbon linear macromolecules having in the chain at least 85 wt% of units with one carbon atom in two carrying a methyl side chain in an isotactic position, and without further substitution.

The availability of raw materials is not a problem, although the supply of propylene may be influenced by the demand for ethylene because ethylene and propylene are coproducts in the cracking of virgin naphtha. Other sources, of lesser importance at the moment, are liquefied refinery gases containing propylene and propane gases.

### 1.2.2. Staple Fiber Production Process

Polypropylene can be produced by two different processes: (1) *conventional spinning* is based on high spinning speeds and spinnerets with a rather limited number of holes; (2) *short spinning* is based on low spinning speeds and spinnerets with a large number of holes. In the first process, because of the high spinning speeds, postspinning operations are performed in a separate step; in the second process, because of the low spinning speeds, spinning and postspinning are carried out continuously. Postspinning operations are basically the same for both processes.

***Conventional Spinning Process.*** The polymer, to which stabilizers and pigments have been added, is introduced into the extruder (a) (Fig. 3). The

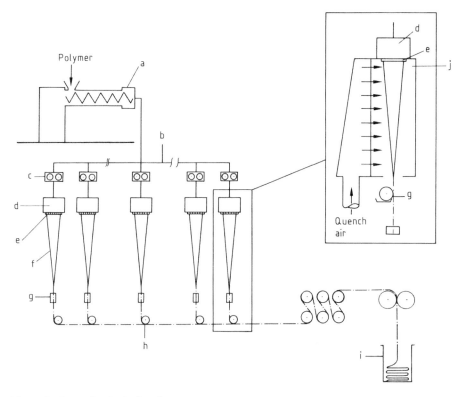

**Figure 3.** Conventional spinning plant
a) Extruder; b) Melt distribution; c) Metering pump; d) Spinning pack; e) Spinneret; f) Filament; g) Oiling device; h) Take-up; i) Can coiler; j) Quench box

molten material is fed to the spinnerets (e) by metering pumps (c), which control the throughput of each spinneret. The extruder is heated electrically, while the melt distribution manifold is heated by means of a diathermic fluid (e.g., Dowtherm and Diphil). Filaments (f) emerging from the spinneret are quenched with cold air (air speed 1–2 m/s) and, when solidified, collected in cans (i). An antistatic and lubricating agent is usually applied to the filaments before they come in contact with any solid surface (g).

The filament fineness depends on the throughput of the metering pump, the number of holes in the spinneret, and the take-up speed, according to the following relationship:

$$d = \frac{1000 \, Q}{3.6 \, n \cdot v}$$

where $d$ is the fineness, decitex; $Q$ is the pump throughput, kilograms per hour; $n$ is the number of holes; and $v$ is the take-up speed, meters per second. The spinning speed of PP fiber is usually 600–1200 m/min, and the temperature of the molten polymer in the spinneret is 220–300 °C, depending on molecular mass of the polymer, spinning speed, and desired characteristics of the fiber being produced. Spinnerets used for staple production are 150–250 mm in diameter and have between 200 and 2000 holes, depending on the fineness of the fiber. In some cases, rectangular spinnerets are used to achieve higher productivity. The number of holes is limited to avoid having the filaments stick together (filament marriage). Polymers with a fairly low $M_r$ (MFI 5–20) are used commonly in the production of standard fibers; higher $M_r$ (MFI ca. 2) polymers are used only for high-tenacity fibers. To obtain satisfactory spinning continuity, the polymer must be filtered before reaching the spinneret; both sand filters and screen filters are used.

*Postspinning Operations* Drawing, crimping, heat setting, cutting, and baling are the steps that follow spinning in the production process.

Several tows coming from the spinning section are processed together to reach a total fineness of 200–300 ktex (0.2–0.3 kg/m). They are stretched between two sets of rolls; the draw ratio depends on the ratio of the roll speeds of these two sets. The draw ratio for standard fibers is between 1:3 and 1:5. To obtain uniformly drawn fibers, the tow is heated in a steam chamber located between the two sets of rolls. Alternatively, heated rolls can be used in the first set. If high-tenacity fibers are desired, the drawing operation is sometimes divided into two steps by means of a third set of rolls. Crimping is intended to impart bulk and cohesion to the fiber for subsequent conversion to final products. The fiber is compressed in the crimper (stuffer box) and subsequently dried and heat-set in a hot-air oven (100–130 °C) to stabilize crimp and reduce thermal shrinkage after drawing. A finish is usually applied to the fiber during the post-spinning operations. Staple cutting and baling operations are the final stages of production. The speed of these operations is limited by the crimping speed and is generally 150–250 m/min.

*Short Spinning Process.* The rheological behavior of PP is such that, when spinning speeds are fairly low (120 m/min), and the stream of quenching air blows through the filaments close to the spinneret, the filaments do not stick together even with close hole spacing (ca. 1 mm). Then, the speed of consecutive operations can be matched, and a continuous, uninterrupted, one-step process can be achieved. A consequence of the high hole density and the reduction in height of the quenching section is that plants can be made more compact for a lower cost. For this reason and reduced manpower requirements, the short spinning process has become very popular.

### 1.2.3. Continuous Filament Yarn Production Process

In principle, spinning continuous filament yarn is similar to spinning staple fiber; the main difference is that in the former, each yarn is collected on an individual bobbin. Drawing of the yarn (Fig. 4) is usually performed by two pairs of rolls (f), which are generally heated electrically. For some applications, continuous filament is used as it comes from the drawing machine (flat yarn), but in most cases, filament bulk must be increased by means of a texturing system. A mechanical texturing system is preferred for the very fine yarns used in the textile industry. In this case, the yarn is highly twisted, heat-set, and detwisted in a continuous false-twisting process to give each individual filament the shape of a spring. A subsequent heat-setting operation can be introduced in the detwisting zone to reduce yarn elasticity. For coarser continuous filament yarns (carpet yarns), systems based on fluid bulking jets are used. Consistent with the continuing trend toward reduced production costs, machines have been developed in which the different operations are performed continuously in a single step.

### 1.2.4. Stabilization

Polypropylene, although relatively stable to heat and light in the absence of oxygen, has poor

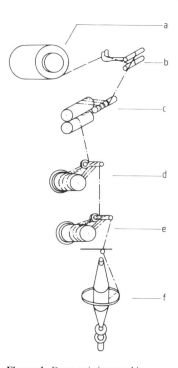

**Figure 4.** Draw-twisting machine
a) Spun yarn; b) Pre-tension guide; c) Pre-tension rolls and guides; d) Feed roll; e) Draw roll; f) Twisting take-up

resistance to thermooxidative and photooxidative degradation [3 pp. 38–93]. Thermooxidative degradation occurs during conversion of the polymer to a fiber. Photooxidative degradation occurs when the end product, which contains PP fiber, is exposed to direct or diffuse solar radiation in the presence of oxygen. As a hydrocarbon, PP is transparent to UV light but it is sensitive to it because of the presence of traces of oxygen-containing groups, such as hydroperoxides and ketones, which absorb at 310–330 nm. Depending on the temperature during exposure, thermooxidative degradation and photooxidative degradation often occur simultaneously. Both processes generate polymer radicals, which catalyze further molecular cleavage.

Therefore, suitable antioxidants and light stabilizers are added to protect the polymer melt at high temperature and to impart sufficient aging resistance to the finished products. (→ Antioxidants, Section 5.4). These stabilizers can be classified according to the mechanism by which they protect the polymer: (1) chain terminators or free-radical scavengers, which terminate the chain reaction by giving rise to nonreactive radicals; (2) hydroperoxide deactivators, which decompose hydroperoxides without formation of radicals; (3) UV absorbers, which screen UV radiation; and (4) quenchers, which quench the excited states of ketone groups; and (5) free-radical scavengers that trap the radicals and regenerate themselves (HALS: hindered amine light stabilizers). The following are some of the chemicals used at present:

| | |
|---|---|
| Chain terminators | phenolic antioxidants |
| Peroxide decomposers | disulfides, trialkyl phosphites |
| UV absorbers | derivatives of hydroxybenzo phenone and benzotriazole |
| Energy quenchers | nickel chelate stabilizers |
| Free-radical scavengers | hindered amines [4] |

The effectiveness of the stabilizer formulation depends on the choice of individual components. This choice must also take into account the pigments used, because they can affect heat and light stability significantly and can even interfere with stabilizers. In addition, attention must be paid to possible concomitant phenomena, such as gas fading, which is the color change caused by nitrogen oxides present in air. These oxides can react with some components of the stabilizing system, resulting in nitration of aromatic rings or addition of nitrogen oxides to double bonds. Nitrogen oxides can also react with certain finishes and cause color changes via some of the reactions mentioned previously.

The stabilizer formulation also includes antacids, typically calcium stearate, which react with the HCl formed by reaction of catalyst residue with substances containing active hydrogen. Stabilizer systems that comply with international regulations for chemicals in food contact applications are also available.

**Stabilizer Evaluation.** The following test methods have been developed to select the most suitable stabilizers and predict actual fiber behavior.

*Oxidation Resistance.* Fiber bundles, hung on a rotating frame under a small weight, are exposed in a forced-draft oven at 110 °C. Resistance to oven aging is expressed as days to failure. For operating conditions, see ASTM D 3045–74 (reapproved 1979).

*Light Resistance.* Accelerated tests with instruments based on a xenon lamp are preferred to those using a carbon-arc apparatus because the xenon lamp gives a spectrum similar to that of summer sunlight, whereas the carbon-arc apparatus is too high in UV radiation. Exposure conditions are selected so that the temperature is comparable to that experienced in use because the rate of photooxidation is a function of temperature. For indoor uses, such as carpets and wall coverings, fiber bundles are exposed in a Xenotest or Weatherometer according to ASTM G 26–84, DIN 54 004, ISO B 02. The resistance to artificial light is expressed as residual tenacity after exposure. Fiber exposed to combined heat and light, e.g., on rear window shelves in cars, is tested in the Xenotest at high temperature.

*Gas-Fading Resistance.* The fiber specimen is exposed to nitrogen oxide containing gases derived from the combustion of butane. Color variation is assessed by comparing the exposed and original samples after four exposure cycles of controlled duration.

### 1.2.5. Pigmentation

Polypropylene cannot be dyed by conventional methods because it lacks polar groups that can provide suitable sites for dyes. The techniques proposed thus far for making PP dyeable by traditional acid or disperse dye systems do not have near- or medium-term commercial potential, either because of the relatively high cost of additives involved or because of the as yet insufficient colorfastness. Only nickel-modified fibers are used commercially for space dyeing or printing.

For these reasons, PP fiber is colored by blending the polymer with suitable pigments prior to extrusion; this system is known as "melt dyeing" or "mass dyeing." Because of various requirements, such as thermal stability at high temperature, the range of pigments that can be used is limited; however, mass-dyed PP fibers are characterized by high colorfastness to light, heat, rubbing, and washing. Colorfastness may, of course, be affected by stabilizer–pigment interaction or by additives that facilitate extraction of the pigments. The main classes of pigments used are the following:

| | |
|---|---|
| Inorganic: | titanium dioxide (rutile or anatase) |
| | iron oxides |
| Organic: | insoluble azo pigments |
| | copper phthalocyanines |
| | vat pigments |
| | carbazole dioxyazine |
| | carbon black |

## 1.3. Uses

***Types of Fibers Produced.*** Polypropylene fiber is produced in a wide range of finenesses, both as staple and continuous filament. Staple fibers range from ca. 1 to 250 dtex per filament. The fiber cross section is primarily circular; the fineness range of continuous-filament yarn is 70–12 000 dtex. The filament cross section is predominantly circular in the low to medium fineness for apparel, upholstery, and industrial uses, and triangular or trilobal for the fibers used in floor covering. For the latter application, the basic fineness range is 1300–4000 dtex; higher fineness up to 12 000 dtex is obtained by comingling.

***Commercial Mass.*** The International Bureau for the Standardization of Man-Made Fibers (BISFA) has agreed on the following definition of commercial mass: "the commercial mass shall be obtained by adding to the oven-dry mass, i.e., after removal of finish, etc., the mass corresponding to the conventional allowance" where "the conventional allowance for polypropylene staple fiber, sliver or top and tow is 2.0 %."

***Properties.*** Polypropylene fiber has physical, chemical, and mechanical properties that make it suitable for a wide range of applications. These properties can be divided into the following groups:

**Good**

Relative density
Resistance to acid and alkali
Abrasion resistance
Moisture absorption
Mechanical properties under dry and wet conditions
Dimensional stability
Stain and soil removal
Easy washing
Quick drying
Static buildup in carpets
Resistance to mildew, microorganisms, and insects
Thermal bondability
Colorfastness of suitably pigmented fibers
Light resistance of suitably stabilized fibers

**Fair**

Flammability
Frictional characteristics
Oil-based soil removal

**Poor**

Dyeability with common dyeing methods (need for pigmentation)
Heat and light resistance (need for stabilization)
Resistance to dry-cleaning solvents
Behavior during ironing

Some properties are inherent in the component material; others are imparted or enhanced by the production process or by suitable additives. A brief survey of the major properties follows, with an indication of their effect on fiber application.

*Relative density* (0.91) is the lowest of all commercial textile fibers; this is an advantage in all applications because it means lightness and high cover (especially in nonwoven construction).

*Chemical resistance,* especially to acid and alkali, is outstanding. Polypropylene fiber was treated at 20 °C with the following chemicals; the residual tenacity after 96 h is given (in percent):

| | |
|---|---|
| Nitric acid (66%) | 80–87 |
| Sulfuric acid (95%) | 95–100 |
| Formic acid (75%) | 100 |
| Sodium hydroxide (40%) | 100 |
| Concentrated aqueous ammonia | 95–100 |
| Trichloroethylene | 80 |
| Perchloroethylene | 80 |
| Xylene | 80 |
| Toluene | 90 |
| Benzene | 90 |
| Acetone | 100 |
| Sodium hypochlorite (5% active chlorine) | 85 |
| Hydrogen peroxide (12 parts) | 90 |

Hot halogenated solvents swell PP fiber, with subsequent shrinkage on drying (trichloroethylene), or even dissolve it (boiling perchloroethylene). Moreover, some boiling aromatic solvents dissolve the fiber (e.g., xylene, decahydronaphthalene), and strong oxidizing agents attack it. In the last case, however, the fiber can be improved considerably by suitable stabilizer formulations. The high chemical resistance is exploited in industrial filtration.

*Abrasion resistance* is comparable to that of polyamide. It increases with increasing crystallinity and molecular mass and is influenced by frictional properties. The high abrasion resistance is exploited in floor covering, upholstery, hosiery, and industrial applications.

*Moisture regain* is virtually nil (0.05% at 65% R.H. and 21 °C). The insensitivity of PP fiber to water brings about the following advantages: dimensional stability, fabric dryness in contact with the skin, quick drying, and no variation in mechanical properties under wet conditions.

*Temperature Behavior.* Polypropylene fiber does not distort when exposed to temperatures of 120–130 °C during conversion processes; however, it is sensitive to pressure at elevated temperature, so ironing can cause the fabric to stiffen. The softening point is 150 °C, and melting occurs at 168 °C. The relatively low melting point is exploited in thermal bonding, which has become important in hygienic and medical applications (calender bonding is generally used, but flow-through bonding with hot air can also be applied). High-temperature tentering treatment is used to produce dense nonwovens suitable as coating substrates.

*Dyeability.* Because PP fiber is not dyeable by common dyeing methods, pigmentation is used. Although a nickel-modified fiber is available for printing or space dyeing with chelatable dyes, its use is limited. The absence of polar groups in the PP molecule makes for easy stain removal; on the other hand, PP is more easily soiled by oil-based stains than other synthetic fibers.

*Resistance to Mildew, Microorganisms, and Insects.* Polypropylene fiber is neither attacked by mildew and molds nor damaged by insects.

*Static Buildup.* Limited charges are separated when a person walks on a PP carpet and, consequently, a low body voltage is generated, lower than that at which uncomfortable shocks can be felt.

*Flammability.* Application of a flame causes PP to shrink, melt, and burn if contact with the flame is sufficiently prolonged. Polypropylene burns slowly and drips like a candle. Burning can continue even if the flame is withdrawn because the limiting oxygen index of PP is 18.5. The self-ignition temperature is 570 °C.

*Frictional Characteristics.* The coefficient of friction of PP fiber, both fiber-to-fiber and fiber-to-metal, is fairly high. Suitable lubricants are applied to keep it low so as to allow conversion by textile processing. Cotton spinning is one of the most critical processes.

*Electrical Properties.* Polypropylene is a good insulator and exhibits an extremely low power loss, even at high frequencies. The dielectric constant is 2.1 at 60 Hz and 2.2 at 1 MHz; the power loss factor is 0.0002 at 60 Hz and 0.0003 at 1 MHz; volume resistivity is $4.9 \times 10^{14}$ Ω cm.

*Thermal Properties.* The thermal conductivity of PP fiber is the lowest among commercial fibers (0.138 W m$^{-1}$ K$^{-1}$). This contributes to high thermal insulation in textiles, although the major contribution comes from air trapped in the structure.

*Allergic Phenomena and Nontoxicity.* Polypropylene itself does not cause skin irritation or sensitization and is not toxic; surface finishes must be selected correctly to retain these advantages.

**Applications.** The three major fields of application are in industrial, home textile, and apparel end uses.

*Industrial applications* include rope, twine, conveyor belts, industrial sewing thread, filter cloth, paper reinforcement, geotextiles, coating substrates, and carpet backing. Diaper facing is an example of a hygienic–medical product in which PP has made rapid inroads.

*Home textile applications* include floor covering (tufted, woven, and needle punched), upholstery fabric, wall covering, fiberfill, blankets, warp yarns in woven carpets, and yarns for secondary backing for tufted carpets. Floor coverings constitute the most important outlet for PP fiber, in both bulked continuous filament (BCF) and staple.

The *apparel uses* of PP are limited by lack of dyeability, difficulty in ironing, and inability to be dry-cleaned. However, the fiber is established in knitwear, especially sportswear. Staple and textured continuous filament yarns are used. Other applications include fleece fabrics, hosiery, and hand knitting yarns.

## 1.4. Economic Aspects [5]

The Oerlikon Textile report "The Fiber Year 2009/2010" estimated the world PP market for 2006 at $2.6 \times 10^6$ t. In the USA, slow demand for carpet yarn and increasing substitution by polyester BCF yarn have put additional pressure on the industry. While staple fiber applications increased by 3.5% to $1.1 \times 10^6$ t, output of filament yarn declined by 12.7% to $1.5 \times 10^6$ t.

## 2. Polyolefin Film Yarns and Split Yarns

The production of polyolefin film yarns and split yarns involves production of a primary film, which is subsequently cut, stretched, and possibly fibrillated mechanically. This method had been developed and patented by the end of the 1930s [6] but was not exploited industrially on a broader basis until the 1960s.

In principle, other thermoplastics can also be produced by this means. However, it is important only for polypropylene and for high-density polyethylene (HDPE).

The usual size of individual PP film ribbons is between 300 and 2500 dtex; split yarns often have sizes up to 10 000 dtex. The smallest size attainable by special processes is ca. 10 dtex.

The literature on PP film yarns and split yarns has been reviewed [7–11].

### 2.1. Raw Materials

The choice of raw materials depends not only on what is required of the finished product but also on the processing properties. Basically, raw materials with higher molecular masses (lower melt flow indices) are used to produce film yarns and monofilaments than to produce melt-spun multifilaments. The choice between PP and HDPE is governed by the following properties:

1. Polypropylene has higher strength, i.e., greater tenacity, than HDPE (see Fig. 5) [12, 13]. In addition, PP exhibits less "cold flow"; i.e., products of PP have lower

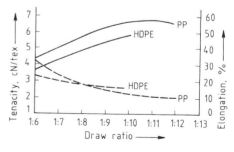

**Figure 5.** Tenacity (———) and break elongation (– – – –) of film yarn of polypropylene (PP) and high-density polyethylene (HDPE) as a function of draw ratio [12]

elongation under continuous mechanical loading than those of HDPE.
2. Tapes and monofilaments of HDPE are more flexible than those of PP and give finished products with a softer hand.
3. Polypropylene can withstand a higher thermal load because of its higher melting temperature (ca. 160 °C vs. ca. 130 °C for HDPE).
4. On the other hand, HDPE has better resistance to cold. Although stretching substantially improves the resistance of PP to cold, HDPE is at present preferred for applications below freezing temperature because it offers a greater margin of safety.
5. Tapes and monofilaments of HDPE have lower coefficients of sliding friction than those of PP. Consequently, they are better suited for processes in which good sliding ability is needed (for example, processing on knitting machines).
6. The tendency of HDPE tapes to split is much lower than that of PP. Depending on end use, this can be an advantage or a disadvantage.
7. The UV stability of HDPE is inherently better than that of PP. Nevertheless, very good weathering resistance has now been achieved with PP tapes and monofilaments by using high-efficiency UV stabilizers of the hindered amine light stabilizer (HALS) type.

The HDPE types used have densities between 0.945 and 0.960 g/cm$^3$ and melt flow index values (MFI 190/2.16, i.e., MFI measured at 190 °C under a load of 2.16 kg/cm$^2$) between 0.3 and 0.7 g/10 min; a broad molecular mass distribution is preferable.

Polypropylene tapes are produced almost exclusively from homopolymers. Depending on the process employed and end uses envisioned, melt flow indices (MFI 230/2.16) between 1.5 and 6 g/10 min, and preferably between 2.5 and 4 g/10 min, are common. If the process involves cooling via a water bath, PP types with low water pickup (carry over) are required.

Coloration is achieved by pigment compounds; to minimize splitting and assist in weaving PP ribbons, inorganic fillers (e.g., chalk) are used.

In special cases, copolymers (i.e., blends with other polymers) or coextrudates can be used to achieve certain properties. Coextruded films permit the production of crimped tapes and yarns [14–16].

## 2.2. Production of Tapes and Split Yarns

The stretching process in which the film strips are drawn to many times their original length (in most cases, six to ten times) is the central processing step. It leads to high molecular orientation in the longitudinal direction and to substantial increase in strength with simultaneous decrease in break elongation. However, strength in the transverse direction clearly decreases. At high draw ratios, this effect can cause the tape to "split," i.e., to fibrillate even at low mechanical stress. This phenomenon, which is more pronounced for PP than for HDPE tapes, is undesirable in some cases, e. g., for bagging and packaging fabrics. In other cases, it is exploited and reinforced intentionally (e.g., by profiling the ribbons or by using fibrillators) to arrive at products with lower fineness and more textilelike properties (split yarns and fibers).

Figure 6 is a generalized scheme for the production of film ribbons and split yarns. The processing steps follow.

*Production of Film.* In general, extruders (a) with screw diameters of 90 or 120 mm and screw lengths $\geq$ 25 times the diameter are used. Screws with a compression ratio of ca. 1:3, with a shear and mixing section, have proved useful. Cooled groove boxes in the feed section lead to more uniform granulate feed and higher throughput. In this case, flat cut screws with combined high-efficiency mixing and shearing sections can be used. The type of film production as well as the conditions of cooling and extrusion affect processing and properties of the end product [12, 17, 18]. Tapes of HDPE are usually produced from tubular films ($b_1$), and PP tapes preferably from cast films. The cast film can be cooled in a water bath ($b_2$) at 15–40 °C. This type of cooling is particularly effective because it permits high operating speed, leads to the formation of small

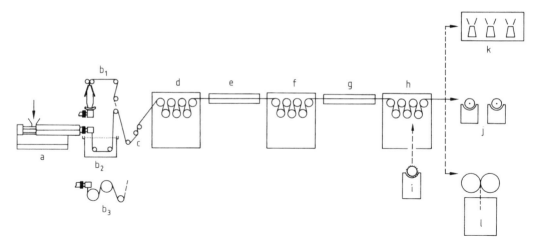

**Figure 6.** Production of film yarns and split fibers
a) Extruder; $b_1$) Tubular film (air cooled); $b_2$) Cast film (water quenched); $b_3$) Cast film (chill roll quenched); c) Knife beam; d) First godet stand; e) Stretching oven; f) Second godet stand; g) Setting oven; h) Third godet stand; i) Fibrillator (optional, only used for split fiber production); j) Windup; k) Twister; l) Staple cutter

crystallites, and is advantageous with respect to stretching behavior and strength of the tapes obtained. Raw materials with low water carry over must be chosen, and wiping, suctioning, and squeezing devices must be provided so that no water is entrained in the stretching section or the edge strip return. Film production by the chill roll process ($b_3$) is used only when exceptionally close thickness tolerance and high tape uniformity are required. Throughput is normally lower in tubular film installations than in cast film units because of lower cooling intensity; in addition, larger variations in film thickness are to be expected.

*Cutting.* Normally the film is cut prior to stretching. This is accomplished with blades arranged at the desired distance on a knife beam (c). In principle, it is also possible to stretch the whole film and cut it only after stretching and setting. The high splitting tendency of PP makes it difficult to obtain a cut completely free of nicks, so only HDPE is processed this way [19, 20].

*Stretching.* Stretching is accomplished by guiding the film strips over mechanically driven cantilevered feed rolls. The stretching unit is made up of two sets of godets (d) and (f) usually consisting of seven godets each. The second set runs at a higher speed than the first; the draw ratio is determined by the difference in speeds. Draw ratios from 1:6 to 1:8 are used to produce tape for weaving and warp knitting; for rope, cable, and twine, draw ratios are ca. 1:9 to 1:10. Operating speeds are $\leq 300$ m/min.

Stretching in the short gap between rolls occurs mainly through reduction in thickness without appreciable loss in width. This leads to some orientation in the transverse direction and, hence, a reduction in splitting tendency. Stretching is generally carried out in hot air ovens (e) or on heated plates.

The overall stretching process is exothermic. However, heat must be supplied to initiate it and to assure uniformity at the high speeds required. The required temperature depends on the raw material, the operating speed, and the desired tape properties; for the most part, it is between 130 and 180 °C. Whereas the temperature of the tape itself cannot be measured, the uniformity of stretching conditions can be checked by measuring yarn tension [21].

*Setting.* After stretching, setting (g), i.e., controlled shrinkage, is accomplished between the second and third godet stand by again supplying heat (in a hot-air oven or with the aid of heated rolls). In this annealing process, the orientation stresses introduced during stretching are relieved. In so doing, subsequent undesired and uncontrolled shrinkage is reduced or prevented.

The setting temperature is either equal to the stretching temperature or 10–20 °C below it. A higher setting temperature leads to less shrinkage but adversely affects tape strength. Setting shrinkage results from the difference in drawoff speeds between the second and third godet stand. For most applications, adjusting the setting shrinkage to ca. 5 % is sufficient. For end uses in which particularly high dimensional stability is required (e.g., carpet backing), a higher setting shrinkage (10–15 %) must be selected.

***Splitting and Fibrillating.*** Polypropylene tape has a relatively high tendency to split, and the propensity for fibrillation increases with increasing degree of stretching. The spontaneous splitting tendency is occasionally exploited in the production of twine and packaging yarns. The tapes split even during twisting (k) and give a fibrous, textilelike end product. The splitting tendency can be reinforced by adding other polymers [22].

Fibrillated yarns or fibers with still lower individual fineness and greater softness are obtained by special fibrillation processes. Stretched film strips are generally guided over a needle roll revolving at controlled speed (i); the direction of rotation of the roll coincides with the running direction of the tapes. The tapes are split into a networklike multiplicity of fibrils (Fig. 7). Depending on the number, fineness, and arrangement of the needles, as well as the speed of rotation of the roll, various fibrillar structures can be obtained [22–26].

Strongly fibrillated tape yarn can also be obtained without needle roll from profiled primary film. Profiling is produced either by a comblike jet lip (Barfilex process [27]) during extrusion or by roll embossing [28, 29]. In both cases, the impressed longitudinal grooves act as preset break points, which tear at the high draw ratios used (ca. 1:10). However, these processes are of minor practical importance compared to needle roll fibrillation.

Split fibers are obtained by cutting the fibrillated tapes in staple cutters (l). This process, however, is used rarely because of the development of short spinning technology to produce PP spun fibers of appropriate deniers at a reasonable cost.

End use for fibrillated tapes (split yarns) include twine, rope, reinforcement and weft yarn in woven carpets, and coarse yarn for geotextiles.

***Coloration and Finishing.*** Tapes and split yarns are almost exclusively bulk-dyed with the aid of pigment concentrates, which are added prior to extrusion. The problem of coloring polyolefins by bath dyeing has not yet been solved satisfactorily.

Inorganic compounds such as chalk are added to reduce splitting and facilitate weaving. Ultraviolet stabilizers, required for many areas, are often provided by the manufacturer of the raw material, but the processor can also blend in concentrates. In general, UV stabilizers of the HALS type are used. They also impart good weathering resistance to PP ribbons [30]. However, HALS stabilizers may interact with certain pigments.

Antistatic agents and incorporated lubricants are generally not effective for ribbons and monofilaments. In addition, they can cause problems during film and ribbon production (e.g., plate out and water carry over). Subsequent external application (finishing) gives better results.

## 2.3. Uses

Tapes and split yarns of polyolefins are used in textile products that require high strength; low weight; chemical, moisture, and abrasion

**Figure 7.** Networklike fibrillated yarn of propylene

resistance; as well as other properties. Because of the ease of processing and the low volume cost compared to other fiber raw materials, their use is especially economical in many cases.

To a large extent, PP and HDPE tapes have replaced jute, hemp, sisal, and other natural fibers. Film yarns and split yarns (tenacities commonly 4–6 cN/dtex) are used particularly in the manufacture of bags, flexible bulk containers and other packaging fabrics and knits; industrial fabric (e.g., tarpaulin, geotextiles, filter fabrics); carpet backing; netting (protection against hail, birds, solar radiation, etc.); awnings; camouflage shields; twine; rope and cable.

***Packaging and Industrial Fabrics.*** Wide width fabrics are usually produced on flat looms [31]. Circular looms are useful for bagging, because in addition to simpler preparation for weaving (no beaming; weaving directly from the reel), elimination of the side seam gives better bag strength and saves labor as well as material [32].

By subsequent extrusion coating, mostly with LDPE or PP, water- and dustproof fabrics are obtained. In addition, the coating prevents ribbon slippage and loosening of the fabric.

***Carpet Backing.*** Because of the relatively high thermal stress during carpet manufacture (dispersion coating and drying), only PP ribbon can be used. The major application is as primary backing in the manufacture of tufted carpets; to a lesser extent, PP is used as secondary backing and as binding warp yarns in conventional woven carpets. The splitting behavior must be precisely controlled, especially in primary backings, and shrinkage must be low (<1 % at 132 °C, 20 s, European Association for Textile Polyolefins method) [18]. Too little splitting leads to deflection of the tufting needles and, hence, defects in the carpet as well as breakage of individual tapes. Too much splitting also affects fabric strength adversely. Tenter setting, previously used to minimize shrinkage of the entire fabric, is employed only occasionally.

***Knit Goods.*** Cloth for bagging or for shade, camouflage, and hail protection can be made on warp knitting machines as well as on looms [19, 33]. In most cases, netlike, coarse-stitch products, e.g., bags and pouches for fruit, vegetables, and other crops, are made.

The raw material is almost exclusively HDPE, frequently via the split-knitting process [19, 20]. Polypropylene tapes are difficult to fabricate on knitting machines because of poor sliding properties and tendency to split and break.

***Twine, String, and Rope.*** Twine and string for packaging are obtained from highly stretched, spontaneously splitting, or fibrillated tapes of PP, by twisting and plying. For high-strength ropes and cables, fibrillated ribbons, PP monofilaments, and long staple fibers are used [34].

***Floor Covering.*** Split fibers can, like other fibers, be processed into needle felts, which are used as floor or wall coverings. The use of split fibers in these areas has, however, decreased considerably and given way to spun fibers.

## 3. Monofilaments

### 3.1. Production

Monofilaments of PP or HDPE usually have a round or elliptical cross section, with a diameter of 0.1–3 mm (mostly 0.1–0.5 mm).

The steps in monofilament production are, for the most part, identical to those in tape manufacture. Monofilament machines consist of extruder, spinning unit (spinning head with spinning pump and spinneret), cooling bath, roll stretching units (usually two sets of seven rolls and one set of three rolls), hot-air ovens or a hot-water bath (for HDPE), and windups [10, 35]. Raw materials are generally propylene homopolymers with melt flow indices between ca. 2 and 6 g/10 min (MFI 230/2.16). If improved abrasion and rubbing resistance as well as better cold toughness are required, propylene copolymers are also used. Among the HDPE types, those with densities of 0.950–0.960 g/cm$^3$, MFI 190/2.16 values between 0.4 and 1.0 g/10 min, and narrow molecular mass distribution have been found to perform best.

***Extruder–Spinneret.*** Extruders with a screw diameter of 45–90 mm are used for the production of monofilaments. Spinneret holes are arranged in concentric circles on the plate. Depending on the size of the unit and the diameter of the monofilaments, ca. 50–100 monofilaments are extruded through one spinneret. In general, the spinneret hole diameter for polyolefins is three to four times the final diameter.

***Cooling Bath.*** Monofilaments are produced exclusively with water bath cooling; the water temperature is 20–60 °C, depending on working material and thickness of the monofilaments.

***Stretching.*** Stretching of PP monofilaments occurs in hot-air ovens between two stretching units with seven rolls each. The usual draw ratio is 1:8 to 1:12, and the temperature is 120–160 °C. Because of their lower softening range, HDPE monofilaments can be stretched in water baths at, or slightly below, 100 °C. Compared to stretching in hot-air ovens, this method has the advantage of better and more uniform heat transfer. Stretch ratios are generally between 1:8 and 1:11.

***Setting.*** As in tape production, setting occurs in hot air after stretching. It is also done under tension, which is controlled by suitable adjustment of drawoff roll speed. Setting temperature is in the range of stretching temperature; setting shrinkage is usually adjusted to 4–8 %.

***Windup.*** in contrast to film yarns (which are wound on packages with traverse laydown), monofilaments are usually wound in parallel on flanged disk spools. Windup speed is ca. 100–180 m/min. Monofilaments for bristles are wound on large-diameter reels or wheels to avoid too high a curvature.

***Coloration and Finish.*** Coloring and finishing are similar to the process steps for tapes. However, addition of mineral fillers to reduce splitting is not required for monofilaments. To avoid fibrillation of the outer layers, surface finishes are applied (e.g., during twisting) to improve lubricity.

## 3.2. Uses

Monofilaments of PP and HDPE are processed into fabrics, knit goods, rope, and cable on the same machines used for film ribbons and split yarns.

Areas of application include fishing nets (HDPE); nets for protection from hail, birds, etc.; packaging nets; sieves and filter fabrics; rope and cable; bristles; reinforcing grids for construction sheeting; and tarpaulin.

Bristles for sweepers, road brooms, and the like are sometimes mechanically split at the ends. Because sufficient weathering resistance can be obtained with little or no UV stabilizer, HDPE is often used for protective netting.

## 4. High-Modulus Polyolefin Yarns

Conventional processes for producing polyolefin tapes, monofilaments, and fibers give a structure that consists of more or less oriented crystals as well as amorphous regions. Considerably higher values of the modulus of elasticity and tensile strength are to be expected if chain folding can be avoided, and as many molecular chains as possible can be oriented in the direction in which the load is applied. This ideal case can be approximated most closely with HDPE. Based on its relatively high packing density in the crystal lattice, HDPE should give the highest theoretical stiffness and strength value [36]. On the other hand, HDPE can be relatively easily prepared with very high molecular mass, so the number of chain ends and defects in the crystal can be minimized. The theoretically attainable stiffness and strength values of polypropylene are lower because of its helical structure. However, PP is also of interest because of its higher heat resistance.

The necessary "ultrahigh" orientation of the molecular chains can be achieved essentially in two ways with high molecular mass polyolefins: (1) drawing in the solid state (at a temperature clearly below the melting range) [37–40], and (2) spinning of highly viscous solutions (gel spinning) and subsequent one- or two-stage drawing [41, 42].

Higher modulus and strength are attainable by drawing in solution than by drawing in the solid state because products with higher molecular mass can be used, and a higher degree of orientation can be achieved.

### 4.1. Production

Two technologies are available; both are still in the process of commercialization. Drawing in

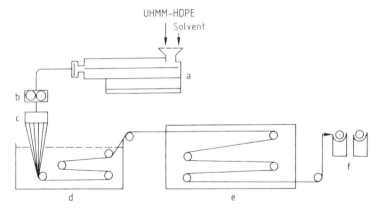

**Figure 8.** Gel-spinning process for the production of high-modulus UHMM HDPE * yarns
a) Extruder; b) Spinning pump; c) Spinneret; d) Precipitating bath – extraction; e) Heating – drawing; f) Windup
* Ultrahigh molecular mass high-density polyethylene.

the solid state is accomplished in draw jets at ca. 100 °C [43]. At draw ratios up to 1:30, moduli of ca. 50 GPa are attained for HDPE and ca. 20 GPa for PP. This process is being commercialized by the British Petroleum (BP) and Celanese companies.

In gel spinning (Fig. 8), a highly viscous solution in decalin, paraffin oil, or paraffin wax is first prepared in a kneading or stirring unit [44]. It is then extruded through a spinneret (c) into a precipitating bath (d). After entering the precipitating and cooling bath, the filaments are in a gel state in which the molecules remain largely disentangled. For orientation in the fiber direction and subsequent crystallization with final removal of solvent, a single- or multiple-step drawing process follows (e). Drawing occurs under heat and with a stretch factor ≥30. Preferably, HDPE with a molecular mass of ca. $2 \times 10^6$ is used (UHMMPE). The yarns can subsequently be cross-linked using radiation.

## 4.2. Properties and Uses

By gel spinning, HDPE yarns with moduli of elasticity of 50–100 GPa, and tenacities of ≤ 35 cN/dtex can be produced [45]. Because of their low density (0.97 g/cm$^3$), these yarns have an extremely high specific strength and are very abrasion, light, chemical, and moisture resistant. In addition, they exhibit high energy absorption and knot strength as well as good resistance to cold. Limitations in use include a relatively high tendency to creep and a relatively low melting temperature (ca. 150 °C). Radiation cross-linking increases heat resistance and reduces the tendency to creep.

End-use possibilities are extremely high-strength rope and cable, fabric for sails, industrial filters, bulletproof vests, and reinforcing layers for composites.

## References

1. Montecatini, IT 535712, 1954 (G. Natta, P. Pino, G. Mazzanti). Montecatini, IT 537425, 1954 (G. Natta, G. Mazzanti, P. Pino). Montecatini, BE 538782, 1954–1955 (G. Natta, P. Pino, G. Mazzanti, K. Ziegler).
2. M. Compostella in A.D. Ketley (ed.): *The Manufacture and Commercial Applications of Stereoregular Polymers*, 1st ed., vol. **1**, Marcel Dekker, New York 1967, pp. 322–337.
3. M. Ahmed: *Polypropylene Fibers–Science and Technology*, Elsevier Scientific Publ. Co., Amsterdam-Oxford-New York 1982, pp. 181–184.
4. A. Tozzi, G. Cantatore, F. Masina, *Text. Res. J.* **48** (1978, Aug.) 433–436.
5. *Chem. Fib. Int.* **2** (2010) 76.
6. I. G. Farbenindustrie, DE 667234, 1936 (H. Jacqué); DE 746593, 1939.
7. H. Krässig, *J. Polym. Sci. Macromol. Rev.* **12** (1977) 321–410.
8. F. Hensen, S. Braun, *Kunststoffe* **68** (1978) 221–229.
9. F. Hensen, *Kunststoffe* **71** (1981) 643–652.
10. F. Hensen, M. Noe in VDI-Gesellschaft Kunststofftechnik (ed.): *Folien Gewebe Vliesstoffe aus Polypropylen*, VDI-Verlag, Düsseldorf 1979, pp. 107–149, 151–174.
11. The Plastics & Rubber Inst. (ed.): *Prepr. Conference on Polypropylene Fibres and Textiles*, The Plastics & Rubber Inst., LondonSept. 30–Oct. 1, 1975;Sept 26–29, 1979; Oct. 4–6, 1983.
12. L.H. Gouw, W. H, Skoroszewski, *Plastica* **22** (1969) no. 10, 438–447.

13 W.H. Skoroszewski, *Plast. Polym.* **40** (1972) no. 147, 142–152.
14 Hercules, US 3470285, 1969 (C.W. Kim, S.D. Samluk).
15 A.R. Freedman, *Mod. Plast.* **47** (1970) no. 6, 108–110.
16 J. Harms, H. Krässig, F. Sasshofer, *Chemiefasern Text.-Anwendungstech./Text.-Ind.* **23** (1973) 845–849, 979–981.
17 D.E. Hanson, F.H. Reed, *Plast. Rubber Process.* **1** (1976) no. 1, 28–32.
18 P.E. Nott, *Kunststoffe* **67** (1977) 490–497.
19 "Das Verarbeiten von Polyolefin-Bändchen auf Raschelmaschinen," *Kettenwirkpraxis* **4** (1970) 9–18.
20 A. Balaz, *Melliand Textilber.* **6** (1971) 664–665.
21 G. Kautz, *Kunststoffe* **63** (1973) 682–686.
22 D. Moorwessel, G. Pilz, *Kunststoffe* **59** (1969) 205–210, 539–544.
23 Phillips Petroleum Co., US 3302501, 1967 (M.E. Greene).
24 Shell International Research M., NL 6511455, 1967.
25 Chevron Research Co., GB 1128274, 1967.
26 J.H.L. Choufoer: *Conference on Polypropylene Fibres in Textiles*, York Sept. 30–Oct. 1, 1975.
27 Barmag, DE-AS 1660230, 1971 (H. Schippers, R. Hessenbruch); DE-OS 2127792, 1972 (E. Lenk, R. Jung, G. Albrecht).
28 Shell International Research M., NL 6905427, 1969; FR 2046406, 1971.
29 G.J. Nichols, *Plast. Polym.* **40** (1972) no. 146, 84–88.
30 F. Guggumus in R. Gächter, H. Müller (eds.): *Plastics Additives Handbook*, 2nd ed., Hanser-Verlag, Munich-Vienna-New York 1984, pp. 152, 158.
31 K.H. Kessels, *Text. Prax. Int.* **31** (1976) no. 3, 250–260; no. 4, 355–358.
32 H. Kirchenberger, *Melliand Textilber.* **7** (1974) 599–603.
33 K.D. Darlington, *Knitting Times* 1975, (Aug.) 12–17.
34 M.R. Parsey: *Conference on Polypropylene Fibres and Textiles*, York Oct. 4–6, 1983.
35 F. Hensen, S. Braun, *Kunststoffe* **64** (1974) 228–233.
36 G.W. Ehrenstein, C. Maertin, B. Pornnimit, D. Scherz, *Plastverarbeiter* **37** (1986) no. 10, 206–219.
37 G. Capaccio, I.M. Ward, *Polymer* **15** (1974) 233–238.
38 A.G. Gibson, I.M. Ward, *J. Mater. Sci.* **15** (1980) 979–986.
39 P.D. Coates, I.M. Ward, *Polymer* **20** (1979) 1553–1560.
40 W.G. Perkins, R.S. Porter, *J. Mater. Sci.* **17** (1982) 1700–1712.
41 P. Smith, P.J. Lemstra, *J. Mater. Sci.* **15** (1980) 505–514.
42 Stamicarbon, BE 881587, 1979; US 4422993, 1983 (P. Smith, P.J. Lemstra).
43 A.G. Gibson, I.M. Ward, *Polymer Eng. Sci.* **20** (1980) no. 18, 1229–1235.
44 *Chemiefasern/Text.-Ind.* **35/87** (1985) no. 1, 33–34.
45 R. Kirschbaum, H. Yasuda, E.H.M. van Gorp, *International Man-Made Fibers Congress*, Dornbirn, Sept. 24–26, 1986.

# Polyacrylonitrile Fibers

ALFRED NOGAJ, Bayer AG, Dormagen, Germany

CARLHANS SÜLING, Bayer AG, Leverkusen, Germany

MICHAEL SCHWEIZER, ITCF Denkendorf, Denkendorf, Germany

| | | | | | |
|---|---|---|---|---|---|
| 1. | Physical Properties | 1514 | 3.4. | After treatment | 1518 |
| 2. | Chemical Properties | 1514 | 3.5. | Special Processes | 1520 |
| 3. | Production | 1514 | 4. | Analysis | 1521 |
| 3.1. | Polymerization | 1514 | 5. | Types of Fibers | 1522 |
| 3.2. | Dissolving | 1516 | 6. | Economic Aspects | 1526 |
| 3.3. | Spinning | 1517 | | References | 1527 |

"Polyacrylonitrile fibers" is a collective name for all fibers that contain at least 85% polymerized acrylonitrile [107-13-1]. These fibers are also referred to as "acrylic fibers." Most polyacrylonitrile [25014-41-9] (PAC) fibers consist of ternary copolymers with 89–95% acrylonitrile, 4–10% of a nonionogenic comonomer, and 0.5–1% of an iogenic comonomer containing a sulfo ($-SO_3H$) or sulfonate ($-OSO_3H$) group. Fibers of 100% acrylonitrile are also produced for industrial use [1]. Because polyacrylonitrile decomposes before melting, it cannot be spun from the melt but only from solution.

*History*. In the early 1930s, H. REIN at I.G. Farbenindustrie carried out experiments to produce fibers from polyacrylonitrile. Because of polyacrylonitrile's insolubility in organic solvents, its nonmelting properties, and its high softening point, advantageous fiber properties were expected. Patents for the use of dimethylformamide (DMF) as a solvent for acrylonitrile polymerization were filed in Germany in 1942 by I.G. Farben and in the United States in 1944 by Du Pont. DMF was the first industrially useful solvent for the production of polyacrylonitrile fibers. Soon thereafter, dimethylacetamide (DMA) was also found to be a solvent. Technology for dissolving the polymer in aqueous solutions of certain organic salts was also developed. The first acrylic fiber was commercialized by Du Pont in 1948. In the Federal Republic of Germany, production of acrylic fibers began in 1954 [2].

The composition of the copolymer determines the dyeing behavior of the fibers. The nonionogenic comonomer serves as an internal plasticizer and increases the rate of dye uptake. The sulfonate group of the ionogenic comonomer makes possible chemically bonding ionogenic dyes, whose color-imparting ion is a cation (basic dyes). The amount of ionogenic comonomer determines the saturation value, i.e., the maximum amount of dye that can be chemically bonded by the fibers. Acidic ionogenic comonomers are replaced by basic comonomers if use of acid dyes is desirable.

Basic-dyeable acrylic fibers are the major type of fiber consumed; those made of 100% acrylonitrile and those dyeable with acid dyes lag far behind.

The term "modacrylic" designates all fibers that contain 50–85% of bound polyacrylonitrile. Fibers with 40–60% copolymerized vinyl chloride and fibers with 30–45% copolymerized vinylidene chloride are important because of their flame-retardant properties.

Polyacrylonitrile fibers are wool-like in nature. An important step in their development was the improvement in dyeability by modification with suitable comonomers. Their wool-like nature and good dyeability have established a secure place for acrylic fibers in the textile market along with polyester and polyamide fibers.

## 1. Physical Properties

Articles of acrylic fibers have a wool-like hand. Lower fineness fibers produce a softer hand; higher fineness makes it harsher. The high bulk and low density (compared to natural fibers) make possible light, airy textiles with good recovery, warmth retention, and covering power. Because acrylic fibers do not swell, articles made from them remain permeable to air even when wet. The tensile strength of acrylic fibers and the abrasion resistance of articles made from them are greater than those of natural fibers.

Commercial fibers have different dye affinities and dye uptake rates. For light shades, dispersion and basic dyes are used; for dark shades, only basic dyes. Dyeing temperature is generally ca. 100°C. Dyeings are distinguished by brilliance and fastnesses. The degree of luster is determined by the type, brilliant or delustered.

Articles of acrylic fibers fall into the easy-care category. They absorb little moisture, can be washed at moderate temperature, and dry rapidly. They are resistant to many organic solvents, and stains can be removed easily. They do not felt. Weathering and lightfastness are excellent and unsurpassed by any natural or synthetic fiber. This makes the fiber especially suitable for outdoor uses such as awnings. The fiber is not attacked by bacteria, mildew, or insects, and does not rot.

## 2. Chemical Properties [3]

Good chemical resistance of the homopolymer to hydrolysis permits industrial applications up to 140°C in a humid atmosphere, e.g., for filter fabrics. The resistance of acrylic fibers to industrial gases, smoke, and soot is very good. Acrylic fibers are resistant to moderate concentrations of mineral acids and to oxidizing agents. They are soluble in concentrated sulfuric acid and concentrated nitric acid. Resistance to weak alkali is sufficient. Acrylic fibers are insoluble in most organic solvents, e.g., benzene, mineral oil, carbon tetrachloride, aliphatic hydrocarbons, aliphatic alcohols (e.g., ethanol), aliphatic esters, and aliphatic ketones (e.g., acetone).

In the burn test, the fibers melt, ignite, and burn with a yellow, sooty flame; a dark, hard, and brittle residue remains. In contrast, modacrylic fibers melt in the flame to dark, hard lumps; burn partially; but extinguish again if the flame is removed.

## 3. Production

Production of polyacrylonitrile fibers includes the following processing steps (Fig. 1) [4]:

1. Polymerization (Section 3.1)
2. Dissolving (Section 3.2)
3. Spinning (Section 3.3)
4. Aftertreatment (Section 3.4)

### 3.1. Polymerization

The fiber raw materials for acrylic or modacrylic fibers are produced by polymerization or copolymerization of acrylonitrile (→ Acrylonitrile). For further details concerning reactivity and material constants see [5].

Polyacrylonitrile can be made by anionic or radical polymerization. Although polymer solutions can be produced in high space–time yields by anionic polymerization, radical-initiated polymerization or copolymerization of acrylonitrile is generally the method of choice in the manufacture of raw materials for acrylic or modacrylic fibers. Either solution polymerization or precipitation polymerization is used.

Like all vinyl monomers, acrylonitrile can polymerize with the formation of isotactic as well as syndiotactic sequences. The processes used to produce acrylonitrile polymers give about equal amounts of isotactic and syndiotactic sequences [6].

For commercially useful acrylic fibers or yarns, homo- or copolymers of acrylonitrile are used as raw materials. The raw materials for acrylic fibers in *textile applications* are acrylonitrile terpolymers.

The same initiator systems and similar reaction conditions can be used for the industrial homopolymerization of acrylonitrile as are used to produce acrylonitrile copolymers, the raw materials for the usual acrylic fibers in textile applications.

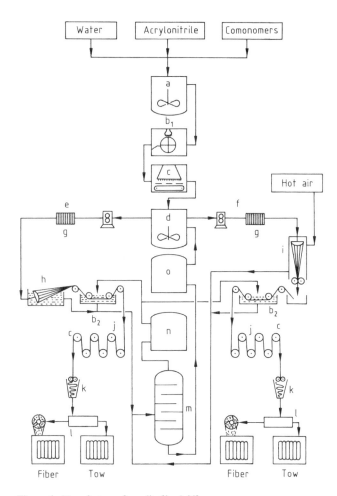

**Figure 1.** Manufacture of acrylic fiber* [4]
a) Polymerization; $b_1$) Washing and filtering of polymer; $b_2$) Washing of filaments; c) Drying; d) Dissolving; e) Wet-spinning process; f) Dry-spinning process; g) Filter; h) Spinning bath; i) Spinning shaft; j) Drawing, drying; k) Crimping; l) Steaming; m) Distillation; n) Water; o) Solvent
* Wet spinning—lower left; dry spinning—lower right.

Interest has developed in very high molecular mass acrylonitrile copolymers that contain small amounts of comonomers. These materials are used to produce high-strength acrylic yarns [7].

To obtain the desired properties in acrylic fibers for textile applications, acrylonitrile terpolymers are used as raw materials. These polymers contain 4–10% comonomers, are free of ionic groups, and are particularly suited for statistical copolymerization with acrylonitrile. Examples are the methyl esters of acrylic or methacrylic acid, vinyl acetate, or acrylamide. Depending on the need for raw materials for cationically or anionically dyeable acrylic fibers, an acidic or basic comonomer is used as an additional component. Monomers containing sulfonic acid groups have proved particularly useful as acidic or anionic comonomers, e.g., methallyl sulfonic acid, styrene sulfonic acid, and 2-methyl-2-sulfopropylacrylamide, or soluble salts of these sulfonic acids, respectively. Examples of basic comonomers are dialkyl aminoalkyl methacrylates or vinylpyridines.

Although the processing technology of the terpolymerization reaction is largely independent of monomer composition, special attention must be given to the structure of the dyeing comonomer when selecting the initiator or initiator system.

***Precipitation Polymerization.*** In the precipitation polymerization of acrylonitrile in water to produce acrylic fiber raw materials, use is made of the relatively good water solubility of the monomer. To introduce end groups into the fiber raw material, which can be utilized to adjust dyeability properties, initiator systems are used. The polymerization is initiated with water-soluble peroxodisulfate redox systems, such as potassium, ammonium, or sodium peroxodisulfate, combined with water-soluble salts of sulfur (IV) or derivatives of sulfur (II). Especially useful in the production of raw materials for cationically dyeable acrylic fibers is the initiator system peroxodisulfate–hydrogen sulfite. The decomposition reaction of this redox system has been thoroughly studied [8].

The relationship between dyeability and the number of different end groups in the copolymers is known [9]. The type and concentration of polymer end groups of the fiber raw material have a major influence not only on the coloring properties but also on the resistance to thermally induced yellowing and the degree of whiteness of the resulting acrylic fibers.

To obtain cationic dyeable acrylic fibers, precipitation polymerization is carried out at 40–70°C and at a pH of <6. Peroxodisulfate–hydrogensulfite is used as initiator of polymerization with a several-fold molar excess of hydrogensulfite relative to peroxodisulfate. To purify and demonomerize the polymer completely and gently, a compact polymer granulate, which can be dried after washing and filtering, is produced.

In the synthesis of fiber raw materials for acrylic fibers that are to be dyed with anionic dyes, redox systems, e.g., $H_2O_2$–thioglycerine [10], that do not result in anionic end groups are used.

A detailed description of redox polymerization of acrylonitrile and a comprehensive review of the use of redox systems can be found in handbooks [11]. The constancy of distribution of molecular mass, chemical composition, and degree of whiteness is particularly significant to the final properties of acrylic fibers.

***Solution Polymerization.*** The copolymerization of acrylonitrile in solution results in polymer solutions that can be spun without additional process steps after the elimination of residual monomers. The polymerization reaction conditions must be optimized so that monomer conversion is almost complete; low molecular mass fractions that could adversely affect the mechanical and thermal properties of the resulting acrylic fibers must be avoided as much as possible. Examples of suitable solvents are dimethylformamide, dimethyl sulfoxide, nitric acid, and concentrated aqueous solutions of ammonium thiocyanate or zinc chloride. In radical polymerization, these solvents—but especially the polymer solutions—are effective chain-transfer agents. For that reason, specially selected initiator systems and special reactors are used for the large-scale commercial solution polymerization of acrylonitrile [12].

The choice of precipitation polymerization or solution polymerization often depends on the prior history of the acrylic fiber producer. Both processes have advantages and disadvantages, and economics—size of installation, type of spinning process in use and raw material situation—is often the deciding factor.

## 3.2. Dissolving

The solvents used in dry and wet spinning are listed in Table 1.

***Commercial Process.*** A dough, which is as free of clumps as possible in order to avoid long dissolution times, is prepared from the powdery polymer and the solvent by growing a compact polymer granulate; intimate mixing of polymer and solvent stream, e.g., in screw mixers, turbines, or paddle mixers; and, providing low solvent temperature, which initially produces a suspension with good distribution of polymer in the solvent.

A two-step process is used for dissolution in aqueous salt solutions. The polymer is first suspended in a salt solution that does not dissolve the polymer; then the salt concentration is increased to the amount necessary for dissolution. This process takes place in agitating pans, either batchwise or continuously. The batchwise process requires several hours to produce dimethylformamide solutions. The dissolving

**Table 1.** Distribution of installed capacity by solvents, 1985

| Solvent | CAS registry number | Formula | Spinning process | Percent of total capacity |
|---|---|---|---|---|
| N,N-Dimethylformamide (DMF) | [68-12-2] | $HCON(CH_3)_2$ | dry / wet | 15 / 17 |
| Sodium thiocyanate* | [540-72-7] | NaSCN | wet | 24 |
| N,N-Dimethylacetamide (DMA) | [127-19-5] | $CH_3CON(CH_3)_2$ | wet | 24 |
| Nitric acid | [7697-37-2] | $HNO_3$ | wet | 12 |
| Zinc chloride* | [7646-85-7] | $ZnCl_2$ | wet | 3 |
| Dimethyl sulfoxide (DMSO) | [67-68-5] | $(CH_3)_2SO$ | wet | 2 |
| Ethylene carbonate | [96-49-1] | (cyclic carbonate) | wet | 3 |

* Aqueous solution.

temperature is between 60 and 100°C. In the continuous process, the holdup time in the agitating pan is shorter (e.g., 5–30 min), but the temperature of dissolution is correspondingly higher (e.g., up to 120°C) and the intensity of agitation is greater [13]. The solution can then be treated thermally in heat exchangers or exposed to shearing forces in screws or turbines.

The concentration of the spinning solutions is higher for dry spinning than for wet spinning. When dimethylformamide is the solvent, 25–30% solutions are normally used for dry spinning and 20–25% solutions for wet spinning. Concentration and viscosity of the spinning solution are monitored constantly.

To keep the spinning solutions free of gel particles or other impurities, which is a prerequisite for long spinneret life, they are filtered in filter presses. Heat, amines, oxygen, and ions of heavy metals (especially iron) damage initial color (degree of whiteness). Therefore, an inert atmosphere, acid, and reducing and complexing stabilizers are needed [14, 15]. Additives have also been developed to protect against saltlike clouding of the solutions (in dimethylformamide) because of the polymerization catalyst [16, 17].

Stable viscosity is desirable so that spinning remains a steady-state process; it is obtained by optimizing the temperature of dissolution (ca. 70–100°C) or by using suitable additives [18, 19].

Gelation occurs at low temperature via secondary valence forces and at high temperature via cross-linking reactions. The higher the viscosity of the solution and the molecular mass of the polymer, the greater is the tendency to gel; more uniform polymers have less tendency to gel [20].

Aqueous zinc chloride solutions and solutions in dimethyl sulfoxide and ethylene carbonate also tend to gel. Because partial hydrolysis of the nitrile groups occurs when nitric acid is used as the solvent, these spinning solutions should not be heated above 5°C.

### 3.3. Spinning

The polymer contained in the spinning solution is shaped into a yarn by either wet or dry spinning. In both processes, the spinning solution is pressed through small holes in the spinneret, thus giving the polymer its threadlike shape. The processes differ in the state of aggregation of the medium that the spinning solution enters on leaving the spinneret. In wet spinning, a precipitating bath of water and the solvent is involved; in dry spinning, hot air or a hot inert gas.

**Wet Spinning.** The 10–25% polymer solutions are deaerated, filtered, and pumped through the 10 000–80 000-hole spinnerets into the precipitating bath, where the solution thread line is coagulated to a highly swollen gel yarn by the exchange of solvent in the yarn for water. Coagulation conditions determine the strength of the sheath–core structure of the yarn, as well as its porosity and cross-sectional shape. Wet-spun yarns are aftertreated immediately without intermediate storage.

***Dry-Jet Process.*** In a variation of the wet-spinning process, the solution thread line leaving the spinneret is first passed through air before entering the precipitating bath. Fibers with higher strength and elongation, smoother surface, and greater uniformity can be made by this process [21].

***Dry Spinning.*** The only solvent used in dry spinning of polyacrylonitrile fibers is dimethylformamide having the following properties:

| | |
|---|---|
| *bp* (101.2 kPa) | 153°C |
| Heat of vaporization | 571.24 kJ/kg |
| Water | <0.3% |
| Amines | <0.01% dimethylamine |
| Iron | <0.05 ppm |

The pH of a 20% aqueous solution at 25°C should be 6.5–9.

The degassed and filtered spinning solution is heated to 100–150°C and spun into vertical spinning shafts through spinnerets (200–2000 spinneret holes; hole diameter, 0.1–0.5 mm). Several spinning shafts make up a spinning machine. The walls of the shafts are heated to 150–220°C; hot air or a hot inert gas (200–240°C), passes through the shaft in the same direction as the yarn. The oxygen and water content of the spinning gas should be as low as possible to avoid discoloration of the spun yarn and decomposition of dimethylformamide. Dimethylformamide evaporates in the spinning shaft and is exhausted together with the spinning gas at the bottom of the shaft. The performance of a spinning shaft is, in the final analysis, determined by the geometry of the engineering apparatus and by the amount of heat transferred to the spinning threads via the hot gas and radiation from the heated walls of the shaft. Of special significance for the quality and performance of the spun material is the air guidance in the immediate vicinity of the spinneret [22–24].

Solidified, dried spun yarns are drawn off at a rate of 200–500 m/min. Spun yarns from the shafts of one spinning machine are collected and deposited in a can. (In the manufacture of continuous filaments, spinnerets with a maximum of 200 holes are used. In this case, an oily finish is applied to the spun material below the spinning shaft, and it is wound up on bobbins.)

The dimethylformamide that has evaporated is condensed, distilled, and recycled. The spun material still contains 7–30% DMF, which is leached out during aftertreatment and recovered by distillation.

The spinning conditions (temperature of solution, temperature of shaft, temperature and amount of spinning gas, and drawoff speed) greatly affect both further processing of the spun yarns (e.g., drawability) and properties of the end product (e.g., strength and elongation).

***Melt Spinning.*** Mixtures of polyacrylonitrile and water (ca. 13–33% relative to the polymer), in contrast to anhydrous polyacrylonitrile, can form a molten, single-phase system without decomposition of the polymer. Fibers from a polyacrylonitrile–water mixture were prepared as early as 1948 and again around 1970–1971 [1, p. 260 ff]. The advantage of such a process could be the absence of solvent and lower costs. However, many of research and development have not yet resulted in a commercial process. Possible reasons for this are the high viscosities of the melts and the difficulty of preventing the melt from becoming two-phase at any point in the process.

## 3.4. After treatment

The aftertreatment steps are as follows:

1. Washing
2. Drawing
3. Finish application
4. Crimping
5. Drying
6. Cutting
7. Steaming
8. Packaging

Cutting is omitted in the production of tow. During aftertreatment, the spun material acquires the necessary physical properties, such as strength, modulus, elongation, shrinkage, crimp, and crimp stability. Large-scale aftertreatment units are capable of processing tow up to 770 ktex (770 g/m) at rates up to 120 m/min in dry spinning and up to 80 m/min in wet spinning.

***Washing.*** Washing removes residual solvent and salts by leaching with hot water (e.g., 80°C) in multilayer or drum washing machines. A closed construction is used to protect the environment. The washing machine consists of many individual zones. The wash water travels in a direction opposite to the direction of travel of the yarn tow. The wash water is distilled or otherwise processed to recover the solvent. Washing usually precedes drawing because of the longer holdup time in the washing baths resulting from the relatively low tow speeds prior to drawing, which makes the washing process correspondingly more intense.

***Drawing.*** Drawing imparts the required strength to the yarn and occurs in two or more steps in drawing tanks between drawing units. The preferred drawing medium is water just below the boil (e.g., 99°C). The draw ratio is between 1 : 2 and 1 : 10. Wet-spun yarns must be drawn more (because of the absence of preorientation) than dry-spun yarns to attain the same strength. The rule "the higher the draw, the higher are the strength and modulus and the lower is the break elongation" applies to both processes.

After being drawn, the yarns have a boil-off shrinkage, which, depending on the draw ratio, is between 15 and 45%. Shrinkage increases as draw temperature decreases and passes through a maximum with decreasing draw ratio. This behavior is applied to production of shrinkable fibers. Fibers already shrunk can be made shrinkable by a postdrawing treatment.

In addition to water drawing, steam drawing or drawing on feed wheels is also possible.

***Finishing.*** Finish application occurs in an immersion bath. Most commonly, active anionic compounds are applied to the fibers from an aqueous medium. Examples are ethoxylated fatty acids, salts of ethoxylated sulfonic acids, or phosphoric acid esters of long-chain aliphatic alcohols. Finish application prevents the buildup of static charges during further processing of the fibers and imparts the proper degree of adhesion and smoothness to the surface. The finish content of the finished fiber is 0.2–0.4%. It can be determined by extracting with ether or a mixture of methanol–benzene, evaporating the solvent, and weighing.

***Crimping.*** In general, crimping is carried out in a stuffer box after drying (stuffer box crimping). A steaming zone, in which saturated steam condenses on the tow, heats the tow and makes it more formable. Two crimping rolls, kept at constant temperature by circulating cooling water, pull tow bands of up to $3 \times 10^6$ dtex into the crimping chamber, stuff them, and push them continuously from the chamber at a predetermined pressure. In certain crimping units, steam can be introduced into the stuffer box to shrink the tow and increase the crimp stability (setting).

The type of crimp is important in processing fibers because of its effect on cohesion and on the hand of the yarns produced from the fibers. Crimp is determined by the conditions of crimping (temperature and stuffing box pressure), the percentage of comonomer in the polymer, and other factors that affect the thermal plasticity or friction, e.g., moisture content or type of finish.

***Drying.*** Screen drum dryers, screen belt dryers, or calenders are used for drying. Drying temperatures of 120–170°C are common.

During tensionless drying, i.e., drying loosely laid tow bands or cut fibers, the shrinkage built into the fiber in the drawing process is released and removed. This process can be enhanced by steaming. When dried under tension, the fibers must subsequently be relaxed, usually by steaming. Drying reduces the moisture content of the fibers to 1–2% and increases their density.

Although polyacrylonitrile does not crystallize like polyamide or polyester, an improvement or increase in regions of near-order by drying or annealing can, nevertheless, be demonstrated by X-ray analysis. Because of the consolidating effect of drying, uniform drying is especially important for uniform dyeability of the fibers.

***Steaming.*** Steaming generally occurs in screen belt steamers or in autoclaves. Steaming with saturated steam shrinks the fibers and increases the crimp stability. To ensure that the shrinking process is complete, a high and uniform steam atmosphere (>95% steam) is required.

Steaming increases the elongation of the fibers and slightly reduces their strength.

Steaming under pressure is especially effective in increasing the rate of dye uptake.

***Cutting and Packaging.*** In the production of endless tow for processing on stretch–break machines, the crimped and dried tow bands are packaged in cartons or bales. The mass of such tows ranges from 50 to 110 g/m. The bundle mass is 100 kg per carton and up to 800 kg per bale.

In the production of staple fibers, the crimped and dried tow bands are cut into staple fibers with slotted wheels or rotary cutters. The usual cut lengths are 40, 60, 80, 100, or 150 mm, depending on end use. Cut fibers are compressed and packaged into bales of ca. 200–300 kg in packing presses.

The aftertreatment of *filament yarns* of polyacrylonitrile involves processes similar to those for tow and fibers: drawing and setting. Equipment is tailored to the particular filament count of the product. In addition, the filaments are twisted to tighten the yarn and packaged on cones, which are shipped at a mass of 1.5–2.0 kg. Commercial counts of industrial filaments are 220 dtex/96 filaments, 440 dtex/192 filaments, or 890 dtex/384 filaments.

## 3.5. Special Processes

***Continuous Dry Spinning Process.*** Continuous dry spinning and aftertreatment processes have eliminated storage following the dry-spinning process [25].

Because of the relatively high spinning speed (compared with wet spinning), the tow speed after drawing must be 500–2000 m/min. A vibrating trough is used to transport the spun tow during washing. Drawing is carried out in steam; crimping, in an aerodynamic crimper with a gaseous medium (at 1.5–1.6 MPa and 50–210 °C). Finish application can occur before, during, or after crimping. The steaming process is carried out in a vibrating trough in a steamer; drying, in a belt dryer.

In this process, products of uniform quality are obtained; solvent emission and waste are reduced; and manufacturing costs are lower [25].

In one variant of this process, washing is also eliminated [26].

***Producer Coloring.*** The manufacture of colored fibers is of economic interest if large amounts of one color are to be produced. Fiber producers can make colored fibers in the following ways:

1. Spin dyeing,
2. Gel dyeing,
3. Dyeing of spun material

*Spin Dyeing [27].* Dyes, either cationic or pigment dyes, are added to the spinning solution prior to spinning.

The dyes are free of adulterants. They are soluble in the solvent used and should be insoluble in water (the subsequent washing medium). Dye solutions also contain polymer to adjust their viscosity to that of the spinning solution. In general, combinations of dye solutions are metered into the polymer solution to achieve a given color; they are homogenized by means of statically or mechanically operating mixers.

Ionic bonding of the color-imparting cation to the sulfite or sulfate groups of the polyacrylonitrile polymer is rapid and quantitative. Uniformly dyed tows with brilliant color and good textile fastness are obtained by both the dry-spinning and the wet-spinning processes.

Organic pigments are used when especially high lightfastness is required, e.g., in textiles used outdoors. The pigments are dispersed in a dilute solution of polymer. In so doing, stabilizers are added to prevent coagulation of the dispersed dyes. This is followed by metering into the spinning solution as with cationic dyes.

*Gel Dyeing.* Dyeing during aftertreatment generally occurs while yarns are in the gel state after wet spinning. Dyeing is possible because cationic dyes rapidly diffuse into the highly swollen yarn. Dyeing can occur in the precipitating bath, the drawing tank, or a separate bath. It must be carried out prior to drying, which collapses the structure of the swollen yarn.

Dyeing by both the extraction method and the padding method has been patented. Important variables are temperature, concentration of dye, and time in the dye bath; these determine the amount of dye required and the wash- and lightfastness of the dyed material [28]. The choice of dyes is also important in the quality of dyeing [29].

Gel dyeing has the following advantages [30]:

1. Good dye uniformity within the batch and from batch to batch
2. Same tow quality as with undyed fibers because all physical perturbations are avoided
3. Economical batch size (≥5 t) because hardly any material is wasted during dye change

*Dyeing of Spun Material.* Dry-spun yarns can also be dyed during aftertreatment, but the rate of dye uptake must be increased by suitable measures [31].

*Uses.* Fibers dyed in the course of fiber manufacture are used in clothing, home furnishings, and outdoor textiles. Yarns, pigment-colored during spinning, are recommended for outdoor textiles (awnings, sunshades, camping and terrace furniture, camping tents, and boat covers).

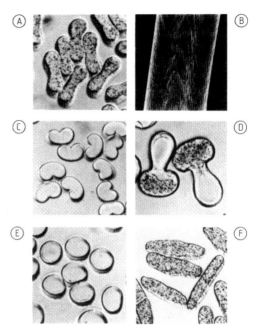

**Figure 2.** Acrylic fibers, cross sections and surfaces
A) Dry spun, delustered; B) Surface of A; C) Wet spun; D) Bifilar, dry spun, single component, delustered; E) Modacrylic, dry spun; F) Modacrylic, wet spun, delustered

## 4. Analysis

*Cross Section* (Fig. 2). The cross-sectional structure of polyacrylonitrile fibers and modacrylic fibers is determined by the composition of the polymer, the spinning process, and spinning conditions. The shape of the cross section determines the luster, moisture regain, and mechanical properties of the fiber [32].

*Stress–Strain [1, pp. 73, 208].* The stress–strain curve of acrylic fibers (Fig. 3) is characterized below their glass transition temperature (50–95°C with decreasing moisture content) by a steep initial region A, followed by zone B, where small changes in stress result in large changes in elongation. Above the glass transition temperature, A no longer exists; the fiber is more extensible, with slowly increasing resistance until a yield stress $\sigma_c$, at which structurally determined resistance to plastic sliding processes are overcome, is attained.

The stress–strain curve is affected by the type and amount of comonomer, the molecular mass and molecular mass distribution and the conditions of spinning and aftertreatment. Figure 4 shows the effect of the draw ratio.

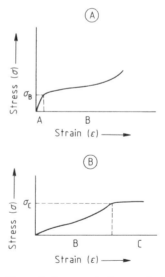

**Figure 3.** Stress–strain diagram of polyacrylonitrile (PAC) fibers [1]
A) Yield points below glass transition temperature $T < T_e$; B) Yield points above glass transition temperature $T > T_e$ $T_e$ = second order transition temperature; $\sigma_B$ and $\sigma_C$ = points at which small increases in stress result in large changes in elongation

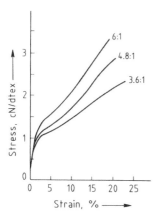

**Figure 4.** Effect of three different draw ratios on elongation and stress

***Thermomechanical Analysis [1, pp. 77, 209].*** In thermomechanical analysis (TMA), the change in length of a fiber under a constant preload is observed at a constant rate of heating. From the changes in length of fibers under low load, conclusions can be drawn concerning their degree of heat-setting. In Figure 5, brands 1 and 2 were more intensively heat-set during manufacture than brand 3, in which extension of the yarns above 110°C is prevented, principally by release of shrinkage tension.

***Analytical Procedure.*** The *cross section* of the fiber material provides a good indication that an acrylic fiber is present. Dumbbell-shaped cross sections indicate a dry-spun acrylic fiber; beanshaped cross sections, a wet spun fiber (cf. Fig. 2).

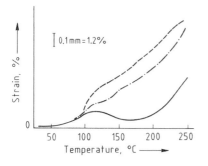

**Figure 5.** Thermomechanical deformation of acrylic fiber tow of three different brands* [1]
—— Brand 1; – · – Brand 2; ——— Brand 3
* Pretension 0.015 cN/dtex; rate of heating 10°C/min.

***Type reactions*** can be used to recognize and distinguish an acrylic fiber from other synthetic fibers. Each of these reactions, must be carried out on a freshly prepared sample; acrylic fibers should be [4]:

1. Soluble in concentrated nitric acid at room temperature
2. Insoluble in concentrated formic acid at room temperature
3. Insoluble in boiling cyclohexanone

Analysis of *polymer composition* is carried out by IR spectroscopy [33]. A first indication of the composition of the copolymer is obtained from analysis of the acrylonitrile content. To determine the brand, the exact composition of the polymer is required. The major neutral comonomers are vinyl acetate, methyl methacrylate, and methyl acrylate. Major ionogenic comonomers are styrene sulfonate, methallyl sulfonate, and itaconate, or in acid-dyeable types, cyclic or aliphatic amines. In addition, the following are present in flame resistant types (modacrylic fibers): vinyl bromide, vinyl chloride, and vinylidene chloride [34].

The number of ionogenic groups can be determined titrimetrically. Acid groups can be determined potentiometrically in DMF solution with 0.1 M NaOH [35]; basic groups, in nitromethane with perchloric acid in dioxane solution [36]. The quantitative determination of sulfur-containing polymer building blocks can also be accomplished by ignition in an oxygen flask and determination of the sulfate formed or by determining the sulfur with X-ray fluorescence analysis [37]. Indications of the "physical history" of an acrylic fiber are obtained by physical methods, e.g., thermomechanical analysis (previously discussed) or by determining the degree of relaxation with a solubility test [38].

The percentage of acrylic or modacrylic fibers in fiber blends can be determined by selective dissolution in DMF or DMA [39]. For comprehensive information on analysis, see [1 pp. 379–423].

## 5. Types of Fibers [40]

***Classification.*** The multiplicity of acrylic fibers can be classified by various systems.

Chemical composition determines certain properties that are used in classification; fibers can be

1. Basic dyeable
2. Dark dyeable
3. Acid dyeable
4. Self-crimping
5. Flame retardant

Characteristic descriptions, such as brilliant, delustered, bright, and spun-dyed, are also used.

Subdivision into spun fiber and tow accounts for various processing possibilities: (1) spun fiber for woolen, cotton, worsted, and open-end (rotor) spinning processes; and (2) tow for processing on stretch–break machines followed by spinning. Classification can be by fineness (denier) range (in decitex):

1. Finest: 0.4, 0.6, 0.9
2. Fine: 1.6, 2.2, 2.6
3. Medium: 3.3, 5.0, 6.7
4. Coarse: 11.0, 15.0, 17.0

Cut lengths are 28, 40, 50, 60, 100, 120, and 150 mm.

Physical modifications important to the quality of the goods are also used in classification. The following distinctions are made with respect to fiber shrinkage:

| | |
|---|---|
| N-fiber | 0–2% shrinkage |
| S-fiber | ca. 20% shrinkage |
| HS-fiber | ca. 40% shrinkage |

The type of crimp is another criterion that affects the quality of the goods. Qualitative data on crimp and crimp stability (low, medium, or high) serve as characterization parameters.

The range of finenesses for tow for stretch-breaking is 1.5–17 dtex. Tow can be bright, delustered, or spun-dyed.

The number of combinations of various characteristics for basic dyeable fibers alone runs into several hundred. Some inkling of the extraordinary diversity of possible applications of the most important synthetic staple fiber can be obtained by considering that the processor has available more than 20 different brands with varying properties, that polyacrylonitrile fiber types can be blended with each other or with other synthetic or natural fibers to achieve certain properties, and that the quality of the goods is substantially affected by the many possible yarn and fabric constructions.

***Industrial Fibers.*** Fibers for industrial use consist mostly of 100% or almost 100% acrylonitrile. Molecular mass is made as high as processibility will permit because the higher the molecular mass, the higher the strength. The fibers absorb very little moisture and, therefore, do not swell. They are resistant to hydrolysis, especially in acid medium, and are heat resistant up to 140°C.

Fibers are used in filter hoses and filter pockets in dry filtration, e.g., in the separation of fly ash in coal-fired power stations. Another end use is in dryer felts in paper machines, which absorb water that could not be squeezed out of the paper sheets. Fibers of 100% acrylonitrile are also used as asbestos replacement. For this end use, particularly high modulus and strength are required [41]. Nonmelting fibers that are nonflammable in air are obtained by heat-treating acrylic fibers in an oxidizing atmosphere at 300°C. These fibers can also replace asbestos [42]. Filaments of 100% polymerized acrylonitrile are good starting materials for high-strength carbon fibers [43].

***Dyeing Modifications.*** Dyeability is generally achieved in polyacrylonitrile fibers by inclusion of a neutral, plasticizing comonomer (5–10%) and an ionogenic comonomer (e.g., methallyl sulfonic, 0.5–1.0%) that has an affinity for dyes. By increasing the amount of comonomer with dye affinity to 3–4%, fibers are obtained with a four- to fivefold increase in rate of dye uptake. Such fibers can be dyed at a temperature as low as 85°C rather than at the boil. Dye migration is especially good in such fibers, and the time required to fix dyes is markedly reduced. Use of a rapidly absorbing fiber is based on the differential dyeing effect, which occurs when the fiber is dyed with a conventional fiber. Special-effect yarns can be produced from homogeneous or inhomogeneous blends of fibers or rovings, e.g., yarns of mixed colors, slub yarns, and flake twist yarns. Combinations of yarns result in twisted yarns with multicolor effects.

Another dye modification is represented by acid-dyeable fibers that contain 5–15% of a basic comonomer with affinity for acid dyes. Possible comonomers are the vinylpyridines, e.g., 2-methyl-5-vinylpyridine [29994-42-1], or even aliphatic amines, e.g., 2-(dimethylamino) ethyl methacrylate [2867-47-2]. Yarns of acid-dyeable acrylic fibers are processed and dyed together with yarns of basic-dyeable acrylic fibers. Dyeing is carried out in the same dye bath containing both acid and basic dyes. In this way, interesting color effects can be achieved such as two-color effects or tone-on-tone effects. If an undyed–colored combination is desired, only acid dyes are used, to which a conventional acrylic yarn is relatively insensitive.

***Modacrylic Fibers [44].*** Modacrylic fibers contain 50–80% bound acrylonitrile. The comonomer that makes up the difference is essentially a halogen-containing monomer. Vinyl chloride [75-01-4] and vinylidene chloride [75-35-4] are the most frequently used halogen-containing comonomers. The main advantage of modacrylic fibers is their reduced flammability. Whereas polyacrylonitrile fibers melt in the flame, ignite, and continue to burn after removal of the flame, modacrylic fibers are self-extinguishing once the flame is removed. In addition, they do not exhibit an undesirable dripping of molten fiber. The burnt residue is solid and black. In case of fire a disadvantage is the formation of hydrogen chloride. Fibers containing vinyl chloride exhibit even less flammability than vinylidene chloride fibers.

Bromine-containing monomers, e.g., vinyl bromide or vinylidene bromide, are also used. The development of polymerizable phosphorus compounds has become known as well.

In addition to acrylonitrile and halogen-containing comonomers, modacrylic fibers generally contain other comonomers that are intended to improve the dyeability of the fibers. These are mostly monomers with acid groups, e.g., sodium styrene sulfonate [2695-37-6] or sodium methallyl sulfonate [149-44-0]. Neutral comonomers, e.g., alkyl acrylamides, which improve dyeability by loosening fiber structure, can also be included. The manufacture of modacrylic fibers in essence follows the same process steps as the manufacture of polyacrylonitrile fibers.

Because of the type and high percentage of comonomer, modacrylic fibers exhibit higher plasticity, which is expressed as lower sticking temperature or higher shrinkage. Attempts were made to eliminate this disadvantage by reducing the halogen content of the polymer and including additives in the spinning solution that act synergistically to reduce flammability. Such additives are antimony trioxide, organic antimony compounds, or halogen compounds of antimony or tin. Halogen compounds can form the oxide and can be dispersed in the fiber so finely that they do not act as delusterants. Even metal compounds (e.g., barium titanate) or phosphorus compounds (e.g., derivatives of phosphorus acids) serve as flame retardants.

Modacrylic fibers tend to lose their luster under hot–wet treatment, even during dyeing. Salts are added to the dye bath or dyeing is done at lower temperature to prevent this. However, the best route is to reluster the dyed fibers with dry heat. Luster stability has been improved by changes in spinning and aftertreatment technology, by specific changes in polymer composition, and by the addition of luster stabilizers.

Reduced flammability favors the use of modacrylic fibers wherever there are fire hazards or where the danger in case of fire is particularly severe. This includes such uses as children's sleepwear; home furnishings, especially drapes; wall coverings in public buildings; and automobile upholstery. A large area of application is in pile fabrics, carpets, and imitation furs. Modacrylics are also used in wigs.

Modacrylic fibers are processed by themselves or in blends with other fibers, e.g., acrylic fibers, polyester fibers, or poly(vinyl chloride) fibers. By blending and proper construction of the finished article, special processing requirements, as well as the requirements of flammability tests, can be met.

***Absorbent Fibers [1].*** To produce fibers with greater wear comfort, special attention has been paid to developing moisture-absorbent fibers. Such fibers must be able to absorb moisture from the air and from water, either by virtue of their ability to swell or by means of a system of pores in the interior of each fiber. With acrylic fibers, the formation of a pore system is especially significant.

The system of pores must be protected by a sheath of proper thickness to ensure that the fiber can be processed without problem. The sheath must have a multiplicity of fine channels that can conduct water into the porous interior of the fiber. The properties of such a fiber include high adsorptivity, no swelling, high threshold of clammy feel, rapid transport of water, rapid evaporation, and low density [45].

The manufacture of absorptive fibers by both dry- and wet-spinning processes has been described. In dry spinning, an appropriate precipitant can be added to the spinning solution. This results in the formation of a sheath around the filament and an internal system of pores during spinning [46]. In wet spinning, highly swollen yarns with many pores are first formed during coagulation in the precipitating bath [47]. By skillful tuning of processing conditions, these pores can be retained in the final product. Water-soluble substances can also be allowed to diffuse into the swollen filaments, which are then dried. By subsequent leaching of water-soluble compounds, a system of pores is created.

Fibers with high swellability can be produced by the inclusion of comonomers with hydrophilic groups. Examples of such comonomers are acrylic acid, methacrylic acid, or 2-(dimethylamino)ethyl methacrylate. The saponification of comonomers on the fiber produces hydrophilic carboxyl groups. A popular route to the production of water-absorbent fibers is the spinning of mixtures of polyacrylonitrile and other polymers. Addition of cellulose acetate favors pore formation. Addition of hydrophilic polymers, e.g., saponified acrylonitrile, favors water absorption.

***Bicomponent Fibers.*** The side-by-side type of bicomponent fibers in which both components are placed eccentrically in the cross section and run along the entire length of the fiber, is commercially important (see Fig. 2D). Differential shrinkage of the components forming the fiber produces three-dimensional crimp when the shrinkage is released. The differential shrinkage of the two components is achieved by a difference in the amount of comonomers. If the concentration of hydrophilic, usually ionic, groups differs in the two components, reversible crimp results. This refers to a crimp that can be removed by washing and regenerated by drying. If the concentration of hydrophilic groups of the two components is essentially equal, the crimp is irreversible, i.e., permanent. The shrinkage of the fiber, and hence the crimp, can be released during fiber manufacture by drying or steaming or during subsequent processing, e.g., dyeing of the fiber or yarn.

Bicomponent fibers are produced both as staple fibers and as tow. Knit fabrics of these yarns are distinguished by good volume, pleasant, wool-like hand, and good stitch definition. Bicomponent fibers are also used in carpets because of covering power [48, 49]. The market for these fibers is primarily in Japan and the United States.

***Shrinkable Fibers.*** Shrinkable fibers can be produced from tow by processing on stretch-break machines. In general, the fiber shrinkage produced on these machines is 15–24%. If suitable measures are taken during tow production and stretch-breaking, fiber shrinkages of 24–40% are attainable.

Shrinkable staple fibers are generally offered at two levels of shrinkage: (1) shrinkable fibers with ca. 20% shrinkage and (2) high-shrink fibers with 35–40% shrinkage. The desired level of shrinkage can be obtained most easily during the manufacture of shrinkable fibers by suitable choice of draw ratio and draw temperature with subsequent mild drying. Care must be taken, however, to produce a shrinkable fiber free of voids so that no electrostatic problems occur in further processing and no changes in coloration or luster occur in the processed fiber. Good technical properties are also desirable. Criteria for a void-free structure are negative birefringence and density $>1.180$ g/cm$^3$ [50].

Shrinkable fibers are processed in blends with nonshrinkable fibers to give high-bulk yarns, which are distinguished by their high volume. High-shrinkage fibers are also used in imitation furs as bottom hair, in the production of multistep plushes, and in felts. Manufacturing methods are described in detail in many patents [1]. Important measures are consolidation of the yarn by removal of solvent, drying, and heat treatment in water or steam, partly prior to drawing and partly between two drawing steps. The rate of drawing is also important.

## 6. Economic Aspects

At its peak in 1983, polyacrylonitrile fibers accounted for 7.1% of total world fiber production, but since then share has been falling such that by 2009 the figure was 2.9% [51]. The downward trend is expected to continue in the future, with Tecnon OrbiChem forecasting that by 2020 acrylic fiber production will account for around 1.7% of the global total. Although share has been falling annually, total worldwide production of polyacrylonitrile staple remained in the range of 2.0 to $2.7 \times 10^6$ t for almost 30 years before falling by more than 20% in 2007 to around $1.9 \times 10^6$ t. However, despite continuing competition from polyester and other fibers as well as rising raw material costs, Tecnon OrbiChem believes that global demand for polyacrylonitrile fiber will stabilize in the future with recent increases in production in parts of Asia and the Middle East/Africa balancing out the continuing decline in Europe, the Americas, and Northeast Asia. Thus by 2020 global production of polyacrylonitrile fibers is expected to be similar to the present level though the regional balance will have changed in favor of Asia.

Output of polyacrylonitrile fibers in China has grown by around 4%/a during the past 10 years, such that since 2003 it has been the largest polyacrylonitrile fiber producing region in the world. In 2009, Chinese output accounted for 32% of the world total, a figure that is expected to rise through to 2020. Fastest growth is expected in South and Southeast Asia and the Middle East, where output is forecast to rise by around 2.5%/a.

Demand in China is expected to rise faster than production, thereby ensuring future opportunities for polyacrylonitrile fiber exports. Western Europe is a region that relies heavily on exports, particularly to China, to maintain utilization levels. China continues to be the biggest importer of polyacrylonitrile staple, accounting for almost a quarter of the global total. Compared with the previous year, the volume in 2008 was down by almost 50% following a similar decrease (40%) the year before. Some recovery was evident in 2009. Turkey continues to be the second largest market for exporters, with Indonesia, the USA, and Iran all importing significant volumes. Together, these five countries account for 50% of world trade in 2008.

Trade names and producers of polyacrylonitrile fibers are listed in Table 2. World production of synthetic polyester fibers (yarn and staple) are listed in Table 3.

Table 2. Trade names and Producers of polyacrylic fibers

| Trade name | Types | Manufacturer |
|---|---|---|
| Acelan | PAN | Tae Kwang Industrial Co./Korea |
| Aksacryl | PAN | AKSA AS/Turkey |
| Beslon | PAN | Toho Tenax Co./Japan |
| Bulana | PAN | Bulana/Bulgaria |
| Crumeron | PAN | Zoltek Magyar Viscosa RT/Hungary |
| Dolan | PAN | Dolan GmbH/Germany |
| Dolanit | PAN | Dolan GmbH/Germany |
| Dralon | PAN | Dralon GmbH/Germany |
| Drytex | PAN | Sudamericana de Fibras SA/Peru |
| Exlan | PAN | Japan Exlan Co./Japan |
| Fisivon | PAN | FISIPE SA/Portugal |
| Hanilon | PAN | Hanil Synthetic Fiber Industrial Co./Korea |
| Jayacrylic | PAN | J.K. Synthetics Ltd./India |
| Kanecaron | MAC | Kaneka Corp./Japan |
| Kanekalon | MAC | Kaneka Corp./Japan |
| Leacril | PAN | Montefibre SpA/Italy |
| Leacril | PAN | Montefibre Hispania SA/Spain |
| Mavilon | PAN | Zoltek Magyar Viscosa RT/Hungary |
| Myoliss | PAN | Montefibre SpA/Italy |
| Nitron-M | MAC | Polymir/Belarus |
| Protex | MAC | Kanebo Corp./Japan |
| Recrylic | PAN | Reliance Industries Ltd./India |
| Recrylon | PAN | Reliance Industries Ltd./India |
| Ricem | PAN | Montefibre SpA/Italy |
| Silpalon | PAN | Mitsubishi Rayon Co./Japan |
| Tairylan | PAN | Formosa Plastics Corp./Taiwan |
| Texlan | PAN | Thai Acrylic Fiber Co./Thailand |
| Toraylon | PAN | Toray Industries, Inc./Japan |
| Vonnel | PAN | Mitsubishi Rayon Co./Japan |
| Vonnel | PAN | FISIPE Sarl/Portugal |
| Yalova | PAN/MAC | Yalova Eliat AS/Turkey |

PAN = acrylic fibers
MAC = modacrylic fibers

Table 3. World Production of synthetic polyacrylonitrile fibers

| Year | Production, t |
|---|---|
| 1980 | 2 060 000 |
| 1990 | 2 270 000 |
| 1995 | 2 420 000 |
| 2000 | 2 669 000 |
| 2005 | 2 632 000 |
| 2009 | 2 005 000* |

* Staple fibers.

# References

1. B. von Falkai (ed.): *Synthesefasern*, Verlag Chemie, Weinheim 1981p. 198.
2. H. Klare: *Geschichte der Chemiefaserforschung*, Akademie-Verlag, Berlin 1985, pp. 180–186.
3. P.A. Koch, "Faserstoff-Tabellen," *Chemiefasern/Text.-Ind.* **27/29** (1977, Jun.) 513–524.
4. *Textil-Praxis Int.* **5** (1985) 549–552.
5. J.F. Kroschwitz (ed.): *Encyclopedia of Polymer Science and Engineering*, 2nd ed., J. Wiley & Sons, New York-Chicester-Brisbane-Toronto-Singapore 1985, p. 426.
6. G. Svegliado, G. Talamini, G. Vidotto, *J. Polym. Sci., Part A* **5** (1967) 2875–2881.
7. Stamicarbon, EPA 0144983, 1985 (R.M.A.M. Schellekens, P.J. Lemstra, G.M.L.M. Leherte et al.).
8. P. Fritsche, J. Ulbricht, *Faserforsch. Textiltech.* **14** (1963) 320–325; **14** (1963) 517–521; **15** (1964) 93–100.
9. L.H. Peebles, Jr., R.B. Thompons, Jr., J.R. Kirby, M.E. Gibson, *J. Appl. Polym. Sci.* **17** (1973) 113. J. Tsuda, *J. Appl. Polym. Sci.* **13** (1961) 104.
10. Du Pont, DE-AS 1174070, 1964 (G.N. Milford, W.K. Wilkinson).
11. *Houben-Weyl–Methoden der organischen Chemie*, 4th ed. vol. 14/1, Georg Thieme Verlag, Stuttgart 1961, 297 ff andp. 973 ff.
12. M. Sittig: *Chem. Process Review No. 62: Acrylic and Vinyl Fibers*, Noyes Data Corp., Park Ridge, N. J. 1972.
13. Hoechst AG, DE 3048059, 1982 (E. Höroldt, H. Vollmüller, H. D. Berndhard, H. Strobel).
14. J. Runge, *Acta Polym.* **33** (1982) no. 12, 708–713.
15. J. Runge, *Acta Polym.* **34** (1983) no. 10, 631–636.
16. VEB Friedrich Engels, DD 215342, 1984 (J. Aurich, H. Ebeling, E. Seidel, P. Leppin).
17. Bayer AG, DE 2832212, 1980 (K.H. David, A. Nogaj, H. Rinkler).
18. Bayer AG, DE 3333145, 1985 (H. Engelhard, A. Nogaj, M. Reichardt, P. Kruchem).
19. Bayer AG, DE 3333146, 1985 (J. König, S. Korte, H. Engelhard).
20. K. Jost, *Rheol. Acta* **1** (1958) 303–315.
21. G.C. East, J.E. Mc Intyre, G.C. Ratel, *J. Text. Inst.* **75** (1984) no. 3, 196–200.
22. Bayer AG, DE 3424343, 1986 (Ch. Pieper, H.K. Burghartz, R. Hirsch, N. Rink).
23. Bayer AG, DE 3141490, 1983 (H.J. Behrens, H. Herold, E. Muschelknautz, R. Vogelsgesang).
24. Du Pont, DE 3339501, 1984 (Berry, Jr., W. Cunningham).
25. Bayer AG, DE 3308657, 1984 (M. Bueb, D. Paulini, E. Muschelknautz, W. Wagner et al.).
26. Bayer AG, DE 3225266, 1984 (M. Reinehr, H. Uhlemann).
27. W. Teige, *Chemiefasern/Text.-Ind.* **33/85** (1983, Sept.) 636–642.
28. R. Detscheva, R. Fltscheva, St. Stojanov, K. Dimov, *Textilveredlung* **9** (1974) 312–316.
29. Hoechst AG, DE 3141082, 1983 (M. Hähnke, W. Teige).
30. D.E. Titheridge, *Text.-Betr.* **100** (1982, Jul./Aug.) 45,46, 58.
31. Bayer AG, DE 2317132, 1974 (M. Reinehr, A. Nogaj, G. Hölzing).
32. J.G. Cook: Man-made Fibers,*Handbook of Textile Fibers*, vol. 2, Merrow Publ. Co. Watford, England, 1968, p. 442.
33. T. Müller, *Chemiefasern/Text.-Ind.* **35/87** (1985, Jun.) 390–392.
34. H. Engelhard, *Chemiefasern/Text.-Ind.* **34/86** (1984, Jun.) 400–401.
35. O. Glenz, W. Beckmann, *Melliand Textilber.* **38** (1957) 296–300, 783–787.
36. C.A. Streuli, *Fresenius Z. Anal. Chem.* **153** (1956) 71.
37. H. Nestler, J. Mai, V. Stepputat, *Acta Polym.* **31** (1980) no. 4, 267–271.
38. J. Glacén, J. Maillo, J.T. Baixauli, *Melliand Textilber.* **66** (1985) 127–132.
39. R.V. Flor, M.J. Prager, *Text. Res. J.* **53** (1983, Jan.) 15–18.
40. "Spezialitäten-Lieferprogramm der Chemiefaser-Hersteller," *Chemiefasern/Text.-Ind.* **33/85**,(1983, Nov.) 782–786.
41. H. Haehne, *Chemiefasern/Text.-Ind.* **33/85** (1983, Dec.) 839,842–844, 846.
42. J. Vogelsgesang, H.D. Gölden, *Chemiefasern/Text.-Ind.* **32/84** (1982, Jun.) 422–432.
43. F. Fourné, *Chemiefasern/Text.-Ind.* **32/84** (1982, Jun.) 433–436.
44. J. Atkinson in P.J. Alvey (ed.): *The Production of Man-made Fibers*,Textile Progress, vol. **8**, no. 1, Textile Institute, Manchester 1976, pp. 61–69.
45. W. Körner, G. Blankenstein, P. Dorsch, U. Reinehr, *Chemiefasern/Textil.-Ind.* **29/81** (1979, Jun.) 452–462.
46. Bayer AG, DE 2554124, 1977 (U. Reinehr, T. Herbertz, H. Jungsverdorben, P. Kleinschmidt et al.).
47. Mitsubishi, GB 1377200, 1974.
48. W.E. Fitzgerald, J.P. Knudsen, *Text. Res. J.* **37** (1967) no. 6, 447–453.
49. O. Heuberger, A.J. Ultee, *Lenzinger Ber.* **38** (1975) 154–163.
50. Du Pont, DE 1435611, 1970 (M.L. Davis, L. Merton, S.C. Camden).
51. N. Bywater, *Chem Fib. Int.* **2** (2010) 60–62.

# Polyvinyl Fibers

JACQUES MENAULT, St-Foy-les-Lyons, France

TETSURO OSUGI, Nippon Chemtec Consulting Inc., Osaka, Japan

OSAMU MORIMOTO, Nippon Chemtec Consulting Inc., Osaka, Japan

| | | | | | |
|---|---|---|---|---|---|
| 1. | Poly(Vinyl Chloride) and Poly(Vinylidene Chloride) Fibers | 1529 | 1.4. | Fibers from Poly(Vinylidene Chloride) and Copolymers | 1532 |
| 1.1. | Fibers from Pure Poly(Vinyl Chloride) Homopolymer | 1529 | 2. | Poly(Vinyl Alcohol) Fibers | 1533 |
| 1.1.1. | Fibers from Pure Atactic Homopolymer. | 1529 | 2.1. | Physical and Chemical Properties | 1533 |
| 1.1.2. | Fibers from Blends of Poly(Vinyl Chloride) Homopolymer | 1531 | 2.2. | Poly(Vinyl Alcohol) for Fiber | 1534 |
| 1.1.3. | Products, Properties, and Uses | 1531 | 2.2.1. | Wet Spinning | 1534 |
| 1.2. | Fibers from Postchlorinated Poly(vinyl chloride) (CPVC) | 1532 | 2.2.2. | Dry Spinning | 1535 |
| 1.3. | Fibers from Vinyl Chloride Copolymers | 1532 | 2.2.3. | Mixed Spinning | 1536 |
| | | | 2.3. | Uses | 1536 |
| | | | 2.4. | Trade Names and Producers | 1537 |
| | | | | References | 1537 |

## 1. Poly(Vinyl Chloride) and Poly(Vinylidene Chloride) Fibers

According to standard ISO terminology, the generic name for poly(vinyl chloride) [9002-86-2] (PVC) and poly(vinylidene chloride) [9002-85-1] (PVDC) fiber is *chlorofiber*; this designates products containing at least 50% vinyl chloride or vinylidene chloride units. In the United States, the Textile Fiber Products Identification Act defines any fiber containing at least 85 wt% of vinyl chloride units as *vinyon* and any fiber composed of at least 80 wt% of vinylidene chloride as *saran*.

In terms of comfort, flame retardancy, chemical resistance, and shrinkability, these fibers may be considered specialty items, essentially having the properties of the homopolymer, e.g., chemical composition, low crystallinity, high dipole moment, and low moisture regain. The obstacles to commercial development of these fibers result mainly from the low softening point, which leads to poor dimensional stability at elevated temperature and poor downstream processability compared to other synthetic fibers.

## 1.1. Fibers from Pure Poly(Vinyl Chloride) Homopolymer

### 1.1.1. Fibers from Pure Atactic Homopolymer

The poly(vinyl chloride) generally employed for fiber production is simply a variant of the standard polymer used in plastic compounds. Bulk polymerization at 60–70°C is generally preferred because contamination from polymerization additives is low; this results in an atactic polymer with a molecular mass of 70 000–85 000. The melting point of pure homopolymer is close to the decomposition point. Dry or wet spinning from solution is therefore the only industrial route available to produce commercial fibers.

*Dry Spinning.* To avoid thermal decomposition with dehydrochlorination, the spinning temperature must be kept relatively low; therefore, the boiling point of the solvent also must be low.

Only a few common solvents are suitable for the homopolymer. Blends of equal volumes of carbon disulfide and acetone (*bp* 38°C) and benzene and acetone (*bp* 64.5°C) were patented and are used in production [1]. The spinning composition consists of polymer, solvent, and additives. Polymer content varies with the molecular mass of the polymer and ranges from 25 to 30 wt%. The additives have definite functions, e.g., improving heat stability, lightfastness, whiteness, and luster [2]. Certain producers of homopolymer chlorofiber overcome the difficulties in conventional dyeing caused by the low shrinkage temperature of the fiber by adding dye to the spinning dope.

Early swelling of the complex polymer grain and high dope viscosity make dope preparation difficult. The three possible ways to overcome this problem are (1) flash mixing of polymer and solvent with a short residence time in the shearing system [3], (2) premixing the polymer with a nonsolvent [4], and (3) kneading under high shearing force with long residence time.

When the solution is processed at ambient temperature, the microcrystalline regions of the polymer are not dissolved and the solution is colloidal. Dissolution at a temperature >100°C and processing the solution above the boiling point of the solvent result in a more homogeneous solution and a large decrease in viscosity, with concomitant advantages in further processing [5]. After the dope is dissolved, it is filtered in one or two steps.

The solution is heated or cooled to the spinning temperature and extruded through a 1000- to 2000-hole spinneret in a long tube with hot air moving concurrently with the filaments. The emerging filaments are cooled, wetted by an aqueous lubricant, and wound up or collected at 300–700 m/min. The spinning parameters are important because they determine characteristics of the as-spun filament, such as cross-sectional shape and specific surface.

***Staple Route.*** After spinning, the tow is preheated at a temperature (80°C) lower than the glass transition temperature and drawn 400–800% at a temperature (85°C) higher than the glass transition temperature. The drawing medium may be hot water or low-pressure steam.

After drawing, the fiber is thermally unstable and has a strong tendency to return to its original length; the shrinkage and shrinkage stress may be as great as 70% and 0.6 cN/dtex, respectively. For this reason, the tow is annealed under tension in water or steam [6]. The residual shrinkage is removed by relaxing the fiber tension-free at the boil or slightly above.

The next steps in the production of standard fiber are oiling, drying, crimping, and cutting, by conventional technology. Some commercial interest has been shown in high-shrinkage (30–65%) chlorofibers with their attendant high shrinkage stress. Shrinkable fibers are obtained by varying the draw ratio and avoiding the annealing and relaxation steps.

***Wet Spinning.*** At the time of writing (1987) one industrial plant (Kustanai, former States of USSR) used the wet process to produce commercial chlorofiber. Suitable common solvents are $N,N$-dimethylacetamide (DMA) and $N,N$-dimethylformamide (DMF). Special additives are required to limit discoloration of the solution. With these solvents, the rate of coagulation is high and voids appear in the coagulated gel, permanently impairing fiber performance. To exert positive control over the coagulation rate, water is added to the solvent before the solution is prepared; the solvent is then heated to improve solubility [7]. High solvent content (up to 85%) and low temperature (5–15°C) are suitable for the coagulating bath.

After the gel has coagulated, it is drawn at a low draw ratio and then washed on a multiple-stage countercurrent machine to recover the solvent as a concentrated aqueous solution. A second draw of 400–600% is carried out at boiling or intermediate temperature; the tow is then dried under tension. Annealing can be carried out during or after drying. The tow is then relaxed tension-free, oiled, cut into staple, and baled.

In wet processing, obtaining a homogeneous structure and avoiding discoloration and residual solvent are inherently difficult. The properties of the fibers are generally inferior to those produced by the dry process.

### 1.1.2. Fibers from Blends of Poly(Vinyl Chloride) Homopolymer

To avoid the inherent limitations of homopolymer fibers without loss of specific and outstanding properties, spinning blends of homopolymer and minor amounts of a polymer with fibrillar structure to act as reinforcement has been proposed. Post-chlorinated PVC (CPVC) or cellulose acetate [8] can be used without modification of the process. Other blends require special processes. Dry-spun fibers, which contain ca. 80–85% PVC homopolymer and 15–20% CPVC, exhibit the structure of a matrix reinforced with fibrils and have greatly improved heat dimensional stability, mechanical properties, and creep [9]. Such fibers are produced on a commercial scale by Rhône-Poulenc. If cellulose acetate or cellulose is used to replace CPVC, the level of heat stability increases [8, 10]. Researchers in the former Soviet Union proposed wet-spinning a blend of homopolymer with 4-phenylbenzimidazoterephthalamide in dimethylacetamide [11]. In all these cases, the PVC homopolymer is the matrix.

A Japanese firm (Kohjin Co.) produces a fiber (Cordelan) in which PVC and PVC–PVA [poly(vinyl alcohol)] graft-polymer particles are dispersed in a PVA matrix. This fiber gains heat dimensional stability, with concomitant reduction in chemical resistance, thermal insulation, and flame retardancy.

### 1.1.3. Products, Properties, and Uses

***Products.*** Three major grades of products are available: (1) shrinkable products with different levels of shrinkage, (2) stabilized products for tow dyeing for apparel, and (3) highly stabilized products for piece dyeing and home furnishings (matrix-fibrils). Table 19 lists the physical characteristics of these grades. The stabilized grades are available in a wide range of fineness from 2.4 to 15 dtex and of cut lengths required for cotton, woolen, or worsted spinning systems.

Chlorofiber can be spun on cotton, woolen, or worsted systems as 100% highly stabilized fiber or as blends of shrinkable and stabilized fibers. The fibers may be dyed at the boiling temperature or slightly below by use of selected disperse dyes. Only highly stabilized fibers can be yarn- or piece-dyed.

***Properties and Related End Uses.*** The main end uses for chlorofibers are related to comfort, flame retardancy, chemical resistance, and surface processing.

*Comfort.* Chlorofibers exhibit high thermal and tactile comfort [12, 13]:

High thermal resistance (0.23–0.25 $m^2$ K $W^{-1}$ $cm^{-1}$)

Constant thermal resistance with humidity transfer

Water transfer

High rate of evaporation of perspiration

Feeling of warmth

No lag time between sweat production and evaporative heat loss

*Flame Retardancy.* Chlorofibers have a very high level of flame retardancy as shown by the limiting oxygen index (LOI) of 46% at 50°C and 49% at 150°C. They do not ignite and spread flame, nor do they melt and form droplets in contrast to many other synthetic fibers. They may be blended with other flammable, staple fibers, such as wool and polyester, and retain their flame retardancy over a specific range of blend ratios. They can also be blended with thermally resistant fibers, such as aramid and preoxidized acrylic fiber, and exhibit outstanding heat resistance in terms of ignition temperature (640–680°C) and ignition flux (6.4 $W/cm^2$).

*Chemical Resistance.* Homopolymer fibers withstand weathering and are resistant to most chemicals with the exception of polar organic compounds (chlorinated hydrocarbons, aromatic hydrocarbons, ketones, and a few others).

*Surface Processing.* Woven or nonwoven fabrics of chlorofibers may be processed by thermal molding. Under the combined action of temperature and pressure, any shape can be achieved with a woven or knitted fabric composed of chlorofibers. These fibers may be heated and welded by high-frequency heating because of their low dielectric loss factor (0.013 at 50 Hz).

*Trade Names.* Retractyl-Fivravyl, Thermovyl L9, and ZC, Rhovyl (Tronville-en-Barrois, France); Teviron, Teijin (Iwakuni, Japan); (Kustanai, former USSR); (Canton, Chunking, Foochow, Shanghai, Sengli, and Tientsin, all P. R. China).

## 1.2. Fibers from Postchlorinated Poly(vinyl chloride) (CPVC)

Poly(vinyl chloride) homopolymer (56.7% Cl) can be postchlorinated to a maximum Cl content of 73.2% by free-radical chlorination. Solubility in common solvents, especially in acetone or blends containing acetone, depends on the chlorine content of different molecular chains and the distribution of blocks of poly(vinyl chloride) and 1,2-dichloroethane in each chain.

The *solution process* is believed to be the most suitable method of obtaining concentrated spinning solutions, with maximum solubility at a Cl content of 63%. Soviet (Chlorin) and German Democratic Republic (Piviacid) fiber producers claimed a solution process using either tetrachloroethane or dichlorobenzene as solvent. The processing steps include solution preparation, chlorination, precipitation of polymer with nonsolvent, polymer washing, and solvent recovery [7].

The *quality specifications* of spinnable polymer are:

| | |
|---|---|
| Chlorine content | 63–65 wt% |
| Specific viscosity at 0.2% in cyclohexane | 0.2 |
| Degree of polymerization | 900–1000 |
| Ash content | 0.03–0.08 wt% |
| Iron content | $1 \times 10^{-3}$ wt% |

Filament or staple fiber based on CPVC is spun by the dry or wet process in acetone. In wet spinning, the coagulation bath consists of water and a small amount of acetone; the spun gel is drawn 300–700% in one or two steps.

The properties of the filament are as follows:

| | |
|---|---|
| Linear mass | 3.6–4.2–5.6 dtex |
| Break strength | >1.6 cN/dtex |
| Break elongation | >40% |
| Density | 1.44 g/cm$^3$ |
| Shrinkage in boiling water | 72% |
| Limiting oxygen index | 51.5% |

Fibers of CPVC have properties similar to homofibers, with slightly higher shrinkage and flame retardancy and lower chemical resistance and ease of dehydrochlorination [14].

Some attempts have been made to improve certain limiting characteristics: (1) grafting methyl methacrylate to increase moisture regain, (2) grafting acrylonitrile to improve thermal stability, and (3) blending CPVC with cellulose acetate at a ratio of 85:15 to increase the shrinkage temperature [7].

## 1.3. Fibers from Vinyl Chloride Copolymers

Copolymers of PVC with a 13–15% vinyl acetate content and a viscosity index of 57–58 exhibit a lower glass transition temperature (73°C) than homopolymers, and the bondability improves at temperatures lower than the melting point.

The process used is *dry spinning*. A solution of copolymer (≤50 wt%) in acetone, containing heat stabilizers and additives, is prepared at ambient temperature, filtered, and extruded at a temperature dependent on the polymer content (50–80°C). A hot aqueous treatment without tension, crimping, or oiling, and low-temperature drying follow if the fiber is to be used in the dry-laid nonwoven process; for the wet-laid nonwoven process, crimping is omitted.

The relevant key *properties* are:

| | |
|---|---|
| Chlorine content | 46% |
| Break strength | 0.6–0.7 cN/dtex |
| Break elongation | 150–200% |
| Softening point | 69°C |
| *mp* | 155–160°C |
| Decomposition temperature | 230°C |
| Shrinkage in water | 18–28% at 80°C |
| Shrinkage tension | 0.003 cN/dtex |
| Shrinkage temperature | 60°C |

Bonding may be carried out between 80 and 150°C or between 150 and 170°C with transformation of the fiber into melt droplets. Thermal bonding is used to produce nonwovens, blended with other natural, regenerated, or synthetic fibers at a ca. 30% maximum content.

## 1.4. Fibers from Poly(Vinylidene Chloride) and Copolymers [7, 15, 16]

The homopolymer of PVDC has a softening point of 185–200°C. Its two outstanding properties, high crystallinity and insolubility, make it difficult to process. Therefore, to ease processing conditions, copolymers with small quantities of vinyl chloride are used.

The usual copolymers contain 85–90 wt% vinylidene chloride (saran). Copolymers are processed by melt spinning or extrusion. Because they are heat sensitive and have a limited thermal life at the extrusion temperature, specially designed extruders and alloys are required. Depending on end use, plasticizers, heat and light stabilizers, and colored pigments may be added to the copolymer. Monofilaments or multifilaments are extruded at 160–175°C, quenched in water at 10–15°C with a short distance between the spinneret and the level of the quenching bath, and then drawn 500–600%.

***Products, Properties, and Uses.*** The range of *products* includes monofilaments (70–120 dtex) and multifilaments (10–78 dtex). Typical properties are:

| | |
|---|---|
| Density | 1.65–1.75 g/cm$^3$ |
| Break strength | 1.5–2.0 cN/dtex |
| Break elongation | 15–25% |
| Initial modulus | 6–10 cN/dtex |
| Moisture regain | >0.1% |
| Softening point | 115–135°C |
| mp | 170–177°C |
| Shrinkage temperature | 70°C |
| Dielectric constant at 60×10$^6$ Hz | 3–5 |
| Dielectric loss (tan δ) | 0.05 |

Poly(vinylidene) chloride has excellent resistance to acid, alkali (except ammonium hydroxide), and weathering but poor heat resistance. The products cannot be dyed easily; disperse dyes may be used but fastness is poor.

## 2. Poly(Vinyl Alcohol) Fibers

W. O. HERMANN and W. HAENEL applied for the first patent on poly(vinyl alcohol) [9002-89-5] (PVA) fiber in 1931. They reported that the fiber could be made from PVA by the well-known wet- and dry-spinning methods, and that the water resistance of the fiber could be improved by physical and chemical aftertreatment. However, they did not investigate further the production of a water-resistant textile fiber because their interest was limited mainly to the solubility of the fiber in water. Water-soluble PVA monofilament, made by dry spinning and called "Synthofil," was manufactured for a short time in Germany.

In 1940, I. SAKURADA and his colleagues improved the hot-water resistance of wet-spun PVA fiber by heat-treating it in hot air and acetalizing it with formaldehyde; they developed the technology for production of a water-resistant textile fiber [17].

Wet-spun staple fiber based on this technology was first available commercially in Japan in 1950; the fiber was called "Vinylon." Initially, Vinylon was distinguished from other synthetic fibers by its hydrophilic properties and it was marketed as a substitute for cotton in both clothing and industrial fibers. Improved production technology led to remarkable improvement in the mechanical properties of the fiber. In 1959, high-tenacity filament produced by dry spinning, and later by wet spinning, became available. The use of Vinylon has become almost entirely industrial. In addition to the water-resistant fiber, a water-sensitive variety has been produced for water-soluble fiber and binder fiber for making paper.

The staple fiber has been also produced commercially in the Democratic People's Republic of Korea since 1961 and the People's Republic of China since 1965.

### 2.1. Physical and Chemical Properties

The mechanical properties of commercial PVA fibers (Vinylon) produced in Japan are shown in Table 2 [18]. The tenacity and modulus are the highest among commercial general-purpose synthetic fibers. Fiber having a tenacity of 12 cN/dtex and a modulus of 274 cN/dtex has been produced for cement reinforcement. Resistance to abrasion and fatigue is also excellent. The fiber absorbs slightly more water than polyamide fiber; and its mechanical properties in the wet state are somewhat lower. Formalization is effective in improving water resistance of the fiber but may lower the mechanical properties. The elastic recovery is slightly lower than that of acrylic fiber but better than cotton or rayon. Acetalization with benzaldehyde is an effective technique for improving the elastic recovery of PVA fibers used in clothing. The resistance of PVA fiber to various chemicals is generally superior to that of polyamide and polyester fibers. In particular, its excellent alkali resistance is important in the reinforcement of cement products [19]. Because of its high polarity, PVA fiber adheres well to matrix materials in fiber-reinforced composites for resin, rubber, and cement. Its weather resistance is the highest of the various synthetic fibers.

The fiber is softened by heating above 230–240°C, but it has no definite melting point. Commercial fibers with sufficient hot-water resistance, and water-soluble fibers that dissolve in 20–90°C water, are produced by applying the appropriate heat or acetalization treatments. For use in wearing apparel, the formalized staple fiber, mainly blended with cotton, is dyed with direct, vat, sulfur, or metallized dyes.

## 2.2. Poly(Vinyl Alcohol) for Fiber

Poly(vinyl alcohol) for fiber is produced by polymerization of vinyl acetate in methanol and subsequent alkaline hydrolysis of poly(vinyl acetate). The material must have a degree of polymerization of 1200–2500 and a narrow molecular mass distribution. These properties are attained by adjusting polymerization conditions, and especially by conducting the polymerization at a lower temperature and keeping the conversion of vinyl acetate in methanol below 70%.

For the production of industrial fiber, PVA with a degree of polymerization of ca. 1700 is generally used. The amount of residual acetyl groups in PVA should be <0.1 mol% for heat treatment to produce a sufficiently water-resistant fiber.

### 2.2.1. Wet Spinning

***Sodium Sulfate Coagulation Bath Method.*** Staple fiber and tow are produced by wet spinning. In ordinary wet spinning, PVA is dissolved in water to a concentration of 14–16% by heating. After filtration and deaeration, the spinning solution is kept above 70°C to avoid gelation. It is extruded through spinnerets into a coagulation bath containing a saturated solution of sodium sulfate at 40–50°C. The coagulation speed is much slower than that of viscose, and a longer residence time in the bath is necessary. For the coagulation operation to run smoothly, and to minimize floor space requirements, a vertical spinning machine is preferred [20]. The spun fiber tow is usually drawn while passing through a second hot coagulation bath and then dried.

The wet-spun and dried fiber is soluble in hot water. It is further drawn and heat-treated in hot air or by hot rollers at 210–240°C. The temperature of the heat-treatment process must be higher than that of the hot drawing process. In heat treatment, a small amount of shrinkage is usually allowed in order to stabilize fiber structure. This treatment orients and crystallizes the polymer chains in the fiber, reducing the fiber solubility in hot water.

The heat-treated fiber is further acetalized in an aqueous solution containing 1–5% formaldehyde, 10–20% sulfuric acid, and 5–20% sodium sulfate at 50–80°C for 10–60 min to reach a degree of formalization of 25–35 mol%. Acetalization is affected not only by reaction conditions but also by conditions of the drawing and heat-treatment processes applied to the fiber. Acetalization greatly improves the boiling-water resistance of the fiber and also changes its mechanical and chemical properties to some extent.

In the production of staple fiber, the heat-treated tow is usually crimped mechanically and cut, and the cut fiber is formalized. For industrial application, the formalized tow is subjected directly to a stretch-break spinning system (Perlock or Converter system) to convert it to spun yarn. This yields spun yarn with high tenacity and high toughness.

Short-cut fiber, destined to be water-swellable binder fiber or water-resistant substrate fiber for making paper, is produced by cutting wet-spun and dried tow, or tow that has been heat-treated and sometimes formalized, to the proper length of 3–12 mm.

***Alkaline Coagulation Bath Method.*** Fiber that has been wet-spun in a sodium sulfate coagulation bath has a cocoon-shaped cross section and skin–core of inhomogeneous structure because coagulation proceeds with rapid dehydration from the outside of the extruded jet of spinning dope. This makes it difficult to obtain a clear fiber with high tensile strength.

When PVA dope is spun into an aqueous solution of concentrated alkali, fiber with a round cross section and a homogeneous, compact structure can be obtained [21]. The fiber is transparent and can be highly drawn to give improved tensile strength and modulus, and better water resistance. In this case, the spinning dope extruded in the coagulation bath is gelated instantaneously by contact with the alkaline coagulant, so dehydration proceeds homogeneously. This wet-spinning method is used in Japan. In commercial practice, the aqueous coagulation bath contains sodium hydroxide at a concentration >200 g/L and at a temperature of 40–50°C.

The alkaline coagulation method was modified to give more highly drawn fiber with a higher modulus by using aqueous acidic spinning dope containing a small amount of boric acid [22]. Because of the presence of boron, cross-linking appears to occur between the polymer chains of PVA during fiber formation. This suppresses entangling of the polymer chains and makes coagulation more homogeneous.

The spinning dope, containing boric acid of 0.5–2.0 wt% (relative to PVA) at pH 3–5.5, is spun into an aqueous alkaline coagulation bath containing 20–100 g/L of sodium hydroxide and 100–300 g/L of sodium sulfate at pH 13.5–14.0. After removal from the coagulation bath, the spun fibers are treated with acid, washed with water, dried, and drawn to a draw ratio >10 to give high tenacity (>9 cN/dtex) and high modulus (>177 cN/dtex).

This technology was developed initially for production of high-tenacity, high-modulus filament yarn for use in belt cord for radial tires. The yarn is modified further to produce high-modulus fiber, which can be substituted for asbestos in cement reinforcement; this fiber has a tenacity of 12 cN/dtex and a modulus of 274 cN/dtex [23].

**Wet Spinning with Organic Solvent.** Poly(vinyl alcohol) has a high crystalline modulus comparable to that of polyethylene. This high modulus has not been exploited extensively to date in the preparation of fiber. Various investigations have been carried out into the preparation of PVA fiber with a much higher modulus. In many cases, this has involved modified wet-spinning methods that use organic solvents.

When a hot, dilute solution of PVA in polyol is cooled slowly, highly crystallized PVA is deposited [24]. This phenomenon has been connected with gel-spinning–ultrahigh-drawing procedures. When the gel fiber from a solution of PVA in glycerin, having an ultra-high degree of polymerization of 38 000, was hot-drawn at high temperature, the resulting fiber had a tenacity of 17 cN/dtex and a modulus of 555 cN/dtex [25].

### 2.2.2. Dry Spinning

Dry spinning of PVA is used in the production of filament yarns. The two classifications of dry spinning are low-draft spinning [26] and high-draft spinning [27].

**Low-Draft Spinning.** Aqueous spinning dope, which has a concentration of 41–45% and an extremely high viscosity, is used in low-draft spinning. It is prepared by dissolving PVA granules, conditioned with water, by heating them in an extruder under pressure.

The high-viscosity spinning dope at a temperature of 130–160°C, is spun in air at ca. 50°C. The cooled and solidified filament is wound in a low-draft ratio of 0.3–1.0 based on the extrusion rate of the spinning dope from the spinneret (i.e., draft ratio = winding rate/extrusion rate), and gradually dried while the drying temperature is raised. The dried filaments have a circular cross section, a homogeneous structure, and low strain; they can be drawn to the highest draw ratio and heat-treated near the melting point. The resulting filament has adequate hot-water resistance and is used commercially without formalization.

This method is suitable for producing heavier filament for industrial applications. It was originally developed for the production of heavier monofilaments of 110–1100 dtex. It is now also applied to the production of multifilaments, such as 1330 dtex/200 filaments.

**High-Draft Spinning.** In high-draft spinning, aqueous spinning dope containing PVA at a concentration of 28–41% is spun at 90–95°C into a spinning tube consisting of a drafting zone at 30–80°C and 55–95% R.H., and a drying zone at >80°C and <50% R.H. While passing through the spinning tube, the filament is highly drafted, dried, and taken off at a winding speed of 200–500 m/min. The dried filaments are hot-drawn and heat-treated, and sometimes further acetalized.

This method can give finer filaments (<5 dtex) for textile applications. Some filament for apparel, made from a mixture of PVA and aminoacetalized PVA is produced commercially and can be dyed with acid dyes. This method is now used primarily to produce water-soluble filament. Filament soluble in water at 20–90°C is produced commercially by using PVA with low degrees of polymerization and saponification, and by adjusting heat-treatment conditions [28].

### 2.2.3. Mixed Spinning

The technology for spinning bicomponent fiber from PVA and PVC in an emulsion was developed in 1961 [29]. Vinyl chloride is emulsion-polymerized in an aqueous solution of PVA, and the emulsion is mixed with PVA to give a spinning dope containing PVC and PVA in ca. 1 : 1 mass ratio. The fiber is produced in essentially the same way as regular PVA fiber, by using the sodium sulfate coagulation bath method. The fiber is characterized by its flame retardancy and is produced commercially in Japan under the name Polychlal.

### 2.3. Uses

In Japan, about 90% of Vinylon fiber production, ca. 30 000 t/a, is consumed in industrial applications that require such advantages as high tenacity, excellent weather resistance, and effective reinforcing effect in matrices. The main industrial uses are fishing nets, ropes, belts, canvasses, and sewing threads. Perlock-type spun yarns from 100% Vinylon tows and filament yarns are used predominantly. The water-soluble filaments are used to produce chemical lace fabrics.

**Table 1.** Trade names and producers of polyvinyl fibers

| Trade name | Type[*] | Producer |
|---|---|---|
| Clevyl | CLF | Rhovyl SA/France |
| Envilon | CLF | Denki Kagaku KK/Japan |
| Fibravyl | CLF | Rhovyl SA/France |
| Krehalon | PVDC | Kureha Gosen Co./Japan |
| Kugafil | PVDC | Fuagafil-saran GmbH/Germany |
| Kuralon | PVAL | Kuraray Co./Japan |
| Mewlon | PVAL | Unitika Ltd./Japan |
| Niti-Vilon | PVAL | Nitivy Co./Japan |
| Quing Wei | PVAL | Ankui Wanwei/China |
| Retractyl | CLF | Rhovyl SA/France |
| Rhovyl | CLF | Rhovyl SA/France |
| Saran | PVDC | Fugafil-saran GmbH/Germany |
| Shuang Lun | PVAL | Fujian Textil & Chemical Fibre/China |
| Solvron | PVAL | Nitivy Co./Japan |
| Teviron | CLF | Teijin Ltd./Japan |
| Thermovyl | CLF | Rhovyl SA/France |
| Toykolon | CLF | Denki Kagaku KK/Japan |
| Viclon | CLF | Kureha Gosen Co./Japan |
| Vilon | PVAL | Nitivy Co./Japan |

[*] PVAL = poly(vinyl alcohol), CLF = chlorofibers = chlorinated polyvinyl fibers, PVC = poly(vinyl chloride), PVDC = poly(vinylidene chloride).

**Table 2.** Mechanical properties of Vinylon fibers [18]

| Property | Staple and tow | | | Filament | |
|---|---|---|---|---|---|
| | Regular tenacity | High tenacity | Regular tenacity | High tenacity | |
| Tenacity, cN/dtex | | | | | |
| Standard | 3.5–5.7 | 6.0–9.3 | 2.6–3.5 | 5.3–10.6 | |
| Wet | 2.8–4.6 | 4.7–7.9 | 1.9–2.8 | 4.4–9.3 | |
| Wet–standard ratio,% | 72–85 | 78–85 | 70–80 | 75–90 | |
| Loop tenacity, cN/dtex | 2.8–4.6 | 4.4–5.1 | 4.0–5.3 | 6.2–11.5 | |
| Knot tenacity, cN/dtex | 2.1–3.5 | 4.0–4.6 | 1.9–2.6 | 2.4–4.4 | |
| Elongation,% | | | | | |
| Standard | 12–26 | 9–17 | 17–22 | 6–22 | |
| Wet | 12–26 | 9–17 | 17–25 | 8–26 | |
| Young's modulus, cN/dtex | 22–62 | 62–221 | 53–79 | 62–221 | |
| Resilience at 3% elongation,% | 70–85 | 72–85 | 70–90 | 70–90 | |
| Relative density | 1.26–1.30 | 1.26–1.30 | 1.26–1.30 | 1.26–1.30 | |
| Commercial regain,% | 5.0 | 5.0 | 5.0 | 5.0 | |
| Water absorbency,% | | | | | |
| 20°C, 20% R.H. | 1.2–1.8 | 1.0–1.5 | 1.2–1.8 | 1.0–1.5 | |
| 20°C, 65% R.H. | 4.5–5.0 | 3.5–4.5 | 3.5–4.5 | 2.5–4.5 | |
| 20°C, 95% R.H. | 10.0–12.0 | 8.0–10.0 | 10.0–12.0 | 8.0–10.0 | |

The development of special grades has increased the demand for Vinylon remarkably. The water-swellable, short-cut fibers are popular as binders in the production of paper and wet-laid nonwoven fabrics. Another short-cut fiber, modified to have the highest tenacity and modulus as well as a special surface structure that enhances its adhesion to matrices, is used as a substitute for asbestos to reinforce cement products. These special grades are exported to the world market.

Only 10% of Vinylon production is used domestically in Japan for apparel, mainly by blending Vinylon with cotton and other fibers. The main apparel application is work clothing that requires durability and chemical resistance.

In Korea and China, significant quantities of PVA fiber are blended with cotton and used for apparel. In these countries too, fiber applications in the future will shift gradually to industrial uses.

## 2.4. Trade Names and Producers

Trade names and producers of polyvinyl fibers are listed in Table 1.

# References

1. Rhodiaceta, FR 913919, 1942.
2. Ch. Heiberger, *Encyclopedia of PVC*, Dekker, New York 1986, pp. 416–429, 417–428,632–636.
3. Rhône-Poulenc, FR 2475418, 1980 (A. Berchoux).
4. Teijin, JP 352668, 1968.
5. Rhovyl, FR 2524475, 1982 (G. Achard, G. Anouilh, J. Menault).
6. Rhône-Poulenc, FR 2495645, 1980 (G. Achard, J. Menault, C. Thouvenot).
7. A. Rogovin, W. Albrecht: *Chemie Fasern*, G. Thieme Verlag, Stuttgart 1981, pp. 314–322.
8. Rhovyl, FR 2524498, 1982 (G. Achard, P. Chion, J. Menault).
9. B. Catoire, G. Gastadi, R. Hagege, *Bull. Sci. Inst. Text. Fr.* **11** (1974) 205–217.
10. Rhône-Poulenc, FR 2483966, 1980 (P. Chion, J. Menault, A. Rodier, J.P. Sacre).
11. E.A. Rassolova et al., *Fibre Chem.* **15** (1983)no. 4, 8.
12. J. Menault, *Ind. Text. (Paris)* **1158** (1985) 829–831.
13. Shirley Institute, Report DS. *1045*,Dec. 12, 1983, unpublished.
14. D. Werner, *Textiltechnik* **35** (1985) 8.
15. B. von Falkai, *Synthesefasern*, Verlag Chemie, Weinheim, Germany 1981, pp. 216–218.
16. *Ullmanns Encyklopädie der Technischen Chemie*, 4th ed., vol. 7, Verlag Chemie, Weinheim,pp. 296–308.
17. Nippon Kagaku Sen'i Kenkyusho, JP 147958, 1939 (I. Sakurada, S. Lee, M. Kawakami); JP 157076, 1940 (S. Lee, I. Sakurada, H. Kawakami, K. Hirabayashi et al.).
18. Japan Chemical Fibers Assn. (ed.): *Kasen Handbook (Handbook of Chemical Fibers)*, Japan Chemical Fibers Assn., Tokyo 1985.
19. A. Mizobe, *Sen'i Gakkaishi* **41** (1985) 180.
20. Kurashiki Rayon Co., US 2642333, 1950 (T. Tomonari, T. Akaboshi, M. Nagai, T. Osugi).
21. Noguchi Laboratory, JP 162708, 1942 (M. Nakanishi, M. Kubota).
22. Kurashiki Rayon Co., US 3660556, 1969 (T. Ashikaga, T. Kosaka).
23. Kuraray Co., JP-Kokai 56125264, 1984 (S. Oka, M. Mizobe, J. Hikasa, M. Okazaki et al.).
24. Kurashiki Rayon Co., US 3427298, 1969 (K. Tsuboi, T. Mochizuki).
25. Allied, US 4440771, 1982 (Y.D. Kwon, S. Kavesh, D.C. Prevosek).
26. Kurashiki Rayon Co., JP 317932, 1958 (S. Nakajo, E. Morita).
27. Kurashiki Rayon Co., JP 232163, 1956 (K. Kawai, S. Miyazaki, K. Tanabe, T. Osugi).
28. M. Uzumaki, *Sen'i Gakkaishi* **33** (1977) 55.
29. Toyo Chemical Co., JP 417267, 1961 (S. Okamura, H. Asakura).

# Polytetrafluoroethylene Fibers

PETER E. FRANKENBURG, E. I. Du Pont de Nemours, Wilmington, Delaware, USA

MICHAEL SCHWEIZER, ITCF Denkendorf, Denkendorf, Germany

1. Physical and Chemical Properties... 1539
2. Production ................... 1539
3. Uses ....................... 1540
References. .................. 1540

Polymerization technology for polytetrafluoroethylene [9002-84-0] (PTFE) has been known since 1941 and has become well established, leading to utilization of the polymer not only as a plastic but also in fiber form [1]. This is undoubtedly because it combines interesting chemical (inertness) and physical (low friction, temperature resistance, and triboelectric characteristics) properties.

Tetrafluoroethylene is polymerized under pressure in the presence of persulfate or peroxide in water; oxygen is excluded. The reaction is exothermic and increases in velocity with increased pressure. The viscosity of the polymer, as well as its molecular mass cannot be measured readily; however, indications are that the polymer chain is linear. At the normal polymerization level, molecular masses are estimated to be between a few hundred thousand and several million.

## 1. Physical and Chemical Properties

Some of the physical properties of polytetrafluoroethylene are as follows:

| | |
|---|---|
| mp | 327°C |
| Glass transition temperature | ca. 120°C |
| ϱ | 2.1 g/cm$^3$ |
| Melt viscosity (380°C) | 10 GPa |

These fibers are similar to the polymer (i.e., highly crystalline 93–98%) and, because of the high energy of the C-F bond, display appreciable resistance to heat and chemicals. As a result, the polymer is difficult to burn. The fibers also are distinctive because of their good light and weathering durability, and their utility at temperatures up to 290°C. Phase changes that occur below 30°C are responsible for the cold flow characteristics of the material, although the fibers are reported to be more dimensionally stable than bulk polymer because of their molecular orientation. Decomposition products are generated above 290°C at a rate of 0.0002%/h. Copolymers, containing minor amounts of other fluorinated monomers, are also available. These are melt-extrudable but, because of their high melt viscosities, only into monofils of =100 dtex. A 50–50 copolymer with ethylene (Tefzel, Du Pont) has a lower melt viscosity and can be extruded at commercially attractive speeds; however, its temperature–chemical characteristics are altered by the presence of the nonfluorinated comonomer.

## 2. Production

Polytetrafluoroethylene can be sintered above its melting point into a tough, viscous melt. However, melt spinning is not feasible with the homopolymer because the temperature required to overcome the viscosity leads to polymer degradation. Because economically attractive solvents for the polymer have not been found (a few fluorinated hydrocarbons dissolve several percent of polymer above 300°C), either paste extrusion below the melting point of the polymer or dispersion spinning to form fibers is used. To promote agglomeration during extrusion of plasticized (with 20 wt% kerosene) homopolymer, extrusion is done under conditions favorable to production of particulates in ribbon form, so that at least 30% have a width of 0.1 nm, an aspect ratio of at least 5:1, and a cross section of <0.001 nm$^2$. Sintering followed by drawing and heat-setting results in solid or expanded films (depending on process conditions), which are then split into filaments of ca. $\geq$100 dtex.

An alternate process is used to produce finer filaments. A 60% solid dispersion of polytetrafluoroethylene polymer, stabilized with aryl or alkyl polyglycol ethers, is blended with viscose solution (5%). The latter acts as a matrix during early phases of the process. The mixture is filtered and then extruded at room temperature through a multihole spinneret (hole diameter 0.15 mm) into a coagulating bath containing 6% sulfuric acid, 16% sodium sulfate, and 0.3% zinc sulfate. The spinning velocity is about 12 m/min.

After coagulation, the yarn is washed with water at 90°C, dried at 190°C, and then the temperature raised to 340–360°C, the temperature at which polymer particles are sintered and the viscose residue degrades and is partially burned. After a brief contact time at 360–390°C, the yarn is stretched four to eight times its original length.

The resultant continuous-filament yarn has a tenacity of 1.8 cN/dtex and an elongation of 20–30%. The filament mass is 6 dtex. The fibers are brown and can be bleached in boiling sulfuric acid containing nitric acid or by exposure for several days in an air oven. The latter must be done cautiously by raising the temperature stepwise to 350°C, to avoid exotherms because of the sudden oxidation of the viscose residues.

## 3. Uses

Polytetrafluoroethylene fibers are utilized in woven and nonwoven forms as filter media for aggressive liquids and gases (i.e., strong acids, bases, and oxidizing agents), including air pollution control fabrics. Fibers are also used in gaskets and pump seals because of their chemical resistance and low-friction characteristics. In electrical systems, they are utilized as insulators for wires and components, and as cleaners–wipers in xerographic devices, where their high-temperature and triboelectric characteristics are important. Because of their great chemical stability, these fibers are used as the binder component in asbestos chlor-alkali cell diaphragms. The low-friction characteristics and high length–diameter ratio of the fibers are important in their use as solid lubricants in engineering plastics. Monofils have found application in demisters in the production of mineral acids.

In addition, dispersion of fibers on woven or nonwoven substrates imparts chemical stability to the surface and results in layered composites with excellent dielectric properties and corrosion resistance for utilization in xerography.

Trade names and producers of polytetrafluoroethylene fibers are listed in Table 1.

**Table 1.** Trade names and producers of polytetrafluoroethylene fibers

| Trade name | Type | Producer |
| --- | --- | --- |
| Gore | PTFE | Gore & Ass. GmbH/Germany |
| Lenzing Profilen | PTFE | Lenzing Plastics GmbH/Austria |
| Lenzing PTFE | PTFE | Lenzing Plastics GmbH/Austria |
| Teflon | PTFE | Toray Fluorofibers (America) Inc./USA |
| Tefzel | PTFE | Albany International, Inc./USA |

## References

1 Du Pont, US 2559750, 1951 (K. L. Berry)

# High-Performance Fibers

VLODEK GABARA, Retired from E.I. Du Pont de Nemours and Company, Richmond, Va., USA

| | | |
|---|---|---|
| 1. | Introduction | 1541 |
| 1.1. | Definition and Classification | 1541 |
| 1.2. | Historical Overview | 1542 |
| 2. | Heat Resistant Fibers | 1542 |
| 2.1. | *meta*-Oriented Aromatic Polyamide Fibers | 1543 |
| 2.1.1. | Solution Polymerization | 1544 |
| 2.1.2. | Interfacial Polymerization | 1545 |
| 2.1.3. | Fiber Formation | 1545 |
| 2.1.4. | Washing, Drawing, Heat Treatment | 1547 |
| 2.1.5. | Fiber Properties | 1547 |
| 2.1.6. | Uses | 1547 |
| 2.2. | Fibers from Heterocyclic Polymers | 1548 |
| 2.3. | Fibers from Melt-Processable Polymers | 1549 |
| 2.4. | Cross-linked Fibers | 1550 |
| 3. | High-Strength and High-Modulus Fibers | 1550 |
| 3.1. | *para*-Oriented Aromatic Polyamide Fibers | 1551 |
| 3.1.1. | Fibers from Lytropic Solutions | 1551 |
| 3.1.2. | Fibers from Isotropic Solutions | 1553 |
| 3.1.3. | Fiber Structure | 1553 |
| 3.2. | Polyazole Fibers | 1554 |
| 3.2.1. | Fiber Structure | 1555 |
| 3.3. | Fibers from Thermotropic Polymers | 1555 |
| 3.4. | High Strength Polyethylene Fibers | 1555 |
| 3.4.1. | Gel Spinning | 1556 |
| 3.4.2. | Extrusion Process | 1556 |
| 3.4.3. | Fiber Structure | 1557 |
| 3.5. | Properties and Uses of High Strength Fibers | 1557 |
| 3.5.1. | Fiber Properties | 1557 |
| 3.5.2. | Uses | 1558 |
| | References | 1560 |

## 1. Introduction

### 1.1. Definition and Classification

Although a strict definition of high-performance fibers does not exist, the term generally denotes fibers that can perform under severe conditions of temperature, chemical attack, mechanical stress as well as flame or offer an unusual functionality e.g., conductivity or magnetic properties. It commonly refers to fibers with some characteristics that differentiate them from commodity fibers such as nylon, polyester, and acrylic fibers. Synonyms are advanced fibers, specialty fibers and, in some cases, high-functional fibers. High-performance fibers can be classified broadly into three categories according to their unique characteristics:

1. Heat-resistant fibers
2. High-strength and high-modulus fibers
3. Other specialty fibers

The unique characteristics of the materials are closely linked with their composition. Thus, it is expected that fibers from ceramic materials have outstanding thermal and compressive properties; many metallic fibers are conductive while unique structure of carbon fiber offers an excellent combination of compressive and tensile properties. This article focuses on how to derive these unique characteristics using organic polymers (both aromatic and aliphatic) based on chemistry of carbon and thus in general less thermally stable. Information on inorganic fibers like carbon, alumina, and boron, can be found in (→ Fibers, 11. Inorganic Fibers; → Fibers, 13. Ceramic Fibers; → Fibers, 15. Carbon Fibers). Testing methods are described in (→ Fibers 16, Testing and Analysis) while some uses of high performance fibers are discussed in (→ Composite Materials).

Organic High-performance fibers can be also classified on the basis of the processes that are used in their production. Obviously the unusual characteristics of these fibers makes the processing conditions very demanding. Most of them do not melt and thus have to be shaped by spinning from solution (*solution spinning*). Solution can be spun into a hot gas to evaporate some of the solvent before coagulation (*dry spinning*) or directly into a coagulating liquid (*wet spinning*). A few examples of polymers which do melt can be spun in that form at very high temperature (*melt spinning*) (→ Fibers, 3. General Production Technology).

As described in greater detail in this article, the commercial examples of wholly aromatic polymers include polyamides, polyesters, and polyketones. They can be classified as *meta*- or *para*-oriented, depending on their chemical structure. Fibers from *meta*-oriented polymers are useful as heat-resistant fibers while *para*-oriented polymers are useful not only as heat-resistant fibers, but also as high-strength (HS) and high-modulus (HM) fibers.

## 1.2. Historical Overview

In the late 1940s, DuPont defined a broad direction for its future research. Substitution of existing materials as an organizing principle led to the search for fibers with high temperature resistance, high strength, high elasticity, etc. In the area of high-performance fibers, *meta*-oriented wholly aromatic polyamide [poly(*m*-phenyleneisophthalamide) – PMIA] was the first example developed in the 1950s. The commercial version of the product was called Nomex. Continuing development in understanding factors that govern high-temperature resistance of polymer shifted the focus on polymers with *para* orientation. This lead to the development of poly(phenyleneterephthalamide) (PPTA) which when properly shaped into fiber (Section 3.1 para-Oriented Aromatic Polyamide Fibers) exhibited not only the expected high temperature performance but also very unique high strength and stiffness. In 1972, the commercial version of this fiber was introduced as Kevlar.

The pioneering nature of this work not only lead to the development of useful commercial products but provided also fundamental underpinning to a development of advanced fibers based on different chemical structure and through very different shaping processes. More than 50 years of research in this field identified a large number of compositions. More than 100 compositions of polymers belonging to this class of materials were already identified by 1989 [1], and the number has increased significantly since then. Most of them were based on aromatic polyamides but other chemistries, such as aromatic polyesters, aromatic polyketones and aromatic polyethers were also investigated and some are being produced on a commercial scale albeit significantly smaller than polyamides. The fundamental understanding of structural requirements for very high strength and stiffness developed during work on Kevlar led to the development of very high strength and stiffness fibers based on chemistry as different from *p*-aramids as aliphatic polyethylene (1980s).

Some of these developments are described below, but it seems useful to stress the rapid development in the field, which allowed to approach in a macroscopic form and on a commercial scale, fiber properties of the same order of magnitude as those of the chemical bonds in just 25 years.

## 2. Heat Resistant Fibers

Early research on resistance of organic polymers to high temperatures focused on their thermal degradation [2, 3] which was linked with the strength of chemical bonds, especially those in the main chain. The bond dissociation energies for aromatic or heterocyclic compounds are higher than those for aliphatic ones. This suggests that aromatic polymers possess better thermal stability than aliphatic polymers (see Table 1). Thus DuPont's early research on heat resistant fibers was focused on aromatic polymers.

The use of fibers under high temperature brings into focus not only polymer decomposition, but also the retention of physical properties under these conditions. Fundamental understanding at the time led scientists to

**Table 1.** Bond dissociation energies

| Bond | Bond dissociation energies, kJ/mol | |
|---|---|---|
| | Aromatic or heterocyclic | Aliphatic |
| C–C | 410 | 347 |
| C–H | 427 | 406 |
| C–O | 448 | 389 |
| C–N | 460 | 343 |

concentrate on compositions with high glass transition temperature ($T_g$) and high melting point ($T_m$), which in turn lead to compositions with relatively high chain rigidity and strong interchain interactions due to dipolar forces or hydrogen bonding. Thus it is not surprising that the first commercial products coming from DuPont (late 1950s and early 1960s) were aromatic polyamides [4–6]. They are still the largest volume commercial products serving thermal and flame protection applications.

Other compositions that found applications in this area include aromatic polymers with different groups linking the rings, such as polyoxadiazole (POD), polyimide (PI), poly(phenylene sulfide) (PPS), and polyetheretherketone (PEEK).

Continuing research in the area of aromatic polymers led to the development of aromatic polyesters while further search for very rigid polymers led to polyazole fibers. The latter, while possessing high temperature properties, were designed primarily for high strength and high stiffness applications and are discussed in Section 3.2 Polyazole Fibers and Section 3.3 Fibers from Thermotropic Polymers.

## 2.1. *meta*-Oriented Aromatic Polyamide Fibers

The unique characteristics of fibers based on aromatic polyamides, which differ significantly from the class of fibers known as polyamides (see → Fibers, 4. Polyamide Fibers), led the US Federal Trade Commission to define the term *aramid* for fibers based on aromatic polyamides in which at least 85% of the amide linkages are attached directly to two aromatic rings. A very detailed description of preparation, properties,

and applications of aramids is beyond the scope of this article but is provided in [7].

In spite of the very large amount of work and exploration of countless aramid compositions only three of them reached commercial significance. Poly(*m*-phenyleneisophthalamide) (PMIA) [24938-60-1] is the basis for most of today's commercial aromatic polyamide fibers used in thermal and flame protection. Poly(*p*-phenyleneterephthalamide) (PPTA, **2**) [24938-64-5] and copoly(*p*-phenylene/3,4-diphenyl ether terephthalamide) (POP, **3**) are most often used in applications requiring high strength and stiffness. The fundamentals of their preparation and uses are discussed in Section 3.1 para-Oriented Aromatic Polyamide Fibers.

The importance of PMIA in heat and flame protection applications is best illustrated by the number of commercial products introduced so far. The unique properties of aromatic polyamides result also in unique difficulties in their synthesis and processing. Technological difficulties lead frequently to subpar fiber quality or too high manufacturing costs and resulted in abandoning production by some of the producers. The trade names and producers are listed below.

| Trade name of PMIA products | Producers |
|---|---|
| Nomex | DuPont |
| Tejinconex | Teijin |
| Newstat | Yantai |
| Chinfunex | Charming |
| X-Fiper | SRO |

| The following products are no longer produced: | |
|---|---|
| Fenilon | Kazakhstan |
| Apyeil | Unitika |
| KM-21 | Mitsui-Toatsu |

Potential solutions to these difficulties in all phases of the manufacturing process are illustrated below.

### 2.1.1. Solution Polymerization

A typical synthesis of aliphatic polyamides by a reaction of diamines with diacid in a melt and removal of water to shift the equilibrium of the reaction towards polymer formation (see → Fibers, 4. Polyamide Fibers) is not applicable to aromatic polyamides. Their very high glass transition (275°C for PMIA) and melting temperature (435°C for PMIA), combined with proximity to the decomposition temperature (450°C for PMIA) [8], make polymerization in melt impossible. The lower reactivity of aromatic amines requires also change from diacids to their more reactive derivatives. The most commonly used process involves solution polymerization but interfacial polymerization or its modifications have also been used. Acid chlorides are most frequently used as diacid derivatives. Both solution and interfacial polymerizations for polyamides in general have been described and their fundamental characteristics have been analyzed [9]. The solution polymerization process for aromatic polyamides is described in [10].

Synthesis of these polymers is based on the Schotten–Baumann reaction. Several unique characteristics of these reactions necessitated creative engineering solutions to produce polymers with high molecular mass in commercial facilities. The high molecular mass is a prerequisite for good physical properties of fibers. These reactions are extremely fast (50% conversion can be achieved in a matter of milliseconds even at very low temperatures) and highly exothermic. The byproduct of the reaction, hydrochloric acid, blocks unreacted amine groups making them nonreactive thus limiting the conversion and molecular mass, leading to formation of oligomeric species only. Limited solubility represents an additional challenge for solvent selection as well as polymerization conditions.

Traditional acid acceptors can be used to reduce the limiting effect of HCl. The effect of different acid acceptors and criteria for selection has been investigated in great detail [9]. High molecular mass aromatic polyamide can be obtained when using tertiary amines as acid acceptors [11]. A basic solvent can be both an acid acceptor and a solvent. Examples of such solvents include dimethylacetamide [*127-19-5*] (DMA) and *N*-methyl-2-pyrrolidone (NMP) [872-50-4] [12]. However, when a separate acid acceptor is used, its amount has to be close to stoichiometric quantities of HCl generated. This frequently requires large quantities of an additional ingredient in the process which is often undesirable. Thus the use of a basic solvent is usually a better option.

The reaction mechanism and kinetics are complex. Contrary to standard melt polymerization the reaction using diamines and diacid chlorides is irreversible. General theoretical considerations for these type reactions are given in [13] and kinetics for the aramids polymerization process is described in [14]. The reaction is divided into two phases: A free amine phase (less than 50% conversion, Eq. 1), and a hydrochloride phase (above 50%, Eq. 2 and Eq. 3). During the free amine phase the reaction is extremely fast (milliseconds) but as unreacted amine groups are blocked by HCl the rate of reaction is controlled by the rate of exchange of amine hydrochloride end groups with the solvent (sol.), which in this case acts as an acid acceptor.

$$2-NH_2 + -COCl \rightarrow -NHCO- + -NH_2 \cdot HCl \quad (1)$$

$$-NH_2 \cdot HCl + sol. \rightleftharpoons sol. \cdot HCl + -NH_2 \quad (2)$$

$$-NH_2 + -COCl \rightarrow -NHCO- + sol. \cdot HCl \quad (3)$$

This very fast reaction, when combined with high heat of reaction, presents significant technological challenges in designing appropriate approaches to minimize temperature rise because high temperature can lead to side reactions of highly reactive species, i.e., acid chloride groups and HCl. These side reactions can produce nonreactive end-groups and thus decrease the attainable molecular mass of the polymer. Selection of the solvent is critical for these systems. The solvent has to dissolve the intermediates but also dissolve or at least swell the growing polymer [9]. The solvent acts as a heat sink and thus limits adiabatic temperature rise. It needs to be stable under reaction conditions. Purity of the solvent plays a key role as several impurities can react with highly reactive ends in this polymerization system. The most common impurity is water. The ratio of monomers in this system plays an as import role in defining molecular mass as in traditional melt polymerization processes.

In general, after polymerization, the polymer remains dissolved in the reaction medium, and the solution can be used directly for fiber spinning. While it is not absolutely necessary, in commercial processes HCl produced during polymerization is usually neutralized [5] to prevent excessive corrosion of the equipment. Calcium hydroxide is one of the possible bases used and the formed calcium chloride remains dissolved in the polymer solution. The presence of salts like $CaCl_2$ improves polymer solution stability [15]. Another approach to neutralization involves the use of liquid ammonia and the formed insoluble $NH_4Cl$ is filtered off, producing a solution without or with a low salt concentration [6].

### 2.1.2. Interfacial Polymerization

Similarly to solution polymerization, interfacial polymerization requires very fast reactions and thus uses the same highly reactive ingredients. Monomers are dissolved in two immiscible liquids, and the solutions are brought into contact. In general, the two layers are mixed vigorously to maximize the contact area. In this process the diamines are dissolved in water whereas diacid chlorides are dissolved in solvents, such as dichloromethane, xylene, or hexane [10, 16, 17]. The reaction occurs close to the interface but in general in the organic layer. Amines are somewhat soluble in the organic solvents used whereas acid chlorides have almost no affinity to the water phase. The reaction of acid chlorides with water has to be significantly slower than that with diamines, and the generated HCl needs to be neutralized to obtain a high molecular mass polymer. In general, an inorganic base (e.g., sodium carbonate) is included as a neutralizing agent in the water phase. Polymerization is carried out at a relatively low temperature since an increase in temperature lowers the molecular mass of the polymer by the increasing rate of side reactions e.g., with water. The molecular mass of the polymer produced via interfacial polymerization is less sensitive to the ratio of monomers than in the solution polymerization because locally their ratio is controlled by diffusion into the reaction zone.

In a modified interfacial polymerization process [18–20], the reaction begins in a homogenous phase with both diacid chlorides and diamine dissolved in tetrahydrofuran (THF). As in other systems, polymerization cannot proceed beyond 50% conversion (no acid acceptor), and it stops at the stage of oligomers. The oligomers are either in solution or in suspension. Addition of a water phase with a dissolved base (e.g., sodium carbonate) allows polymerization to proceed to completion with the polymer precipitating as fine particles, which are washed and dried for further use.

In general, the heterogeneous nature of the interfacial polymerization process results in a polymer with significantly broader molecular mass distribution than the one produced in solution polymerization.

### 2.1.3. Fiber Formation

Fibers based on aromatic polyamides are spun from solution because their very high melting points are too close to their decomposition temperatures to allow for melt spinning. The limited solubility of these polymers requires highly polar solvents, which are frequently used in conjunction with salts [15]. Two fiber spinning processes are used to commercially produce these fibers, i.e., dry spinning and wet spinning. Both of these processes include extrusion of a viscous polymer solution through a spinneret and coagulation (precipitation of the polymer) through contact with a nonsolvent. The main difference between the two is that in dry spinning a fraction of the solvent is

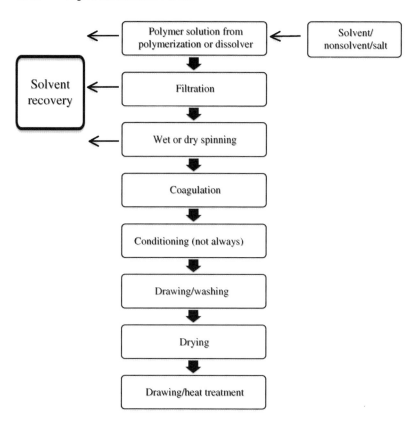

**Figure 1.** A typical fiber formation process [21].

evaporated in a heated column and thus the coagulation occurs at a higher polymer concentration in the "liquid" filament than in wet spinning where the solution is extruded directly into the nonsolvent. A good example of a fiber formation process is shown in Figure 1 with potential variants of the process.

***Dry Spinning.*** A version of the dry spinning process useful for PMIA solutions has been developed [10, 22]. In this process the polymer solution is extruded at high temperature through a spinneret into a column of hot inert gas. For dry spinning it is advantageous to use a spinning solution containing salts (e.g., $CaCl_2$) to improve solution stability at elevated temperatures [22]. As the fiber moves through the column, the solvent is evaporated. In general, for *m*-aramids solutions, solvent evaporation during spinning is not complete and complete removal is not necessary when the spinning is followed by a wet drawing process. As the solution is extruded into a gas, a low drag environment, the spinning process can be reasonably fast, in the order of a hundred meters per minute.

***Wet Spinning.*** In the wet spinning process, the solution is extruded directly into a coagulating bath. The coagulation process is critical for producing good fiber. During coagulation the polymer solution undergoes spinodal decomposition (separation into polymer rich and polymer poor phases), which, if not controlled well, can lead to a highly voided fiber structure that is unsuitable for the subsequent drawing process. Factors affecting this include solution composition (polymer concentration, presence of salts), coagulation bath composition, and temperature. In general, this process requires that the polymer spinning solution contains little or no salt and high salt concentration is needed in the coagulation medium [23]. A wet spinning process has been developed that allows the use of a spinning solution with a relatively high salt concentration [21]. As in wet spinning the solution is extruded into a liquid (water solutions), which is a relatively high drag environment; the process has to be slow, in the order of ten meters per minute.

### 2.1.4. Washing, Drawing, Heat Treatment

To obtain good mechanical properties, fibers spun via either of these processes need to be oriented. Orientation is accomplished in a drawing process in which the spun fiber is extended several times in its original length. The process is carried out above glass transition temperature. If the drawing is to be done in wet form, i.e., in water, the addition of a plasticizer (in general the solvent from which the fiber was spun) is required to lower the glass transition temperature below the boiling point of water. The drawing process is combined with washing of the fiber to remove the residual solvent before drying. As these polymers are very thermally stable, the fiber can be drawn in a dry state as well but at temperatures significantly above $T_g$. This drawing process may be linked with the heat treatment process above the glass transition temperature if higher crystallinity of the final fiber is desired.

### 2.1.5. Fiber Properties

PMIA-based fibers can be produced at a substantial level of crystallinity. The unit cell is triclinic with a calculated density of 1.45 g/cm$^3$ (observed for heat treated fibers is 1.38 g/cm$^3$) [24]. Both density and crystallinity vary with fiber orientation and heat treatment conditions. Hydrogen bonding and relative chain rigidity are responsible for the high glass transition (275°C) and high melting point of the crystalline phase (435°C) whereas high bond dissociation energy leads to a high decomposition temperature (450°C). This combination of properties results in good dimensional stability (low shrinkage) when exposed to high temperature as well as very good retention of mechanical properties at elevated temperatures. The high degree of aromaticity is responsible for the limiting oxygen index (LOI), which is a measure of flame retardance and represents the percent concentration of oxygen needed for self-supporting combustion. PMIA-based fibers reach a LOI of ~ 30%. However, the aromatic structure is also responsible for UV absorption leading to color change and, at longer exposure, to a decrease of mechanical properties.

Fiber structure is also responsible for the good chemical resistance of these fibers. They are only soluble in very aggressive solvents (e.g., NMP, DMA, DMF, strong acids like $H_2SO_4$ or HF). The chemistry of the chain (amide groups) is responsible for susceptibility to hydrolytic attack in the presence of acids or bases as well as to oxidation by strong oxidizing agents. Amide groups are also the cause of relatively high water absorption (5–7% depending on the humidity).

Table 2 summarizes fiber properties of some commercial PMIA fibers.

### 2.1.6. Uses

The unusual properties of these fibers open a very broad range of applications. This breadth of uses is predicated on understanding how these fibers work in final structures and on developing special ways to use them. To maximize the value of these fibers it was necessary to create a special field of research, i.e., application research.

***Protection.*** The best-known application of *m*-aramid fibers is in protective garments designed to protect workers from the hazards of flame and heat. These include industrial workers in the oil industry, electrical workers, and firemen. The

**Table 2**. Selected properties of commercial PMIA fibers [25, 26]

| Property | Nomex 430 | Teijinconex std |
|---|---|---|
| Density, g/m$^3$ | 1.38 | 1.38 |
| Equilibrium moist content, % | 5.2 | 5.2 |
| Tensile strength, GPa | | |
| at rt* | 0.59 | 0.61–0.68 |
| at 200°C | 0.39 | 0.41 |
| at 250°C | 0.32 | 0.35 |
| Tensile modulus, GPa | | |
| at rt* | 11.5 | 7.9–9.8 |
| at 200°C | 9.9 | 6.1 |
| at 250°C | 9.4 | 4.3 |
| Tensile elongation at rt, % | 31 | 35–45 |
| LOI, % | 28 | 29–32 |

*rt = room temperature

first two are designed to protect people in case of a rare event whereas the latter is designed to protect people who actually go into a flame environment. Low shrinkage of fabrics is critical in this application. The shrinkage forces are high enough to "break open" a garment when exposed to flame and destroy the protective barrier. While $m$-aramids exhibit low shrinkage under most usage conditions, to meet the requirement of fire-protection it is important to further reduce shrinkage at flame temperatures. This can be accomplished by blending the fibers with $p$-aramid fibers [27]. Conditions of use of these garments are obviously aggressive and thus, in addition to inherent flame retardency, good durability of the fibers is critical.

In addition to the functional requirements of protection, it is important to be able to produce garments in a broad range of colors. Some of the products available on the market can be "piece" dyed (dyeing of the textile fabric), others are based on producer dyed fibers (by including pigments in the spinning solution). The first one offers a broader range of colors while the second one has better light fastness. In this application most of the fibers are used as cut fiber (staple products), which are processed into yarns on equipment used for cotton or wool. Only some use continuous filament yarns, e.g., car racer or pilot suits.

*Industrial Applications.* In addition to protecting people, these fibers are used in the protection of structures. Aircraft seat covers block fire from spreading to more flammable cushions.

**Industrial Materials.** The combination of good chemical and thermal resistance opened a broad field of application in hot gas filtration. The durability of these fibers is an additional characteristic governing their selection for bag filters.

Very good fatigue properties allow the use of the continuous filament yarns as mechanical reinforcement of elastomeric hoses and belts in automotive applications.

A very unique application is in the preparation of aramid papers. In these structures short fibers are combined with an $m$-aramid binder (small high surface particles called "fibrids"). The papers are used as electrical insulation with a constant use temperature of 220°C. More detailed description of their properties and applications can be found in DuPont product literature [28]. The papers are also used in the production of honeycomb structures, which combine high stiffness and low mass with $m$-aramid flame retardant characteristics and thus are ideal for aerospace and aircraft applications [29].

## 2.2. Fibers from Heterocyclic Polymers

Fibers from heterocyclic polymers are based on different chemistries but share a common characteristic; in spite of a good and at times excellent level of some properties, their applications are limited primarily because of high cost or a very specific deficiency. The group includes polybenzimidazole fiber (PBI, produced by PBI Performance Products Inc.); polyamide-imide fiber Kermel (developed by Rhône-Poulenc and now owned by Kermel); polyimide fiber P84 (produced by Evonik Industries); Arimid, a polyimide fiber (produced by a Russian company that has been shut down); and polyoxadiazoles fibers (Oxalon and Arselon produced in Belarus and Podlon in China).

*Polybenzimidazole Fiber* (PBI, **4**) [25734-65-0] polymer is produced by reaction of diphenylisophthalate with 3,3′,4,4′ tetraaminobiphenyl. The details of its manufacturing process are given in [30]: Whereas the polymerization is carried out in solid state, the fiber is formed by dry spinning from a DMA solution. The final step of the process involves treatment with sulfuric acid to improve thermal dimensional stability. The fiber was developed in response to the Apollo 1 fire. Today it is used primarily for personal protection in the fire service industry. In comparison to other high temperature resistant fibers PBI is characterized by a high LOI of 41% and a high moisture regain of 15% (at 65% relative humidity) but relatively low strength (tenacity of 2.7 g/d) [31]. PBI is frequently used in blends with other fibers (most often with $m$-aramids or $p$-aramids) at less than 50% to both improve fabric strength and lower its cost.

**4**

***Polyamide-Imide Fiber.*** Kermel fiber is currently available in two basic grades: Kermel (**5**) and Kermel Tech (**6**) [32]. Although both are produced by reaction of diisocyanates and diamines and by a similar process (solution polymerization and wet spinning [33]), they differ in the amine used. Kermel is used mainly in protective apparel whereas Kermel Tech is used in industrial applications (hot gas filtration).

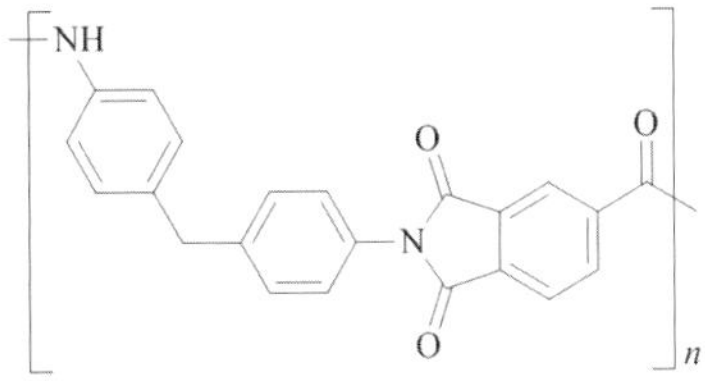

**5**

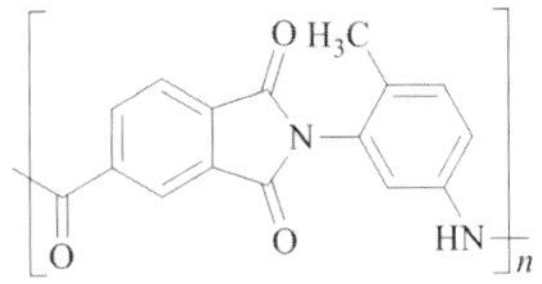

**6**

***Polyimide Fiber*** P84 (**7**) [65328-60-1], developed by Lenzing in Austria, is based on a polyimide copolymer (3,3′,4,4′benzophenone tetracarboxylic dianhydride with 20% of methylphenylenediamine and 80% toluenediamine) [34] and is shaped by dry spinning with DMA as a solvent [35]. Like most polyimides, it has very good high temperature resistance leading to its main application in hot gas filtration. The fiber has a trilobal cross-section which the producers claim is helpful in this application.

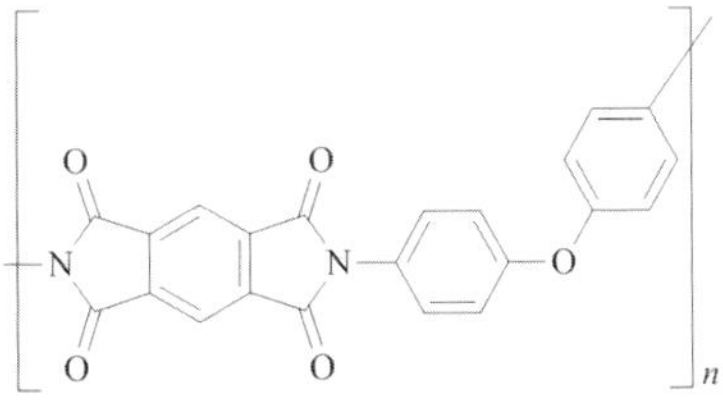

**7**

***Polyoxadiazoles Fibers*** (POD, **8**). The literature on POD is very rich, even though the volume of the product in the market place is small. The commercial (actually semi-commercial) products include Oxalon and Arselon from Belarus and Podlon from China. The polymer chemistry goes back to early work at DuPont and later Monsanto. A good summary of reactions used to produce these polymers is given in [36]. The most frequently used process on the commercial scale involves the reaction of hydrazine sulfate with diacids (most often aromatic) in oleum. Oleum acts as a solvent as well as dehydration agent. A critical fact is that the cyclization rate is much faster than the rate of polymerization [37]. Therefore, the yield of cyclization is very high (97–98%) [37]. POD fibers possess good thermal properties but UV stabilization needs to be provided [38]. The relatively low LOI (~ 21%) of unmodified fibers is a disadvantage. All commercial spinning processes use wet spinning for fiber production. The spinning process as well as fiber morphology and properties are described in [39].

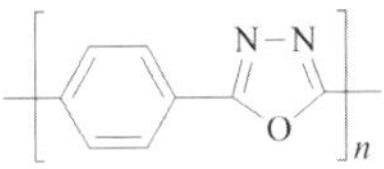

**8**

## 2.3. Fibers from Melt-Processable Polymers

The performance of melt-processable polymers is less desirable for protective apparel applications than that of aramids or heterocyclic fibers. Melting is not desirable because molten droplets are likely to add to injury severity and lead, in an extreme case, to the elimination of a protective barrier. Most of these fibers are also very difficult to dye which further decreases their utility in this application. This directs their use to industrial application, such as flame barriers and hot gas filtration.

The group includes fibers based on polymers like PEEK (**9**) [31694-16-3] and poly(phenylene sulfide) (PPS) [9016-75-5].

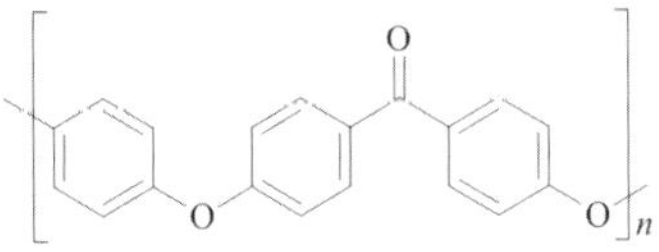

**9**

For both polymers, engineering resins applications are the main market. Use as industrial fibers constitutes only a very small fraction of their production. Vitrex produces PEEK polymer whereas Zyex produces the fiber. PEEK has good thermal properties although a $T_g$ of 143°C and $T_m$ of 334°C are significantly below those of *m*-aramid. High melting temperature and processing of high molecular mass polymer required for fiber applications result in filaments with higher dtex numbers (usually above 3) as thinner fibers are too difficult to produce. This further points toward industrial applications as high filament diameter decreases the comfort of the fabric (higher stiffness). Chemical composition and spinning process are responsible for a very low moisture regain and good chemical stability (oxidative stability may be the exception). The relationship between fiber spinning conditions, structure and properties of these fibers are described in [40]. A higher molecular mass allows for higher maximum draw ratio and thus better mechanical properties [41].

As for PEEK, fiber applications of PPS are only a very small fraction of its uses. Phillips [42] was the original developer of the polymer, which is now also produced by Ticona and Toray. Of these Toray produces also fibers and films. More recently a substantial manufacturing capacity of PPS (both polymer and fiber) has been created in China. In general the polymer is produced by reaction of dichlorobenzene with sodium disulfide in a solvent [43]. The fibers are formed by a melt spinning process (see → Fibers, 3. General Production Technology). The largest application of these fibers is hot gas filtration because of their very good chemical resistance to acids. Again, the weak point is its relatively low resistance to oxidation at high temperature.

## 2.4. Cross-linked Fibers

In the scope of this article it is not possible to describe all thermally stable fibers. The last group of fibers worth mentioning is based on cross-linked phenol-formaldehyde resin (Kynol) and melamine-formaldehyde resin (Basofil). Kynol (**10**) [9003-55-8] [44] is melt spun as novolak and cross-linked by reaction with formaldehyde.

Basofil [45] is spun via centrifugal spinning, which produces short fibers of variable diameter and variable length. Basofil is easy to dye while the deep color of Kynol makes dyeing difficult. Both are used in very limited (mostly in personal protection) volumes as they are relatively brittle and weak. As a result to be commercially usable they have to be processed in blends with other fibers.

## 3. High-Strength and High-Modulus Fibers

Realization that almost perfect orientation and full chain extension of the polymer are critical to approaching theoretical bond strength and stiffness in practical fibers goes as far back as the 1930s [46, 47]. Calculations of modulus based on bond force constants [48] and of strength using the weakest link theory [49] showed values significantly higher than those observed. For example, theoretical modulus of 240 GPa has been calculated for polyethylene whereas the measured modulus is in the range of 1–4 GPa. The issue faced by this field of science was whether one could achieve in a practical environment the level of orientation and chain extension required to obtain these properties.

Theoretical work on solution properties of rigid polymers for relatively low polymer concentration [50] and for a full range of concentrations [51] showed that above a critical concentration the solutions separate into two phases, i.e., anisotropic (locally organized) and isotropic (not organized). The materials where the change from isotropic to anisotropic phase occurs as a result of change of polymer concentration in solution are called lyotropic. In thermotropic materials this change occurs in a melt as a result of change of temperature. The phenomenon is based on packing considerations for the rigid particles. The effect of polymer concentration and the molecular mass of the polymer on the critical point of phase separation for lyotropic systems was calculated in [51]. In the domains of anisotropic phase the local orientation of chains is very good. This then makes it easier to obtain near perfect global orientation in the fiber.

The initial interest in aromatic polyamides was due to the higher bond strength and expected better thermostability. DuPont's success with work on *m*-aramids, culminating in commercial introduction of Nomex fiber, focused further work in this basic area of polymers. The chemistry indicated that movement from *meta*- to *para*-configuration would further improve fiber thermal properties. The science developed during work on *m*-aramids was fully applicable to *p*-aramids, but clearly additional advances were needed to deal with even less tractable polymers. The work on solution polymerization [12] and on polymer solutions [15] was critical for these more difficult to dissolve polymers. While the work began as normative extension of the *m*-aramid effort to improve thermal stability, it led to the unexpected result of producing a fiber now known as Kevlar, with very high strength and stiffness.

This was a clear breakthrough not only because of the utility of these properties, but also because it demonstrated that properties approaching theoretical ones can be achieved. It spawned research on other aramid and non-aramid chemistries yielding not only a family of products with very impressive properties but also a new field in fiber science. Chemistry developments moved towards super rigid molecules of polyazoles whereas processing work led to drawable systems of semirigid polymers. The final step was the development of gel spinning, which allowed transformation of flexible molecules into oriented and chain extended configurations.

## 3.1. *para*-Oriented Aromatic Polyamide Fibers

Scientific literature contains hundreds of compositions studied but only three have reached commercial reality and only poly(*p*-phenyleneterephthalamide), (PPTA, **2**) has reached a very large scale of operation with more than one producer.

PPTA fibers are based on spinning a lytropic solution while fibers based on polyamides from 2-(4-aminophenyl)-1*H*-benzimidazol-6-amine (DAPBI, **11**) and copoly(*p*-phenylene/3,4'-diphenyletherterephthalamide) (POP, **3**) are based on spinning isotropic solutions followed by fiber drawing to achieve desired orientation. The breakthrough began with KWOLEK spinning an organic solvent solution of poly(1,4-benzamide) which is more tractable than PPTA (compare nylon 6 with nylon 6,6) [52, 53]. A more complex monomer synthesis, the need to block the amine groups before polymerization (usually with HCl), and the need for an additional acid acceptor were the major reasons for higher costs. They became the driver for future transition to PPTA.

### 3.1.1. Fibers from Lytropic Solutions

There are several producers offering fibers based on PPTA: The original producer DuPont (Kevlar), Teijin (Twaron; developed by AKZO), and Kolon (Heracron). Several very small producers in China are at early development stages.

Typical polymerization and fiber preparation processes are shown in Figure 2 [54]. In all cases, the process is split between polymerization and fiber formation, because the polymer is not soluble in the polymerization mixture.

***Polymerization.*** The chemistry of the polymerization process of PPTA is the same as for PMIA, but it starts with the *para* monomers. The technological solutions for production of the polymer are significantly different because the polymer is much less tractable. It is not soluble in the reaction mixture. Hence, the best possible solvent is required to obtain high molecular mass polymer. Even with the best solvent, the polymer molecular mass is a result of competition between the rate of

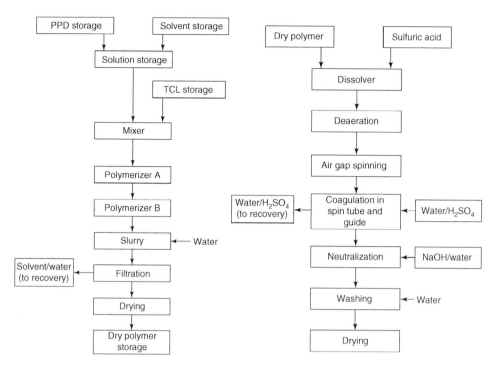

**Figure 2.** Polymerization (A) and spinning (B) of PPTA fibers [54].

polymerization and the rate of precipitation. Original work used hexamethylphosphoramide (HMPA) as a solvent, which was later abandoned due to toxicity concerns. Different solvent systems have been explored. Most of them are based on using some salts (e.g., $CaCl_2$, LiCl) to improve polymer solubility in amide solvents [15]. DMA was used with LiCl [55] and NMP with $CaCl_2$ [56]. Efforts to accelerate the polymerization reaction focused on the use of acid acceptors (beyond the basic solvents). Their use with DMA as solvent systems was described in [57].

A description of the polymerization process is provided in [58]. The reaction starts in a solution, but in a very short time, in the order of seconds, the materials convert into a friable gel that is broken into particles by mechanical action. This material exhibits flow behavior similar to Bingham type plastics. When the yield stress exceeds a critical value, the material begins to flow, but almost immediately solidifies when sheer stress is removed. Design of the reaction system, which can handle such a material, is not trivial. A twin-screw fully wiped extruder was proposed [58], which allow for continuous polymerization process whereas others described a batch reactor vessel [56, 59].

***Solution Preparation and Spinning.*** The anisotropic nature of PPTA and of poly(1,4-benzamide) solutions was a critical factor in obtaining unique properties of fibers based on these polymers. This is due to the chain rigidity leading to highly extended chain configuration. The conformation of PPTA and poly(1,4-benzamide) chains is nearly as extended as in a crystal with persistence length of 175 and 500 Å, respectively [60–62]. These values are about 25–50% of the chain length of molecular mass of 20 000.

Results of viscosity measurement as a function of concentration of the PPTA polymer with moderate molecular mass in sulfuric acid solution (Fig. 3) [63] are in agreement with those calculated by FLORY [51]. This fundamental characteristic of poly(1,4-benzamide) permitted KWOLEK to spin high strength and high stiffness fibers for the first time. The second breakthrough was BLADES' use of air-gap spinning. The principles of air-gap spinning are shown on Figure 4 [64]. The concentrated spinning

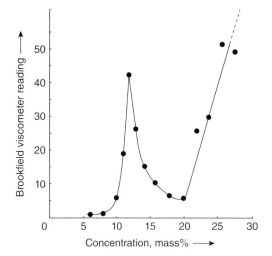

**Figure 3.** Viscosity versus concentration for PPTA polymers [63].

solution is extruded into a non-coagulating medium (usually air) and then exposed to a coagulating environment (usually water). This allows spinning at very high polymer concentration, which requires higher solution temperature and allows coagulation at a much lower temperature [64]. While wet spinning can be used, the air-gap allows for an orientation via

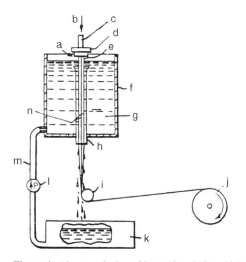

**Figure 4.** Air-gap spinning of lyotropic solutions [64] a) Spinneret; b) Spin dope; c) Transfer line; d) Spinning block; e) Air gap; f) Container; g) Coagulating liquid; h) Spin tube; i) Guide; j) Rotating bobbin; k) Container; l) Pump; m) Tube; n) Filaments.

elongational flow as well as permitting higher spinning speeds because of significantly lower drag forces (similar to a difference between dry and wet spinning).

The fibers exhibited high strength (2–4 times that of wet spinning) and stiffness immediately after spinning. Further improvement of fiber modulus can be achieved by heat treatment of the fiber at high tension and high temperature [65]. Modulus of 150 GPa for Kevlar 149 obtained under commercial condition represents 80% of the theoretical value. This is a testimonial of perfection of orientation achieved under large scale production conditions. The principles of air-gap spinning are shown on Figure 4 [64].

### 3.1.2. Fibers from Isotropic Solutions

The best-known fiber from isotropic solution is Technora (Teijin) [66]. It is a copolymer of 3,4'-diaminodiphenyl ether and $p$-phenylenediamine (reacted with terephthaloyl chloride). Polymerization is carried out in an amide solvent (likely NMP) with $CaCl_2$. The fact that it is a copolymer improves its solubility, and the polymer remains in solution after polymerization. It is spun from the neutralized polymerization mixture. The solution is isotropic. Both wet and air-gap spinning of the fiber is possible [66]. The polymer composition and the essentially noncrystalline character of the polymer, combined with good thermal stability of aromatic material, permit it to be drawn as much as 10 times at a high temperature of the order of 500°C. Under these conditions a very good orientation with very good properties is obtained.

Another example of isotropic spinning is based on polymers and copolymers of 2-(4-aminophenyl)-1$H$-benzimidazol-6-amine (**11**) and terephthaloyl chloride [67]. The technology was developed originally in Russia and several fibers have been described (SVM, Armos, Rusar etc.) but all are at very low volume.

### 3.1.3. Fiber Structure

Fibers based on PPTA are highly crystalline. In general, this structure is considered as essentially 100% crystalline, with some defects, thus

Table 3. Tensile properties of p-aramid fibers

| Property | Kevlar 29 [76] | Kevlar 49 [76] | Kevlar AP (K29) [77] | Twaron [78] | Twaron [78] HM | Technora [79] |
|---|---|---|---|---|---|---|
| Tenacity, GPa | 3.0 | 3.0 | 3.4 | 2.8 | 2.8 | 3.07 |
| Modulus, GPa | 70 | 112 | 79 | 80 | 125 | 70 |
| Elongation, % | 3.6 | 2.4 | - | 3.3 | 2.0 | 4.4 |

it is difficult to talk about glass transition for this polymer. Some attempts to measure $T_g$ on materials with relatively low crystallinity yielded values as different as 295°C and 492°C [68, 69]. The melting point of PPTA exceeds its decomposition temperature of ~ 550°C. Commercial fibers are very well oriented, e.g., Kevlar fiber shows orientation angle as measured by WAXD of 12° which decreases to 9° after heat treatment for Kevlar 49 [70]. It has been shown that Kevlar has an unusual radial orientation of hydrogen-bonded sheets (Fig. 5) [71]. The sheets exhibit regular, cooperative missorientation of $\pm 5°$, which is retained in Kevlar 49 but is absent in the product with the highest modulus (and thus the highest orientation), Kevlar 149 [72].

Aramid fibers spun from isotropic solutions, such as Technora and Armos, are copolymers and thus noncrystalline. The structure of Technora fiber (initially called HM-50) was studied [73–75]. X-ray results indicate a nematic array of copolymer chains that have completely random monomer sequences and highly extended conformations. The disagreement between predicted and observed peak profiles indicates a nonlinearity of the chain conformation. This is likely to be due to some distortion of the bond and torsion angles required to align the random sequences.

Comparison of mechanical properties of these fibers is given in Table 3 while a full discussion of properties and uses are described in Section 3.5 Properties and Uses of High Strength Fibers.

## 3.2. Polyazole Fibers

Continuing expansion of work on the anisotropic solution led to the synthesis of a group of rigid rod polymers, i.e., polyazoles.

These polymers can be synthesized to a very high molecular mass [80]. Polymers synthesized in polyphosphoric acid remain in solution. The very high rigidity of these polymers and their very high molecular mass result in extremely high viscosity solutions, necessitating air-gap spinning at very high temperature. While early work concentrated on poly(p-phenylene benzobisthiazole) (PBZT, **12**) [69794-31-6], progress in obtaining high molecular mass of poly(p-phenylene-2,6-benzobisoxazole) (PBO, **13**) [60871-72-9] and its better properties redirected research to this polymer. The work was initially carried out at Dow Chemical and later at Toyobo, which commercialized the fiber under the trademark Zylon. Following a somewhat similar fundamental polymerization path, using again air-gap spinning, M5 fiber (poly(2,6,-diimidazo[4,5-b:4',5'-e]pyridinylene-1,4,[2,5-dihydroxy]phenylene, **14**) was developed [81]. In 2006, it has been reported that Zylon fiber loses a significant fraction of its strength under atmospheric conditions at elevated temperature and high humidity. The mechanism of this unexpected phenomenon has been proposed [82].

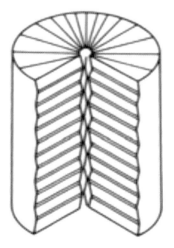

Figure 5. Radially oriented pleated sheets of Kevlar.

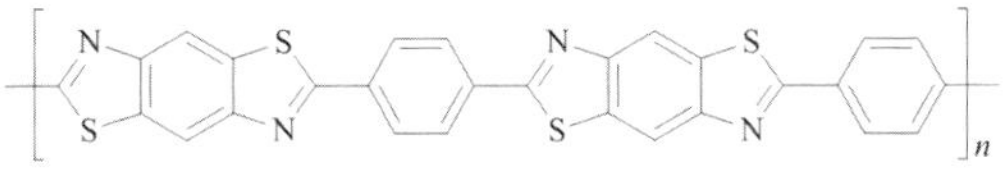

**12**

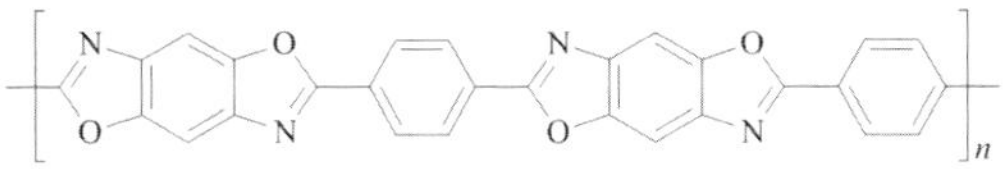

**13**

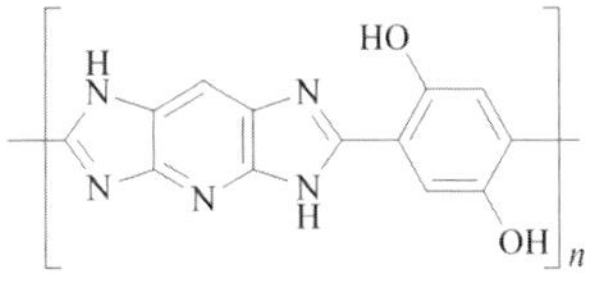

**14**

### 3.2.1. Fiber Structure

The PBO fiber is crystalline, with monoclinic unit cells [83]. The crystallinity of PBO improves with heat treatment, which is also accompanied by a significant improvement in fiber modulus [84]. The M5 fiber with its hydroxyl groups exhibits some unique characteristics. The as spun fiber forms a hydrate with a substantial amount of water being released upon heat treatment of the fiber. It has been proposed that during heat treatment the crystal hydrate transforms into a bidirectional hydrogen-bonded structure [85] (expected by some to improve the compressive properties of the fiber). Their mechanical properties are summarized in Table 4 while the properties and uses are discussed in Section 3.5 Properties and Uses of High Strength Fibers in more detail.

### 3.3. Fibers from Thermotropic Polymers

Development of liquid crystalline polyesters (LCP) followed the development of aramids.

In spite of the substantial amount of work, success in the fiber field was limited. Much more successful was the development of engineering resins for injection molding where excellent flow characteristics of LCPs offer special advantages. Initial work focused on compression moldable poly-$p$-benzoate [88]. Injection molding resins based on LCP were developed soon after [89]. The path from compression molding to injection molding was made possible by improvement of tractability by copolymerization.

To produce fibers further improvements in tractability were necessary. A very large number of publications both in patent and refereed journals evaluated hundreds of compositions [90, 91] but at this point only one found commercial realization (Vectran, **15**). The typical process involves melt spinning of a polymer of relatively low molecular mass. The fiber is oriented, but has low strength because of inadequate molecular mass. To increase the fiber strength, the fiber is subjected to heat treatment at temperatures very close to the melting temperature of the polymer [92] during which a solid phase polymerization increases the molecular mass of the polymer.

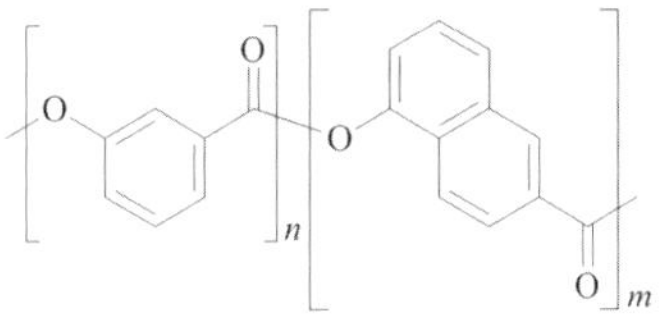

**15**

The typical properties of these fibers are given in Table 5.

### 3.4. High Strength Polyethylene Fibers

High-strength polyethylene (PE) fiber can be obtained via two very different technologies. The first one uses high-molecular-mass polyethylene (HMPE) that is extruded and then

**Table 4.** Tensile properties of polyazole fibers

| Property | Zylon Regular [86] | Zylon high mod. [86] | M5 (2001) [87] | M5 estimated [87] |
|---|---|---|---|---|
| Tenacity, GPa | 5.8 | 5.8 | 3.96 | 8.5 |
| Modulus, GPa | 180 | 270 | 271 | 300 |
| Elongation, % | 3.5 | 2.5 | 1.4 | 2.5 |

Table 5. Tensile properties of Vectran

| Property | Vectran HT [93] |
|---|---|
| Tenacity, GPa | 3.2 |
| Modulus, GPa | 75 |
| Elongation, % | 3.8 |

drawn in a solid state. The second one, commonly referred to as gel spinning, uses solution spinning of ultra-high-molecular-mass polyethylene (UHMWPE) and solid-state drawing. The second technology produces significantly stronger and stiffer fiber (about three times) but further developments could change that.

### 3.4.1. Gel Spinning

Polyethylene has a very low cross-sectional area of its chain and consequently the theoretical modulus and theoretical strength are highest among all organic high performance fibers. It is interesting that the first very high tenacity for polyethylene was reported even before work on Kevlar began. "Multifibrous strands" of polyethylene were developed by flash spinning which, when twisted, had a tenacity exceeding 2 GPa (23 gpd) [94]. Development of high strength polyethylene fiber was based on initial work on crystallization from solution and then spinning of the fiber [95, 96]. This work lead to a conclusion that fiber strength was almost linearly related to drawing stress, and the number of chain entanglements had a larger effect than the number of chain ends. The next step was the development of a gel spinning process which is solution spinning during which the fiber becomes a gel upon cooling [97].

Fundamentally, the technology is aimed at obtaining almost a perfectly oriented and fully extended chain configuration. To accomplish this with a flexible chain like polyethylene one needs to use a very high draw ratio of ~ 40× or more (semi rigid molecules like Technora are drawn up to 10 times). Thus, the morphology of as-spun fiber has to allow large extension without breaking. This is controlled by the level of entanglements in the fiber. The level of entanglement has to be small enough to allow such a degree of translational chain mobility and be large enough to maintain the integrity of the structure. This is accomplished by spinning solutions with a very low polymer concentration (few percent). In turn, for the solution to be spinnable at this low polymer content, the polymer has to be of a very high molecular mass (order of millions). Therefore the fiber is frequently referred to as based on UHMWPE (ultrahigh-molecular-mass polyethylene). This very high molecular mass also reduces structure defects by lowering the number of chain ends. The additional requirement is that the polymer has to be linear.

There are essentially two type of processes used today to gel spin UHMWPE. The first one involves dry spinning (not that different from the process described for *m*-aramids) of solution in a volatile solvent (e.g., decalin) [97]. In general, a very large fraction of the solvent is removed during that process and the fiber is quenched to below gelation temperature (most often with water). The fiber is then drawn at temperatures very close to its melting point. The draw is often accomplished in multiple stages.

In the second process, the polymer is dissolved in an essentially nonvolatile solvent (e.g., mineral oil) and is spun via a process very similar to air-gap spinning for lytropic systems [98]. The fiber is quenched below the gel point by cooling with a solvent, which is immiscible with the spinning solvent (e.g., water). The nonvolatile solvent is extracted from the solid fiber using a volatile solvent (e.g., hydrocarbon). The dry fiber is then drawn under conditions similar to the dry-spinning process. An example of an apparatus used for dry spinning and the multi stage drawing processes is given in [99]. The parameters affecting fiber drawability and properties obtained are discussed in [100]. Patent literature indicates that DSM and Toyobo produce Dyneema fiber via a process similar to that described in [100] while Honeywell uses a process to produce Spectra as described in [98]. There are number of small producers in China. Most of them use the two solvent system.

UHMWPE polymer (see → Polyethylene) is available on the open market (e.g., Ticona) and most fiber producers purchase it.

### 3.4.2. Extrusion Process

Understanding gained during the development of the gel spinning process for UHMWPE fibers redirected research on processing of UHMWPE via the extrusion process. Drawing of "virgin" polyethylene films produced by direct polymerization (no melting or dissolution) was described [101]. Aggregates of molecules in these films were disentangled directly during polymerization when they formed as a solid. This allowed them to be drawn as much 60 times with resulting strength of 3.5 GPa and modulus of 116 GPa, which are similar to results obtained via gel spinning (Table 4). This work clearly demonstrated that to attain very high draw ratios, the polymer needs to be shaped directly after polymerization without melting. Other than films, precursors to high temperature draw were also evaluated. Powder rolling was studied in [102] resulting in fiber properties of 1.3 and 81 GPa for tenacity and modulus, respectively. Solid-state coextrusion of mats of polymeric powder from solution crystallization was studied in [103]. The effect of the molecular mass and its distribution on maximum draw and the resulting properties were investigated, and tensile strength approaching 5 GPa and modulus of 220 GPa have been obtained. These are just some illustrations of the very large amount of work in this area. The work led so far to three commercial products: Tensylon (DuPont), Endumax (Teijin), Dyneema Ballistic Tape (DSM), and Tsunooga (Toyobo). All of them possess middle of the range properties.

### 3.4.3. Fiber Structure

A very high draw ratio used in the manufacturing of UHMWPE fibers is responsible for high orientation. What clearly distinguishes it from other polyethylene fibers (see → Fibers, 7. Polyolefin Fibers) is that they are made of essentially fully extended chains. A transition from shish-kebab to fibrillar structures during fiber drawing and development of chain extended conformation is described [104] and two melting points of two different polymorphs are observed [105, 106]. The higher melting peak at 423 K is associated with the orthorhombic crystal structure. The fibrils in fully drawn fibers were composed of chain-extended orthorhombic crystallites with an average length 70 nm. They were interrupted by defect regions containing trapped entanglements or chain ends.

**Table 6.** Tensile properties of UHMWPE fibers

| Property | Spectra 1000 [107] | Dyneema SK 71 [108] |
|---|---|---|
| Tenacity, GPa | 3.2 | 3.4 |
| Modulus, GPa | 113 | 119 |
| Elongation, % | 2.9 | 3.5 |

The defect regions were approximately 4 nm long. The mechanical properties of these fibers are compared in Table 6.

### 3.5. Properties and Uses of High Strength Fibers

While this group of fibers is classified as high strength and high stiffness, other properties (e.g., thermal stability, flame resistance) are also important in view of the applications in which these fibers are used. In some cases these additional characteristics become dominant, as for example in applications, which use staple fibers or pulp.

### 3.5.1. Fiber Properties

There is a broad range of commercially high-performance fibers currently available with excellent tensile properties (Table 3–6). The density of the different fibers varies from 0.97 g/cm$^3$ for Spectra and Dyneema to ~1.4 g/cm$^3$ for Vectran, 1.44 g/cm$^3$ for Kevlar and 1.55 g/cm$^3$ for Zylon. All of them are lower than 1.8 g/cm$^3$ for carbon fiber or 2.5 g/cm$^3$ for S-glass. However, these excellent tensile properties are not matched by compressive properties. PPTA-based fibers exhibit a yield stress at 0.5% [109]. At the yield point a significant change in the morphology of the fiber is observed. Kink bands at 45° angle to fiber direction are formed. But even at the compressive strain rate of 3% the loss of tensile properties is very small (~ 10%) [110]. The fact that compressive properties are significantly lower than tensile properties is connected to the highly anisotropic character of these materials (excellent longitudinal properties with relatively weak lateral properties). The coil test was used to determine the compressive strength of several fibers [111]. Whereas Kevlar was measured at 25.8 N/tex;

PBO fiber had strength of 13.3 N/tex and high strength polyethylene 7.5 N/tex. This sequence corresponds to the decrease of lateral properties in these fibers. Hydrogen bonding in aramids is responsible for the better lateral properties of these fibers. Consequently one would expect compressive properties for M5 fiber to be better than other polyazole fibers [112]. Fatigue measurements for these fibers show a similar picture: Tension–tension fatigue is significantly better than the compressive fatigue. Significant differences exist between these fibers as far as creep is concerned. UHMWPE fibers exhibit the highest creep. The lifetime of PPTA and UHMWPE fibers was calculated using a kinetic model of failure [113]. The results show that at a load of 50% of their breaking strength UHMWPE fiber fail after 2.5 min whereas Kevlar supports the load for 100 years.

Additional reduction of creep for PPTA fiber is achieved by improvement of its structure after heat treatment e.g., Kevlar 49 has a creep about half of that of Kevlar 29. Technora exhibits higher creep than fibers based on PPTA [114] whereas the creep of Zylon HM and Vectran is a little lower. This significant deficiency of UHMWPE fibers attracted a lot of work searching for improvement. One of the most frequently considered approaches was cross-linking using UV [115] and electron irradiation [116], but at this point none of them have entered commercial operation.

In addition to mechanical properties, thermal properties and flammability of these fibers control their applications. UHMWPE fiber is the worst in both areas. Its melting point is the lowest (below 150°C), and it is flammable. LOI of PPTA-based fibers is about 28%, a little lower for Technora and Vectran whereas Zylon and M5 with their highly aromatic structure have LOI above 65%. Among these high temperature performing fibers only Vectran melts. All other polymers decompose before melting. As far as thermal stability is concerned both Zylon and M5 are more thermally stable than aramids. All these fibers (except UHMWPE) retain their mechanical properties well at elevated temperatures e.g., at 300°C Kevlar still retains about 50% of its strength and 70% of its modulus when compared with room temperature data [76]. At the same temperature Zylon retains 60% of its strength and 80% of its modulus [86]. Excellent orientation and chain extension of these fibers result in a negative expansion coefficient for all of them, which is of value in some applications.

In addition to very good mechanical and thermal properties, high-performance fibers also have good chemical resistance. The chemistry of UHMWPE fiber dictates its very good chemical resistance to hydrolytic attack and many solvents (except under some conditions to hydrocarbons). The same properties are responsible for good UV stability (with use of UV stabilizers), but poor adhesion to resins [108]. Chemical properties of *p*-aramids are determined by the presence of the amide group, which is subject to hydrolytic attack. The rate of hydrolysis is influenced to some degree by the fiber morphology, with Technora having somewhat better stability than PPTA-based fibers and heat treated PPTA fibers better stability than standard varieties. Melt spinning of Vectran results in a very well consolidated morphology giving better chemical resistance than aramids. While having, in general, better chemical resistance than *p*-aramids, Zylon exhibits unusual degradation under humid and hot conditions. UV stability is reversed with PPTA-based fibers, which have better UV stability than other aromatic fibers. These properties are summarized in technical bulletins [76–79, 86, 87, 93, 107, 108].

### 3.5.2. Uses

When Kevlar was being developed the original expectation at DuPont was that it would be used mostly as tire cord. It seems that forecasting market developments is as difficult as forecasting breakthroughs in technology. In most cases the applications are driven by the unique combination of mechanical and thermal properties with low density. At this point, applications for high strength fibers are very diverse and development of new uses frequently requires different fiber forms beyond the continuous filament yarn envisioned in the original tire cord application. They are addressed in discussions of the specific uses. The review of applications is brief, but it should be remembered that these new materials frequently require the development of a very sophisticated understanding of the way they work in the final structures. Consequently, design of structures is optimized specifically for these materials. One critical difference versus traditional materials like steel and concrete is the anisotropy of the properties of high-performance fibers. A more complete (but still brief) review of applications can be found elsewhere [117, 118].

**Ballistic Protection.** This is the most broadly known application of these materials though not the largest one by volume. Fiber strength is obviously important for this application, but it is also very important to design structures to maximize the amount of material participating in stopping a projectile. The wave propagation during the event is influenced by modulus and, as a result, the amount of material involved in the event is 3–4 times greater for Kevlar than for nylon. The vests are multilayer structures, and the number of layers depends on the threat they are designed to protect from. Fabric design plays an important role, as the transverse deflection of the fabric is responsible for a large fraction of energy absorption. In addition to woven structures, there are soft composites in which fibers within a layer are arranged unidirectionally (UD structures), and the layers are stacked so that the fibers are not oriented in the same directions. UD structures minimize some negative effects of the anisotropy of the fibers (two orders of magnitude difference of properties in the fiber direction versus that of cross direction) and decrease the importance of fiber–fiber friction. Ballistic application include helmets, spall liners, etc., in which fibers are used in the form of hard composites. In these cases interaction with the resin has to be considered (good adhesion is not always good for good protection). Many of the fibers considered here are used in this application: PPTA-based *p*-aramids, PBO, and UHMWPE fibers.

**Composites.** High specific strength and stiffness offered by these fibers are important values for applications as composites. The range of composite materials is very broad, from hard composites with resins, such as epoxy to soft ones e.g., reinforcement of rubber. Aerospace and aircraft are the most important applications for hard composites as here the value of mass reduction is highest. While relatively poor compressive properties limit applications in structural materials, the good impact damage tolerance opens some uses as hybrids with carbon fibers.

Rubber reinforcement includes high performance automotive tires but also tires for aircrafts, heavy duty machinery, bicycles as well as in a variety of transmission belts and hoses. Increased temperatures under the hood of automobiles increase value of these products. Tires represent a good example where anisotropic properties of the fiber affected tire design. The design has to avoid putting these fibers under compression during operation. Aramids are the only materials used in this application because processing and use temperatures are too high for polyethylene and cost is too high for other fibers.

A unique application is in honeycomb structures (see Section 2.1.6 Uses for *m*-aramid honeycombs). In this case *p*-aramid is used as a floc (short fiber) from which paper is produced for manufacturing the honeycomb composite. High stiffness, low mass and self-extinguishing characteristics drive applications in aircraft.

More information can be found in → Composite Materials.

**Ropes and Cables.** The high specific strength of these high-performance materials opens a substantial market in underwater applications. PPTA-based aramid fibers find use in anchoring oil rigs (especially in deep water exploration) or mooring lines for ships. Again, it is mostly *p*-aramids and LCP fibers that are used in this application as too high creep for UHMWPE is a limitation.

Reinforcement of optical cables is one of the best-known uses among the cable applications.

**Protective Clothing.** The use of *p*-aramid staple in blends with *m*-aramids to control fabric shrinkage and break open in flame is described in Section 2.1.6 Uses. Another area of high volume use of these fibers is cut protection. Gloves and chain saw chaps made from *p*-aramids and UHMWPE (if high temperature or flame protection is not needed) are used in various industries.

**Asbestos Replacement and Other Applications.** The unique morphology of *p*-aramid fibers based on PPTA makes it possible to produce short, fibrillated particles by mechanical treatment of short fibers. The high aspect ratio of the fibers offers mechanical reinforcement. The very high surface area combined with good thermal properties allows them to be used in brake and clutch plates.

A long and growing list of smaller applications includes sport and recreation equipment: Boat sails, parachute fabrics, golf clubs, tennis rackets, etc.

## Abbreviations used in this article:

| | |
|---|---|
| DAPBI | 2-(4-aminophenyl)-1*H*-benzimidazol-6-amine |
| DMA | dimethylacetamide |
| HMPA | hexamethylphosphoramide |
| LCP | liquid crystalline polyester |
| LOI | limiting oxygen index |
| NMP | *N*-methyl-2-pyrrolidone |
| PBI | polybenzimidazole |
| PBO | poly(*p*-phenylene-2,6-benzobisoxazole) |
| PBZT | poly(*p*-phenylene benzobisthiazole) |
| PE | polyethylene |
| PEEK | polyetherether ketone |
| PI | polyimide |
| PMIA | poly(*m*-phenyleneisophthalamide) |
| POD | polyoxadiazole |
| POP | copoly(*p*-phenylene/3,4-diphenyl ether terephthalamide) |
| PPS | poly(phenylene sulfide) |
| PPTA | poly(*p*-phenyleneterephthalamide) |
| UD | unidirectional |

## References

1. H. H. Yang (ed.): *Aromatic High-Strength Fibers*, J. Wiley & Sons, New York 1989.
2. N. Grassie (ed.): *The Chemistry of High Polymer Degradation Processes*, Butterworths, London 1955.
3. A. H. Frazer (ed.): *High Temperature Resistant Polymers*, Wiley-Interscience; New York, 1968.
4. E. I. du Pont de Nemours, BE 569760, 1958.
5. E. I. du Pont de Nemours, US 3006899, 1961 (H. W. Hill, S. L. Kwolek, P. W. Morgan).
6. E. I. du Pont de Nemours, US 3063966, 1962 (S. L. Kwolek, P. W. Morgan, W. R. Sorenson).
7. V. Gabara, J. D. Hartzler, K. S. Lee, D. J. Rodini, H. H. Yang in *Handbook of Fiber Chemistry*, 3rd ed., CRC Press, Boca Raton, FL 2007, p. 975.
8. R. Takatsuka, K. Uno, Y. Iwakura, *J. Polym. Zi.* **15** (1977) 1905.
9. P. W. Morgan (ed.): *Condensation Polymers: By Interfacial and Solution Methods*, Wiley-Interscience, New York 1965.
10. E. I. du Pont de Nemours, US 3287324, 1966 (W. Sweeny).
11. P. W. Morgan, S. L. Kwolek, *J. Polym. Sci A* **2** (1964) no. 1, 181–208.
12. E. I. du Pont de Nemours, US3063966, 1962 (S. L. Kwolek, P. W. Morgan).
13. H. Kilkson, *Ind. Eng. Chem. Fundam.* **3** (1964) no. 4, 281.
14. V. Gabara, H. Kilkson: "Aramids Polymerization Kinetics" *Proceedings from AIChE Annual Meeting '03*, San Francisco, CA, 2003.
15. E. I. du Pont de Nemours, US 3068188, 1962 (L. F. Beste, C. W. Stephens).
16. E. L. Wittbeckerand, P. W. Morgan, *J. Polym. Sci.* **40** (1959) 289.
17. P. W. Morgan, S. L. Kwolek, *J. Polym. Sci.* **49** (1959) 299.
18. Teijin, JP47010863, 1972 (S. Ozawa, H. Fujie).
19. Teijin, JP49047276, 1974 (S. Ozawa. A. Aoki, H. Fujie).
20. Teijin, JP49061287, 1974 (H. Fujie, T. Noma, S. Ozawa).
21. E. I. du Pont de Nemour, US5667743, 1997 (T. M. Tai, D. J. Rodini, J. C. Masson, R. I. Leonard).
22. E. I. du Pont de Nemour, US3360598, 1967 (C. R. Earnhart).
23. K. Kouzai, K. Matsuda, Y. Tabe, H. Hond, K. Mori, *Sen-I Gakkaishi* **48** (1992) 55.
24. K. Tashiro, M. Kobayashi, H. Tadokoro, H., *Macromolecules* **10** (1977) no. 2, 413.
25. Technical Guide for Nomex® Brand Fiber, H-52703, E.I. du Pont de Nemours & Co., Inc. Wilmington Del., Apr. 1999.
26. Teijinconex® - Technical Information, CN02/91.2, Teijin Ltd., Japan.
27. E.I. du Pont de Nemours, US4198494, 1980 (W. G. Burckel).
28. DuPont Technical Bulletin H0949 (2000).
29. R. Pinzelli, H. Loken, *JEC Composites* **8** (2004) 133.
30. A. B. Conciatori, B. Alan, D. E. Stuets: "Polybenzoimidazole Fibers" in J. Preston (ed.): *High Technology Fibers, Part A*, Marcel Dekker, New York 1985, Chap. 6, pp. 221–267.
31. PBI Performance Products, http://www.pbiproducts.com (accessed 18 March 2014).
32. Technical Brochures of Kermel® and Kermel Tech® by Kermel.
33. J. Chambion, *Chim. Ind. Genie Chim.* **106** (1973) 453.
34. M. S. Fragoso Da Silva: "*Polyimide and Polyetherimide Organic Solvent Nanofiltration Membranes*", Dissertation at Department of Science and Technology in New University of Lisbon (2007).
35. Chemiefaser Lenzing, EP119185 A2, 1984 (T. Jeszenszky, H. Schmidt, S. Bauman, J. Kalleitner).
36. D. Gomes, S. P. Nunes, J. C. Pinto, C. Borges, *Polymer* **44** (2003) 3633.
37. N. P. Volkhina, L. M. Levites, *Vysokomolekularnye Soedineniya, Seryia B* (1979) 116.
38. K. L. Perepelkin, R. A. Makaraova, *Chem. Fibers Int.* (2006) 224.
39. Z. Zhang, G. Ye, W. Li, T. Li, J. Xu *J. App. Polym. Sci.* **114** (2009) 1485.
40. J. Shimizu, T. Kikutani, Y. Ookoshi, A. Takaku, *Sen'i Gakkaishi* **43** (1987) 507.
41. Y. Ohkoshi, A. Konda, H. Ohshima, K. Toriumi, Y. Shimizu, M. Nagura, *Sen'i Gakkaishi* **49** (1993) no. 4, 151–156.
42. Phillips Petroleum, US3354129, 1967 (J. T. Ednonds, H. W. Hill).

43 Dainippon Ink and Chemicals, Inc., Japan JP62252430 A, (1987) (T. Mine, S. Hasegawa, T. Sugie).
44 Nippom Kynol, US3972959, 1976, (H. Koyama).
45 BASF, DE4315609 A1, 1994 (H. Voelker, H. D. Zettler, W. Fath, H. Berbner).
46 H. Mark (ed.): *"Physik und Chemie Der Cellulose"* Julius Springer, Berlin 1932.
47 H. Mark, *Trans. Faraday Soc.* **32** (1936) 143.
48 L. R. G. Treloar, *Polymer* **1** (1960) 95–103.
49 I. Sakurada, Y. Nukushina, T. Ito, *J. Poly. Sci.* **57** (1962) 651.
50 L. Onsager, *Ann. N. Y. Acad. Sci.* **51** (1949) 627.
51 P. J. Flory, *Proc. Royal Soc. London* **73** (1956) no. A234.
52 E.I. du Pont de Nemours, US3951914, 1976 (S. L Kwolek).
53 S. L. Kwolek, P. W. Morgan, J. R. Schaefgen, L. W. Gulrich, *Macromolecules* **10** (1977) 1390.
54 J. B. Gallini, *Kirk Othmer Encyclopedia of Chemical Technology*, 5th ed., vol. 19 John Wiley & Sons, Hoboken, NJ 2006, pp. 713–738.
55 V. M. Savinov, L. B. Sokolov, SU402289 A1, 1987.
56 Akzo Nobel, US43084521, 1981 (T. J. Veerman, L. Volbracht).
57 Y. B. Rotenberg, T. I. Tarasova, V. M. Savinov, *Khimicheskie Volokna* (1986) no. 2, 26–27.
58 E.I. du Pont de Nemours, US3850888, 1974, (J. A. Fitzgerald, K. K. Likhyani).
59 Akzo Nobel, WO9521883 A1, 1995 (A. E. M. Bannenberg-Wiggers, J. A. Van Omme, J. M. Surquin).
60 M. Arpin, C. Strazielle, A. Skoulios, *J. Phys. (Paris)* **38** (1977) 307.
61 M. Arpin, C. Strazielle, *Polymer* **18** (1977) 591.
62 M. Arpin, C. Strazielle, G. Weill, H. Benoit, *Polymer* **18** (1977) 262.
63 E.I. du Pont de Nemours, US3673143, 1974 (T. I. Bair, P. W. Morgan).
64 E.I. du Pont de Nemours, US3767756, 1973 (H. H. Blades).
65 E.I. du Pont de Nemours, US3869429, 1975 (H. H. Blades).
66 Teijin, US4075172 A, 1978 (S. Ozawa, Y. Nakagawa, K. Matsuda, T. Nishihara, H. Yunoki).
67 National Research Council, *High Performance Structural Fibers for Advanced Polymer Matrix Composites*, The National Academies Press, Washington, DC 2005.
68 S. M. Aharoni, S. A. Curran, N. S. Murthy, *Macromolecules* **25** (1992) 4431.
69 A. S. Badaev, AI. I. Perepelkin, V. E. Sorokin, *Vyskomol. Soed, Ser A.* **24** (1986) 277.
70 D. Tanner, V. Gabara, J. Schaefgen, in M. Lewin (ed.). *Polymers for Advanced Technologies, IUPAC, International Symposium*, VCH Publishers, New York 1988.
71 M. G. Dobb, D. J. Johnson, B. P. Saville, *J. Polymer Sci. Phys. Ed.* **15** (1977) 2201.
72 S. R. Allen, E. J. Roche, *Polymer* **30** (1989) 996.
73 J. Blackwell, A. Biswas, H. M. Cheng, R. A. Cageao, *Mol. Cryst. Liq. Cryst. Pt. B* **155** (1988) 299.
74 R. A. Cageao, A. I. Schneider, A. Biswas, J. Blackwell, *Macromolecules* **23** (1990) 2843.
75 J. Blackwell, R. A. Cageao, A. Biswas, *Macromolecules* **20** (1987) 667.
76 Technical Guide – Kevlar® Aramid Fiber, H-77848 4/00.
77 Technical Guide – DuPont™ Kevlar® AP, K-24368 4/11.
78 Twaron® Technical Information 1987.
79 Technora® Technical Information 1987.
80 SRI International, US4533692, 1985; US4533693, 1985 (J. F. Wolfe, P. D. Sybert, J. R. Sybert).
81 Akzo Nobel, US5674969, 1997 (V. L. Lishinsky, D. J. Sikkema).
82 P. J. Walsh, X. Hu, P. Cunniff, P. A. J. Lesser, *J. Appl. Polym. Sci.* **102** (2006) 3517.
83 A. V. Fratini, P. G. Lenhert, T. J. Resch, W. W. Adams, *Mater. Res. Soc. Symp. Proc. (Mater. Sci. Eng. Rigid-Rod Polym.)* **134** (1989) 431.
84 R. J. Young, R. J. Day, M. Zakikhani, *J. Mater Sci.* **25** (1990) no. 1A, 127.
85 M. Lammers, E. A. Klop, M. G. Northolt, D. J. Sikkema, *Polymer* **39** (1998) no. 24, 5999.
86 Zylon® Fiber Technical Information, Toyobo (2001).
87 P. Cunniff, M. Auerbach, E. Vetter, D. Sikkema: "High Performance M5 Fiber for Ballistics/Structural Composites" *23$^{rd}$ Army Science Conference*, Orlando, Florida, AO-04, (2004).
88 J. Economy, B. E. Nowak, S. G. Cottis, *Polym. Prepr. (Am. Chem. Soc., Div. Polym. Chem.* **11** (1970) no. 1, 332–333.
89 J. W. Jackson, H. F. Kuhfuss, *J. Polym. Sci.* **14** (1976) 2043.
90 E.I. du Pont de Nemours, US4118372, 1978 (J. R. Schaefgen).
91 E.I. du Pont de Nemours, US4500699, 1984 (R. S. Irwin, F. M. Logullo).
92 E.I. du Pont de Nemours, US4183895, 1980 (R. R. Luise).
93 Vectran® Technical Information, Kuraray (2014).
94 E.I. du Pont de Nemours, US3081519, 1963 (H. Blades, J. R. White).
95 A. J. Pennings, K. E. Melhulzen, *Ultra-High Modulus Polym., Lect. Semin.* **3** (1979) 117.
96 B. Kalb, A. J. Pennings, *Polym. Bull. (Berlin, Ger.)* **1** (1979) no. 12, 871.
97 Stamicarbon, US4344908, 1982; US4422993, 1983; US4436689, 1984; US4430383, 1984 (P. Smith, P. J. Lemstra).
98 Allied Corporation, US4413110, 1983 (S. Kavesh, D. C. Prevorsek).
99 Y. Sun, et al., *J. Appl. Polym. Sci.* **98** (2005) no. 1, 474.
100 P. Smith, P. J. Lemstra, *Colloid. Polym. Sci.* **258** (1980) no. 7, 891; *Polymer* **21** (1980) no. 11, 1341.
101 P. Smith, H. D. Chanzy, B. P. Rotzinger, *Polym. Commun.* **26** (1985) 258.
102 L. H. Wang, R. S. Porter, *J. Appl. Poly. Sci.* **43** (1991) 1559.
103 D. Sawai, K. Nagai, M. Kubota, T. Ohama, T. Kanamoto, *J. Polym. Sci. Part B: Polym. Phys.* **44** (2006) 153.
104 W. Hoogsteen, G. Ten Brinke, A. J. Pennings, *Colloid Polym. Sci.* **266** (1988) 1003.
105 Y. K. Kwon, B. Wunderlich, *Proc. NATAS Annu. Conf. Therm. Anal. Appl.* **27** (1999).
106 J. Smook, J. Pennings, *Colloid Polym. Sci.* **262** (1984) 712.
107 Honeywell Spectra® Fiber Technical Information (2013).
108 Dyneema® Fiber Technical Information, Nippon Dyneema Co., Ltd (2008).
109 D. Tanner, in M. Lewin, J. Preston (eds.): *High Technology Fibers, Part B*, Marcel Dekker, New York 1989 chap. 2.
110 S. J. Deteresa, R. J. Farris, R. S. Porter *Polym. Compos.* **3** (1982) 57.
111 S. R. Allen, *J. Mater. Sci.* **22** (1987) 857.
112 M. Lammers, E. A. Klop, M. G. Northolt, D. J. Sikkema, *Polymer* **39** (1998) 5999.
113 Y. Termonia, P. Smith, *Polymer* **27** (1986) 1845.
114 I. P. Giannopoulos, C. J. Burgoyne, *J. Applied Polym. Sci.* **126** (2012) 91.

115. M. Jacobs, N. Heijnen, C. Bastiaansen, P. Lemstra, *Macromol. Mater. Eng.* **283** (2000) 120.
116. R. Hikmet, P. J. Lemstra, A. Keller, *Colloid Polym. Sci.* **265** (1987) 185.
117. D. C. Pervorsek in H. Bordy (ed.): *Synthetic Fibre Materials*, Longman Scientific & Technical Harlow, UK 1994, chap. 10.
118. V. Gabara, in H. Bordy (ed.): *Synthetic Fibre Materials*, Longman Scientific & Technical, Harlow, UK 1994, chap. 9.

# Further Reading

H. Mera, T. Takata: "High-Performance Fibers", in *Ullmanns Encyclopedia of Industrial Chemistry*, 7th edn., Wiley-VCH Weinheim, 2011, p. 573–597

# Foamed Plastics

HEINZ WEBER, BASF Aktiengesellschaft, Ludwigshafen, Federal Republic of Germany

ISIDOOR DE GRAVE, BASF Aktiengesellschaft, Ludwigshafen, Federal Republic of Germany

ECKHARD RÖHRL, BASF Aktiengesellschaft, Ludwigshafen, Federal Republic of Germany

| | | |
|---|---|---|
| 1. | Introduction | 1563 |
| 2. | Production | 1564 |
| 2.1. | Principles | 1564 |
| 2.2. | Blowing Agents | 1564 |
| 2.3. | Additives | 1565 |
| 3. | Properties | 1565 |
| 4. | Specific Foamed Plastics | 1568 |
| 4.1. | Polyolefin (PO) Foams | 1569 |
| 4.1.1. | Low-Density Polyolefin Foams (25 – 250 kg/m$^3$) | 1569 |
| 4.1.2. | High-Density Polyolefin Foams (250 – 700 kg/m$^3$) | 1572 |
| 4.2. | Polystyrene (PS) Foams | 1572 |
| 4.2.1. | Expandable Polystyrene (EPS) Foams | 1572 |
| 4.2.2. | Extruded Polystyrene (XPS) Foams | 1574 |
| 4.2.3. | Structural Polystyrene Foams | 1576 |
| 4.3. | Foams from Phenol–Formaldehyde (PF) Resins | 1576 |
| 4.3.1. | Rigid Poly(Vinyl Chloride) Foams | 1577 |
| 4.3.2. | Flexible Poly(Vinyl Chloride) Foams | 1578 |
| 4.4. | Foams from Urea–Formaldehyde (UF) Resins | 1578 |
| 4.5. | Foams from Melamine–Formaldehyde (MF) Resins | 1580 |
| 4.6. | Foams from Melamine–Formaldehyde (MF) Resins | 1581 |
| 4.7. | Silicone (SI) Foams | 1581 |
| 4.8. | Epoxy (EP) Foams | 1582 |
| 4.9. | Polyimide (PI) Foams | 1583 |
| 4.10. | Polymethacrylimide (PMI) Foams | 1584 |
| 4.11. | Poly(Methyl Methacrylate) (PMMA) Foams | 1585 |
| 4.12. | Polyamide (PA) Foams | 1585 |
| 4.13. | Poly(Phenylene Oxide) (PPE) Foams | 1585 |
| 4.14. | Polysulfone (PSU) Foams | 1586 |
| 4.15. | Polycarbonate (PC) Foams | 1586 |
| 5. | Testing | 1586 |
| 5.1. | Mechanical Properties | 1586 |
| 5.1.1. | Rigid Foams | 1586 |
| 5.1.2. | Flexible Foams | 1587 |
| 5.2. | Thermal Properties | 1588 |
| 5.3. | Miscellaneous Properties | 1589 |
| 6. | Economic Aspects | 1589 |
| 7. | Toxicology and Environmental Aspects | 1589 |
| | Abbreviations | 1590 |
| | References | 1591 |

## 1. Introduction

Foamed plastics are materials that have cells distributed throughout their entire mass. A cell is a small void formed during the manufacture of the foam; it is enclosed partly or completely by *cell walls* or *struts*. In contrast to an open cell, a closed cell is totally surrounded by its walls and, therefore, is not interconnected with other cells via the gas phase.

The apparent density of a foamed plastic depends on the fraction of the total volume occupied by the polymer phase that forms the skeleton; it is always lower than the density of the skeletal substance [12].

*Structural foams* consist of a light, closed-cell core that is sandwiched between solid integral skins. *Syntactic foamed plastics* are gas-filled polymers that are not, however, foamed. They contain completely closed cells that are formed by hollow spheres distributed in the polymer matrix (see Syntactic foams).

The constituents of the polymer skeleton can be produced by polymerization, polyaddition, or polycondensation.

**History.** The first foamed polymers were produced in the 1920s from rubber (Dunlop). These were followed in the 1930s by foamed plastics from urea resin (BASF), flexible poly (vinyl chloride) (I.G. Farben), and polyurethane

(Bayer). Polyethylene foams (Du Pont), extruded polystyrene foams (Dow), rigid poly(vinyl chloride) foams (Lonza), and the first phenolic resin foam (Dynamit Nobel) were developed during World War II. Foams from polystyrene beads (BASF), epoxy resins (Shell), and silicones (Dow Corning) were announced in the late 1940s and early 1950s. In the 1960s, polyamide (BASF), polycarbonate (General Electric), polysulfone (Pechiney-Saint-Gobain, ICI), polyimide (Du Pont), and polymethacrylimide (Röhm) foams were described. Poly(phenylene oxide) (General Electric), and melamine resin (BASF) foams have been known since the 1970s.

## 2. Production (→ Plastics, Processing 1. Processing of Thermoplastics)

Production processes for individual foams are described in Chapter 4. Only a brief overview is given here.

### 2.1. Principles

The three key starting materials for a foam are polymer, blowing agent, and additives.

In the manufacture of *thermoplastic foams*, the starting material is generally an existing polymer. Foams are produced in both continuous and batchwise processes which frequently involve an extrusion step. Thus, boards or films as well as profiled and shaped parts with various apparent densities can be produced (e.g., Section 4.2.2).

Foamed thermoplastic beads are fused with steam to form a variety of shaped parts, including blocks (Sections 4.1.1 and 4.2.1).

In the manufacture of *thermoset foams*, the macromolecules are cross-linked during the foaming process. The foams are produced in various ways. The pourable charge is distributed on a conveyor belt where the mixture foams and hardens continuously (Sections 4.4 and 4.6). The resulting paste can also be poured into block molds (Section 4.4) or other cavities where the reaction is completed (Sections 4.7 and 4.8). Whipping (Section 4.5) and spraying (Sections 4.4 and 4.8) processes may also be used.

### 2.2. Blowing Agents

In almost all foamed plastics, the volume fraction of the gas phase (a function of density) is far greater than that of the skeletal substance. Thus, for example, at an apparent density of 10 kg/m$^3$, the gas phase accounts for ca. 99 % of the volume. The gas phase is produced by using a blowing agent.

***Physical blowing agents*** may be compressed gases (air, nitrogen) or low-boiling hydrocarbon or chlorofluorocarbon (CFC) liquids, e.g., butane, pentane, trichlorofluoromethane (CFC 11). They do not leave undesirable decomposition products.

In the oldest method, originally developed for rubber, air is whipped mechanically into the latex [13]. The resulting froth is fixed by cross-linking. This process is commonly used for urea–formaldehyde resin foams (Section 4.5) and flexible poly(vinyl chloride) (PVC) foams (Section 4.3.2).

In an alternative method, compressed nitrogen is dissolved in a polymer melt. The cellular structure is then obtained by reducing the pressure. Examples are the batchwise production of cross-linked polyethylene foam (Section 4.1.1) and injection molding.

Incorporation of the blowing agent pentane [109-66-0] into polystyrene beads (Section 4.2.1) renders them heat-expandable.

Low-boiling chloro- and chlorofluorocarbons are incorporated into polyethylene (Section 4.1.1) or polystyrene melts (Section 4.2.2) in an extruder. Trichlorofluoromethane [75-69-4] (CFC 11) and 1,2-dichloro-11, 22-tetrafluoroethane [76-14-2] (CFC 114) are used as blowing agents for polyethylene. Methyl chloride [74-87-3], ethyl chloride [75-00-3], and trichlorofluoromethane [75-69-4] (CFC 11) are used for expandable polystyrene. The blowing agents evaporate and the polymer expands. Cooling increases the viscosity and sets the skeleton of the foam.

Thermoset foams (Sections 4.4 and 4.6) are also produced with low-boiling liquid blowing agents which are activated by the heat of the cross-linking reaction or by externally supplied energy.

The choice of blowing agent is based on a knowledge of physical and chemical properties [14].

***Chemical blowing agents*** are solids which decompose at elevated temperature with the formation of gas. For optimum action, they must be distributed very uniformly in the substrate. Their decomposition temperature must be suitable for the processing temperature of the polymer. Numerous organic, nitrogen-containing substances with decomposition temperatures of 100 – 250 °C [15], [16] or higher [17] are known (e.g., azo compounds, nitroso compounds, and hydrazides). The most frequently used compound is azodicarbonamide [123-77-3] which splits off nitrogen and carbon monoxide at about 200 °C. The decomposition temperature can be lowered by various additives such as zinc oxide [18].

The compounds, which contain polymers and blowing agents, are then subjected to extrusion processing.

If chemical blowing agents are used, thermoplastic foam injection molding (TFI) and thermoplastic foam extrusion (TFE) are usually employed to yield heavy, foamed plastics known as *structural* (*integral*) *foams* (Section 4.2.3). The injection mold (TFI) or the die (TFE) is cooled so that the foam collapses at the wall of the mold or die and forms a compact surface layer. The compacted surface has nearly the same density as the polymer matrix, whereas the internal layers are foamed.

Inorganic blowing agents, such as carbonates, are of minor importance. A mixture of sodium bicarbonate [144-55-8] and citric acid [77-22-9] serves as an additional, nucleating blowing agent in combination with pentane (Section 4.2.2). Sodium borohydride [16940-66-2], which splits off hydrogen in the presence of water and is adsorbed on fumed silica, has been reported to be an effective blowing agent [19].

Blowing agents can also be formed as reaction side products. For example, in certain polyurethane foams and in thermoset-modified PVC foams, carbon dioxide is formed from isocyanate and water (Section 4.3.1). Polyimide foam is blown by the alcohol (usually ethanol) formed in the imidization reaction (Section 4.9).

## 2.3. Additives

***Nucleating agents*** are important additives in foam manufacture, especially in physically blown, thermoplastic foams. They are finely dispersed, often silicate-like solids but they may also be waxlike. They act as boiling chips in the polymer melt which is supersaturated with dissolved blowing gas.

Minute, nucleating gas bubbles, called hot spots, are also formed during the initial decomposition of chemical blowing agents [20]. Gaseous side products of the reaction act in the same manner.

***Surfactants*** are required in thermosetting plastics to emulsify the liquid blowing agent. They reduce interfacial tension and thus facilitate nucleation. The surfactant also stabilizes the foam during expansion until the foam is set by the simultaneous hardening process.

***Other additives*** include dyes, inert or reinforcing fillers and fireproofing agents. Antioxidants and UV stabilizers are sometimes needed.

## 3. Properties

The properties of foamed plastics are determined by their polymer substrate, apparent density, and morphology.

All three parameters, but especially apparent density, affect mechanical behavior. Resistance to elevated temperature and to chemical agents, as well as fire behavior, are primarily determined by the polymer. Morphology is responsible for the acoustical and thermal insulating properties. Both morphology and polymer substrate determine the water balance (water absorption, diffusion of water vapor).

***Compressive Strength.*** One the basis of their compressive strengths determined in a compression test at 10 % deformation (DIN 53 421, ISO 844), foamed plastics are classified as *rigid* (> 0.08 MPa), *semirigid* (0.015 – 0.08 MPa) or *flexible* (< 0.015 MPa) [12]. The measured compressive strength depends primarily on the apparent density of the foamed plastic but also on the modulus of elasticity of the polymer [21].

Figure 1 provides an overview of the ranges of apparent density and compressive strength for individual foams. A linear relationship exists between the logarithms of the compressive strength and of the apparent density for all rigid plastic foams [15, p. 411], [22]. Polyethylene foams lie parallel to the scatter band; flexible plastic foams and some semirigid are outside it.

***Cell Structure.*** Rigid and semirigid plastic foams generally have *closed cells* which do not interconnect via the gas phase. They can, therefore, also contain gases other than air, e.g., fluorocarbons. Flexible foams, on the other hand, are usually *open-celled* and filled with

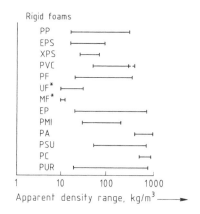

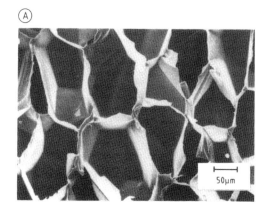

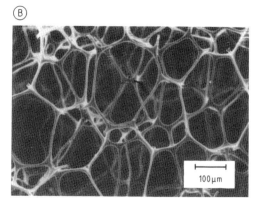

Figure 2. Scanning electron micrographs of (A) a closed-cell, extruded polystyrene foam and (B) an open-cell melamine–formaldehyde foam

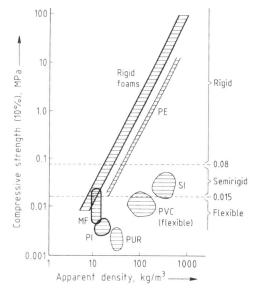

**Figure 1.** Compressive strength and apparent density of foamed plastics
*Also include flexible foams
Abbreviations are defined in the abbreviation list at the beginning of this article

air. Exceptions to this rule are, however, found. Figure 2 shows scanning electron micrographs of a closed-cell (A) and an open-cell (B) plastic foam.

The *cell size* of most plastic foams is 0.05–1 mm. Thermoset-modified PVC foams, and noncross-linked, extruded polyethylene and poly(methyl methacrylate) foams have cell sizes > 1 mm. Foams with smaller cells usually exhibit better mechanical behavior.

If the cells are oriented, e.g., stretched in the rising direction of the foam, the mechanical properties are no longer isotropic.

The thickness of the cell wall of closed-cell foams, the strut thickness of open-cell foams, and the cell size affect the radiant fraction of thermal conductivity. Each material has an optimum value [23], [24].

***Thermal Insulation.*** The thermal insulation of a foamed plastic depends primarily on the thermal conductivity of the filling gas, followed by the apparent density and then morphology. Foams filled with heavy chlorofluorocarbon gases, therefore, have better thermal insulation characteristics than air-filled foams. Figure 3 shows the density dependence of the thermal conductivities of several foams measured at 10 °C [25], [26].

Depending on prevailing conditions, open-cell products can absorb water and thus increase their thermal conductivity [27]. Since they are also open to diffusion, they must be provided with a barrier layer in some cases.

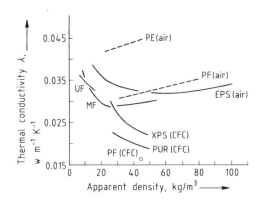

**Figure 3.** Thermal conductivity of foamed plastics as a function of apparent density
Values were measured at 10 °C according to DIN 52 612, ISO-DIS 8302, ASTM C 177. The gases used to fill the foams, air or a chlorofluorocarbon (CFC), are indicated in parentheses. Dashed lines indicate that values vary widely. EPS = expandable polystyrene; MF = melamine–formaldehyde resin; PE = polyethylene; PF = phenol–formaldehyde resin; PUR = polyurethane; UF = urea–formaldehyde resin; XPS = extruded polystyrene

*Sound Insulation.* Open-cell foams are well suited for airborne sound insulation. Closed-cell products are useful for impact sound insulation, provided they have sufficient resilience [28]. Rigid foams are less suitable for acoustic applications.

*Fire Behavior.* Polymer-dependent properties (i.e., maximum long-term service temperature and oxygen index) are shown in Figure 4 for the individual foams discussed in Chapter 4. Thermoplastics differ from thermosets: when exposed to a flame, thermoplastic foams sinter above their softening range and melt, whereas thermosets form a carbon skeleton.

The oxygen index provides some indication of the flammability of a polymer. It measures the minimum oxygen concentration necessary to support combustion of the material [29]. Thus, higher values are better. The oxygen index of a foamed plastic may often be only slightly higher than that of the compact polymer, exceptions are

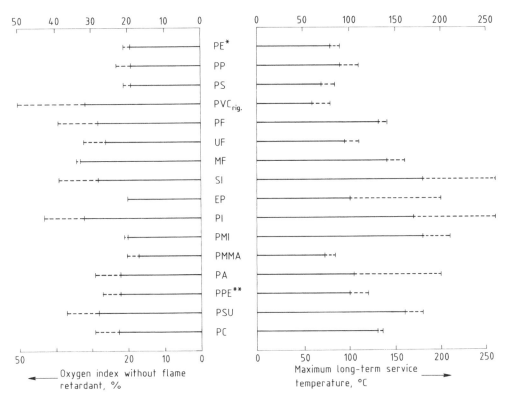

**Figure 4.** Oxygen index and maximum long-term service temperature of foamed plastics
The dashed lines indicate the range of variation (i.e., minimum and maximum values)
*Abbreviations are defined at the beginning of this article
**Ratio of PPE:PS = 1:1

urea–formaldehyde resins and poly(phenylene oxide).

Other important parameters for judging fire behavior are the smoke density and the toxicity of the combustion gases.

In compliance with the flammability requirements specified in guidelines for use in construction or transportation, some foams contain flame retardants. These are often halogen compounds, examples include chlorinated paraffins and hexabromocyclodecane for polyethylene and polystyrene, as well as boron trioxide for phenol-–formaldehyde resins (cf. Chap. 5).

## 4. Specific Foamed Plastics

Polyurethane (PUR) foams are treated elsewhere (→ Polyurethanes). For a description of foamed PVC coatings, see → Leather Imitates. Trade names of individual foams are listed in Table 1.

**Table 1.** Trade names of foamed and foamable plastics

| Trade name and raw material | Producer |
|---|---|
| **Polyolefin (PO)**[*] | |
| Alveolit (PE) | Alveo (Switzerland) |
| Apro (PP) | Arco Chemical (USA) |
| Arpak (PE) | Arco Chemical (USA) |
| Eperan (PE) | Kanegafuchi Chemical Industry (Japan) |
| Ethafoam (PE) | Dow Chemical (USA) |
| Evazote (EVM) | Bakelite Xylonite Ltd. (UK) |
| Foam ace (PE) | Furukawa (Japan) |
| Frelen (PE) | Freudenberg (FRG) |
| Neopolen (PE) | BASF (FRG) |
| Opcel (PE) | Sentinel (France) |
| Plastazote (PE) | Bakelite Xylonite Ltd. (UK) |
| Softlon (PE) | Sekisui (Japan) |
| To-re-pepu (PE) | Toyo Rayon (Japan) |
| Trocellen (PE) | Dynamit Nobel (FRG) |
| **Polystyrene (PS)** | |
| Afcolene | Société Chimique des Charbonnages (France) |
| Arcel | Arco Chemical (USA) |
| Dylark | Arco Chemical (USA) |
| Dytherm | Arco Chemical (USA) |
| Eslen | Sekisui (Japan) |
| Extir | Montedison (Italy) |
| Kane Pearl | Kanegafuchi Chemical Industry (Japan) |
| Rigipore | British Petroleum (UK) |
| Roofmate | Dow Chemical (USA) |
| Styro-lite | 3M (USA) |
| Styrocell | Shell Chemical (UK) |
| Styrodur | BASF (FRG) |
| Styrofoam | Dow Chemical (USA) |
| Styropor | BASF (FRG) |
| Vestypor | Hüls (FRG) |
| **Poly(vinyl chloride) (PVC)** | |
| Airex | Airex (Switzerland) |
| Conticell | Continental-Gummi (FRG) |
| Divinycell | Diab-Barracuda (Sweden) |
| Ensolite | Uniroyal (UK) |
| Foamcell | Pelf (Italy) |
| Klegecell | Kleber-Colombes (France) |
| Kömacell | Kömmerling (FRG) |
| Pliovic | Goodyear (USA) |
| PVC Foam | Shin-Etsu (Japan) |
| Rigicell | Goodrich (USA) |
| Solvic | Solvay (Belgium) |
| Vestolit | Hüls (FRG) |
| Vinoflex | BASF (FRG) |
| **Phenol-formaldehyde (PF) resins** | |
| Cellobond | British Petroleum (UK) |
| Dynapor | Dynamit Nobel (FRG) |
| Eurothane xtra | Recticel (Belgium) |
| Exeltherm xtra | Koppers (USA) |
| Fenomo | Isomo (Belgium) |
| Mosy | Compo (FRG) |
| Oasis | Smithers (USA) |
| Phenexpan | Société Chimique des Charbonnages (France) |
| Phenobil | Roland-Werke (FRG) |
| Phenocomp | Sekisui (Japan) |
| Phenolite | Bridgestone (Japan) |
| Pyrofoam | Union Carbide (USA) |
| Resiphen | Texaco (USA) |
| **Urea–formaldehyde (UF) resins** | |
| Aerolite | Ciba-Geigy (Switzerland) |
| Beetle UF Foam | British Industrial Plastics (UK) |
| Cascofoam | Borden Inc. (UK) |
| Hygromull | BASF (FRG) |
| Isoschaum | Schaumchemie (FRG) |
| Kobosan | Vorwerk (FRG) |
| **Melamine–formaldehyde (MF) resins** | |
| Basotect | BASF (FRG) |
| **Silicones (SI)** | |
| Cohrlastic | Connecticut Hard Rubber (USA) |
| Foamega | Bisco Prod. (USA) |
| Polyvoltac | Polyvoltac (UK) |
| Rau-Sik | Rehau (FRG) |
| **Epoxy resins (EP)** | |
| Araldit | Ciba-Geigy (Switzerland) |
| Scotchplay | 3M (USA) |
| **Polyimide (PI)** | |
| Chem-Lon | Chem-Tronics (USA) |
| LMB 1907 | Ciba-Geigy (Switzerland) |
| Polyimid 2080 | Upjohn (USA) |
| Skybond | Monsanto (USA) |
| Solimide | International Harvester (USA) |
| **Polymethacrylimide (PMI)** | |
| Rohacell | Röhm (FRG) |
| **Poly(methyl methacrylate) (PMMA)** | |
| Isocryl | IMC Acrylguß (FRG) |
| **Polyamide (PA)** | |
| Maranyl | ICI (UK) |
| Rilsan | Société Nationale des Pétroles d'Aquitaine (France) |
| **Poly(phenylene oxide) (PPE)** | |
| Noryl | General Electric (USA) |
| Prevex | Borg-Warner (USA) |
| **Polycarbonate (PC)** | |
| Lexan | General Electric (USA) |
| Makrolon | Bayer (FRG) |
| Merlon | Mobay Chemical (USA) |

[*] PE = polyethylene; PP = polypropylene; EVM = ethylene–vinyl acetate copolymer.

## 4.1. Polyolefin (PO) Foams

Polyolefin (PO) foams are manufactured by many producers in a variety of processes and are classified as low density (25 – 250 kg/m$^3$) or high density (250 – 700 kg/m$^3$). The market has expanded relative to other foamed plastics, especially in Japan.

Most polyolefin foams are made from low-density polyethylene [9002-88-43] (LDPE). Copolymers such as the more flexible ethylene–vinyl acetate [24937-78-8] (EVM) and, more rarely, high-density polyethylene [9002-88-4] (HDPE) are also found. Finally, foams may be produced from linear low-density polyethylene (LLDPE), a copolymer of ethylene and other olefins (1-butene to 1-octene) which is synthesized with a Ziegler catalyst. Polypropylene [9003-07-0] (PP), which is more resistant to heat deformation, stiffer, and tougher than polyethylene, was first processed into foamed sheets in the 1960s. Today, a polypropylene copolymer with a few percent ethylene is also utilized. For a detailed description of these polyolefins, see → Polyethylene, → Polypropylene. Trade names of polyolefin foams are listed in Table 1.

*Fire Behavior.* Like their starting materials, polyolefin foams are flammable. The smoke density is low. The combustion gases contain only carbon dioxide and, in case of oxygen deficiency, carbon monoxide.

### 4.1.1. Low-Density Polyolefin Foams (25 – 250 kg/m$^3$)

Boards, sheets, or profiled shapes are manufactured by continuous processes; boards are also produced by batch processes. In addition, beads can be made continuously or batchwise.

***Continuous Process for Non-Cross-Linked Foam.*** Non-cross-linked LDPE foam is produced by applying the technique used for polystyrene (Styrofoam, see Section 4.2.2) which involves the gassing of a polymer melt with low-boiling halogenated hydrocarbons in an extruder. The blowing agent evaporates as it emerges from the nozzle; the melt then cools and its viscosity increases [30]. A strand of closed-cell foam (cell size 0.5 – 1.5 mm) is formed, with a thickness of 80 mm and an apparent density of 30 kg/m$^3$. Thin sheets of foam up to 5 mm thick are produced in the same manner or via the film blowing process (→ Films, Section 2.2.2.0.0.2.,). An example of such a film is Ethafoam produced by Dow Chemical.

***Cross-Linked Foams.*** As shown in Figure 5, LDPE melts in a very narrow range [31]. Cross-linked LDPE exhibits a less precipitous drop in viscosity with increasing temperature. Thus, a broader temperature range for processing is obtained; in addition, the foamed plastic has smaller cells.

Cross-linking is accomplished either chemically (e.g., with dicumyl peroxide [80-43-3]) or with beams of electrons at doses up to 20 Mrad.

The gel content (degree of cross-linking) should be ca. 30 – 60 % [32]. Occasionally, divinylbenzene [1321-74-0] [33] or oximes [34] are added to the peroxide. Polyolefins grafted with silanes can be cross-linked catalytically in contact with water [35], [36]. For polypropylene, cross-linking agents with higher decomposition temperatures are used, e.g., polysulfone azides [37].

***Batch Process for Molded, Cross-Linked Foam.*** In the in-mold, molded foam (batch) method, preformed polyethylene boards containing peroxide cross-linking agents are placed in a mold and exposed to nitrogen at ca. 38 MPa and 180 °C. After cooling and careful pressure reduction, the nitrogen remains dissolved in the melt. When heated to 100 – 130 °C, the partially cross-linked board foams to yield a foamed plastic. Its cells are very fine (0.1 mm), and

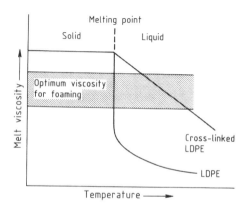

**Figure 5.** Melt viscosity of low-density polyethylene (LDPE) as a function of temperature

its density is 24 – 70 kg/m$^3$ [38], [39]. This process is used by Bakelite Xylonite Ltd. to manufacture Plastazote.

Instead of nitrogen, chemical blowing agents such as azodicarbonamide can also be used. The gel is removed from the mold rapidly while it is still hot and somewhat larger cells result.

*Continuous Process for Cross-Linked Sheets.* For the continuous manufacture of foam sheeting cross-linked with peroxides, the polyolefin (e.g., LDPE) is premixed carefully with up to 10 parts by weight of blowing agent (azodicarbonamide) and < 1 part by weight of cross-linking agent (dicumyl peroxide). The mixture is extruded to form a thin matrix which is then cross-linked by passing it through a heating tunnel at 180 – 230 °C. The sheet is finally foamed to a thickness of 4 – 20 mm and an apparent density of 25 – 250 kg/m$^3$. This process was developed by Furukawa and licensed to other companies (Bayer, Dynamit Nobel, Freudenberg) [32], [40], [41].

In a similar process, the matrix containing the blowing agent is cross-linked by irradiation with electrons. Subsequently, it is foamed in a shaft by exposure to IR heaters and hot air at 200 – 250 °C (Fig. 6). Free expansion can also be carried out in a hot fluid bath. The depth of penetration of the electron beam limits the maximum thickness of the foamed sheet to 5 mm. Apparent densities are 30 – 200 kg/m$^3$. This process was developed by Sekisui and Alveo, it is used to produce Alveolit [42], [42], [43].

*Continuous Process Using Cross-Linked Beads.* Strands of foam, produced by extruding LDPE with added butane, are comminuted to spherical beads. After the polyethylene has been cross-linked by an electron beam, the beads are fused continuously with hot air to form foam boards with thicknesses up to 120 mm and apparent densities in the range 28 – 45 kg/m$^3$. Cell size is about 0.5 mm. This process is used to manufacture Neopolen (BASF) [44], [45].

*Batch Processes Using Beads.* Pellets of weakly cross-linked LDPE, propylene–ethylene copolymers, or LLDPE in an aqueous dispersion are impregnated with low-boiling (halogenated) hydrocarbons as blowing agents. Impregnation is performed at a temperature near the softening range in the presence of a

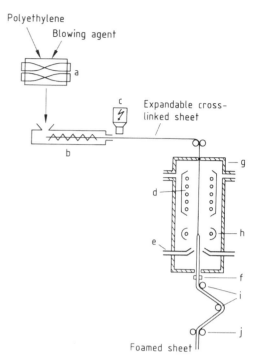

**Figure 6.** Production of radiation-cross-linked polyethylene by using a chemical blowing agent
a) Mixer; b) Extruder; c) Ionization irradiation unit; d) Preheater; e) Hot-air blower; f) Wrinkle-removing unit; g) Oven; h) Hot-air heater for foaming; i) Chill rolls; j) Take-up roll

suspension stabilizer (e.g., tricalcium phosphate, aluminum oxide).

When the pressure is reduced, the particles expand into foam beads with bulk densities of 20 – 70 kg/m$^3$, depending on the amount of blowing agent and the temperature (Fig. 7) [46], [47]. Uniform cell structure and low bulk densities can be obtained, for example, by post-treatment with nitrogen under pressure [48], [49]. To ensure that the final molded parts have a satisfactory surface, the original pellets should be small (mass, 1 mg) and the beads should have a diameter of 3 – 6 mm. The beads can be processed into molded parts without further cross-linking. This process is used by Kanegafuchi Chemical Industries, Japan Styrene Paper, and Yuka Badische [44], [47].

***Processing.*** Polyolefin foam boards can be cut or split readily with band saws or slicers. They can also be milled, cut thermally, or punched.

By heating and subsequent pressing (hot-melt lamination), cross-linked sheets can be

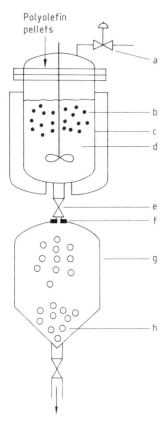

**Figure 7.** Production of expanded foam beads from polyolefin pellets
a) Pressure control valve; b) Polyolefin pellets; c) Pressure-resistant container; d) Water; e) Discharge valve; f) Orifice plate; g) Discharge tank; h) Expanded foam beads

fused to each other or to film, fabric, or sheet metal. They can also be cemented together with rubber-based contact adhesives or elastic melt adhesives. Thermoforming can be carried out effectively on commercial equipment [50–52].

Foam beads can be fused into parts of any desired shape, even boards, without waste by using steam or hot air; the process is similar to that used for expandable polystyrene beads (EPS) (see Section 4.2.1). Additional steps are required to completely fill the interstices between the spheres with foam:

1. In the Neopolen process for molded parts (Fig. 8), beads are placed in the mold cavity at elevated pressure and allowed to expand into the interstices by reducing the pressure. Only then are they exposed to steam, cooled, and removed from the mold [44], [53], [54].

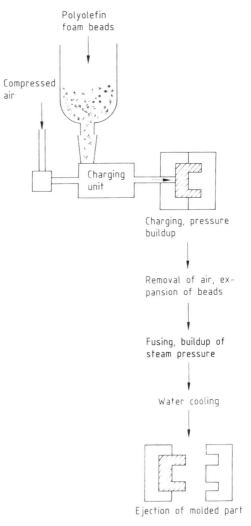

**Figure 8.** Manufacture of molded parts from polyolefin beads

2. The interstitial volume can also be reduced by a movable mold face which compresses the contents of the mold [55].
3. The internal pressure, and thus the expandability of the beads, can be increased immediately prior to mold foaming by using a compressed gas, e.g., nitrogen [56].

***Properties and Uses.*** Polyolefin foams are predominantly of the closed-cell type. They become semirigid as the crystallinity of the substrate and the apparent density decrease. Because of the air trapped in the cells, PO foams display excellent shock absorption properties which barely deteriorate under repeated impact. During prolonged loading, some of the air escapes from the cells. However, when the load is removed, air diffuses back into the cells and a slow recovery from deformation occurs, which depends on the apparent density and the load. Consequently, polyethylene foams are especially suitable for packaging impact-sensitive goods, e.g., electronic products. In packaging heavy goods, polypropylene foams can support higher loads and thus permit smaller support surfaces than the softer polyethylene foams. Polypropylene is particularly useful when greater dimensional stability under heat is required, e.g., in the automotive sector for items such as bumper cores [57].

The low moisture permeability of polyolefin foams combined with their low water absorption and relatively good weather resistance make them suitable for marine applications (e.g., ship fenders, life buoys, and life jackets) [54]. Protection against UV radiation is recommended for outdoor uses.

Because of their low thermal conductivity, which is only slightly greater than that of other foamed plastics, polyolefin foams are used as flexible pipe covers and insulating bags. More flexible types are used for impact soundproofing and resilient flooring in sports halls [50], [51], [58], [59].

### 4.1.2. High-Density Polyolefin Foams (250 – 700 kg/m$^3$)

Low-density polyethylene is foamed with chemical blowing agents (e.g., azodicarbonamide) to give apparent densities up to 450 kg/m$^3$. This material is used to insulate wires in telephone cables and coaxial cables; its dielectric constant is $3 \times 10^{-4}$ [60].

Larger, shaped parts with apparent densities of 500 – 700 kg/m$^3$ can be produced from HDPE and polypropylene by the thermoplastic foam injection molding process. The integral skins of these structural foams gives a higher stiffness than is expected from their mean density [10, pp. 1 – 16].

Polypropylene is also foamed in extruders with chemical blowing agents to produce films and tapes in the density range of 400 – 700 kg/m$^3$. These products have high surface luster and are used in packaging [10, p. 15], [61].

## 4.2. Polystyrene (PS) Foams (→ Polystyrene and Styrene Copolymers)

Polystyrene [9003-53-6] (PS) foamed plastics are important because of their physical properties, their broad spectrum of applications, and their favorable price–performance relationship. They can be produced by various processes and with different apparent densities; accordingly, they differ in properties and in possible uses. Trade names of foamed and expandable polystyrene are listed in Table 1.

### 4.2.1. Expandable Polystyrene (EPS) Foams

*Production.* The techniques used for the production of expandable polystyrene (EPS) beads and their processing to expanded polystyrene foams were developed at the end of the 1940s by BASF [62] who marketed the new raw material under the trade name Styropor. Due to licensing and the expiry of patents, other raw material manufacturers and trade names have appeared (see Table 1).

Expandable polystyrene is produced by suspension polymerization of styrene with addition of blowing agents; the resulting polymer beads are then sieved into fractions with various ranges of bead size. Depending on end use different coatings may be applied (→ Plastics, Processing, 3. Machining, Bonding, Surface Treatment).

In addition to pure polystyrene, foams are also made from polystyrene copolymers (e.g., with acrylonitrile or maleic anhydride). These foams possess special properties such as increased solvent or heat resistance. Incorporation of surface coating with organobromine compounds (e.g., hexabromocyclododecane [25637-99-4], pentabromophenyl allyl ether [3555-11-1]) act as flame retardants and result in flameproofed foams, e.g., Class B 1 building materials according to DIN 4102, M 1 according to NF P 92–505 (Norme Française), or MF-1 according to UL-94 (Underwriter's Laboratories). Blowing agents include low-boiling hydrocarbons, particularly pentane, but also butane, propane, and their mixtures; EPS beads contain 4 – 7 wt % of these substances.

Two other production methods for EPS consist of impregnating standard polystyrene with blowing agents in pressurized vessels and extrusion of EPS or standard polystyrene with added blowing agents. Cylindrical pellets are formed instead of beads.

***Processing into Foam.*** Foam is obtained from EPS in three steps—prefoaming, intermediate storage, and final foaming (or molding) — as illustrated in Figure 9.

During *prefoaming*, the particles are heated and the polystyrene is thereby softened, the blowing agent in the polystyrene evaporates to form rapidly growing bubbles until either the heat supply is shut off or the expandability exhausted [63]. The bead diameter increases about threefold and the bead volume (bulk volume) about thirtyfold. The apparent density of the foamed product is determined by the bulk density obtained during prefoaming.

*Intermediate storage* is required prior to final foaming into blocks, boards, or molded parts to allow air to diffuse into the individual cells. Air is needed in subsequent molding: it acts as a supplementary blowing agent and also enables the soft cellular structure to withstand the external atmospheric pressure when the finished part is removed from the mold.

*Final foaming* is usually fully automated [64], [65]: perforated molds are completely filled with prefoamed beads and exposed to steam. The beads expand to fill the residual voids and are fused together. In modern installations, a vacuum is also applied. Some of the water condensed in the foam then evaporates and internal cooling reduces the pressure of the foamed plastic more rapidly, so that the parts can be quickly removed from the mold (short cycle times).

Boards are manufactured batchwise with block molds or with board molding machines; they may also be produced continuously on special moving-belt machines.

The glass transition temperature of polystyrene is reduced if pentane, which acts as a plasticizer, is used as blowing agent. The temperature range used during prefoaming and final foaming is then 85 – 120 °C. The most widely used heat-transfer medium is steam, which has a large heat capacity; it diffuses rapidly through the cell walls, and thus acts as a supplementary

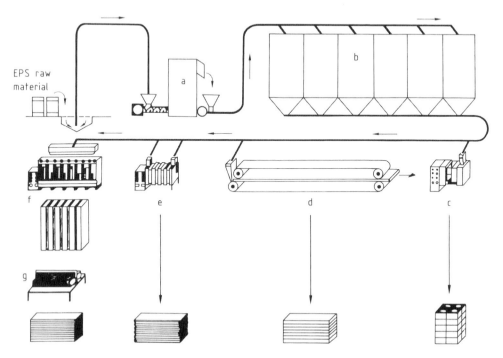

**Figure 9.** Processing of expandable polystyrene (EPS) foam into molded parts and insulation board
a) Prefoamer; b) Intermediate storage; c) Automatic molding machine; d) Continuous foaming unit; e) Automatic board molding machine; f) Block mold; g) Cutter

blowing agent. In some special cases, hot air is used instead of steam.

Foam boards for heat insulation are made primarily from foamed blocks 1 – 6 m$^3$ in volume. After stabilization for several hours to complete cooling and air uptake, the blocks are trimmed and cut into boards on cutters using electrically heated or oscillating wires. Post-treatment such as grooving or slotting is carried out on milling machines. Sheets of any desired length and 1 – 6 mm thick can be produced from special cylindrical blocks by using peeling machines with rotary blades. These peeled sheets are used under wallpaper and in packaging.

The resilience of boards for impact soundproofing is increased by briefly compressing the blocks prior to cutting or by rolling (or squeezing) the individual boards.

In a number of applications, the foamed plastic must be stored prior to use. This helps to remove internal moisture (steam condensate and residual cooling water) and to reduce postshrinkage of the foam.

***Properties.*** Most of the physical properties of EPS foams depend strongly on the apparent density, the most important are summarized in Table 2. Boards for thermal insulation generally have densities between 15 and 30 kg/m$^3$, whereas the density of boards used for impact soundproofing is ca. 12 kg/m$^3$. The density of shape-molded parts is generally 20 – 30 kg/m$^3$, for special applications it may be much higher.

The chemical resistance of the foams corresponds largely to that of polystyrene.

***Uses.*** Expandable polystyrene foams have many applications including thermal insulation, impact soundproofing [66], and formwork elements for concrete in the building industry; insulation of cold-storage depots and cooling cells, storage and transport tanks, pipes, etc.; and molded parts for packaging [67], promotion articles, toys, and sporting goods. It is also used for producing molded articles by lost-foam casting, a process that is similar to wax molding. A molded EPS form is embedded in sand and molten metal is poured into this composite. The foam shape evaporates and is replaced by metal.

Considerable amounts of nonfused, loose beads are also used as additives in bricks; (during firing they decompose to leave uniformly distributed pores). Further uses are as additives in light-weight concrete [68] and as renderings for thermal insulation or as drainage material.

Foamed EPS waste that accumulates during processing can be recycled; it is ground, mixed with preformed beads, and reused during molding. Ground EPS foam has also proved useful for loosening soil, for drainage, and as a filling material. These foams are inert and do not harm ground water, flora, or fauna.

### 4.2.2. Extruded Polystyrene (XPS) Foams

Foamed polystyrene boards and films are produced in large amounts by extrusion [69].

***Production.*** In the original DOW process [70], polystyrene and a blowing agent were fed into a

**Table 2.** Physical properties of expandable polystyrene (EPS) foams as a function of apparent density

| Property | Testing method | Apparent density, kg/m$^{3*}$ | | | | |
|---|---|---|---|---|---|---|
| | | 15 | 20 | 30 | 60 | 100 |
| Compressive stress at 10% deformation, N/mm$^2$ | DIN 53 421 | 0.06 – 0.10 | 0.10 – 0.14 | 0.18 – 0.25 | 0.40 – 0.65 | 0.8 – 1.3 |
| Dimensional stability/ compressive creep, % | DIN 18 164 | < 5 | < 5 | < 5 | | |
| Thermal conductivity at 10 °C, W m$^{-1}$ K$^{-1}$ | DIN 52 612 | 0.036 – 0.039 | 0.033 – 0.036 | 0.031 – 0.035 | 0.031 – 0.035 | 0.036 – 0.039 |
| Specific heat, kJ kg$^{-1}$ K$^{-1}$ | | 1.21 | 1.21 | 1.21 | 1.21 | 1.21 |
| Coefficient of linear thermal expansion, K$^{-1}$ | | $7 \times 10^{-5}$ | $7 \times 10^{-5}$ | $7 \times 10^{-5}$ | $7 \times 10^{-5}$ | $7 \times 10^{-5}$ |
| Water vapor transmission rate, g m$^{-2}$ h$^{-1}$ | DIN 53 429 | 1.7 | 1.6 | 0.8 | 0.3 | 0.15 |
| Water absorption after 7 days, wt % | DIN 53 434 | 0.4 – 2 | 0.4 – 1.5 | 0.3 – 1.0 | 0.3 – 0.7 | 0.3 – 0.7 |

*Density determined according to DIN 53 420.

pressurized autoclave and melted. The molding batch was then cooled to about 100 °C; upon emerging into the atmosphere or into a mold chamber, it foamed and formed a uniform cell structure.

The melting of the polystyrene pellets and the gassing, homogenizing, and cooling of the melt were then performed in extruders [71]. The continuously emerging strand foams, is formed into a flat sheet, solidified in a cooling zone, and cut into boards.

Blowing agents are mainly halogenated hydrocarbons which increase the foam's thermal resistance (decrease thermal conductivity) because of their high molecular mass.

In addition to Styrofoam (Dow Chemical), similar XPS boards with apparent densities of ca. 25 – 60 kg/m³ are also available, e.g. Styrodur (BASF).

Foam boards are generally dyed during extrusion and provided with flame retardants (e.g., hexabromocyclododecane).

Depending on the end use, the boards are produced with different apparent densities, thicknesses, surface textures (foam skin, milled), and edges (flat, step groove, or groove and tongue).

*Properties.* The most important physical properties of extruded polystyrene (XPS) foams at different densities are summarized in Table 3. The requirements for use as thermal insulation material in building applications are specified in ISO 4898. The chemical resistance corresponds to that of polystyrene.

*Uses.* Extruded polystyrene foam boards are expecially suitable for thermal insulation in special constructions where high compressive strength, good appearance in exposed locations, and very low water adsorption are required.

**Extruded polystyrene films** are produced exclusively by the film blowing process (→ Films). A tubular film is extruded from an annular spinneret. It is then expanded three- to sixfold by using a high internal gas pressure and is thus stretched transverse to the direction of drawoff. The film is then calibrated via a cooling mandrel, cut into one or more strips, flattened, and either wound or processed immediately by thermoforming or lamination.

In the past films were produced in smaller, simple extruders from expandable polystyrene containing a blowing agent. However, the growing demand for such films required the use of large, direct-gassing installations that use standard polystyrene and an added blowing agent as in the extrusion of boards [72]. Low-boiling, preferably nonflammable, hydrocarbons are usually used as blowing agents. Chemical agents (e.g., citric acid plus sodium bicarbonate) and nucleating agents are also added to achieve a uniform cell structure.

Depending on the end use, film thicknesses range from several millimeters to a fraction of a millimeter and apparent densities from 100 kg/m³ for thick films to 200 kg/m³ for thin ones.

During thermoforming or lamination with cardboard and other layers, the films are often postfoamed; their apparent density is thus reduced and their thickness is increased. In this way, items such as egg cartons and packaging trays for fruit and meat are obtained with apparent densities of 60 kg/m³ or less.

Polystyrene film laminated with cardboard is being tested as a replacement for corrugated

**Table 3.** Physical properties of extruded polystyrene (XPS) foams as a function of apparent density

| Property | Testing method | Apparent density, kg/m³* | | | |
|---|---|---|---|---|---|
| | | 25 | 30 | 40 | 50 |
| Compressive stress at 10 % deformation, N/mm² | DIN 53 421 | 0.1 – 0.2 | 0.2 – 0.4 | 0.4 – 0.6 | 0.5 – 0.7 |
| Compressive creep, % | DIN 18 164 | < 5 | < 5 | < 5 | < 5 |
| Thermal conductivity at 10 °C, W m⁻¹ K⁻¹ | DIN 52 612 | 0.029 – 0.033 | 0.025 – 0.030 | 0.025 – 0.030 | 0.025 – 0.030 |
| Specific heat, kJ kg⁻¹ K⁻¹ | | 1.45 | 1.45 | 1.45 | 1.45 |
| Coefficient of linear thermal expansion, K⁻¹ | | $7 \times 10^{-5}$ | $7 \times 10^{-5}$ | $7 \times 10^{-5}$ | $7 \times 10^{-5}$ |
| Water absorption after 28 days, wt % | DIN 53 434 | < 0.1 | < 0.1 | < 0.1 | < 0.1 |

*Density determined according to DIN 53 420.

cardboard where outstanding wet strength is required. Thin films are used for decorative purposes and as nonskid interlayers.

### 4.2.3. Structural Polystyrene Foams

Special starting materials are available for structural foams derived from polystyrene, its copolymers, and polymer blends. They already contain chemical blowing agents and are specially formulated for thermoplastic foam injection molding (TFI) and thermoplastic foam extrusion (TFE).

Products from butadiene-modified, impact-resistant standard polystyrene and from acrylonitrile–butadiene–styrene (ABS) polymers have the highest usage, followed by structural foams produced from standard polystyrene and styrene–acrylonitrile (SAN) polymers. For a detailed description of these polymers, see → Polystyrene and Styrene Copolymers.

*Production.* Physical blowing agents in the form of halogenated hydrocarbons are in principle suitable for structural polystyrene foams. However, they have not been widely used because more sophisticated processing technology is required than for chemical agents and control of foam structure is more difficult.

The following chemical blowing agents are commonly used: azo compounds (especially azodicarbonamide, sometimes modified), hydrazines (e.g., benzenesulfohydrazide), and sodium bicarbonate with citric acid.

Cell structure is affected not only by the blowing agents and the processing conditions, but also by the amount and type of nucleating additives such as talc or pigments.

*Thermoplastic Foam Injection (TFI) Molding.* Molded TFI parts can be produced on conventional injection molding machines with shut-off nozzles. However, low-pressure TFI machines are more economical. The mold is filled in two steps: (1) injection of the melt containing the blowing agent and (2) postexpansion to fill the mold cavity. Thus, the clamping unit experiences only the expansion pressure which is about one-tenth of the injection pressure required for compact polystyrene. In this two-step process, removal of heat from the melt starts with evaporation of the blowing agent and surface streaks are created on the molded parts. Streaking can be prevented by various techniques, some of which are patented [8].

*Thermoplastic Foam Extrusion (TFE).* The formulations of raw materials used in TFE are similar to those used in TFI. However, since the temperature of the extrusion mass is generally lower, blowing agents with a lower decomposition temperature are used or catalysts are added to promote decomposition of the blowing agent. Although a number of patented extrusion processes exist (e.g., the Celuka process for very smooth surfaces, the Woodlite process for woodlike grain), conventional devices, profiled nozzles, and subsequent calibration are commonly employed [73]. Prior to use, the parts or profiles are frequently subjected to surface treatment such as spraying, printing, or lacquering.

*Properties.* The physical and chemical properties of structural polystyrene foams depend highly on the raw materials used. With their generally very high apparent densities and virtually compacted surface layers, the mechanical properties of structural foams are only slightly inferior to those of solid parts.

*Uses* [8]. The major end user of structural polystyrene foam is the furniture industry. The product is not used simply to imitate wood, but to construct structural parts for furniture of modern, functional design that exploits the properties of the raw material. The hi-fi and television industries are also important customers for housings and other TFI components. Significant quantities are also used in the building and transport industries, in the manufacture of sports equipment, and in the automotive sector.

### 4.3. Poly(Vinyl Chloride) (PVC) Foams [→ Poly(Vinyl Chloride)]

Depending on the content of plasticizers, polymer processing aids, and thermosetting binders, poly(vinyl chloride) [9002-86-2] foams can be flexible, semirigid, or even rigid. Because of their range of properties, PVC foams have numerous uses. Frequently, they form part of structurally complex, multicomponent articles such as floor coverings or window frames.

The composition of PVC foams varies widely. Some consist mainly of PVC and contain only minor amounts of additives necessary for smooth processing and trouble-free application (stabilizers, lubricants, fillers, pigments, or blowing agents and their decomposition products). Others contain larger amounts of modifiers (plasticizers, cross-linking agents, processing aids, or impact modifiers.

Dry blends, granules and fluid or paste mixtures containing blowing agents are used to manufacture PVC foam. The most important processing technologies for dry blends and granules are molding, extrusion, calendering, and injection molding. Fluid or paste mixtures are usually processed by spread coating.

The PVC molding masses are foamed with both chemical and physical blowing agents (see Section 2.2) [74]. The foaming rate can be influenced by kickers or inhibitors. Poly(vinyl chloride) foams cannot be consistently classified because of the different manufacturing processes involved and their broad spectrum of properties and uses. However, division into rigid and flexible PVC foams has proved useful. Semirigid PVC foams are classified as rigid PVC foams although they contain plasticizers.

Trade names of PVC foams are listed in Table 1.

### 4.3.1. Rigid Poly(Vinyl Chloride) Foams

High mechanical strength, along with resistance to moisture, rot, and chemicals, is common to all rigid PVC foams. Thermal conductivity is 0.036 – 0.040 W m$^{-1}$ K$^{-1}$. These foams have a high oxygen index and their combustion gases contain carbon dioxide, carbon monoxide, and hydrogen chloride.

Most rigid PVC structural foams are produced by extrusion of PVC compound (dry blend or granules containing blowing agents) in the form of pipes, shaped parts, or boards [75–85]. Metal pipes or shapes can also be coated with foamed PVC by extrusion [86], [87]. Furthermore, rigid PVC foam shapes can be reinforced with glass fibers. Very few rigid PVC foam components are produced by injection molding [88].

In the extrusion of molded rigid PVC parts, processing conditions are adjusted so that the blowing agent foams the molding mass only after it leaves the nozzle. Foamed parts have an outer skin whose thickness and density vary depending on how the melt foams and how rapidly the boundary phase is cooled. Complex shapes and boards of PVC foam can be produced by the so-called Celuka process [89] in which appropriate machine design allows foaming to proceed "inwards".

In the coextrusion process [90], [91], a compact PVC layer (which may be stabilized and pigmented as desired) is melted onto the foam during extrusion as an outer skin. Such profiles are preferred for outdoor use, e.g., as panels.

Rigid PVC foams range from predominantly closed-cell to open-cell, depending on the type and amount of blowing agent and the process used. These foams are used in the building and furniture industries and in the do-it-yourself sector for thermal insulation and as wood substitutes.

***Thermoset-modified rigid PVC foams*** are generally manufactured by a two-step process [92–95]. In the first, so-called high-pressure stage, a paste containing PVC, plasticizer, isocyanate cross-linking agent, stabilizers, and a chemical blowing agent is heated in closed molds to about 180 °C under high pressure. The PVC and the plasticizer then form a gel, and the blowing agent decomposes to form tiny closed cells. After about 1 h, the molds are cooled while the pressure is maintained. In the second, so-called expansion stage, the molded articles containing the blowing agent are exposed to water vapor for several hours at 90 – 100 °C and reduced pressure. The material is still thermoplastic and thus softens; the expansion pressure of the occluded blowing agent increases the volume of the molded article. Some of the isocyanate additive reacts with water to form carbon dioxide which acts as an additional blowing agent; the remainder, together with a special auxiliary system, cross-links the PVC and significantly consolidates the foam structure.

The closed-cell foams produced are used primarily as core materials of high mechanical quality for composites in automotive, ship, and airplane construction. Important manufacturers are Kleber Colombes (France), Airex (Switzerland), and Conti-Gummiwerke (Federal Republic of Germany).

*Rigid thermoplastic PVC foams* can be produced by the same mold foaming process used for thermoset-modified foams. They are manufactured from polymer granules without using plasticizers, isocyanate, or cross-linking agents. These foams are used as core materials for composites and, because of their closed outer skin, are also used in marine buoyancy or flotation items.

### 4.3.2. Flexible Poly(Vinyl Chloride) Foams

Flexible PVC foams can be manufactured by various processes, either with or without a carrier.

*Flexible PVC foams without support* can be manufactured in the same way as rigid PVC foams, by a two-step process using dry blends, granules, or pastes containing blowing agents. The PVC foams obtained are predominantly closed cell; their flexibility and elasticity depend on the cell structure and the plasticizer content. Flexible PVC foams can also be produced continuously but with limited cross section by paste spreading or extrusion.

Flexible foams are used for sealing and insulation in the building trade, for heat insulation and buoyancy in clothing, and for impact protection in sports equipment.

*Flexible PVC foams with support* are produced by spreading pastes containing blowing agents on various supporting materials. Synthetic PVC leather, for example, consists of a textile carrier, a base layer of flexible PVC, and a layer of flexible PVC foam with a thin, compact surface. The flexible foam layer imparts a soft feel to synthetic PVC leather (→ Leather Imitates).

Modern, chemically embossed, color-printed floor and wall coverings have a similar structure. Inhibitors in the printing ink retard or prevent decomposition of the chemical blowing agent and thus create an embossed pattern that matches the printed pattern [96]. These materials are especially important as floor coverings (Congoleum process) because they are pleasant to walk on (→ Floor Coverings).

A spreadable whipped foam is obtained by intensively whipping gases (nitrogen, carbon dioxide, air) into PVC pastes. This foam is used mainly to back textile or rubber floor covering with an elastic layer that contains largely open pores after gelling [97–99].

*Miscellaneous Flexible PVC Foams.* Other flexible PVC foam items are produced from pastes containing blowing agents. They are used predominantly as sealing materials. Thus, for example, seals in crowned cork bottle caps used for beverages are produced by gelling and foaming a small amount of PVC paste applied directly to the cap. Similarly, flexible PVC foam seals for industrial components (e.g., air filters for automobiles) can be foamed in situ.

## 4.4. Foams from Phenol–Formaldehyde (PF) Resins (→ Phenolic Resins)

*Production.* Phenolic resin foams are produced by the acid-catalyzed, exothermic hardening of a phenol–formaldehyde resin [9003-35-4] condensed under alkaline conditions [100], [101] (→ Phenolic Resins). Evaporation of an added (halogenated) hydrocarbon blowing agent by the heat of the reaction produces foaming. Surfactants emulsify the blowing agent and stabilize the expanding foam. The foam is hardened with catalysts such as sulfuric acid, phosphoric acid, or arylsulfonic acids. Hydrochloric acid is no longer used [102]. For complete and thorough hardening, heat treatment may follow [103].

Hydrogen produced by reaction of powdered aluminum with an acid catalyst can also act as a blowing agent. This method is employed in the former Soviet Union where PF foams account for about one quarter of foamed plastic production [104].

Production is predominantly a batch process in which the foaming mixture is poured into block or other cavity molds. The use of spray guns has also been reported [105].

The well-known continuous production of laminated sheets [106] is used by Koppers [108], [108] and British Petroleum Chemicals [109], [110]. Closed-cell foams are presumably created. The starting materials are special resols with a high resin solids content. The blowing agent is a mixture of trichloromonofluoromethane [75-69-4] (CFC 11) and 1,1,2-trichloro-1,2,2-trifluoroethane [76-13-1] (CFC 113). Alkoxylated silicones [108] or glycerol ricinoleates [110] are used as surfactants.

For hardening, strong organic acids such as an anhydrous mixture of tolyl- and xylylsulfonic acids [108] or 50 % sulfuric acid [110] are used. Special mixing heads have been developed for intensive mixing of components [111].

The mixture is sandwiched between two surface layers of, for example, aluminum-coated cardboard; the foaming mass is shaped and hardened, while still plastic, in a heated, double-belt press (Fig. 10). The pressure exerted in the press is reported to have an important effect on foam quality. It reduces the open-cell content [107] and void formation underneath the upper layer [109].

The resulting laminates are between 30 and 55 mm thick and have apparent densities of 35 – 60 kg/m$^3$.

***Properties.*** The hard, brittle, thermoset nature of cured PF resin is shared by the foam. It is, however, less pronounced in foams with smaller cells (cell size < 80 μm) or with highly perforated cell walls [112].

All PF foams in existence at the end of the 1970s were open-cell types. Here, the cells are interconnected by holes in their walls which are several micrometers wide. Because the holes also allow communication with the surroundings, the cells are air-filled [113–115]. Like polyurethane and polystyrene foams, closed-cell PF foams can now be produced, whose cell walls are intact and free of pores [116], [117]. These cells can thus retain fluorohydrocarbon blowing agents. Because the thermal conductivity of these heavy gases is lower than that of air, the resulting foams are better insulators than air-filled foams (thermal conductivity < 0.02 W m$^{-1}$ K$^{-1}$ and 0.033 W m$^{-1}$ K$^{-1}$, respectively).

Water absorption naturally increases with the number of open cells. Light, largely open-cell PF foams (apparent density ca. 20 kg/m$^3$) may even be used as floral foam in flower arrangements [118].

If the foam is only partially closed-cell, expensive water-vapor barriers may be needed as, for example, in underground long-distance heating lines insulated with PF foam having an apparent density of 100 kg/m$^3$ [119]. Infiltration of water would not only impair heat insulation but also initiate corrosion in combination with the acidic hardening agents. Despite the generally high acid content (up to 20 wt %), corrosion problems are seldom encountered, especially with closed-cell products. Corrosion inhibitors such as calcium carbonate can be added.

Over short periods PF foams can resist temperatures up to 180 °C and, over longer periods, up to 130 °C. They are resistant to many solvents and attacked only by concentrated acids or bases.

Compared to other organic insulators, PF foam has a favorable fire behavior primarily because of the low smoke density of its combustion gases and its low flammability. Like other thermosets, the material does not melt. When ignited, a carbon skeleton is formed. With respect to the evolution of toxic gases (carbon monoxide), however, PF foams do not have any advantage over other foamed plastics [100], [101].

The afterglow of open-cell foams (punking) can be suppressed by the addition of flame retardants such as boron trioxide [1303-86-2] or aluminum hydroxide [21645-51-2]. Closed-cell PF foams hardened with sulfuric acid do not exhibit punking; when burnt, they "pop"; due to

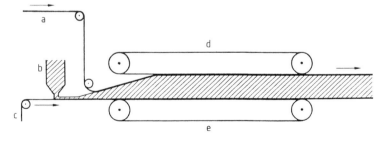

**Figure 10.** Production of laminated phenolic foam board
a) Upper surface layer; b) Dispenser for foamable phenolic resin; c) Lower surface layer; d) Upper conveyor belt; e) Lower conveyor belt

the noisy bursting of small particles of material [120].

**Uses.** In the building industry, for example, PF foam boards covered with glass-fiber webs are used as thermal insulation for flat roofs. They can be processed with hot bitumen because of their high heat resistance. When reinforced with an impregnated paper honeycomb, PF foam serves as the core of structural sandwich elements for walls and roofing, which are capable of bearing static loads. Sheet metal, polyester–glass fiber, plywood, or plaster may be used as surface layers [120–122].

Hemispheres cut from blocks are used for pipe insulation. A pour-in-place process with special molds has also been described [119].

In the former Soviet Union, PF foam is preferred for insulation in transport (motor vehicles, railroad cars, ships) as well as in aeronautic and space applications [104].

**Special PF Foams.** Inorganic additives (fillers) are added to PF foams to improve their mechanical strength and fire behavior. Depending on the type and amount of filler, *filled PF foams* with apparent densities of 100 – 300 kg/m$^3$ are obtained. Fillers include expanded silicates such as perlite [7344-61-8] [123]. Products that contain phenolic and furan resins are highly filled with materials such as aluminum hydroxide and are used as noncombustible building materials [124].

Phenolic foam with an apparent density of 20 – 30 kg/m$^3$, which is filled with polystyrene foam beads, can be obtained by dielectric heating [125].

*Hybrid foams* with good mechanical properties and low combustibility can be produced from PF resins and polyisocyanates [126].

## 4.5. Foams from Urea–Formaldehyde (UF) Resins (→ Amino Resins)

**Production.** A surfactant solution containing a hardener (e.g., phosphoric acid) is foamed with compressed air and blended with an aqueous 30 – 40 % solution of urea–formaldehyde resin [9011-05-6]. The resin is distributed throughout the surfactant foam, and cross-linking is initiated by the acidic hardener. A light, whipped-cream-like foam is formed which can be shaped easily and used to fill any cavity. The foam hardens rapidly to a water-insoluble, rot-resistant, foamed plastic without creating expansion pressure [127], [128]. A hardening time of 15 – 40 s is generally selected, which can be controlled within certain limits by changing external variables (temperature, hardener concentration).

Newly formed foam contains ca. 70 – 75 wt % of water. During drying, cracks are formed in the lamellae and the foam becomes open-celled [129]. This is accompanied by a ca. 8 % decrease in volume which can be reduced to 3 % by additives (e.g., polyols) [128], [130], [131]. Any free formaldehyde remaining is lost along with the water. Problems caused by formaldehyde release can be avoided largely by using more urea or improved resin formulations [132–134].

Stationary and portable equipment is available for foaming. Insulating layers can thus be introduced in cavities that are not easily accessible (e.g., in double walls).

**Properties.** Apparent densities of UF foam range from 10 to 30 kg/m$^3$ depending on the application. It can support little or no mechanical load and is, therefore, not used as semifinished product.

Urea–formaldehyde foam can withstand continuous exposure to temperatures up to 100 °C; at higher temperature, gradual shrinkage occurs. This material decomposes above 200 °C [128].

The thermal conductivity of UF foam (ca. 0.035 W m$^{-1}$ K$^{-1}$) is in the same range as that of other air-filled insulating foams (cf. Fig. 3).

The material has a relatively favorable oxygen index (cf. Fig. 4). Although UF foam does not meet the flammability requirements of DIN 4102, it performs better in other flammability tests [ASTM D 1692–59 T, BS 2782 (1985) 508 D] [128].

**Uses.** The apparent density of in situ UF foam used in buildings must be at least 11 kg/m$^3$ (DIN 18 159). The foam serves mainly as thermal insulation in wall cavities [129], [135] and in plumbing ducts [136].

In mining, cavities (e.g., gallery roofs) are filled with UF foam to prevent accumulation of dangerous firedamp [137], [138]. The foam is also employed in the construction of airtight dams to seal off sections of mines (e.g., abandoned shafts) [139].

In agriculture and horticulture, UF foams with apparent densities > 20 kg/m$^3$ and preferably in the form of flocks (e.g., Hygromull, BASF) are mixed into the soil to improve heat balance. Because of their open-cell structure, the foams also store water and dissolved nutrients.

Kindling is obtained by emulsifying liquid petroleum products with UF foam during manufacture [140].

Foamed plastic with an apparent density of ca. 30 kg/m$^3$, ground to a particle size of 0.1 – 6 mm and containing a surfactant and an antistatic agent, is used in carpet cleaning [141].

Urea–formaldehyde foam is suitable for cushioning impacts only if no more than one violent impact is anticipated (parachute deployment) [142]. For trade names of UF foams, see Table 1.

## 4.6. Foams from Melamine–Formaldehyde (MF) Resins (→ Amino Resins)

The rigid, brittle melamine–formaldehyde resin [9003-08-1] foams known until the late 1970s had no practical significance [143]. However, a flexible, elastic foam can now be produced from the same resin. This foam (Basotect, BASF) has a completely different morphology [144], [145]; it is open-celled and has slender, flexible struts (cf. Fig. 2B). The length : diameter ratio of the struts is crucial to its elastic behavior. For strut diameters from 7 to 15 µm, the ratio must exceed 10.

***Production.*** A surfactant is used to emulsify pentane with a melamine–formaldehyde precondensate (molar ratio 1 : 3). Hardening occurs upon addition of formic acid. Controlled expansion in a continuous foaming tunnel is initiated by heating [146], [147].

Pentane and water from the reaction mixture act as blowing agents [148]. After removal of water and pentane, the foam is perfused continuously with hot air at 220 – 250 °C, and any cleavable formaldehyde is completely eliminated [149]. Flexible foams result only if the solids content exceeds 70 % [146]. Brittle products are obtained if the solids content is lower.

***Properties.*** Density ranges from ca. 10 to 12 kg/m$^3$. Morphology and, hence, mechanical properties can be modified by pressing, which slightly bends the originally straight struts. The material becomes more flexible, with a negligible increase in density. For example, the compressive stress at 40 % deformation drops from 1.1 to 0.8 MPa [150]. Extreme mechanical–thermal treatment produces a highly consolidated, weblike material [151].

Because of the large proportion of IR radiation scattered by the thin struts, the thermal conductivity of the foam is in the lower range of air-filled insulators (Fig. 3).

Melamine resin foam can withstand short exposure up to 250 °C and long exposure up to 150 °C. No solvents are known for the foam; it is attacked by concentrated acids and bases.

Water absorption and diffusion properties are similar to those described in Chapter 3 for open-cell foams. The open, extremely fine structure and the flexibility of the foam result in especially favorable acoustical properties; despite the low apparent density, these properties are similar to those of open-cell polyurethane foams and mineral fibers.

The material meets the requirements of flammability class B 1 building materials, specified in DIN 4102 ("schwer entflammbar," i.e., of low flammability), without an added flame retardant. According to the tunnel test of ASTM E 84–82 A, it is a class A material as defined in the Standard Building Codes. The smoke density of the combustion gases is low but does not reach the minimum values of PF foam. The material carbonizes during burning, without melting, and decomposes above 350 °C [145].

***Uses.*** Thermal insulation, heat resistance, and ease of processing have opened up many applications for MF foams in industrial heat insulation (e.g., pipe insulation).

The material is used for sound insulation in the form of decorative ceiling and wall paneling, as well as for casings in automotive engines. Because of its low dynamic stiffness, MF foam is also employed for impact sound insulation. Laminates with plasterboard provide good soundproofing in addition to the desired thermal insulation.

## 4.7. Silicone (SI) Foams

***Production.*** Silicone foams are produced from hot- or cold-vulcanized silicone rubber (→ Rubber, 5. Technology, → Silicones).

*Heat-activated vulcanization systems* contain mostly high molecular mass polysiloxanes incorporating methyl as well as vinyl groups. Additional components are chemical, heat-activated blowing agents, peroxide cross-linking agents, and silicate fillers [152]. These "single"-component blends are stable at room temperature and begin to foam when vulcanized by the action of heat.

Generally, preforms of the mixture or previously formulated dry blends are foamed in compression molds or extruders at 150 – 200 °C. The particles of silicone foam are then annealed at ca. 200 °C to remove any decomposition products derived from the cross-linking agent and to complete residual cross-linking.

*Room-Temperature Vulcanization (RTV) Systems.* Pourable mixtures that can be vulcanized and foamed at room temperature are required for producing molded articles and for filling cavities or coating parts with silicone foam.

Two-component systems are used in which short-chain polysiloxanes containing Si — H groups are allowed to react with low-viscosity polysiloxanes containing Si — OH groups. The reaction is performed at room temperature in the presence of a condensation catalyst:

$$\equiv Si-H + \equiv Si-OH \xrightarrow[\text{Room temp.}]{\text{Catalyst}} \equiv Si-O-Si\equiv + H_2$$

The hydrogen evolved acts as a blowing agent, but additional blowing agents may be necessary. Organotin compounds are used as catalysts [153], [154]; platinum catalysts yield products with improved heat and fire resistance [155–157].

The degree of cross-linking and the morphology of the foam can be controlled by the ratio of silanol to silane groups, especially by their functionality. Foams that range from rigid to flexible and from largely open-cell to closed-cell can be produced this way [158].

The above reaction takes only a few minutes but can be retarded by using acetylene alcohols or rhodium compounds as catalyst. The system is then stable for several hours and can be reactivated with heat [159].

Foam systems that can be cross-linked with atmospheric moisture have also been described [160].

A froth produced from a silicone emulsion can be converted to a light, elastomeric silicone foam in a microwave oven; no additional blowing agent is required [161].

*Properties.* Silicone foams are produced with apparent densities of 60 – 300 kg/m$^3$, but highly filled products are heavier. The heat distortion temperature is at least 200 °C; the material generally decomposes above 350 °C although higher limits have been recorded [162]. Thermal conductivity is reported to be 0.035 – 0.07 W m$^{-1}$ K$^{-1}$.

Silicone foams display largely constant cushioning properties over a wide temperature range (− 80 to + 200 °C), and compression set is low. These materials exhibit good dielectric properties; are resistant to hydrolysis, radiation, aging, and weathering; and are physiologically inert.

The materials have a high flame resistance, with a favorable oxygen index of about 30 and a low smoke density. Combustion gases contain only carbon monoxide and carbon dioxide.

*Uses* [157]. Examples of the uses of silicone foams are

- sealing (shaped profiles, strings, rings) including heavy-duty applications in nuclear power plants;
- filling cracks or cavities, and sealing cable inlets and outlets in construction work;
- thermal and electrical insulation of fragile components in aerospace applications, with simultaneous vibration damping;
- production of carbonizing heat shields used in space vehicles (filled foams) [163];
- cushioning in airplane seats [162]; and
- fitting artificial limbs and producing inert implants for medical use [164].

## 4.8. Epoxy (EP) Foams (→ Epoxy Resins)

*Production.* The starting components for epoxide foams are epoxide oligomers, cross-linking agents, blowing agents, surfactants, and possibly also fillers. Silicone–glycol copolymers are described as surfactants [165]. The preferred epoxide oligomers are reaction products of 2,2-bis(4-hydroxyphenyl) propane [80-05-7] (bisphenol A) and epichlorohydrin [106-89-8] (→ Epoxy Resins). Cross-linking agents include fast-acting aliphatic amines (e.g., diethylenetriamine [111-40-0]) or slower acting aromatic amines (e.g., 1,3-phenylenediamine [108-45-2]) that are effective only at elevated temperature. The reaction is highly exothermic and thus activates the blowing agent which is often a fluorinated hydrocarbon or a compound that splits off nitrogen (e.g., sulfonyl hydrazides). Precise temperature control during foaming is important. Too high a temperature causes cracking and deformation of the foam.

Foams with apparent densities of 20 – 40 kg/m$^3$ are produced from fast-acting systems by a two-component spray process. However, mold-pouring or injection processes are more common and can also be used with slow-acting systems that require external heat. The resulting foams have apparent densities of 60 – 300 kg/m$^3$ [166]. Trade names are listed in Table 1.

*Syntactic foams* (see also 1) [167] often consist of epoxy resins and, therefore, are discussed in this section. Their cells are formed by hollow spheres (microballoons) that are distributed in the polymer matrix as a filler. The hollow spheres are composed of glass or phenolic resin and have a diameter of ca. 100 µm and a wall thickness of ca. 2 µm. Larger "macrospheres" composed of epoxy resin and with a diameter of over 1 mm have also been described [168].

The hollow spheres are mixed with a low-viscosity epoxy resin in the same way as fillers. When the fraction of spheres is 40 vol % or less, mold pouring is used for further processing; extrusion or compression molding can be employed for fractions up to 67 vol %. A vacuum is applied for degassing. The products have apparent densities of 200 – 800 kg/m$^3$.

Phenolic resins, polyimides [169], silicones [170], and polyether sulfones are also used as matrix resins.

*Properties and Uses.* Epoxy foams are usually rigid and primarily closed-cell materials. They have favorable mechanical properties, absorb very little water, and are not hydrolyzable. They are resistant to many chemicals but not to ketones.

The good adhesion of EP foams to many materials makes them suitable as laminate cores. They also serve as edge seals for honeycombs made from materials such as the aramid Nomex that are used in airplane construction.

Because of their low dielectric constant (< 2), EP foams are used in protective housings for radar equipment (radomes). Materials with a higher dielectric constant are obtained by addition of metal flocks (e.g., aluminum) and are used as radar lenses.

Sensitive electronic components are protected by encapsulation in EP foam [165].

The heat distortion temperature is reported to be 100 – 200 °C. The thermal conductivity of EP foams containing heavy gases is ca. 0.02 W m$^{-1}$ K$^{-1}$, whereas that of foams manufactured with other blowing agents exceeds 0.035 W m$^{-1}$ K$^{-1}$, depending on the density. Therefore, these foams are also used as insulation for hot-water pipes.

Epoxy foams that do not contain fire retardants are flammable. The oxygen index is 20.

Syntactic foams are very rigid and able to withstand very high hydrostatic pressures. They are employed in laminates for underwater vehicles, including a shuttle for deep-sea mining to a depth of 6000 m. Other uses include flotation devices with a density of 0.5 g/cm$^3$ for steel pipes in offshore oil exploration [168], and ablation materials for space flight.

## 4.9. Polyimide (PI) Foams (→ Polyimides)

The increasing importance of polyimides as temperature-resistant materials has led to the development of PI foams. Trade names are listed in Table 1.

*Production.* In a typical polyimide synthesis, the hemiester (**2**) is prepared from 3,3′,4,4′-benzophenonetetracarboxylic acid dianhydride [2421-28-5] (**1**) and ethanol. The polyimide (**4**) is obtained by reacting (**2**) with a diamine, in this case 4,4′-diaminodiphenylmethane [101-77-9] (**3**) (and if necessary with 2,6-diaminopyridine [141-86-6]), in the presence of suitable surfactants.

The powder obtained after spray-drying the precursor solution is foamed into large blocks in a microwave oven. The water and alcohol liberated upon imidization act as blowing agents.

Flexible, open-cell foams are formed [171]. Morphology and especially thermal properties can be modified by using other diamines containing, for example, hexafluoropropyl groups [173], [173].

Rigid panels can be produced by impregnation of fibrous webs or honeycombs with precursor solution and subsequent foaming. Syntactic PI foams (cf. Syntactic foams) and rigid PI foams [174] are also known.

*Properties.* The range of apparent densities of flexible PI foams is 6 – 18 kg/m$^3$; the thermal conductivity is ca. 0.035 – 0.040 W m$^{-1}$ K$^{-1}$ at 10 °C.

The materials are resistant to solvents. They do not undergo hydrolysis and can be used between − 195 and + 260 °C. Their fire behavior is particularly favorable: they have a very low smoke density and a high oxygen index [175–177].

*Uses.* Lower density PI foam is used for acoustic and thermal insulation. Its low mass and fire behavior make it highly suitable for aerospace applications [178] where material cost is not a primary concern.

Examples of uses are as a heat and vibration protection for sensitive military and aerospace electronics equipment, a lightweight fire barrier for airplane cabins, and cryogenic insulation for fuel tanks. The higher density foams can replace polyurethane foams in airplane seats because of their cushioning properties [179]. Uses in shipbuilding have also been reported.

## 4.10. Polymethacrylimide (PMI) Foams (→ Polyimides)

Methacrylic acid [79-41-4] and methacrylonitrile [126-98-7] are subjected to radical copolymerization after the addition of blowing agents (e.g., formamide, or 1-propanol) that will come into play later. A transparent sheet is formed which is heated in a second step to 175 – 220 °C.

Copolymer of methacrylic acid and methacrylonitrile → Polymethacrylimide [53112-45-7]

Heating results in imide formation and the production of a rigid, closed-cell foam (Rohacell, produced by Röhm, Federal Republic of Germany) [180–182].

*Properties.* The relatively rigid material has apparent density of 30 – 200 kg/m$^3$. The heat distortion temperature (180 °C) can be increased to ca. 205 °C by annealing, which increases the degree of imidization. While warm, the foamed plastic can be reshaped and compacted. The thermal conductivity is 0.030 – 0.034 W m$^{-1}$ K$^{-1}$.

Polymethacrylimide foam is resistant to organic solvents but attacked by alkaline media. The material burns, giving off little smoke; its oxygen index is 20.

*Uses.* Polymethacrylimide foam serves as a core material for lightweight laminates with high flexural rigidity which can be loaded dynamically. Because the foam is thermoplastic and can be compacted, complex forms can be produced with precision from a preform in a press. Therefore, the product frequently simplifies existing fabrication techniques. It is used in skis, tennis racquets, helicopter rotor blades, and load-bearing parts in airplanes.

Components with a large surface area (e.g., for railroad car construction) can be produced from sheet-molding compound sandwiches in short cycles at a molding pressure of ca. 1 MPa [183–185].

## 4.11. Poly(Methyl Methacrylate) (PMMA) Foams (→ Polymethacrylates)

*Production.* Monomeric methyl methacrylate [80-62-6] is subjected to radical polymerization in the form of a thin layer between glass plates. It contains trichlorofluoromethane or pentane as a blowing agent and butyl methacrylate as a plasticizer. A board is formed (typical thickness 4 mm) which is subsequently foamed at 110 °C [186].

*Properties.* The foamed sheets (typical thickness 16 mm) have very large closed cells (diameter 2 – 10 mm) and thus absorb very little visible light. Material with an apparent density of 30 kg/m$^3$ has a thermal conductivity of 0.035 – 0.044 W m$^{-1}$ K$^{-1}$. It is highly weather resistant.

The maximum long-term service temperature of the PMMA foam Isocryl (IMC Acrylguß) is 80 °C.

*Uses.* The transparent insulating material is used for double-glazing improvements to increase the thermal insulation of windows in large buildings and greenhouses. It is also used in ceiling construction [187].

## 4.12. Polyamide (PA) Foams (→ Polyamides)

*Production.* A structural foam with only slightly lowered apparent core density (1.1 → 1.0 kg/m$^3$) is produced by the reaction injection molding (RIM) process. Caprolactam [105-60-2], for example, is injected into a mold with sodium lactamate [2123-24-3] as catalyst and an acyl lactam as activator [188–190]. Nitrogen serves as blowing agent [191]. Reaction injection molding permits good filling of the mold with low postshrinkage of the molded part. No sink marks are formed.

Foams with apparent densities down to ca. 400 kg/m$^3$ are made by conventional thermoplastic injection molding from polyamide (nylon 6) and a second batch containing blowing agents (e.g., azocarbonimide) [192], [193].

Polyamide foams with apparent densities below 200 kg/m$^3$ can also be obtained by alkaline lactam polymerization with low-boiling hydrocarbons as blowing agents [192–195]. However, they have no commercial significance.

*Properties and Uses.* Polyamide foams with higher apparent densities are used more widely than those with low densities, and their properties resemble more closely those of solid polyamides. Their toughness, rigidity, and abrasion resistance are worthy of mention. The high heat resistance, solvent resistance, and relatively favorable fire behavior are additional advantages. The inherent water absorption of polyamides can alter the mechanical and thermal behavior of the foam.

Some applications of structural PA foams are as housings for electrical apparatus, lamp sockets, furniture parts, pallets, containers, and bowling pins. Trade names are listed in Table 1.

## 4.13. Poly(Phenylene Oxide) (PPE) Foams [→ Poly(Phenylene Oxides)]

Blends of poly(phenylene oxide) [25667-40-7] with high-impact polystyrene are fabricated into structural foams by thermoplastic foam injection molding. Trade names are listed in Table 1.

The material is distinguished by good mechanical properties and a high heat resistance. The polystyrene content has an adverse effect on the favorable fire behavior of PPE (oxygen index = 30).

Applications in the automobile sector (crash pads) and in computer manufacture have been described [10, p. 16].

In the patent literature, the manufacture of extruded, lighter weight foams from PPE – PS blends, by using chlorinated hydrocarbons as blowing agents, is described [196].

## 4.14. Polysulfone (PSU) Foams
(→ Specialty Plastics)

*Production.* In the *dough-foam process*, particles of polysulfone [25135-51-7] are allowed to swell and form a gel in a solvent such as methylene chloride or a ketone. A cold-extruded dough sheet is comminuted to dough pellets, which can then be processed into foam-molded parts by using EPS technology (cf. Section 4.2.1). The gel can also be pressed directly to give a preform which foams when heated. Heating causes the solvent or blowing agent to escape and the low glass transition temperature of the gel increases to the high polysulfone value of ca. 200 °C [197–199].

*Properties and Uses.* The apparent densities of PSU foams are between 40 and 500 kg/m$^3$, depending on the process used. Unfilled products made by the dough-foam process, for example, have apparent densities of 390 kg/m$^3$; fiber-reinforced foams may have apparent densities as high as 700 kg/m$^3$.

Polysulfone foams display high heat resistance and a favorable fire behavior.

Possible applications include self-lubricating bearings, as well as special laboratory and medical products.

Highly filled syntactic polyether sulfone foams have been proposed for radomes [200].

## 4.15. Polycarbonate (PC) Foams
(→ Polycarbonates)

*Production.* Structural polycarbonate [24936-68-3] foams with apparent densities of 560 – 900 kg/m$^3$ are produced by low-pressure thermoplastic foam injection molding. Special chemical blowing agents [201] are admixed prior to processing which occurs at about 300 °C [202].

The manufacture of lighter weight foams by extrusion with chlorinated hydrocarbon blowing agents has also been described [203].

*Properties and Uses.* Polycarbonate foams, like the solid material, exhibit high impact resistance, rigidity, and heat resistance. They are flame-resistant and have a high electrical resistance. They are resistant to acid, but are attacked by bases, and by aromatic or halogenated hydrocarbons.

Examples of applications are housings for electrical apparatus, sports equipment (water skis), and furniture parts. Trade names are listed in Table 1.

# 5. Testing (→ Plastics, Properties and Testing)

Almost all major industrialized countries have adopted their own standards for testing foamed plastics (e.g., ASTM, United States; BS, Great Britain; DIN, Federal Republic of Germany; Association Française de Normalisation AFNOR, France). International standards (ISO) also exist. In some countries, test methods are specified in end-use-related standards. Thus, for example, foams for the building industry and their test methods are specified in ASTM C 578 "Standardized Specification for Preformed, Cellular Polystyrene Thermal Insulation."

In this overview, only the principles of the most important, generally valid test methods are described. Detailed information necessary for actual testing must be obtained by referring to the standards listed.

## 5.1. Mechanical Properties

*Density* (ASTM D 1622, ASTM C 303, ISO 845, DIN 53 420). The density is determined by using cubical test specimens or slabs; it is measured in kilograms per cubic meter.

*Open- and Closed-Cell Content* (ASTM D 2856). The content of open and closed cells is determined by measuring the apparent and true volumes with a special apparatus. The fraction of open or closed cells is reported in volume percent.

### 5.1.1. Rigid Foams

*Compressive Properties* (ASTM D 1621, ISO 844, DIN 53 421). The *compressive stress* is the compressive load per unit area of minimum original cross section carried by the test specimen at any given moment (Fig. 11). It is expressed in force per unit area, e.g., newtons per square meter or pascals. The *compressive strength* is the stress at the yield point (the first point on the stress–strain curve at which an

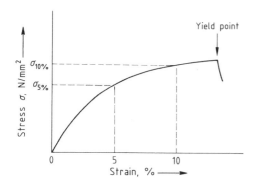

**Figure 11.** Typical stress–strain curve for a foamed plastic

increase in strain occurs without an increase in stress). In the absence of such a yield point, the compressive strength is the stress at 10% deformation.

***Tensile Properties*** (ASTM D 1623, ISO 1962, DIN 53 430). The *tensile strength* $\sigma_B$ is obtained by dividing the breaking load in newtons by the original minimum cross-sectional area of the specimen in square centimeters. The *elongation* is the extension of the material at the moment of rupture expressed as a percentage of its original length measured between gauge marks. The shape of the test specimen used depends on the amount of sample available, as well as the stiffness of the material. It may be round or square and have tapered ends or rounded shoulders; length varies from 10 cm to 1 m. A typical shoulder shape is shown in Figure 12. Speed of testing strongly influences the results; a standard speed is chosen such that rupture occurs in 1 – 6 min.

***Breaking Load and Flexural Properties*** (ASTM C 203, ISO 1209, DIN 53 423). The breaking load and flexural properties are tested by different procedures on a bar of rectangular cross section which can be loaded in two ways:

1. The bar rests on two supports and is loaded midway between the supports.
2. The bar rests on two supports and is loaded at the two quarter points.

Normally, values are measured at the maximum load or breaking point. The *flexural strength* is equal to the maximum stress in the outer layers at break; it is calculated from the load applied at the break point by means of

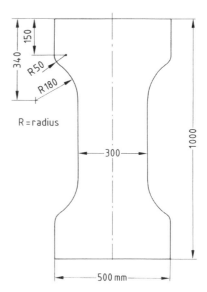

**Figure 12.** Specimen for tensile testing
All lengths are given in millimeters

equations that depend on the testing procedure used. Flexural properties may vary with specimen span-to-thickness ratio, temperature, difference in rate of straining, and testing procedure followed. In comparing results, these parameters must be consistent.

***Dynamic Stiffness*** (DIN 52 214). The dynamic stiffness of elastified rigid foams is easily measured and is used to estimate their potential as impact soundproofing material. A metal plate that applies a 2 kPa (kN/m$^2$) load uniformly to the test specimen is affixed to it with plaster. An alternating force is applied perpendicular to the metal plate. The resonance frequency $f_r$ of the system is measured. The dynamic stiffness $s'$ is proportional to $f_r^2$.

### 5.1.2. Flexible Foams

Several procedures are used to test the compressive behavior of flexible foams.

***Compression Load Deflection Test*** (ASTM D 1564, ASTM D 2406, DIN 53 577). This test involves measuring the load necessary to produce a defined compression (20, 40, or 60%) over the entire surface area of the foam specimen. The compressive stress at the specified compression is given in pascals or newtons per square meter. The test is also used to measure the relaxation of flexible foams after removal of the compressive load.

***Indentation Load Deflection*** (ILD) (ASTM D 1564, ASTM D 2406, DIN 53 576). This test measures the load necessary to produce a defined indentation (e.g., 25, 65, or 70 %) in the foam product. A flat, circular indentor foot of specified diameter is connected to a load-measuring device by means of a swivel joint and is pressed into the specimen. The force in newtons required for 25 and 65 % indentation is known as the 25 and 65 % ILD value, respectively.

The ASTM tests also describe a procedure for measuring indentation on loading. These values are known as indentation residual gage load (IRGL) values. The load deflection is determined by measuring the thickness of the foam under fixed loads on a circular indentor foot.

***Compression Set Test*** (ASTM D 1564, ASTM D 2406, DIN 53 572). This test involves compressing the foam specimen under specified conditions of time and temperature, and noting the effect on the thickness of the specimen. The compression device consists of two or more flat plates arranged so that the plates are held parallel to each other by bolts or clamps; the space between the plates can be adjusted to the required thickness by means of spacers.

Standard testing conditions are a compression of 50 or 75 %, a temperature of 70 °C for 22 h, and a relaxation time of 30 min. The constant-deflection compression set is calculated and expressed either as a percentage of the original thickness or as a percentage of the original deflection.

***Tensile Properties*** (ASTM D 1564, ASTM D 2406, DIN 53 571). Tensile stress, tensile strength, and elongation at break are measured when a tensile load is applied to the foam. The test specimen is stamped out with a die of a specified shape (shoulder shape, see Fig. 12). The *tensile stress*, in pascals or newtons per square millimeter, at a given elongation is the load applied at this elongation divided by the original cross-sectional area of the specimen. The *tensile strength* is the tensile stress at the breaking point.

***Resilience (Ball Rebound) Test*** (ASTM D 1564). The ball rebound test is used to evaluate the resilience of foams; a steel ball is dropped on a foam specimen, and the height of rebound is noted. Values are given as a percentage of the dropping height and are known as ball rebound resilience (BRR) values.

## 5.2. Thermal Properties

***Dimensional Stability*** (ISO 2796, DIN 53 431) consists of measuring the dimensions of a test specimen at a standard test temperature (e.g., −25, 0, or +70 °C). The change in linear dimensions is expressed as a percentage of the values measured at room temperature.

***Compressive Creep*** (ISO/TR 2799, ISO 4898, DIN 18 164) consists of measuring the change in thickness of a test specimen when it is subjected to a load at elevated temperature. The change is given as a percentage of the original thickness. Several testing procedures are used; common testing conditions follow.

| $t$, °C | Load, kPa | Time, h |
|---|---|---|
| 70 | 0 | 48 |
| 80 | 20 | 48 |
| 70 | 40 | 168 |

***Deflection Temperature*** (ASTM D 648, DIN 53 424) consists of determining the temperature at which an arbitrary deformation occurs when a specimen is subjected to an arbitrary set of testing conditions. Either a compressive or a flexural load may be used.

*Flexural Load.* A bar of rectangular cross section is tested as a simple beam; the load is applied at its center, or the bar is held at one end while the load is applied at the other. The loaded specimen is immersed in a heat-transfer medium whose temperature can be increased at a constant rate (e.g., 2 °C/min or 50 °C/h). The temperature of the medium is measured when a certain deflection (e.g., 2 mm) has been reached and is recorded as the deflection temperature under flexural load.

*Compressive Load.* A test specimen of $50 \times 50 \times 20$ mm is loaded with 25 kPa (0.025 N/mm$^2$). Then, as with flexural load, the temperature of the test specimen is raised at a constant rate (e.g., 50 °C/h). The temperature at which a certain deflection is reached is recorded as the deflection temperature under compressive load.

*Thermal Conductivity* (DIN 52 612, ASTM C 177, ASTM C 518, ASTM C 236). The thermal conductivity, expressed in watts per meter per kelvin, is the rate of heat flow through a unit thickness of an infinite slab of a homogeneous material in a direction perpendicular to the surface, induced by a unit temperature difference.

Thermal conductivity is one of the most important properties of foamed plastics and is generally low due to their cellular structure.

Because of the technical importance of this property and the complicated nature of the measurements, different test methods exist. In principle, a steady-state heat flux is measured through flat slabs of the specimen using a guarded hot-plate apparatus. Single-sided and double-sided modes of measurement are used. An idealized double-sided system is shown in Figure 13. The specimen is sandwiched between two isothermal cold plates and a guarded hot plate. The guard portion of the isothermal hot plate provides the power necessary to create the proper thermal conditions within the test volume.

A good review of terms, definitions, symbols, and units pertaining to thermal insulating materials is given in ASTM C 168.

*Fire Behavior.* The fire behavior of foamed plastics depends on many parameters. Therefore, many different tests exist, each of which emphasizes different aspects of burning, i.e. ease of ignition, flame propagation, heat release, combustion products (smoke density and toxicity). The most important tests are DIN 4102, DIN 53 438, ASTM E 84, NF P92 501, NF P92 504, NF P92 505, NF P92 506, UL 94, ASTM D 2863–77.

## 5.3. Miscellaneous Properties

*Water absorption* (ASTM D 2842, ISO 2896, DIN 53 433) of cellular plastics is determined by measuring the change in buoyancy resulting from immersion under a head of water for a certain time (2 h, 4 d, 7 d, 28 d, or longer). Water absorbed by the specimen lowers buoyancy by reducing the volume of water displaced. Water absorption may be expressed in volume percent or weight percent.

*Thermal and Humid Aging* (ASTM D 2126). Because of the wide variety of potential uses of cellular plastics, artificial aging tests must be based on the intended application. Normally, responses to temperature and humidity are selected. Temperature ranges from − 70 to + 200 °C and humidity ranges from ambient to 97 % are used. The specimen is examined visually; changes in density and dimensions are also measured.

*Electrical properties* of foamed plastics are seldom measured. Testing methods are identical with those used for standard plastics (see → Plastics, Properties and Testing).

## 6. Economic Aspects

Polyurethanes are the most important foams as far as consumption is concerned and are described elsewhere (→ Polyurethanes). The consumption of other major foamed plastics in Western Europe, the United States, and Japan is summarized in Table 4. The most important of these are foams derived from polystyrene beads and extruded polystyrene. Significant amounts of foam from polyolefins, poly(vinyl chloride), phenol–formaldehyde resins, and urea–formaldehyde resins are also used.

## 7. Toxicology and Environmental Aspects

The polymeric materials used for foamed plastic production generally do not present a

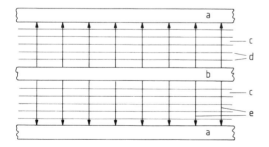

**Figure 13.** Schematic of the guarded hot-plate apparatus for determining thermal conductivity
a) Cold isothermal surface; b) Hot isothermal surface; c) Specimen; d) Isothermal planes; e) Heat flux lines

**Table 4.** Consumption of foamed plastics in 1986

| Foamed plastics[*] | Consumption, $10^3$ t | | |
|---|---|---|---|
| | Western Europe | United States | Japan |
| Expanded polystyrene (EPS) | 485 | 283 | 190 |
| Extruded polystyrene (XPS) | 135 | 293 | 300 |
| Poly(vinyl chloride) (PVC)[**] | 25 | 150 | |
| Polyolefin (PO) | 20 | 25 | 37 |
| Urea–formaldehyde (UF) resin | 12 | | |
| Phenol–formaldehyde (PF) resin | 10 | 8 | 3 |

[*] Foamed plastics with an apparent density of < 200 kg/m$^3$.
[**] Including foams with apparent density > 200 kg/m$^3$.

toxicological hazard because they are not absorbed by the human body. Of greater concern are residual monomers and additives, i.e., materials of lower molecular mass. All legal requirements pertaining to such constituents must be observed. For the benefit of processors and end users, the manufacturers of synthetic foams provide pamphlets, safety bulletins, and technical information on the processing and proper handling of synthetic foams.

Foamed plastic waste is bulky and is therefore often compacted before disposal. Since toxic substances are not released on storage, contamination of ground water does not occur. After being dumped the foamed plastics are slowly degraded, the rate depending on the nature of the plastic [204]. During the controlled aerobic degradation of noncomminuted thermoplasts and thermosets (composting), soil-inhabiting Actinomycetes were shown to attack aminoplasts most effectively [205].

Although foamed plastics have a high calorific value, they do not play an important role in refuse incineration plants because of their low mass. However, thermoplastic foams must be mixed with sufficient quantities of nonmelting materials with a low calorific value.

The combustion products of most foamed plastics are carbon dioxide and steam which do not pose any environmental problems. However, chlorine-containing products release hydrogen chloride which must be removed by washing the flue gas [206].

Chlorofluorocarbons (CFCs) are used in the production of many foamed plastics. They remain enclosed within the cells of the plastic and increase the thermal insulation of air-filled foams.

The ozone depletion hypothesis [207] suggests that CFCs damage the stratospheric ozone layer that prevents shortwave UV radiation from reaching the earth. Models of stratosphere chemistry, propose as many as 200 possible reactions that involve CFCs, carbon dioxide, nitrogen oxides, and methane. Chlorine atoms seem to play an important role in initiating ozone-consuming reactions [208–210].

"Soft" CFCs (e.g., chlorodifluoromethane [75-46-6]) are being introduced as alternative blowing agents. Since they are less stable than the customery CFCs, they are reported to decompose in the troposhere and thus have a much lower ozone depletion potential [211], [212].

## 8. Abbreviations used in this article:

| | |
|---|---|
| ABS | acrylonitrile–butadiene–styrene |
| BRR | ball rebound resilience |
| EP | epoxy |
| EPS | expandable polystyrene |
| EVM | ethylene–vinyl acetate copolymer |
| HDPE | high-density polyethylene |
| ILD | indentation load deflection |
| LDPE | low-density polyethylene |
| LLDPE | linear low-density polyethylene |
| MF | melamine–formaldehyde |
| NF | Norme Française |
| PA | polyamide |
| PC | polycarbonate |
| PE | polyethylene |
| PF | phenol–formaldehyde |
| PI | polyimide |
| PMMA | poly(methyl methacrylate) |
| PMI | polymethacrylimide |
| PO | polyolefin |
| PP | polypropylene |
| PPE | poly(phenylene oxide) |
| PS | polystyrene |
| PSU | polysulfone |
| PUR | polyurethane |
| PVC | poly(vinyl chloride) |
| RIM | reaction injection molding |
| RTV | room-temperature vulcanization |
| SAN | styrene–acrylonitrile |
| SI | silicone |
| TFE | thermoplastic foam extrusion |
| TFI | thermoplastic foam injection molding |
| UF | urea–formaldehyde |
| UL | Underwriter's Laboratories |
| XPS | extruded polystyrene |

# References
## General Reference

1. C. J. Benning: *Plastic Foams*, vols. 1, 2, Wiley-Interscience, New York 1969.
2. A. A. Berlin, F. A. Shutov, A. K. Zhitinkina: *Foam based on Reactive Oligomers*, Technomic Publ. Co, Westport 1982.
3. K. C. Frisch, J. H. Saunders: *Plastic Foams*, vols. 1, 2, Marcel Dekker, New York 1972/73.
4. N. C. Hilyard: *Mechanics of Cellular Plastics*, Applied Science Publ., London 1982.
5. D. Homann: *Kunststoff-Schaumstoffe*, Hanser Verlag, München 1966.
6. T. H. Ferrigno, *Rigid Plastic Foams*, Reinhold Publ. Co., New York 1967.
7. W. Knappe in R. Vieweg, D. Braun (ed.): *Kunststoff-Handbuch*, vol. 1, Chap. 5.1.4.6, Hanser Verlag, München 1975.
8. H. Piechota, H. Röhr, *Integralschaumstoffe*, Hanser Verlag, München – Wien 1975.
9. F. A. Shutov: *Syntactic Polymer Foams*, Springer Verlag, Berlin – Heidelberg 1986.
10. G. Trausch in VDI-Ges. Kunststofftechnik (ed.): *Schäume aus der thermoplastischen Schmelze*, VDI-Verlag, Düsseldorf 1981.
11. B. R. Wendle: *Structural Foam*, Marcel Dekker, New York 1985.

## Specific Reference

12. DIN 7726, (1982): Cellular Materials: Definition of Terms, Classification.
13. F. Lober, *Angew. Chemie* **64** (1952) 65 – 92.
14. R. A. Gorski, R. B. Ramsey, K. T. Dishart, *J. Cell. Plast.* **22** (1986) 21 – 52.
15. J. H. Saunders, R. H. Hansen in K. C. Frisch, J. H. Saunders (ed.): *Plastic foams, Part I, Chap. 2*, Marcel Dekker, New York 1972.
16. W. S. Fong, *Foamed Plastics, Suppl. A*, Process Economics Program Report 97A, SRI Int., Menlo Park, California, Jan. 1984, pp. 29 – 36.
17. *Mod. Plast. Int.*, **6** (1976) no. 9, 56 – 57. Ciba Geigy, EP 7 499, 1978.
18. R. J. Bathgate, K. T. Collington, *39th Annu. Tech. Conf.-Soc. Plast. Eng.* **27** (1981) 856 – 862.
19. J. A. Gribens, R. C. Wade, *39th Annu. Tech. Conf.-Soc. Plast. Eng.* **27** (1981) 852 – 855.
20. R. H. Hansen, W. M. Martin, *Ind. Eng. Chem.* **3** (1964) no. 2, 137 – 141.
21. G. Menges, F. Knipschild in N. C. Hilyard (ed.): *Mechanics of Cellular Plastics*, Applied Science Publ., London 1982, p. 27 – 72.
22. D. J. Doherty, R. Hurd, G. R. Lester, *Chem. Ind. London* **30** (1962) 1340 – 1356.
23. B. Koglin, *Kältetech. Klima.* **21** (1969) 122 – 125.
24. N. E. Hager, R. C. Steere, *J. Appl. Phys.* **38** (1967) 4663 – 4668.
25. W. Küster, *Kunststoffe* **60** (1970) 249 – 255.
26. H. Zehendner, 12. Int. Fachtagung für Schaumkunststoffe, Düsseldorf 1983, pp. 153 – 158.
27. H. Mittasch, *Plaste und Kautsch.* **16** (1969) 268 – 271.
28. DIN 18 164/2–1979: Insulating Materials for Impact Sound Insulation.
29. ASTM D 2863–77: Oxygen Index.
30. Dow Chemical, US 3 067 147, 1957.
31. A. Osakada, *Jpn. Chem. Q.* **5** (1969) 55.
32. N. Shiina, M. Tsuchiya, H. Nakae, *Jpn. Plast. Age* **10** (1972) no. 12, 37 – 48.
33. Toray Ind., US 3 542 702, 1967.
34. Sekisui, JP-Kokai 61–143 450, 1986.
35. Mitsubishi Petr., JP-Kokai 58–134 131, 1983.
36. Y. D. Lee, L. F. Wang, *J. Appl. Polym. Sci.* **32** (1986) 4639 – 4647.
37. Hercules, US 3 341 481, 1964.
38. M. B. Johnson, *Polym. Age* **6** (1975) 88 – 89.
39. Dynamit Nobel, DE 3 430 108, 1984.
40. Furukawa, DE-AS 1 694 130, 1967.
41. F. W. Alfter, *Plastverarbeiter* **29** (1978) 129 – 132.
42. Sekisui, DE 1 947 589, 1969.
43. *GAK Gummi, Asbest + Kunstst.* **36** (1983) 356 – 357.
44. BASF, DE-OS 2 542 452, 1975.
45. F. Stastny, R. Gäth, U. Haardt, *Kunststoffe* **61** (1971) 745 – 749.
46. Kanegafuchi, US 4 448 901, 1983.
47. Japan Styrene Paper, US 4 399 087, 1981.
48. Yuka Badische, DE-OS 3 431 245, 1983.
49. Japan Styrene Paper, US 4 440 703, 1982.
50. N. Shiina, M. Tsuchiya, H. Nakae, *Jpn. Plast. Age* **11** (1973) no. 3, 47 – 53.
51. F. Stastny, R. Gäth, U. Haardt, *GAK Gummi, Asbest + Kunstst.* **25** (1972) 398 – 402.
52. F. Kleiner, *Kautsch. + Gummi, Kunstst.* **28** (1975) 149 – 153.
53. BASF, DE 1 629 316, 1966.
54. U. G. Haardt, *Kunststoffe* **68** (1978) 468 – 469.
55. BASF, DE-AS 1 629 281, 1966.
56. Kanegafuchi, US 3 953 558, 1973.
57. Japan Styrene Paper, EP 097 504, 1983.
58. N. Shiina, M. Tsuchiya, H. Nakae, *Jpn. Plast. Age* **11** (1973) no. 1, 49 – 53, **11** (1973) no. 2, 45 – 50.
59. I. Tamai, K. Yamaguchi, *Jpn. Plast. Age* **10** (1972) no. 12, 22 – 27.
60. K. Leilich, K. E. Büscher, in R. Vieweg (ed.): *Kunststoff-Handbuch*, vol. IV, Hanser Verlag, München 1969, p. 709.
61. J. Helberg, W. Löw, T. Riedel, *Kunststoffe* **75** (1985) 342 – 345.
62. BASF, DE 845 264, 1951; DE 941 389, 1951.
63. K. Hinselmann, D. Stockburger, *Kunststoffe* **61** (1971) 152 – 157.
64. I. De Grave, *Kunststoffe* **70** (1980) no. 10, 625 – 629.
65. K. Stange, *Kunststoffe* **74** (1984) no. 10, 556 – 559.
66. E. Neufert: *Styropor-Handbuch*, 2nd ed., Bauverlag GmbH, Wiesbaden – Berlin 1971.
67. W. Prankel, *Tech. Rundsch.* **39** (1979) 9 – 15;**42** (1979)5 – 7.
68. G. Baum, *Beton Information* **5** (1974) 58 – 67.
69. W. Albert, H. Gümpel, H. Röhr, *Kunststoffe* **61** (1971) 386 – 396.
70. Dow Chemical, US 2 450 436, 1947.
71. Dow Chemical, US 2 669 751, 1950.
72. A. Krämer, *Plastverarbeiter* **25** (1974) 73 – 78.
73. A. Krämer, *Kunststoffe* **64** (1974) 350 – 353.
74. R. Städter, *Kunststoff J.* **8** (1974) no. 7/8, 22 – 28.
75. R. P. Sacha, 29th Annu. Techn. Conf. – Soc. Plast. Eng., May 10 – 13 1971, pp. 326 – 328.
76. Gebr. Anger, DE 1 963 811-Q, 1971.
77. Dynamit Nobel, DE 2 158 673-Q, 1973.
78. *Kunstst. Berat.* **18** (1973) no. 11, 812.
79. *Kunstst. Berat.* **19** (1974) no. 4, 180 – 182.
80. K. Wetzel, *Plaste Kautsch.* **24** (1977) no. 3, 197 – 199.
81. H. Barth, *Kunststoffe* **67** (1977) no. 3, 130 – 135.
82. Dow Chemical, US 4 146 563, 1979.

83. *Mod. Plast. Int.* **12** (1982) no. 4, 18.
84. H. W. Matill, A. Schaefer, *Kunstst. Bau* **20** (1985) no. 2, 81 – 83.
85. P. Zingsheim in W. Becker, D. Braun (eds.): *Kunststoff Handbuch*, vol. **2/2**, Carl Hanser Verlag, p. 977 –1000.
86. Vallourec, FR 2033 555-Q, 1970.
87. E. Morianz, *"Schäume aus der thermoplastischen Schmelze,"* in Reference 10 pp. 85 – 93.
88. *Plast. Ind. News* **24** (1978) no. 2, 26.
89. Ugine Kuhlmann, DE-OS 1 919 921, 1969.
90. L. F. Fow, *Soc. Plast. Eng. (Tech. Pap.)* **24** (1978) 29 – 33.
91. *Mod. Plast. Int.* **13** (1983) no. 6, 34 – 35.
92. Kleber Colombes, DE-OS 1 569 238, 1963.
93. *Elastomerics* **114** (1982) no. 4, 18.
94. M. Bouzim, *Rev. Gen. Caoutch. Plast.* **49** (1972) no. 6, 541.
95. Y. J. Landler, *J. Cellular Plast.* **3** (1967) no. 9, 400 – 404.
96. H. Boelle, *GAK Gummi, Asbest + Kunstst.* **34** (1981) no. 9, 546 – 552.
97. UCC, DE-AS 10 10 271, 1955.
98. Israel Dennis, US 2 763 475, 1954.
99. Vanderbilt, DE-AS 1 719 311, 1967.
100. K. Hillier, *Plast. Rubber Process Appl.* **1** (1981) 39 – 42.
101. C. J. Hilado, P. A. Huttlinger, *J. Cell. Plast.* **16** (1980) 285 – 292.
102. A. J. Lowe, A. Barnatt, *Chimica Ind. (Milan)* **61** (1979) 656 – 660.
103. M. Cherubim in *Schaumkunststoffe, Berichte von den FSK-Tagungen 1971 – 1975*, Hanser Verlag, München – Wien 1976, pp. 402 – 406.
104. F. A. Shutov, *Cell. Polym.* **3** (1984) 95 – 104.
105. V. D. Valgin, *Europlast. Mon.* **46** (1973) no. 7, 57 – 60.
106. K. Brüning in *Schaumkunststoffe*, cf. Reference 103, pp. 407 – 410.
107. Koppers, DE-OS 3 139 890, 1981.
108. Koppers, DE-OS 3 324 431, 432, 433, 1983.
109. BP Chemicals, EP-A 154 452, 1985.
110. BP Chemicals, EP-A 170 357, 1985.
111. Isomo, EP-A 108 167, 1982.
112. BP Chemicals, EP-A 068 673, 1982.
113. A. J. Lowe, A. Barnatt, E. F. Chandley, R. Dyke, K. Hillier, *Eur. J. Cell. Plast.* **1** (1978) no. 1, 42 – 49.
114. A. Vazquez, R. J. Williams, *Cell. Polym.* **5** (1986) 123 – 140.
115. F. A. Shutov, *Cell. Polym.* **1** (1982) 73 – 75.
116. G. C. Marks, 12. Int. Schaumkunststofftagung, Düsseldorf 1983, pp. 59 – 67.
117. A. S. Wood, *Mod. Plast. Int.* **11** (1981) no. 10, 48 – 50.
118. Smithers, US 2 753 277, 1953.
119. F. A. Shutov, V. V. Ivanov, *Cell. Polym.* **3** (1984) 369 – 382.
120. E. G. Norman: "Plastics and Rubber," *Processing* **4** (1979) no. 3, 41 – 46.
121. H. D. Schmidt, *Kunstst. Bau* **10** (1975) no. 2, 8 – 10.
122. *Mod. Plast. Int.* **6** (1976) no. 6, 8 – 10.
123. Sekisui Plast., JP-Koaki 59–155 016, 1983.
124. E. Rühl, EP-A 154 874, 1985.
125. B. Mosier, US 4 596 682, 1984.
126. Thermocell Dev., US 4 390 641, 1982.
127. L. Unterstenhöfer, *Kunststoffe* **57** (1967) 850 – 855.
128. H. Baumann, *Plastverarbeiter* **27** (1976) 235 – 243.
129. E. K. H. Wulkan, 9. Int. Fachtagung für Schaumkunststoffe, Düsseldorf 1979, pp. 59 – 82.
130. Schaumchemie, DE 1 054 232, 1956.
131. BASF, DE-OS 2 064 102, 1970.
132. H. Baumann, *Cell. Polym.* **3** (1984) 249 – 261.
133. Schaumchemie, DE 3 216 897, 1982.
134. Schaumchemie, DE-OS 3 108 279, 1981.
135. E. K. H. Wulkan, *Kunstst. Bau* **14** (1979) no. 4, 167 – 171.
136. L. Gerhardy, H. Lottner, *Baugewerbe* (1980) no. 3, 30 – 34.
137. R. Grossmann, *Glückauf* **112** (1976) no. 14, 803 – 807.
138. Bauer, DE-AS 1 014 508, 1953.
139. Schaumchemie, DE 1 298 488, 1967.
140. BASF, DE-AS 1 794 131, 1968.
141. Vorwerk, DE 2 732 011, 1977.
142. J. Miltz, G. Gruenbaum, *J. Cell. Plast.* **17** (1981) no. 7/8, 213 – 219.
143. CIBA, DE 870 027, 1948.
144. H. Weber, F. Krückau, *Kunststoffe* **75** (1985) 843 – 848.
145. H. Weber, 14. Schaumkunststofftagung, Düsseldorf 1986, pp. 61 – 77.
146. BASF, EP 074 593, 1979.
147. BASF, EP 037 470, 1981.
148. BASF, DE-OS 3 109 929, 1981.
149. BASF, DE-OS 3 138 862, 1981.
150. DIN 53 577 (1976); ISO 3386/1: Determination of Compression Stress/strain Characteristics and Compression Stress Value.
151. BASF, EP-A 111 860, 1983.
152. General Electric, US 2 875 163, 1956.
153. Dow Corning, US 3 070 555, 1959.
154. Wacker-Chemie, DE 1 224 040, 1963.
155. General Electric, US 3 425 967, 1965.
156. Dow Corning, US 3 923 705, 1974.
157. C. L. Lee, R. G. Niemi, K. M. Kelly, *J. Cell. Plast.* **13** (1977) no. 1/2, 62 – 66.
158. C. L. Lee, G. M. Ronk, *J. Cell. Plast.* **18** (1982) no. 5/6, 178 – 182.
159. C. L. Lee, G. M. Ronk, S. Spells, *J. Cell. Plast.* **19** (1983) no. 1/2, 29 – 33.
160. Dow Corning, US 4 567 212, 1985.
161. Dow Corning, EP-A 099 002, 1983.
162. Polyvoltac Ltd. (UK), Information sheet, Southall, England, May 1986.
163. US Air Force, US 3 429 838, 1966.
164. H. L. Vincent in K. C. Frisch, J. H. Saunders (ed.): *Plastic Foams, Part I, "Silicone Foams,"* Marcel Dekker, New York 1972, pp. 385 – 396.
165. M. Hajimichel, A. Lewis, D. Scholey, C. Simmonds, *Br. Polym. J.* **18** (1986) no. 5, 307 – 311.
166. F. A. Shutov, A. A. Berlin: *Epoxy Oligomer Based Foams in Foam based on Reactive Polymers*, Technomic Publishing Co., Westport 1982, pp. 215 – 261.
167. F. A. Shutov: "Syntactic Polymer Foams," in *Advances in Polymer Science*, vol. 73/74, Springer Verlag, Berlin – Heidelberg 1986, pp. 63 – 123.
168. *Shell Polymers* **7** (1983) no. 3, 80 – 81.
169. M. Narkis, S. Kenig, M. Putermann, *Polym. Compos.* **5** (1984) 159 – 164.
170. S. Kenig, I. Raiter, M. Narkis, *J. Cell. Plast.* **20** (1984) 423 – 429.
171. Int. Harvester, US 4 296 208, 1980.
172. IMI-Tech Corp., US 4 535 101, 1984.
173. P. B. Rand, *J. Cell. Plast.* **9** (1973) 130 – 133.
174. *Mod. Plast. Int.* **10** (1980) no. 3, 21.
175. *Plast. Technol.* **28** no. 3, March 1982, pp. 15 – 17.
176. W. Frank, 14. Schaumkunststofftagung, Düsseldorf 1986, pp. 9 – 22.
177. J. Gagliani, *NASA Tech Briefs*, Fall 1980, pp. 384 – 387.
178. *Chem. Eng. News* **59** (1981) no. 1, 21.
179. *Kunststoffe* **77** (1987) 57.
180. Röhm, DE 1 817 156, 1968.
181. Röhm, DE 2 726 259, 1977.
182. Röhm, DE 2 726 260, 1977.

183 W. Pip, *Kunststoffe* **64** (1974) 23 – 27.
184 W. Pip, *Plastverarbeiter* **34** (1983) 629 – 632.
185 Röhm: *Rohacell-Handbuch*, Darmstadt 1986.
186 Imchemie, EP 068 439, 1982 (A. Hohnholz).
187 *Kunstst. Bau* **20** (1985) 75.
188 J. D. Gabbert, R. M. Hedrick, *Kunststoffe* **75** (1985) 416 – 420.
189 P. Wagner, *Kunststoffe* **73** (1983) 588 – 590.
190 J. Härting, *Kunstst. Berat.* **20** (1984) no. 11/12, 30 – 32.
191 Rogers Corp., US 4 464 491, 1983.
192 M. Kuga, K. Matsukura, *Jpn. Plast. Age* **10** (1972) 32 – 39.
193 UBE Industries, JP-Kokai 57 000 144, 1982.
194 BASF, DE-AS 1 159 643, 1961.
195 BASF, DE-AS 1 177 340, 1962.
196 Mobil Oil, US 4 598 101, 1985.
197 W. H. Smarook, *Soc. Plast. Ind., 32nd Annu. Tech. Conf.*, 1977 Section 11-G, 1 – 8.
198 D. A. Blackadder, A. Gupta, W. D. James, *Chem. Eng. Sci.* **38** (1983) no. 1, 181 – 182.
199 H. R. Smith, WO 82/00649 (1981).
200 ICI, EP 16 544, 1980.
201 Bayer, DE 1 252 891, 1961.
202 W. Cohnen, W. Meyer, *Kunstst. Rundsch.* **20** (1973) 425 – 434.
203 Mobil Oil, US 4 579 874, 1985.
204 E. Küster, *J. Appl. Polym. Sci. Appl. Polym. Symp.* **35** (1979) 395 – 404.
205 P. Chrometzka, F. X. Kneer, *Forum Umwelt Hyg.* **26** (1975) no. 8, 225 – 230.
206 H. Reimer, *Müll Abfall* **4** (1972) no. 1, 19 – 21.
207 F. S. Rowland, M. J. Molina, *Rev. Geophys. Space Phys.* **1** (1975) 1 – 35.
208 J. F. Mills, *Polyurethanes World Congress 1987*, Aachen, pp. 54 – 58.
209 ICI Brochure: *Chlorofluorocarbons and the Ozone Layer*, ICI, Runcorn, Ch, England, 1986.
210 L. R. Ember, P. L. Layman, W. Lepkowski, P. S. Zurer, *Chem. Eng. News* **64** (1986) no. 47, 14 – 64.
211 K. T. Dishart, J. A. Creazzo, M. R. Ascough in Reference 208, pp. 59 – 66.
212 *Mod. Plastics Int.* **17** (1987) no. 12, 57 – 59.

# Further Reading

C. A. Harper: *Handbook of Plastic Processes*, Wiley, Hoboken, NJ 2006.

K. W. Suh: *Foamed Plastics*, "Kirk Othmer Encyclopedia of Chemical Technology", 5th edition, John Wiley & Sons, Hoboken, NJ, online DOI: 10.1002/0471238961.06150113192108.a01.

M. Tolinski: *Additives for Polyolefins*, William Andrew, Oxford 2009.

# Part 4

# Resins

# Alkyd Resins

FRANK N. JONES, Coatings Research Institute, Eastern Michigan University, Ypsilanti, MI 48197, USA

| | | | |
|---|---|---|---|
| 1. | Introduction . . . . . . . . . . . . . . . . . . 1597 | 3. | Film Formation by Coatings Based on Alkyd Resins . . . . . . . . . . . . . . . |
| 2. | Production . . . . . . . . . . . . . . . . . . . 1599 | 4. | Modified Alkyds . . . . . . . . . . . . . . 1607 |
| 2.1. | Raw Materials. . . . . . . . . . . . . . . . . 1599 | 5. | Low-Solvent Coatings . . . . . . . . . . 1609 |
| 2.2. | Polymerization . . . . . . . . . . . . . . . 1601 | 6. | Markets and End Uses. . . . . . . . . . 1611 |
| 2.3. | Formulation and Process Types; Laboratory Processing. . . . . . . . . . 1602 | | References. . . . . . . . . . . . . . . . . . . 1612 |
| 2.4. | Manufacture . . . . . . . . . . . . . . . . . 1604 | | |

## 1. Introduction

Alkyd resins are, literally, polyesters. They are made by condensation polymerization of three types of monomers:

1. Polyols
2. Polybasic acids
3. Fatty acids or triglyceride oils

Because of their distinctive properties, alkyds often are classified separately from other polyesters, as they are in this encyclopedia.

The largest use of alkyds is for surface coatings (paints, enamels, lacquers, and varnishes) in which the resins function as binders, forming a tough, continuous film that adheres to the object coated. Alkyds are the most versatile of coating binders, and they are used extensively in all major categories of coatings: architectural, industrial, and special purpose (see Chap. 6). Despite gradual replacement of alkyds with other types of resins, a substantial fraction of all organic coatings applied worldwide use alkyds as a primary binder or as an additive.

Alkyds are also important ink binders. Other uses include caulks, adhesives, and plasticizers.

Relative to most commercial polymers, alkyd resins often have low glass transition temperatures $T_g$. Consequently, they are viscous, tacky materials that are difficult to handle. Most often, the handling problems are overcome by dissolving the resins in organic solvents to give solutions that can be poured and pumped.

Because of their low $T_g$, most alkyds are useful as coating binders only after they are chemically cured to harden the film after the liquid coating has been applied. Curing is accomplished by cross-linking chemical reactions. Cross-linking may result from chemical reaction of the coating film with oxygen in the air or from chemical reaction with a second resin or substance. The second resin, called a cross-linker, is blended with the alkyd at the time of coating manufacture or application. Certain alkyds are added to acrylic and nitrocellulose lacquers, where their role is to plasticize the brittle thermoplastic resins. Alkyds used in this way do not cross-link.

*History.* Paints containing vegetable oils have been used for at least 600 years. A major goal in the early development of alkyd resins was to combine vegetable oil technology with polyester technology to make improved materials. Probably the first polyester resin was made by BERZELIUS [18] in 1847. It was an intractable product of glycerol and tartaric acid. In 1901 W. SMITH [19] reported a brittle glycerol – phthalic anhydride polymer, and by 1910 workers at General Electric were attempting to develop more tractable materials for electrical end uses. Work was probably hampered by lack of an accepted theory of polymerization, but several patents were obtained.

The General Electric program eventually led to preparation in 1921 of glycerol – phthalic anhydride – fatty acid polymers. By 1926 these polymers were being used commercially as adhesives, and research on their use in coatings was underway [20]. The first account of this work was published in 1929 [21]. A U.S. patent was issued in 1933 [22] but was declared invalid in 1935. Technical progress presumably was aided by emergence of polymerization theory in the late 1920s [23], [24].

An important early development was the monoglyceride process for making alkyds directly from vegetable oils, eliminating the expense of preparing and purifying glycerine and fatty acids. In the 1930s, linseed oil was the most common binder for architectural paints, but alkyds were beginning to replace it. Alkyds offered faster drying, less yellowing, and superior protective properties. By 1940 U.S. consumption of alkyds exceeded $4 \times 10^4$ t/a.

By the early 1950s, alkyds had become the preeminent binder for coatings. U.S. consumption had reached $2 \times 10^5$ t/a and was rising. However, water-based architectural paints were being introduced. These paints generally used polyvinyl or polyacrylic emulsion resins as binders and contained little or no alkyd. During the 1950s, 1960s, and 1970s, water-based paints gradually captured most of the architectural coatings market from alkyd- and oil-based paints. The newer paints were popular because they could be cleaned up with water and were economical to produce because of low petroleum prices. Despite declining use in architectural paints, total use of alkyds continued to increase during most of this period because of rapidly growing use in a huge variety of industrial and special purpose coatings and in inks. Alkyd consumption in the USA reached a plateau at about $3 \times 10^5$ t/a in 1960 – 1980 and then began declining at a rate of about 2 %/a, reaching about $2.3 \times 10^5$ t/a in 1996 [25]. Worldwide consumption of alkyds is probably declining at a slower rate and perhaps holding steady.

Alkyd technology is being adapted to meet contemporary needs. Concern about air quality led to development of alkyds for use in high-solids and water-reducible coatings. Rising petrochemical prices have increased interest in all polymers made partly from renewable resources. Besides, alkyds are entrenched in innumerable specialized end uses, so wholesale replacement would be difficult.

***Definitions and Classifications.*** Terminology relating to alkyd resins can be confusing because it has never been uniform and because it has changed considerably over the years. The ASTM defines an alkyd as "a synthetic resin made from polyhydric alcohols and polybasic acids; generally modified with resins, fatty oils, or fatty acids" [26]. The DIN definition is similar [27].

For this article, slightly narrower and more exact definitions have been adopted. They are conveniently represented as follows:

| Types of monomer | | Terminology | |
|---|---|---|---|
| Polyols | ⎫ | ⎫ | |
| Polybasic acids | ⎬ Polyester resin ⎬ Alkyd resin | ⎬ Modified alkyd resin |
| Fatty acids | ⎭ | | |
| Others | | ⎭ | |

In this article the word "alkyd" is used as both a noun and an adjective. Therefore, "alkyd" and "alkyd resin" are synonymous. This usage is widespread, although some authors feel that only the latter usage is correct.

The scope of this article includes alkyds and modified alkyds. Polyesters are treated under separate keywords (see → Polyester Resins, Unsaturated, → Polyesters).

A brief retracing of the history of resin terminology may help clarify some of the terms that may be encountered. "Alkyd" is a respelling of "alcid", a word apparently coined at General Electric for the reaction product of polyfunctional alcohols and organic acids. Therefore, in the 1920s "alkyd" meant what is now meant by "polyester". The resins now called alkyds were called "oil-modified alkyds" during the 1930s. For a time, almost all commercial alkyds contained oil, so the phrase "oil-modified" was redundant and fell out of use. As polyester resins gained popularity in coatings, they began to be called "oil-free alkyds", a term that is now becoming obsolete. In the older literature, the term "glyptal" is found. It is synonymous with "alkyd". It was originally a General Electric trade name.

Today, the term "modified alkyd" usually refers to a resin in which a fourth type of monomer (a modifier) is incorporated in the polymer molecules. This definition is used in this article.

Because alkyds are a large and diverse family of resins, several *subclassifications* are often used. These groupings describe the features that most strongly affect resin properties and end uses, such as the type and amount of fatty acid and the identity of other components.

*Fatty Acid Type.* There are two broad groupings:

1. *Drying alkyds* contain enough unsaturated fatty acids to make curing by oxygen possible.
2. *Nondrying alkyds* contain lower levels of unsaturated fatty acids and are not polymerized appreciably with oxygen.

To be more specific, the particular oil used in preparing the resins is often stated.

*Fatty Acid Amount.* A classification called "oil length" arbitrarily groups alkyds by the amount of oil or fatty acid present. For alkyds made directly from oils, the oil length is given by Equation (1).

$$\text{Oil length} = \frac{\text{(mass of oil used to make alkyd resin} \times 100)}{\text{(mass of reactants} - \text{mass of water evolved in process)}} \quad (1)$$

For alkyds made from fatty acids, the formula is amended to correct for the extra mass of water lost (usually about 4 %) during esterification of the fatty acids, i.e., the factor of 100 in Equation (1) is replaced by 104.

Many authors classify alkyds by "oil length", a term apparently derived from terminology in varnish production in which oils were blended with other ingredients such as rosin or phenolic resins. Oil-length classifications are approximate, mainly because the ranges tabulated in Table 1 are standardized in Europe but not, apparently, in

**Table 1.** Percentage of oil in alkyd classes

| Oil length | Europe (DIN 55 945) | Common U.S. usage |
|---|---|---|
| Short oil alkyds | < 40 | < 45 |
| Medium oil alkyds | 40 – 60 | 45 – 55 |
| Long oil alkyds | 60 – 70 | 55 – 70 |
| Very long oil alkyds ("alkyd oils") | >70 | > 70 |

the rest of the world. Different authors assign somewhat different dividing points between short and medium, medium and long, etc. Despite being inexact, oil length classifications are used widely. The ranges used in Europe and the USA are listed in Table 1 [28].

Other authors, however, give different ranges for the USA; in one case they are the same as STOYE gives for Europe [29]. In view of the ambiguities involved, it is safest to work with the actual oil length numbers, not the verbal ranges.

*Other Constituents.* When the polyol and polybasic acid are not specified in describing an alkyd, it probably contains glycerol and phthalic anhydride, the most common ingredients. Other ingredients often are mentioned in describing a resin, e.g., "an isophthalic alkyd".

Modified alkyds usually, but not always, are classified according to the generic type of modifer, e.g., "a styrenated alkyd" or "a silicone-modified alkyd".

## 2. Production

### 2.1. Raw Materials

Well over one hundred raw materials are used in commercial alkyd manufacture. Each resin contains a minimum of three monomers and often five to seven. Only the most common raw materials are listed here.

**Polyols** (→ Alcohols, Polyhydric, → Ethylene Glycol, → Glycerol). Alkyds produced by processes that start with triglyceride oils (see Section 2.3) always contain at least some glycerol. Other polyols often are used, leading to alkyds of mixed polyol content.

Processes starting with fatty acids (see Section 2.3) allow the use of any polyol or polyol mixture to achieve the desired spectrum of properties. Glycerol [56-81-5] can be used, but less expensive synthetic polyols are a common choice. 1,1,1-Trimethylpropane [77-99-6] is a popular synthetic alternative to glycerol. Pentaerythritol [115-77-5], with four hydroxyl groups per molecule, is popular because it builds highly branched polymers and because it is relatively inexpensive. Mixtures of pentaerythritol and ethylene glycol [107-21-1] may be used as a low-cost substitute for glycerol.

***Polybasic Acids and Their Anhydrides.*** Phthalic anhydride [85-44-9] (→ Phthalic Acid and Derivatives) is used more than any other polybasic acid for reasons of economy and easy processing. Isophthalic acid [121-91-5] is often used where superior weatherability and corrosion resistance are required. Cycloaliphatic diacids, such as 1,4-cyclohexanedicarboxylic acid [1076-97-7], are also said to give better weatherability than phthalic acid. Trimellitic anhydride [552-30-7] (→ Carboxylic Acids, Aromatic) is used mainly in water-reducible alkyds. Many alkyds contain a small percentage of maleic anhydride [108-31-6] (→ Maleic and Fumaric Acids), which is said to reduce resin color and may enhance oxidative cross-linking.

[Structure: trimellitic anhydride — benzene ring with HOOC group and cyclic anhydride]

1,2-Polybasic acids are never used as such. Instead, their internal cyclic anhydrides are used. The anhydrides form nominally the same polymer composition as the acids, but they are easier to handle, easier to polymerize, more stable, and less costly. In large-scale operations phthalic anhydride and maleic anhydride are kept hot and are handled in liquid form, an economy not feasible with the high-melting acids.

***Triglyceride Oils and Fatty Acids*** (→ Fats and Fatty Oils, → Fatty Acids, → Drying Oils and Related Products). *Triglycerides* containing unsaturated and mixed fatty acids can remain liquid at temperatures well below 0 °C. For example, linseed oil contains a mixture of $C_{18}$ fatty acids with zero, one, two, and three double bonds per $C_{18}$ chain; it does not freeze at the very cold temperatures to which flax seeds may be exposed. A typical molecule might have the structure:

$H_2C-O-\overset{O}{\overset{\|}{C}}(CH_2)_7(CH=CHCH_2)_2(CH_2)_3CH_3$ (linoleic)

$HC-O-\overset{O}{\overset{\|}{C}}(CH_2)_7(CH=CHCH_2)_3CH_3$ (linolenic)

$H_2C-O-\overset{O}{\overset{\|}{C}}(CH_2)_7CH=CH(CH_2)_7CH_3$ (oleic)

The total number of double bonds in a fatty acid, the Wijs iodine number, is measured by titration with iodine [30]. This number, however, correlates poorly with the reactivity of a triglyceride oil toward oxygen, a critical factor for coatings based on drying alkyds. Fatty acid residues containing only one double bond are almost unreactive. Those containing more than one double bond separated by a single methylene group, e.g., $-CH=CHCH_2CH=CH-$, react readily with oxygen and impart drying capablities to the oil or alkyd. Linoleic and linolenic acids have two and three such double bonds, respectively. Conjugated double bonds, e.g., $-CH=CHCH=CH-$, within the fatty acid are even more reactive.

Nondrying alkyds, for which reactivity with oxygen is unwanted, are made from triglyceride oils that contain a high proportion of saturated fatty acids, such as coconut oil. They can also be made from synthetic fatty acids.

While the fatty acid makeup of each type of oil is somewhat characteristic, the exact composition of the fatty acid mixture may vary widely with soil, climate, strain of plant, extraction process, and other factors. For example, the iodine number of soybean oil may vary from 103 to 151 [30]. Commercial oils are reliably within a much narrower range, for example 128 – 135 for soybean oil [31].

Usually the double bonds of the fatty acids are in the relatively unreactive *cis*, nonconjugated configuration. Certain oils, such as tung, oiticica, and dehydrated castor, contain conjugated and/or *trans* double bonds and are used to make alkyds for fast air-drying and low-bake coatings.

Some of the oils frequently used in alkyds are listed in Table 2 along with their iodine numbers and an indication of the degree of unsaturation of the fatty acids present. The numbers are averages of published data and, of course, are approximate. A wide variety of other oils is used, including tung oil, whale oil, and fish oils from various sources.

The *fatty acids* used in alkyd production are usually made by hydrolyzing (saponifying or "splitting") triglyceride oils to obtain fatty acid mixtures. Linseed oil fatty acid and soybean oil fatty acid [68308-53-2] are used widely. The fatty acid composition of the oil is not substantially changed by the hydrolysis process. Some

Table 2. Makeup of oils used in alkyds

| Oil | Iodine no. | Typical fatty acid composition, % * | | | | |
|---|---|---|---|---|---|---|
| | | 0 | 1 | 2 | 3 | 4 – 6 |
| Linseed | [8001-26-1] | 190 | 11 | 18 | 16 | 55 | – |
| Manhaden | [8002-50-4] | 180 | 37 | 30 | 2 | 1 | 30 |
| Safflower | [8001-23-8] | 145 | 10 | 12 | 78 | tr. | – |
| Sunflower | [8001-21-6] | 135 | 10 | 25 | 65 | <1 | – |
| Dehydrated castor | [64147-40-6] | 135 | <1 | 7 | 92 ** | – | – |
| Soybean | [8001-22-7] | 132 | 15 | 25 | 52 | 8 | – |
| Coconut | [8001-31-8] | 8 | 91 | 7 | 2 | – | – |

* Indicates percentage of fatty acid chains containing zero, one, two, etc., double bonds per fatty acid chain
** Contains a high proportion of conjugated double bonds

fatty acid producers offer fractionated products that are enriched in particular acids.

An important fatty acid mixture not derived from a seed oil is tall oil fatty acid [61790-12-3] (→ Tall Oil). It is a byproduct of kraft pulping. The name derives from the Swedish word for pine. The composition and performance are fairly similar to those of soybean oil fatty acid. The presence of about 10 % of fast-reacting conjugated double bonds compensates for its lower iodine number of about 125. Tall oil fatty acid always contains some rosin as an impurity; it is sold in grades specifying the maximum amount of rosin. The 1.5 % and 4 % grades are used widely in alkyds. Tall oil produced in Europe has a somewhat different composition than tall oil produced in North America; resin production formulas developed with one type are likely to fail if the other type is substituted.

Synthetic fatty acids are used in alkyds to a modest extent.

***Modifiers*** (see Chap.).

***Solvents.*** Immediately after production almost all alkyds are thinned in organic solvent to make them pumpable and pourable. Generally, these solvents are not removed at the time of coating manufacture, and they become part of the coating formula. Medium and long oil types usually are thinned with mineral spirits. Short oil alkyds require more polar solvents, often blends of aromatic hydrocarbon solvents. Alkyds to be used in high-solids coatings are sometimes thinned with ester or ketone solvents, which are efficient viscosity reducers. Water-reducible alkyds are thinned with water-miscible solvents such as 2-butoxyethanol or 1-propoxy-2-propanol.

Many alkyd resin and paint formulas were changed during the 1980s and 1990s to avoid solvents classified in the USA as "hazardous air pollutants" (HAPs). Similar changes were made in other countries.

## 2.2. Polymerization Theory

When a difunctional acid or anhydride (AA) and a difunctional alcohol (BB) are heated together, a stepwise reaction occurs that builds up linear chain molecules:

$$AA \xrightarrow[-H_2O]{+BB} AA-BB \xrightarrow[-H_2O]{+AA} AA-BB-AA \xrightarrow[-H_2O]{+BB} $$

$$\downarrow {+AA-BB \atop -H_2O} \qquad BB-AA-BB-AA$$

$$AA-BB-AA-BB \rightarrow \text{etc.} \qquad \downarrow {+AA-BB-AA \atop -H_2O}$$

$$AA(BB-AA)_3 \rightarrow \text{etc.}$$

Compositions containing short chains (3 to about 20 monomer units per chain) are called "oligomers". Those containing long chains are "polymers". Under ideal conditions polymer chains containing thousands of monomer units can be made. Conditions necessary for formation of very high molecular mass chains include the following:

1. The mixture must be heated long enough to complete the reaction.
2. Equimolar quantities of AA and BB must be present.
3. No alternative reactions may be possible.
4. AA and BB must contain no monofunctional reactive impurities.

The average molecular mass can be deliberately held at lower levels by altering any of the above ideal conditions, for example:

$$9\,AA + 8\,BB \xrightleftharpoons{-16\,H_2O} AA\text{-}(BB\text{-}AA)_8$$

$$9\,AA + 10\,BB + 2\,A \xrightleftharpoons{-20\,H_2O} A\text{-}(BB\text{-}AA)_9\text{-}BB\text{-}A$$

where A represents a monofunctional acid. The actual products of such reactions are mixtures of various chain lengths.

If a trifunctional monomer BBB is added, chain branching occurs:

$$4\,AA + 2\,BB + BBB \xrightarrow{-6\,H_2O} \begin{array}{c} AA\text{-}BBB\text{-}AA\text{-}BB\text{-}AA \\ | \\ AA\text{-}BB \end{array}$$

If substantial amounts of BBB are present, continued polymerization leads to formation of a three-dimensional network, or gel:

One of the goals in alkyd resin synthesis is to prepare structures that are branched but are not quite gelled. A fairly close approach to gelation is needed to form a polymer which readily forms cross-linked networks after application as a coating. However, gelation in the reactor causes a total loss of materials and a difficult clean-up job.

Much research has been directed toward understanding polymerization and developing general principles to help the alkyd formulator achieve the desired degree of polymerization and to optimize properties. Early theories of polymerization were developed by CAROTHERS [23], [29] and by KIENLE et al. [20], [22], [24], [34], [35]. CAROTHERS hypothesized a uniform, stepwise polymerization process. In his model, gelation occurs as the polymer molecular mass approaches infinity. The condition for gelation would be

$$p_g = \frac{2}{F_{av}}$$

where

    $p_g$    is the extent of reaction at the gel point and

    $F_{av}$    is the average functionality of the monomer mixture.

According to this equation pure difunctional monomers can never gel, but a mixture of glycerol and phthalic anhydride ($F_{av} = 2.5$) would gel at $p_g = 0.8$; that is, it would gel when 80 % of the functional groups have reacted.

In practice gelation usually occurs at lower $p_g$ than predicted by the Carothers theory. To explain this fact FLORY et al. [32], [33] proposed a probabilistic theory: gelation occurs when a relatively small proportion of the monomer molecules first becomes bonded into a three-dimensional network. FLORY's theory is satisfactory for predicting polymerizations of the chemically straightforward monomers used to make polyesters. However, alkyd formulators must deal with monomers that are far from straightforward. Complications include

- Variable starting materials
- Competing side reactions
- Reaction reversibility
- Differential reactivity of the functional groups
- Differential solubility effects
- Nonuniform process conditions

Gradual progress has been made toward a general and reliable theory of alkyd formation by several authors; see, e.g., [6], [38–44]. A completely precise theory is unlikely, mainly because alkyd polymerizations probably are not homogeneous during the late stages of reaction. Most alkyds contain small, soft gel particles as well as soluble material [44], and these gel particles are in some cases essential to the attainment of optimum properties [45], [46]. Despite this difficulty, certain alkyd polymerizations can be accurately modeled with the aid of computers and advanced statistical techniques [47]. Such theoretical methods offer considerable help and insight to the alkyd formulator. They can reduce reliance on empirical trial and error, but they will not completely eliminate it for the foreseeable future.

## 2.3. Formulation and Process Types; Laboratory Processing

Alkyds are made in batch processes, of which several types have evolved.

Table 3. Composition of long oil alkyd

|  | Equivalents COOH | OH | g | % in alkyd |
|---|---|---|---|---|
| Soybean fatty acids | 1.246 | – | 349.0 | 60.0 |
| Pentaerythritol | – | 3.545 | 125.5 | 21.5 |
| Phthalic anhydride | 2.000 | – | 148.0 | 25.4 |
| Totals | 3.246 | 3.545 | 622.5 | 106.9 |
| H$_2$O |  |  | –40.4 | –6.9 |
| Theoretical yield of resin |  |  | 582.1 | 100.0 |

***Fusion Process Using Fatty Acids.*** The fusion process starting with fatty acids is the oldest and most straightforward. A representative formulation is given in Table 3 [48].

Like nearly all alkyd formulations, it contains an excess of hydroxyl groups, 9.2 % in this case. This excess is essential to assure that the rate of viscosity increase will be controllable as the polymerization process approaches completion. To visualize the consequences of not using an excess of OH groups, suppose that exactly 3.246 equivalents of pentaerythritol have been used. The $F_{av}$ would have been 2.07, and CAROTHERS' theory predicts an exponential increase in viscosity as the reaction approaches completion, ending in gelation at $p_g = 0.96$. As FLORY's theory and practical experience demonstrate, gelation would actually occur at lower $p_g$. Therefore in the late stages of reaction, viscosity would be increasing rapidly and exponentially. However, the 9.2 % excess of OH groups assures that the number average chain length will remain modest even when nearly all of the carboxyl groups have reacted, limiting viscosity at the desired level. Chain lengths of individual molecules in the resin will vary from the oligomer to the high-polymer range.

Viscosity could be controlled by using an excess of acid, but inferior coatings would result because of moisture sensitivity imparted by the unconverted COOH groups in the film. Unconverted OH groups impart less moisture sensitivity, and they play a useful role in certain cross-linking reactions (see Chap. 3).

To prepare the above resin on a *laboratory scale* the ingredients are placed in a round-bottom reaction flask, heated at about 240 °C, and stirred mechanically. Inert gas (N$_2$ or CO$_2$) is passed through the reaction mass to facilitate water removal; the water is condensed and measured to estimate the extent of reaction. As the reaction approaches completion (5 – 10 h), samples are removed and the acid value and/or viscosity are measured. When the desired degree of reaction has been reached, the reaction mass is cooled slightly and is diluted with about an equal volume of mineral spirits.

***Fusion Process Using Triglyceride Oils.*** The fusion process starting from triglyceride oils is often called the *alcoholysis* or the *monoglyceride process*. It is popular because the oil provides an economical source of all of the fatty acid and part of the polyol. It is a two-stage, batch process:

1)

$$\text{Triglyceride} + 2\text{ Glycerol} \xrightleftharpoons[\text{catalyst}]{220-260\,°C} 3\text{ "Monoglyceride"}$$

2)

$$\text{Monoglyceride} + \text{Polybasic acid} \xrightleftharpoons[-H_2O]{220-250\,°C} \text{Alkyd resin}$$

The first, or alcoholysis, stage is an ester interchange reaction. Equilibrium is reached within an hour or two at sufficiently elevated temperature when the reaction is catalyzed by litharge (PbO) or by certain Li or Ca salts. Although the product is called a monoglyceride, it is actually a mixture of isomers of mono-, di-, and triglycerides and of glycerol. Polyols other than glycerol often are used and yield even more complex mixtures. In the second stage polybasic acids or anhydrides are polymerized with the monoglyceride mixture to form the resin. Other ingredients often are added during the second stage.

Procedures and equipment for laboratory preparation are generally similar to those used for the fatty acid process.

Both fusion processes suffer from several practical difficulties. To assure reasonable process rates the water of reaction must be removed by "sparging" the reaction mass with a stream of inert gas. Without sparging, process rates are slow because the esterification reaction is highly reversible. With sparging, however, the gas stream sweeps vapors of phthalic anhydride and polyol out of the reaction mass. These materials may be lost altogether or they may condense in the upper parts of the apparatus from which they may fall back into the reactor, causing dark or nonuniform product.

***Solvent Processes.*** To overcome these problems solvent processes are used widely, both with fatty acid and triglyceride starting materials. In these processes a few percent of unreactive, water-immiscible solvent is added to the formulation. Commercial xylene, also called xylol (mixed dimethylbenzenes and ethylbenzene, *bp* 138 – 140 °C), is used commonly. Typically a stirred alkyd polymerization mass containing 4 % xylene will boil vigorously at about 250 °C. The xylene vapors form an azeotrope with water and carry it out of the reactor. Usually the vapor stream is passed through a "partial condenser" operated at about 100 °C; this allows water vapor to escape, but it condenses most of the xylene and almost all of the monomers and returns them to the reactor. Whether or not a partial condenser is used, the vapors are condensed. Water and xylene separate into two layers, and the xylene is returned to the reactor.

Laboratory equipment for solvent processing has been described in detail [49]. Losses of xylene and of monomers can be minimized with proper operation, and uniform products can be produced.

## 2.4. Manufacture

A variety of equipment for alkyd production has been described [5], [8], [50]. Equipment for *solvent processing* will be described here because the trend is in that direction. However, fusion processes, which require simpler equipment, often produce satisfactory products. The fusion process is preferable for isophthalic alkyds.

Figure 1 shows a versatile configuration designed for the solvent process. The reactor typically would be constructed of no. 316 stainless steel. Capacity of commercial-scale reactors ranges from 2 to 35 $m^3$. A turbine agitator driven by a powerful electric motor is necessary. Heating and cooling generally are accomplished by a heat-transfer fluid flowing either through a jacket around the reactor (as shown in Figure 1) or through coils within the reactor. High-pressure steam and gas-flame heating have been used but are less satisfactory and more hazardous.

Whereas the reactor and overhead shown in Figure 1 are the heart of an alkyd plant, other equipment is needed for efficient, large-scale production. Such equipment includes the following:

1. Raw material storage and metering equipment (phthalic anhydride often handled in hot, molten form)
2. A manhole in the reactor top for adding solids and to provide access for cleaning
3. A heater for heat-transfer fluid
4. A thin tank about 2.5 times as large as the reactor, equipped with a condenser and agitator
5. Product storage tanks and drumming equipment
6. Environmental control equipment, including incinerators for gaseous and liquid waste
7. Filters
8. Safety and fire protection equipment, including rupture disks, blowout vents, and sprinklers

Automated alkyd reactors have been installed in Europe, North America, and Asia [51]. Improved safety, quality, and productivity are claimed. However, several factors have discouraged retrofitting of reactors to automate them. Although the necessary microprocessors are relatively inexpensive, automation requires costly feedback systems and servomotor driven valves. Detailed process studies are necessary to program the microprocessor to deal with all contingencies.

## 3. Film Formation by Coatings Based on Alkyd Resins

There are two mechanisms by which a liquid coating can be transformed into a tough, protective film after it has been applied:

1. Lacquer drying by simple solvent evaporation
2. Chemical cross-linking by reaction of the polymeric binders to form a cross-linked polymer network

Coatings based on alkyd resins often depend on both mechanisms to some degree. Chemical

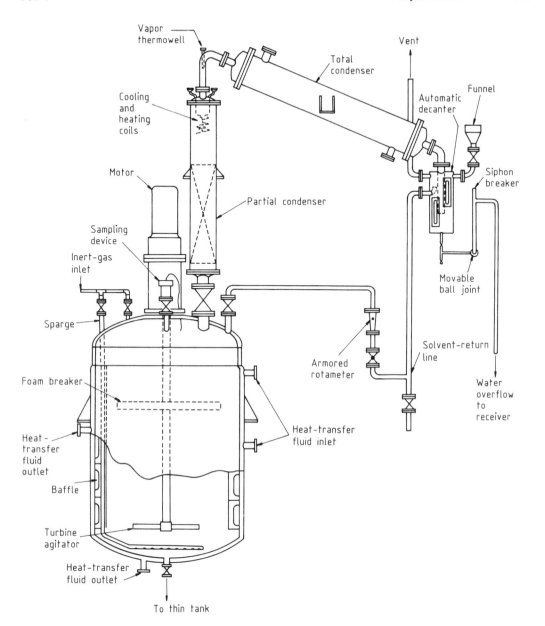

**Figure 1.** Alkyd solvent-processing equipment with partial condenser (adapted from [30])

cross-linking plays a vital role in converting a soft, tacky alkyd into a hard, tough film.

**Drying Alkyds.** The majority of alkyd coatings are designed for film formation at ambient or slightly elevated temperatures. The chemical mechanism of curing involves autoxidation of the unsaturated fatty acid residues in the resin. This chemically complex process has been studied extensively [52–58], but it is not yet completely understood. Most of the studies have involved model compounds such as monoesters of fatty acids rather than the more complex alkyd resins.

Autoxidation of alkyds is a chain reaction that proceeds by a free-radical mechanism. A key step is the formation of a hydroperoxide on an allylic methylene group:

$$\underset{H}{\overset{CH_2}{\underset{|}{C=C}}}\underset{H}{\overset{|}{\phantom{C}}} \xrightarrow[\text{catalyst}]{O_2} \underset{H}{\overset{CH-OOH}{\underset{|}{C=C}}}\underset{H}{\overset{|}{\phantom{C}}}$$

The hydroperoxide is capable of several further reactions, some of which lead to formation of stable, covalent bonds with sites on other resin molecules. This bonding builds up a three-dimensional polymeric network, and a tough film is formed.

Small amounts of catalysts (driers) that promote autoxidation and cross-linking are added to drying-type alkyd coatings. Driers are generally mixtures of oil-soluble salts of Co, Mn, Zr, Ca or other metals [59]: soluble Pb salts are also effective but have been banned in many countries.

Fatty acid chains containing only one double bond are autoxidized slowly. Those containing allylic methylene groups activated by two double bonds react over 100 times faster. Reaction sequences probably involve rearrangement and isomerization of the double bonds, for example:

[reaction scheme showing progression of double bond isomerization with R and R' substituents, ending with HOO group]

Studies on model compound [56–58] confirm the mechanism shown above for the early stages of cross-linking, but reveal that subsequent film-forming reactions are extremely complex and beyond the scope of this article.

Useful drying rates are achieved only if appreciable concentrations of polyunsaturated fatty acid residues are present. Soybean, safflower, sunflower, and tall oil alkyds are about the least unsaturated that dry well. These oils have high linoleic acid (two double bonds) content but relatively low levels of fatty acids with three or more double bonds. They are preferred for decorative coatings where resistance to yellowing is desired and moderate cross-linking is adequate.

Higher levels of cross-linking and therefore better protective properties are achieved by use of fatty acids or oils with substantial contents of triply unsaturated fatty acid residues (linseed or tung) or of those containing multiple unsaturated fatty acid residues (fish oils). Blends of fatty acids or oils are used to reduce cost and sometimes to minimize premature cross-linking during alkyd manufacture.

Drying rate is often an important consideration. Oils containing two conjugated double bonds dry faster than those containing similar levels of nonconjugated double bonds. Thus, chemically modified oils and fatty acids with two conjugated double bonds (dehydrated castor oil and conjugated tall oil fatty acids) find considerable use. Fastest of all are oils that contain fatty acids with three conjugated double bonds, such as tung oil.

Rates of autoxidation are useful above about 0 °C and sharply increase with increasing temperature. A significant advantage of drying oil alkyd-based paints over latices is that they can form satisfactory films when applied at 0 °C or even a bit below.

Drawbacks of all alkyd paints paints include a tendency of the films to turn yellow and become brittle with age. Their low permeability to water can also lead to blistering when painted on wood. Use of high-linoleic, low-linolenic oils such as safflower and sunflower can minimize these problems, but can not eliminate them. Short oil alkyds are used in applications where rapid partial drying is needed. Although they have less cross-linking capability than longer oil types, they quickly develop some degree of hardness by lacquer drying as the solvent evaporates. Quick drying is desirable in many manufacturing situations, as it permits gentle handling within 15 – 30 min after coating. Film properties improve for several days or weeks as the slower cross-linking process occurs.

*Nondrying Alkyds.* Alkyds are used extensively in industrial coatings that are designed to be cured at elevated temperatures (baking or stoving enamels). Such enamels are usually blends of alkyds with aminoplasts, e.g., alkylated melamine – formaldehyde or urea – formaldehyde resins. Cross-linking occurs by acid-catalyzed condensation of the OH groups of the alkyd with the aminoplast:

$$4 \; (RE)\!\!\begin{array}{c}OH\\-OH\\OH\end{array} \;+\; O\!=\!C\!\!-\!\!N\!\!\begin{pmatrix}OCH_3\\OCH_3\end{pmatrix}_2 \; \xrightleftharpoons[-CH_3OH]{\sim 100\,°C}$$

Hydroxy functional resin    Urea-formaldehyde resin

$$O\!=\!C\!\!-\!\!N\!\!\begin{pmatrix}HO \quad OH\\-O\!-\!(RE)\\-O\!-\!(RE)\\HO \quad OH\end{pmatrix}_2 \longrightarrow \text{etc.}$$

Other chemical cross-linkers can be blended with alkyds. For example, alkyd – polyisocyanate blends are used as binders for two-package, ambient temperature curing polyurethane coatings.

Alkyds for baking enamels do not need to be autoxidizable and often are made from oils containing only saturated fatty acids, coconut, for example. Soybean and tall oil alkyds often are used in alkyd – amino enamels; such enamels cure by a mixture of condensation and autoxidative cross-linking.

Polyvalent metal alkoxides, such as Al$(OC_4H_9)_3$, can serve as trifunctional cross-linkers. They react with hydroxyl groups on the alkyd resin to form Al(O–alkyd)$_3$ structures within the film. This expedient can enhance hardness and through-cure of the paints.

Nondrying alkyds are used as plasticizers in lacquers. Their role is to soften high molecular mass thermoplastic resins that otherwise would be too hard and brittle for coatings use. Examples are nitrocellulose – alkyd blends for furniture finishes and acrylic – alkyd blends for automotive refinish lacquers. Note that in Europe the term "lacquers" is frequently applied to all types of organic coatings, while in North America the term tends to be limited to coatings made from thermoplastic, non-cross-linking binders.

## 4. Modified Alkyds

Roughly one-fourth of commercial alkyd usage is of modified alkyds, resins in which modifiers are at least partly interpolymerized with the usual alkyd monomers. The more important types are described in this section.

*Chain-stopped alkyds* contain aromatic monobasic acids (see Section 2.2), such as benzoic acid [65-85-0] or 4-(1,1-dimethylethyl)benzoic acid (*p-tert*-butylbenzoic acid) [98-73-7]. These modifiers improve lacquer-dry characteristics and harden films. However, because they do not participate in covalent cross-linking, the films tend toward brittleness.

Inclusion of monobasic acids during polymerization reduces $F_{av}$; to compensate, highly functional polyols such as pentaerythritol are used. Because pentaerythritol and benzoic acid are relatively inexpensive, low-cost resins with reasonably good properties are attainable. Uses are diverse, a major one being traffic-control striping. Chain-stopped alkyds are useful in high-solids coatings (see Chapter 5).

For premium-quality chain-stopped resins, 4-*tert*-butylbenzoic acid affords an excellent combination of fast dry and film plasticity. However, use of this modifier is limited by concerns about its potential for long-term health effects.

*Styrenated alkyds* are made in a two-stage process [54], [60]. First a drying oil alkyd is produced using a formulation designed to limit molecular mass. The resin is then thinned, usually with xylene, and styrene [100-42-5] is added to the solution. Addition of a free-radical initiator starts the second stage of the process, the addition polymerization of the styrene. This step is effected at higher temperature (about 140 °C) than most addition polymerizations, conditions that favor grafting of polystyrene segments onto the alkyd. The product is a mixture of graft copolymer, ungrafted alkyd, and homopolystyrene [61]. Apparently sufficient grafting occurs to "compatibilize" the mixture.

Styrene content ranges up to 60 %. The homopolystyrene ($T_g \approx 100$ °C) segments impart hardness, chemical resistance, and very fast lacquer dry characteristics. However, high levels of styrene reduce weatherability and toughness.

*Acrylated alkyds* are made by a similar process using acrylic or methacrylic monomers in place of styrene [62]. These monomers are more costly than styrene, but their presence improves resistance to sunlight. For example, an alkyd modified with methyl methacrylate [80-62-6] (homopolymer $T_g \approx 105$ °C) would be similar in most properties to a styrenated alkyd but would be more weatherable. Alkyds modified with lower $T_g$ acrylic or methacrylic monomers can be made readily. They are less important because they offer few advantages over unmodified alkyds.

*Other olefinic monomers,* such as acrylonitrile [*107-13-1*], α-methylstyrene [*98-83-9*], and mixed 3- and 4-vinyltoluene (VT), are used commercially to modify alkyds. The latter mixture affords the commercially important "VT alkyds", which lacquer dry nearly as fast as styrenated alkyds but have better film properties. Uses include fast-dry industrial enamels and aerosol paints.

*Silicone alkyds* are prepared by interpolymerizing hydroxyfunctional organosilicone oligomers with alkyd resins [63]. Methyl/phenyl-substituted silicones are used. The proportion of silicone ranges from 15 to 55 %. These premium quality resins have superior resistance to heat and weather. Drying oil silicone alkyds are used in heat-resistant coatings such as stove enamels and in exterior applications in which prolonged retention of properties and appearance saves substantial cost by reducing the frequency of repainting, e.g., petroleum storage tanks. Blends of nondrying silicone alkyds with melamine resins have been used in coil-applied coatings for exterior siding.

*Thixotropic alkyds* are modified at high temperatures with an oil-miscible polyamide resin, for example, a polymer [*51178-84-8*] of octadecadienoic acid dimer [*6144-28-1*] and 1,2-diaminoethane [*107-15-3*]. Proportions and reaction conditions must be carefully controlled [64]. Paints containing these resins are viscous, but their viscosity drops under mechanical shear and then recovers when the shear is removed. It is unclear from published information whether their behavior is simple shear thinning (near instantaneous recovery when shear is removed) or genuine thixotropy (time-delayed recovery). Regardless of mechanism, the paints resist running and sagging when brushed or sprayed in thick coats. Applications include bridge and marine paints.

*Uralkyds* are usually air-drying compositions in which part of the polybasic acid is replaced by a polyfunctional isocyanate, most commonly toluene diisocyanate [*26471-62-5*] (→ Isocyanates, Organic) to form a polyester – polyurethane interpolymer. The resins are usually made in a two-stage process, the reactive toluene diisocyanate being added in the second stage at relatively low temperature.

An excess of OH groups is used so that all of the isocyanate reacts to form urethane linkages. Cross-linking is accomplished by the usual autoxidative mechanism. Presence of the urethane linkages imparts outstanding toughness, water resistance, and abrasion resistance; thus, urethane alkyds are used in premium-quality varnishes, for example, for gymnasium floors. A drawback is that resins containing toluene diisocyanate are brownish and yield yellowish varnishes that darken with age. This problem can be almost eliminated by use of the more costly aliphatic polyisocyanates in place of toluene diisocyanate.

A second class of uralkyds is prepared with a deficiency of OH groups so that unconverted isocyanate groups are present in the resin. After application these resins cross-link by reaction of atmospheric moisture with the isocyanate groups. They form exceptionally tough films. However, the coatings are difficult to apply and may cause skin irritation.

*Epoxy esters* do not fit the strict definition of alkyd resins, but they are closely related. These resins usually are formed by condensing an unsaturated fatty acid with an epoxy resin of low to intermediate molecular mass (→ Epoxy Resins). Epoxy ester coatings combine part of the toughness, adhesion, and corrosion resistance of epoxies with the easy application and cure of alkyds. They also share the main shortcomings of epoxies: surface chalking when exposed to sunlight and discoloration with age. Epoxy esters are used where their adhesion and corrosion resistance are needed and where surface weatherability is unimportant, for example, in auto body primers and in industrial maintenance paints.

*Phenolic modified alkyds* are used in applications in which hardness and moisture resistance are important. High-solids formulations have been developed [65].

*Rosin and rosin esters* (→ Resins, Natural) are used as modifiers mainly to reduce cost. Coating properties are similar to those of chain-stopped alkyds, but films yellow with age and have inferior weatherability. These modifiers are used in wall primers.

## 5. Low-Solvent Coatings

In most coating processes, the solvent in the coating evaporates into the atmosphere. Formerly the amounts were thought harmless, but by the 1960s it was recognized that the solvents used in alkyd and other coatings contribute substantially to air pollution. In this context, the solvents are called the "volatile organic content" (VOC) of the coating. In air, organic solvents react in a complex photochemical process to produce ozone, one of the most troublesome constituents of photochemical smog.

In the late 1960s and early 1970s governments in many industrial areas adopted "technology forcing" regulations designed to compel coatings users to find ways to reduce VOC emissions. These regulations stimulated research on alkyds during the 1970s and 1980s. Two major approaches were used (→ Paints and Coatings, 1. Introduction): replacing most of the solvent with water (water-borne coatings), or sharply increasing the proportion of coatings solids to solvent (high-solids coatings). It is estimated that 10 % of alkyd usage is now of the water-borne and high-solids types. The most common approach to water-borne alkyds is to make special alkyds (see below), dissolve them in water-miscible organic solvents, and then add water. These are called "water-reducible" resins. Additional approaches have been introduced more recently, such as solvent-free dispersions of alkyd resins in water. All approaches require substantial changes in conventional alkyd resin technology.

***Water Reducible Resins.*** To make alkyd resins suitable for use in water-reducible paints, it is necessary to chemically change them so they are miscible with blends of water and limited amounts of organic solvent. This is accomplished by attaching hydrophilic sites to the polymer molecules. Numerous procedures for doing so have been published [66–68]. The most common method is to place acidic carboxy groups on the molecule and then to "salt" the resin with a base, usually ammonia or an amine.

A representative method uses a two-stage process involving trimellitic anhydride [69] (mass parts in parenthesis):

Trimethylol propane (277)
+ tall oil fatty acid (456)   $\xrightarrow{250\,°C}$   alkyd (913)
+ isophthalic acid (264)                              + H$_2$O (84)

Alkyd (913)
+ trimellitic anhydride (87)   $\xrightarrow{193\,°C}$   [alkyd-trimellitate structure] (1000)

Reduced temperature is used in the second stage to encourage selective reaction of the anhydride group. However, there are several competing side reactions: at 193 °C the anhydride reaction is not completely selective, and acid – ester interchange reactions occur to some degree. For this reason, the process requires careful control to assure consistent product.

If enough carboxyl groups are present to give an acid value of about 45 mg of KOH per gram of resin or higher, the resin forms a transparent mixture when salted and dissolved in 20:80 solvent: water medium. Such resins are sometimes called "water-soluble" alkyds, but their abnormal solution rheology, light scattering, and other evidence indicate that they are seldom, if ever, true solutions. The terms "water-reducible", "water-thinnable", or "water-dilutable" are more descriptive.

Similar resins having acid numbers below about 45 form hazy or milky dispersions when salted and thinned with water. They are often termed "water-dispersible" as well as "water-reducible" alkyds. Such resins are gaining commercial importance because lowering the acid number seems to improve corrosion resistance of the coatings.

Most types of modified alkyds can be made in water-reducible form. For example, water-reducible acrylated [70], silicone-modified [71], and urethane-modified [72] alkyds have been developed.

A disadvantage of this approach is that the ester bond that connects trimellitic anhydride residues to the alkyd resin is more vulnerable to hydrolysis than the ester bonds in most polyesters. For this reason, stability of water-borne paints containing such resins is limited to 3 – 6 months. If a substantial fraction of the critical ester bond is hydrolyzed, the paint becomes unstable and unusable.

Numerous other approaches to formulating alkyds in aqueous media have been studied, often in efforts to improve package stability. Among the more promising are to copolymerize hydrophilic poly(ethylene oxide) segments into the resin [73], to emulsify hydrophobic alkyds with conventional surfactants [74], to prepare acrylated alkyds in which the acrylic polymer segments contain –COOH or other hydrophylic groups, and to prepare transparent microemulsions [75]. These approaches can be combined. The emulsification approach has been widely studied in recent years [76], [77], and is the basis for commercial products. Careful selection of emulsifiers, pigments, and additives is critical to assure stability and acceptable application characteristics and to avoid adverse interactions that deactivate driers [77].

***High-Solids Coatings.*** There seems to be no standard definition of the term "high-solids coatings". The level of 70 vol % solids has been suggested as a minimum for high solids, but such a definition is arbitrary and does not conform to general usage. In this article, the term is used in a vague but useful way: a high-solids coating is one that is substantially more concentrated than similar coatings that were in general use in the 1960s. For example, an appliance finish that is sprayed at 62 vol % solids is called a high-solids coating because earlier generations were sprayed at 34 %. To emphasize the relative aspect of the definition, some authors prefer the term "higher-solids coatings".

The essential technical problem in formulating high-solids coatings is to find ways to achieve low viscosity without using very much volatile organic solvent. Many formulation and application variables can be manipulated in pursuit of this goal, but the most important is resin viscosity [78]. Therefore, the challenge to the alkyd resin chemist is to reduce the resin viscosity in solution without serious adverse effect on coating properties. To accomplish this difficult task it is necessary to reduce the molecular mass of the resin while increasing its functionality for cross-linking. Especially important is reducing the proportion of soluble high molecular mass molecules in the polymer mixture. These molecules contribute disproportionately to the viscosity [79]; unfortunately, their removal may harm film properties [46]. These difficulties have, to a degree, been overcome, and high-solids alkyd coatings are gaining commercial acceptance.

In air-dry, high-solids alkyds, the most difficult problem is to attain commercially acceptable drying rates without excessive yellowing. Resins that autoxidize very rapidly can be made by use of conjugated or highly unsaturated fatty acids. However, coatings made with such resins tend to turn yellow quickly and to cure too quickly at the surface of the coating film. Surface cure creates an oxygen barrier film that unacceptably slows cure in the interior of the coating. Solutions proposed to solve this problem include use of reactive diluents [80–82], use of auxiliary cross-linkers [83], and use of special autoxidation catalysts [84]. In each case, specially tailored alkyd resins are required.

Alkyds also are used in high-solids baking enamels, where they usually are cross-linked with amino resins, and in two-package alkyd – isocyanate enamels. In both of these applications, alkyds compete with polyester and acrylic high-solids resins.

RYER presents an overview of new alkyd chemistry and gives several specific formulas [85]. For example, a chain-stopped alkyd suitable for coatings for agricultural and heavy equipment is composed of 72.8 parts (by weight) of soya fatty acid, 44.9 parts of pentaerithritol, 40.0 parts of phthalic anhydride, and 24.4 parts of benzoic acid. Given that the water of reaction is 13.1 parts, one can use Equation (1) to calculate the oil length of this alkyd as 44.8.

***Compliance Solvents.*** In the USA, three solvents, acetone, methyl acetate, and 4-chlorobenzotrifluoride, have been exempted from regulations limiting VOC emissions because of their very low photochemical reactivity. Exemption of one or two more solvents, including *tert*-butyl acetate, is expected. These regulatory changes are very favorable for the future of alkyds, since resourceful formulators can use them to help satisfy air-pollution regulations that limit VOC. Two chlorinated organic solvents, $CH_3CCl_3$ and $CH_2Cl_2$, were previously exempted [87] but are little used because of concerns about their safety, corrosiveness, and greenhouse effects.

***Latex Modifiers.*** A few percent of alkyd resin is often added to water-borne architectural paints, the principal binder of which is a polyvinyl or polyacrylic emulsion resin. The alkyd improves adhesion to chalky or oily surfaces. Addition of much larger proportions of alkyd, up to 50 %, is feasible [74], but incentives to commercialize this technology have been lacking because latices have been less expensive than alkyd resins most of the time since 1970.

## 6. Markets and End Uses

U.S. Tariff Commission states that alkyd resin production in the USA was about $(3-3.5) \times 10^5$ t/a between 1960 and 1980 and then began a slow decline, reaching about $2.3 \times 10^5$ t/a by 1996 [25]. The decline is thought to be continuing at a rate of about 2 %/a. Worldwide production figures are not readily available. If the proportion of alkyd used in the rest of the world is the same as in the USA, gobal production would be roughly three times as great as the U.S. figures. The proportion appears to be higher in some major markets, and this suggests that global production may be on the order of $10^6$ t/a.

A very high percentage of alkyd production is used in coatings (→ Paints and Coatings, 1. Introduction). In this market alkyds compete with many other generic types of resins. While they are gradually being displaced by other types of resins, alkyds continue to compete effectively in many end uses. The continuing popularity of alkyds seems to be based on three factors:

1. *Versatility.* Whereas alkyds cannot match the outstanding features of some competing resin types (hardness and weatherability of acrylics, corrosion resistance and adhesion of epoxies, and toughness and abrasion resistance of polyurethanes), they can be formulated to provide a satisfactory balance of properties for many purposes.
2. *Ease of Use.* Well formulated alkyd coatings tend to be easier to apply by spray, brush, or roller than other generic types. They seem less susceptible to popping, sagging, cratering, and other appearance defects. They wet oily or dirty surfaces relatively well. Alkyds are package stable and do not have to be formulated in two-package systems. Air-dry formulations dry quickly, and baking enamels cure at low temperatures. An important but often overlooked factor in the continuing popularity of alkyds is that many existing finishing facilities and procedures have been designed for alkyds, and it has proved difficult to substitute other types of coatings without expensive capital investments.
3. *Economy.* During the 1960s and early 1970s petrochemical monomers, such as acrylics and vinyls, were less costly than the mixture of petrochemical and biomass ingredients used in alkyds. By 1983 the situation had reversed; soybean oil and tall oil fatty acids were selling at less than half the price of basic acrylic monomers. The cost advantage of alkyds is diluted by the petrochemical content of alkyd resins, which averages roughly 50 %, and the need, in most cases, to use petrochemical-based solvents. Since 1983, the relative costs of alkyd resins and 100 % petrochemical-based resins have fluctuated roughly in proportion to the prices of petroleum and of vegetable oils. Soybean oil currently sell for about 0.55 $/kg in the USA, while key vinyl and acrylic monomers (including styrene) sell for 1.5 to 3.5 times as much.

A summary of where alkyds fit in each of the three major coatings market segments follows.

*Trade sales,* or *architectural coatings*, comprise roughly 52 % of the physical volume and 42 % of the value of coatings sold in the USA. It is estimated that 55 % of the total alkyd usage in paints is in this sector [17], [25]. Water-borne formulations based on polyacrylic and polyvinyl emulsion resins have captured over three-fourths of this market in the USA. Their chief advantages are easier cleanup, and better exterior durability. Alkyd- and alkyd – vegetable oil-based paints remain popular with many coating contractors because they can be applied faster and at lower temperatures; emulsion resin paints do not form films properly below about 5 – 8 °C, but alkyds can be used at 0 °C or even lower.

Alkyd-based paints retain a major share of certain trade sales markets, for example, stains, varnishes, high-gloss enamels, and primers. In these market segments, the hard, tough, cross-linked films of alkyds offer advantages over latices, and their superior adhesion is important.

As noted above, a few percent of alkyd is often added to acrylic emulsion paint formulations to improve adhesion.

*Industrial coatings* are, by definition, coatings applied at the factory to manufactured goods. They account for roughly 31 % of the volume and 36 % of the value of coatings sold in the USA About 21 % of the alkyds used in paints go into these markets [17], [25].

Alkyd – melamine automotive topcoats have been almost completely replaced by more weatherable acrylic – melamine – formaldehyde and acrylic – polyurethane topcoats. Alkyds still find uses in auto and truck production, for example, in coatings for under-hood parts.

Factory wood finishing is a large user of coatings. Alkyds are used extensively both in alkyd – aminoplast thermosetting enamels and as plasticizers for nitrocellulose lacquers.

Other large-scale manufacturing operations that use alkyds include the following:

1. Metal containers – exterior can finishes
2. Coil coatings – flexible enamels for light fixtures, siding, Venetian blinds, etc.
3. Metal office furniture – hard alkyd – aminoplast enamels
4. Farm implements – fast air-drying enamels

Alkyd coatings are popular in smaller scale factory finishing operations, often because of their ease of application. Examples are tools, fixtures, drums, toys, and shelving. These diverse operations often are called the *general industrial* coatings market; it is a substantial consumer of alkyds.

*Special purpose coatings* include all organic coatings that do not fit into the above categories. It is a diverse market that is often characterized by sophisticated, specialized coatings. It accounts for the remaining 17 % of the volume and 22 % of the value of U.S. coatings. About 24 % of U.S. alkyd production goes to this segment [17], [25]. Alkyds and modified alkyds share this market with polyacrylic, epoxy, and polyurethane resins. Alkyds are used wherever they are satisfactory because in most cases they are the least expensive and easiest to apply. There are many examples:

1. Highway maintenance – alkyds, thixotropic alkyds, and epoxy esters for bridges, etc.
2. Traffic-control striping – usually very fast drying chain-stopped or styrenated alkyds blended with chlorinated rubber resin; a large market
3. Auto and truck refinish – epoxy ester primers and chain-stopped alkyd topcoats; alkyds losing share to acrylics
4. Marine – alkyds used mainly above the water line; fish oil alkyds popular for rust penetration, silicone alkyds for gloss retention
5. Industrial maintenance – linseed and silicone alkyds for refineries, tanks, factories, etc.

*Noncoatings uses.* The ink industry is a significant user of alkyd binders, mainly of the longoil isophthalic type. Other uses include caulks, adhesives, sealants, and linoleum. At one time alkyds were used as binders in fiberglass-reinforced plastics, but they have been supplanted almost completely by unsaturated polyesters.

# References

# General References

1 R. H. Kienle: "Alkyd Resins, Development of and Contribution to Polymer Theory", *Ind. Eng. Chem.* **41** (1949) 726 – 729.
2 *The Chemistry and Processing of Alkyd Resins*, Monsanto Co., St. Louis, Mo., 1952.
3 P. J. Flory: *Principles of Polymer Chemistry*, Cornell University Press, Ithaca, N.Y., 1953.
4 A. R. H. Tawn: "Some Problems in the Fundamental Study of Alkyd Resins", *J. Oil Colour Chem. Assoc.* **39** (1956) 223 – 252.
5 C. R. Martens: *Alkyd Resins*, Reinhold Publ. Corp., New York 1961.
6 T. C. Patton: *Alkyd Resin Technology: Formulating Techniques and Allied Calculations*, Wiley-Interscience, New York 1962.
7 R. P. A. Sims, W. H. Hoffman, W. J. Steward in W. O. Lundberg (ed): *Autooxidation and Antioxidants*, vol. **II**, J. Wiley & Sons, New York 1962, pp. 683 – 694.
8 J. B. Blegen, W. R. Fuller: *Alkyd Resins*, Unit Five of Federation Series on Coatings Technology, Federation of Societies for Paint Technology, Philadelphia 1967.
9 R. A. Myers, J. S. Long (eds.): *Treatise on Coatings*, vol. **1**, pt. 1, Marcel Dekker, New York 1967.
10 F. W. Billmeyer, Jr.: *Textbook of Polymer Science*, 2nd ed., J. Wiley & Sons, New York 1971.
11 M. W. Formo in D. Swern (ed.): *Bailey's Industrial Oil and Fat Products*, 4th ed., vol. **I**, J. Wiley & Sons, New York 1979, pp. 687 – 816.
12 T. C. Patton: *Paint Flow and Pigment Dispersion*, 2nd ed., J. Wiley & Sons, New York 1979.

13. *Annual Book of ASTM Standards, Paints, Related Coatings and Aromatics*, vol. **6**.01, 6.02, and 6.03, ASTM, Philadelphia (revision issued annually).
14. K. Holmberg, *High Solids Alkyd Resins*, M. Dekker, New York, 1987.
15. A. R. Marrion (ed.): *The Chemistry and Physics of Coatings*, Royal Society of London, London, 1994
16. D. Stoye, W. Freitag, *Resins for Coatings*, Hanser, Munich, 1996, pp. 60 – 80.
17. Z. W. Wicks, Jr., F. N. Jones, S. P. Pappas, *Organic Coatings Science and Technology*, 2nd Ed., Wiley, New York, 1999.

## Specific References

18. J. Berzelius, *Rapp. Annu. Inst. Geol. Hong.* **26** (1847).
19. W. Smith, *J. Soc. Chem. Ind. London* **20** (1901) 1075 – 1076.
20. R. H. Kienle, *Ind. Eng. Chem.* **41** (1949) 726 – 729.
21. General Electric, *US 1893873*, 1933 (R. H. Kienle).
22. R. H. Kienle, C. S. Ferguson, *Ind. Eng. Chem.* **21** (1929) 349 – 352.
23. W. H. Carothers, *J. Am. Chem. Soc.* **51** (1929) 2548 – 2559.
24. R. H. Kienle, A. G. Hovey, *J. Am. Chem. Soc.* **51** (1929) 509 – 519; **52** (1930) 3636 – 3645.
25. K. McNally et al., *The Coatings Agenda America 1999/2000*, Campden, London, 1999, pp. 83 – 84.
26. *Annual Book of ASTM Standards, Paints, Related Coatings and Aromatics*, vol. **6**.01 § D 16 – 82 a ASTM, Philadelphia 1983, p. 13.
27. *DIN 55945*(1968).
28. D. Stoye, W. Freitag, *Resins for Coatings*, Hanser, Munich, 1996, pp. 60 – 80.
29. *Organic Coatings Science and Technology*, 2nd Ed., Wiley, New York, 1999, p. 268.
30. N. O. V. Sonntag in D. Swern (ed.): *Bailey's Industrial Oil and Fat Products*, 4th ed., vol. **I**, J. Wiley & Sons, New York 1979, pp. 289 – 477.
31. *Annual Book of ASTM Standards, Paints, Related Coatings and Aromatics, D*, vol. **6**.03, ASTM, Philadelphia 1959, pp. 282 – 285
32. C. Williams, *Inform* **11** (2000) 580 – 588.
33. W. H. Carothers, *Trans. Faraday Soc.* **32** (1936) 38.
34. R. H. Kienle, P. A. Vander Muelen, F. E. Petke, *J. Am. Chem. Soc.* **61** (1939) 2258, 2268 – 2271.
35. R. H. Kienle, F. E. Petke, *J. Am. Chem. Soc.* **62** (1940) 1053 – 1056; **63** (1941) 481– 484.
36. P. J. Flory, *J. Am. Chem. Soc.* **63** (1941) 3083, 3091, 3096 – 3100.
37. P. J. Flory, *Chem. Rev.* **39** (1946) 137 – 197.
38. W. H. Stockmayer, *J. Chem. Phys.* **11** (1943) 45; **12** (1944) 125.
39. W. H. Stockmayer, *J. Polym. Sci.* **9** (1952) 69; **11** (1953) 424.
40. L. C. Case, *J. Polym. Sci.* **26** (1957) 333 – 350.
41. M. Jonason, *J. Appl. Polym. Sci.* **4** (1960) 129 – 140.
42. H. S. Lilley, *J. Appl. Polym. Sci.* **5** (1961) no. 16, S16 – S18.
43. E. G. Babolek, *Am. Paint J.* **42** (15 Sept. 1958).
44. E. G. Babolek, E. R. Moore, S. S. Levy, C. C. Lee, *J. Appl. Polym. Sci.* **8** (1964) 625 – 657.
45. H. Hata, J. Kumoanotani, Y. Nishizawa, H. Tomita, *XIVth FATIPEC Congr.* 1978, 358 – 364.
46. J. Kuomanotani, H. Hata, H. Masuda, *Proc. VIIIth Intl. Conf. Org. Coatings Sci. Technol.*, Athens 1982, 123 – 144.
47. C. M. Bruneau, *XVth FATIPEC Congr.*, Amsterdam 1980, vol. I, 123 – 144.
48. *Alkyd Report No. 1.2*, Hercules, Wilmington, Del.
49. *Alkyd Report No. 1.3*, Hercules, Wilmington, Del.
50. A. G. Hovey, *Ind. Eng. Chem.* **41** (1949) 730 – 737.
51. W. S. Armstrong, B. F. Coe, *Chem. Eng. Prog.* **79** (1983) 56 – 61.
52. H. Wexler, *Chem. Rev.* **64** (1964) 591 – 611.
53. J. H. Hartshorn, *J. Coat. Technol.* **54** (1982) 53 – 61.
54. J. Scheiber, *Farbe Lack* **63** (1957) 443.
55. N. A. Falla, *J. Coat. Technol.* **64** (1992) no. 815, 55 – 60.
56. J. C. Hubert, R. A. M. Venderbosch, W. J. Muizebelt, R. P. Klaasen, K. H. Zabel, *J. Coat. Technol.* **69** (1997) no. 869, 59 – 64.
57. W. J. Muizebelt, J. J. Donkerbroek, M. W. F. Nielen, J. B. Hussern, M. E. F. Biemond, R. P. Klaasen, K. H. Zabel, *J. Coat. Technol.* **70** (1998) no. 876, 83 – 93.
58. W. J. Muizebelt, J. C. Hubert, M. W. F. Nielen, R. P. Klaasen, K. H. Zabel, *Prog. Org. Coat.* **40** (2000) 121 – 130.
59. R. W. Hein, *J. Coat. Technol.* **71** (1999)no. 898, 21 – 25.
60. Lewis Berger & Sons, *DE 975352*, 1949.
61. E. F. Redknap, *J. Oil Colour Chem. Assoc.* **43** (1960) 260.
62. Rohm & Haas, *DE 912752*, 1951; *DE 1002 381*, 1953.
63. Dow Corning, Midland, Mich., Bulletin 03-003.
64. T. F. Washburn Co., *US 2633649*, 1952.
65. J. S. Fry, C. N. Meriam, G. J. Misko, *Am. Chem. Soc. Div. Org. Coat. Appl. Polym. Sci. Preprints* **47** (1982) 540 – 548.
66. R. R. Engelhardt, *Double Liaison Chim. Peint.* **27** (1980) no. 296, 207 – 213; *Pigm. Resin Technol.* 1979 (Aug.) 5 – 15.
67. F. Sheme, S. Belote, L. Gott, *Polym. Paint Colour J.* 1975, 784 – 790.
68. Bayer, *DE 2837552*, 1979; *US 4321 170*, 1982 (K. Zabrocki et al.).
69. Amoco Chemicals Corp., Tech. Bull. TMA-1086 (1981).
70. E. Levine, E. J. Kuzma, *J. Coat. Technol.* **51** (1979) no. 657, 35 – 48.
71. K. Fey, W. Finzel, *Am. Paint Coat. J.* 1982 (June 28)41 – 45.
72. J. C. Laout, J. F. May, C. Nicaud, *XVth FATIPEC Congr.*, Amsterdam 1980, vol. II, 227-246.
73. Ashland Oil & Refining Co., *US 3442835*, 1969 (G. M. Curtice, D. D. Taft).
74. B. G. Bufkin, J. R. Grawe, *J. Coat. Technol.* **50** (1978) no. 647, 65 – 96.
75. Perstorp AB, Report 1978 (S. E. Friberg, E. G. Gillberg-LaForce, K.-H. Falklin).
76. B. Bergenstahl, A. Hofland, G. Ostberg, A Larsson in C. Salamone (ed.): *Polymeric Materials Encyclopedia*, CRC Press, Boca Raton, FL, 1996, pp. 154 – 160.
77. P. K. Wissenborn, A. Motiejauskaite, *Prog. Org. Coat.*, **40** (2000) 253 – 266.
78. L. W. Hill, Z. W. Wicks, Jr., *Prog. Org. Coat.* **10** (1982) no. 1, 55 – 89.
79. K. Holmberg, J.-A. Johansson, *Proc. VIIIth Int. Conf. Org. Coatings Sci. Technol.*, Athens 1982, 255 – 275.
80. D. B. Larson, W. D. Emmons, *Proc. VIIIth Int. Conf. Org. Coatings Sci. Technol.*, Athens 1982, 193 – 216.
81. S. Enomoto et al., *J. Appl. Polym. Sci.* **22** (1978) 252 – 265.
82. E. Levine, E. J. Kuzma, M. T. Nowak, *Mod. Paint Coat.* **66** (1976) no. 8, 23 – 30.
83. D. J. Love, *J. Coat. Technol.* **53** (1981) no. 680, 55 – 58.
84. V. Verkholantsev, *Eur. Coat. J.* (2000) no. 1/2, 76 – 83.

85 D. Ryer, *Paint and Coatings Industry* **14** (1998) no. 1, 76 – 83.
86 A. Heitkamp, J. Hamre, *J. Water Borne Coat.* **6** (1983) no. 2, 20 – 28.
87 F. N. Jones, C. P. Alexander, J. M. Larson, *Mod. Paint Coat.* **71** (1981) no. 8, 54 – 56.

# Further Reading

Z. W. Wicks: *Alkyd Resins*, "Kirk Othmer Encyclopedia of Chemical Technology", 5th edition, John Wiley & Sons, Hoboken, NJ, online DOI: *10.1002/0471238961.01121125120914.a01.pub2*.

# Amino Resins

HANS DIEM, BASF Aktiengesellschaft, Ludwigshafen, Germany

GÜNTHER MATTHIAS, BASF Aktiengesellschaft, Ludwigshafen, Germany

ROBERT A. WAGNER, Dynea Austria GmbH, Krems/Donau, Austria

| | | |
|---|---|---|
| 1. | Introduction | 1615 |
| 2. | Physical Properties | 1616 |
| 3. | Chemical Properties | 1616 |
| 4. | Raw Materials | 1617 |
| 5. | Production | 1619 |
| 5.1. | Principle Aspects | 1619 |
| 5.1.1. | Hydroxymethylation | 1620 |
| 5.1.2. | Condensation | 1621 |
| 5.1.3. | Principal Components of an Amino Resin Solution | 1622 |
| 5.2. | Production Processes | 1622 |
| 5.2.1. | Batchwise Production | 1622 |
| 5.2.2. | Continuous Production | 1625 |
| 5.2.3. | Production of Special Products and Foams | 1627 |
| 6. | Environmental Protection | 1627 |
| 7. | Types of Resins and Their Properties | 1628 |
| 7.1. | Urea Resins (Urea–Formaldehyde Resins) | 1628 |
| 7.1.1. | Pure Urea–Formaldehyde Resins | 1628 |
| 7.1.2. | Mixed Condensates | 1629 |
| 7.1.3. | Condensation of Urea with Other Aldehydes | 1629 |
| 7.1.4. | Resins from Urea Derivatives | 1629 |
| 7.1.5. | Modified Urea Resins | 1630 |
| 7.2. | Melamine Resins | 1632 |
| 7.2.1. | Unmodified Melamine Resins | 1632 |
| 7.2.2. | Modified Melamine Resins | 1632 |
| 7.3. | Urethane Resins | 1633 |
| 7.4. | Cyanamide and Dicyanodiamide Resins | 1633 |
| 7.5. | Sulfonamide Resins | 1634 |
| 7.6. | Aniline Resins | 1634 |
| 8. | Analysis | 1634 |
| 9. | Storage and Transportation | 1635 |
| 10. | Uses | 1636 |
| 11. | Economic Aspects | 1639 |
| | References | 1639 |

## 1. Introduction

Amino resins are condensates formed when carbonyl compounds react with compounds containing amino, imino, or amide groups, liberating water. Before they have hardened, these products consist principally of oligomers, which are also called prepolymers.

A very large number of carbonyl- and nitrogen-containing compounds yield primarily condensates of relatively low molecular mass. Among such compounds, formaldehyde, urea, and melamine are particularly important, because of their relatively low price. The type of chemical reaction yielding these prepolymers is determined by the functional groups. The amino resins are obtained by large-scale production in modern industrial plants and offer technical advantages in many applications.

**History.** The history of amino resins began with the synthesis of the two main raw materials, urea and formaldehyde; the former was prepared by F. WÖHLER in 1824, and the latter by A. BUTLEROV in 1859. However, the first investigations of urea–formaldehyde resins were not carried out until about 1880. The first reaction products were reported by B. TOLLENS (1884) and C. GOLDSCHMIDT (1887), and the basic chemistry of the amino resins was established about 1908 [16].

F. POLLACK and K. RIPPER (1921) synthesized a clear transparent material, which they called "Pollopas." This product never became commercially important, but it did give rise to a number of experiments. The first commercial products, manufactured by E. C. ROSSITER of British Cyanides Co. in 1924 and 1925, led to molding materials still in use today.

A major step forward in the industrialization of amino resins became possible by the patent of A. SCHMID and M. HIMMELHEBER (1932), in which they laid the basis for resin-bonded particle board in order to increase the yield of sawn wood for furniture production from 40–80% [17].

The melamine resins were introduced in Germany by Henkel in 1935 [18]. Their properties are similar to those of the urea–formaldehyde resins, but enable the manufacture of products of higher quality. For example, they are chemically stable after the resin has hardened and become insoluble.

An ever-increasing number of mixtures of urea–formaldehyde and melamine–formaldehyde, or of urea–melamine–formaldehyde resins are produced for the most diverse uses. Lower costs, processing advantages, and the better performance of the products made from these mixtures account for their popularity.

Amino resins find their major use as adhesives. They are used in the production of particle board, medium density fiber board (MDF), high density fiber board (HDF), plywood, waferboard, and strandboard. Other types are employed for the production of laminated beams and parquet flooring, and for furniture assembly. Amino resins are also used to strengthen paper, to produce molding compounds and the flexible backing of carpets, as leather auxiliaries and soil conditioners, and for protective surface coatings. Another important application of amino resins is to saturate decorative papers for their final use in laminate manufacture.

## 2. Physical Properties

Amino resins are available commercially as concentrated solutions or solids, e.g., powders. The solutions generally contain water as a solvent, but in the surface-coating resins the solvents are alcohols. The solutions range from colorless and clear to milky and cloudy, and are tacky to the touch. Their viscosities are substantially higher than that of water, from 20–70 000 mPa·s—in general from 40–2000 mPa·s—measured at 20°C. The amino resins can be odorless, possess a characteristic smell, or, depending on the content of free formaldehyde, have a more or less sharp odor. The densities of the solutions at the commercially available concentrations of 50–80% solids are from 1.15–1.31 g/cm$^3$. An important property of the resins is the amount of water that can be added without precipitating solids, the so-called water tolerance or water dilutability. At relatively low temperatures, i.e., below about 20°C, pronounced cloudiness and an unusually large increase in viscosity are observed in many cases.

Some resin solutions exhibit the phenomenon of pseudoplasticity (thixotropy), which may or may not be desirable. Thixotropy means that the viscosity decreases as the shearing force increases [19, 20].

In powder form amino resins are white. In the spray-dried, commercially available form they have a particle size generally of 15–70 μm, but always < 200 μm. This particle size is a result of the spray-drying process [21]. The bulk density of the commercially available products is 0.5–0.8 kg/L. The powder resins dissolve in water to usually give milky liquids, however, in some cases clear solutions can be achieved. Flour-like additives can be mixed with some resins, which then assume the color of the additive to a greater or lesser extent.

## 3. Chemical Properties

Even when they are prepared from only two components, the amino resins are mixtures of various substances. For example, in the case of urea–formaldehyde resins the major components of the mixture are various hydroxymethyl compounds and oligomers composed of different combinations of basic units and substituted basic units. These oligomers differ in size. As an approximation, the basic unit can be described as methylene–urea (**1**), which can carry various substituents at the nitrogen atoms:

$$\left[ -CH_2-N-\underset{R^1}{\overset{O}{\underset{\|}{C}}}-\underset{R^2}{N}- \right] \qquad R^1, R^2 = -CH_2-N-\overset{O}{\underset{\|}{C}}-\\ -CH_2OH\\ -H$$

**1**

Hence, it is not possible to attribute specific chemical properties to these substances, as in the case of a pure compound.

However, some chemical properties are common to all or most amino resins.

The chemical properties given below are important with regard to processing the resins. The equations given for urea–formaldehyde also apply in principle to other reactants employed for these classes of products.

- Formaldehyde is slowly split off from hydroxymethyl compounds and undergoes addition reactions with other amino groups in accordance with the equilibrium equation [22, 23]:

$$H_2N-CO-NH_2 + CH_2O \rightleftharpoons H_2N-CO-NH-CH_2OH$$

- Hydroxymethyl compounds undergo slow condensation in the absence of heat or react with urea to give methylene compounds, such as methylene–diurea or methylene–oligourea:

$$H_2N-CO-NH-CH_2OH + H_2N-CO-NH_2 \rightarrow$$
$$H_2N-CO-NH-CH_2-NH-CO-NH_2 + H_2O$$

$$2\,H_2N-CO-NH-CH_2OH \rightarrow$$
$$H_2N-CO-NH-CH_2-NH-CO-NH_2 + CH_2O + H_2O$$

These reactions are responsible for the gradual increase in viscosity that amino resins undergo during storage [24].

- It is known [25] that bis-methylene–urea ethers (**2**) are formed [13, p. 324, [23, 26]], but amino resin solutions contain rather small amounts of these ethers. This is due to the fact that the formation of the structural units (**1**) and (**2**) depends on the pH. (**1**) is formed in acidic media, while (**2**) is formed at both alkaline and acidic pH. Moreover, (**1**) is rather stable and basically no equilibrium with the monomers exists, while (**2**) allows a retro reaction into urea and formaldehyde. As a consequence, the amount of ether bridged diurea decreases, while the concentration of methylene bridges increases by time. Under certain production conditions cyclic ethers, so-called urones (**3**), also occur. Their formation is favored by acidic pH values (pH < 3) and at elevated temperatures (> 80 °C)

$$H_2N-CO-NH-CH_2-O-CH_2-NH-CO-NH_2$$
**2**

$R = -H, -CH_2OH, -CH_2OCH_3$

**3**

- Amino resin solutions are generally rendered slightly alkaline, i.e., pH 7.5–9, in order to stabilize them, but the pH slowly decreases as formic acid is formed. This effect is more pronounced for urea–formaldehyde resins and much less significant in melamine–formaldehyde resins.

These reactions, especially condensation with the formation of methylene bridges, are accelerated by acid catalysts and elevated temperatures. Hence, in processing amino resins, acids or acid-forming substances are important as hardeners that accelerate the condensation. Alkali-catalyzed condensation [13, pp. 331–332, 27] is of no practical importance.

The hydroxymethyl groups can be converted with alcohols into the corresponding ethers. This reaction is utilized in surface-coating resins and impregnation resins for the manufacture of paper-based edge-banding material and, to a lesser extent, in textile and paper auxiliaries.

The free formaldehyde, which is present in concentrations from 0.03% to about 2%, is of practical importance – not only because of classifications with respect to transport and storage of the resins – but in many cases because it reacts with ammonium salts of acids to give hexamethylenetetramine and the free acid, which acts as a hardener.

Resin powders also condense over time to give relatively high molecular mass products, but in this case the reaction is slower by a factor of at least five.

## 4. Raw Materials

The principal and most important carbonyl component is formaldehyde. Furfural and glyoxal are also employed, but these play a very minor role.

The most important amine-containing components are urea and melamine, but thiourea, guanidine, cyanamide, dicyanodiamide, and to a lesser extent acid amides, guanamines, and aniline, are also employed. Further components include the derivatives of these compounds, e.g., substituted ureas.

Relevant information on raw materials used to produce amino resins can be found in [1, pp. 3–85].

Products that are added in relatively small amounts to effect slight but often important changes in the properties are called modifiers. These include alcohols, amines, aminoalcohols, and a very large variety of other organic and inorganic substances [1, pp. 7–122].

Finally, it should also be mentioned that inert inorganic salts sometimes are added during the production or processing of glue resins [8, p. 34, [28, 29]].

***Formaldehyde [50-00-0]***, $CH_2O$, $M_r$ 30.03, (production and properties → Formaldehyde). In the production of amino resins, formaldehyde is used either in the form of solid paraformaldehyde, in aqueous solution, or as a urea–formaldehyde precondensate that contains from 4–5 mol of formaldehyde per mole of urea. Paraformaldehyde is a mixture of hydrated polyoxymethylene molecules, with a degree of polymerization of 6–100 units. The melting point ranges from 120–170°C, depending on the molecular mass.

$HO - [CH_2O]_n - H$, where $n = 6 - 100$

The most frequently used commercial form is a 30–55% aqueous solution of formaldehyde. These solutions are clear, caustic, and inflammable, have a sharp odor, and are infinitely miscible with water. Formaldehyde is present in aqueous solutions predominantly as the hydrate ($HO-CH_2-OH$), which is in equilibrium with its oligomers:

$n\, CH_2(OH)_2 \rightleftharpoons HO(CH_2O)_nH + (n-1)H_2O$

Depending on the method of production, the solutions are contaminated with a few hundredths of a percent of formic acid and from 0.4–2% of methanol. The content of formic acid and methanol increases during storage, as a result of the Cannizzaro reaction:

$2\, CH_2O + H_2O \rightarrow HCOOH + CH_3OH$

Moreover, on long-term storage at low temperatures, paraformaldehyde separates as a hazy suspension or a precipitate. This process is retarded by the addition of stabilizers, e.g., up to 12% of methanol or triazine derivatives.

Depending on the concentration and the methanol content, the flash point is about 50–85°C. The regulations governing flammable liquids do not apply to standard aqueous formaldehyde solutions with low methanol content.

Formaldehyde solutions can be stored and transported in containers made of stainless steel, aluminum, polyester resin, or appropriately lined iron.

Finally, aqueous solutions of urea – formaldehyde precondensates also are employed. Commercially available solutions contain from 70–85% of the active ingredients, calculated as urea and formaldehyde. The molar ratio of formaldehyde to urea ranges from 4 : 1 to 5.5 : 1, and the mass ratio is therefore 2 : 1 to 2.75 : 1. These solutions are substantially more resistant to polymer precipitation than are pure formaldehyde solutions. In the urea–formaldehyde precondensates, the major part of the formaldehyde is present in the form of urea–hydroxymethyl compounds, a smaller part as formaldehyde or its oligomers in the hydrated form, and a very small part as condensates. In this last case, the urea molecules are linked mainly by methylene bridges.

***Biological Effect.*** Since 2000, the maximum allowable concentration (MAK value) recommended by the Senate Committee of the Deutsche Forschungsgemeinschaft (DFG) in Germany is 0.3 ppm. The corresponding values for the United States of America (given by OSHA) and Japan are 0.75 ppm (TWA 8 h) and 0.5 ppm (OEL (operational exposure limit)), respectively. At present, regulations vary from country to country (→ Formaldehyde, Chap. 10).

***Other Aldehydes.*** Other aldehydes play a minor role. *Glyoxal* [107-22-2] is available commercially in the form of 30–40% aqueous solutions. The purity is 97–99%. Depending on the production method, the solutions contain 0.1–0.2% acid constituents. The color number should have a maximum value of 15 (APHA). Glyoxal is hydrated in the aqueous solution.

***Urea*** [57-13-6], carbamide, diamide of carbonic acid, $H_2NCONH_2$, $M_r$ 60.06 (→ Urea).

Urea is available commercially as a white crystalline product or as prills. For amino resin production, a technical-grade product is usually adequate.

It should be noted that fertilizer urea is frequently contaminated with ammonium salts, and these may act as hardeners and cause the glue mixture to solidify even in the production stage. Furthermore, the prilled urea can be mixed with sulfur or formaldehyde to keep it free flowing. Sulfur-coated urea is not recommended for resin production. The formaldehyde forms insoluble condensates on the surface of the prills, and these condensates can lead to a slight cloudiness when the prills are dissolved.

The dissolution of urea in water is an endothermic process. The enthalpy of solution ($\Delta H$) is + 257.5 J/g, measured at 10°C in a 20-fold amount of water.

In the production of amino resins, urea is employed in solid form, in solution, or as a urea–formaldehyde precondensate.

***Thiourea*** [62-56-6], thiocarbamide, $NH_2CSNH_2$, $M_r$ 76.12 ($\rightarrow$ Thiourea and Thiourea Derivatives).

Thiourea is a relatively expensive raw material and is, therefore, not very important industrially. Apart from its use in special products, it serves only as a modifier, for example, in impregnating resins. Thiourea improves the resin properties, e.g., hardness, surface gloss, and water resistance.

The melting point of thiourea is 180°C. Its solubility in water is 9% at 100°C, substantially lower than that of urea.

***Substituted Ureas.*** Cyclic ethylene ureas and propylene ureas (2-oxoimidazolidine and 2-oxohexahydropyrimidine) are employed for the preparation of textile auxiliaries. Because of their high price and their limited ability to form condensates, they are not used in large amounts for any other purposes.

***Melamine*** [108-78-1], 2,4,6-triamino-1,3,5-triazine, cyanurotriamide, $M_r$ 126.13 ($\rightarrow$ Melamine and Guanamines). Melamine (**4**) is a white powder that sublimes on heating. The particle diameter depends on the production process and varies from 5 to 300 µm. The rate of solubility in formaldehyde solutions depends on the particle size and is an important parameter given in the information leaflets of melamine producers. The color number of a 50% melamine–formaldehyde solution may not exceed 20 (APHA). This is important with respect to raw materials for surface coatings and to impregnating resins.

Melamine derivatives, e.g., benzoguanamine (**5**), are employed as modifiers.

***Modifiers.*** A very large number of additives from all areas of organic and inorganic chemistry have been used to modify amino resins; most of these additives are no longer important [1, pp. 107–122]. Aliphatic alcohols, glycols, sugars, sugar alcohols, polysaccharides, and caprolactam are additives for special purposes and are still used today.

## 5. Production

### 5.1. Principle Aspects

Amino resin production consists of two stages: *hydroxymethylation* and *condensation* [30]. Depending on the reaction conditions, hydroxymethylation is accompanied to a greater or lesser extent by condensation. The following equations illustrate the process in a simple case, for example, for formaldehyde plus urea:

Hydroxymethylation:

$$NH_2CONH_2 + 2\,CH_2O \rightleftharpoons HOCH_2-NHCONH-CH_2OH$$

Condensation:

$$HOCH_2-NHCONH-CH_2OH + NH_2CONH_2 \rightarrow$$
$$HOCH_2-NHCONH-CH_2-NHCONH_2 + H_2O$$

Further condensation leads to longer chains of the basic unit

$$-N-CO-N-CH_2-.$$

Hydroxymethylation is carried out in slightly alkaline or slightly acidic media, whereas the condensation requires a more strongly acidic solution. Exceptions are the condensation conditions of melamine–formaldehyde resins, which require an alkaline pH [31], and the manufacture of melamine–urea–formaldehyde resins, which are condensed in slightly alkaline media. The process for the production of the resin is continued until the resulting product is an oligomeric mixture that is still soluble or fusible. It is protected against further condensation by rendering the medium alkaline. Condensation of the resin by means of acid catalysis, i.e., cross-linking to give a substantially infusible product, is not carried out until the resin is put to use. This process is called *curing*.

However, free hydroxymethyl groups are present even in this state.

The monomolecular and hydroxymethyl compounds with low degrees of polymerization and up to six urea–formaldehyde units are virtually all known and can be prepared in pure form [1, p. 20]. Aqueous solutions of the oligomers with low degrees of polymerization do not have a sufficiently long shelf life. Only when the concentrations of hydroxymethyl and $NH_2$ groups are reduced by partial condensation or when the products are present in solid form do they possess an industrially useful shelf life, i.e., of several weeks or more.

Condensations carried out in an excess of an alcohol, such as methanol or butanol, with the addition of acid and removal of water as necessary, produce curable etherified (i.e., alkylated) amino resins. These are soluble in nonaqueous solvents and can be mixed with alkyd resins, epoxy resins, etc. They are used as starting materials in the production of surface coatings.

Resin glues based on urea, melamine, or both have been subjected to partial etherification to stabilize them against further condensation [32–34]. Historically, these resins have been used for the manufacture of plywood, but are practically not used any longer.

Preventing the resins from solidifying during the production process and selectively achieving certain properties – such as long shelf life, low formaldehyde content and, low formaldehyde emissions from final products, strength, and resistance to swelling of particle board after curing – are the basic difficulties encountered. Frequently, novel properties are required on short notice by the market.

### 5.1.1. Hydroxymethylation

Hydroxymethylation can be carried out using either alkaline or acid catalysis, but the former is faster. The reaction mechanisms in acidic and alkaline media differ. In acids, condensation takes place at the same time. Hence, hydroxymethylation is carried out industrially in an alkaline or slightly acidic solution.

In hydroxymethylation, a semiaminal bond is formed:

$$H_2C-N(H)-CO-NH_2$$
with OH on the carbon

Monohydroxymethylurea

This is cleaved readily, and hydroxymethylation reactions are reversible and establish an equilibrium [35]. Although mono-, di-, and trihydroxymethylurea are formed and isolated easily, tetrahydroxymethylurea is virtually nonexistent in aqueous solution.

The complete hydroxymethylation of all amino groups to form hexahydroxymethylmelamine can be carried out readily. The hydroxymethylation of urea is slightly exothermic:

$$H_2N-CO-NH_2 + CH_2O \rightleftharpoons H_2N-CO-NH-CH_2OH$$

$$\Delta H = -23 \text{ kJ/mol}$$

The $\Delta H$ data given in the literature fluctuate about this value by $\pm 5$ kJ/mol. The enthalpy of hydroxymethylation of melamine, relative to the amount of formaldehyde, is $-15$ kJ/mol [36, 37].

Reaction rates have been measured frequently, but they are sensitive to many parameters [22, 36].

Industrial processes produce variable mixtures of oligomers with mixed repeat units, so the overall reaction rate is also variable and cannot be defined. Moreover, the important resin properties have not been correlated with the individual reaction rates. In general, the hydroxymethylation rate increases with increased concentration, temperature, and pH. Substituents also have an effect; adjacent electron-withdrawing groups reduce the reaction rate, as do bulky substituents.

### 5.1.2. Condensation

The condensation is catalyzed by acids. Catalysis by strong alkalis also has been described, but the resulting amino resins are not important industrially [27]. The condensation can take place by the following routes:

Although the presence of methylene compounds in amino resins can be established unambiguously by means of $^{13}C$ NMR spectroscopy, bis(ureamethylene) ethers can be detected only by an indirect procedure [38]. $^{13}C$ NMR spectroscopy permits the elucidation of many of the structures but does not distinguish between the **C** atoms shown in boldface in (**6**) and (**7**), although their third nearest neighbors are different.

$$-N-CH_2-O-CH_2-N-\qquad -N-CH_2-O-CH_2-OH$$
$$\quad\quad\quad\quad 6 \quad\quad\quad\quad\quad\quad\quad\quad\quad\quad 7$$

This problem also occurs in the case of melamine resins [39].

The methylene bonds between two urea molecules are very stable and can be hydrolyzed only by strong acids at elevated temperatures (see Chap. 8). The condensation reaction is slightly exothermic; the enthalpy is $-28$ kJ/mol [36].

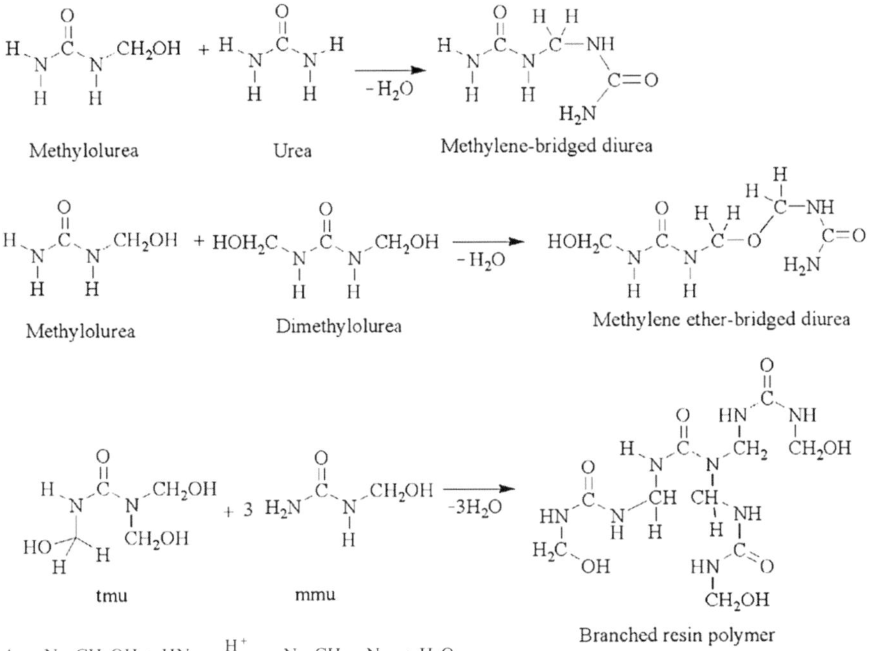

Methylolurea + Urea → Methylene-bridged diurea

Methylolurea + Dimethylolurea → Methylene ether-bridged diurea

tmu + 3 H₂N– → mmu → Branched resin polymer

1. $-N-CH_2OH + HN- \xrightarrow{H^+} -N-CH_2-N- + H_2O$

2. $-N-CH_2OH + HOCH_2-N- \xrightarrow{H^+} -N-CH_2-N- + CH_2O + H_2O$

3. $-N-CH_2OH + HOCH_2-N- \xrightarrow{H^+} -N-CH_2-O-CH_2-N- + H_2O$

The stability of the methylene–ether bond is between that of the hydroxymethyl bond and that of the methylene bond.

It is impossible to give a single reaction rate for the condensation process because there are many different starting materials (hydroxymethyl compounds) and many end products (methylene compounds).

The type of solvent also has an effect on the condensation rate [40]. The dependence of the reactivity of hydroxymethyl compounds on the chemical composition of the amino compounds can be interpreted on the basis of several qualitative rules [41].

### 5.1.3. Principal Components of an Amino Resin Solution

It is possible to identify the most important structural moieties of the oligomers in amino resin solutions by means of $^1$H and $^{13}$C NMR spectroscopy [26, 38, 42–44].

The hydrates of the formaldehyde oligomers HO(CH$_2$O)$_n$H, where $n$ is about 5, can be detected along with the following hydroxymethyl structures:

```
-N-CO-N-CH₂OH        -N-CO-N-CH₂OH
 |      |             |      |
 H                    R
-N-CO-N-CH₂OCH₂OH    -N-CO-N-CH₂OCH₂OH
 |      |             |      |
 H                    R
```

Ethers, such as hydroxymethyl methyl ether, also can be identified in the spectrum by distinguishing the C atoms of the parent alcohol.

Finally, the structures (**8**) and (**9**) and ring compounds, including cyclic ethers of the urone type, also appear.

```
-N-CH₂-N-          -N-CH₂-N-
 |     |            |     |
 H     R            R     R
Asymmetrical       Symmetrical
     8                  9
```

In a typical particle board glue, the $^{13}$C NMR spectrum might show, for example, the following distribution of the carbon atoms of the original formaldehyde (mole fractions):

| | |
|---|---|
| formaldehyde | 0.03 |
| hydroxymethyl compounds | 0.72 |
| methylene compounds | 0.15 |
| cyclic compounds | 0.10 |

It is worth being mentioned that other analytical methods can be and are applied for further structure elucidation, too; for example:

- HPLC, mainly for low molecular mass constituents, like urea, monomethylolurea, $N,N$-dimethylolurea, $N,N'$-dimethylolurea, trimethylolurea, and methylene diurea
- Size exclusion chromatography, mainly for higher molecular mass species, like dimers and oligomers.

## 5.2. Production Processes

Amino resins usually are prepared in 5–100 m$^3$ batches. However, continuous processes are also employed, particularly for the production of resin glues. The patent literature describes processes where a particular stage is operated continuously while other stages are operated batchwise.

Because the reaction rate and the composition of the end product are influenced by a large number of parameters, the reaction must be controlled carefully. Important reaction parameters are:

- purity or composition of the starting materials
- molar ratio in each of the reaction stages
- type and amount of modifier
- reactant concentrations
- the pH value at each stage of the reaction
- temperature at each stage of the reaction
- type and concentration of catalyst
- amount of buffer salt
- reaction time in each stage

The large number of reaction parameters and resultant adaptations of the process lead to a very extensive body of relevant literature, particularly patent literature. Only the important processes are described here, but a comprehensive review can be found in [1, pp. 101–105].

### 5.2.1. Batchwise Production

The batchwise procedure is the most widely used method for the industrial production of amino resins, but it has the disadvantage of a relatively small production capacity. Nevertheless, this process permits an extensive variety of products and frequent changes of product.

The reactions are carried out in stirred kettles, in two or more stages, at 70–100°C. In the first stage, which is carried out in a slightly acidic to alkaline solution, the principal reaction

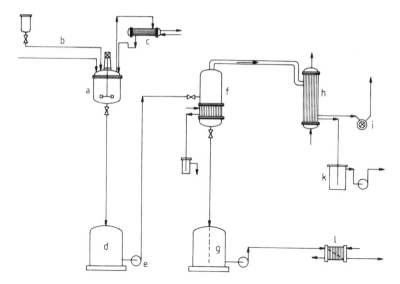

**Figure 1.** Plant for the batchwise production of aqueous amino resin solutions
a) Stirred kettle; b) Raw material feed; c) Reflux condenser; d) Temporary container; e) Pump; f) Vaporizer; g) Container for finished product; h) Condenser; i) Vacuum pump; k) Vessel for condensed vapors; l) Cooler

is hydroxymethylation. In subsequent stages, the condensation is carried out, water is split off, and higher molecular mass products of increasing viscosity are obtained. The duration of the condensation depends on the properties desired in the product. In general, the products are concentrated to 60–70% aqueous solutions by evaporation. The starting materials, urea or melamine, are employed in powder form. Urea can also be used in solution. Formaldehyde is employed as a 37–55% aqueous solution. In a few special cases, formaldehyde is used in the solid form, paraformaldehyde. Finally, it is possible to employ urea–formaldehyde precondensates in concentrations of not more than 80% by mass of the active ingredients.

Figure 1 shows a diagram of a plant for the batchwise production of aqueous solutions.

**Aqueous Solutions.** The reactor consists of a kettle equipped with heating and cooling devices, a stirrer, inlets for feeding in the raw materials and for measuring devices, a valve at the bottom for discharging the kettle, and a manhole for cleaning.

The material used is mainly stainless steel, material no. St. 1.4541 or St. 1.4571. Internal coils or external, welded-on, semicircular pipes enable heating with steam or cooling with cold water. Stirring is carried out by means of a powerful unit consisting of disk or anchor stirrers. Nozzles are welded on in order to meter in the raw materials and to permit measurement of the temperature and pH.

When the condensation is complete, the product can be evaporated. This is carried out in the stirred kettle, or more economically in a single- or multistage tubular vaporizer, with the pressure reduced to protect the resin. The quality of the resin is then compared to its specifications and it is discharged into a storage container.

The pipelines can be of aluminum or steel because the medium is slightly alkaline, the product tanks can also be of glass fiber reinforced polyester. It is advantageous to provide lined iron tanks.

A typical, simple resin glue for particle board production is made as follows [45]: 85 parts (by mass) of solid urea are introduced into 158 parts of a 50% formaldehyde solution that has been rendered neutral. The mixture is kept at 80°C for 10 min, during which time hydroxymethylation takes place. The pH is then brought to 4–5 by adding a 10% formic acid solution. The condensation causes the temperature to rise to 90–100°C within a few minutes. This temperature is maintained until a mixture of 1 part of reaction solution and 5 parts of water gives a white precipitate at a certain temperature (sharp endpoint) or a certain viscosity is achieved. The reaction mixture is then brought to a slightly alkaline pH with a 25% sodium hydroxide solution. This terminates the condensation. A further 60 parts (by mass) of urea are then added, so that the molar ratio of formaldehyde to urea in the end product is 1.1. The resulting solution has a solids content of ~60%. The resin solution can be evaporated under reduced pressure to the commercial concentration of 66%.

This 66% solution of amino resin forms a milky solution and has a viscosity of about 500 mPa·s (at 20°C) and a density of 1.290 g/cm$^3$.

A typical resin for the impregnation of paper webs used in producing decorative laminates can be made as follows [12, p. 258]:

126 parts (by mass) of melamine are introduced into a stirred solution of 120 parts of a 40% formaldehyde solution and 70 parts of water at room temperature. The reaction mixture is brought to pH 9 by adding aqueous sodium hydroxide solution and is then heated very rapidly, i.e., in 20–30 min for an industrial-scale mixture, to a temperature of 100°C. After the melamine has dissolved, a process that takes place exothermally, sodium hydroxide solution is added continuously to maintain the pH value of the reaction mixture at 8.5–8.8 throughout the condensation. Condensation is carried out, with gentle refluxing and stirring, until the reaction mixture is compatible with water when mixed in the ratio 1:1.5. This is the case when a sample of the reaction mixture exhibits slight cloudiness when water, at 1.5 times the amount of solution, is added at 20°C. The reaction solution is then cooled rapidly and brought to pH 9.5–10 at room temperature by the addition of sodium hydroxide solution. The resultant impregnating resin solution has a solids content of about 55% and is usually clear.

***Amino Resin Powders.*** In the production of amino resin powders, an aqueous resin solution is first produced (Fig. 2) and then fed into a spray drier, where it is atomized by a spray disk or nozzle. The resultant droplets are dried in the stream of hot gas produced by heating air indirectly in a heat exchanger or by mixing hot exhaust gas with air. The powder is collected in a tower and in cyclone separators or filters. It passes through a vibrating sieve into a mixing bin and then pours into sacks or drums.

***Etherified Products Containing Solvent.*** Products of this type, which are used as starting materials for making surface coatings, can be produced in a reactor like that described in the production of aqueous solutions of amino resins (Fig. 3). The reactor must be equipped with a sufficiently large heat exchanger to vaporize the mixture of water and excess alcohol, e.g.,

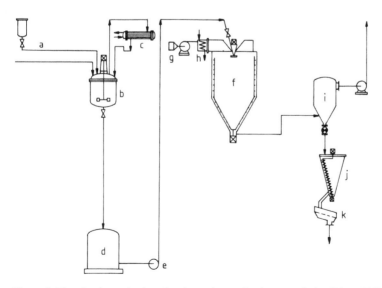

**Figure 2.** Plant for the production of amino resin powders by spray drying [21, p. 501]
a) Starting material feed; b) Stirred kettle; c) Reflux condenser; d) Temporary container; e) Pump; f) Spray drier; g) Blower; h) Air heater; i) Filter; j) Mixing bin; k) Vibrating sieve

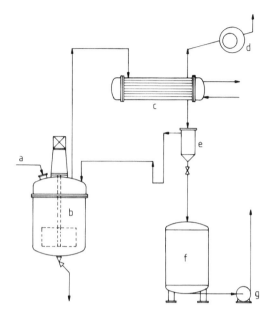

**Figure 3.** Plant for the batchwise production of etherified amino resins
a) Raw material feed; b) Stirred kettle; c) Condenser; d) Vacuum pump; e) Separating vessel (water separator); f) Product vessel; g) Pump

butanol. Downstream from the heat exchanger is a water separator. The separated aqueous phase still contains from 1–2% formaldehyde and 20% alcohol and, therefore, must be worked up. At the appropriate alcohol content, this solvent phase is recycled to the production process.

A butylated melamine–formaldehyde resin for surface coatings can be produced in the following way [11, p. 69]:

60 parts (by mass) of melamine, 60 parts (by mass) of formaldehyde (calculated at 100%) in the form of a 40% aqueous solution, and 220 parts (by mass) of butanol are refluxed for 30 min in a stirred kettle equipped for heating. Distillation is then carried out until water no longer evaporates. A further 20–25 parts (by mass) of anhydrous butanol are then distilled. The mixture is cooled, the resin content is determined, and the concentration is then brought to 50% with butanol.

## 5.2.2. Continuous Production

The continuous industrial production of amino resins was begun in order to meet rising demand. The disadvantages of continuous production are that the amount produced per unit time in a particular plant can vary only within relatively narrow limits because prolonged residence under otherwise constant conditions increases the proportion of molecules with high degrees of condensation. Changing the product is also not simple. If the plant is not cleaned out beforehand, a product having a composition between the old product and the new will be formed for a time. For this reason, the continuous production process is employed where the variety of products is restricted to a relatively narrow range. On the other hand, continuous procedures give a very uniform product quality.

The patent literature describes a large number of continuous processes. They differ only in process technology for the product flow. In general, the differences include variations in the temperature, pH, concentration, or modifiers. The process and product flow otherwise remain unchanged. In many cases the publications only describe laboratory- or pilot-scale plants that have not been scaled up for industrial use. Of the continuous processes that have been described, only those with basic differences in type are discussed here.

*Aqueous Solutions.* The apparatuses employed for the continuous production of aqueous amino resin solutions are tube reactors, stirred kettle cascades, or combinations of the two. The danger of incrustation by highly condensed, insoluble products is greater in the continuous process than in the batch process. In the latter, the residues are dissolved by fresh resin solutions; this is not the case in the continuous process because very little backmixing occurs.

A continuous two-reactor process was described by the Girdler Corp., United States of America, in 1943 [46]. The reactants were mixed thoroughly in the first reactor and heated to about 150°C under pressure. In the second stage, the solution under pressure was maintained at this temperature and condensation was carried out. The pressure of the reactor was then released, with the result that volatile constituents, especially water, escaped from the resin and the solution cooled.

According to a 1949 patent granted to Sherwood Paints, UK, the reaction is carried out in a packed column [47] and water vapor is expelled by means of a countercurrent gas stream.

A process of Allied Chemical Corp., United States of America (1951), operates in a similar manner and employs a tubular coil as the reactor [48]. The resin solution is subsequently evaporated. Another process of Allied Chemical Corp. is carried out under pressure, using injected steam (1952) [49].

According to an application by Spumalit-Anstalt, Liechtenstein, the starting materials are passed over a solid catalyst (polycarbamide resin) under a pressure of 400 kPa at 140–150°C to form the resin [50].

A process of Rütgerswerke, Germany, from 1955 also uses a fixed-bed catalyst, an ion exchanger [51]. The resins obtained are used as starting materials for making molding compounds.

A Swedish application by Skanska Attifabriken AB from 1955 describes a process in which starting materials are heated separately and then sprayed into a chamber to undergo partial condensation [52]; the reaction is completed in another chamber. The advantage claimed is that none of the final products participates in the initial stages of the condensation.

An application by Du Pont, United States of America, in 1956, describes a continuous process for producing etherified formaldehyde–urea resins [53]. The stepwise reaction takes place in a cascade of two to four stirred kettles. The free alcohols (e.g., butanol, isobutanol, and others) are distilled in the last stage, and some of them are recycled.

In a Meissner process from 1966, the hydroxymethylation stage is carried out continuously and the subsequent condensation stage is operated batchwise [54]. The continuously flowing product stream from the first stage passes through a stirred kettle, which is one of a large number of kettles, so that the product flow need not be interrupted.

Other continuous processes have been worked out by Rheinpreussen AG [55] and the St. Regis Paper Co. [56], but these processes predominantly relate to phenoplasts.

A BASF process (1971) is shown diagrammatically in Figure 4 [57]. The reaction is carried out in a series of three or more stirred kettles. At the beginning of the cascade predominantly hydroxymethylation takes place. Condensation occurs later in the cascade. The reaction is terminated in the last section by

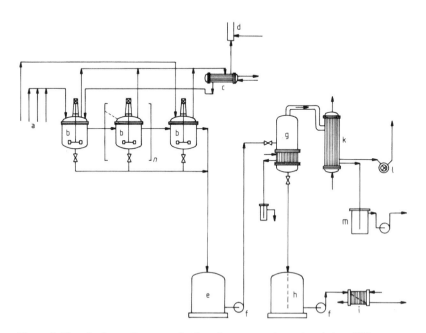

**Figure 4.** Plant for the continuous production of aqueous amino resin solutions [57]
a) Starting material feeds; b) Stirred kettle; c) Reflux condenser; d) Flare; e) Temporary container; f) Pump; g) Vaporizer; h) Product container; i) Cooler; k) Condenser; l) Vacuum pump; m) Container for condensed vapors

the addition of sodium hydroxide solution. The resin solution passes into a temporary container and is subsequently evaporated under reduced pressure in a single- or multi-stage vaporizer. The vapors, which contain about 2% formaldehyde, are worked up and the effluent is discharged for treatment. The end product is cooled and pumped into a tank.

Other cascade processes have been described in applications of SIR, Milano, Italy, (1972) [58] and Montedison (1980) [59].

Finally, mention should be made of the process in which formaldehyde is absorbed directly in urea solutions to produce highly concentrated, formaldehyde-rich solutions that are used as starting materials for glue resins. Examples of conventional commercial products are Kaurit-Konzentrat 244 fl. or 255 fl. (BASF), Prefere 30K180 (Dynea Austria GmbH), and Prefere 30K080 (Dynea NV, Gent, Belgium).

Stamicarbon [60] and American Cyanamid [61] have described the continuous production of concentrated, formaldehyde-rich melamine–formaldehyde solutions. These too can be used as starting materials for amino resins.

According to DSM the condensation resin is prepared in an aqueous medium at elevated temperatures (70–200°C) and pressure (0.2–20 MPa), where the monomers of the condensation resin are fed continuously to a tubular reactor provided with static mixer elements [62].

***Resins in Powdered Form.*** Resins in powdered form also can be produced by a continuous procedure if the spray solutions, which in general are produced batchwise, are fed continuously from a temporary container into a spray tower [21].

### 5.2.3. Production of Special Products and Foams

In addition to impregnating resins, special products include foamed resins and resins for paper auxiliaries, label adhesives, concrete liquefiers, leather auxiliaries, carpet coatings, microcapsules, etc. Because of the relatively small individual amounts and the large variety of products, these resins are generally produced in batches. The principle of the production is the same as that for the manufacture of resin glues. The proprietary know-how relates to modifiers and small but important variations in the manner in which the process is carried out.

The foamed resins (→ Foamed Plastics) are an interesting type of amino resin. They are used to produce solid foams, which have a wide variety of applications. The production of the foamed resin solution and the end product, i.e., the foam, are discussed together.

In 1927, E. C. BAYER discovered that porous, heat-insulating materials are obtained if a foam or a foam-forming substance is added to a glue solution and the mixture is foamed and allowed to harden. In 1933, a process for the production of a foamed urea–formaldehyde resin was developed by I. G. Farbenindustrie. In this process, a foaming agent and air were used to produce a primary foam that itself contained the acid, e.g., phosphoric acid, for curing the amino resin. Curing begins after a few seconds and is complete after 12–24 h at room temperature. After drying, the foamed urea formaldehyde resins have bulk densities of 5–70 kg/m$^3$. More than 60% of the foam is in the form of open cells.

Whereas urea–formaldehyde foams previously could be prepared only in factories of fixed location, W. BAUER in 1953 described a transportable apparatus that permitted on-site production of the foam by means of compressed air [63]. A number of other apparatuses subsequently were developed. The state of the art is reported in [64] with respect to the design of foam-producing apparatus and a novel high-performance apparatus of this type is described. The development of foam-producing apparatuses led to much wider applications: soil conditioning in agriculture, heat insulation, sound insulation, filling of mining cavities, carrier material for cleaning fluids, and covering for landfills.

## 6. Environmental Protection

***Effluents [66].*** The condensed vapors or effluents from amino resin production may not be introduced into water without biological treatment. It should be further ensured that these effluents cannot penetrate into ground water.

Waste waters or partially hardened residues inevitably result when machines and containers are cleaned. The amount of wash water or solvent should be kept as small as possible. If the waste water contains predominantly urea or formaldehyde, it can be reused in the production process. Otherwise, it is eliminated, for example, by incineration.

The biological treatment of industrial effluents from amino resin production is carried out primarily because of the formaldehyde, which is toxic. The usual effluents contain formaldehyde in concentrations that can be degraded readily by bacteria, which are always present. In general, a critical factor in the biological treatment of any effluent is the concentration in which the various toxic components are present.

***Air.*** The treatment of waste gas by washing with water, catalytic combustion, or thermal combustion is used in industry. The waste gases should be very concentrated to minimize treatment costs.

***Solid Wastes.*** Amino resins can be solidified by hardening them with ammonium chloride or phosphoric acid, or they can be absorbed in sawdust or sand. In this form, they may be deposited in a landfill in accordance with the relevant provisos and legal requirements.

## 7. Types of Resins and Their Properties

In addition to the simple types of resin that are produced in the largest amounts, i.e., the urea–formaldehyde and melamine–formaldehyde resins, other types of resin with different properties have been developed. These are modifications, i.e., versions of a resin obtained by use of relatively small amounts of additive or by changing the aldehyde or amino component. However, even where the composition is fixed the resin type can be varied by changing the reaction procedure. This leads to different distributions of oligomers or to different substitutions of the repeat units. The distribution of hydroxymethyl groups is the determining feature.

### 7.1. Urea Resins (Urea–Formaldehyde Resins)

All the urea–formaldehyde condensates [9011-05-6] containing the repeat unit **1** (see Chap. 3), from monomolecular to high molecular mass

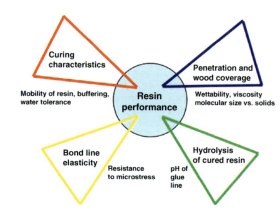

**Figure 5.** Interaction of resin properties

product, are useful substances. In the 1970s, a number of discoveries have been made concerning the composition of urea–formaldehyde resins. Systematic investigations into the process of condensation and the structural composition had long been impossible because of the lack of suitable analytical methods. Industrial development took place largely on an empirical basis. However, using a combination of analytical methods, e.g., $^1$H NMR and $^{13}$C NMR spectroscopy, C, H, N, O elemental analysis, determination of hydroxymethyl groups, and determination of total formaldehyde, it is possible in principle to determine the structural units and their concentrations [26]. However, the relationship between the technical properties of the amino resins or the materials produced from these resins and the distribution of the structural units is still unclear in most cases, so development is still largely empirical, but, the complexity of the interaction between substrate, resin, curing, processability, and final properties can be depicted as shown in Figure 5.

#### 7.1.1. Pure Urea–Formaldehyde Resins

Important macroscopic parameters are the molar ratio of formaldehyde to urea, the viscosity, and the gelling time. Urea–formaldehyde condensates that have a molar formaldehyde-to-urea ratio of 1 usually give insoluble polymethylene ureas that do not possess any adhesive properties. If, however, the molar ratio is reduced from a relatively high value gradually or stepwise during the course of the reaction, the resultant resin possesses adhesive properties. These properties improve as the molar ratio increases, but the undesirable release of formaldehyde increases simultaneously.

The highly viscous resin solutions obtainable by prolonged industrial condensation of urea and formaldehyde are low molecular mass products. They contain about six to eight urea units, as can be determined by gel chromatographic [24] and cryoscopic measurements and by endgroup determinations. This also applies to the completely insoluble powdered or glassy condensates [67]. During storage, the concentration of methylene groups increases at the expense of the hydroxymethyl groups. The viscosity also increases, initially slowly and virtually linearly, and then, after a critical time, rather steeply. Some or all of the hydroxymethyl groups generally can be blocked by etherification, and the shelf life of the amino resin is thus extended [33, 68, 69]. In principle the same reactions take place when the resin is hardened by the user, but they are substantially more rapid. However, a residual amount of hydroxymethyl groups remains after hardening [24]. Because they are susceptible to hydrolysis, even the cured urea resins are more water sensitive than the melamine or phenol resins.

The gelling time [11, p. 174] or reactivity of the resin is an important parameter for the user. This time can be determined relatively precisely, and is the time in which the resin changes from a fluid to a virtually immobile solid following the addition of a certain amount of hardener at a particular temperature. The reactivity increases with increasing molar ratio, content of hydroxymethyl groups, and content of free formaldehyde.

### 7.1.2. Mixed Condensates

Industrial amino resins contain mixtures of various condensates and cocondensates. Amino resin solutions are always polymeric mixtures having various degrees of condensation and differing distributions of functional groups. Resin-forming monomers employed as textile auxiliaries constitute an exception, but these in themselves are not resins. The term mixed condensates is rather loosely applied to those resins which are formed in the condensation of urea and formaldehyde when the reaction is carried out using substances that also form resins with formaldehyde. These are, for example, melamine, phenol, organic acid amides, sulfonamides, and carbamates. Therefore, in the presence of melamine or phenol, it is possible to prepare urea resins with improved strength and swelling properties for the manufacture of particle board, i.e., weather-resistant particle board. Modification with melamine gives resin glues that can be used to produce good-quality particle board even when the molar ratio of formaldehyde to urea is low.

Mixed resins with other resin formers, e.g., acrylic acid, acrylates [70, 71], acrylamide [1, p. 109] and their derivatives, are also sometimes advantageous. These are also mixtures of polymers, of cocondensates, or of homologous polymers and cocondensates.

### 7.1.3. Condensation of Urea with Other Aldehydes

Condensation of urea with isobutyraldehyde, acetaldehyde, or crotonaldehyde is carried out industrially to obtain sparingly soluble nitrogen fertilizers, i.e., controlled release fertilizers, rather than resins [72, 73]. The condensation of urea or melamine with glyoxal gives products that can be employed as textile auxiliaries. Products obtained by the subsequent condensation of urea–formaldehyde resins with glyoxal (e.g., dimethylol dihydroxy ethylene urea, DMDHEU [1854-26-8]) also are used for textile finishing [74, 75].

### 7.1.4. Resins from Urea Derivatives

Condensation with urea derivatives or substitution products are well known, but these reactions are of industrial importance only for making textile and paper auxiliaries. Reactions with aldehydes can take place depending on the hydrogen atoms still present at the nitrogen, but the ability to undergo condensation decreases sharply with the size and number of substituents.

The reaction of a symmetrically disubstituted urea derivative, e.g., $N,N'$-dimethylurea or dihydroxymethylurea dimethyl ether, with formaldehyde in an acidic medium gives cyclic urea derivatives of the urone type [76, p. 9]:

$$RN\text{-}CO\text{-}NR\text{H} + 2\,CH_2O \longrightarrow RN(CH_2OH)\text{-}CO\text{-}NR(CH_2OH) \xrightarrow{H^+} \text{cyclic urone} + H_2O$$

Dihydroxymethylurone dimethyl ether, bis(methoxymethyl)urone [7388-44-5] is used industrially to provide cellulose fabrics with a wrinkle-resistant finish. It has middling reactivity and stability to hydrolysis [77, pp. 164–174]. The urones do react with cellulose but form only minor amounts of self-crosslinked resins, such as the urea–formaldehyde compounds.

When S is substituted for O in urea, thiourea is obtained. The latter undergoes all of the principal condensation reactions of urea. In a weakly alkaline medium, $N$-hydroxymethyl compounds and $S$-hydroxymethyl compounds are formed. The tautomeric structure $NH_2\text{-}C(SH)=NH$ may be assumed initially for the latter. The thiourea–formaldehyde condensates are more water resistant than the corresponding urea condensates. Some of the thiourea condensates are assumed to be cyclic compounds [78].

### 7.1.5. Modified Urea Resins

There is very extensive literature on the possible methods of modifying amino resins [1, pp. 95] and only the most important modifications are discussed here.

***Reactions with Alcohols.*** The hydroxymethyl groups can be etherified with alcohols in an acidic medium [13, pp. 334].

$$-NH\text{-}CH_2\text{-}OH + ROH \xrightleftharpoons{H^+} -NH\text{-}CH_2OR + H_2O$$

The equilibrium constants for the etherification of monohydroxymethylurea and dihydroxymethylurea with methanol and ethanol have been measured [68]. The enthalpy of reaction, estimated from the temperature dependence of the equilibrium constant, is $\Delta H = -6$ to $-8$ kJ/mol. The equilibrium constants ($K$) differ depending on the substitution:

$K_{30\,°C,\,R=CH_3}$

$H_2N\text{-}CO\text{-}NH\text{-}CH_2OH + ROH$
$\rightleftharpoons H_2N\text{-}CO\text{-}NH\text{-}CH_2OR + H_2O$    0.6

$HOCH_2\text{-}HN\text{-}CO\text{-}NH\text{-}CH_2OH + ROH$
$\rightleftharpoons ROCH_2\text{-}HN\text{-}CO\text{-}NH\text{-}CH_2OH + H_2O$    7.6

$ROCH_2\text{-}HN\text{-}CO\text{-}NH\text{-}CH_2OH + ROH$
$\rightleftharpoons ROCH_2\text{-}HN\text{-}CO\text{-}NH\text{-}CH_2OR + H_2O$    2.0

$$K = \frac{[\text{Ether}] \cdot [H_2O]}{[ROH] \cdot [\text{Hydroxymethyl}]}$$

As the number of C atoms in the alcohol increases, the equilibrium is shifted to the left and the reaction rate reduced so that the etherification becomes more difficult. The ethers of the low-boiling alcohols can be transetherified with higher boiling alcohols by heating the mixture in weakly acidic solution and distilling the more volatile alcohols [13, p. 336].

Frequently only some of the hydroxymethyl groups of resin glues and impregnating resins undergo etherification with monohydric or polyhydric alcohols of less than five carbon atoms. As a result, some of the hydroxymethyl groups are protected against condensation, and the resins have a longer shelf life [1, p. 115, [79, 80]]. In the starting materials for surface coatings, 20–100% of the hydroxymethyl groups undergo etherification, depending on the type. The amino resins etherified with butyl groups are the most important for wood coatings, e.g., parquet coating. Where complete etherification takes place, the reaction is carried out using an entraining agent – the alcohol itself, toluene, or xylene – and the water is distilled azeotropically.

Ethers of one- to three-carbon alcohols with low molecular mass hydroxymethyl compounds of urea, ethylene urea (imidazolidinone [28906-87-8]), propylene urea (1,3-bis(hydroxymethyl) tetrahydro-2-(1$H$)-pyrimidinone [65405-39-2]), and melamine have achieved some importance in the field of textile auxiliaries [3, p. 168, [68, 77]]. These compounds are precondensates having relatively low molecular masses of about 600.

Urea resins etherified with butyl groups can be employed as starting materials for making surface coatings. However, when baked alone, they give brittle and water-sensitive films. Therefore, they cannot be used alone as binders. In combination with nitrocellulose, or with alkyd, epoxy, or acrylate resins, they give excellent baked finishes.

## Reactions with Ammonia and Amines.

Amines are very reactive toward hydroxymethyl compounds obtained from urea and formaldehyde. Ammonia reacts with the hydroxymethyl group to form an amino or imino group, eliminating water:

$$-NH-CH_2OH + NH_3 \rightarrow -NH-CH_2NH_2 + H_2O$$

$$\underset{H}{-N}-CH_2OH + NH_3 \rightarrow \underset{H}{-N}-CH_2NH_2 + H_2O$$

$$\underset{H}{-N}-CH_2OH + \underset{H}{-N}-CH_2NH_2 \rightarrow \underset{H}{-N}-CH_2\diagdown_{NH}^{} + H_2O$$
$$\phantom{xxxxxxxxxxxxxxxxxxxxxxx} \underset{H}{-N}-CH_2\diagup$$

$$\left(\underset{H}{-N}-CH_2-\right)_2 =NH + -N-CH_2NH_2 \rightarrow \left(\underset{H}{-N}-CH_2-\right)_3 \equiv N + NH_3$$

The reaction of ammonia with formaldehyde is exothermic ($\Delta H = -57$ kJ/mol) [1, p. 49] and gives hexamethylenetetramine. This reaction has been used industrially to trap formaldehyde from waste gases, for example, from particle board production [81–83]. The reaction of urea and aldehydes with primary amines gives cyclic compounds (triazines) [84] of the type **10**, where R' can be hydroxymethyl and its methyl and ethyl ethers.

```
       O
       ‖
   R'-N   N-R'
      |   |
    H₂C   CH₂
       \ /
        N
        |
        R
       10
```

Ammonia, primary or secondary amines, or polyamines containing two or more primary amino groups are used for the production of amino-modified resins [3, p. 206]. They are more stable towards water than the unmodified resins produced by a similar procedure.

As shown in the equation above, ammonia and amines can react with the hydroxymethyl groups in urea, in a weakly acidic or alkaline medium, to give N-aminomethyl groups. In strongly acidic solution, these eliminate the amine substituents to form condensates. This condensation is similar to that starting with hydroxymethyl groups. Even very highly condensed urea resins are able to bind large amounts of ammonia and amines because a substantial part of the formaldehyde is still present in the form of hydroxymethyl groups. This reaction is utilized industrially, for example, to deodorize resins or particle board [81].

Condensation of the resin molecule with an amine produces a resin that has a cationic character; this increases the resin's affinity for cellulose fibers when it is used as an auxiliary for textiles or paper. These basic urea resins are therefore mainly important because they improve the wet tensile strength of paper.

## Reaction with Sulfites.

Just as formaldehyde hydrate reacts with a sulfite, so the hydroxymethyl groups of the amino resins react with a hydrogen or other sulfite to give an aminomethanesulfonate [1, p. 120]:

$$R-NH-CH_2OH + [HSO_3]^- \rightarrow [R-NH-CH_2SO_3]^- + H_2O$$

This imparts an anionic character to the resins [3, p. 206]. The pH of the solution controls its sulfite content.

$$[SO_3H]^- \rightleftharpoons [SO_3]^{2-} + H^+$$

To produce resins of this type, the reaction of, e.g., urea and formaldehyde is begun in a neutral or alkaline medium in order to achieve hydroxymethylation. When the desired degree of conversion has been achieved, condensation is carried out using sulfite or hydrogen sulfite in a slightly acid solution [85]. The resin can react with hydrogen sulfite because many hydroxymethyl groups are still present. Water-insoluble resins can even be rendered soluble by posttreatment with hydrogen sulfite. Although, like unmodified resins, the amino resins produced using sulfites exhibit an increase in viscosity on aging, they remain water soluble because of their hydrophilic ionic groups.

The anionic resins are of particular industrial importance in the field of water-soluble or water-dispersible adhesives. Other uses include textile auxiliaries [86], paper auxiliaries [87], leather auxiliaries, and deodorants [88] for urea–formaldehyde resins [89].

In addition to modification with sulfite, the relevant patents mention many other modifiers, such as aldehydes and triazines. The affinity of the anionic resins for cellulose (which carries a slight negative charge) is lower than that of the cationic resins; therefore, the fibers must be posttreated with aluminum sulfate to fix the resins.

The modification of urea–formaldehyde resin glues with ligninsulfonic acid should also be mentioned [90, 91]. The products are dark resin solutions that have been used from time to time in the woodworking industry. The mechanism involved in the incorporation of the nonuniform lignin is still unclear [1, p. 120]. Lignin contains phenolic groups. Incorporation of the lignin components imparts relatively high water resistance, among other things. The lignins are of different origins and compositions, so that the production process must be adapted empirically to the different lignin varieties. This fact and the dark coloration have prevented more extensive use of lignin wastes from wood pulp production.

## 7.2. Melamine Resins

### 7.2.1. Unmodified Melamine Resins [9003-08-1]

Melamine, like urea, can react with formaldehyde [9, p. 65]. The addition of formaldehyde at the amino group is carried out at pH 9–10 and takes place very rapidly. One mole of melamine combines with 6 mol of formaldehyde to form hexahydroxymethylmelamine. Melamine in itself is only slightly soluble even in hot water [9, p. 60]. However, it dissolves readily in (alkaline) aqueous formaldehyde solutions, forming hydroxymethylmelamines. A small part of the melamine remains physically dissolved. When heated, hydroxymethylmelamines undergo cross-linking [9, p. 66] and form resins that become more insoluble the longer they are heated. Melamine condenses with formaldehyde in acid solution to form hydroxymethyl compounds. This is followed by a more rapid polycondensation, yielding cross-linked products, than in the case of urea–formaldehyde [13, p. 359]. Melamine–formaldehyde resins can be hardened thermally in the absence of a hardener or they can be hardened in the absence of heat, using a very strongly acidic hardener. The hardening processes, as in the case of the urea resins, are not entirely clear. Cross-linking takes place principally by methylene bridges, but it also has been suggested that methylene–ether bridges play a role. The melamine formaldehyde condensates possess good thermal stability, optical properties, and water resistance [13, p. 350].

### 7.2.2. Modified Melamine Resins

*Reaction with Alcohols.* Melamine hydroxymethyl compounds can be etherified with alcohols in the same manner as hydroxymethylurea compounds [13, p. 361]. The most important use of the products is as surface-coating resins, but they are also used as textile auxiliaries and even as melamine–urea–formaldehyde resins and impregnating resins. As in the case of the urea resins, etherification is carried out at relatively high temperatures with acid catalysis. Where etherification is carried out to a substantial degree, it is important to remove the water or excess alcohol from the weakly acidic solution, because cross-linking otherwise competes with etherification. The mono- to hexaalkyl ethers of the $C_1$ to $C_4$ alcohols can be prepared readily. Good conversions are obtained for higher alcohols only by means of transetherification [92].

Methyl ethers of melamine hydroxymethyl compounds are readily soluble in water even at room temperature and crystallize readily [3, p. 361]. The aqueous solutions of the methyl ethers prepared from higher hydroxymethyl compounds, i.e., those with a higher molar ratio of hydroxymethyl groups, become cloudy on heating and clarify again when cooled. Most of the methyl ethers are soluble not only in water but also in benzene and similar solvents. The products etherified with higher alcohols, e.g., butanol or benzyl alcohol, are water insoluble but are readily soluble in alcohols and some hydrocarbons. In addition to monoalcohols, polyhydric alcohols, e.g., ethylene glycol, glycerol, 3-hydroxymethyl-1,5-pentanediol, and pentaerythritol, can be employed as etherification agents. This etherification must be carried out in neutral solution because the competing cross-linking reaction otherwise takes place more rapidly and forms insoluble products.

Like the corresponding urea resins, the etherified melamine–formaldehyde resins used as starting materials for surface coatings are condensates having low molecular masses of 600–1400 [93]. These are presumably triazine structures linked by methylene bridges and contain from two to four melamine groups.

The ether groups are distributed randomly over the triazine rings.

***Reaction with Amines.*** Amines and aminocarboxylic acids react readily with the hydroxymethyl group of hydroxymethylmelamines in the same way as with urea resins. Modification with amines or amino acids improves the hydrophilic character of the melamine resins. Condensation with amino acids gives alkali-soluble melamine resins that are used to produce impregnating or molding materials [94]. Basic products, which form salts with acids, are obtained by modification with aminoalcohols, e.g., aminoethanol [95], and are used as textile auxiliaries. These products are not of major industrial importance.

***Reaction with Hydrogen Sulfite.*** Melamine resins, like urea resins, react with hydrogen sulfites. Industrially, the cheapest product, i.e., sodium sulfite, sodium hydrogen sulfite, or pyrosulfite is used, and the reaction is carried out in an acidic or basic medium. The products are readily water-soluble melamine resins that can be precipitated with acids. Hydrogen sulfite-modified melamine–formaldehyde resins are used industrially as concrete liquefiers in the building materials industry. In addition to sulfite, other modifiers described in the patent literature are dithionite and formaldehyde sodium sulfoxylate ($NaHSO_2 \cdot CH_2O$) [96, 97].

***Other Modifiers.*** The modification of hydroxymethylmelamine resins with polyalcohols, sugars, toluenesulfonamides, lactams, nitriles [98–100], acrylates, and polyacrylonitrile [1, p. 109, 101] is of great importance in the production of impregnating resins. Such modifications improve the large number of laminating properties required and stabilize the resin solution against clouding and an excessively rapid increase in viscosity.

Mixed condensates and cocondensates of melamine with other compounds – e.g., phenol, urea, cyanamide, and dicyanodiamide – that react with formaldehyde have also been described in a large number of publications. For example, phenol-modified melamine–urea–formaldehyde resins are used as resin glues for weather-resistant particle board [102, 103]. These resins, too, can be modified in a large variety of ways.

## 7.3. Urethane Resins

Condensation of a urethane (carbamate) of the formula $H_2N–COOR$ with an aldehyde gives a distinct compound or a resin, depending on the production conditions [8, p. 220], [13, p. 371]. In an alkaline medium, carbamates react with formaldehyde to give hydroxymethyl compounds that split back into the starting materials at higher temperatures. In an acidic medium, they cross-link to form methylene bridges and produce resins that are used to manufacture molding materials. A patent has been applied regarding the use of urethane resins derived from long-chain aliphatic alcohols as plasticizers [3, p. 220]. Urethane–urea–formaldehyde mixed resins have also been described. The reaction of cyanuric acid with polyhydric alcohols and formaldehyde gives resins that can be used as paper and textile auxiliaries and as starting materials for surface coatings.

In another procedure, urea is converted into the appropriate carbamate by heating in the presence of an alcohol and eliminating ammonia. The subsequent reaction with formaldehyde and cross-linking are carried out as described above [13, p. 374]. The resultant highly viscous solutions can be used as label adhesives and for the adhesive bonding of cardboard boxes.

The properties of urethane resins vary depending on the components and method of production. They range from oily to brittle depending on the degree of condensation. Lower aliphatic and cycloaliphatic alcohols yield hard resins, whereas higher alcohols give wax-like resins.

## 7.4. Cyanamide and Dicyanodiamide Resins

Cyanamide and dicyanodiamide react with formaldehyde both in acid and in alkaline media [8, p. 199]. Cyanamide reacts with an equimolar amount of formaldehyde at pH 6–7 to give a white precipitate, which loses water to form a crystalline compound. In general, cyanamide and dicyanodiamide are used as modifiers [1, p. 118, 13, p. 382]. Cyanamide–formaldehyde and dicyanodiamide–formaldehyde resins have been used as mixed resins for starting materials in making molding compositions, as dyeing and tanning auxiliaries, for textile finishing, and for preserving urea–formaldehyde foams.

## 7.5. Sulfonamide Resins

The most commonly used basic component for this type of resins is *p*-toluenesulfonamide:

$$H_3C-\langle\text{C}_6H_4\rangle-SO_2NH_2$$

With formaldehyde, *p*-toluenesulfonamide forms resins that are fusible but not curable, whereas disulfonamides or amides possessing other groups capable of condensation give insoluble products that are infusible [104]. The special properties of *p*-toluenesulfonamide resins include compatibility with cellulose surface coatings, pale natural color (in contrast to that of phenol resins), and a low tendency to yellow.

Only the *p*-toluenesulfonamide–formaldehyde condensates have become important industrially, and these only as cocondensates or modifiers. Only the combinations with urea–melamine resins, phenol resins, and phthalate resins are of importance [105]. They also are added to melamine–formaldehyde impregnating resins as plasticizers in amounts of 3–8% [1, pp. 109].

## 7.6. Aniline Resins

Aniline resins include the condensates of formaldehyde with aromatic amines, such as aniline [25214-70-4], toluidines, xylidines, naphthylamines, and phenylenediamines [13, p. 192].

In alkaline solution, the first stage, i.e., hydroxymethylation, leads to a simple condensate:

$$2\,\langle\text{Ph}\rangle\text{-NH}_2 + CH_2O \rightarrow \langle\text{Ph}\rangle\text{-NH-CH}_2\text{-NH-}\langle\text{Ph}\rangle + H_2O$$

Condensation to give a cyclic product has been reported also.

$$\text{Ar-N}\begin{array}{c}\text{Ar}\\|\\\text{N}\\H_2C\diagup\;\diagdown CH_2\\|\quad\quad|\\\diagdown\,C\diagup\\H_2\end{array}\text{N-Ar}\qquad Ar = aryl$$

The reaction with formaldehyde in an acid medium gives resins in which methylene groups link the aromatic rings directly, as in the case of the phenoplasts (→ Phenolic Resins). The para position is activated by salt formation at the amino group, so formaldehyde can add at this position to form *p*-aminobenzyl alcohol.

$$CH_2O + \langle\overset{NH_3Cl}{\text{C}_6H_5}\rangle \rightarrow \langle\overset{NH_3Cl}{\underset{CH_2OH}{\text{C}_6H_4}}\rangle \rightarrow \langle\overset{NH_2}{\underset{CH_2OH}{\text{C}_6H_4}}\rangle + HCl$$

Polymer chains are formed and water is eliminated in this process. In the presence of an excess of formaldehyde, three-dimensional networks also form.

The amino groups have a directing function, so tertiary aromatic amines can also form resins. Unlike the phenol resins, aniline resins do not harden further. The aniline resins are therefore thermoplasts in contrast to the resins discussed above, which are duroplasts. The aniline resins are not very important industrially. They have been used in small amounts in the production of molding materials [1, p. 120] and ion-exchange resins and, because of their high resistance to tracking currents, they have been mixed with epoxy resins and used for wire enamels and insulating coatings [106].

## 8. Analysis

Analysis, insofar as it is used for product control during production, storage, and use, amounts to testing various types of resins for characteristic parameters or selected specifications. Complete elucidation of the chemical structure, including the sizes and amounts of the oligomers, would entail an unacceptably great expense.

Characterization of the products usually includes the following determinations [11, p. 171, [107–109]]:

- density in g/cm$^3$ (in the case of foams, in kg/m$^3$)
- solids content in % by mass
- viscosity in mPa · s, usually at 20°C
- free formaldehyde
- total formaldehyde
- free urea
- total urea

- melamine
- methoxy content
- ash content (incombustible modifiers)

The urea–formaldehyde or amino groups–formaldehyde molar ratio is calculated from the analytical data.

Further tests may be employed if special properties are required of the resins. The development of novel products and applications increases the need for analytical effort.

***Solids Content.*** The determination of the solids content in resin solutions is a determination of the dry matter content. The value is useful only for comparison. It is determined by evaporating the volatile components at 100–130°C over a fixed period of time. This procedure removes not only the water present as a solvent and the free formaldehyde, but also the water and formaldehyde formed during the drying process in the condensation. Therefore the dry matter content is by no means to be equated with the active ingredient content. More exact data are obtained by the determination of water by the Karl Fischer method [110], by freeze drying, or, in organic solvents, by the Aufhauser method. In industry, however, the water content is estimated by the drying method detailed above because this method is simple and quick.

***Formaldehyde*** [111–113]. The free formaldehyde can be determined directly and the total formaldehyde can be determined after the product has been hydrolyzed completely. The hydroxymethyl content can be determined by hydrolysis under relatively mild conditions. The formaldehyde bonded in the form of methylene groups is calculated by difference.

***Urea.*** The total urea can be determined by hydrolysis and the Kjeldahl procedure. The free urea is degraded to ammonia by the enzyme urease [114]. If no other form of nitrogen is present in the resin, the urea content can be calculated from the nitrogen content, which is determined by combustion analysis. The *molar ratio* of urea to formaldehyde can be calculated from the total content of formaldehyde. This ratio provides information on the formaldehyde elimination to be expected.

***Melamine*** in amino resins is determined by hydrolysis of the product under mild conditions followed by titration with perchloric acid. Resins having a low degree of condensation can be titrated directly using a computerized automatic apparatus.

***Methanol*** is generally present in the form of methoxy groups bonded to hydroxymethyl groups by an ether bond; it is determined by the Zeisel's procedure.

***Other components*** are important, particularly where the resins are used as paper or textile auxiliaries or as starting materials for making surface coatings. Some of these are maleic acid, glyoxal, phenol, amines, and sulfites. The methods of determination are discussed in textbooks on analysis or, for example, in specific articles. These substances are not determined during routine product control.

For research and development purposes, paper chromatography, thin-layer chromatography [115], column chromatography [116], gel permeation chromatography, $^{13}C$ NMR spectroscopy [26, 38, 42], $^{15}N$ NMR spectroscopy [117], $^{1}H$ NMR spectroscopy (using dimethyl sulfoxide as the solvent) [38, 118], and infrared spectroscopy are used.

## 9. Storage and Transportation

Aqueous or alcoholic solutions of amino resins are stored in steel tanks with or without internal epoxide coatings, in stainless steel tanks, or in containers made of fiberglass-reinforced polyester resin. Temperatures above 20–25°C must be avoided because they bring on rapid condensation, which increases the viscosity and soon renders the resin unprocessible. Rotary pumps or – in the case of highly viscous liquids – positive-displacement pumps, e.g., gear pumps, are employed.

Powdered amino resins are packaged in airtight, plastic-lined paper sacks that are sealed against moisture. Certain types of paper bags can withstand tropical climates. Under high pressure, the materials may solidify. Spilled product is taken to a waste dump.

Liquid resin loads of up to about 250 kg are dispatched in small polyethylene barrels, whereas greater loads, up to about 1 t, are dispatched in polyethylene containers encased in sheet metal. Larger amounts are transported in containers, railroad tank cars, tank trucks, or marine tankers. Amino resins must be disposed of if they are damaged or become unstable during transport in tanks.

The relevant regulations governing the disposal of amino resins differ from country to country and even within some countries. A common feature of all the regulations, however, is that the resin must be hardened by the addition of acid curing agents and must be delivered to a licensed landfill. Where leakage occurs during transport, it is advisable to absorb the resin solution with any appropriate agent that is available (e.g., sand or sawdust) and then to deliver the resulting material for incineration. In the Federal Republic of Germany, amino resins are not included among substances regarded as endangering water.

## 10. Uses

The properties of amino resins have been adapted by industry to a very large number of applications. New uses are being developed constantly and any list is bound to be incomplete.

Uses include the following: adhesives; impregnating resins; molding materials; starting materials for making surface coatings; auxiliaries for paper, textiles, leather, and flotation; strengtheners for building materials; concrete liquefiers; binders for glass fibers and foundry sand casting; fire lighters; emery papers; flame retardant coatings; flameproofed combustible items; foamed resins for many purposes; grinding wheels; ion-exchange resins; sewage flocculants; and microcapsule production.

The largest amount of amino resins is consumed by the woodworking industry. These resins, along with the phenol resins, have largely replaced the natural glues based on proteins (glutin, casein, soybean, and blood albumin glues). Resins are used as resin glues in the particle board, MDF, HDF, plywood, and furniture industries. Impregnating resins also are used to impregnate papers for decorative laminates and for coating wood particle board.

***Resin Glues for the Woodworking Industry [120].*** The amino resins used as glues in the woodworking industry are available commercially predominantly in the form of 55–70% aqueous solutions or as solids, for example, powders. The powder resins entail lower transport costs but are more expensive because of the costs involved in drying them and must be dissolved by the user. Whereas the powder resins have shelf lives of one year or longer if stored correctly, the liquid products, depending on the type, have shelf lives of several weeks to six months, typically four weeks. During this period the viscosity remains within limits that permit the resin to be pumped with a conventional pump. The amino resins are more or less stable intermediates; condensation is completed by the addition of a curing agent and/or by heat treatment. A curing agent, e.g., ammonium chloride, is virtually always required for the urea resins, whereas melamine resins can be cured at 120°C. Compared to the melamine or phenol resin glues, particle board made from urea resin glue has a lower water resistance and is not resistant to boiling. If these properties are required, melamine resin glues or mixed condensates of formaldehyde, urea, melamine, and/or phenol are employed.

In addition to water resistance, the tendency to release formaldehyde is very important. If a low level of formaldehyde release is desired, the formaldehyde content of the resin glue must be correspondingly low. In some countries, and for structural applications, the glue must receive the approval of the authorities.

Since the 1980s, MDF resins have become increasingly important. The MDF resins are pure or melamine-modified urea–formaldehyde resins that have a low viscosity, about 200 mPa·s. Other modifiers may be used also. Medium-density fiberboard is used in the production of furniture with structured surfaces [1, p. 179, 119, pp. 266–275].

From an economic point of view, the urea resins are the most important amino resin glues because they are inexpensive, enable short press times, and possess technical properties adequate for most purposes.

The main use is in the production of wood particle board. These resins also play an important role as glues for plywood and blockboard and for furniture construction. The gluing of load-bearing wood components and shutter panels and the production of shaped articles are further examples of the large variety of uses of amino resin wood glues.

***Impregnating Resins [121].*** Urea–formaldehyde or melamine–formaldehyde impregnating resins are used to treat papers for decorative purposes. The resins themselves are brittle products and it may be necessary to modify them with a plasticizer. Because of their good properties, i.e., hardness, abrasion resistance, scratch resistance, and stability to light, the amino resins are particularly suitable for this purpose. The impregnated papers are used either to produce decorative laminates or for coating wood particle board. Papers impregnated with melamine resin are more heat stable and water resistant than those impregnated with urea resin [122].

The paper webs impregnated with the amino resin are dried in a furnace at 70–200°C and are stored under air-conditioning until they are used.

Various types of amino resin-impregnated papers are required for decorative laminates. The most important are papers that impart a decorative pattern to the board which itself consists of several layers of phenol resin-impregnated kraft paper (high pressure laminate (HPL) and, continuously pressed laminate (CPL)).

However, the major share of laminates manufactured today is produced by directly pressing amino resin-impregnated decorative papers onto a suitable substrate, such as particle board, MDF, or HDF. These laminates are abbreviated as DPL, (direct pressed laminate) or LPM (low pressure melamine).

Ground films and decorative films are likewise amino resin-impregnated papers and are used in the same way as wood veneers in the furniture industry. These films generally are produced from urea resins and to a certain extent also from melamine–urea resin mixtures.

***Molding Materials.*** Molding materials consist of a filler, e.g., cellulose or sawdust, and a binder, which can be a urea resin or a melamine resin. The molding materials obtained from melamine resins are superior to those obtained from urea resins because of their better hardening properties, higher mechanical strength, and greater resistance to tracking currents and moisture [3, pp. 68–100].

The molding materials are employed for objects in daily use; e.g., white electrical plugs, switches, covers, telephone accessories, and buttons are made from urea resin. Melamine resin is used for camping, hospital, and canteen tableware, and for junction boxes and insulating components. Mixed resins that combine some of the properties of the individual resins are also used.

***Raw Materials for Surface Coatings.*** Pure amino resins are not very useful as surface coatings. The majority of the commercially available amino resins used as starting materials for surface coatings are etherified with butanol or isobutanol and dissolved therein [3, pp. 124–148]. Special brands are also modified with methanol, ethanol, or propanols.

Because the surface coatings obtained using amino resins are too brittle, these resins cannot be used alone as binders. However, they are very useful in combination with other binders. Urea resins and melamine resins are the most important cross-linking components for alkyd resins.

Physically drying finishes are obtained by adding amino resins to nitrocellulose finishes. The addition of these resins improves the hardness, body, and stability to light. The principal field of use is in furniture finishing.

Acid-curable finishes are obtained by adding alkyd resins or saturated polyester resins. These finishes can be hardened by adding an acid, e.g., hydrochloric acid, *p*-toluenesulfonic acid, or a phosphoric acid derivative, at an elevated temperature. The resulting coatings are very hard and resistant to scratches, solvents, and the effects of light [3, p. 130]. Urea resins are more important than melamine resins, particularly in furniture finishing and parquet sealing.

Coatings that can be cross-linked by heat treatment alone are obtained by combining amino resins with alkyd resins, saturated polyester resins, heat-curable acrylate resins, or epoxy resins. Because of their good mechanical properties, their gloss, and their resistance, these baked finishes are extremely important in industrial metal finishing, for example, for automobile bodywork, household appliances, metal furniture, and containers [123, p. 177].

Because they are etherified and essentially free of water, the amino resins used in the surface-coating industry have shelf lives of six months to several years at room temperature. They are dispatched in plain or galvanized sheet metal containers. Only stainless steel or aluminum tanks are used for transport by tanker trucks and railroad tank cars.

***Paper Auxiliaries.*** In the paper industry, amino resins are used to increase the dry tensile strength and especially the wet tensile strength of the product [3, pp. 205–209, [124, 125]]. In contrast to the procedure followed in the textile industry (see Textile Auxiliaries below), the resin components are added to the paper stock before condensation. They are adsorbed onto the fiber and then undergo condensation in an acid medium. The maximum strength of the paper is reached only after a maturing period of two to four weeks. Based on the fiber, 1–5 wt% of resin is sufficient to achieve good wet strength. The amino resins also are used for the surface finishing of papers. They impart water resistance to starch-containing paper-coating compositions.

The urea resins, which constitute the most important group, are generally hydroxymethyl compounds that have been modified by the addition of amines to give them a cationic character or by the addition of sulfites to provide an anionic character. After modification, these compounds are subjected to a condensation reaction in an acid medium. The resultant ionic resins are water soluble. The ionic character also enables them to be adsorbed onto anionic paper pulp, but anionic resins must be used with aluminum sulfate [124, p. 43].

Melamine resins are also important to the wet strength of paper. In the absence of other modifiers, they react with acids to form strongly acid cationic colloids that are stabilized by an electric charge and consist of partially condensed hydroxymethyl melamines. Melamine resins modified with aminocaproic acid are also suitable substitutes, either alone or in combination with other polymers, for some or all of the colophony soaps used in the internal sizing of paper [124, p. 42].

A further field of use is the production of moisture-resistant starch pastes for the adhesive bonding of corrugated board [126].

***Textile Auxiliaries (→ Textile Auxiliaries, 7. Finishing Agents).*** The textile auxiliaries based on derivatives of urea, melamine, and formaldehyde [3, pp. 166–181] are predominantly monomeric, and therefore should not be regarded as resins. They contain only unavoidable, small, random amounts of higher condensates. The molecules react with the primary hydroxyl groups of the cellulose on contact with the cellulose fibers. The relatively weak hydrogen bridges of the cellulose molecules are thereby supplemented by additional, stronger covalent bonds. Self-condensation takes place on the surface of synthetic fibers, which do not react with the "amino resin" monomers. This type of auxiliary is discussed here, as are others that stiffen the finish of synthetic fabrics.

The pure urea–formaldehyde adducts, mono- and dihydroxymethylurea, are water soluble and crystallize readily. In crystalline form, they possess the largest share of the market for easy-care and wrinkle-resistant finishes for cellulose textiles and mixtures of cellulose fibers with synthetics and wool. The pastes also consist principally of these basic substances, but some of them are etherified with lower alcohols and exhibit a minor amount of condensation. They are therefore used to provide wrinkle-resistant and wash-and-wear finishes as well as stiffened finishes on synthetic fabrics. The fabrics treated in this manner possess improved wash resistance. Hydroxymethyl compounds of urea that exhibit a high degree of condensation and possess a high concentration of ether groups are used to produce highly elastic stiffening effects and good wash resistance in cellulose fabrics. Because of the relatively low concentration of "amino resin" monomers (about 1 mol of resin per kg of cellulose), cross-linking between the cellulose groups is strongly favored over self-cross-linking. A disadvantage of this class of products is the relatively low stability of the acid treatment liquors in the bath, where condensation can take place easily between the hydroxymethylurea molecules.

An important group is constituted by hydroxymethyl compounds of cyclic urea derivatives; examples are dihydroxymethylethylene urea (**11**), dihydroxymethylpropylene urea (**12**), and dihydroxymethylurone (**13**).

Acyclic compounds, such as various alkyl carbamates, are also typical finishing agents. In contrast to the pure urea–formaldehyde compounds, these exhibit little tendency to form self-cross-linked resins and react predominantly with cellulose to crosslink the fibers. The other advantages, over and above the mechanical properties, are resistance to chlorinated water, lightfastness, and shade constancy. These are closely related to the chemical structure and the resistance of the C–N and N–O bonds to hydrolysis.

Hydroxymethyl melamine derivatives and their ethers are of some importance for providing high-grade finishes, i.e., embossed finishes, Schreiner finishes, chintz finishes, and plisse finishes, but they have largely been replaced by the reactant types. The finishes obtained using hydroxymethylmelamines are more wash resistant than those made with hydroxymethylureas. Hydroxymethylmelamine compounds are preferred for embossed finishes (Everglaze) where the fabric can be damaged mechanically.

***Leather Auxiliaries.*** Resin retannage of chrome-tanned leather is carried out using hydroxymethyl compounds of urea, thiourea, melamine, dicyanodiamide, and ethers of these [3, p. 211, 89]. Aqueous solutions of the resins should penetrate the hide before condensation begins. The acid condensation catalyst (pH 3.9–4.5) should not reduce the period of activity of the tanning liquor or damage the hide. As in the case of the paper auxiliaries, the resins are rendered cationic using amines or anionic by means of a bisulfite. The anionic resins based on melamine or dicyanodiamide make leather very strong and supple and leave the color unchanged. The hydroxymethyl groups react with the active groups of the collagen fiber, which has a high content of glycine, proline, and oxyproline.

***Other Uses of Amino Resins.*** The number of possible uses is increasing constantly.

The importance of many uses, e.g., flameproofing agents, has declined greatly. Some applications, e.g., casting resins, have proved unsatisfactory in practice, whereas others are only at the development stage and their industrial suitability cannot yet be assessed.

## 11. Economic Aspects

Amino resins have become increasingly important after World War II. The most important amino resins are urea-formaldehyde, melamine-urea-formaldehyde, and melamine-formaldehyde resins [127]. The other resins mentioned in this keyword reflect only a very minor share in terms of volume and economic value. The predominant application of urea- and melamine-fortified urea-formaldehyde resins is for production of panel boards, whereas the major uses for melamine-formaldehyde resins are manufacture of impregnated paper for laminates and surface coatings.

Almost two thirds of the UF and MUF resins are consumed by particle board production; about one third by production of MDF, the rest of about 5% is used by all other end applications. For these applications, an overall European consumption of about $5.5 \times 10^6$ t can be assumed for 2004. A very small share of UF resins is also used as saturating resins in the production of laminates, for this a total volume of about 200 000 t (2009) can be estimated.

For the melamine-formaldehyde resins, the major end use is the production of laminates (HPL, CPL, DPL; about two thirds of the total volume) leaving the use of MFs as constituents of surface coatings behind (about 30%). For 2009, the total European market for MF saturating resins was about 520 000 t (2007: 525 000 t; 2003: 480 000 t).

Overall, and very simplifying, for all these aforementioned resins used in the wood based panels industry about 55% of the annual quantity is produced by the consumers themselves being backward integrated into resin production (like Kronospan Group, Sadepan Chimica, Egger, Pfleiderer, Sonae), about 45% of the annual resin consumption are covered by nonintegrated resin producers (Dynea Chemicals, BASF, Hexion, Ercros, Ineos, PA Resins).

In the application of surface finishings MF resins used in melamine films clearly dominate over other surface finishing materials such as thermoplastic films [128].

## References

## General References

1 B. Meyer: *Urea-Formaldehyde Resins*, Addison-Wesley, London 1979.
2 H. Baumann: *Leime und Kontaktkleber*, Springer Verlag, Berlin–Göttingen–Heidelberg–New York 1967.

3. A. Bachmann, T. Bertz: *Aminoplaste*, VEB-Verlag f. Grundstoffindustrie, Leipzig 1970.
4. R. Houwink, A.J. Staverman: *Chemie und Technologische Kunststoffe*, 4th ed., vol. **I**: pp. 102–103, vol. II/1: pp. 623–639, vol. II/2: pp. 681–685, Akadem. Verlagsgesellschaft, Leipzig 1962.
5. H. Petersen: *Grundzüge der Aminoplastchemie*, W. Pansegrau Verlag, Berlin 1968(Reprint from Kunstoffjahrbuch, 10th Issue).
6. E. Plath, L. Plath: *Taschenbuch der Kitte und Klebstoffe*, Wissenschaftl. Verlags-GmbH, Stuttgart 1963.
7. J.F. Blais: *Amino Resins*, Reinhold Publ. Co., New York 1959.
8. P. Talet: *Aminoplastes*, Dunod, Paris 1951.
9. C.P. Vale: *Aminoplastics*, Cleave-Hume Press Ltd., New York 1950.
10. R. Vieweg, E. Becker: "Duroplaste", in *Kunststoff-Handbuch*, vol. X, Hanser Verlag, München 1968, p. 134.
11. G.F. D'Alelio: "Aminoplasts," in *Kunstoff-Praktikum*, Hanser Verlag, München 1952, pp. 50–69.
12. D. Braun, H. Cherdron, W. Kern: *Praktikum der makromolekularen organischen Chemie*, 2nd ed., Hüthig Verlag, Heidelberg 1971.
13. E. Müller (ed.) in "Houben-Weyl", *Methoden der organischen Chemie*, vol. XIV/2: Makromolekulare Stoffe, "part 2", Thieme Verlag, Stuttgart 1963, pp. 319–400.
14. British Ind. Plastics, GB 524350, 1939.
15. ICI, DE 2726617, 1977 (D.A. Hubbard).

# Specific References

16. E. Einhorn, A. Hamburger, *Ber. Dtsch. Chem. Ges.* **41** (1908) 24.
17. A. Schmid, M. Himmelheber,CH 182058, 1932.
18. Henkel, DE 647303, 1935 (W. Hentrich, R. Köhler).
19. H. Umstätter: *Einführung in die Viskosimetrie und Rheometrie*, Springer Verlag, Berlin–Göttingen–Heidelberg 1952, pp. 5,17.
20. A. Eucken: *Lehrbuch der chemischen Physik*, 2nd ed., vol. II, Akad. Verlagsges., Leipzig 1944, pp. 1119–1120.
21. K. Masters: *Spray Drying*, Leonhard Hill Books, London 1972, p. 489.
22. A.F. Price, A.R. Cooper, A.S. Meskin: "Urea-Formaldehyde Reaction System, an Experimental Investigation," *J. Appl. Polym. Sci.* **25** (1980) 2597–2611.
23. G. Zigeuner, *Kunstoffe* **41** (1951) no. 7, 221–224.
24. D. Braun, F. Bayersdorf, *Angew. Makromol. Chem.* **89** (1980) 183–200.
25. M.L. Scheepers, P.J. Adriaensens, J.M. Gelan, R.A. Carleer, D.J. Vanderzande, N.K. De Vries, P.M. Brandts, *J. Polym. Sci., Part A: Polym. Chem.* **33** (1995) 915–920.
26. V. Horn, G. Benndorf, K.-P. Rädler: "Strukturgruppenanalyse von Harnstoff-Formaldehyd-Leimen", *Plaste Kautsch.* **25** (1978) no. 10, 570–575.
27. K. Kumlin, R. Simonson: "Urea-Formaldehyde Resins," *Angew. Makromol. Chem.* **93** (1981) 27–42.
28. Teukros, DE 2745951, 1976.
29. Borden Inc., GB 1319110, 1970.
30. A. Pizzi: in *Advanced Wood Adhesives Technology*, "Urea formaldehyde resins;" "Melamine formaldehyde resins" Marcel Dekker, New York 1994.
31. R.A. Wagner, J. Binder: *Proceedings Xth Austrian Chemical Days* 2002, ISBN 3-900-554-358.
32. Hitachi Chem. K. K., JP 56024415, 1979.
33. Evidenzbüro Österreichischer Zuckerfabriken, DE 2928003, 1979 (G. Greber, H. Andres, W. Pichler).
34. BASF, EP 52211, 1980 (W. Weiss, H. Petersen, H. Etling, V.M. Duda, L. Lelgemann, W. Pfalzgraf).
35. S.H. Nam, J.I. Ha, *Hwahak Kwa Hwahak Kongop* **19** (1976) 180–182, 194; *Chem. Abstr.* **86** (1977) 44 266 w.
36. A. Šebenik, U. Osredkar, M. Žigon, I. Vizovišek: "Study of the reaction between Urea and Formaldehyde by DSC and $^{13}$C NMR Spectroscopy," *Angew. Makromol. Chem.* **102** (1982) 81–85.
37. T.J. Perelpekova, F.G. Igranova, V.D. Moiseev: "Calorimetric study of the methylolation of urea by formaldehyde," *Khim. Promst. Ser. Proizvod. Pererab. Plastmass Sint. Smol.* 1981 no. 5, 12–14; *Chem. Abstr.* **96** (1982) 35762j.
38. B. Tomita, S. Hatono: "Urea-Formaldehyde Resins. III. Constitutional Characterisation by $^{13}$C Fourier Transform NMR Spectroscopy," *J. Polym. Sci.* **16** (1978) 2509–2525.
39. W.J. Blank, *J. Coat. Technol.* **51** (1979) 62.
40. H. Petersen, *Chem. Ztg.* **95** (1971) 695.
41. H. Petersen, *Textilveredlung* **5** (1970) 570–588.
42. M. Chiavarini, N. Del Fanti, R. Bigatto: "Compositive Characterization of Urea-Formaldehyde Adhesives by NMR Spectroscopy," *Angew. Makromol. Chem.* **46** (1975) 151–162.
43. K. Siimer, T. Pehk, P. Christjanson, *Proc. Estonian Acad. Sci. Chem.* **55** (2006) 212–225.
44. A. Pizzi, T.A. Mercer, *J. Appl. Polymer Sci.* **61** (1996) 1687–1695.
45. W. Kantner, Dynea Austria GmbH, private communication, 2010.
46. Girdler Corp., US 2456192, 1943 (H. G. Houlton).
47. Sherwood Paints, DE 863417, 1949.
48. Allied Chem. and Dye Corp., US 2658054, 1951 (G.A. Coleman, R.B. Greene, J.H. Merriam, S.P. Miller, R.F. Shannon).
49. Allied Chem. and Dye Corp., US 2729616, 1952 (M.H. Bigelow, H.N. Spurlock).
50. Spumalit-Anstalt, FR 1104018, 1954.
51. Rütgerswerke, DE 1029150, 1955 (H. Sauer, A. Tusch).
52. Skanska Attifabriken Aktiebolag, GB 800201, 1951.
53. Du Pont, US 2849421, 1956 (T.D. Weldin).
54. J. Meissner, DE 1570998, 1965 (L. Brucker-Voigt).
55. Rheinpreussen AG für Bergbau und Chemie, BE 711431, 1967.
56. St. Regis Paper Comp., US 2688606, 1951 (G.P Schmitt, C. Werberig).
57. BASF, DE 2109754, 1971 (F. Brunnmüller, H. Schatz, J. Mayer, O. Grabowsky).
58. Società Italiana Resine, DE 2263125, 1972 (S. Vargiu, G.S. Sesto, G. Mazzolani, U. Nistri).
59. Montedison, EP 38180, 1980 (D. Cannaloni, N. Conti).
60. Stamicarbon, BE 635085, 1963.
61. American Cyanamid Corp., FR 955810, 1947 (K.L. Lynch, A.J. Grossman).
62. DSM IP Assets BV, WO 2008128908 (J.J. Nusselder, W.J. Cramer, et. al).
63. W. Bauer, DE 1043628, 1953.
64. L. Unterstenhöfer, *Kunststoffe* **57** (1967) no. 11, 850.
65. W. Kantner, private communication, 2009.
66. *Lehr- und Handbuch der Abwassertechnik*, Verlag W. Ernst & Sohn, Berlin–München 1969.
67. H. Staudinger, K. Wagner, *Makromol. Chem.* **12** (1954) no. 168, 194.
68. H. Sobue, K. Murakami, C.Y. Tae, *Kobunshi Kagaku* **12** (1955) 240–257, *Chem. Abstr.* **51** (1957) 734 a.
69. BASF, EP 52211, 1980 (W. Weiss, H. Petersen, H. Etling, U.M. Duda, L. Lelgemann, W. Pfalzgraf).
70. H. Scheibler, *Z. Angew. Chem.* **53** (1940) 303.
71. Th. Goldschmidt, DE 2734628, 1977 (A. Laqua, U. Holtschmidt, E. Schamberg, D. Hellwig).

72. I. G. Farbenind., DE 733496, 1938 (K. Ott, K. Hamann).
73. Mitsubishi Kasei Kogyo, DE 1146080, 1951 (M. Hamamoto, S. Yasuhiro).
74. BASF, DE 889152, 1952 (B. von Reibnitz).
75. Phrix Werke, US 2876062, 1953 (E. Torke, J. König).
76. H. Petersen, *Text. Rundsch.* **16** (1961) no. 9/10, 2–11.
77. H. Petersen, *Text. Res. J.* **38** (1968) 156–176.
78. H. Petersen, *Text. Res. J.* **40** (1970) 335–344.
79. BASF, EP 62900, 1982 (C. Dudeck, E. Weber, H. Diem, O. Wittmann).
80. BASF, EP 53762, 1982 (C. Dudeck, E. Weber, H. Diem, J. Mayer, O. Wittmann, G. Lehmann).
81. Verkor N. V., Belgien, DE-OS 2903254, 1979.
82. Fraunhofer-Gesellschaft, DE 2829021, 1978 (E. Roffael, L. Mehlhorn).
83. Aktiebolaget Statens, EP 27583, 1980 (A.W. Westling).
84. M. Paquin, *Angew. Chem.* **60** (1948) 267–271.
85. BASF, DE 889225, 1949 (H. Scheuermann, J. Lenz).
86. Böhme Fettchemie, US 2978359, 1954 (H. Wedell).
87. The Resinous Products and Chem. Comp., US 2407599, 1946 (R.W. Auten, J.L. Rainey).
88. West Point Manufacturing Comp., US 2870041, 1958 (H.M. Waddle, J.F. Cotton, R.E. Hudson).
89. Nopco Chem. Comp., US 3063781, 1959 (C.A. Fetscher, S.L. Lipowski).
90. E. Roffael: "Fortschritte in der Verwendung von Sulfitablaugen als Binde- und Zusatzmittel bei der Herstellung von Holzspanplatten", *Adhäsion* (1979) no. 11, 334–336; (1979) no. 12, 368–370.
91. Champion Int. Corp., US 3994850, 1976 (W.N. Willegger).
92. A. Gams, G. Widmer, W. Fisch, *Helv. Chim. Acta* **24** (1941) 302E–319E.
93. D. Braun, W. Pandjojo: "Gelchromatographische Untersuchungen vonungehärteten Kondensaten aus Melamin und Formaldehyd", *Angew. Makromol. Chem.* **80** (1979) 195–205.
94. BASF, DE 888169, 1949 (H. Scheuermann).
95. I. G. Farbenind., FR 880189, 1942.
96. BASF, DE 2424379, 1974 (U. Hampel, R. Petri, H. Petersen, J. Lenz, G. Matthias, W. Reuther).
97. Süddeutsche Kalkstickstoffwerke, DE 1671017, 1966 (H. Aignesberger, H. Michaud).
98. Cassella, DE 3104420, 1981 (H. Hönel, W. Michel, S. Piltsch, K. Schlüter, A. Wolf).
99. Société Chimique des Charbonnages, FR 80/09 193, 1980.
100. BASF, DE 1595368, 1966 (H. Etling, H. Henkel, F. Meyer).
101. Torbjörn Reitberger, SE 8000–663, 1980.
102. BASF, DE 2020481, 1970 (J. Mayer, C. Schmidt-Hellerau).
103. BASF, DE 2951957, 1979 (J. Mayer, C. Schmidt-Hellerau, C. Dudeck, H. Diem, H. Schatz).
104. L. McMaster, *J. Am. Chem. Soc.* **56** (1954) 204–206.
105. The Resinous Products and Chem. Comp., US 2160196, 1937 (H.A. Bruson, J.W. Eastes).
106. Dr. Beck and Co., DE 888294, 1949 (H. Beck, A. Wiegand).
107. H. Schindlbauer, B. Holzhöfer, J. Schuster, *Kunststoffe* **69** (1973) no. 3, 158–162.
108. W.G.K. Taylor: "Aminoplastics", *Rep. on the Progress of Appl. Chem.*,vol. XLIX, Society of Chemical Industry, London 1964, pp. 572–581.
109. G. Widmer: "Quantitative Bestimmung von Harnstoff und Melamin in Gemischen von Harnstoff- und Melamin-Formaldehyd-Kondensationsprodukten", *Kunststoffe* **46** (1956) no. 8, 359–362.
110. DIN 51777: *Bestimmung des Wassergehaltes nach Karl Fischer*, Beuth-Vertrieb GmbH, Berlin–Köln.
111. F. Käsbauer, D. Merkel, O. Wittmann: "Bestimmung von freiem Formaldehyd und N-Methylol-Formaldehyd in Harnstoff-Formaldehyd-Kondensaten", *Fresenius Z. Anal. Chem.* **281** (1976) 17–21.
112. H. Petersen, L. Klug, H. Hahn, W. Huber, *Farbe + Lack* **87** (1981) no. 8, 647–652.
113. G. Groh, H. Petersen, L. Klug, *Farbe + Lack* **87** (1981) no. 9, 744–748.
114. AOAC-Methods (Association of Official Analytical Chemists), 13th ed., Washington 1980, p. 17.
115. D. Braun, P. Günther: "Methoden zur Verfolgung der Harnstoff-Formaldehyd-Harz-Synthese", *Kunststoffe* **72** (1982) no. 12, 785–790.
116. K. Kumlin, R. Simonson: "Urea-Formaldehyde Resins, 1. Separationof Low Molecular Weight Components in Urea-Formaldehyde Resins by means of Liquid Chromatography," *Angew. Makromol. Chem.* **68** (1978) 175–184.
117. J.R. Ebdon, P.E. Heaton, W.T.S. O'Rourke, J. Parkin, *Polymer* **25** (1984) 821–825.
118. M. Chiavarini, R. Bigatto, N. Conti: "Synthesis of Urea-Formaldehyde Resins: NMR Studies on Reaction Mechanisms," *Angew. Makromol. Chem.* **70** (1978) 49–58.
119. H.J. Deppe, K. Ernst: *Taschenbuch der Spanplattentechnik*, 2nd ed., DRW-Verlag, Leinfelden 1982.
120. M. Dunky, P. Niemz, *Holzwerkstoffe und Leime*, Springer Verlag, Berlin 2002.
121. J. Binder, K. Lepedat, R.A. Wagner, in K. Fischer (ed.): *Proceedings of the Asia Pacific Laminates Workshop*, 2003.
122. T. Götze, P. Dörries, *Plastverarbeiter* **33** (1982) no. 9, 1118–1122.
123. H. Kittel: *Lehrbuch der Lacke und Beschichtungen*, vol. IV, Heenemann, Berlin 1976.
124. F. Wultsch: *Hilfsmittel und ihre Anwendung in der Papiererzeugung*, G. Staib Verlag, Biberach 1966.
125. "Wet Strength Paper and Paperboard," *TAPPI Monogr. Ser.* **29** (1965) 1–32.
126. K. Holmberg: "Alkali Catalysed Curing of Starch-Melamine Resin Systems," *Polym. Bull. (Berlin)* **6** (1982) 553–558.
127. "Socio-Economic Benefits of Formaldehyde to the European Union (EU 25) and Norway," released June 2007, http://www.formaldehyde-europe.org/fileadmin/formaldehyde/PDF/Socio-Economic-Benefits-Study.pdf
128. K. Lepedat, J. Lang, R. A. Wagner, in L. Pilato (ed.): *Phenolic Resins: A Century of Progress*, Springer, Heidelberg 2010

## Further Reading

M. Manea: *High Solid Binders*, Vincentz Network, Hannover 2008.
E. Petrie: *Handbook of Adhesives and Sealants*, 2nd ed., McGraw-Hill, New York, NY 2007.
A. A. Tracton (ed.): *Coatings Materials and Surface Coatings*, CRC Press, Boca Raton, FL 2007.
J. Tulla-Puche, F. Albericio (eds.): *The Power of Functional Resins in Organic Synthesis*, Wiley-VCH, Weinheim 2008.
Z. W. Wicks Jr., F. N. Jones, S. P. Pappas, D. A. Wicks: *Organic Coatings*, 3rd ed., Wiley-Interscience, Hoboken, NJ 2007.
L. L. Williams: *Amino Resins and Plastics*, Kirk Othmer Encyclopedia of Chemical Technology, 5th edition, vol. 2, p. 618–652, John Wiley & Sons, Hoboken, NJ, 2005, online: DOI: 10.1002/0471238961.0113091423091212.a01.pub2.

# Epoxy Resins

HA Q. PHAM, Dow Chemical, Freeport, Texas

MAURICE J. MARKS, Dow Chemical, Freeport, Texas

*ULLMANN'S ENCYCLOPEDIA OF INDUSTRIAL CHEMISTRY*

| | | |
|---|---|---|
| 1. | Introduction | 1644 |
| 2. | History | 1645 |
| 3. | Industry Overview | 1647 |
| 4. | Classes of Epoxy Resins and Manufacturing Processes | 1649 |
| 5. | Liquid Epoxy Resins (DGEBA) | 1649 |
| 5.1. | Caustic Coupling Process | 1651 |
| 5.2. | Phase-Transfer Catalyst Process | 1651 |
| 6. | Solid Epoxy Resins Based on DGEBA | 1652 |
| 6.1. | SER Continuous Advancement Process | 1657 |
| 6.2. | Phenoxy Resins | 1657 |
| 6.3. | Epoxy-Based Thermoplastics | 1658 |
| 7. | Halogenated Epoxy Resins | 1658 |
| 7.1. | Brominated Bisphenol A Based Epoxy Resins | 1658 |
| 7.2. | Fluorinated Epoxy Resins | 1659 |
| 8. | Multifunctional Epoxy Resins | 1659 |
| 8.1. | Epoxy Novolac Resins | 1659 |
| 8.1.1. | Bisphenol F Epoxy Resin | 1660 |
| 8.1.2. | Cresol Epoxy Novolacs | 1660 |
| 8.1.3. | Glycidyl Ethers of Hydrocarbon Epoxy Novolacs | 1661 |
| 8.1.4. | Bisphenol A Epoxy Novolacs | 1661 |
| 8.2. | Other Polynuclear Phenol Glycidyl Ether Derived Resins | 1661 |
| 8.2.1. | Glycidyl Ether of Tetrakis(4-hydroxyphenyl)ethane | 1662 |
| 8.2.2. | Trisphenol Epoxy Novolacs | 1662 |
| 8.3. | Aromatic Glycidyl Amine Resins | 1662 |
| 8.3.1. | Triglycidyl Ether of *p*-Aminophenol | 1663 |
| 8.3.2. | Tetraglycidyl Methylenedianiline (MDA) | 1663 |
| 9. | Specialty Epoxy Resins | 1663 |
| 9.1. | Crystalline Epoxy Resins Development | 1663 |
| 9.2. | Weatherable Epoxy Resins | 1664 |
| 9.2.1. | Hydrogenated DGEBA | 1664 |
| 9.2.2. | Heterocyclic Glycidyl Imides and Amides | 1664 |
| 9.2.3. | Hydantoin-Based Epoxy Resins | 1665 |
| 9.3. | Elastomer-Modified Epoxies | 1665 |
| 10. | Monofunctional Glycidyl Ethers and Aliphatic Glycidyl Ethers | 1665 |
| 11. | Cycloaliphatic Epoxy Resins and Epoxidized Vegetable Oils | 1666 |
| 12. | Epoxy Esters and Derivatives | 1668 |
| 12.1. | Epoxy Esters | 1668 |
| 12.2. | Glycidyl Esters | 1669 |
| 12.3. | Epoxy Acrylates | 1669 |
| 12.4. | Epoxy Vinyl Esters | 1670 |
| 12.5. | Epoxy Phosphate Esters | 1670 |
| 13. | Characterization of Uncured Epoxies | 1671 |
| 14. | Curing of Epoxy Resins | 1673 |
| 15. | Coreactive Curing Agents | 1675 |
| 15.1. | Amine Functional Curing Agents | 1675 |
| 15.1.1. | Primary and Secondary Amines | 1675 |
| 15.1.1.1. | Aliphatic Amines | 1677 |
| 15.1.1.2. | Cycloaliphatic Amines | 1678 |
| 15.1.1.3. | Aromatic Amines | 1679 |
| 15.1.1.4. | Arylyl Amines | 1679 |
| 15.1.2. | Polyamides | 1679 |
| 15.1.3. | Amidoamines | 1682 |
| 15.1.4. | Dicyandiamide | 1682 |
| 15.2. | Carboxylic Functional Polyester and Anhydride Curing Agents | 1683 |
| 15.2.1. | Carboxylic Functional Polyesters | 1684 |
| 15.2.2. | Acid Anhydrides | 1684 |
| 15.3. | Phenolic-Terminated Curing Agents | 1687 |
| 15.4. | Melamine–, Urea–, and Phenol–Formaldehyde Resins | 1687 |
| 15.5. | Mercaptans (Polysulfides and Polymercaptans) Curing Agents | 1688 |
| 15.6. | Cyclic Amidines Curing Agents | 1689 |
| 15.7. | Isocyanate Curing Agents | 1689 |

Ullmann's Polymers and Plastics: Products and Processes
© 2016 Wiley-VCH Verlag GmbH & Co. KGaA, Weinheim
ISBN: 978-3-527-33823-8 / DOI: 10.1002/14356007.a09_547.pub2

| | | | | | | |
|---|---|---|---|---|---|---|
| 15.8. | Cyanate Ester Curing Agents. | 1689 | | 20.1.2. | High Solids Solventborne Coatings | 1709 |
| 16. | Catalytic Cure | 1689 | | 20.1.3. | Solvent-Free Coatings (100 % Solids) | 1709 |
| 16.1. | Lewis Bases | 1690 | | 20.1.4. | Waterborne Coatings | 1709 |
| 16.2. | Lewis Acids | 1691 | | 20.1.5. | Powder Coatings | 1709 |
| 16.3. | Photoinitiated Cationic Cure | 1691 | | 20.1.6. | Radiation-Curable Coatings | 1711 |
| 17. | Formulation Development With Epoxy Resins | 1692 | | 20.2. | Epoxy Coatings Markets | 1712 |
| 17.1. | Relationship Between Cured Epoxy Resin Structure and Properties | 1692 | | 20.2.1. | Marine and Industrial Maintenance Coatings | 1712 |
| 17.2. | Selection of Epoxy Resins | 1694 | | 20.2.2. | Metal Container and Coil Coatings | 1714 |
| 17.3. | Selection of Curing Agents | 1694 | | 20.2.3. | Automotive Coatings | 1716 |
| 17.4. | Epoxy/Curing Agent Stoichiometric Ratios | 1695 | | 20.3. | Inks and Resists | 1717 |
| 17.5. | Catalysts | 1696 | | 21. | Structural Applications | 1718 |
| 17.6. | Accelerators | 1697 | | 21.1. | Structural Composites | 1718 |
| 18. | Epoxy Curing Process | 1697 | | 21.1.1. | Epoxy Composites | 1719 |
| 18.1. | Characterization of Epoxy Curing and Cured Networks | 1699 | | 21.1.2. | Epoxy Vinyl Ester Composites | 1719 |
| 19. | Formulation Modifiers | 1702 | | 21.1.3. | Mineral-Filled Composites | 1720 |
| 19.1. | Diluents | 1702 | | 21.2. | Civil Engineering, Flooring, and Construction | 1720 |
| 19.2. | Thixotropic Agents | 1703 | | 21.3. | Electrical Laminates | 1721 |
| 19.3. | Fillers | 1703 | | 21.4. | Other Electrical and Electronic Applications | 1723 |
| 19.4. | Epoxy Nanocomposites | 1704 | | 21.4.1. | Casting, Potting, and Encapsulation | 1723 |
| 19.5. | Toughening Agents and Flexiblizers | 1706 | | 21.4.2. | Transfer Molding | 1724 |
| 20. | Coatings Applications | 1707 | | 21.5. | Adhesives | 1725 |
| 20.1. | Coatings Application Technologies | 1709 | | 21.6. | Tooling | 1725 |
| 20.1.1. | Low Solids Solventborne Coatings | 1709 | | 22. | Health and Safety Factors | 1726 |
| | | | | 23. | Acknowledgments | 1727 |
| | | | | | Bibliography | 1727 |

## 1. Introduction

Epoxy resins are an important class of polymeric materials, characterized by the presence of more than one three-membered ring known as the epoxy, epoxide, oxirane, or ethoxyline group.

The word "epoxy" is derived from the Greek prefix "ep," which means over and between, and "oxy," the combining form of oxygen [1]. By strict definition, epoxy resins refer only to uncross-linked monomers or oligomers containing epoxy groups. However, in practice, the term epoxy resins is loosely used to include cured epoxy systems. It should be noted that very high molecular weight epoxy resins and cured epoxy resins contain very little or no epoxide groups.

The vast majority of industrially important epoxy resins are bi- or multifunctional epoxides. The monofunctional epoxides are primarily used as reactive diluents, viscosity modifiers, or adhesion promoters, but they are included here because of their relevance in the field of epoxy polymers.

Epoxies are one of the most versatile classes of polymers with diverse applications such as metal can coatings, automotive primer, printed circuit boards, semiconductor encapsulants, adhesives, and aerospace composites. Most cured epoxy resins provide amorphous thermosets with excellent mechanical strength and toughness; outstanding chemical, moisture, and corrosion resistance; good thermal, adhesive, and electrical properties; no volatiles emission and low shrinkage upon cure; and dimensional stability—a unique combination of properties generally not found in any other plastic material. These superior performance characteristics, coupled with outstanding formulating versatility and reasonable costs, have gained epoxy resins wide acceptance as materials of choice for a multitude of bonding, structural, and protective coatings applications.

Commercial epoxy resins contain aliphatic, cycloaliphatic, or aromatic backbones and are available in a wide range of molecular weights from several hundreds to tens of thousands. The most widely used epoxies are the glycidyl ether derivatives of bisphenol A (> 75 % of resin sales volume). The capability of the highly strained epoxy ring to react with a wide variety of curing agents under diverse conditions and temperatures imparts additional versatility to the epoxies. The major industrial utility of epoxy resins is in thermosetting applications. Treatment with curing agents gives insoluble and intractable thermoset polymers. In order to facilitate processing and to modify cured resin properties, other constituents may be included in the compositions: fillers, solvents, diluents, plasticizers, catalysts, accelerators, and tougheners.

Epoxy resins were first offered commercially in the late 1940s and are now used in a number of industries, often in demanding applications where their performance attributes are needed and their modestly high prices are justified. However, aromatic epoxies find limited uses in exterior applications because of their poor ultraviolet (UV) light resistance. Highly crosslinked epoxy thermosets sometimes suffer from brittleness and are often modified with tougheners for improved impact resistance.

The largest use of epoxy resins is in protective coatings (> 50 %), with the remainder being in structural applications such as printed circuit board (PCB) laminates, semiconductor encapsulants, and structural composites; tooling, molding, and casting; flooring; and adhesives. New, growing applications include lithographic inks and photoresists for the electronics industry.

## 2. History

The patent literature indicates that the synthesis of epoxy compounds was discovered as early as the late 1890s [2]. In 1934, Schlack of I.G. Farbenindustrie AG in Germany filed a patent application for the preparation of reaction products of amines with epoxies, including one epoxy based on bisphenol A and epichlorohydrin [3]. However, the commercial possibilities for epoxy resins were only recognized a few years later, simultaneously and independently, by the DeTrey Fréres Co. in Switzerland [4] and by the DeVoe and Raynolds Co. [5] in the United States.

In 1936, Pierre Castan of DeTrey Fréres Co. produced a low melting epoxy resin from bisphenol A and epichlorohydrin that gave a thermoset composition with phthalic anhydride. Application of the hardened composition was foreseen in dental products, but initial attempts to market the resin were unsuccessful. The patents were licensed to Ciba AG of Basel, Switzerland, and in 1946 the first epoxy adhesive was shown at the Swiss Industries Fair, and samples of casting resin were offered to the electrical industry.

Immediately after World War II, Sylvan Greenlee of DeVoe and Raynolds Co. patented a series of high molecular weight (MW) epoxy resin compositions for coating applications. These resins were based on the reaction of bisphenol A and epichlorohydrin, and were marketed through the subsidiary Jones-Dabney Co. as polyhydroxy ethers used for esterification with drying oil fatty acids to produce alkyd-type epoxy ester coatings. Protective surface coatings were the first major commercial application of epoxy resins, and they remain a major outlet for epoxy resin consumption today. Concurrently, epoxidation of multifunctional olefins with peroxy acids was studied by Daniel Swern as an alternative route to epoxy resins [6]. Meanwhile, Ciba AG, under license from DeTrey Fréres, further developed epoxy resins for casting, laminating, and adhesive applications, and the Ciba Products Co. was established in the United States.

In the late 1940s, two U.S. companies, Shell Chemical Co. and Union Carbide Corp. (then Bakelite Co.), began research on bisphenol A based epoxy resins. At that time, Shell was the only supplier of epichlorohydrin, and Bakelite was a leading supplier of phenolic resins and bisphenol A. In 1955, the four U.S. epoxy resin manufacturers entered into a cross-licensing agreement. Subsequently, The Dow Chemical Co. and Reichhold Chemicals, Inc. joined the patent pool and began manufacturing epoxy resins.

In the 1960s, a number of multifunctional epoxy resins were developed for higher temperature applications. Ciba Products Co. manufactured and marketed *o*-cresol epoxy novolac resins, which had been developed by Koppers Co. Dow developed the phenol novolac epoxy resins, Shell introduced polyglycidyl ethers of tetrafunctional phenols, and Union Carbide developed a triglycidyl p-aminophenol resin. These products continue to find uses today in highly demanding applications such as semiconductor encapsulants and aerospace composites where their performance justifies their higher costs relative to bisphenol A based epoxies.

The peracetic acid epoxidation of olefins was developed in the 1950s by Union Carbide in the United States and by Ciba AG in Europe for cycloaliphatic structures. Ciba Products marketed cycloaliphatic epoxy resins in 1963 and licensed several multifunctional resins from Union Carbide in 1965. The ensuing years witnessed the development of general-purpose epoxy resins with improved weathering characteristics based on the five-membered hydantoin ring and also on hydrogenated bisphenol A, but their commercial success has been limited because of their higher costs. Flame-retardant epoxy resins based on tetrabromobisphenol A were developed and commercialized by Dow Chemical for electrical laminate and composite applications in the late 1960s.

In the 1970s, the development of two breakthrough waterborne coating technologies based on epoxy resins helped establish the dominant position of epoxies in these markets: PPG's cathodic electrodeposition automotive primer and ICI-Glidden's epoxy acrylic interior can coatings.

While epoxy resins are known for excellent chemical resistance properties, the development and commercialization of epoxy vinyl ester resins in the 1970s by Shell and Dow offered enhanced resistance properties for hard-to-hold, corrosive chemicals such as acids, bases, and organic solvents. In conjunction with the development of the structural composites industry, epoxy vinyl ester resin composites found applications in demanding environments such as tanks, pipes and ancillary equipment for petrochemical plants and oil refineries, automotive valve covers, and oil pans. More recently, epoxy and vinyl esters are used in the construction of windmill blades for wind energy farms. Increasing requirements in the composite industries for aerospace and defense applications in the 1980s led to the development of new, high performance multifunctional epoxy resins based on complex amine and phenolic structures. Examples of those products are the trisphenol epoxy novolacs developed by Dow Chemical and now marketed by Huntsman (formerly Ciba).

The development of the electronics and computer industries in the 1980s demanded higher performance epoxy resins. Faster speeds and more densely packed semiconductors required epoxy encapsulants with higher thermal stability, better moisture resistance, and higher device reliability. Significant advances in the manufacturing processes of epoxy resins led to the development of electronic-grade materials with lower ionic and chloride impurities and improved electrical properties. Dow Chemical introduced a number of new, high performance products such as hydrocarbon epoxy novolacs based on dicyclopentadiene. The 1980s also witnessed the development of the Japanese epoxy resin industry with focus on specialty, high performing and high purity resins for the electronics industry. These include the commercialization of crystalline resins such as biphenol diglycidyl ether.

More recently, in order to comply with more stringent environmental regulations, there has been increased attention to the development of epoxy resins for high solids, powder, and waterborne and radiation-curable coatings. Powder coatings based on epoxy–polyester and epoxy–acrylate hybrids have continued to grow in the global markets, including new applications such as primer-surfacer and topcoats for automotive coatings. Radiation-curable epoxy–acrylates and cycloaliphatic epoxies showed tremendous growth in the 1990s in radiation-curable applications. These include important and new uses of epoxy resins such as the photoresists and lithographic inks for the electronics industry. Waterborne epoxy coatings are projected to grow substantially.

The continuing trend of device miniaturization in the computer industry, and the explosive growth of portable electronics and communications devices such as wireless cellular telephones in the 1990s demanded new, high performance resins for the PCB market. This has led to the development of new epoxies and epoxy hybrid systems having lower dielectric constants ($D_k$), higher glass-transition temperatures ($T_g$), and higher thermal decomposition temperatures ($T_d$) for electrical laminates. Environmental pressures in the PCB industry have fueled the development of a number of new bromine-free resin systems, but their commercialization is limited because of higher costs.

Significant efforts have been directed toward performance enhancements of epoxy structural composites. Advances have been made in the epoxy-toughening area. Epoxy nanocomposites and nanotube systems have been studied and are claimed to bring exceptional thermal, chemical, and mechanical property improvements. However, commercialization has not yet materialized.

In 1999, Dow Chemical introduced a new epoxy-based thermoplastic resin, BLOX*, for gas barrier, adhesives, and coatings applications.

## 3. Industry Overview

From the first commercial introduction of diglycidyl ether of bisphenol A (DGEBA) resins in the 1940s, epoxy resins have gradually established their position as an important class of industrial polymers. Epoxy resin sales increased rapidly in the 1970s and continued to rise into the 1980s as new applications were developed (annual growth rate > 10 % in the U.S. market, Table 1). More recently, the slower growth rates (3–4 %) of the U.S., Japanese, and European markets in the 1990s were made up for by the higher growth rate (5–10 %) in the Asia-Pacific markets outside of Japan, particularly in Taiwan and China. Epoxy resin growth has historically tracked well with economic developments and demands for durable goods, and so the growth of the epoxy markets in Asia-Pacific is expected to continue into the next decade.

The global market for epoxy resins is estimated at approximately 1.15 million metric tons (MT) for the year 2000 [8]. This is an increase of 5 % over 1999 demands. The North American market consumed over 330,000 MT of epoxy resins, the European market is estimated at more than 370,000 MT, and the Asian market has surpassed both the North American and European markets by consuming 400,000 MT of epoxy resins. About 50,000 MT of epoxies were consumed in the South American markets. Imports of epoxy resins from Asia into North America has steadily grown to about 120,000 MT in 2000. Epoxy resins were used with over 400,000 MT of curing agents to produce an estimated 3 million MT of formulated compounds, worth over $20 billion.

Up until the mid-1990s, the major worldwide producers of epoxy resins were Dow Chemical, Shell, and Ciba-Geigy. However, both Shell and Ciba-Geigy have recently divested their epoxy resins businesses. Shell sold their epoxy business to Apollo Management LP (based in New York City) in the year 2000 and the company was renamed Resolution Performance Products. Similarly, Ciba's epoxy business was sold in 2000 to Morgan Grenfell, a London (U.K.)-based private equity firm, and the new company name was Vantico. More recently, in June 2003, the Vantico group of companies joined Huntsman. The Vantico business units are now named Huntsman Advanced Materials. The cycloaliphatic epoxy business of Union Carbide became part of The Dow Chemical Company after their merger in the year 2001. Together, these three producers continue to dominate the world market for epoxy resins, accounting for almost 65 % of the global market. However, this is a reduction from over 70 % of market shares owned by the three largest producers in the 1980s. Smaller producers of epoxy resins for the North American markets are Reichhold (owned by Dainippon Ink and Chemicals), CVC Specialty Chemicals, Pacific Epoxy Polymers, and InChem (phenoxy thermoplastic resins). Suppliers of epoxy derivatives include Ashland Specialty Chemical, UCB Chemicals (Radcure), AOC LLC, Eastman Chemical, and Interplastic Corp.

The market in Europe is similarly dominated by the three big producers: Dow, Resolution, and Huntsman and their affiliated joint ventures. Other smaller epoxy producers include Bakelite AG, LEUNA-Harze, Solutia, SIR Industriale, and EMS-CHEMIE. Imports from Asia have become significant in recent years.

Table 1. History of U.S. epoxy resin annual production[a]

| Year | Production, $10^3$ MT |
|---|---|
| 1955 | 10 |
| 1960 | 30 |
| 1965 | 55 |
| 1970 | 79 |
| 1975 | 100 |
| 1980 | 201 |
| 1985 | 347 |
| 1990 | 475 |
| 1994 | 433 |

[a] Data from U.S. International Trade Commission, *Synthetic Organic Chemicals*. Data include modified and unmodified epoxy resins. Modified epoxy resins include solid epoxy resin (SER), vinyl ester resins, epoxy acrylates, etc. There appear to be some discrepancies in epoxy resin production and market data as reported by different publications and organizations [7]. This is primarily due to the fact that some epoxy resins such as liquid DGEBA resins and epoxy novolacs are used as raw materials to produce modified or advanced epoxy resins, which may be further converted to end-use products. Some publications report only unmodified epoxies.

The last two decades marked the emergence of the Asian epoxy industry. In the 1980s, the Japanese epoxy industry was transformed from a number of joint venture companies with Dow, Shell, and Ciba into independent producers and the emergence of a high number of new producers. This coincides with the development of Japan as a world-class manufacturing base. The Japanese epoxy industry is known for their special focus on high performance, high purity resins for the electronics industry. According to data from the Japan Epoxy Resin Manufacturers Association, the total Japanese market demand is estimated at approximately 200,000 MT for the year 2000. The production capacity is estimated at 240,000 MT annually. Exports accounted for an estimated 40,000 MT in 2000. Major Japanese epoxy resin producers are Tohto Kasei, Japan Epoxy Resins Corp. (formerly Yuka-Shell), Asahi Kasei, Dai Nippon Ink and Chemicals, Dow Chemical Japan, Mitsui Chemicals, Nihon Kayaku, Sumitomo Chemical, and Asahi Denka Kogyo. In Japan, Tohto Kasei is a leading resin producer, with epoxy technology licensing arrangements with numerous resin producers in Asia.

Outside of Japan, there have been significant increases in epoxy market demands and capacity in the 1990s. This is due to the migration of many PCB, electronic, computer, and durable goods manufacturing plants into the region, which has considerably lower manufacturing costs. Nan Ya, a subsidiary of the Formosa Plastics Group based in Taiwan, is emerging as a major epoxy resin producer with some import presence in North America and Europe. Similarly, Kukdo of Korea also exports to the North American and European markets. The output of these two companies now account for an estimated 15% of the world market. In China, there are numerous (more than 200) small domestic producers of epoxy resins. Recently, a number of major epoxy producers have announced joint ventures or plans to build manufacturing plants in China. These include a number of companies with integrated capacity into electrical laminates and PCB manufacturing, following the business model pioneered by the Formosa Plastics Group. Other notable Asian producers include Asia Pacific Resins, Chang Chun, and Eternal Chemical of Taiwan; Thai Epoxy of Thailand; Kumho, LG Chemical, and Pacific Epoxy Resins of Korea; and Guangdong Ciba Polymers, Sinopec Baling Petrochemical, Jiangsu Sanmu, and Wuxi DIC Epoxy Resin of China. The LG Chemical epoxy business was purchased by Bakelite in late 2002. A significant amount of resin produced in Taiwan and China is directed toward electrical laminates applications. The aggressive buildup of epoxy capacity in Asia has put significant pressures on resin prices, particularly the high volume products such as liquid epoxy resins based on bisphenol A (Table 2). But as of January 2004, the epoxy market demand in China alone has increased to more than 500,000 MT (Chinese Epoxy Industry Web site).

Estimated average prices for epoxy resin products in North America are given in Table 3. As with other petrochemical-based products, they depend on crude-oil prices. Prices of

Table 2. Epoxy production capacity in Asia-Pacific[a] (2001)

| Country | Existing capacity, 1000 MT/year | Announced capacity, 1000 MT/year |
|---|---|---|
| Japan | 240 | |
| Taiwan | 239 | 70 |
| China | 100 | 255 |
| Korea | 180 | |
| Thailand | 30 | |
| Malaysia | 10 | |
| Philippines | 10 | |
| Total | 809 | 325 |

[a]Compilation of published data by Dow Chemical.

Table 3. U.S. average epoxy resin prices and applications (2000)

| Resin | $/kg | Applications |
|---|---|---|
| Liquid epoxy resins (Diglycidyl ether of bisphenol A, DGEBA) | 2.2 | coatings, castings, tooling, flooring, adhesives, composites |
| Solid epoxy resins (SER) | 2.4 | powder coating; epoxy esters for coatings; can, drum, and maintenance coatings |
| Bisphenol F epoxy | 4.4 | coatings |
| Multifunctional | | |
| Phenol epoxy novolac | 4.8 | castings, coatings, laminates |
| Cresol epoxy novolac | 8.8 | electronics encapsulants, powder coatings, laminates |
| Other multifunctional epoxies | 11–44 | composites, adhesives, laminates, electronics |
| Cycloaliphatic epoxies | 6.6 | electrical castings, coatings, electronics |
| Brominated epoxies | 3.3–5.5 | printed wiring boards, composites |
| Epoxy vinyl esters | 3.3 | composites |
| Phenoxy resins | 11–17 | coatings, laminates, glass sizing |
| Epoxy diluents | 4–11 | |

**Table 4.** List of some epoxy resin producers and their product trade names

| Company | Trade name |
| --- | --- |
| Resolution Performance Products | Epon, Eponol, Eponex, Epi-Cure, Epikote |
| Dow Chemical | D.E.R., D.E.N., D.E.H., Derakane, E.R.L |
| Hunstman Advanced Materials (formerly Ciba, Vantico) | Araldite, Aralcast |
| Reichhold Chemical | Epotuf |
| Nan Ya | NPEL, NPES |
| Kukdo Chemical | YD |
| Dainippon Ink & Chemical (DIC) | Epiclon |
| Tohto Kasei | Epotohto |
| Japan Epoxy Resin (JER) | Epikote |
| Asahi Kasei | A.E.R. |
| Mitsui Chemical | Eponik |
| Sumitomo Chemical | Sumiepoxy |
| Thai Epoxy | Epotec |
| Chang Chun | |
| InChem | Paphen |
| Pacific Epoxy Polymers | PEP |
| CVC Specialty Chemicals | Erysis, Epalloy |

multifunctional resins are typically higher. They are based on more expensive raw materials than DGEBA resins and involve more complex manufacturing procedures. A listing of some major epoxy resin producers and the trade names of their products is shown in Table 4.

There are numerous suppliers of epoxy curing agents. Some of the major producers are Air Products and Chemicals, Cognis, Degussa, DSM, Huntsman, and Resolution.

## 4. Classes of Epoxy Resins and Manufacturing Processes

Most commercially important epoxy resins are prepared by the coupling reaction of compounds containing at least two active hydrogen atoms with epichlorohydrin followed by dehydrohalogenation:

$$R-H + \overset{O}{\underset{}{\triangle}}\!\!\text{CH}_2\text{Cl} \longrightarrow R\text{-CH}_2\text{-CH(OH)-CH}_2\text{Cl}$$

$$R\text{-CH}_2\text{-CH(OH)-CH}_2\text{Cl} \xrightarrow{-\text{HCl}} R\text{-CH}_2\text{-CH}\overset{O}{\underset{}{\triangle}}\text{CH}_2$$

These included polyphenolic compounds, mono and diamines, amino phenols, heterocyclic imides and amides, aliphatic diols and polyols, and dimeric fatty acids. Epoxy resins derived from epichlorohydrin are termed glycidyl-based resins.

Alternatively, epoxy resins based on epoxidized aliphatic or cycloaliphatic dienes are produced by direct epoxidation of olefins by peracids:

$$\text{RHC=CHR'} + \text{R''COOOH} \longrightarrow \text{RHC}\overset{O}{-}\text{CHR'} + \text{R''COOH}$$

Approximately 75% of the epoxy resins currently used worldwide are derived from DGEBA. This market dominance of bisphenol A based epoxy resins is a result of a combination of their relatively low cost and adequate-to-superior performance in many applications. Figure 1 shows U.S. consumption of major epoxy resin types for the year 2000.

## 5. Liquid Epoxy Resins (DGEBA)

The most important intermediate in epoxy resin technology is the reaction product of epichlorohydrin and bisphenol A. It is often referred to in the industry as liquid epoxy resin (LER), which can be described as the crude DGEBA where the degree of polymerization, $n$, is very low ($n \cong 0.2$):

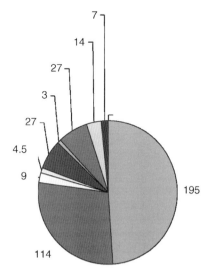

**Figure 1.** Major epoxy resin and derivatives markets ($10^3$ MT). ■ LER; ■ SER; □ epoxy novolacs; ▨ other multifunctional epoxies; ■ brominated epoxies; ■ cycloaliphatic; ■ vinyl esters; and ▨ epoxy acrylates.

Pure DGEBA is a crystalline solid (*mp* 43 °C) with an epoxide equivalent weight (EEW) of 170. The typical commercial unmodified liquid resins are viscous liquids with viscosities of 11,000–16,000 mPa·s (= cP) at 25 °C, and an epoxide equivalent weight of ca 188.

EEW is the weight of resin required to obtain one equivalent of epoxy functional group. It is widely used to calculate reactant stoichiometric ratios for reacting or curing epoxy resins. It is related to the epoxide content (%) of the epoxy resin through the following relationship:

$$\text{EEW} = \frac{43.05}{\%\text{Epoxide}} \times 100 \qquad (1)$$

where 43.05 is the molecular mass of the epoxide group, $-C_2H_3O$. Other equivalent terminologies common in the industry include weight per epoxide (Wpe) or epoxide equivalent mass (EEM).

The outstanding performance characteristics of the resins are conveyed by the bisphenol A moiety (toughness, rigidity, and elevated temperature performance), the ether linkages (chemical resistance), and the hydroxyl and epoxy groups (adhesive properties and formulation latitude; reactivity with a wide variety of chemical curing agents).

LERs are used in coatings, flooring and composites formulations where their low viscosity facilitates processing. A large majority of LERs are used as starting materials to produce higher molecular weight (MW) solid epoxy resins (SER) and brominated epoxy resins, and to convert to epoxy derivatives such as epoxy vinyl esters, epoxy acrylates, etc. The bisphenol A derived epoxy resins are most frequently cured with anhydrides, aliphatic amines, phenolics, or polyamides, depending on desired end properties. Some of the outstanding properties are superior electrical properties, chemical resistance, heat resistance, and adhesion. Cured LERs give tight cross-linked networks having good strength and hardness but have limited flexibility and toughness.

Epichlorohydrin, or 3-chloro-1,2-epoxy propane (*bp* 115 °C), is more commonly prepared from propylene by chlorination to allyl chloride, followed by treatment with hypochlorous acid. This yields glycerol dichlorohydrin, which is dehydrochlorinated by sodium hydroxide or calcium hydroxide [9].

In industrial practices, epichlorohydrin is produced by direct chlorohydroxylation of allyl chloride in chlorine and water [10–13]. Alternatively, a new epichlorohydrin process has been developed and commercialized by Showa Denko [14] in Japan in 1985. It involves the chlorination of allyl alcohol as the precursor and is claimed to be more efficient in chlorine usage.

Bisphenol A (*mp* 153 °C), or 2,2-bis(*p*-hydroxyphenyl)propane, is prepared from 2 moles of phenol and 1 mole of acetone [15, 16]

Bisphenol A based liquid epoxy resins are prepared in a two-step reaction sequence from epichlorohydrin and bisphenol A. The first step is the base-catalyzed *coupling* of bisphenol A and epichlorohydrin to yield a chlorohydrin.

Bases that may be used to catalyze this step include sodium hydroxide, lithium salts, and quaternary ammonium salts. Dehydrohalogenation of the chlorohydrin intermediate with a stoichiometric amount of base affords the glycidyl ether. Manufacturing processes can be divided into two broad categories according to the type of catalyst used to couple epichlorohydrin and bisphenol A [17, 18].

## 5.1. Caustic Coupling Process

In this process, caustic is used as a catalyst for the nucleophilic ring-opening (coupling reaction) of the epoxide group on the primary carbon atom of epichlorohydrin by the phenolic hydroxyl group and as a dehydrochlorinating agent for conversion of the chlorohydrin to the epoxide group:

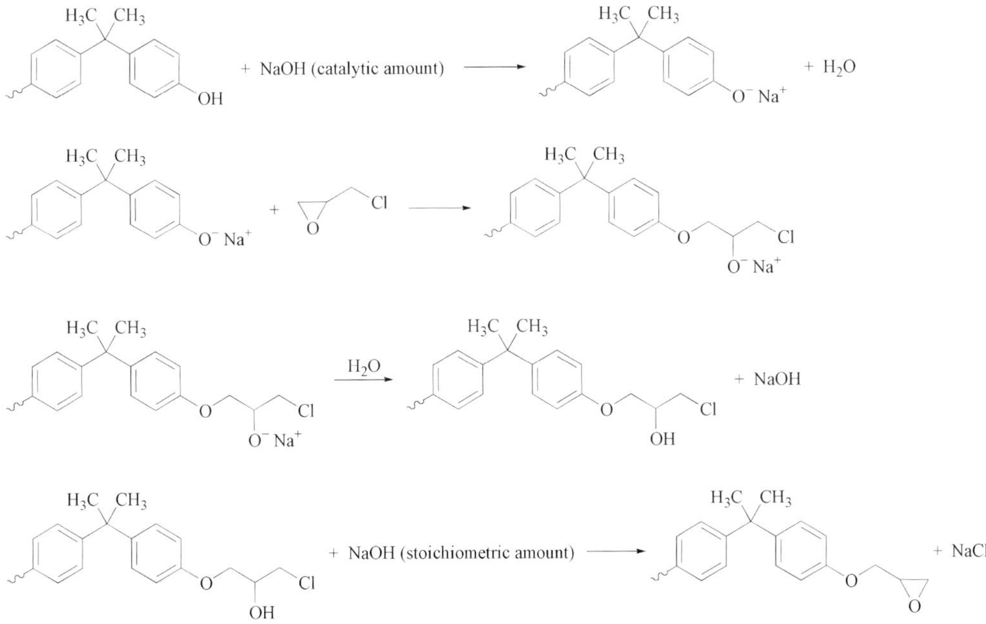

In caustic coupling processes, caustic (20–50 % sodium hydroxide in water) is slowly added to an agitated mixture of epichlorohydrin and bisphenol A. The highly exothermic coupling reaction proceeds during the initial stages. As the coupling reaction nears completion, dehydrochlorination becomes the predominant reaction. A high ratio (usually 10:1) of epichlorohydrin/bisphenol A is charged to the reactor to maximize the yield of monomeric ($n = 0$) DGEBA. At a 10:1 level of epichlorohydrin/bisphenol A, the $n = 0$ monomer comprises $>85\%$ of the reaction product mixture.

## 5.2. Phase-Transfer Catalyst Process

Alternatively, the coupling reaction and dehydrochlorination can be performed separately by using phase-transfer coupling catalysts, such as quaternary ammonium salts [19], which are not strong enough bases to promote dehydrochlorination. Once the coupling reaction is completed, caustic is added to carry out the dehydrochlorination step. Higher yields of the $n = 0$ monomer ($>90\%$) are readily available via this method.

Many variations of these two basic processes are described in process patents [20, 21], including the use of co-solvents and azeotropic removal of water to facilitate the reactions and to minimize undesirable byproducts such as insoluble polymers. The original batch methods have been modified to allow for continuous or semicontinuous production. New developments have been focused on improving manufacturing yield and resin purity.

The description of liquid DGEBA resins presented so far is oversimplified. In reality, *side reactions* result in the formation of low levels of impurities that both decrease the epoxide content from the theoretical amount of 2 per molecule and affect the resins properties, both before and after curing [22]. The five common side reactions are as follows:

1. *Hydrolysis of epoxy groups.* Unavoidable hydrolysis of the epoxide ring gives a small amount (0.1–5 %) of monohydrolyzed resin (MHR) or α-glycol. It has been reported that dispersability of pigments are enhanced and rates of epoxy resin curing with diamines can be dramatically increased by higher levels of MHR [23].

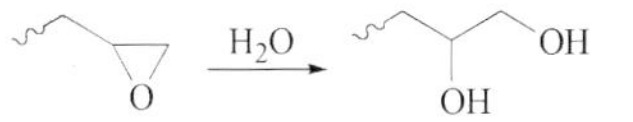

   α-Glycol formation

2. *Incomplete dehydrochlorination* results in residual saponifiable or hydrolyzable chloride:

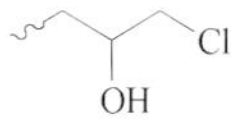

   Incomplete dehydrochlorination increases the level of hydrolyzable chloride in the resin, which affects its suitablity for applications requiring superior electrical properties. In addition, hydrolyzable chlorides can affect reactivity by neutralizing basic catalysts such as tertiary amines. Many formulators adjust their formulations according to resin hydrolyzable chloride content. Typical hydrolyzable chloride contents of LERs range from < 100 ppm for electronic grade resin to 200–1000 ppm for standard grade resins.

3. *Abnormal addition of epichlorohydrin,* i.e., abnormal phenoxide attack at the central carbon of epichlorohydrin results in an end group that is more difficult to dehydrochlorinate:

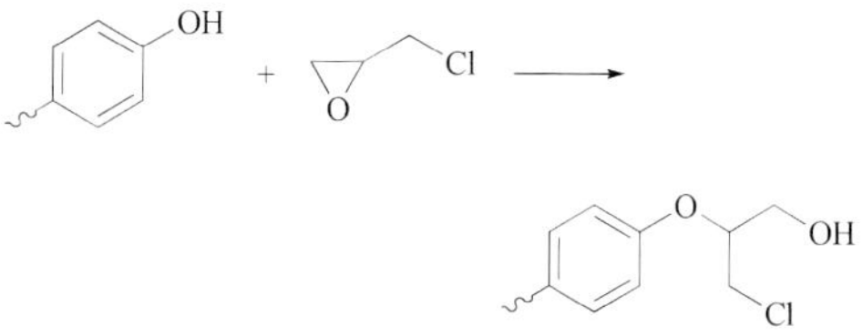

4. *Formation of bound chlorides* by reaction of epichlorohydrin with hydroxy groups in the polymer backbones:

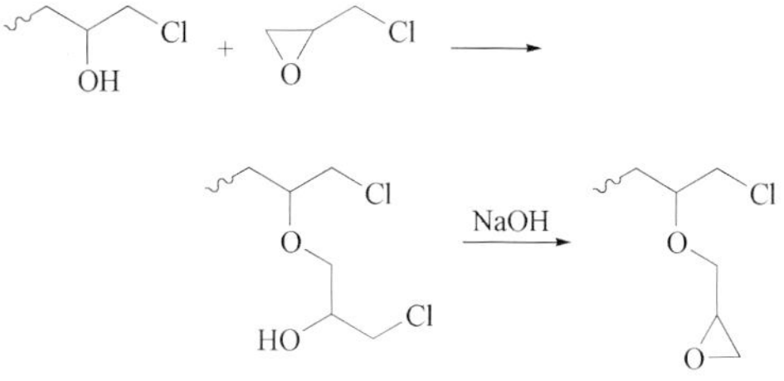

   The bound chloride is not readily saponified with metal hydroxide solutions and is analyzed as part of the total chloride of the resin. Typical total chlorides values are 1000–2000 ppm.

5. *Higher oligomer formation.* Reaction of a phenolic terminal group with another epoxy resin molecule instead of an epichlorohydrin molecule gives epoxy resins with broader oligomer distribution and increased viscosity ($n = 1$ and higher oligomers). Typical LERs contain 5–15 % of the higher oligomers, mostly $n = 1$ and $n = 2$ compounds.

Pure DGEBA is a solid melting at 43 °C. The unmodified commercial liquid resins are supercooled liquids with the potential for crystallization, depending on purity and storage conditions. This causes handling problems, particularly for ambient cure applications. Addition of certain reactive diluents and fillers can either accelerate or retard crystallization. Crystallization-resistant, modified resins are available. A crystallized resin can be restored to its liquid form by warming.

## 6. Solid Epoxy Resins Based on DGEBA

High molecular weight (MW) SERs based on DGEBA are characterized by a repeat unit containing a secondary hydroxyl group with degrees of polymerization, i.e., $n$ values ranging from 2 to about 35 in commercial resins; two terminal epoxy groups are theoretically present.

epoxy resins. Lower epichlorohydrin to bisphenol A ratios are used to promote formation of high MW resins. The term taffy is derived from the appearance of the advanced epoxy resin prior to its separation from water and precipitated salts.

The epoxy industry has adopted a common nomenclature to describe the SERs. They are called type "1," "2" up to type "10" resins, which correspond to the increased values of $n$, the degree of polymerization, EEW, MW, and viscosity. Examples of SERs are D.E.R. 661, 662, 664, 667, 669 resins from Dow Chemical, and Epon 1001 to 1009 series from Resolution. A comparison of some key properties of LERs and SERs is shown in Table 5.

SERs based on DGEBA are widely used in the coatings industry. The longer backbones give more distance between cross-links when cross-linked through the terminal epoxy groups, resulting in improved flexibility and toughness. Furthermore, the resins can also be cured through the multiple hydroxyl groups along the backbones using cross-linkers such as phenol–formaldehyde resoles or isocyanates to create different network structures and performance.

SERs are prepared by two processes: the *taffy* process and the *advancement* or *fusion* process. The first is directly from epichlorohydrin, bisphenol A, and a stoichiometric amount of NaOH. This process is very similar to the *caustic coupling* process used to prepare liquid epoxy resins.

In the taffy process, a calculated excess of epichlorohydrin governs the degree of polymerization. However, preparation of the higher molecular weight species is subject to practical limitations of handling and agitation of highly viscous materials. The effect of epichlorohydrin–bisphenol A (ECH–BPA) ratio for a series of solid resins is shown in Table 6.

In commercial practice, the *taffy* method is used to prepare lower MW solid resins, i.e., those with maximum EEW values of about 1000 (type "4"). Upon completion of the polymerization, the

Table 5. DGEBA-based epoxy resins

| Resin type | $n$ value[a] | EEW | Mettler softening point, °C | Molecular weight ($M_w$)[b] | Viscosity at 25 °C, mPa·s (= cP) |
|---|---|---|---|---|---|
| Low viscosity LER | <0.1 | 172–176 | | ∼350 | 4,000–6,000 |
| Medium viscosity LER | ∼0.1 | 176–185 | | ∼370 | 7,000–10,000 |
| Standard grade LER | ∼0.2 | 185–195 | | ∼380 | 11,000–16,000 |
| Type 1 SER | ∼2 | 450–560 | 70–85 | ∼1,500 | 160–250[c] |
| Type 4 SER | ∼5 | 800–950 | 95–110 | ∼3,000 | 450–600[c] |
| Type 7 SER | ∼15 | 1,600–2,500 | 120–140 | ∼10,000 | 1,500–3,000[c] |
| Type 9 SER | ∼25 | 2,500–4,000 | 145–160 | ∼15,000 | 3,500–10,000[c] |
| Type 10 SER | ∼35 | 4,000–6,000 | 150–180 | ∼20,000 | 10,000–40,000[c] |
| Phenoxy resin | ∼100 | >20,000 | >200 | >40,000 | |

[a] $n$ value is the number-average degree of polymerization which approximates the repeating units and the hydroxyl functionality of the resin.
[b] Molecular weight is weight average ($M_w$) measured by gel-permeation chromatography (GPC) using polystyrene standard.
[c] Viscosity of SERs is determined by kinematic method using 40% solids in diethylene glycol monobutyl ether solution.

Table 6. Effect of epichlorohydrin–bisphenol a ratio on resin properties of taffy SERs

| Mole ratio ECH/BPA | EEW | Softening point, °C |
|---|---|---|
| 1.57:1.0 | 450–525 | 65–75 |
| 1.22:1.0 | 870–1025 | 95–105 |
| 1.15:1.0 | 1650–2050 | 125–135 |
| 1.11:1.0 | 2400–4000 | 145–155 |

mixture consists of an alkaline brine solution and a water–resin emulsion. The product is recovered by separating the phases, washing the taffy resin with water, and removing the water under vacuum. One disadvantage of the taffy process is the formation of insoluble polymers, which create handling and disposal problems. Only a few epoxy producers currently manufacture SERs using the taffy process. A detailed description of a taffy procedure follows [24].

A mixture of bisphenol A (228 parts by weight) and 10 % aqueous sodium hydroxide solution (75 parts by weight) is introduced into a reactor equipped with a powerful agitator. The mixture is heated to ca 45 °C and epichlorohydrin (145 parts by weight) is added rapidly with agitation, giving off heat. The temperature is allowed to rise to 95 °C, where it is maintained about 80 min. for completion of the reaction. Agitation is stopped, and the mixture separates into two layers. The heavier aqueous layer is drawn off and the molten, taffy-like product is washed with hot water until the wash water is neutral. The taffy-like product is dried at 130 °C, giving a solid resin with a softening point of 70 °C and an EEW of ca. 500.

Alternatively, epichlorohydrin and water are removed by distillation at temperatures up to 180 °C under vacuum. The crude resin/salt mixture is then dissolved in a secondary solvent to facilitate water washing and salt removal. The secondary solvent is then removed via vacuum distillation to obtain the taffy–resin product.

Resins produced by this process exhibit relatively high α-glycol values, i.e., ca 0.5 eq/kg, attributable to hydrolysis of epoxy groups in the aqueous phase. Although detracting from epoxide functionality, such groups act as accelerators for amine curing. Resins produced by the taffy process exhibit $n$ values of 0, 1, 2, 3, etc., whereas resins produced by the *advancement* process (described below) exhibit mostly even-numbered $n$ values because a difunctional phenol is added to a diglycidyl ether of a difunctional phenol.

An alternative method is the chain-extension reaction of liquid epoxy resin (crude DGEBA) with bisphenol A, often referred to as the advancement or fusion process [25] which requires an advancement catalyst:

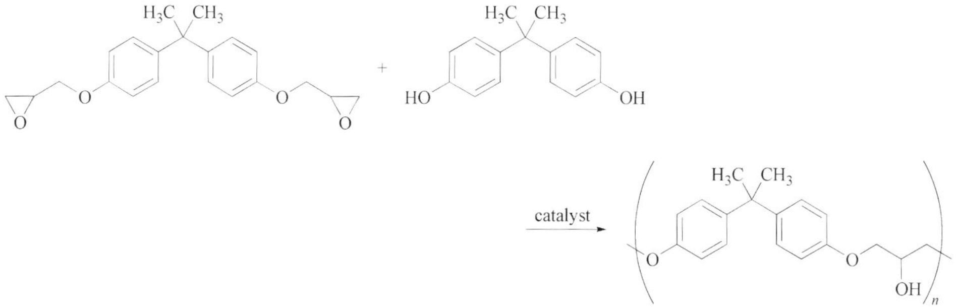

The advancement process is more widely used in commercial practice. Isolation of the polymerized product is simpler, since removal of copious amounts of NaCl is unnecessary. The reaction can be carried out with or without solvents. Solution advancement is widely practiced by coatings producers to facilitate handling of the high MW, high viscosity epoxy resins used in many coating formulations. The degree of polymerization is dictated by the ratio of LER to bisphenol A; an excess of the former provides epoxy terminal groups. The actual MW attained depends on the purity of the starting materials, the type of solvents used, and the catalyst. Reactive monofunctional groups can be used as chain terminators to control MW and viscosity build.

The following formula can be used to calculate the relative amount of bisphenol A that must be reacted with epoxy resin to give an advanced epoxy resin of predetermined EEW:

$$\text{Bis A} = \frac{\text{EEW}_i^{-1} - \text{EEW}_f^{-1}}{\text{EEW}_i^{-1} + \text{PEW}^{-1}}$$

where Bis A is the mass fraction of bisphenol A in the mixture prior to advancement, $\text{EEW}_i$ is the EEW of the epoxy resin that is to be advanced, $\text{EEW}_f$ is the EEW of the advanced epoxy resin, and PEW is the phenol equivalent mass of the bisphenol, which is 114 g per equivalent for bisphenol A.

In a typical advancement process, bisphenol A and a liquid DGEBA resin (175–185 EEW) are heated to ca 150–190 °C in the presence of a catalyst and reacted (i.e., advanced) to form a high MW resin. The oligomerization is exothermic and proceeds rapidly to near completion. The exotherm temperatures are dependent upon the targeted EEW and the reaction mass. In the cases of higher MW resins such as type "7" and higher, exotherm temperatures of > 200 °C are routinely encountered.

Advancement reaction catalysts facilitate the rapid preparation of medium and high MW linear resins and control prominent side reactions inherent in epoxy resin preparations, e.g., chain branching due to addition of the secondary alcohol group generated in the chain-lengthening process to the epoxy group [26, 27]. Nuclear Magnetic Resonance (NMR) spectroscopy can be used to determine the extent of branching [28].

Conventional advancement catalysts include basic inorganic reagents, e.g., NaOH, KOH, $Na_2CO_3$, or LiOH, and amines and quaternary ammonium salts. One mechanism proposed for the basic catalysts involves proton abstraction of the phenolic compound as the initiation step:

The phenoxide ion then attacks the epoxy ring, generating an alkoxide, which immediately abstracts a proton from another phenolic OH group. This is called the propagation step. Regeneration of the phenoxide ion repeats the cycle. The potential for side reactions increases after the phenolic OH groups have been consumed, particularly in melt (i.e., fusion) polymerization reactions.

One key disadvantage of catalysts based on inorganic bases and salts is the increased ionic impurities added to the resin, which is not desirable in certain applications.

Imidazoles, substituted imidazoles, and triethanolamine have been patented as advancement catalysts [29]. However, most of the inorganic bases, salts, and amines produce resins with broad MW distribution and viscosity instability. This is due to poor catalyst selectivity and the continuing activity of the catalyst after completion of the advancement reaction.

Alternatively, a broad class of catalysts derived from aryl or alkyl phosphonium compounds were developed. Extensive patent literature claims a high order of selectivity [30, 31]. Selections of the phosphonium cation and counter ion have been shown to affect initiation rate, catalyst selectivity, catalyst lifetime, and, consequently, product quality and consistency. Some of the phosphonium salts are deactivated at high temperatures by the reaction exotherm, and are claimed to give better resin stability in terms of viscosity, EEW, and MW during the subsequent finishing steps [32–35].

Few mechanistic studies have been published on the selectivity of phosphonium compounds, but one publication describes the role of triphenylphosphine in advancement catalysis [36]. Nucleophilic attack by triphenylphosphine opens the epoxy ring, producing a betaine:

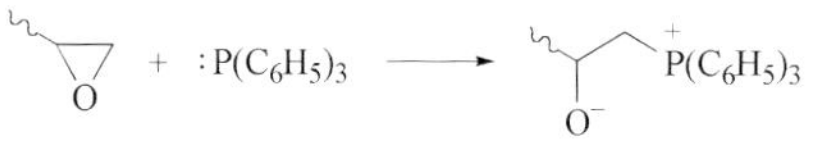

Proton abstraction from bisphenol A yields the phenoxide anion, forming a phosphonium salt. The phenoxide reacts with the electrophilic carbon attached to the positive phosphorus regenerating the catalyst:

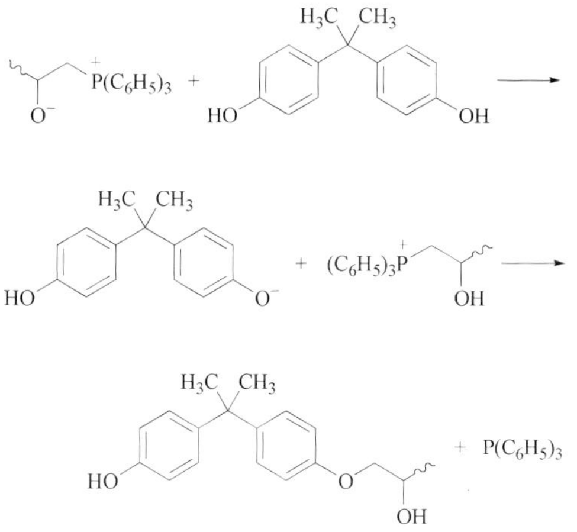

When the bisphenol A is consumed, the betaine decomposes into a terminal olefin and triphenylphosphine oxide:

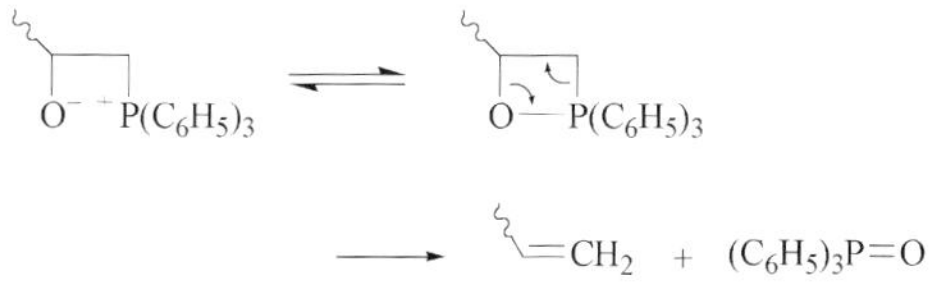

Branched epoxies [37] are prepared by advancing LER with bisphenol A in the presence of epoxy novolac resins. Such compositions exhibit enhanced thermal and solvent resistance.

SERs are available commercially in solid form or in solution. MW distributions of SERs have been examined by means of theoretical models and compared with experimental results [38]. Taffy-processed resins were compared with advancement-processed resins by gel-permeation chromatography (GPC) and high performance liquid chromatography (HPLC) [39] in conjunction with statistical calculations. The major differences are in the higher $\alpha$-glycol content and the repeating units of oligomers. Resin viscosity and softening points are also lower with taffy resins. In addition, certain formulations based on taffy resins exhibit different behavior in pigment loading, formulation rheology, reactivity, and mechanical properties compared to those based on advancement resins.

## 6.1. SER Continuous Advancement Process

The recent literature review indicates efforts to develop continuous advancement processes to produce SERs. Companies seek to improve process efficiencies and product quality. One of the major deficiencies of the traditional batch advancement process is the long reaction time, resulting in EEW and viscosity drift, variable product quality, and gel formation. In addition, it is difficult to batch process higher MW, higher viscosity SERs such as types "9" and "10" resins. Shell patented several versions of the continuous resin advancement processes using modified reactor designs [40]. Dow Chemical received patents covering the uses of reactive extrusion [41] (REX) to produce SERs and other epoxy thermoplastic resins [42]. The latter process makes use of a self-wiping twin-screw extruder. LER, bisphenol A, and catalyst are fed directly to the extruder to complete the resin advancement reaction in several minutes compared to the traditional several hours in a batch process. The process is claimed to be very efficient and is particularly suitable for the production of high molecular weight SERs, phenoxy resins, and epoxy thermoplastic resins. Compared to the traditional taffy processes used to produce phenoxy resins, the chemistry is salt-free, and the resins made via the REX process are fully converted in a matter of minutes, significantly reducing manufacturing costs. Additional benefits include reduced lot-to-lot variations in MW distribution, the flexibility to make small lots of varying molecular weights with minimal waste, and the ability to make custom resins with a variety of additives such as pigments and flow modifiers.

## 6.2. Phenoxy Resins

Phenoxy resins are thermoplastic polymers derived from bisphenol A and epichlorohydrin. Their weight-average molecular weights ($M_w$) are higher (i.e., > 30,000) than those of conventional SERs (i.e., 25,000 maximum). They lack terminal epoxides but have the same repeat unit as SERs and are classified as polyols or polyhydroxy ethers:

Phenoxy resins were originally developed and produced by Union Carbide (trade names PKHH, PKHC, PKHJ) using the taffy process. The process involves reaction of high purity bisphenol A with epichlorohydrin in a 1:1 mole ratio. Alternatively, phenoxy resins can be produced by the fusion process which uses high purity LER and bisphenol A in a 1:1 mole ratio. High purity monomers and high conversions are both needed to produce high MW phenoxy resins. The effects of monomer purity on phenoxy resin production are significant: monofunctional components limit MW, and functionality > 2 causes excess branching and increased polydispersity. Solution polymerization may be employed to achieve the MW and processability needed [43]. This, however, adds to the high costs of manufacturing of phenoxy resins, limiting their commercial applications.

The phenoxies are offered as solids, solutions, and waterborne dispersions. The majority of phenoxy resins are used as thermoplastics, but some are used as additives in thermoset formulations. Their high MW provide improved flexibility and abrasion resistance. Their primary uses are in automotive zinc-rich primers, metal can/drum coatings, magnet wire enamels, and magnetic tape coatings. However, the zinc-rich primers are being phased out in favor of galvanized steel by the automotive industry. Smaller volumes of phenoxy resins are used as flexibility or rheology modifiers in composites and electrical laminate applications, and as composite honeycomb impregnating resins. A new, emerging application is fiber sizing, which utilizes waterborne phenoxies. Literature references indicate their potential uses as compatibilizers for thermoplastic resins such as polyesters, nylons, and polycarbonates because of their high hydroxyl contents.

Current producers of phenoxy resins include the Phenoxy Specialties division of InChem

## 6.3. Epoxy-Based Thermoplastics

Some of the new epoxy products developed in the past few years are the thermoplastic resins based on epoxy monomers. Polyhydroxy amino ether [44, 45] (PHAE) was commercialized by Dow Chemical in 1999 and trade named BLOX (Dow Chemical Co.). It is produced by the reaction of DGEBA with monoethanol amine using the reactive extrusion process. The high cohesive energy density of the resin gives it excellent gas-barrier properties against oxygen and carbon dioxide. It also possesses excellent adhesion to many substrates, optical clarity, excellent melt strength, and good mechanical properties. The product has been evaluated as a barrier resin for beer and beverage plastic bottles, as thermoplastic powder coatings, and as a toughener for starch-based foam [46]. Another epoxy thermoplastic resin under development by Dow is the polyhydroxy ester ether (PHEE). It is a reaction product of DGEBA with difunctional acids. The ester linkage makes it suitable for biodegradable applications [47].

## 7. Halogenated Epoxy Resins

A number of halogenated epoxy resins have been developed and commercialized to meet specific application requirements. Chlorinated and brominated epoxies were evaluated for flame retardancy properties. The brominated epoxy resins were found to have the best combination of cost/performance and were commercialized by Dow Chemical in the late 1960s.

### 7.1. Brominated Bisphenol A Based Epoxy Resins

Many applications of epoxy resins require the system to be ignition-resistant, e.g., electrical laminates for PCBs and certain structural composites. A common method of imparting this ignition resistance is the incorporation of tetrabromobisphenol A (TBBA), 2,2-bis(3,5-dibromophenyl)propane, or the diglycidyl ether of TBBA, 2,2-bis[3,5-dibromo-4-(2,3-epoxypropoxy)phenyl]propane, into the resin formulation. The diglycidyl ether of TBBA is produced via conventional liquid epoxy resin processes. Higher MW resins can be produced by advancing LERs or diglycidyl ether of TBBA with TBBA. The lower cost, advanced brominated epoxies based on LERs and TBBA containing ca. 20 wt.% Br are extensively employed in the PCB industry. The diglycidyl ether of TBBA (ca. 50 wt.% Br) is used for critical electrical/electronic encapsulation where high flame retardancy is required. Brominated epoxies are also used to produce epoxy vinyl esters for structural applications. Very high MW versions of brominated epoxies are used as flame-retardant additives to engineering thermoplastics used in computer housings.

## 7.2. Fluorinated Epoxy Resins

In order to meet increased requirements of the PCB industry for higher glass-transition temperature ($T_g$), higher thermal decomposition temperature ($T_d$), and lower dielectric constant ($D_k$) products, a number of new epoxy resins have been developed [48, 49].

Fluorinated epoxy resins have been researched for a number of years for high performance end-use applications [50]. Fluorinated epoxies are highly resistant to chemical and physical abuse and should prove useful in high performance applications, including specialty coatings and composites, where their high cost may be offset by their special properties and long service life. The following fluorinated diglycidyl ether, 5-heptafluoropropyl-1,3-bis[2-(2,3-epoxypropoxy) hexafluoro-2-propyl] benzene, illustrates an example of fluoroepoxy resins [51] under development.

This resin is a viscous, colorless liquid (bp 118 °C at 20 Pa · s) that contains 52 wt. % fluorine. It has a low surface tension, which makes it a superior wetting agent for glass fibers. The reactivity of this resin with amine or anhydride curing agents is comparable to epoxy resins based on bisphenol A and results in a thermoset that has a low affinity for water and excellent chemical resistance. Another fluorinated epoxy resin derived from hexafluorobisphenol A was introduced to the marketplace aiming at the anticorrosion coatings market for industrial vessels and pipes. The key disadvantages of fluorinated epoxies are their relatively high costs and low $T_g$, which limit their commercialization [52].

## 8. Multifunctional Epoxy Resins

The multifunctionality of these resins provides higher cross-linking density, leading to improved thermal and chemical resistance properties over bisphenol A epoxies.

## 8.1. Epoxy Novolac Resins

Epoxy novolacs are multifunctional epoxies based on phenolic formaldehyde novolacs. Both epoxy phenol novolac resins (EPN) and epoxy cresol novolac resins (ECN) have attained commercial importance [53]. The former is made by epoxidation of the phenol–formaldehyde condensates (novolacs) obtained from acid-catalyzed condensation of phenol and formaldehyde (see → Phenolic Resins). This produces random *ortho*- and *para*-methylene bridges.

EPN, R = H
ECN, R = CH$_3$

An increase in the molecular weight of the novolac increases the functionality of the resin. This is accomplished by changing the phenol or cresol to formaldehyde ratio. Epoxidation with an excess of epichlorohydrin minimizes the reaction of the phenolic OH groups with epoxidized phenolic groups and prevents branching. The epoxidation is similar to the procedure described for bisphenol A. EPN resins range from a high viscosity liquid of $n = 0.2$ to a solid of $n > 3$. The epoxy functionality is between 2.2 and 3.8. Properties of epoxy phenol novolacs are given in Table 7. When cured with aromatic amines such as methylenedianiline, the heat

Table 7. Typical properties of epoxy phenol novolacs

| Property | D.E.N. 431,[a] EPN 1139[b] | D.E.N. 438,[a] EPN 1138[b] | D.E.N. 439[a] |
|---|---|---|---|
| $n$ | 0.2 | 1.6 | 1.8 |
| EEW[c] | 175 | 178 | 200 |
| Viscosity, mPa·s (= cP) | 1,400[d] | 35,000[d] | 3,000[e] |
| Softening point[f] | | | 53 |
| Color, Gardner | 1 | 2 | 2 |

[a] The Dow Chemical Co.
[b] Huntsman.
[c] Epoxide equivalent weight.
[d] Temperature of measurement = 52 °C.
[e] Temperature of measurement = 100 °C.
[f] Durran's mercury method.

distortion temperatures (HDT) of EPN-based thermosets range from 150 °C to 200 °C, depending on cure and postcure schedules

Curing agents that give the optimum in elevated temperature properties for epoxy novolacs are those with good high temperature performance, such as aromatic amines, catalytic curing agents, phenolics, and some anhydrides. When cured with polyamide or aliphatic polyamines and their adducts, epoxy novolacs show improvement over bisphenol A epoxies, but the critical performance of each cure is limited by the performance of the curing agent.

The improved thermal stability of EPN-based thermosets is useful in elevated temperature services, such as aerospace composites. Filament-wound pipe and storage tanks, liners for pumps and other chemical process equipment, and corrosion-resistant coatings are typical applications which take advantage of the chemical resistant properties of EPN resins. However, the high cross-link density of EPN-based thermosets can result in increased brittleness and reduced toughness.

### 8.1.1. Bisphenol F Epoxy Resin

The lowest MW member of the phenol novolacs is bisphenol F, which is prepared with a large excess of phenol to formaldehyde; a mixture of $o,o'$, $o,p'$, and $p,p'$ isomers is obtained:

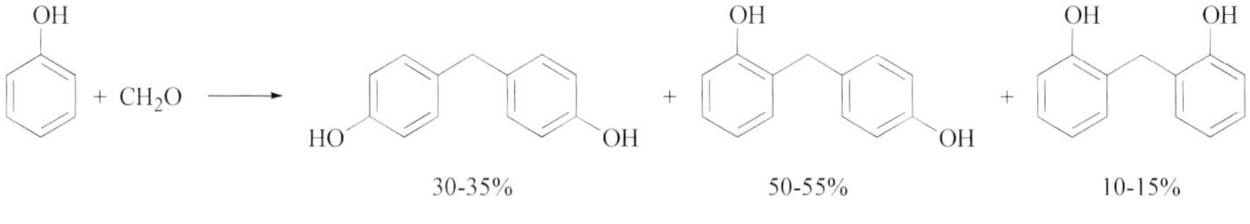

Epoxidation yields a liquid bisphenol F epoxy resin with a viscosity of 4000–6000 mPa·s (= cP), an EEW of 165, and $n \cong 0.15$.

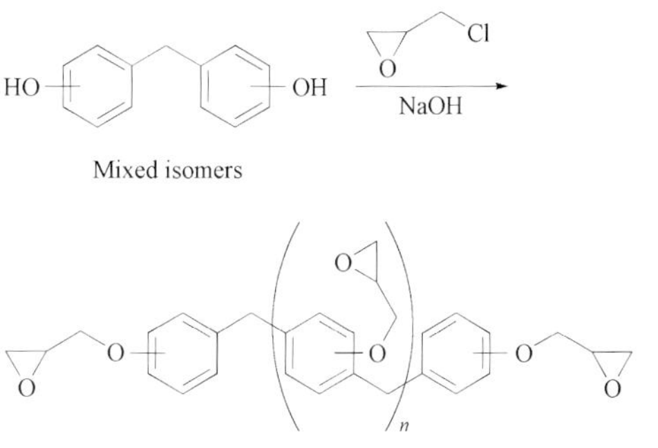

Bisphenol F resin
(3)

This unmodified, low viscosity liquid resin exhibits slightly higher functionality than unmodified bisphenol A liquid resins. Crystallization, often a problem with liquid bisphenol A resins, is reduced with bisphenol F resin. Consequently, noncrystallizing LERs which are blends of DGEBA and bisphenol F epoxy are available. Epoxy resins based on bisphenol F are used primarily as functional diluents in applications requiring a low viscosity, high performance resin system (e.g., solvent-free coatings). Higher filler levels and faster bubble release are possible because of the low viscosity. The higher epoxy content and functionality of bisphenol F epoxy resins provide improved chemical resistance compared to conventional bisphenol A epoxies. Bisphenol F epoxy resins are used in high solids, high build systems such as tank and pipe linings, industrial floors, road and bridge deck toppings, structural adhesives, grouts, coatings, and electrical varnishes.

### 8.1.2. Cresol Epoxy Novolacs

The o-cresol novolac epoxy resins (ECN) are analogous to phenol novolac resins. ECNs exhibit better formulated stability and lower moisture adsorption than EPNs, but have higher costs. ECN resins are widely used as base components in high performance electronic

Table 8. Typical properties of epoxy cresol novolac resins[a]

| Property | ECN 1235 | ECN 1273 | ECN 1280 | ECN 1299 |
|---|---|---|---|---|
| Molecular weight | 540 | 1080 | 1170 | 1270 |
| EEW[b] | 200 | 225 | 229 | 235 |
| Softening point, °C | 35 | 73 | 80 | 99 |
| Epoxide functionality | 2.7 | 4.8 | 5.1 | 5.4 |

[a] Huntsman.
[b] Epoxide equivalent weight.

(semiconductors) and structural molding compounds, high temperature adhesives, castings and laminating systems, tooling, and powder coatings. Increasing demands by the semiconductor industry has led to significant advances in ECN resin manufacturing technologies to reduce impurities, mainly the ionic content, hydrolyzable chlorides, and total chlorides. The use of polar, aprotic solvents, such as dimethyl sulfoxide (DMSO), as a co-solvent to facilitate chloride reduction has been patented [54]. Typical high purity ECN resins contain <1000 ppm total chlorides and <50 ppm hydrolyzable chlorides.

The melt viscosity of these resins, which are solids at room temperature, decreases sharply with increasing temperature (Table 8). This affords the formulator an excellent tool for controlling the flow of molding compounds and facilitating the incorporation of ECN resins into other epoxies, e.g., for powder coatings. While Ciba-Geigy was the first producer of ECN resins, many Japanese companies (Nippon Kayaku, Sumitomo Chemical, DIC, and Tohto Kasei) supply the majority of high purity ECN resins for the semiconductor industry today. Other suppliers are based in Korea and Taiwan.

### 8.1.3. Glycidyl Ethers of Hydrocarbon Epoxy Novolacs

In response to the increased performance demands of the semiconductor industry, hydrocarbon epoxy novolacs (HENs) were developed by Dow Chemical in the 1980s. HENs exhibit a much lower affinity for water compared to cresol or phenol epoxy novolacs. This translates directly into increased electrical property retention, which is important in the reliability of an electronic device encapsulated in the resin. An epoxy resin that is typical of this class is based on the alkylation product of phenol and dicyclopentadiene [55] ($n = 0.1$), 2,5-bis[(2,3-epoxypropoxy) phenyl]octahydro-4,7-methano-5$H$-indene (272 EEW; softening point 85 °C; η at 150 °C 0.4 Pa·s). The product is available from Huntsman as TACTIX* 556. Similar products based on $o$-cresol are commercialized in Japan by DIC (EPICLON HP-7200L).

### 8.1.4. Bisphenol A Epoxy Novolacs

Bisphenol A novolacs are produced by reacting bisphenol A and formaldehyde with acid catalysts. Epoxidation of the bisphenol A novolacs gives bisphenol A epoxy novolac (BPAN) with improved thermal properties such as $T_g$, $T_d$ of the epoxy-based electrical laminates.

Bisphenol A epoxy novolac

### 8.2. Other Polynuclear Phenol Glycidyl Ether Derived Resins

In addition to the epoxy novolacs, there are other epoxy resins derived from phenol–

aldehyde condensation products. New applications that require increased performance from the epoxy resin, particularly in the electronics, aerospace, and military industries, have made these types of resins more attractive despite their relatively high cost.

### 8.2.1. Glycidyl Ether of Tetrakis(4-hydroxyphenyl)ethane

One of the first polyfunctional resins to be marketed (by Shell) is based on 1,1,2,2-tetrakis[4-(2,3-epoxypropoxy)phenyl]ethane [56]. It is used primarily as additives in standard epoxy resin formulations for electrical laminates, molding compounds, and adhesives in which increased heat distortion temperature and improved chemical resistance are desired. Tetrakis(4-hydroxyphenyl)ethane is prepared by reaction of glyoxal with phenol in the presence of HCl. The tetraglycidyl ether (*mp* ca. 80 °C, and an EEW of 185–208) possesses a theoretical epoxide functionality of 4.

The commercial products Araldite 0163 (Huntsman) and Epon 1031 (Resolution) are tan-colored solids. They are widely used in high temperature resistance electrical laminates for high density PCBs and military applications. Their costs are typically higher than those of phenol and cresol epoxy novolacs.

### 8.2.2. Trisphenol Epoxy Novolacs

In the 1980s, new trifunctional epoxy resins based on tris[4-(2,3-epoxypropoxy)phenyl] methane isomers were introduced by Dow Chemical to help close the performance gap between phenol and cresol epoxy novolacs and high performance engineering thermoplastics [57]. These products were later sold to Ciba-Geigy and continued to be marketed under the TACTIX* 740 and XD 9053 trade names by Huntsman.

The resins are prepared via acid-catalyzed condensation of phenol and a hydroxybenzaldehyde, e.g., salicylaldehyde, to afford the trifunctional phenol, which is epoxidized with epichlorohydrin. These resins range from semisolids (162 EEW; Durran softening point 55 °C; η at 60 °C 11.5 Pa·s, at 150 °C 0.055 Pa·s) to nonsintering solids (220 EEW; Durran softening point 85 °C; η at 150 °C 0.45 Pa·s).

The semisolid resins are used in advanced composites and adhesives where toughness, hot-wet strength, and resistance to high temperature oxidation are required. Their purity, formulated stability, fast reactivity, and retention of electrical properties over a broad temperature range make the solid resins suitable for use in the semiconductor molding powders industry. The trisphenol-based epoxies command significant high prices ($28–48/kg), limiting their uses.

## 8.3. Aromatic Glycidyl Amine Resins

Among the multifunctional epoxy resins containing an aromatic amine backbone, only a few have attained commercial significance. Their higher costs limit their uses to critical applications where their costs are justified. Glycidyl amines contain internal tertiary amines in the resin backbone, hence their high reactivity. Epoxy resins with such built-in curing catalysts are less thermally stable than nitrogen-free multifunctional epoxy resins.

## 8.3.1. Triglycidyl Ether of p-Aminophenol

The triglycidyl derivative of p-aminophenol was originally developed by Union Carbide [58] and is currently marketed by Huntsman under the designation MY 0500 and 0510. Epoxidation of p-aminophenol is carried out with a large excess of epichlorohydrin under carefully controlled conditions, since the triglycidylated resin exhibits limited thermal stability and polymerizes vigorously under the influence of its tertiary amine moiety.

Triglycidyl-p-aminophenol

The resin exhibits a low viscosity, 2500–5000 mPa·s (= cP) at 25 °C, and an EEW of 105–114; a molecularly distilled version (0510) has a viscosity of 550–850 mPa·s (= cP) at 25 °C and an EEW of 95–107. It is considerably more reactive toward amines than standard bisphenol A resins. The trifunctional resin permits curing at low temperatures, i.e., 70 °C, and rapidly develops excellent elevated-temperature properties. Used as additives to increase cure speed, heat resistance, and $T_g$ of bisphenol A epoxy resins, it has utility in such diverse applications as high temperatures adhesives, tooling compounds, and laminating systems.

## 8.3.2. Tetraglycidyl Methylenedianiline (MDA)

These resins are used as binders in graphite-reinforced composites and are the binders of choice for many military applications. Epoxidation of MDA is carried out with stoichiometric excess of epichlorohydrin and under carefully controlled conditions to avoid rapid polymerization side reactions.

The tetrafunctional glycidylated MDA resins range in viscosity from 5000 to 25,000 mPa·s (= cP) at 50 °C and have an EEW of 117–133; they are commercially available as Araldite MY 720 (Huntsman) and Epiclon 430 (DIC). When used in combination with the curing agent 4,4′-diaminodiphenylsulfone (DADS), it is the first system to meet the performance requirements set by the aerospace industry and is the standard against which other resin systems are judged [59]. Because of its outstanding properties, this resin is often used as the primary resin in high heat resistance formulations for military applications, despite its high costs (~ $22/kg). Among its attributes are excellent mechanical strength, high $T_g$, good chemical resistance, and radiation stability.

N,N,N′,N′-Tetraglycidyl-4,4′-diaminodiphenylmethane

Another commercially important aromatic glycidyl amine resin is triglycidyl isocyanurate (TGIC), which is discussed in the "Weatherable Epoxy Resins" section.

## 9. Specialty Epoxy Resins

### 9.1. Crystalline Epoxy Resins Development

A number of new epoxy resins used in epoxy molding compounds (EMC) have been developed by Japanese resin producers in response to the increased performance requirements of the semiconductor industry. Most notable are the commercialization of crystalline epoxies based on biphenol by Yuka-Shell [60]:

Diglycidyl ether of tetramethyl biphenol

The very low viscosity of these crystalline, solid epoxies when molten allows very high filler loading (up to 90 wt. %) for molding compounds. The high filler loading reduces the coefficient of thermal expansion (CET) and helps manage thermal shock and moisture and crack resistance of molding compounds used in new, demanding semiconductor manufacturing processes such as Surface Mount Technology (SMT). It should be noted that cured thermosets derived from these crystalline resins do not retain crystallinity.

Recently, a number of capacity expansions were announced for biphenol epoxies (sold as YX-4000 resin by Japan Epoxy Resins Corporation, formerly Yuka-Shell). DIC has developed dihydroxy naphthalene based epoxies [61] as the next generation product for this high performance market. Prices for crystalline epoxies are generally high ($22–26/kg), limiting their uses to high end applications.

Diglycidyl ether of 1,5-dihydroxy naphthalene

Tetraglycidyl ether based on dihydroxy naphthalene

Dow Chemical developed liquid crystalline polymers (LCP) based on diglycidyl ether of 4-4′-dihydroxy-α-methylstilbene in the 1980s [62, 63]. Liquid crystal thermoplastics and thermosets based on this novel chemistry showed excellent combinations of thermal, mechanical, and chemical properties, unachievable with traditional epoxies. However, commercialization of these products has not materialized.

Diglycidyl ether of 4,4'-dihydroxy-α-methylstilbene

## 9.2. Weatherable Epoxy Resins

One of the major deficiencies of the aromatic epoxies is their poor weatherability, attributable to the aromatic ether segment of the backbone, which is highly susceptible to photoinitiated free-radical degradation. The aromatic ether of bisphenol A absorbs UV lights up to about 310 nm and undergoes photocleavage directly. This in turn produces free radicals that lead to oxidative degradation of bisphenol A epoxies, resulting in chalking. Numerous efforts have been devoted to remedy this issue, resulting in a number of new weatherable epoxy products. However, their commercial success has been limited, primarily because of higher resin costs and the fact that end users can topcoat epoxy primers with weatherable coatings based on other chemistries such as polyesters, polyurethanes, or acrylics. The following epoxy products when formulated with appropriate reactants can provide certain outdoor weatherability.

### 9.2.1. Hydrogenated DGEBA

In 1976, Shell Chemical Co. introduced epoxy resins based on the diglycidyl ether of hydrogenated bisphenol A, 2,2-bis[4-(2,3-epoxypropoxy)cyclohexyl]propane (232–238 EEW; η at 25 °C 2–2.5 Pa · s).

These resins resist yellowing and chalking because of their aliphatic structure. Epoxy resins based on hydrogenated bisphenol A are made by the epoxidation of the saturated diol, 2,2-bis(4-hydroxycyclohexyl)propane or 2,2-bis(4-hydroxycyclohexyl)propane with epichlorohydrin, or by the hydrogenation of a low molecular weight DGEBA resin [64]. Commercially available products include an Epalloy 5000 resin from CVC. One disadvantage is their much higher costs, and consequently, the products have not found broad acceptance in the industry. Furthermore, cross-linked networks based on hydrogenated bisphenol A epoxies lose some of the characteristic temperature and chemical resistances inherent with the bisphenol A backbone.

### 9.2.2. Heterocyclic Glycidyl Imides and Amides

In the 1960s, considerable work was devoted to preparing triglycidyl isocyanurate, 1,3,5-tris (2,3-epoxypropyl)-1,3,5-perhydrotriazine-2,4,6-trione [65]. The epoxidation of cyanuric acid with epichlorohydrin gives triglycidyl isocyanurate

(TGIC), marketed as PT 810 by Huntsman. It is a crystalline compound (*mp* 85–110 °C) with an EEW of ca. 108. Miscibility with organic compounds is limited. Because of its excellent weatherability, TGIC is widely used in outdoor powder coatings with polyesters [66], despite its higher cost (~ $12/kg).

Triglycidyl isocyanurate

### 9.2.3. Hydantoin-Based Epoxy Resins

These resins were commercialized by Ciba-Geigy. Hydantoins are prepared from carbon dioxide, ammonia, hydrogen cyanide, and ketones via the Bucherer reaction and can be epoxidized with epichlorohydrin [67]. Cured and uncured resin properties depend greatly on the nature of the substituents R and R'. The hydantoin derived from acetone furnishes a low viscosity, water-dispersable epoxy resin, 5,5-dimethyl-1,3-bis(2,3-epoxypropyl)-2,4-imidazolidinedione (R = R' = $CH_3$; 145 EEM; η at 25 °C 2.5 Pa · s). A nonsintering solid epoxy resin is obtained if R = R' = $-(CH_2)_5-$.

When cured with aromatic amines or anhydrides, these resins show high heat distortion temperatures and excellent adhesion and weatherability. A variety of applications are suggested for these new resins, particularly in applications in which a non-yellowing epoxy resin is desirable.

### 9.3. Elastomer-Modified Epoxies

Epoxy thermosets derive their thermal, chemical, and mechanical properties from the highly cross-linked networks. Consequently, toughness deficiency is an issue in certain applications. To improve the impact resistance and toughness of epoxy systems, elastomers such as BF Goodrich's CTBN rubbers (carboxyl-terminated butadiene nitrile) are often used as additives or pre-reacted with epoxy resins [68]. Most commonly used products are reaction adducts of liquid epoxy resins (DGEBA) with CTBN in concentrations ranging from 5 wt. % to 50 wt. %. They have been shown to give improved toughness, peel adhesion, and low temperature flexibility over unmodified epoxies. Primary applications are adhesives for aerospace and automotive and as additives to epoxy vinyl esters for structural composites. Formation of adducts of epoxy resins and CTBN is promoted by triphenylphosphine or alkyl phosphonium salts. Other elastomers used to modify epoxies include amine-terminated butadiene nitrile (ATBN), maleated polybutadiene and butadiene–styrene, epoxy-terminated urethane prepolymers, epoxy-terminated polysulfide, epoxy–acrylated urethane, and epoxidized polybutadiene.

## 10. Monofunctional Glycidyl Ethers and Aliphatic Glycidyl Ethers

A number of low MW monofunctional, difunctional, and mutifunctional epoxies are used as reactive diluents, viscosity reducers, flexiblizers, and adhesion promoters. Recent trends toward lower VOC, higher solids and 100 % solids epoxy formulations have resulted in increased utilization of these products. Most of these epoxies are derived from relatively compact hydroxyl-containing compounds, such as alcohols, glycols, phenols, and epichlorohydrin. Epoxidized vegetable oils, such as epoxidized linseed oils, are also used as reactive diluents. They are produced using a peroxidation process and are discussed in more detail in the cycloaliphatic epoxies and epoxidized vegetable oils section. Typically, these products have very low viscosity (1–70 cP at 20 °C) relative to LERs (11,000–16,000 cP). They are often used in the range of 7–20 wt. % to reduce viscosity of the diluted system to 1000 cP. However, the uses of reactive diluents, especially at high levels, often result in decreased chemical resistance and thermal and mechanical properties of the cured epoxies.

**Table 9.** Some common commercial glycidyl ether reactive diluents

| Name | Structure |
|---|---|
| n-Butyl glycidyl ether | H₃C–(CH₂)₃–O–CH₂–CH(–O–)CH₂ |
| C12–C14 Aliphatic glycidyl ether | C₁₂–C₁₄–O–CH₂–CH(–O–)CH₂ |
| o-Cresol glycidyl ether | (o-cresol)–O–CH₂–CH(–O–)CH₂ |
| Neopentylglycol diglycidyl ether | CH₂(–O–)CH–CH₂–O–CH₂–C(CH₃)₂–CH₂–O–CH₂–CH(–O–)CH₂ |
| Butanediol diglycidyl ether | CH₂(–O–)CH–CH₂–O–(CH₂)₄–O–CH₂–CH(–O–)CH₂ |

Important products include butyl glycidyl ether (BGE), alkyl glycidyl ethers of C8–C10 (Epoxide 7) and C12–C14 (Epoxide 8), o-cresol glycidyl ether (CGE), p-tert-butyl glycidyl ether, resorcinol diglycidyl ether (RDGE), and neopentyl glycol diglycidyl ether (Table 9). While BGE is the most efficient viscosity reducer and has been widely used in the industry for many years, it has been losing market share because of its volatility and obnoxiousness. Phenyl glycidyl ether (PGE) is no longer used by many formulators because of its toxicity. The industry trend is moving toward longer chain epoxies such as Epoxide 8 or neopentyl glycol diglycidyl ether.

Major suppliers of these products are Resolution, Air Products, Ciba Specialty Chemicals, Huntsman, CVC Specialty Chemicals, Pacific Epoxy Polymers, and Exxon.

An example of multifunctional aliphatic epoxies is the triglycidyl ether of propoxylated glycerine (Heloxy 84) from Resolution. A similar product is based on epoxidized castor oil (Heloxy 505). These products are used primarily as viscosity reducers while increasing functionality and cross-linking density of the cured systems.

Epoxy resins based on long-chain diols, such as the diglycidyl ether of polypropylene glycol [α,ω-bis(2,3-epoxypropoxy)poly(oxypropylene)] (305–335 EEW; η at 25 °C 0.055–0.10 Pa·s), are used as flexibilizing agents to increase a thermoset's elongation and impact resistance. Because of the low reactivity of the aliphatic diols toward epichlorohydrin, these epoxies are produced by first coupling the diols with epichlorohydrin using phase-transfer catalysts such as ammonium salts or Lewis acid catalysts (boron trifluoride, stannic chloride), followed by epoxidation with caustic [69, 70]. A prominent side-reaction is the conversion of aliphatic hydroxyl groups formed by the initial reaction into chloromethyl groups by epichlorohydrin. The resultant epoxy resins are known to have lower reactivity toward conventional amine curing agents relative to bisphenol A epoxies. Dow Chemical manufactures D.E.R. 732 and D.E.R. 736 aliphatic epoxy resins. They are derived from polyglycols with different chain lengths.

CH₂(–O–)CH–CH₂–[O–CH(CH₃)–CH₂]ₙ–O–CH₂–CH(–O–)CH₂

Polyglycidyl ethers of sorbitol, glycerol, and pentaerythritol are used as adhesion promoters for polyester tire cords. Their high chloride content improves adhesion to rubber.

## 11. Cycloaliphatic Epoxy Resins and Epoxidized Vegetable Oils

Resins based on the diepoxides of cycloaliphatic dienes were first commercialized in the 1950s by Union Carbide Corp. The combination of aliphatic backbone, high oxirane content, and no halogens gives resins with low viscosity, weatherability, low dielectric constant, and high cured $T_g$. This class of epoxy is popular for diverse end uses including auto topcoats, weatherable high voltage insulators, UV coatings, acid scavengers, and encapsulants for both electronics and optoelectronics. A comparison of some properties of two common aliphatic epoxies with those of LER (DGEBA) is shown in Table 10.

The preferred industrial route to cycloaliphatic epoxy resins is based on the epoxidation of cycloolefins with peracids, particularly peracetic acid [71]. Few side reactions are encountered. Some properties of various commercial products are given in Table 11. The peracid cannot be made *in situ* because the cyclic olefins are sensitive to impurities generated in this process.

**Table 10.** Comparative Viscosities of cycloaliphatic epoxies, epoxidized oils, and DGEBA

| Epoxy type | Viscosity, mPa·s(= cP) | EEW |
|---|---|---|
| Cyclo diepoxy ERL-4221[a] (3′,4′-epoxycyclohexylmethyl-3,4-epoxycyclohexanecarboxylate) | 400 | 135 |
| DGEBA | 11000 | 190 |
| Linseed oil epoxy | 730 | 168 |

[a]Trademark of the Dow Chemical Co.

3′-Cyclohexenylmethyl 3-cyclohexenecarboxylate

3′,4′-Epoxycyclohexylmethyl 3,4-epoxycyclohexanecarboxylate

An important secondary reaction is the acid-catalyzed hydrolysis of the epoxide groups. The reaction is minimized at low temperatures and strongly depends on the constituents and the reaction medium.

**Table 11.** Cycloaliphatic epoxy resins

| Chemical name | Structure | Commercial products | EEW[a] | Viscosity, mPa·s (= cP) at 25 °C[b] |
|---|---|---|---|---|
| 3′,4′-Epoxycyclohexylmethyl-3,4-epoxycyclohexanecarboxylate | | ERL-4221[c] UVR-6110 CY-179[d] | 131–143 | 350–450 |
| 3,4-Epoxycyclohexyloxirane | | ERL-4206[c] | 70–74 | 15 |
| 2–(3′,4′-Epoxycyclohexyl)-5,1″-spiro-3″,4″-epoxycyclohexane-1,3-dioxane | | ERL-4234[c] | 133–154 | 7,000–17,000[e] |
| Vinyl cyclohexene monoxide | | VCMX | 124 | 5 |
| 3,4-Epoxycyclohexanecarboxylate methyl ester | | ERL-4140 | 156 | 6 |
| Bis(3,4-epoxycyclohexylmethyl) adipate | | ERL-4299[c] UVR-6128 | 180–210 | 550–750 |

[a]Epoxide equivalent weight.
[b]Unless otherwise stated.
[c]Union Carbide, division of The Dow Chemical Co.
[d]Huntsman.
[e]At 38 °C.

The cycloaliphatic epoxides are more susceptible to electrophilic attack because of the lower electronegativity of the cycloaliphatic ring relative to the bisphenol A aromatic ether group in DGEBA resins. Consequently, cycloaliphatic epoxies do not react well with conventional nucleophilic epoxy curing agents such as amines. They are commonly cured via thermal or UV-initiated cationic cures. In addition, cycloaliphatic epoxy resins are low viscosity liquids that can be thermally cured with anhydrides to yield thermosets having a high heat distortion temperature. They are often used as additives to improve performance of bisphenol A epoxies. Their higher prices ($6.60–8.80/kg) have limited their commercial applications to high end products.

The largest end uses of cycloaliphatic epoxies in order of volume are electrical, electronic components encapsulation, and radiation-curable inks and coatings. A potentially large volume application is UV-curable metal can coatings for beer can exterior and ends, but the market has not been growing significantly in recent years. Other uses include acid scavengers for vinyl-based transformer fluids and lubricating oils; filament winding for aerial booms and antennas; and as viscosity modifier for bisphenol A LERs in tooling compounds. An epoxy silicone containing cycloaliphatic epoxy end groups and a silicone backbone is used as radiation-curable release coatings for pressure-sensitive products. Dow Chemical is the largest producer of cycloaliphatic epoxies. Daicel of Japan has entered the cycloaliphatic epoxy resin market.

Epoxidation of α-olefins, unsaturated fatty acid esters, and glycerol esters is affected readily by peracids including *in situ* peracids generated from hydrogen peroxide and carboxylic acids.

α-Olefin epoxies find utility as reactive diluents for coatings and as chemical intermediates for lubrication fluids. Larger volume epoxidized soybean and linseed oils are most frequently used as secondary plasticizers and co-stabilizers for PVC.

## 12. Epoxy Esters and Derivatives

### 12.1. Epoxy Esters

The esterification of epoxy resins with commercial fatty acids is a well-known process that has been employed for industrial coatings for many years. The carboxylic acids are esterified with the terminal epoxy groups or the pendant hydroxyls on the polymer chain.

A wide variety of saturated and unsaturated fatty acids are utilized to confer properties useful in air-dried, protective, and decorative coatings. Typical fatty acids include tall oil fatty acids, linseed oil fatty acid, soya oil fatty acid and castor oil fatty acid. A medium molecular weight SER, a so-called 4-type, is commonly used. Catalysts such as alkaline metal salts ($Na_2CO_3$) or ammonium salts are essential to prevent chain branching and gelation caused by etherification of the epoxy groups.

Esterification is generally conducted in an inert atmosphere at 225–260 °C, with sparging to remove byproduct water. The course of the reaction is monitored by acid number to a specified end point and by viscosity. The product is then dissolved in a solvent [72].

Metallic driers are incorporated in unsaturated ester solutions to promote cure via air-drying, i.e., oxidative polymerization of the double bonds of the fatty acids. Chemical resistance is generally lower than that of unmodified epoxy resins cured at ambient temperatures with amine hardeners. Epoxy esters are also used to produce anodic electrodeposition (AED) coatings by further reaction with maleic anhydride followed by neutralization with amines to produce water-dispersable coatings. Epoxy esters were widely used as automotive primer-surfacer and metal can ends coatings for many years, but are being replaced by other technologies. Their high viscosity limited their uses in low solids, solventborne coatings. Waterborne epoxy esters are now available and are used in flexographic inks for milk cartons.

## 12.2. Glycidyl Esters

Glycidyl esters are prepared by the reaction of carboxylic acids with epichlorohydrin followed by dehydrochlorination with caustic:

Hexahydrophthalic acid

Diglycidyl ester of hexahydrophthalic acid

The viscosity of these esters is low, i.e., ca. 500 mPa · s (= cP), and their reactivity resembles that of bisphenol A resins. Similar epoxy resin derived from dimerized linoleic acid is also commercially available. They are often used as flexiblizing agents instead of epoxidized long-chain diols.

The *glycidyl ester of versatic acid* or neodecanoic acid is an example of high MW monoglycidyl aliphatic epoxy. The molecule is highly branched, thus providing steric effects to protect it from hydrolysis, resulting in good weatherability and water resistance. On the other hand, it suffers from high viscosity, and therefore it is not an effective diluent. It is often used to improve scrubability, chemical resistance, and weatherability of coatings. The product is commercially available from Resolution (Cardura E-10) and Exxon (Glydexx N-10).

$R^1$, $R^2$ = alkyl or substituted alkyl
Glycidyl ester of versatic acid

A commercially important glycidyl ester is *glycidyl methacrylate (GMA)*, a dual functionality monomer, containing both a terminal epoxy and an acrylic C=C bond. It is produced by the reaction of methacrylic acid with epichlorohydrin. The dual functionality of GMA brings together desirable properties of both epoxies and acrylics, e.g., the weatherability of acrylics and chemical resistance of epoxies, in one product. GMA is useful as a comonomer in the synthesis of epoxy-containing polymers via free-radical polymerization. The resultant epoxy-containing polymers can be further cross-linked. An example of such polymers is GMA acrylic, which is an acrylic copolymer, containing about 10–35 % by weight of GMA. Cure is by reaction with dodecanedioic acid. Its primary uses are in automotive powder coatings. GMA-containing polymers are also used as compatiblizers for engineering thermoplastics, in adhesives and latexes, and as rubber and asphalt modifiers [73]. Dow Chemical and Nippon Oils & Fats are two major producers of GMA.

Glycidyl methacrylate (GMA)

## 12.3. Epoxy Acrylates

Epoxy resins are reacted with acrylic acid to form epoxy acrylate oligomers, curable via free-radical polymerization of the acrylate C=C bonds initiated by light [74]. UV lights are most commonly used, but electron beam (EB) curing is becoming more common because of its decreasing equipment costs. This is a fast-growing market segment for epoxy resins because of the environmental benefits of the UV cure technology: low to zero VOC, low energy requirements. Major applications include coatings for overprint varnishes, wood substrates, and plastics. Radiation-cured epoxy acrylates are also growing in importance in inks, adhesives, and photoresists applications. The 2001 global market for epoxy acrylates was estimated at 40,000 MT with an annual growth rate projected to be 8–10 %.

Epoxy acrylate based on DGEBA

Liquid epoxy resins such as DGEBA are most commonly used to produce epoxy acrylates. When higher thermal performance is required, multifunctional epoxies such as epoxy novolacs are used. Epoxy acrylates from epoxidized soybean oil and linseed oil are often used as blends with aromatic epoxy acrylates to reduce viscosity of the formulations. Major producers of UV-curable epoxy acrylates are UCB, Radcure, Dow Chemical, Sartomer, and Henkel.

## 12.4. Epoxy Vinyl Esters

A major derivative of epoxy resins is the epoxy vinyl ester resin. Originally developed by Dow Chemical [75, 76] and Shell Chemical in the 1970s, it is considered a high performance resin used in glass-reinforced structural composites, particularly for its outstanding chemical resistance and mechanical properties. The resins are made by reacting epoxy resins with methacrylic acid and diluted with styrene to 35–40 % solvent by weight. Liquid epoxy resins (DGEBA) are commonly used. Epoxy novolacs are used where higher thermal or solvent resistance is needed. Brominated epoxies are also used to impart flame retardancy for certain applications. In the final formulation, peroxide initiators are added to initiate the free-radical cure reactions of the methacrylic C=C bonds and styrene to form a random copolymer thermoset network.

The vinyl ester functionality of the epoxy vinyl esters provides outstanding hydrolysis and chemical resistance properties, in addition to the inherent thermal resistance and toughness properties of the epoxy backbone. These attributes have made epoxy vinyl esters a material of choice in demanding structural composite applications such as corrosive chemicals storage tanks, pipes, and ancillary equipment for chemical processing. Other applications include automotive valve covers and oil pans, boats, and pultruded construction parts. Significant efforts have been devoted to improve toughness and to reduce levels of styrene in epoxy vinyl ester formulations because of environmental concerns. In addition to Dow Chemical, other major suppliers of epoxy vinyl esters include Ashland, AOC, DSM, Interplastic, and Reichhold (DIC).

Vinyl ester network

## 12.5. Epoxy Phosphate Esters

Dow Chemical developed the epoxy phosphate ester technology in the 1980s [77]. Epoxy phosphate esters are reaction products of epoxy resins with phosphoric acid. Depending on the stoichiometric ratio and reaction conditions, a mixture of mono-, di- and triesters of phosphoric acid are obtained. Subsequent hydrolysis of the esters is used as a way to control the distribution of phosphate and glycol end groups and product viscosity. Epoxy phosphate esters can be made to disperse in water to produce waterborne coatings. They are used primarily as modifiers to improve the adhesion property of nonepoxy binders in both solventborne and waterborne systems for container and coil coatings.

## 13. Characterization of Uncured Epoxies

Most industrial chemicals and polymers are not the 100 % pure, single chemicals as described in their general chemical structures. In the case of epoxy resins, they often contain isomers, oligomers, and other minor constituents. As a first requirement, one would need to know the epoxy content or EEW of the epoxy resin so the proper stoichiometric amount of cross-linker(s) can be calculated. However, a successful thermoset formulation must also have the proper reactivity, flow, and performance. Consequently, other epoxy resin properties are required by the formulators and supplied by the resin producers.

Liquid epoxy resins are mainly characterized by epoxy content, viscosity, color, density, hydrolyzable chloride, and volatile content [78]. Less often analyzed are α-glycol content, total chloride content, ionic chloride, and sodium. Solid epoxy resins are characterized by epoxy content, solution viscosity, melting point, color, and volatile content. Less often quoted are phenolic hydroxyl content, hydrolyzable chloride, ionic chloride, sodium, and esterification equivalent. Table 12 lists analytical methods adopted by ASTM [79] as standard testing methods for epoxy resins.

In addition, gel-permeation chromatography (GPC), high performance liquid chromatography (HPLC) [39, 80], and other analytical procedures such as nuclear magnetic resonance (NMR) [28] and infrared spectroscopy (IR) [81] are performed to determine MW, MW distribution, oligomer composition, functional groups, and impurities.

Resin components such as α-glycol content and chloride types and levels are known to influence formulation reactivity and rheology, depending on their interactions with the system composition such as basic catalysts (tertiary amines) and/or amine curing agents. Knowing the types and levels of chlorides guides formulators in the adjustment of their formulations for proper reactivity and flow.

***Epoxide Equivalent Weight.*** The epoxy content of liquid resins is frequently expressed as epoxide equivalent weight (EEW) or weight per epoxide (WPE), which is defined as the weight in grams that contains 1 g equivalent of epoxide. A common chemical method of analysis for epoxy content of liquid resins and solid resins is titration of the epoxide ring by hydrogen bromide in acetic acid [82]. Direct titration to a crystal violet indicator end point gives excellent results with glycidyl ethers and cycloaliphatic epoxy resins. The epoxy content of glycidyl amines is determined by differential titration with perchloric acid. The amine content is first determined with perchloric acid. Addition of tetrabutylammonium iodide and additional perchloric acid generates hydrogen iodide, which reacts with the epoxy ring. The epoxy content is obtained by the second perchloric acid titration to a crystal violet end point.

In another procedure, a halogen acid is generated by the reaction of an ionic halide salt, e.g., tetraethylammonium bromide in acetic acid with perchloric acid with subsequent formation of a halohydrin; the epoxy group is determined by back-titration with perchloric acid using crystal violet indicator [83]. The end point can be determined visually or potentiometrically. A monograph on epoxide determinations was published in 1969 [84]. This is the method adopted by ASTM and is currently used by most resin producers.

***Viscosity*** of epoxy resins is an important characteristic affecting handling, processing, and application of the formulations. For example, high viscosity LERs impede good mixing with curing agents, resulting in inhomogeneous

**Table 12.** Uncured epoxy resin test methods

| Test item | Unit | Condition | ASTM method |
|---|---|---|---|
| EEW | | | D1652-97 |
| Viscosity, neat | cP$^a$ | 25 °C | D445-01 |
| Viscosity, solution | cSt$^b$ | 25 °C | D445-01 |
| Viscosity, melt | cSt$^b$ | 150 °C | D445-01 |
| Viscosity, ICI Cone and Plate | Pa · s | | D4287-00 |
| Viscosity, Gardner–Holdt | | | D1545-98 |
| Color, Co-Pt | | | D1209-00 |
| Color, Gardner | | | D1544-98 |
| Color, Gardner in solution | | | D1544-98 |
| Moisture | ppm | | E203-01 |
| Softening point | °C | | D3104-99 |

$^a$cP = mPa · s.
$^b$cSt = mm$^2$/s.

mixtures, incomplete network formation, and poor performance. On the other hand, too low viscosity would affect application characteristics such as coverage and appearance.

Viscosities of liquid resins are typically determined with a Cannon–Fenske capillary viscometer at 25 °C, or a Brookfield viscometer. The viscosity depends on the temperature, as illustrated in Figure 2. Viscosities of solid epoxy resins are determined in butyl carbitol (diethylene glycol monobutyl ether) solutions (40 % solids content) and by comparison with standard bubble tubes (Gardner–Holdt bubble viscosity). The Gardner color of the same resin solution is determined by comparison with a standard color disk. Recently, data have been reported for solid epoxy resins using the ICI Cone and Plate viscometers, which are much more time-efficient because they do not require sample dissolution.

***Hydrolyzable Chloride (HyCl)*** content of liquid and solid epoxy resins is determined by dehydrochlorination with potassium hydroxide solution under reflux conditions and potentiometric titration of the chloride liberated by silver nitrate. The solvent(s) employed and reflux conditions can influence the extent of dehydrochlorination and give different results. The "easily hydrolyzable" HyCl content, which reflects the degree of completion of the dehydrochlorination step in the epoxy resin manufacturing process, is routinely determined by a method using methanol and toluene as solvents. This is the method most commonly used to characterize LER and SER.

For epoxy resins used in electronic applications, such as cresol epoxy novolacs, more powerful polar aprotic solvents such as dioxane or dimethyl formamide (DMF) have been used to hydrolyze the difficult-to-hydrolyze HyCls,

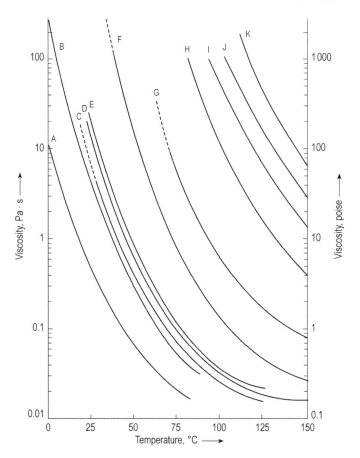

**Figure 2.** Viscosity–temperature profiles for bisphenol A epoxy resins with the following EEWs: A) 175–195 and 195–215 diluted resins; B) 172–178; C) 178–186; D) 185–192; E) 190–198; F) 230–280; G) 290–335; H) 450–550; I) 600–700; J) 675–760; K) 800–975.

such as the abnormal chlorohydrins and the organically "bound" chlorides. The issue here is the inconsistency in results obtained by different methods [78]. The presence of ionic hydrolyzable chlorides and total chlorides has been shown to affect electrical properties of epoxy molding compounds used in semiconductor encapsulation [85]. For these applications, producers offer high purity grade epoxy resins with low ionic, hydrolyzable and total chloride contents.

*Total chloride* content of epoxy resins can be determined by the classical Parr bomb method in which the sample is oxidized in a Parr bomb, followed by titration with silver nitrate [78]. The major disadvantage of this method is that it is time-consuming. Alternatively, X-ray fluorescence has been used successfully as a simple, nondestructive method to determine total chloride of epoxy resins. The method, originally developed by Dow Chemical, has been under consideration for adoption by ASTM.

*The "ball and ring" and Durran's methods* traditionally measure the softening point of SERs, which is important in applications such as powder coatings. The Durran's method involves heating a resin sample topped with a certain weight of mercury in a test tube until the resin reaches its softening point and flows, allowing the mercury to drop to the bottom of the test tube. The method is accurate but involves handling of highly hazardous mercury at elevated temperatures. The Mettlers' softening point method is more widely used recently because of its simplicity.

**The Esterification Equivalent** of solid resins is defined as the weight in grams esterified by one mole of monobasic acid. This value includes both the epoxy and hydroxyl groups of the solid resin. It is determined by esterification of the sample with acetic anhydride in the presence of pyridinium chloride, followed by titration with sodium methoxide to a thymol blue–phenolphthalein end point.

**Molecular Structure of Epoxy Resins.** Infrared spectroscopy (IR) is used to determine the epoxide content of resins as well as their structure. A compilation of IR spectra of uncured resins has been published [86] and their use in quality control and identification of components of resin blends has been described. Recently, near IR (NIR) has emerged as a useful tool to characterize epoxy resins [87].

NMR has been utilized to characterize epoxy resins, formulations and cured networks. It has been shown to be useful in determining the level of branching in epoxy resins and isomers distribution in epoxy novolacs [88, 89].

GPC and HPLC are utilized to characterize both liquid and solid epoxy resins [90]. MW and MW distributions are obtained from GPC measurements, but differences in chemical composition of resin samples are more apparent from HPLC chromatograms because of better resolution [91].

HPLC has proven to be a good fingerprinting tool to characterize LERs and SERs. Chromatograms of liquid epoxy resins (crude DGEBA) indicate a homologue distribution of $n = 0$, 85 %, and, in a specific case, $n = 1$, 11.5 %, although the values obtained depend on the source of the liquid resin. HPLC analysis of both liquid and solid epoxy resins has been studied in some detail using normal-phase and reversed-phase columns, respectively [39].

The difference between taffy-processed and fusion advancement solid resin can be noted in HPLC chromatograms. In the advancement process, the even-membered oligomers predominate, whereas taffy-produced resins exhibit both even- and odd-numbered oligomers. Compounds that contribute to hydrolyzable chloride and α-glycol content can be quantified by HPLC. The presence of branched chain components is detectable in studies using an improved reversed-phase gradient HPLC method [92, 93]. Excellent reviews of applications of chromatographic techniques to the analysis of epoxy resins are available [94].

## 14. Curing of Epoxy Resins

With the exception of the very high MW phenoxy resins and epoxy-based thermoplastic resins, almost all epoxy resins are converted into solid, infusible, and insoluble three-dimensional thermoset networks for their uses by curing with cross-linkers. Optimum performance properties are obtained by cross-linking (qv) the right epoxy resins with the proper cross-linkers, often called hardeners or curing agents. Selecting the proper curing agent is dependent on the requirements of the application process techniques, pot life, cure conditions, and ultimate physical properties. Besides affecting viscosity and reactivity of the formulation, curing agents determine both the types of chemical bonds formed and the degree of cross-linking that will occur. These, in turn, affect the chemical resistance, electrical properties, mechanical properties, and heat resistance of the cured thermosets.

Epoxy resins contain two chemically reactive functional groups: epoxy and hydroxy. Low MW epoxy resins such as LERs are considered difunctional epoxy monomers or prepolymers and are mostly cured via the epoxy group. However, as the MW of SERs increases, the epoxy content decreases, whereas the hydroxyl content increases. High molecular weight SERs can cross-link via reactions with both the epoxy and hydroxyl functionalities, depending on the choice of curing agents and curing conditions. Reaction of the epoxy groups involves opening of the oxirane ring and formation of longer, linear C–O bonds. This feature accounts for the low shrinkage and good dimensional stability of cured epoxies. The polycondensation curing via hydroxyl groups is often accompanied by generation of volatilebyproducts, such as water or alcohol, requiring heat for proper cure and volatiles removal.

It is the unique ability of the strained epoxy ring to react with a wide variety of reactants under many diverse conditions that gives epoxies their versatility [95]. Detailed discussions on the probable electronic configurations, molecular orbitals, bond angles, and reactivity of the epoxy ring are available in the literature [96].

Compared to noncyclic and other cyclic ethers, the epoxy ring is abnormally reactive. It has been postulated that the highly strained bond angles, along with the polarization of the C–C and C–O bonds account for the high reactivity of the epoxide. The electron-deficient carbon can undergo nucleophilic reactions, whereas the electron-rich oxygen can react with electrophiles. It is customary in the epoxy industry to refer to these reactions in terms of anionic and cationic mechanisms. The terminology was attributed to the fact that an anionic intermediate or transition state is involved in a nucleophilic attack of the epoxy while a cationic intermediate or transition state is formed by an electrophilic curing agent [97]. For the sake of clarity, the nucleophilic and electrophilic mechanism terminology is used in this article.

Curing agents are either catalytic or coreactive. A catalytic curing agent functions as an initiator for epoxy resin homopolymerization or as an accelerator for other curing agents, whereas the coreactive curing agent acts as a comonomer in the polymerization process. The majority of epoxy curing occurs by nucleophilic mechanisms. The most important groups of coreactive curing agents are those with active hydrogen atoms, e.g., primary and secondary amines, phenols, thiols, and carboxylic acids (and their anhydride derivatives). Lewis acids, e.g., boron trihalides, and Lewis bases, e.g., tertiary amines, initiate catalytic cures.

The functional groups surrounding the epoxide resin also affect the curing process. Steric factors [98, 99] can influence ease of cure. Electron-withdrawing groups adjacent to the epoxide ring often enhance the reactivity of the epoxy resin to nucleophilic reagents, while retarding its reactivity toward electrophilic reagents [98, 100, 101]. In general, aromatic and brominated aromatic epoxy resins react quite readily with nucleophilic reagents, whereas aliphatic and cycloaliphatic epoxies react sluggishly toward nucleophiles [102].

Figure 3 shows the pseudo first-order kinetic response for the disappearance of the epoxy in buffered methanol solutions (lines are for clarity only).

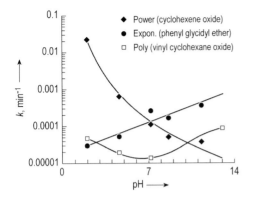

**Figure 3.** Effect of pH on reaction rates of epoxies

Clearly the epoxy structure dramatically influences the cure response of the epoxy as a function of pH. Cycloaliphatic epoxies are fast-reacting under low pH conditions. Aromatic glycidyl ethers are faster under high pH conditions. These results generally agree with "practical" cures: aromatic epoxies are easily cured with amines and amidoamines. Cycloaliphatics [102] are cured with acids and superacids. The behavior of the aliphatic epoxies is more complex but on balance is similar to that of cycloaliphatics.

**Table 13.** U.S. consumption of curing agents for epoxy resins (2001)

| Curing agents | Consumption, $10^3$ MT | Market percentage |
|---|---|---|
| Amine functional compounds | 50 | 48 |
| Aliphatic amines and adducts | 16 | |
| Polyamides | 14 | |
| Amidoamines | 9 | |
| Cycloaliphatic amines | 6.8 | |
| Phenalkamines | 1.8 | |
| Dicyandiamide (DICY) | 1.8 | |
| Aromatic polyamines | 0.9 | |
| Carboxylics | 37 | 36 |
| Polycarboxylic polyesters | 22 | |
| Anhydrides | 15 | |
| Resole resins | 9 | 9 |
| Amino formaldehydes | 4.5 | |
| Phenol formaldehyde | 4.5 | |
| Novolacs and other phenolics | 2.7 | 2.6 |
| Polysulfides and polymercaptans | 14 | 1.3 |
| Catalysts | 3.2 | 3 |
| Anionic | 3.1 | |
| Cationic | 0.1 | |
| Others | 0.9 | < 1 |

In 2001, the U.S. market for epoxy curing agents was estimated at $165 \times 10^3$ MT (see Table 13), while approximately $318 \times 10^3$ MT of epoxy resins was consumed. The most commonly used curing agents are amines, followed by carboxylic-functional polyesters and anhydrides.

A description of advantages, disadvantages and major applications of typical curing agents is given in Table 14.

## 15. Coreactive Curing Agents

Commercially, epoxy resins are predominantly cured with coreactive curing agents. Following are important classes of epoxy coreactive curing agents.

### 15.1. Amine Functional Curing Agents

This section describes one of the most important classes of epoxy coreactive curing agents.

#### 15.1.1. Primary and Secondary Amines

Primary and secondary amines and their adducts are the most widely used curing agents for epoxy resins, accounting for close to 50 % of all the epoxy curing agents used in the United States in 2001. The number of amine hydrogen atoms present on the molecule determines the functionality of an amine. A primary amine group, one which has two hydrogens bound to it, will react with two epoxy groups while a secondary amine will react with only one epoxy group. A tertiary amine group, which has no active hydrogen, will not react readily with the epoxy group, but will act as a catalyst to accelerate epoxy reactions. Reactions of a primary amine with an oxirane group or an epoxy resin are shown in the following [103].

It has been reported that primary amines react much faster than secondary amines [101, 104]. Reaction of an epoxy group with a primary amine initially produces a secondary alcohol and a secondary amine. The secondary amine, in turn, reacts with an epoxy group to give a tertiary amine

Table 14. Curing agents for epoxy resins

| Type | Advantages | Disadvantages | Major applications |
| --- | --- | --- | --- |
| Aliphatic amines and adducts | low viscosity; ambient cure temperature; little color; low cost | short pot life; rapid heat evolution; critical mix ratio; some are moderately toxic; high moisture absorption; blush; carbonation; limited high temperature performance (< 100 °C) | flooring; civil engineering; marine and industrial coatings; adhesives; small castings |
| Cycloaliphatic amines | low viscosity; long pot-life; room temperature (RT) cure and heat-curable; adhesion to wet cement; good color; low toxicity; good electrical, mechanical, thermal properties (high $T_g$) | slower reactivity; high costs | flooring; paving; aggregate; industrial coatings; adhesives; tooling; composites; castings |
| Aromatic amines | excellent elevated temperature performance (150 °C); good chemical resistance; long pot life; low moisture absorption | solids; incompatibility with resins; long cure cycles at high temperature (150 °C); toxicity | high performance composites and coatings; adhesives; electrical encapsulation |
| Amidoamines | low viscosity; reduced volatility; good pot life; ambient cure temperature; convenient mix ratios; good toughness | poor performance at high temperature (< 65 °C); some incompatibility with epoxies | high solids, solvent-free coatings; floorings; concrete bonding; troweling compounds |
| Polyamides | good mix ratios; pot life; RT cure; good concrete wetting; flexibility; low volatility and toxicity | high viscosity; low temperature performance; poor color; higher cost | marine and maintenance coatings; civil engineering; castings; adhesives |
| Anhydrides | low exotherm; good thermal (high $T_g$), mechanical, electrical properties; low shrinkage and viscosity; long pot life; little color | long cure cycles at high temperature (200 °C) | composites; castings; potting; encapsulation |
| Catalytic | long pot life; high temperature resistance | brittle; moisture-sensitive | adhesives; prepregs; electrical encapsulation; powder coatings |
| Dicyandiamide | good electrical properties; high temperature resistance; latent systems | incompatibility with epoxy resins | electrical laminates; powder coatings; single-package adhesives |
| Carboxylic-terminated polyesters | good weatherability, corrosion resistance, and mechanical properties; low cost | poor chemical resistance | powder coatings |
| Isocyanates | fast cure at low temperature; good flexibility and solvent resistance | moisture-sensitive; toxic | powder coatings; maintenance coatings |
| Phenol–formaldehyde, novolacs | good chemical resistance, electrical properties, shelf stability, and compatibility with epoxies; high temperature resistance | high melting solids; high temperature cure; poor UV stability | molding compounds; powder coatings; electrical laminates |
| Polysulfides and polymercaptans | RT rapid cure times; flexible systems; moisture insensitive | poor performance at high temperature; odorous | consumer adhesives; sealants; traffic paints |
| Melamine–formaldehyde | good color and hardness; stable one-component systems | high temperature cure | stove paints; can coatings |
| Urea–formaldehyde | stable one-component systems; little color; low cost | high temperature cure; formaldehyde emission | fast-bake enamels; stove primers; can and drum coatings |
| Phenol–formaldehyde resoles | stable one-component systems; excellent chemical resistance | high temperature cure; brittle; gold color | baked enamels; can, drum and pail coatings; high temperature service coatings |

and two secondary hydroxyl groups. Little competitive reaction is detectable between a secondary hydroxyl group in the backbone and an epoxy group to afford an ether [100], provided a stoichiometric equivalent or excess amine is maintained. However, with excess epoxy, the secondary hydroxyl groups formed gradually add to the epoxide groups [105]. This reaction can be catalyzed by tertiary amines.

Hydroxyl compounds accelerate the rate of amine curing. A mechanism has been proposed [100] in which the hydroxyl group hydrogen bonds to the oxygen atom on the epoxy group, rendering the methylene group more susceptible to attack by the nucleophilic amine. Reactivity is proportional to the hydroxyl acidity and functionality; phenolics, aryl alcohols, and polyfunctional alcohols afford the best results.

$$R'R\text{NH} + \text{epoxide} + \text{HOR''} \longrightarrow$$

$$\longrightarrow R'R\text{N-CH}_2\text{-CH(OH)-R'} + \text{HOR''}$$

In general, reactivity of amines toward aromatic glycidyl ethers follows their nucleophilicity: aliphatic amines > cycloaliphatic amines > aromatic amines. Aliphatic amines cure aromatic glycidyl ether resins at room temperature (RT) without accelerators, whereas aromatic amines require elevated temperatures. However, with the help of accelerators, the cure rates of aromatic amines can approach those of some aliphatic amines. In general, the steric and electronic effects of substituents of the epoxy and the amine influence the reaction rate of an amine with an epoxy resin.

### 15.1.1.1. Aliphatic Amines

The liquid aliphatic polyamines such as polyethylene polyamines (PEPAs) were some of the first curing agents used with epoxies. They give good RT cures with DGEBA-type resins. The low equivalent weights of the ethylene amines give tightly cross-linked networks with good physical properties, including excellent chemical and solvent resistance but limited flexibility and toughness. Good long-term retention of properties is possible at temperatures up to 100 °C. Short-term exposure to higher temperatures can be tolerated. Certain aliphatic amines cured epoxies will blush (or bloom) under humid conditions. This undesirable property has been attributed to the incompatibility of some amine curing agents with epoxy resins. Incompatible amines can exude to the surface during cure and react with atmospheric carbon dioxide and moisture to form undesirable carbamates (carbonation). This, in turn, leads to gloss reduction and intercoat adhesion and recoatability problems in coating applications [106].

$$H_2O + CO_2 \longrightarrow H_2CO_3$$
Water + Carbon dioxide → Carbonic acid

$$H_2CO_3 + RNH_2 \longrightarrow RNHCOOH + H_2O$$
Carbonic acid + Amine → Carbamic acid + Water

$$RNHCOOH + R'NH_2 \longrightarrow R'NH_3^+ \, {}^-OCONHR$$
Carbamic acid + Amine → Carbamate

Mixing ratios with epoxy resin are very critical, and working pot lives are too short for some applications. Aliphatic polyamines are hygroscopic and volatile, have bad odor, and cause dermatitis if improperly handled. Another disadvantage is high exotherm in thick sections or large mass parts that can lead to thermal decomposition. Consequently, significant efforts have been devoted toward remedying these shortcomings by modifications of the polyethylene polyamines. Adducts with epoxy resins (resin adducts), carboxylic acids (polyamides, amidoamines), ketones (ketimines), and phenols/formaldehyde (Mannich bases) [107] are widely used commercially. Longer chain alkylenediamines such as hexamethylenediamine (HMD) and polyetheramines (polyglycol-based polyamines) have also been developed. Currently, very small amounts of unmodified polyamines are used as curing agents for epoxies. They are primarily used to produce epoxy adducts (up to 90 %). Chemical modification by reaction with epoxy groups to yield epoxy adducts affords products with better handling properties, lower vapor pressure, reduced tendency to blush, and less critical mix ratio. For example, diethylenetriamine (DETA) readily reacts with ethylene oxide in the presence of water to give a mixture of mono- and dihydroxyethyl diethylenetriamine with a longer pot life and fewer dermatitic effects than free DETA.

$$H_2N\text{-CH}_2\text{CH}_2\text{-NH-CH}_2\text{CH}_2\text{-NH}_2 + \text{ethylene oxide} \longrightarrow$$

$$H_2N\text{-CH}_2\text{CH}_2\text{-NH-CH}_2\text{CH}_2\text{-N(H)-CH}_2\text{CH}_2\text{OH} +$$

$$HO\text{-CH}_2\text{CH}_2\text{-N(H)-CH}_2\text{CH}_2\text{-N(H)-CH}_2\text{CH}_2\text{-N(H)-CH}_2\text{CH}_2\text{OH}$$

Resinous adducts are produced by reaction of excess diamine with epoxy resins.

$$\text{epoxide-R-epoxide} + H_2N-R'-NH_2 \text{ (excess)} \longrightarrow$$

$$H_2N-\underset{H}{\overset{R'}{N}}-\overset{R}{\underset{OH}{C}}-\underset{OH}{\overset{R}{C}}-\underset{H}{\overset{R'}{N}}-NH_2$$

The higher molecular weight of the adduct affords a more desirable, forgiving ratio of resin to curing agent, lower water absorption, and better resin compatibility.

**Ketimines.** Ketimines are the reaction products of ketones and primary aliphatic amines. In the absence of reactive hydrogens, they do not react with epoxy resins. They can be considered blocked amines or latent hardeners, since they are readily hydrolyzed to regenerate the amines. They have low viscosity, cure rapidly when exposed to atmospheric humidity, and are useful in high solids coatings. Similar products have been obtained with acrylonitrile.

$$2\ R-\overset{O}{\underset{}{C}}-R + H_2N-R'-NH_2 \rightleftharpoons$$

$$\overset{R}{\underset{R}{C}}=N-R'-N=\overset{R}{\underset{R}{C}} + 2\,H_2O$$

**Mannich Base Adducts.** Mannich base adduct is the reaction product of an amine with phenol and formaldehyde.

phenol-OH + $H_2CO$ + $NH_2RNH_2$ ⟶ 2-(NHRNH$_2$-methyl)phenol

Mannich base

The resultant product has an internal phenolic accelerator. Compared to unmodified amines Mannich base adducts have lower volatility, less blushing and carbonation, and, despite their higher MW, faster reactivity.

**Polyetheramines.** Polyetheramines are produced by reacting polyols derived from ethylene oxide or propylene oxide with amines. The more commercially successful adducts are based on propylene oxide and are available in different MWs (JEFFAMINE* from Huntsman). The longer chain backbone provides improved flexibility but slower cure rate. Chemical and thermal resistance properties are also reduced. Polyetheramines are often used in combination with other amines in flooring, and adhesive and electrical potting applications.

### 15.1.1.2. Cycloaliphatic Amines

Cycloaliphatic amines were originally developed in Europe, where their use as epoxy curing agents is well established. Compared to aliphatic amines, cycloaliphatic amines produce cured resins having improved thermal resistance and toughness. Glass-transition temperatures ($T_g$) approach those of aromatic amines ($> 150\,°C$), while percent elongation can be doubled. However, chemical resistance is inferior to that of aromatic amines. Because cycloaliphatic amines are less reactive than acyclic aliphatic amines, their use results in a longer pot life and in the ability to cast larger masses. Unmodified cycloaliphatic amines require elevated temperature cure, but modified systems are RT-curable. Properly formulated, they can give an excellent balance of properties: fast cure, low viscosity, low toxicity, good adhesion to damp concrete, and excellent color stability. They are, however, more expensive than other types of curing agents.

Isophorone diamine (IPDA), bis(4-aminocyclohexyl)methane (PACM), and 1,2-diaminocyclohexane (1,2-DACH) are the principal commercial cycloaliphatic polyamine curing agents. IPDA is the largest volume cycloaliphatic amine. Commercial cycloaliphatic amines are formulated products. In addition to the cycloaliphatic amines, other components such as aliphatic amines and plasticizers are also included to improve RT cure speed and end-use properties. One popular formulation consists of IPDA used in combination with trimethylhexamethylenediamines (TMDA) or *meta*-xylenediamine (MXDA), and plasticizers/accelerators such as nonyl phenol or benzyl alcohol. In some ambient cure coating applications, cycloaliphatic amines can be reacted with phenol and formaldehyde to form the Mannich base products, which have an internal phenol accelerator and cure readily at ambient temperatures.

The largest market for cycloaliphatic amines is in flooring, followed by high solids coatings, composites, adhesives, castings, and tooling. Cycloaliphatic amines experienced significant growth in the early 1990s as replacements for more toxic aromatic amines such as MDA. However, anhydrides have been more successful at replacing aromatic amines in composite applications.

### 15.1.1.3. Aromatic Amines

Because of conjugation, aromatic amines have lower electron density on nitrogen than do the aliphatic and cycloaliphatic amines. Consequently, they are much less reactive toward aromatic epoxies. They have longer pot-lives and usually require elevated temperature cures. Aromatic amines are usually solid at room temperature. These hardeners are routinely melted at elevated temperatures and blended with warmed resins to improve solubility. Eutectic mixtures of *meta*-phenylenediamine (MPD) and methylenedianiline (MDA or DDM) exhibit a depressed melting point resulting in an aromatic hardener that remains a liquid over a short period of time. MDA or 4,4′-diaminodiphenylmethane (DDM), 4,4′-diaminodiphenyl sulfone (DDS or DADPS), and MPD are the principal commercial aromatic amines. A new aromatic amine, diethyltoluenediamine (DETDA) has gained more significant uses in recent years.

Epoxies cured with aromatic amines typically have better chemical resistance and higher thermal resistance properties than products cured with aliphatic and cycloaliphatic amines. Their best attribute is their retention of mechanical properties at long exposures to elevated temperatures (up to 150 °C). Consequently, they are widely used in demanding structural composite applications such as aerospace, PCB laminates, and electronic encapsulation. 4,4′-DDS is the standard curing agent used with a multifunctional amine epoxy (MY 720) for high performance aerospace and military composite application. 3,3′-DDS is used in aerospace honeycomb for its excellent peel strength. MDA, which has excellent mechanical and electrical properties, is the most widely used aromatic curing agent, but recently has been classified as a potential human carcinogen and its volume has been declining.

Alkyl-substituted MDAs such as tetraethyl-MDA have been developed with lower toxicity and improved performance [108, 109]. However, none of the replacement products has the performance/cost combination of MDA. Anhydrides and cycloaliphatic amines have been used to replace aromatic amines in a number of composite applications. Efforts have been made to develop ambient-curable aromatic amines by adding accelerators such as phenols to MDA.

### 15.1.1.4. Arylyl Amines

These amines have cycloaliphatic or aromatic backbones, but the amine functional groups are separated from the backbone by methylene groups (benzylic amines and hydrogenated derivatives). Consequently, arylyl amines are much more reactive toward epoxies than aromatic amines while having improved thermal and chemical resistance over aliphatic amines. Fast cures at ambient and sub-ambient are possible with arylyl amines. These amines are more widely used in Japan and Europe than in North America *meta*-Xylylene diamine (MXDA) and its hydrogenated product, 1,3-bis(aminomethyl cyclohexane) (1,3-BAC) are popular arylyl amines.

The commercial polyamine curing agents are given in Table 15.

The stoichiometric quantity of polyamine used to cure an epoxy resin is a function of the molecular weight and the number of active hydrogens of the polyamine (amine equivalent weight, AEW) and the EEW or equivalent weight of epoxy resin; it is expressed as follows:

$$\left(\frac{AEW}{EEW}\right) \times 100 = \text{parts by weight polyamine}$$
$$\text{per 100 parts by weight epoxy resin}$$

### 15.1.2. Polyamides

Polyamides are one of the largest volume epoxy curing agents used. They are prepared by the reaction of dimerized and trimerized vegetable-oil fatty acids with polyamines. Dimer acid is made by a Diels–Alder reaction between 9,12- and 9,11-linoleic acids. Subsequent reaction with diethylenetriamine or other suitable

**Table 15.** Commercial amine curing agents

| Formula | Name | Abbreviation |
|---|---|---|
| *Aliphatic* | | |
| $NH_2CH_2CH_2NHCH_2CH_2NH_2$ | diethylenetriamine | DETA |
| $NH_2CH_2CH_2NHCH_2CH_2NHCH_2CH_2NH_2$ | triethylenetetramine | TETA |
| (structure) | Poly(oxypropylene diamine) | |
| (structure) | poly(oxypropylene triamine) | |
| $NH_2(CH_2)_3O(CH_2)_2O(CH_2)_3NH_2$ | poly(glycol amine) | |
| (structure) | *N*-aminoethylpiperazine | AEP |
| *Cycloaliphatic* | | |
| (structure) | isophorone diamine | IPDA |
| (structure) | 1,2-diaminocyclohexane | DACH |
| (structure) | bis(4-aminocyclohexyl)methane | PACM |
| *Aromatic* | | |
| (structure) | 4,4′-diamino-diphenylmethane | MDA, DDM |
| (structure) | 4,4′-diaminodiphenyl sulfone | 4,4′-DDS |
| (structure) | *m*-phenylenediamine | MPD |
| (structure) | diethyltoluenediamine | DETDA |

*Arylyl amines*

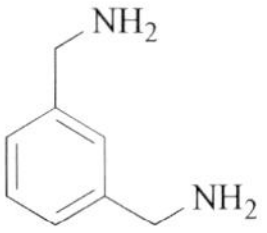

meta-xylene diamine      MXDA

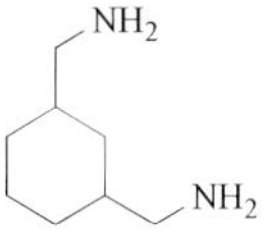

1,3-bis(aminomethyl cyclohexane)      1,3-BAC

multifunctional amines yields the amine-terminated polyamides. They are available in a range of molecular weights and compositions.

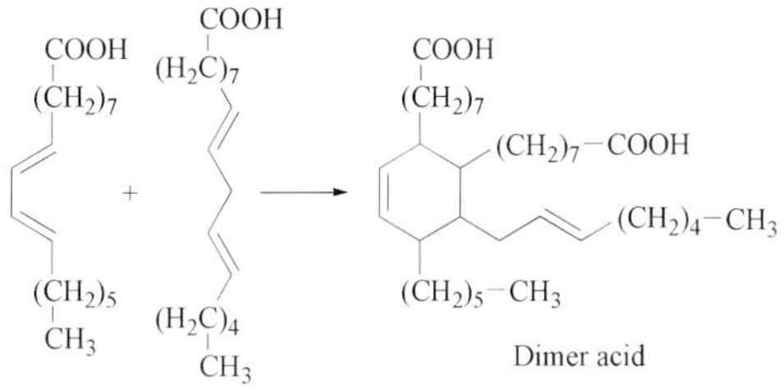

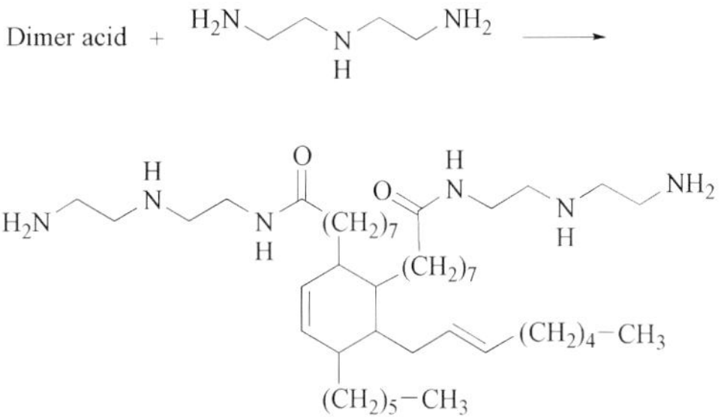

Polyamides are extremely versatile curing agents. The polyamides react with the epoxide group through the terminal amine functionality. The unreacted amide NH groups in the backbone provide good intercoat adhesion and the fatty acid structures provide good moisture resistance and mechanical properties. Wetting of cement surfaces is excellent. As a result of their relatively higher molecular weight, the ratio of polyamide to epoxy is more forgiving than with low MW polyamines. They are inexpensive, less toxic to handle; give no blushing; exhibit readily workable pot lives; and cure under mild conditions. Polyamides are mainly used in coating and adhesive formulations, mostly in industrial maintenance and civil engineering applications. The various MW polyamides exhibit different degrees of compatibility with epoxy resins. To ensure optimum properties, the polyamide/epoxy mixture must be allowed to react partly before being cured. This partial reaction assures compatibility and is known as the induction period.

Disadvantages of polyamides include slower cure speeds and darker color than polyamine-cured epoxies. Polyamide-cured epoxies lose structural strength rapidly with increasing temperature. This limits their use to applications not subjected to temperatures above 65 °C. Formulations with tertiary amines, phenolic amines, or co-curing agents help to speed up cures at low temperatures. Alternatively, polyamides derived from polyamines with phenolic-containing carboxylic acids are called phenalkamines [110]. These curing agents have low viscosity and fast ambient cure speed and are widely used in on-site marine coatings and concrete deck applications.

The high viscosity of polyamides limits their uses primarily to low solids coatings, which have been losing ground to higher solids coatings. Waterborne polyamides have been developed for use with waterborne epoxies, but their growth has been modest over the past decade because the conversion to waterborne epoxy coatings has been slower than expected. Commercial polyamides include the Versamid resins from Cognis, Ancamide resins from Air Products, and Epicure resins from Resolution.

### 15.1.3. Amidoamines

Amidoamines have all the properties of polyamides, except for a significantly lower viscosity, which make them useful in high solids and solvent-free coating formulations. They are prepared by the reaction of a monofunctional acid like tall-oil fatty acid with a multifunctional amine such as DETA, resulting in a mixture of amidoamines and imidazolines.

Imidazoline is formed by intramolecular condensation at high reaction temperatures. Commercial amidoamines are produced with different imidazoline contents to regulate reactivity and cured product performance. The pot life/reactivity of amidoamines varies with imidazoline content. High imidazoline content offers longer pot life and semi-latent curing activated by moisture. They are particularly useful in wet concrete applications. Like the polyamides, amidoamines can be used over a range of levels to enhance a specific property. However, amidoamines offer several advantages over aliphatic amines and polyamides. They offer more convenient mix ratios, increased flexibility, and better moisture resistance than aliphatic polyamines, and they offer lower color and viscosity than polyamides. Consequently, the volume of amidoamines has grown significantly in the past decade.

### 15.1.4. Dicyandiamide

Dicyandiamide (DICY) is a solid latent hardener ($mp$ 208 °C). Its latent nature is due to its insolubility in epoxy resins at RT. DICY can be mixed with epoxy resins to provide a one-package formulation with good stability up to 6 months at ambient temperatures. Cure of epoxies with DICY occurs with heating to 150 °C. It is often used with imidazoles as catalysts. DICY offers the advantage of being latent (reacts with epoxy resin upon heating and stops reacting temporarily when the heat is removed). This partially cured or "B-staged" state is ideal for prepreg applications. Typically, DICY is used at levels of 5–7 parts per 100 parts of liquid epoxy resins and 3–4 parts per 100 parts of solid epoxy resins.

DICY is one of the first curing agents to be used with epoxy resins. It cures with epoxies to give a highly cross-linked thermoset with good mechanical strength, thermal properties, and chemical resistance, and excellent electrical properties. Because of its latency, low quantity requirements and excellent balance of properties, DICY is a widely used curing agent in powder coating and electrical laminate applications. These two applications account for 85 % of DICY consumption as epoxy curing agent.

The curing mechanism is rather complex, involving several simultaneous reactions. There are a number of conflicting proposed mechanisms in the literature. One study proposed the initial reaction of all four active hydrogens with epoxy resin catalyzed by tertiary amine catalysts followed by epoxy homopolymerization. The last step involves reactions between the hydroxyl groups of the epoxy resin with the cyano group [108, 109]. One of the more recent and plausible mechanisms of DICY cure with epoxies is that of Gilbert and co-workers [111]. The Gilbert mechanism is summarized in Figure 4. Gilbert and co-workers investigated the reaction of DICY with methyl glycidyl ether of bisphenol A (MGEBA). Products were analyzed using HPLC, NMR, and FTIR. On the basis of products that were isolated and characterized, Gilbert and co-workers proposed the mechanism shown in Figure 4.

The first step in the mechanism is the reaction of DICY with epoxy to form the alkylated DICY. This was confirmed by the imide IR peak at 1570 cm$^{-1}$. The second step involves further alkylation of the nitrogen that reacted in step 1, to form the N,N-dialkyldicyandiamide. No alkylation of the other amino group was

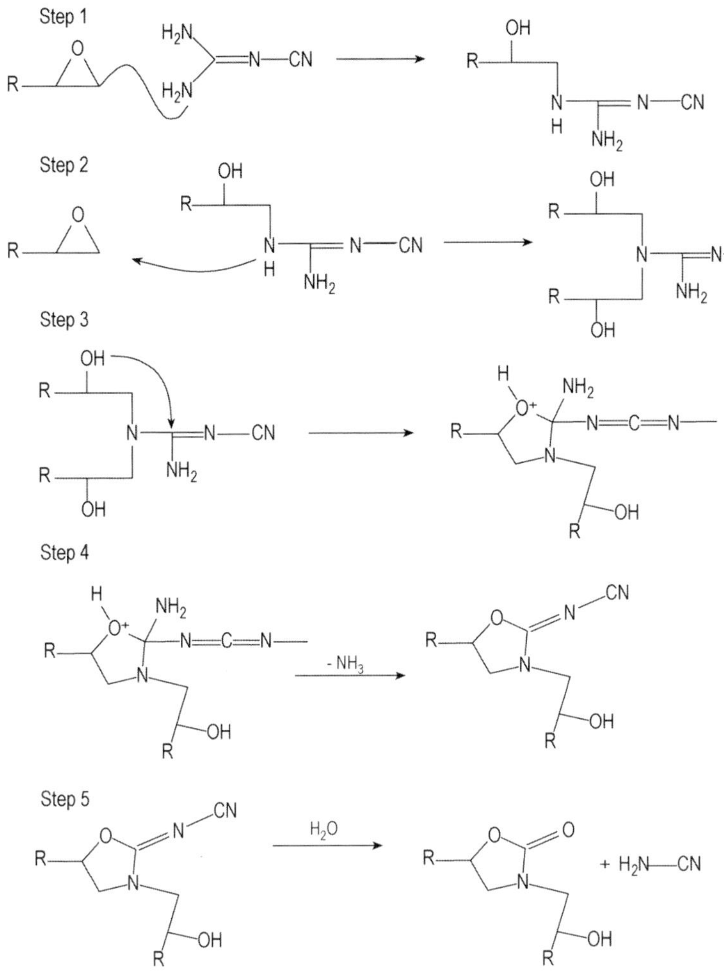

**Figure 4.** The Gilbert mechanism for the DICY curing of epoxy. From Ref. [111].

suggested. The third step is the intramolecular cyclization step to form a zwitterionic five-membered intermediate. This involves the intramolecular reaction of the secondary alcohol formed in step 2 with the imide functionality (−C=N−). This is in contrast with the Zahir mechanism [112] where the intramolecular cyclization involves the hydroxy and the nitrile groups. The fourth step involves the elimination of ammonia and the formation of 2-cyanoimidooxazolidine. The formation of this heterocycle is consistent with the observed bathocromic IR shift from 1570 $cm^{-1}$ to 1650 $cm^{-1}$. The ammonia that is eliminated can then react with epoxy to form a trifunctional crosslink. The last step involves the hydrolysis of the oxazolidine to form the oxazolidone and cyanamide. The hydrolysis step accounts for the formation of the carbonyl group.

## 15.2. Carboxylic Functional Polyester and Anhydride Curing Agents

Carboxylic polyesters and anhydrides are the second most important class of epoxy curing agent. Together, they constitute 36% of the total curing agent volume used in the U.S. market (2001 data). Polyesters have been growing rapidly in powder coatings formulations with epoxy resins, consuming the highest tonnage of epoxy curing agents. This is driven in part by the conversion to the more environmentally friendly powder coating technologies, and in part by the versatility and cost efficiency of polyester–epoxy hybrid powder coatings. Anhydrides have been successfully replacing more toxic aromatic amines in composites. They account for 70 % of the volume of curing agents used in structural composite applications. Both polyesters and anhydrides are used in heat-cured applications only.

### 15.2.1. Carboxylic Functional Polyesters

The reaction of polyacids with polyalcohols produces polyesters. The terminal functionality is dictated by the ratio of the reactants. By virtue of their relatively cheap, widely available raw materials and good flexibility and weatherability, acid functional polyesters are used in hybrid epoxy powder coatings for a wide range of applications. For applications requiring good weatherability, triglycidyl isocyanurate (TGIC) is often used as curing agent for acid functional polyesters.

Terephthalic acid, trimellitic anhydride, and neopentyl glycol are commonly used raw materials to produce polyesters. Other acids, anhydrides, and glycols can also be used to modify functionality, MW, viscosity, and mechanical properties (after curing) of the polyesters. This versatility of the polyester building blocks allows many useful combinations of epoxy–polyester hybrid systems to be developed for a wide range of applications [113]. Major applications include coatings for metal furniture, general metal finishing, appliances, machinery and equipment, automotive, and wood. Automotive is a new, large, and fast growing market with many car makers converting to primer-surfacer based on epoxy–polyester powder coatings. Wood coatings are a new, emerging market.

The curing mechanism of epoxy–polyester thermosets involves reaction of the acid functionality with epoxy followed by esterification of the epoxy hydroxyl groups with the acids [114]. Compounds such as amines and phosphonium salts catalyze these reactions. Water is a condensation reaction byproduct that must be allowed to escape during the curing process to avoid coating defects.

The first product is a β-hydroxypropyl ester, which reacts with a second mole of carboxylic acid to yield a diester. The hydroxyl ester can also undergo polymerization by reaction of its secondary hydroxyl group with an epoxy group.

### 15.2.2. Acid Anhydrides

Anhydrides are some of the very first epoxy curing agents used, and they remain a major class of curing agents used in heat-cured structural composites and electrical encapsulation. Their consumption volume equals that of all aliphatic amines and adducts in 2001 in the United States. While the carboxylic-terminated polyesters find widespread uses in coatings, anhydride use in coatings is minimal.

Epoxy–anhydride systems exhibit low viscosity and long pot life, low exothermic heats of reaction, and little shrinkage when cured at elevated temperatures. The low exotherm heat generation is a unique attribute of anhydrides, making them suitable for uses in large mass epoxy cures. Curing is slow at temperatures below 200 °C and is often catalyzed by Lewis bases or acids. Postcure is often needed to develop optimum properties. Tertiary amines such as benzyldimethylamine, dimethylaminomethylphenol, tris(dimethylaminomethyl)phenol, boron trihalide amine complexes, stannic chloride, ammonium salts, phosphonium salts, and substituted imidazoles are effective catalysts. Proper catalyst concentration (0.5–2.5 % of resin weight) is critical, depending on the types of anhydrides and resins used and the cure schedules, and is known to affect high temperature performance.

Cured epoxy–anhydride systems exhibit excellent thermal, mechanical, and electrical properties, and are used in filament-wound epoxy pipe, PCB laminates, mineral-filled composites, and electrical casting and encapsulation applications. Anhydride-cured epoxies also have better aqueous acid resistance than similar amine-cured systems. Anhydrides are the principal curing agents for cycloaliphatic and epoxidized olefin resins in electrical casting and potting. Some key physical properties of exemplary epoxy resins cured with hexahydrophthalic anhydride are shown in Table 16.

The mechanism of anhydride cure is complex and controversial because of the possibility of several competing reactions. The uncatalyzed reaction of epoxy resins with acid anhydrides proceeds slowly even at 200 °C; both esterification and etherification occur. Secondary alcohols from the epoxy backbone react with the anhydride to give a half ester, which in turn reacts with an epoxy group to give the diester. A competing reaction is etherification of an epoxy with a secondary alcohol, either on the resin backbone or that formed during the esterification, resulting in a β-hydroxy ether. It has been reported that etherification is a probable reaction since only 0.85 equivalents of anhydrides are required to obtain optimum cross-linked density and cured properties [103].

Lewis bases such as tertiary amines and imidazoles are widely used as epoxy–anhydride

Table 16. Formulation and properties of epoxy resins cured with hexahydrophthalic anhydride

|  | DGEBA | 3′,4′-Epoxycyclohexylmethyl 3,4-epoxycyclohexanecarboxylate | Hexahydrophthalic acid diglycidyl ester |
|---|---|---|---|
| Formulation |  |  |  |
| Resin, pbw[a] | 100 | 100 | 100 |
| Hexahydrophthalic anhydride, pbw[a] | 85 | 105 | 100 |
| Accelerator type | tertiary amine | metal alkoxide salt | quaternary ammonium salt |
| pbw[a] | 3 | 12 | 4 |
| Cure schedule, h at °C | 2 at 100 | 4 at 120 | 4 at 80 |
|  | 1 at 150 |  | 4 at 140 |
| Typical cured properties at 25 °C |  |  |  |
| Tensile strength, MPa[b] | 65 | 68 | 83 |
| Tensile modulus, MPa[b] | 3400 | 3300 | 3000 |
| Flexural strength, MPa[b] | 131 | 89 | 127 |
| Flexural modulus, MPa[b] | 3400 | 3000 | 3000 |
| Elongation, % | 5.0 | 2.7 | 3.5 |
| Compressive strength, MPa[b] | 124 | 151 | 124 |
| Heat-deflection temperature, °C | 120 | 150 | 105 |
| Water absorption, % weight gain[c] | 0.5 | 0.4 | 0.4 |
| Dielectric constant at 60 Hz | 3.4 | 3.3 | 3.5 |
| Dissipation factor at 60 Hz | 0.006 | 0.005 | 0.007 |
| Volume resistivity, $\Omega \cdot cm \times 10^{16}$ | 2.0 | 10.0 | 3.0 |

[a] Parts by weight.
[b] To convert MPa to psi, multiply by 145.
[c] After boiling for 1 h.

catalysts. Conflicting mechanisms have been reported for these catalyzed reactions [115]. The more widely accepted mechanism [103] involves the reaction of the basic catalyst with the anhydride in the initiation step to form a betain (internal salt). The propagation step involves the reaction of the carboxylate anion with the epoxy group, generating an alkoxide. The alkoxide then further reacts with another anhydride, propagating the cycle by generating another carboxylate which reacts with another epoxy group. The end result is the formation of polyester-type linkages. In practice, it has been observed that optimum properties are obtained when stoichiometric equivalents of epoxy and anhydride are used with high temperature cures, which is consistent with this mechanism and does not involve etherification reactions. At lower anhydride/epoxy ratios (0.5:1) and lower cure temperatures, some etherifications can take place by reaction of the alkoxide with an epoxy group.

Table 17. Commercially important anhydride curing agents

| Name | Structure |
|---|---|
| Phthalic anhydride (PA) | |
| Tetrahydrophthalic anhydride (THPA) | |
| Methyltetrahydrophthalic anhydride (MTHPA) | |
| Methyl hexahydrophthalic anhydride (MHHPA) | |
| Hexahydrophthalic anhydride (HHPA) | |
| Nadic methyl anhydride or methyl himic anhydride (MHA) | |
| Benzophenonetetracarboxylic dianhydride (BTDA) | |
| Tetrachlorophthalic anhydride (TCPA) | |

Numerous structurally different anhydrides can be used as epoxy curing agents, but the most widely used are liquids for ease of handling. The most important commercial anhydrides are listed in Table 17. Methyltetrahydrophathalic anhydride (MTHPA) is the largest volume anhydride, used in filament-winding composites. Phthalic anhydride (PA) is the next largest volume and is inexpensive; so it is used widely in mineral-filled laboratory bench top manufacturing, which requires low exotherm heat generation to avoid cracking. Dodecylsuccinic anhydride (DDSA) has a long aliphatic chain in the backbone and is used as blends to improve flexibility. Benzophenonetetracarboxylic dianhydride (BTDA) is a relatively new, multifunctional anhydride developed for high temperature applications, capable of achieving a high cross-linking density with a heat distortion temperature (HDT) of 280 °C. It has been

## 15.3. Phenolic-Terminated Curing Agents

Phenolics form a general class of epoxy curing agents containing phenolic hydroxyls capable of reacting with the epoxy groups. They include phenol-, cresol-, and bisphenol A terminated epoxy resin hardener. More recent additions include bisphenol A based novolacs. Cure takes place at elevated temperatures (150–200 °C) and amine catalysts are often used.

The bisphenol A terminated hardeners are produced using liquid epoxy resins and excess bisphenol A in the resin advancement process. They are essentially epoxy resins terminated with bisphenol A. They are popular in epoxy powder coating applications for rebar and pipe, providing more flexible epoxy coatings than the novolacs.

The novolacs are produced via the condensation reaction of phenolic compounds with formaldehyde using acid catalysts. They are essentially precursors to epoxy novolacs. Novolacs are multifunctional curing agents and can impart higher cross-link density, higher $T_g$, and better thermal and chemical resistance than other phenolics. Cresol novolacs provide higher solvent and moisture resistance, but are more brittle than their phenol novolac counterparts. Recently, bisphenol A based novolacs have been used in electrical laminate formulations to improve thermal performance ($T_g$ and $T_d$) [116]. Novolacs are widely used in composites, PCB laminates, and electronic encapsulation applications. Their uses in coatings are limited to high temperature applications such as powder coatings for down-hole oil-field pipe coatings.

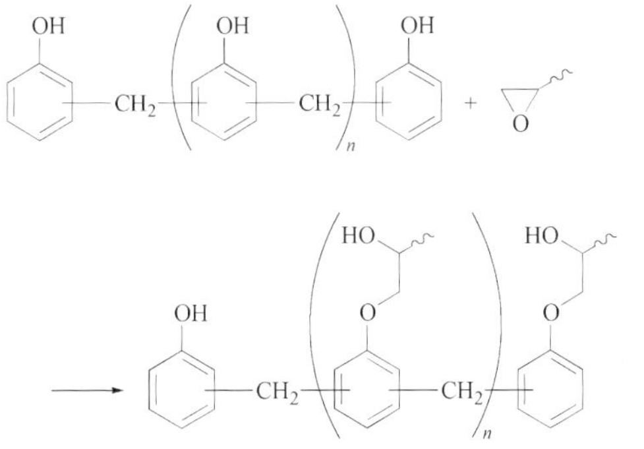

used as a replacement for more toxic aromatic amines. Tetrachlorophthalic anhydride (TCPA) is used in epoxy powder coatings for small electronic components with flame-retardancy requirements.

## 15.4. Melamine–, Urea–, and Phenol–Formaldehyde Resins

Melamine–formaldehyde, urea–formaldehyde, and phenol–formaldehyde resins react with hydroxyl groups of high MW epoxy resins to afford cross-linked networks [72].

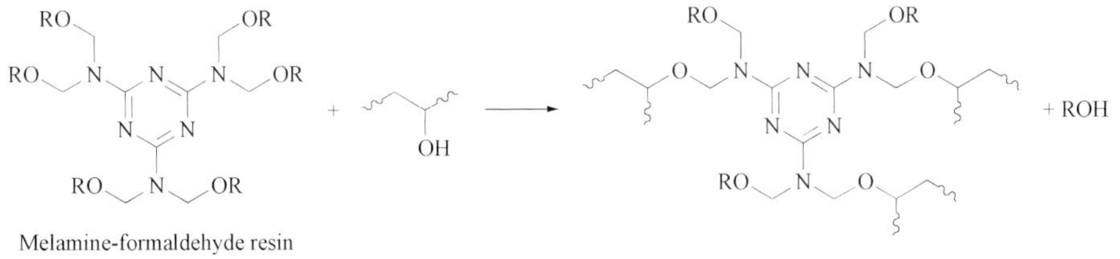

Melamine-formaldehyde resin

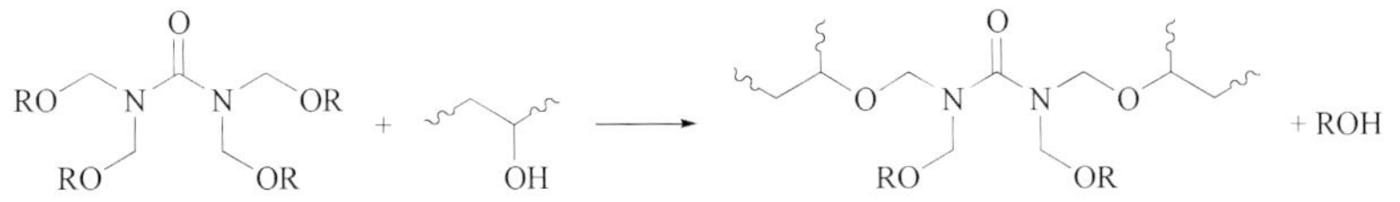

Urea-formaldehyde resin

The condensation reaction occurs primarily between the methylol or alkylated methylol group of the formaldehyde resin and the secondary hydroxyl group on the epoxy resin backbone. The high bake temperatures used in these applications drive off the condensation byproducts (alcohol or water). Acids such as phosphoric acid and sulfonic acids are often used as catalysts.

There are two types of phenol–formaldehyde condensation polymers: resoles and novolacs [117]. Phenol–formaldehyde polymers prepared from the base-catalyzed condensation of phenol and excess formaldehyde are called resoles. In most phenolic resins commonly used with epoxies, the phenolic group is converted into an ether to give improved alkali resistance. At elevated temperatures ($> 150\,°C$), resole resins react with the hydroxyl groups of the epoxy resins to provide highly cross-linked polymers.

The melamine- and urea–formaldehyde resins are also called amino resins [118]. The phenol–formaldehyde resoles are often called phenolic resins, which is rather easily confused with phenolic-terminated cross-linkers such as novolacs and bisphenol A terminated resins.

These formaldehyde-based resins are widely used to cure high MW solid epoxy resins at elevated temperatures (up to $200\,°C$) for metal can, drum, and coil coatings applications. The resultant coatings have excellent chemical resistance, good mechanical properties, and no effects on taste (adding or taking away taste from packaged foods or drinks). The vast majority of the food and beverage cans produced in the world today are coated internally with epoxy–formaldehyde resin coatings. The phenol–formaldehyde resoles are also used with epoxies in coatings for high temperature service pipes and to protect against hot, corrosive liquids.

## 15.5. Mercaptans (Polysulfides and Polymercaptans) Curing Agents

The mercaptan group of curing agents includes polysulfide and polymercaptan compounds which contain terminal thiols.

In the language commonly used in this industry, "polysulfides" typically have a functionality of 2, while "polymercaptans" have an average functionality of 3. By itself, the thiol or mercaptan group (SH) reacts very slowly with epoxy resins at ambient temperature. However, when converted by a tertiary amine to a mercaptide ion, they are extremely reactive [119].

$$NR_3 + RSH \longrightarrow RS^- + \overset{+}{N}HR_3$$

Increasing the basic strength of the amine increases the reaction rate. Polar solvents are also known to speed up these reactions. Fast curing at ambient conditions is the primary attribute of this class of curing agent, lending themselves to applications such as the "5-minutes" consumer adhesives, concrete road repairs, and traffic marker adhesives. In practice, they are often formulated with co-curing agents such as amines or polyamides to achieve a balance of fast cure with improved mechanical properties. The tertiary amine accelerated polymercaptan/epoxy systems exhibit good flexibility and tensile strength at ambient temperature. They are used in high lap-shear adhesion applications such as concrete patch repair adhesives. One disadvantage of polymer-catans is their strong odor. Aliphatic amine/polysulfide co-curing agent systems yield improved initial elevated temperature performance and are widely used as building adhesives for their excellent adhesion to both glass and concrete. However, both systems lose some flexibility on aging.

### 15.6. Cyclic Amidines Curing Agents

Cyclic amidine curing agents are typically used in epoxy powder coating formulations and in decorative epoxy–polyester hybrid powder coatings to produce matte surface for furniture and appliance finishes. 2-Phenyl imidazoline has been used successfully to produce low gloss epoxy powder coatings. It is highly reactive, capable of curing at relatively low temperatures (140 °C) and is suitable for curing of coatings on temperature-sensitive substrates such as wood and plastics. Other curing agents in this group include salts of polycarboxylic acids and cyclic amidines. Their volume is currently small but is expected to grow as the markets for low gloss and low temperature cure powder coatings develop. They can also be used as tertiary amine catalysts similar to imidazoles.

### 15.7. Isocyanate Curing Agents

Isocyanates react with epoxy resins via the epoxy group to produce an oxazolidone structure [120, 121] or with a hydroxyl group to yield a urethane linkage. The urethane linkage provides improved flexibility, impact, and abrasion resistance. The oxazolidone products have been successfully commercialized in high temperature resistance coating and composite applications. Blocked isocyanates are used as cross-linkers for epoxy in PPG's cathodic electrodeposition (CED) coatings. Isocyanates are also used to cure epoxies in some powder coatings, but their toxicity has limited their use.

### 15.8. Cyanate Ester Curing Agents

Cyanate esters can be used to cure epoxy resins to produce highly cross-linked thermosets with high modulus and excellent thermal, electrical, and chemical resistance properties. They are used in high performance electrical laminate and composite applications. Cure involves oxazoline formation catalyzed by metal carboxylates in addition to homopolymerization of both cyanate ester and epoxy [122]. The high costs of cyanate esters however limit their uses.

Bisphenol A based dicyanate ester

## 16. Catalytic Cure

The catalytic curing agents are a group of compounds that promote epoxy reactions without being consumed in the process. In some of the epoxy literature, catalysts are referred to as "accelerators"; the distinction of these two types of additives is discussed in later sections.

## 16.1. Lewis Bases

Lewis bases contain an unshared pair of electrons in an outer orbital and seek reaction with areas of low electron density. They can function as nucleophilic catalytic curing agents for epoxy homopolymerization; as co-curing agents for primary amines, polyamides, and amidoamines; and as catalysts for anhydrides. Tertiary amines and imidazoles are the most commonly used nucleophilic catalysts. Several different mechanisms are possible:

1. The catalytic curing reactions of tertiary amines with epoxy resins follow two different pathways, depending on the presence or absence of hydrogen donors, such as hydroxyl groups. In the absence of hydrogen donors [123], tertiary amines react with the electron-poor methylene carbon of the epoxy group to form an intermediate zwitterion. The zwitterion then attacks another epoxy group to continue homopolymerization via an anionic mechanism. In the presence of hydrogen donors such as alcohols, the zwitterion abstracts the proton from the alcohol to generate an alkoxide,

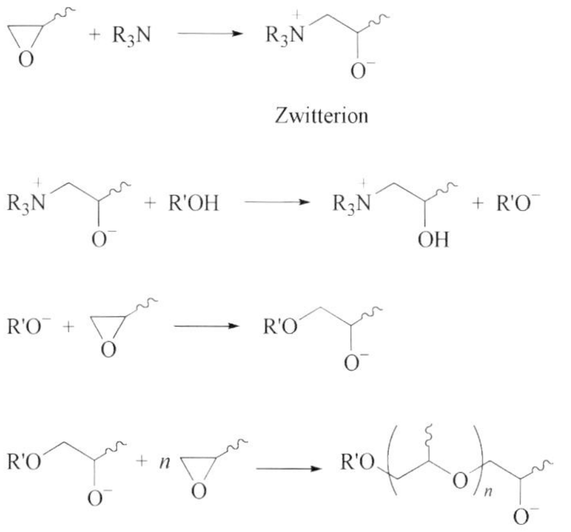

which further reacts with an epoxide group. Chain propagation continues by way of a polymeric anion mechanism.

2. With more acidic hydrogen donors such as benzyl alcohol, phenols, or mercaptans, the tertiary amine acts as a co-curing agent by first abstracting the proton from the hydrogen donor:

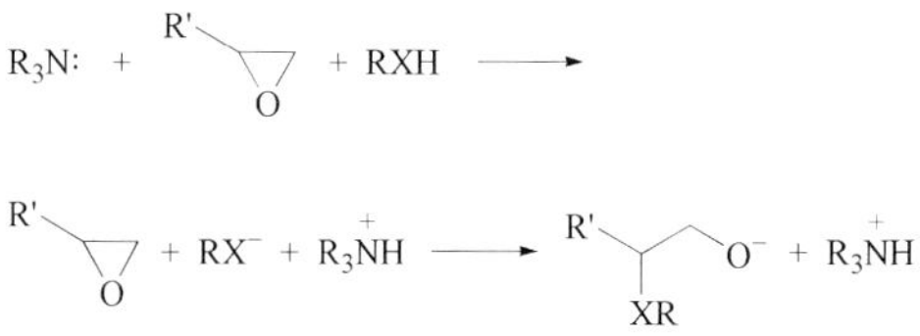

3. With anhydrides, the catalyst facilitates the anhydride ring opening:

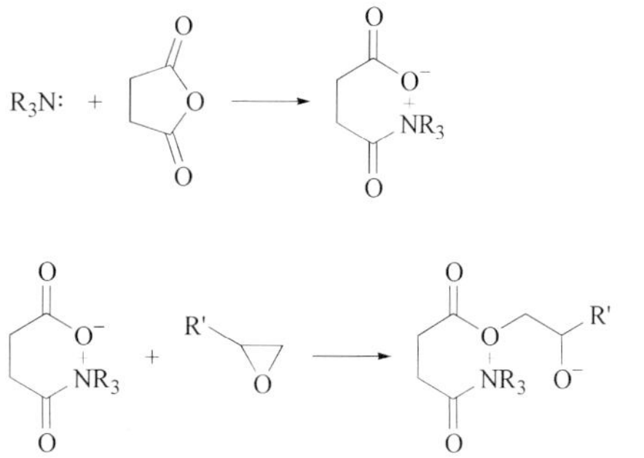

Commonly used tertiary amines include 2-dimethylaminomethylphenol (DMAMP) and 2,4,6-tris(dimethylaminomethyl)phenol (TDMAMP, trade name DMP-30 of Rohm and Haas), which contain built-in phenolic hydroxyl groups and can be used as a good catalysts and co-curing agents for room temperature cure of epoxies.

The rate of cure of epoxy resins with tertiary amines depends primarily upon the extent to which the nitrogen is sterically blocked. The homopolymerization reaction depends on the temperature as well as the concentration and type of tertiary amine. Benzyldimethylamine (BDMA) and TDMAMP are mainly used as accelerators for other curing agents, in the curing of anhydride- and dicyandiamide-based systems. Other tertiary amine catalysts include 1,4-diazabicyclo(2,2,2)octane (DABCO) and diazabicycloundecene (DBU).

*Imidazoles* such as 2-methylimidazole (2-MI) and 2-phenylimidazole (2-PI) contain both a cyclic secondary and a tertiary amine functional groups and are used as catalysts, catalytic curing agents, and accelerators [124, 125]. They are widely used as catalysts for DICY-cured epoxies in electrical laminates. For powder coatings, 2-MI adducts of LER are often used to facilitate dispersion of the components in powder coating formulations and to enhance shelf-life. Other modified imidazoles are also commercially available. The main advantage of imidazoles is the excellent balance of pot life and fast cure. 2-PI is used to increase $T_g$ and thermal resistance.

*Cyclic amidines* such as 2-phenylimidazoline have also been used as a catalyst and co-curing agent in epoxy–polyester and epoxy powder coatings.

*Substituted ureas* are another group of epoxy nucleophilic catalytic curing agent, derived by blocking of isocyanates with dimethylamine. They are commonly used as catalysts for DICY cure of epoxies in adhesives, prepregs, and structural laminates. The ureas exhibit outstanding latency at room temperature and are widely used in one-pack adhesives. The catalytic mechanism of ureas is not well understood, but it has been postulated that DICY assists in deblocking of the urea to generate a tertiary amine, which in turns acts as epoxy curing catalyst. Commercially important substituted ureas are 3-phenyl-1,1-dimethyl urea (Amicure UR by Air Products), a reaction product of phenyl isocyanate with dimethylamine; and Amicure UR 2T, a reaction product of toluene diisocyanate (TDI) with dimethylamine.

*Quaternary phosphonium salts* such as the tetraalkyl and alkyl-triphenylphosphonium halides have been used as fast catalysts for curing of phenolics, carboxylic acid-terminated polyesters, or anhydrides with epoxies [126]. Used in powder coatings, they showed good latency and fast cure rates at moderate temperatures.

Air Products is a major epoxy catalyst supplier. Others include Huntsman, Cognis, and SKW Chemicals.

## 16.2. Lewis Acids

Lewis acids, e.g., boron trihalides, contain an empty outer orbital and therefore seek reaction with areas of high electron density. Boron trifluoride, $BF_3$, a corrosive gas, reacts easily with epoxy resins, causing gelation within a few minutes. Complexation of boron trihalides with amines enhances the curing action. Reasonable pot lives using these complexes can be achieved because elevated temperatures are required for cure. Reactivity is controlled by the choices of the halide and the amine. The amine choice also affects other properties such as solubility in resin and moisture-sensitivity. Boron trifluoride monoethylamine ($BF_3 \cdot NH_2C_2H_5$), a crystalline material which is a commonly used catalyst, cures epoxy resins at 80–100 °C. A chloride version is also commercially available. Other Lewis acids used in epoxy curing include stannic chloride and tin octanate.

Different mechanisms have been proposed for curing epoxy resins with $BF_3$ complexes or salts. In general, it is assumed that complexation with the oxirane oxygen is involved, facilitating proton transfer and ether formation. Thermal dissociation of the $BF_3$–amine complex may form a proton that further reacts with the epoxy group to initiate the curing process [127]. Another mechanism assumes an amine adduct or salt is solvated by the epoxy groups, resulting in an oxonium ion [128]. The curing reaction is initiated and propagated by attack of other epoxy groups on the oxonium ion.

## 16.3. Photoinitiated Cationic Cure

Photoinitiated cationic curing of epoxy resins is a rapidly growing method for the application of coatings from solvent-free or high solids systems. This technology allows the formulation of epoxy coatings and adhesives with essentially "infinite" shelf life, but almost "instantaneous" cure rates. Cycloaliphatic epoxies are widely cured using photoinitiated cationic initiators.

Photoinitiators used for epoxy curing include aryldiazonium salts ($ArN_2^+X^-$), diaryliodonium salts ($Ar_2I^+X^-$), and onium salts of Group VIa elements, especially salts of positively charged sulfur ($Ar_3S^+X^-$). The anions must be of low nucleophilicity, such as tetrafluoroborates or hexafluorophosphates, to promote polymer chain growth rather than chain termination. Upon UV irradiation, photoinitiators yield a "super" acid, which polymerizes the epoxy resins by a conventional electrophilic mechanism.

The photolysis of diaryliodonium and triarylsulfonium salts may proceed via formation of a radical cation, which abstracts a hydrogen atom from a suitable donor.

$$Ar_3S^+ PF_6^- \xrightarrow{UV} Ar_2S^{+\cdot} + Ar\cdot + PF_6^-$$

$$Ar_2S^{+\cdot} + R-H \longrightarrow Ar_2SH^+ + R\cdot$$

$$Ar_2SH^+ + PF_6^- \rightleftharpoons Ar_2S + HPF_6$$

Subsequent loss of a proton yields the Brønsted acid HPF$_6$. Catalytic curing of the epoxy resin proceeds through an onium intermediate:

In the presence of triarylsulfonium and diaryliodonium salts, polymerization continues even if UV irradiation is terminated. This phenomenon is called "dark cure" and is due to the "living" nature of the "superacid" generated cation. The cure regime can be thought of as UV-initiated but "thermally cured." Thermally initiated cationic catalysts are also available [129].

In contrast, dialkylphenacyl sulfonium salts undergo reversible dissociation upon photolysis with formation of an ylid and a Brønsted acid. Cessation of UV activation results in termination of epoxy homopolymerization, since the acid is consumed in a reverse reaction with the ylid.

This type of behavior provides a means of controlling the degree of cure. Dialkylphenacyl sulfonium salts are thermally stable in epoxy resins at room temperature and up to 150 °C for 1–2 h. Significant interest in thermal cationic cure of epoxies, especially cycloaliphatic epoxies, has developed [130].

## 17. Formulation Development With Epoxy Resins

The most important step in using epoxy resins is to develop the appropriate epoxy formulation since most are used as precursors to a three-dimensional cross-linked network. With the exception of the very high MW phenoxy resins and the epoxy-based thermoplastics, epoxy resin is rarely used by itself. It is usually formulated with modifiers such as fillers and used in composite structures with glass fiber or metal substrates (coatings). To design a successful epoxy formulation that will give optimum *processability* and *performance*, the following factors must be carefully considered:

1. Selection of the proper combination of epoxy resin(s) and curing agent(s) structures
2. Epoxy/Curing agent stoichiometric ratio
3. Selection of catalyst/accelerator
4. Curing/postcuring processes and conditions
5. Formulation modifiers such as fillers, diluents, toughening agents, etc
6. Interactions among the formulation ingredients and with the composite materials (fibers, metals, etc) on the system chemistry, adhesion, rheology, morphology, and performance

The development of an epoxy formulation containing a high number of components can be very resource and time-consuming. Techniques such as design of experiments (DOE) are useful tools to facilitate the formulation development process and to obtain optimum performance [131, 132]. Future developments should include application of high throughput techniques to epoxy formulation development and optimization.

## 17.1. Relationship Between Cured Epoxy Resin Structure and Properties

The following diagram illustrates the formation of cured epoxy networks using different ratios of a difunctional epoxy and a tetrafunctional hardener. The structures formed are significantly different, depending on the ratio used. Consequently, it is expected that performance of these networks will be quite different despite the fact that they are derived from identical building blocks (Fig. 5).

The structure between the cross-linking position and the distance between any two of these points are important characteristics. Molecular weight between cross-links ($M_c$) and cross-link density are terms developed to describe "distance" between cross-link points. The concept originated with the rubber elasticity theory developed for the lightly cross-linked elastomers and has been adopted for use with epoxy thermosets with mixed success [133, 134]. The cured epoxy system derives its properties mostly from a combination of cross-link density, monomer structure and the curing process.

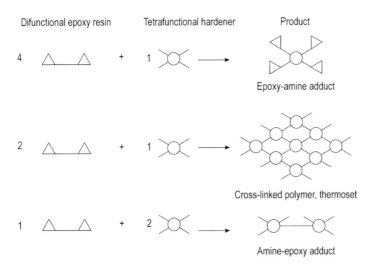

**Figure 5.** Formation of resin–hardener networks

The two-dimensional schematic network structures do not represent spatial reality but have been devised to help understand the nature of the various structures [135]. A good understanding of the structure/property relationship is critical in designing the appropriate epoxy/curing agent combination. For example, cross-linking with dicarboxylic anhydrides yields polyesters that are resistant to oxidation, but less so to moisture, especially in the presence of basic components. Amine cross-linked systems are resistant to saponification but not to oxidation. There is a large body of specific structure/property relationship knowledge in the epoxy industry and literature, but only a few systematic treatments are available [136–138].

Cross-link density increases with degree of cure up to its limit at full conversion of the (limiting) functional groups. The curing temperature and process strongly influence cross-link density, molecular architecture, network morphology, residual stress, and the ultimate performance. The effects of degree of cure and subsequent cross-link density on the chemical resistance of a cured DGEBA–aromatic polyamine adduct system are depicted in Figure 6.

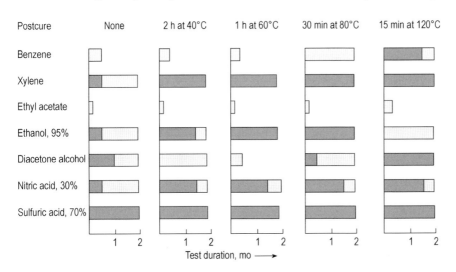

**Figure 6.** Chemical resistance of a DGEBA–aromatic polyamine adduct. Postcured substrate: sandblasted mild steel; film thickness: 300–350 μm; cure: 7 d at 20 °C. ▨ degraded; ■ resistant.

The increase in chemical resistance properties after postcure also demonstrates the effects of increased cross-link density. The cross-link density of a cured epoxy system can be estimated by a number of different techniques as described in the characterization of cured epoxy section.

## 17.2. Selection of Epoxy Resins

Successful performance of epoxy-based systems depends on proper selection and formulation of components. The components that have the most significant influences are the epoxy resins and the curing agents. As discussed in earlier sections, there are numerous choices of epoxy resins and curing agents presenting a wide variety of structure and functionality. Figure 7 shows the general attributes of common types of epoxy resins.

Epoxy resins can be used separately or in combination, such that formulations can be designed to take advantage of the desirable characteristics of several components. Because combining resins from different families can result in certain trade-offs, a careful balance of components should be investigated to produce optimal performance for specific applications. Table 16 shows effects of different resin backbones on cured properties with formulations based on hexahydrophthalic anhydride as curing agent.

The difunctional DGEBA resins are offered commercially in a wide range of molecular weights. As the molecular weight increases, so does the chain length between the epoxy end groups. Table 5 shows the effects of increasing EEW and MW of bisphenol A based epoxy resins on resin properties. The cross-link density of a difunctional resin cured by way of the epoxy group decreases as the resin molecular weight increases. High molecular weight resins are frequently cured via the secondary hydroxyl group, chemistry that results in a different set of structure–property relationships.

Multifunctional epoxy resins are available with functionalities ranging from above 2 to about 5. When cured to the same degree using a given curing agent at stoichiometric ratios, they produce a higher cross-link density, higher glass-transition temperature, and better thermal and chemical resistance compared with difunctional epoxy resins.

## 17.3. Selection of Curing Agents

The selection of curing agents is just as critical as the selection of resins. As discussed in the Curing Agents section, there are numerous types of chemical reagents that can react with epoxy resins. Since coreactive curing agents become part of the network structure, careful consideration must be paid to their contributions. Besides affecting viscosity and reactivity of the formulation, curing agents determine both the types of chemical bonds formed and the functionality of the cross-link junctions that are formed. Table 18 show performance examples of a liquid epoxy resin cured with different curing agents. Several authors have attempted to rationalize the curing agent selection process for different applications [106, 139].

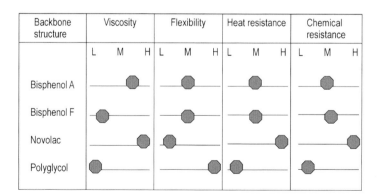

**Figure 7.** Comparison of relative properties of common epoxy resins
L, low; M, medium; H, high.

**Table 18.** Typical properties, chemical resistance, and thermal degradation of liquid DGEBA resin (185 EEW) cured with common hardeners (Dow Chemical data)

| | Curing agent | | | | |
|---|---|---|---|---|---|
| | TETA[a] | MDA[b] | Polyamide[c] | Anhydride[d] | BF$_3$–MEA[e] |
| *Property* | | | | | |
| phr[f] | 13 | 26 | 43 | 87.5 | 3 |
| Formulation viscosity, Pa·s[g] (°C) | 2.25 (25) | 0.110 (70) | 1.25 (50) | 0.038 (80) | 0.040 (100) |
| Cure schedule, h (°C) | 16 (25) | 16 (55) | 16 (25) | 4 (100) | 4 (100) |
| | 3 (100) | 2 (125) | 3 (100) | 4 (165) | 16 (150) |
| | | 2 (175) | | 16 (200) | |
| Heat distortion temperature, °C | 111 | 160 | 101 | 156 | 168 |
| Strength, MPa[h] | | | | | |
| Compression | 112 | 116 | 85.6 | 126 | 114 |
| Flexural | 96 | 93 | 67 | 97 | 100 |
| Tensile | 79 | 70.4 | 57.3 | 69 | 39.4 |
| Modulus, GPa[i] | | | | | |
| Compression | 3.05 | 2.6 | 2.6 | 3.04 | 2.3 |
| Flexural | 3.05 | 2.7 | 2.14 | 3.05 | 3.1 |
| Textile elongation,% | 4.4 | 4.4 | 3.9 | 2.5 | 1.6 |
| Dielectric constant at 10$^3$ Hz | 3.90 | 4.06 | 3.19 | 3.14 | 3.45 |
| Dissipation factor at 10$^3$ Hz | 0.020 | 0.015 | 0.0070 | 0.0054 | 0.0053 |
| Resistivity at 25 °C, 10$^{-17}$ Ω·m | | | | | |
| Volume | 6.1 | 12.2 | 12.2 | 6.1 | 8.6 |
| Surface | 7.8 | > 7.9 | 5.5 | > 7.3 | > 7.9 |
| *Chemical resistance* | | | | | |
| % Weight gain after 28 d | | | | | |
| 50 % NaOH | 0.04 | −0.05 | 0.07 | −0.12 | −0.02 |
| 30 % H$_2$SO$_4$ | 1.8 | 1.6 | 1.9 | 0.83 | 1.1 |
| Acetone | 2.1 | 4.6 | 7.3 | 15.0 | 1.2 |
| Toluene | 0.07 | 0.13 | 3.7 | 0.09 | 0.17 |
| Water | 0.86 | 1.1 | 1.3 | 0.82 | 1.2 |
| *Thermal degradation* | | | | | |
| % Weight loss after 300 h at 210 °C | 6.8 | 5.5 | 5.0 | 1.5 | 4.9 |

[a] Triethylenetetramine.
[b] 4,4′-Methylenedianiline.
[c] Versamide 140 (Henkel Corp.).
[d] Methylbicyclo[2.2.1]heptene-2,3-dicarboxylic anhydride catalyzed with 1.5 phr benzyldimethylamine.
[e] Methylethylamine.
[f] Parts per hundred epoxy resin.
[g] To convert Pa·s to P, multiply by 10.
[h] To convert MPa to psi multiply by 145.
[i] To convert GPa to psi, multiply by 145,000.

The effect of hardener structure on the heat resistance of a cross-linked DGEBA resin is shown in Table 19 [140]. Thermal stability is affected by the structure of the hardener. The heat resistance of aliphatic amine cured epoxy is low as measured by TGA. The nitrogen atoms are oxidized by atmospheric oxygen to amine oxides, which attack the polymer backbone. Anhydride systems tend to split off the anhydride at temperatures well below their decomposition point at about 390 °C. The ether segments formed by 2-MI and phenolic cured epoxies have the highest thermal stability.

## 17.4. Epoxy/Curing Agent Stoichiometric Ratios

In addition to the choices of epoxy resins and curing agents, the stoichiometric ratio of epoxy/curing agent is another factor that has significant effects on the network structure and performance. A variety of products are obtained from different ratios. Network formations for a difunctional epoxy resin and a tetrafunctional amine are illustrated in Figure 5. The products range from an epoxy–amine adduct with excess epoxy to an amine–epoxy adduct with excess amine.

Table 19. Effect of hardener structure on reactivity and heat resistance of a cross-linked bisphenol a diglycidyl ether

| Curing agent | $T_{rmax}$, °C | $E_a$, J/mol[a] | TGA, 4 °C/min Weight loss before decomposition | $T_{dec}$, °C |
|---|---|---|---|---|
| hexahydrophthalic anhydride + catalyst | 125 | 92 | 12 | 392 |
| 4,4'-methylenedianiline | 154 | 50 | 0 | 390 |
| 2,2,4(or 2,4,4)-trimethylhexamethylenediamine | 90 | 58 | | 320 |
| 2-methylimidazole + catalyst | 126 | 67 | 3.2 | 420 |
| pyrogallol (1,2,3-trihydroxybenzene) + catalyst | 185 | 54 | 2.9 | 400 |
| cyanoguanidine (dicyandiamide) | 207 | 125 | 0 | 373 |

[a] To convert J to cal, divide by 4.184.

Theoretically, a cross-linked thermoset polymer structure is obtained when equimolar quantities of resin and hardener are combined. However, in practical applications, epoxy formulations are optimized for *performance* rather than to complete stoichiometric cures. This is especially true when curing of high MW epoxy resins through the hydroxyl groups.

In primary and secondary amines cured systems, normally the hardener is used in near stoichiometric ratio. Because the tertiary amine formed in the reaction has a catalytic effect on reactions of epoxy with co-produced secondary alcohols, slightly less than the theoretical amounts should be used. However, if substantially less than the theoretical amount of amine is used, the epoxy resin will not cure completely unless heat is applied (postcure). The use of excess amine will result in unreacted amine terminated dangling chain ends and reduced cross-linking, yielding a polymer that can be somewhat tougher but which is considerably more susceptible to attack by moisture and chemicals. In formulations containing anhydrides, less than stoichiometric ratios of curing agents normally are used (0.50 to 0.85 of anhydride to 1 epoxy stoichiometric ratio) because of significant epoxy homopolymerization.

Ladder studies are often conducted varying the stoichiometric ratios and other factors to determine the optimum formulations. Statistical design of experiment (DOE) methodology has been used to efficiently carry out ladder studies [141]. Information concerning network structures can be obtained using dynamic mechanical analysis (DMA) [142, 143] and chemorheology to guide formulation development [144, 145].

## 17.5. Catalysts

The choice of a catalyst and of its amount is important. As discussed in previous sections, some tertiary amine catalysts can play multiple roles in the curing reaction. Anhydride cure in particular is highly sensitive to catalyst amount. Nucleophilic catalysts, used with acidic curing agents such as anhydrides and novolacs, can greatly reduce the gel time. In the case of anhydrides, a nucleophilic catalyst attacks the

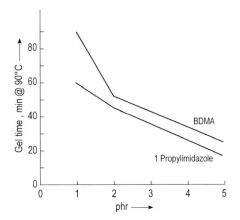

**Figure 8.** Effects of catalyst on epoxy/nadic methyl anhydride cure

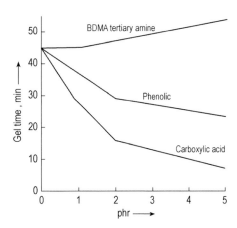

**Figure 9.** Effects of accelerator on epoxy/triethylene triamine cure

anhydride ring, causing the ring to open and promote bonding to the epoxy ring. Figure 8 shows the effect of BDMA and 1-propylimidazole levels on the pot life of a system combining D.E.R. 331 resin and nadic methyl anhydride at 90 °C (194 °F) [146]. Imidazoles are more efficient catalysts than tertiary amines; only half the concentration is required to produce the same catalytic effect.

## 17.6. Accelerators

Accelerators are commonly added to epoxy systems to speed up curing. This term should be used to describe compounds which increase the rate of catalyzed reactions but which by themselves are not catalysts. However, the term accelerator is often used synonymously with catalyst in some of the literature. Hydrogen donors such as hydroxyl groups facilitate epoxy reactions via hydrogen bonding or reaction with the oxygen on the epoxide ring. More acidic donors such as phenols and benzyl alcohols increase the rate of acceleration. However, very strong acids can interfere with amine curing agents by protonation of the amine to form an amine salt, resulting in increased pot life. Figure 9 shows the effects of different accelerators on the rate of a DGEBA/triethylenetriamine formulation.

## 18. Epoxy Curing Process

The epoxy curing process is an important factor affecting the cured epoxy performance. Consequently, it is imperative to understand the curing process and its kinetics to design the proper cure schedule to obtain optimum network structure and performance. Excellent reviews on this topic are available in the literature [147, 148].

The curing of a thermoset epoxy resin can be expressed in terms of a time–temperature-transformation (TTT) diagram (Fig. 10) [149, 150]. Later, a CTP (cure-temperature–property) diagram was proposed as a modification of the TTT diagram [151]. For nonisothermal cure, the conversion-temperature-transformation (CTT) diagram has been shown to be quite useful [152]. In the TTT diagram, the time to gellation and vitrification is plotted as a function of

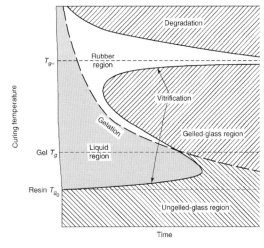

**Figure 10.** Time–temperature-transformation diagram

isothermal cure temperature. Important features are the gel point and the onset of vitrification. The gel point is defined as the onset of the formation of insoluble, cross-linked polymer (gel fraction) in the reaction mixture. However, a portion of the sample may still be soluble (sol fraction). The onset of vitrification is when the glass-transition temperature ($T_g$) of the curing sample approaches the curing temperature $T_c$. Ideally, a useful structural thermoset would cure until all monomers are built into the network, resulting in no soluble fraction.

The S-shaped vitrification curve and the gelation curve divide the time–temperature plot into four distinct states of the thermosetting-cure process: liquid, gelled rubber, ungelled glass, and gelled glass. $T_{g0}$ is the glass-transition temperature of the unreacted resin mixture; $T_{g\infty}$ the glass-transition temperature of the fully cured resin; and gel $T_g$ the point where the vitrification and gellation curves intersect.

In the early stages of cure prior to gelation or vitrification, the epoxy curing reactions are kinetically controlled. When vitrification occurs the reaction is diffusion controlled, and the reaction rate is orders of magnitude below that in the liquid region. With further cross-linking of the glass, the reaction rate continues to decrease and is eventually quenched. In the region between gelation and vitrification (rubber region) the reaction can range from kinetic to diffusion control. This competition causes the minimum in vitrification temperature seen in the TTT diagram between gel $T_g$ and $T_{g\infty}$. As the cure temperature is raised the reaction rate increases and the time to vitrification decreases until the decrease in diffusion begins to overcome the increased kinetic reaction rate. Eventually, slower diffusion in the rubbery region decreases the overall reaction rate and thus the increase in time to vitrify is seen. Below $T_{g\infty}$, the reaction does not go to completion. As curing proceeds, the viscosity of the system increases as a result of increasing molecular weight, and the reaction becomes diffusion-controlled and eventually is quenched as the material vitrifies [153]. After quenching, the cure conversion can be increased by raising the temperature. This is often practiced as postcure for certain epoxy systems to achieve maximum cure and performance. Postcure is only effective at temperatures higher than $T_{g\infty}$. However, it must be noted that at temperatures sufficiently above $T_{g\infty}$, onset of network degradation can also be seen if sufficient time is involved. Thus one must be careful about potential "over-curing."

The TTT diagram is useful in understanding the cure kinetics, conversion, gelation, and vitrification of the curing thermoset. Gelation and vitrification times can be determined from the intersections of the storage and loss moduli and the maxima in the loss modulus of an isothermal dynamic mechanical spectrum, respectively. Recently, techniques have been developed using rheological and dynamic mechanical analysis instruments to determine the gel point and vitrification [154]. Understanding the gelation and vitrification characteristics of an epoxy/ curing agent system is critical in developing the proper cure schedule/process to achieve optimum performance.

One important application is the management of cure temperatures ($T_c$) and heating rate: if $T_c$ is too low, vitrification may occur before gelation and further reactions may not be completed, resulting in an incomplete network structure and poor performance. This is of particular relevance in ambient cures and radiation cures [155]. Furthermore, attention must be paid to the relationship between mixing of reactants and gel point. Epoxy resins and curing agents must be thoroughly mixed prior to the gel point since the rapid viscosity buildup at gel point inhibits homogeneous mixing of reactants, resulting in potential network and morphological inhomogeneities and defects [156].

Curing and quenching processes of epoxies have been reported to affect performance of certain epoxy coatings and composites. These effects have been attributed to phenomena known as internal or residual stress and physical aging of cured epoxies [147].

Internal stresses arise mainly because of the diminishing capacity of the cross-linked polymer to expand or contract to the same extent (volume) with the solid substrate to which it is adhered. This phenomenon is caused by mismatches of coefficients of thermal expansion (CTE) of the substrates (metal, glass, etc) and the cross-linked epoxies during nonisothermal cures; and cure shrinkage (solvent loss, cross-linking). The effect often contributes to adhesion failures and is more prominent in metal coatings and large composite parts manufacturing, especially when the $T_g$ of the cross-linking polymer approaches $T_c$. As discussed previously, while curing of epoxy functional groups via polycondensation reaction results in relatively low shrinkage, failures attributable to internal stresses such as delamination have been observed in certain epoxy coatings of metal substrates, epoxy encapsulants for electronic devices and glass-fiber-reinforced composites [157]. The effect can be very severe in the case of photoinitiated curing of epoxy acrylates as well as free-radical curing of epoxy vinyl esters. Shorter bonds are formed during these free-radical curing processes, which result in significant shrinkage. Postcure with heat is often required to release some of the internal stresses and to improve adhesion. Efforts have been focused on understanding the mechanism of stress development, and stress minimization by modifications of the cure and postcure cycles [158].

Physical aging is a well-known phenomenon in glassy polymers and has been studied quite extensively in amorphous thermoplastics [159]. The term physical aging refers to the gradual changes in polymer physical properties with time after a glassy polymer is heated above its $T_g$ and rapidly cooled (quenching) to temperatures below $T_g$. The physical aging process differs from chemical aging processes, in which breakage or formation of chemical bonds are involved such as continuing cure, hydrolytic aging, and photochemical and thermal degradation. The phenomenon has been attributed to the non-equilibrium state of the glassy polymer at temperatures below its $T_g$, in which the polymer contains excessive free volume as it is quenched. As the polymer recovers gradually over time to approach equilibrium, a reduction in free volume and an increase in density results. Consequently, the term densification is sometimes used to describe physical aging. For certain epoxy systems, physical aging has been reported to cause increases in stiffness and decreases of toughness [160, 161]. Hardening of certain baked epoxy coatings with time and failures of the coatings due to loss of ductility have been observed. However, physical aging has been reported to be reversible (erasable) by postheating above polymer $T_g$. Proper selection of the cure and postcure schedules including quenching cycle is important to minimize the potential detrimental effects of physical aging [162]. In some epoxy systems, it is difficult to distinguish physical aging from the effects of residual solvent loss and/or continuing cross-linking. They all can contribute to increases in stiffness of the system.

To develop a proper curing process, it is important to understand the reactivity of different curing agents toward the epoxy structure of interest. The effect of hardener structure on reactivity of a cross-linked DGEBA resin (determined by DSC) is shown in Table 19. Aliphatic amines show a maximum reaction rate, called $T_{rmax}$, at 90 °C (heating rate 10 °C/min). The same epoxy resin is somewhat less reactive ($T_{rmax}$ = 126 °C) when homopolymerized via initiators. Aromatic amines and phenols cure considerably more slowly, requiring higher curing temperatures. The highest temperatures are required for dicyandiamide curing, which can, however, be accelerated by basic components.

Relative reaction rates are often expressed in terms of the activation energy $E_a$ (Arrhenius type relationship). $E_a$ allows comparisons of reaction rates at different temperatures and is influenced by the type of chemical reactions involved in the cure. Curing of epoxy resins with phenols or aromatic and aliphatic amines proceeds with a fairly low activation energy of 50–58.5 kJ/mol (12–14 kcal/mol). Activation energies are higher when epoxy compounds having low hydroxyl content are cured alone in the presence of catalysts (92 kJ/mol = 22 kcal/mol) or with dicyandiamide (125.5 kJ/mol = 30 kcal/mol).

## 18.1. Characterization of Epoxy Curing and Cured Networks

Cured thermoset polymers are more difficult to analyze than thermoplastics since they are insoluble and generally intractable. Their properties are influenced by factors at the molecular level, such as backbone structures of epoxy resin and curing agent; nature of the covalent bond developed between the epoxy resin and the curing agent during cross-linking; and density and extent of cross-linking, i.e., degree of cure.

Epoxy resin formulators are concerned with formulation reactivity and flow during application. Reactivity tests or gel time tests are used to determine the proper reactivity of the formulations. Formulators also developed flow tests to check for the formulation rheology profile. The coatings industry widely uses MEK (methyl ethyl ketone) double rubs as an indication of cure. While the test does give a relative indication of cure for a certain system, caution must be exercised when comparing different systems, which may have very different inherent resistance against MEK. In general, these end-use tests do not provide insights on the structure–property relationship of the system.

Epoxy curing process can be monitored by a number of different techniques:

1. Analysis of the disappearance and/or formation of functional groups
2. Indirect estimation of cure conversion
3. Measurements of changes in thermal, physical, and mechanical properties of the system

Comprehensive reviews of different techniques for epoxy cure monitoring are available [86, 94]. Wet chemical or physical analysis methods, such as solvent swell [163], titration of functional groups, IR, near IR [164], or NMR spectroscopy, are commonly used to monitor epoxy cure.

The thermal properties of the system reflect the degree of cure, and thermal analysis of polymers (DSC, DMA, TGA) has been used extensively in studies of epoxy resins [156]. Correlation between $T_g$ and degree of cure has been well established for many systems.

Viscosity build is observed with increased reaction conversion in epoxy curing. More recently, chemorheology, which utilizes rheological measurement (qv) and thermal analysis such as DSC, has been applied to study epoxy cure [166, 167].

Since epoxy curing involves epoxy ring opening and the generation of polar groups, which have a high dipole moment, dielectric measurements have been applied to monitor cures. Dielectric methods [168, 169] encompass both macroscopic and microscopic features: the dipoles being oriented during dielectric measurements are on a microscopic scale, whereas the degree and rate of orientation may depend on macroscopic properties such as viscosity and density.

The mechanical properties of a resin system can also be used to estimate the degree of cure [170]. The methods range from hardness evaluation to complex static measurements or sensitive dynamic mechanical analysis (DMA). Table 20 gives ASTM standard procedures for measuring the properties of cured or partially cured epoxy resin systems.

Direct measurement of the cross-link density of thermoset polymers including those from

Table 20. ASTM[a] procedures for cured or partially cured epoxy resin systems

| Test | ASTM standard |
|---|---|
| Chemical | |
| Density by displacement | D792 |
| Water absorption in plastics | D570 |
| Moisture absorption properties in composites | D5229 |
| Void content in composites | D2734 |
| Electrical | |
| Volume resistivity | D257 |
| Surface resistivity | D257 |
| Dielectric strength | D149 |
| Dielectric constant and dissipation factor | D150 |
| Insulation resistance | D257 |
| Thermal | |
| Heat-deflection temperature | D648 |
| Glass-transition temperature | D696 |
| Dynamic mechanical properties of plastics | D4065 |
| Coefficient of thermal linear expansion | D296 |
| Coefficient of linear thermal expansion by thermomechanical analysis | E831 |
| Coefficient of thermal conductivity | C177 |
| Mechanical | |
| Tensile strength | D638 |
| Compressive strength (plastic) | D695 |
| Compressive testing (composite) | D3410 |
| Flexural strength | D790 |
| Impact strength | D256 |
| Fracture strength in cleavage of adhesives in bonded metal joints | D3433 |
| Fracture strength in "T" peel of adhesives in bonded joints | D903 |
| Fracture testing in 180° peel of adhesives | D5528 |
| Mode I interlaminar fracture toughness of composites | D2344 |
| Apparent interlaminar shear strength of composites | D5045 |
| Plane strain fracture toughness of plastics | D4255 |
| On-plane shear response of composites | |
| Hardness, Barcol | D2583 |
| Hardness, Rockwell M | D785 |

[a]From Ref. [171].

epoxy resins remains one of the most difficult analytical challenges in the field. A far too common approach simply relates the rubbery modulus ($G_r$), the thermoset modulus above $T_g$, to the molecular weight between cross-links ($M_c$) using the theory of rubbery elasticity [133, 134]. Unfortunately thermoset networks have much more complex features than do true elastomers, including non-Gaussian chain behavior, interchain interactions, and entanglements [172]. These factors render rubbery elasticity theory inadequate as an absolute measure of $M_c$ from $G_r$, and doing so can lead to totally erroneous conclusions on the network structure [173]. In a given family of thermosets, changes in $G_r$ can be considered to reflect *relative* changes in $M_c$. Estimates of the expected $M_c$ can be calculated from monomer MW and functionality for stochiometric systems [174]. More extensive network structure calculations including $M_c$ are done using statistical relations developed by MILLER and MAKOSCO [175].

In many applications, epoxy systems derive their high thermal and mechanical performance characteristics from highly cross-linked network structures. However, this often results in brittleness of the epoxy thermosets and loss of end-use properties such as impact resistance. Elongation at break (% elongation) has been a popular test used in the industry for many years to measure toughness, ability to resist failure under tensile stress. While useful in certain applications, good correlations between elongation at break and end-use properties of cured epoxies are not always possible. The failure envelope concept has been useful in looking at the entire time–temperature failure spectrum of epoxies [176]. More recently, progress in the field of fracture mechanics [177, 178] has led to advanced fracture toughness tests that are more useful in characterizing cured epoxy performance. Examples of such tests are critical elastic strain release rate ($G_{IC}$) and critical stress intensity factor ($K_{IC}$) [179].

Dynamic mechanical analysis (DMA) of cross-linked epoxy resins typically shows, in order of decreasing temperature, an α transition corresponding to $T_g$, a β transition associated with relaxation of the glyceryl groups, and a γ transition due to methylene group motions [180]. Both the β and the γ transitions, which are typically observed at −30 to −70 °C and at about −140 °C, respectively, are attributed to crankshaft motions of the polymer chain segments. The appearance of transitions between the α and β transitions is highly variable and has been attributed to segmental motions due to particular curing agents [181]. No definitive correlations between the appearance of sub-$T_g$ relaxations and mechanical properties have been observed [182]. Like many other plastics, cross-linked epoxy resins undergo a change in fracture mechanism from brittle to ductile ($T_b$) with increasing temperature. The window between $T_g$ and $T_b$ has been shown to correlate well with the formability of epoxy can coatings in the draw-redraw (DRD) process [183, 184].

Adhesion is an important issue in epoxy applications since epoxy is almost always used as part of a composite system. Examples are epoxy coatings on metal substrates, epoxy adhesives for metal surfaces, and matrix resin in fiber-reinforced composites such as PCB laminates and aerospace composites. Consequently, optimum epoxy adhesion to the substrate is a prerequisite for good system performance in terms of static and dynamic mechanical properties and environmental durability. In rubber-toughened composite systems, it has been reported that a threshold of interfacial adhesion between both phases (rubber and resin matrix) is needed for maximum toughening by promoting the cavitation mechanism and by activating the crack-bridging mechanism [185]. Excellent review papers are available on the issue of adhesion of epoxy in composites [186], coatings, and adhesives [187]. Effects of internal stresses on coating adhesion failures including the role of coating defects and pigments as potential stress concentrators have been reported [188].

Surface properties such as dynamic contact angle and surface tension are used to ensure proper wetting of epoxy and the substrate. Microscopic techniques, such as scanning electron microscopy (SEM), transmission electron microscopy (TEM), and atomic force microscopy (AFM), are widely used to study morphology, fracture, and adhesion issues of cured epoxy systems. Chemical analysis techniques, such as micro-IR, X-ray photoelectron spectrometry (XPS), and secondary ion mass spectrometry (SIMS), are useful in providing functional group analysis at the interfaces.

Consumers of products which use epoxy resins have developed increasing expectations for longer and more reliable performance. In automobiles, for example, the coating is expected to maintain its initial "Class A" finish for 10 years and the composite leaf spring is designed to last for the life of the vehicle. To meet these expectations, the long-term durability of epoxy thermosets is a key material-specific and application-specific consideration. The durability of polymeric materials in general depends on phenomena such as physical aging, environmental exposure (such as weathering), and mechanical experience (such as impact and load). A detailed discussion of this topic is beyond the scope of this review; interested readers are referred to a leading reference [189].

In addition, the processing of epoxy formulations into their final thermoset structure and form has a major effect on ultimate performance. Material properties such as rheology and reaction kinetics interplay with processing variables such as temperature and shear rate to affect key properties of extent of cure, orientation, and residual stress. Design of the final form of the material also should incorporate fundamental thermoset properties using finite element analysis methods. Optimization of any given epoxy thermoset application is therefore very specific to formulation, processing conditions, and final form and use of the material, and involves the contributions from chemistry, engineering, and material science disciplines to be fully successful.

## 19. Formulation Modifiers

The processing behavior (mainly viscosity and substrate wetting) and other properties of an epoxy system can be modified by diluents, fillers, toughening agents, thixotropic agents, etc. Most commercial epoxy resin systems contain modifying agents.

### 19.1. Diluents

Diluents affect the properties of the cured resin system and, in particular, lower the viscosity in order to improve handling and wetting characteristics. They are often used in the range of 2–20 wt. % based on the epoxy resin. Diluents can be classified into *reactive* and *nonreactive* types.

The *reactive* diluents are products with low viscosity (1–500 cP at 25 °C) used to lower the viscosity of standard epoxy formulations. The effect of reactive diluents on DGEBA viscosity is illustrated in Figure 11. Lower viscosity allows higher filler loading, lower costs, and/or improved processability. Because of the epoxy functionality, the diluents become part of the cured network. However, the reactive diluents can negatively impact properties, and so balancing of viscosity reduction and property loss is an important consideration. Decreases in tensile strength, glass-transition temperature, chemical resistance, and electrical properties are usually observed. Toxicity is another concern, particularly the aromatic mono glycidyl ethers such as phenyl glycidyl ether (PGE) and *o*-cresol glycidyl ether (CGE). *n*-Butyl glycidyl ether (BGE) is one of the most efficient viscosity reducers, but it has been losing favor because of its volatility

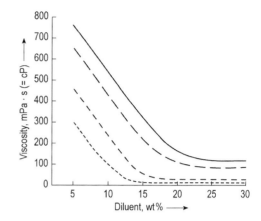

**Figure 11.** Reduction of DGEBA viscosity by reactive diluents: ——, *o*-cresol glycidyl ether; – – –, butanediol diglycidyl ether; ----------, $C_{12}$–$C_{14}$ aliphatic glycidyl ether (Epoxide 8); ··········, *n*-butyl glycidyl ether.

and noxiousness. Longer chain alkyls, polyfunctional or aromatic glycidyl ethers such as bisphenol F epoxy, neopentylglycol diglycidyl ether, and triglycidyl ether of propoxylated glycerine are gaining popularity as epoxy reactive diluents. Cycloaliphatic epoxies and glycidyl esters of acids such as neodecanoic acid are also used as reactive diluents.

Acrylics such as 1,6-hexanediol diacrylate and trimethylolpropane triacrylate are nonepoxy multifunctional diluents, which react readily with primary and secondary amines by means of Michael addition of the the amine to the acrylic double bond [190]. They have been used to increase cure speed or to lower cure temperature of epoxy–amine systems. Caprolactone acrylates have also been used for this application [191].

Solvents and plasticizers are *nonreactive* diluents. The most common nonreactive diluents are nonyl phenol, furfuryl alcohol, benzyl alcohol, and dibutyl phthalate. These materials have the advantage of being able to add to the amine side of the system to better balance mix ratios. Nonyl phenol and furfuryl alcohol also improve wet-out and accelerate cure slightly. They are also capable of reacting with the epoxy group under high temperature cure conditions. Benzyl alcohol is a popular diluent used with amine-cured systems. In addition to viscosity reduction, it is also known to increase cure speed. Benzyl alcohol can be used up to 10 wt. % level without significant effects on cured properties. Dibutyl phthalate is widely used as a nonreactive diluent for liquid resins. However, performance properties will drop off more quickly with increasing levels of nonreactive diluents than with increasing levels of reactive diluents.

Aromatic hydrocarbons, such as toluene or xylene, significantly reduce the viscosity of liquid DGEBA resins, but their use can be accompanied by a 15–25 % decrease in compressive yield strength and a 10–20 % reduction in compressive modulus. If the solvent is trapped in the cured system, solvent resistance is reduced and cracks develop if the resin is used in heat-cured castings. The use of solvents and reactive diluents in epoxy systems is reviewed in References [192] and [193].

## 19.2. Thixotropic Agents

Thixotropy is the tendency of certain colloidal gels to flow when subjected to shear, and then to return to a gel when at rest. A thixotropic gel can be produced through the addition of either high surface area fillers such as colloidal silicas and bentonite clays or of chemical additives. Thixotropy is desirable in applications such as encapsulation where the coating is applied by dipping. The resin will wet out and coat the object being dipped, but will not run off when the object is removed from the dipping bath.

## 19.3. Fillers

Fillers are incorporated in epoxy formulations to enhance or obtain specific desired properties in a system. The type and amount of filler used are determined by the specific properties desired. Fillers can also reduce the cost of epoxy formulations. Inert commercial fillers (qv) can be organic or inorganic, and spheroidal, granular, fibrous, or lamellar in shape. The properties of commercial fillers are given in Table 21, and some effects on epoxy resins are shown in Tables 22 and 23. Some formulations contain up to 90 wt. % fillers. For certain applications, fillers can have significant effects on thermoset morphology, adhesion, and the resulting performance.

Table 21. Typical properties of fillers

| Name | Composition | Particle shape | Surface area volume | Bulk density, kg/m$^3$ | Characteristics and main use |
|---|---|---|---|---|---|
| Marble flour, dolomitic | magnesium–calcium carbonate | granular | medium | 1120–1300 | general-purpose fillers, particularly recommended for castings requiring machining |
| Chalk powder | precipitated calcium carbonate | crystalline | high | 800–880 | |
| Sand | quartz, feldspar, and subsidiary minerals | spheroidal | low | 1500–1700 | bulk filler giving high compressive strength and abrasion resistance; difficult to machine |
| Silica flour | ground quartz | granular | medium | 1100–1150 | standard filler for large electrical castings; high abrasion resistance; difficult to machine |
| Mica flour | muscovite$^a$ | lamellar | high | 300–400 | filler giving high crack resistance to castings exposed to mechanical and thermal shock |
| Slate powder | slate$^a$ | mainly lamellar | medium | 700–900 | general-purpose filler giving high abrasion resistance; difficult to machine |
| Vermiculite$^b$ | vermiculite$^a$ | exfoliated laminae | high | 100–150 | fillers giving lightweight bulk in cores or thick backing to increase the rigidity of thin sections |
| Phenolic microballoons | phenolic resins | hollow spheres | medium | 100–150 | |
| Zircon flour | zircon$^a$ | granular | medium | 1700–1900 | filler giving high abrasion resistance; difficult to machine |
| Aluminum powder | metallic aluminum | granular | medium | 1000–1100 | filler imparting thermal conductivity, e.g., to prevent excessive temperature buildup in electrical components or in tools for hot-forming plastics |
| Chopped glass strand$^c$ | low alkali glass | fibrous | medium | 100–250 | fillers improving the mechanical strength of prominent edges and thin sections |
| Hydrated aluminum oxide | alumina trihydrate | granular | medium | 700–1300 | filler improving wet and dry arc-track resistance and flame retardance |

$^a$Silicate.
$^b$Grain size = 0.15–0.32 cm.
$^c$Length = 0.60 cm.

**Table 22.** Effect of fillers

| Advantages | Disadvantages |
|---|---|
| Lower cost of product | increased weight[a] |
| Reduced shrinkage upon curing | loss of transparency |
| Decreased exothermic temperature rise on curing[a] | tendency to entrap air |
| Increased thermal conductivity[a] | difficulty of machining hard fillers |
| Reduced expansion and contraction with temperature change | decreased impact and tensile strengths |
| Higher deflection temperature | increased dielectric constanta and power factor[a] |
| Improved heat-aging properties[a] | |
| reduced water absorption[a] | |
| Improved abrasion resistance[a] | |
| Increased surface hardness[a] | |
| Increased compressive strength[a] and Young's modulus[a] | |
| Increased electric strength[a] | |

[a]Certain fillers, such as vermiculite and phenolic microballoons, have the reverse effect.

Filler loading is often limited for a given application by the maximum viscosity allowable and/or the reduction in some mechanical properties such as tensile and flexural strength in the cured material. Viscosity can be modified by heat or by addition of a reactive diluent; heating is preferred since diluents affect overall system properties. Some of the major property enhancements affected by fillers are described below.

*Pot life and exotherm.* Fillers can increase pot life and lower exotherm of epoxy systems. Fillers reduce the reactant concentration in the formulation and act as a heat sink. Generally, they have higher heat capacities than the epoxy resins. They are also better heat conductors than the resins, and thus help to dissipate exotherm heat more readily. Commonly used fillers are silica, calcium carbonate, alumina, lithium aluminum silicate, and powdered metals.

*Thermal shock resistance.* Fillers help to increase thermal shock resistance and to decrease the thermal expansion coefficient of an epoxy system by replacing part of the resin with a material that does not change its volume as significantly with temperature variations. Such fillers are clay, alumina, wood flour, sawdust, silica, and mica. Epoxy molding compounds (EMC) can contain up to 90 % of fused silica to manage the thermal stress experienced by encapsulated semiconductors. Powdered metals are used when bonding metals together to better match the coefficient of thermal expansion of the bond with that of the metal, thus minimizing thermal stress.

*Shrinkage.* Using fillers as a partial replacement for a reactive resin that shrinks on curing can reduce shrinkage of the system. Any inert filler will decrease shrinkage, but the most commonly used are silica, calcium carbonate, alumina talc, powdered metals, and lithium aluminum silicate.

*Machinability and abrasion resistance.* The addition of fillers can increase the machinability and abrasion resistance of an epoxy resin system by increasing the hardness of the thermoset. Greater hardness leads to a higher energy required to scratch but cleaner cuts upon machining. Fillers used for this purpose are powdered metals, wood flour, calcium carbonate, sawdust, clay, and talc.

*Electrical conductivity.* In certain applications, conducting fillers are added to epoxy formulations to reduce the good insulating properties of the epoxy systems. The most commonly used fillers are graphite and powdered metals.

Other properties that can be affected with the proper choice of fillers for a specific application include compressive strength, adhesion, arc and tracking resistance, density, and self-lubricating properties.

## 19.4. Epoxy Nanocomposites

Significant recent developments in polymer property enhancement involve polymer nanocomposites. This is a special class of fillers (mostly clay derivatives) in which the nanoscale, highly oriented particles are formed in the polymer matrix through monomer intercalation and particle aggregate exfoliation [194, 195]. The objective is to combine the performance attributes of both hard inorganic and plastic materials. Significant efforts have been dedicated to develop epoxy nanocomposites in the past decade. Improvements in electrical and

**Table 23.** Influence of fillers on epoxy resin properties[a]

| Filler | Ease of dispersion | Antisettling properties | Cost | Shrinkage | Exotherm | Thermal conductivity | Coefficient of expansion | Deflection temperature | Weight | Machinability | Abrasion resistance | Impact strength | Tensile strength | Flexural strength | Compressive strength | Thixotropy | Dielectric constant | Power factor | Electric strength | Protection against Mineral acids | Protection against Caustic alkalies |
|---|---|---|---|---|---|---|---|---|---|---|---|---|---|---|---|---|---|---|---|---|---|
| Marble flour, dolomitic | P | N | − | − | − | + | − | + | + | O | O | − | − | − | + | O | + | O | O | − | − |
| Chalk powder | N | N | − | − | − | + | − | O | + | O | O | − | − | − | O | O | + | O | O | − | − |
| Sand | P | P | − − | − | − | + | − − | + | + | − | + + | − | O | O | + | O | + | + | + | O | O |
| Silica flour | P | N | − | − | − | + | − | + | + | − | + + | + | O | O | + + | O | + | + | + | O | O |
| Mica flour | N | N | − | − | − | + | − | + | + | − | + | − | O | O | O | O | + | + | + | O | O |
| Slate powder | P | P | − | O | + + | + | − | O | + | + + | + + | + | O | O | + | O | + | + | + | O[b] | O[b] |
| Vermiculite | N | N | − − | − − | + + | − | O | O | − − | + + | − − | − | − − | − − | − − | O | − | − | − − | O[b] | O[b] |
| Phenolic microballoons | P | P | − | O | − | − | O | O | − − | − | − − | − | − | − | − − | O | − | − | − − | − | − |
| Zircon flour | P | N | − | − | − | + + | − | + | + + | + | + + | − | O | O | + | O | + + | + | O | O | O |
| Aluminum powder | P | N | O | − | − | + + | − | + | + | − | − | + + | + | + | + | O | + | O | − | − | − |
| Chopped glass strand | N | P | O | − | − | + | − | O | + | − | + | + + | + | + | + | O | + | + | O | O | O |
| Asbestos fiber | P | P | O | − | − | + | − | + | + | − | O | + | + | O | + | O | + | + | O | O | O |
| Hydrated aluminum oxide | N | P | − | − | − | + | − | + | + | O | O | − | − | − | + | O | + | + | O | O | O |

[a] Key: P = positive effect; N = negative effect; O = no significant effect; − = significant decrease; − − = large decrease; + = significant increase; + + = large increase; · = fillers taken for arbitrary standard for comparison of dispersibility and setting.
[b] Porosity of filler reduces protection provided by resin.

Table 24. Effect of flexiblizers

| Flexiblizers | Concentration, % | Advantages | Disadvantages |
|---|---|---|---|
| Poly(propylene glycol) diglycidyl ether | 10–60 | low viscosity, good flexibility | poor water resistance fair impact resistance |
| Polyaminoamides | 30–70 | good abrasion resistance, good flexibility | fair chemical resistance |
| Liquid polysulfides | 10–50 | good corrosion resistance, excellent flexibility | odor Poor heat resistance tendency to cold flow |
| Aliphatic polyester adducts | 10–30 | good water resistance fair flexibility over a range of temperatures | high viscosity |

mechanical properties, chemical resistance, high temperature performance, and flame retardancy have been reported. Other silica-based organic hybrids have been developed [196] for military and aerospace applications.

The emerging field of nanotechnology has produced new materials such as the carbon nanotubes, which are filaments of carbon with atomic dimensions. Recent publications claimed exceptional property enhancement from nanotube-laced epoxy (hardness, electrical and heat-conducting properties) [197]. However, cost remains a barrier for commercialization.

## 19.5. Toughening Agents and Flexiblizers

Some cross-linked, unmodified epoxy systems exhibit brittleness, poor flexibility, and low impact strength and fracture resistance. Modifiers can be used to remedy these shortcomings. However, there usually will be some sacrifices of properties. In general, there are two approaches used to modify epoxies to improve these features.

1. *Flexiblization.* Aliphatic diepoxide reactive diluents enhance the flexibility or elongation by providing chain segments with greater free rotation between cross-links. Polyaminoamide hardeners, based on aliphatic polyamines and dimerized fatty acids, perform similarly. Liquid polysulfide polymers possessing terminal mercaptan functionality improve impact properties in conjunction with polyamine hardeners.

    Flexible chain segments are incorporated in an epoxy resin by many means [189]. One approach is the incorporation of oligomeric aliphatic polyesters containing carboxylic acid end groups, forming an epoxy resin adduct. This is one of the reasons that epoxy–polyester hybrid powder coatings have become very popular. The effects of flexiblizers are shown in Table 24. Flexiblization can enhance elongation of the system but is often accompanied by a reduction of glass-transition temperature, yield stress, and elastic modulus. Other properties (e.g. water absorption and thermal and chemical resistance) may also be affected.

2. *Toughening* refers to the ability to increase resistance to failure under mechanical stress. Epoxies derive their modulus, chemical, and thermal resistance properties from cross-link density and chain rigidity. Increasing cross-link density to meet higher thermal requirements ($T_g$) often comes at the expense of toughness. Toughening approaches for epoxies [199–202] include the dispersion of preformed elastomer particles into the epoxy matrix and reaction-induced phase separation of elastomers or thermoplastic particles during cure.

    Elastomers such as carboxyl-terminated poly(butadiene-*co*-acrylonitrile)s (CTBN) have been popular tougheners for epoxies. Toughening by elastomers can be attributed to the incorporation of a small amount of elastic material as a discrete phase of microscopic particles embedded in the continuous rigid resin matrix. The rubbery particles promote absorption of strain energy by interactions involving craze formation and shear deformation. Craze formation is promoted by particles of 1–5 µm size, and shear deformation by particles > 0.5 µm. Systems possessing both small and large particles, i.e., bimodal distribution, provide maximum toughness [203]. The rubber is incorporated in the epoxy resin in a ratio of 1 : 8 in the presence of an esterification catalyst. The product is an epoxy ester capped with epoxy groups. The adduct is then formulated with unmodified resin and cured with standard hardeners and accelerators. Phase separation, of the adduct occurs during the curing process, resulting in the formation of segregated domains of elastomer-like particles covalently bound to the epoxy resin matrix. Optimum particle size and particle-size distribution, phase separation, and phase morphology are crucial for the development of desirable properties of the system. If the elastomer remains soluble in the epoxy matrix, it serves as a flexiblizer and reduces the glass-transition temperature significantly. Some reductions in $T_g$ and modulus are typical of CTBN-modified epoxies. Amine-terminated poly(butadiene-*co*-acrylonitrile)s (ATBN) are also available [68].

Elastomer-modified epoxy resins are used in composites and structural adhesives, coatings, and electronic applications. A similar approach to toughen epoxy vinyl esters using other elastomeric materials has been reported [204]. Other elastomer-modified epoxies include epoxy-terminated urethane prepolymers, epoxy-terminated polysulfide, epoxy–acrylated urethane, and epoxidized polybutadiene. Preformed dispersions of epoxy-insoluble elastomers have been developed and reported to achieve toughening without $T_g$ reduction [205, 206].

Other epoxy toughening approaches include chemical modifications of the system either through the epoxy backbone and/or crosss-linker. Dow Chemical developed a cross-linkable epoxy thermoplastic system (CET) [207]. The concept involves introducing stiffer polymer segments into the network structure to maintain the glass-transition temperature while allowing cross-link density reduction to improve toughness. Thermoplastics, core-shell rubbers (CSR), and liquid crystal polymers (LCP) have also been used. Semi-interpenetrating network (IPN) approaches involve formation of a dispersed, cross-linked epoxy second phase in a thermoplastic matrix. The systems were reported to have good combinations of toughness, high $T_g$, high modulus, and processability.

Incorporation of block copolymers has been shown to improve toughness of certain epoxy systems [208]. More recently, nanocomposites and self-healing epoxy systems [209] represent new approaches to develop more damage-tolerant epoxies.

Through the proper selection of resin, curing agent, and modifiers, the cured epoxy resin system can be tailored to specific performance characteristics. The choice depends on cost, processing and performance requirements. Cure is possible at ambient and elevated temperatures. Cured epoxies exhibit good combinations of outstanding properties and versatility at moderate cost: excellent adhesion to a variety of substrates; outstanding chemical and corrosion resistance; excellent electrical insulation; high tensile, flexural, and compressive strengths; good thermal stability; relatively low moisture absorption; and low shrinkage upon cure. Consequently, epoxies are used in diverse applications.

## 20. Coatings Applications

Commercial uses of epoxy resins can be generally divided into two major categories: protective coatings and structural applications. U.S. consumption of epoxy resins is given in Figure 12. The largest single use is in coatings (> 50 %), followed by structural composites. Among the structural composite applications, electrical laminates contribute the largest epoxy consumption. A similar trend is observed for the European market, but the Asian consumption is heavily tilted toward electrical laminate and electronic encapsulant applications [210]. Electrical and electronic applications account for the largest consumption of epoxy resins in Japan (> 40 %). In 2000, it is estimated that the Asia-Pacific region consumed up to 70 % of all epoxies used in electrical laminate production worldwide. While the overall epoxy markets continue to grow at a steady pace over the past two decades, more rapid growth has occurred in powder coatings, electrical laminates, electronic encapsulants, adhesives, and radiation-curable epoxies.

The majority of epoxy coatings are based on DGEBA or modifications of DGEBA. Chemical and corrosion-resistant films are obtained by curing at ambient and/or elevated temperatures. Ambient temperature cured coatings primarily involve cross-linking of the epoxy groups in mostly two-package systems, while elevated

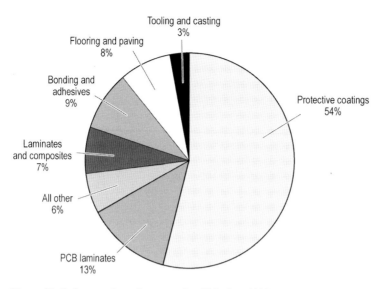

**Figure 12.** End-use markets of epoxy resins (U.S. data, 2000)

temperature cured coatings in one-package systems take advantage of the reactivity of both the epoxy and the secondary hydroxyl groups. As a class, epoxy coatings exhibit superior adhesion (both to substrates and to other coatings), chemical and corrosion resistance, and toughness. However, epoxy coatings have been employed mainly as primers or undercoats because of their tendency to yellow and chalk on exposure to sunlight.

Epoxy-based coatings are the preferred and dominant choices for cathodic electrodeposition of automotive primers, marine and industrial maintenance coatings, and metal container interior coatings. Use of epoxy flooring for institutions and industrial buildings has been growing at a steady rate as the industry becomes more aware of its benefits.

Solvents are commonly used to facilitate dissolution of resins, cross-linkers, and other components, and for ease of handling and application. Although most of the epoxy coatings sold in the 1970s were solventborne types, they made up only 40 % of epoxy coating consumption in 2001 [211]. Economic and ecological pressures to lower the volatile organic content (VOC) of solventborne coatings have stimulated the development of high solids, solvent-free systems (powder and liquid), and waterborne and radiation-curable epoxy coatings technologies [212]. These environmentally friendly coating technologies have experienced rapid growth in the past decade. For example, epoxy powder coatings have been growing at rates exceeding those of other coating technologies as new applications such as automotive primer-surfacer and low temperature cure coatings for heat-sensitive substrates are developed. Radiation-curable liquid coatings based on epoxy acrylates and cycloaliphatic epoxies have also been growing significantly over the last decade. The current distribution of coating technologies is summarized in Figure 13.

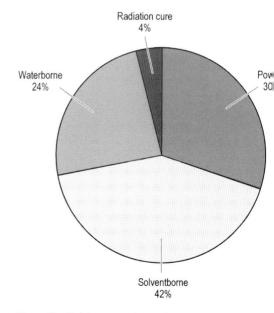

**Figure 13.** Global epoxy coating application technologies [211]

## 20.1. Coatings Application Technologies

### 20.1.1. Low Solids Solventborne Coatings

These traditional low solids coatings contain less than 60 % solids by volume (typically 40 %). Their advantages include established application equipment and experience, fast drying and cure at ambient temperatures and excellent film formation at extremely fast cure conditions like those used in coil coatings (< 30 s, > 200 °C). However, because of stricter VOC regulations, solvent-based coatings have been losing market share steadily to more environmentally friendly technologies.

### 20.1.2. High Solids Solventborne Coatings

High solids coatings contain 60–85 % by volume of solids. They are mostly based on standard LERs or low molecular weight SERs modified by reactive diluents, low viscosity multifunctional aliphatic epoxies, or bisphenol F epoxy resins. High film build is one key advantage of high solids coatings. Examples include the coal-tar epoxy coatings that contain up to 85 % solids used in industrial protective coatings.

### 20.1.3. Solvent-Free Coatings (100 % Solids)

Ecological concerns have led to increasing uses of these materials. Low viscosity LERs based on bisphenol A and bisphenol F epoxies are often used in combination with reactive diluents. The advantages include high buildup in a single application, minimization of surface defects owing to the absence of solvents, excellent heat and chemical resistance, and lower overall application costs. Disadvantages include high viscosity, difficulties to apply and produce thin films, poor impact resistance and flexiblity, short pot life, and increased sensitivity to humidity. Weatherable cycloaliphatic epoxies can be used to formulate solvent-free thermally curable coatings because of their low viscosities [213].

### 20.1.4. Waterborne Coatings

In the switch from solventborne to waterborne systems, epoxies are successfully bridging the gap largely by adaptation of conventional resins. Waterborne coatings accounted for almost 25 % of epoxy coating consumption in 2001.

In addition to the waterborne epoxy dispersions which are typically supplied by epoxy resin producers, significant advances in waterborne coatings have been made by coatings producers such as PPG Industries, ICI Paint, and others utilizing modified epoxies. PPG coatings are used in cathodic electrodeposition systems that are widely accepted for automobile primers. Many patents have been issued for this important technology [214]. The Glidden Co. (now ICI Paint) developed a waterborne system for container coatings based on a graft copolymerization of an advanced epoxy resin and acrylic monomers [215]. These two waterborne epoxy coatings were significant breakthroughs in the coatings industry in the 1970s and are still widely used today.

For ambient temperature cure applications such as industrial maintenance and marine coatings, LERs or low molecular weight SERs (type 1 resin) are dispersed in water with a surfactant package and small amounts of co-solvents [216]. Some producers offer waterborne curing agents that, typically, are salts of polyamines or polyamides. Key disadvantages include higher costs, slow cure at ambient and humid conditions, and tendency to cause flash rush. In addition, expensive stainless steel equipment are required for application. Recent developments include the elimination of co-solvents in some epoxy dispersions [217]. Custom synthesized acrylic latexes have shown promise when thermally cured with cycloaliphatic epoxies [218]. While the overall volume is still relatively modest (estimated at < 20,000 MT in 2000 for the global market), it is expected that future growth rate for this segment will be much higher than standard epoxy resins, particularly in Europe where environmental pressures are stronger.

### 20.1.5. Powder Coatings

Epoxy-based powder coatings exhibit useful properties such as excellent adhesion, abrasion resistance, hardness, and corrosion and chemical resistance. The application possibilities are diverse, including metal finishing, appliances, structural rebars, pipes, machinery and

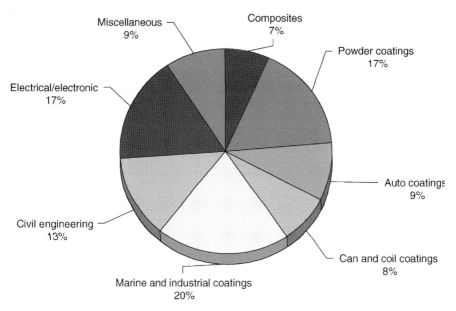

**Figure 14.** Major global epoxy resin markets (Dow Chemical data, 2001)

equipment, furnitures, and automotive coatings. Together, these applications accounted for 30 % of epoxy coatings (Fig. 13) and 17 % of epoxy resin consumption globally in 2001 (Fig. 14). This is a high growth segment of epoxy coatings (see → Paints and Coatings, 1. Introduction).

The development of highly reactive powder systems which cure using low energy (150 °C) and the possibility of economical thin films (30–40 µm) have made powder coatings competitive with waterborne and high solids systems. Powder coatings can be applied by fluidized-bed (thick films, 50–150 µm) or electrostatic spray (thin films, 30–40 µm).

In powder coatings, epoxies are continuing to grow at rates exceeding other technology segments mainly because of the 100 % solids feature, improved coverage, and recyclability of overspray materials. Pipeline projects, important because of worldwide energy problems, are significant consumers of epoxy powder coatings. The value of improved service life is being increasingly accepted even at the somewhat higher material cost of epoxy systems. Four types of epoxy resin systems are commonly used as powder coatings.

1. *Epoxy powder coatings* are based on SERs of intermediate molecular weight (800–2000 EEW). They provide good flow and reactive terminal epoxy functionality. The properties of these thermoset coatings depend on the curing agents, which are friable solids such as dicyandiamide (DICY), phenol-terminated epoxy hardener, and anhydrides. Epoxy powder coatings are generally employed for interior or undercoat uses. Functional epoxy powder coatings are thick films (0.1–0.5 mm, 5–20 mil) used to protect automotive and truck parts, pipe, and concrete reinforcing bars. Fusion-bonded-epoxies (FBE), first developed by 3M Co., are epoxy powder coatings used to protect oil and gas pipelines where long-term corrosion protection under adverse conditions (for example, the Alaska oil and North Sea underwater pipelines) is critical. The performance requirements for FBEs are challenging as the hard, cross-linked coatings are expected to survive the pipe bending/unwinding processes and handling abuses in the field. FBEs are also used to protect rebars embedded in critical concrete structures such as bridges, tunnels, and highways. Their primary function is to extend the lifetime of the concrete structures (5–10 years when built with uncoated rebars) to 20–30 years, reducing maintenance and repair costs. Epoxy powder coatings also serve as electrical insulation for bus bars, motor

armatures, and similar articles. Decorative epoxy coatings are applied as a thin film (0.02–0.1 mm, 1–5 mil) and used mainly in appliance and general metal product applications. Coating for heat-sensitive substrates is an emerging market for epoxy powders.

2. *Epoxy–polyester hybrids* are mixtures of solid epoxy resins based on bisphenol A and acid-terminated polyester solid resin (25–85 acid equivalent weight). These hybrids are typically less expensive than the epoxy-based powder coatings and offer improved weatherability, and better resistance to overbake yellowing while retaining many of the properties of the standard epoxies. Corrosion resistance is equivalent to epoxy powders in most cases, although solvent and alkali resistance is inferior. One significant new application of the epoxy–polyester hybrids is the primer-surfacer coating for automobiles. Primer-surfacer coatings based on epoxy–polyester hybrid is applied in between the epoxy primer and the topcoats. Its functions are to provide intercoat adhesion and to improve the chip resistance properties of the coatings. Automakers have also found that the epoxy–polyester hybrid primer-surfacer give a smoother surface under the top coats, resulting in a better quality appearance [219]. Epoxy–polyester hybrids have experienced exceptional growth in the global market in the past decade.

3. *Polyester–TGIC*, a third type of epoxy powder coating, is based on a mixture of polyester polycarboxylic acids (18–37 acid equivalent mass) and triglycidyl isocyanurate (TGIC). The TGIC-cured powder coatings have excellent UV resistance, good gloss and color retention, as well as good adhesion and mechanical properties. They were originally developed in Europe for coating metal window frames and buildings, exterior siding, outdoor hardware, high quality outdoor furniture, and other articles requiring superior outdoor durability. These polyester–TGIC powder coatings have gained popularity worldwide.

In recent years, there have been concerns over toxicity of TGIC, and a number of potential replacement compounds have been developed. Among these, β- hydroxyalkylamide (HAA), trade named Primid by EMS-Chemie is gaining in popularity, particularly in Northern Europe.

Hydroxy alkylamide (HAA)

4. *Glycidyl methacrylate–acrylic.* These powder coatings are based on copolymers of glycidyl methacrylate (GMA) and acrylic monomers. They are often thermally cured with dodecanedioic acid (DDDA) and are used in automotive primer-surfacer and clear coats of luxury automobiles such as BMW [220, 221].

Recent developments in powder coatings based on epoxy include UV-curable GMA–acrylic coating for automotive parts; lower temperature cure coatings for heat-sensitive substrates such as wood [222, 223] and plastics; and dual cure (thermal/UV) systems.

## 20.1.6. Radiation-Curable Coatings

UV and electron beam (EB) radiation curable coatings [74] is a fast-growing segment of epoxy coatings, increasing at 8–10% annual growth rate. The technology is environmental-friendly. No solvent is used, and volatile emission is essentially eliminated. Cure is highly effective and energy-efficient at ambient temperatures, lending the technology highly applicable to heat-sensitive substrates such as wood, plastics, and paper. Capital cost requirements are low especially when compared to new thermal ovens for solventborne and waterborne coatings. EB cure is a relatively new technology which initiates cure via highly energetic electron beams, and unlike UV cure, no photoinitiator is needed. However, EB capital cost is higher.

Epoxy acrylates are widely used as the base resin in many UV-initiated free-radical cure varnish formulations. Epoxy acrylates provide varnishes with excellent scuff resistance, high gloss, and good adhesion. Major markets are overprint varnishes for papers (books, magazines, cards, labels, etc) and exterior can coatings. Wood furniture and particle board are new but growing markets for this technology. Alternatively, cycloaliphatic epoxies can be UV-cured via a photo-initiated cationic mechanism. They are used in metal container exterior overprint varnish and inks, and high performance electronic applications. While cycloaliphatic epoxies are more expensive than epoxy acrylates, they offer several advantages: better adhesion to metals, fewer hazards in handling, and continued curing in the dark (which is important in certain applications). A related and high value market is inks and resists, where radiation and heat-curable epoxies and epoxy acrylates are used. As discussed in the powder coatings section, radiation-curable epoxy powder coatings are being developed for a number of applications.

## 20.2. Epoxy Coatings Markets

The major global market segments of epoxies are represented in Figure 14. The marine and industrial protective coating is the largest market for epoxy coatings, followed by powder coatings, automotive, and container coatings.

### 20.2.1. Marine and Industrial Maintenance Coatings

The combined marine and industrial maintenance coatings application constitutes the largest epoxy coating market segment globally. While the end-use markets and application requirements are different, the basic epoxy systems utilized in these two markets are quite similar. The basic function of these coatings is to protect metal and concrete structures from degradation in aggressive environments for extended periods of time. The long service life of the coating and/or extended intervals between repairs are critical requirements, especially for marine applications because of the high costs of dry-docking of ships for re-painting. The excellent corrosion, abrasion, and chemical resistance properties of epoxy coatings allow their dominant position in these markets. They are used in new construction as well as in maintenance and repair works. Examples are corrosion-resistant coatings for ships, shipping containers, offshore oil rigs and platforms, transportation infrastructures such as bridges, rail car coatings, coatings for industrial storage tanks, and primers for light industrial and agricultural equipment.

Most coatings used in these markets are two-component systems applied and cured at ambient conditions. LERs and low molecular weight SERs based on bisphenol A and bisphenol F epoxies are commonly used [224]. Aliphatic polyamines, adducts of epoxy resins with aliphatic and aromatic amines, ketimines, phenalkamines, amidoamide, and polyamide resins are employed as curing agents. The working pot life of the amine–epoxy resin systems depends on the curing agent, solvents, catalysts, and temperature. High solids solventborne coatings are most popular. Tighter VOC regulations have facilitated the development of lower VOC, 100% solids, and waterborne epoxy coating systems. Important types of epoxy coatings in this segment include the following.

1. *Two-component epoxy–amine coatings* are used primarily as a primer or mid-coat over the inorganic zinc-rich primer coating. High solids epoxy mastics can be applied over contaminated substrates and form thick, good barrier coatings. These coatings account for the majority of epoxies used in marine and industrial protective coatings markets.

2. *Organic (epoxy) zinc-rich primers* are used in place or to repair imperfections of the inorganic zinc-rich primer. Their advantages over inorganic zincs include improved adhesion to the epoxy primer coating and better tolerance to poor surface cleaning.

3. *Coal-tar epoxies* are historically some of the most popular high solids epoxy coatings, having excellent water barrier, chemical resistance properties, and low costs. They are typically cured with polyamides and are used as ship bottom or primer coatings for tanks, pipes, and steel pilings. However, their use has been declining or banned in certain countries because of concerns and regulations over the toxicity of coal-tar as a suspected carcinogen.
4. *Epoxy esters* are used as primers in less demanding applications. Their performance is inferior to epoxy–amine systems but their costs are lower.
5. *Waterborne coatings.* These are based on two-component epoxy–polyamine/polyamidoamine or epoxy–acrylic latex hybrids. One limitation of the waterborne systems is their poor cure in high humidity conditions. They have made some penetration in industrial maintenance coatings and are expected to grow more significantly in the future.

Epoxy coatings are known to have poor weatherability and often chalk when exposed to sunlight for long periods of time. Over the years, significant efforts have been dedicated to develop weatherable epoxy resin systems such as hydrogenated bisphenol A epoxy cured with siloxane-modified epoxy curing agent [225], but higher costs and compromises in curing characteristics and performance limited their commercialization. Today, most industrial structures are only coated with epoxy coatings which can last up to 10–15 years. When appearance is critical, epoxy primers are often top-coated with aliphatic isocyanate based polyurethane coatings. Marine coatings have very diverse requirements depending on the specific functions of the parts of the ship being coated. For example, ship decks are coated with antislip, abrasion- and corrosion-resistant epoxy coatings; the cargo tanks require highly chemical-resistant coatings; ship exteriors above the water line are coated with epoxy primers followed by urethane top coats for appearance; underwater ship bottoms are coated with multilayer coatings including a zinc-rich or epoxy primer, epoxy intermediate coats and antifouling top coats based on vinyls or acrylates.

Concerns over the safety of large tankers have led to regulations and construction of double-hull ships, increasing epoxy coatings consumption. While Japan was the center of the ship-building industry since the 1980s, Korea and China have emerged as major players in this market because of their low cost advantages. According to data from the Japanese Ship Building Industry Association, Korea has overtaken Japan as the global leader in ship building in the year 2000. The combined market shares of these three Asian countries now account for more than 80% of the global ship-building business. In addition, China already owns 80% of the world shipping container construction business.

The migration of ship building yards to Korea and China has led to significant increases in marine epoxy coatings consumption in that region, and has resulted in increased demands for lower temperature cure (LTC) epoxy systems. Traditional ambient-cure epoxy coatings do not cure well at temperatures below 10 °C (50 °F). They often require excessive cure time, affecting productivity and performance, and shorten the painting season in colder climates such as in Korea. A number of LTC epoxy coating technologies have been developed. Uses of accelerators such as tertiary amines, organic acids or alkyl-substituted phenols have allowed cures at 4.45 °C (40 °F) temperature range. However, shorter pot-life is a limitation of these systems. Newer epoxy coatings utilizing cycloaliphatic amines, phenalkamines as curing agents can cure at temperatures of about 0 °C (30 °F). For industrial protective coatings, systems developed by Ameron International (Amerlock 400) and ICI-Devoe (Bar-Rust 235) are claimed to achieve LTC down to −18 °C (0 °F).

Other new technology developments in this market segment include surface-tolerant epoxy coatings for aged or marginally-prepared surfaces, interval-free epoxy coatings to extend coating service life, mineral spirit-soluble epoxy coatings for shipping containers repairs, and styrene-free coatings to replace foul-smelling and regulated organic solvents such as toluene and xylene in coatings for new shipping containers.

### 20.2.2. Metal Container and Coil Coatings

Metal container and coil coatings represent a major outlet for epoxy resins considering there are more than 100 billion beverage cans and 30 billion food cans produced annually in the United States. Globally, the metal can market is estimated at over 300 billion cans. While the majority of metal containers coated with epoxy coatings are aluminum and steel food and beverage cans, coatings for drums, pails, and aerosol spray cans are included in this market segment. Coil coating is a highly efficient, automated coating process used to produce precoated metal coils, which are subsequently stamped and fabricated to parts. The majority of epoxy coil coatings are used to produce metal can ends and can bodies with smaller amounts going to building products, appliance panels, transportation, and metal furniture applications.

Higher molecular weight SERs (EEW = 2000–4000), which contain predominantly secondary hydroxyl groups, are used in these coatings where maximum resistance to chemicals, good flexibility, freedom from off-taste, good thermal stability, blush (hydrolysis) resistance, and the ability to hold corrosive foods and beverages are needed. In addition, compliance with food regulations such as the Food and Drugs Administration (FDA) rules is required for food and beverage interior can coatings. This application is where the unique combination of properties of epoxy resins stand out.

The can and coil coatings, generally, are cross-linked with phenol, melamine, or urea–formaldehyde condensation products at elevated temperatures (150–200 °C) with acid catalysts. Normal epoxy–amino resin weight ratios are epoxy–urea, 70 : 30; epoxy–benzoguanamine, 70 : 30; epoxy–melamine, 80 : 20 and 90 : 10. Increasing crosslinker levels give improved thermal and chemical resistance at the sacrifice of coating flexibility and adhesion.

Phenol–formaldehyde resole cured epoxies have excellent chemical resistance and hardness and are the popular choices for drum coatings. Their golden color is affecting their uses in food can coatings because of the increasing popularity of the water-white coatings based on melamine–formaldehyde resins, which are perceived to be "cleaner" by the consumers and the food industry. Melamine–formaldehyde resins are the primary crosslinker for beer and beverage interior can coatings. Urea–formaldehyde resins can be cured at lower temperatures and faster speed than phenol and melamine–formaldehyde resins and are widely used in the coil coatings industry where cure schedules are extremely short. However, their use has been declining because of concerns over the release of formaldehyde fumes. Recently, there have been regulatory issues in the can industry concerning worker exposure to volatile formaldehyde emissions from the formaldehyde resins. A number of new, formaldehyde-free coating formulations have been introduced by coating suppliers [226].

High solids binders for metal can coatings have been developed on the basis of dimer acid modification of epoxy resins, whereby a flexible $C_{34}$ difunctional acid is used to esterify a conventional diepoxide resin [227]. The resultant epoxy ester possesses a sufficiently lower viscosity to provide binders with solids contents > 70 vol. %. Curing is accomplished by a melamine–formaldehyde resin (Cymel 303, from Cytec) in conjunction with phosphoric acid catalyst.

In the 1970s, a waterborne coating system for aluminum beverage can coatings was developed by the Glidden Company (ICI Packaging Coatings) on the basis of a graft copolymerization of an advanced epoxy resin and acrylic monomers [228, 229]. The acrylic–vinyl monomers are grafted onto preformed epoxy resins in the presence of a free-radical initiator; grafting occurs mainly at the methylene group of the aliphatic backbone on the epoxy resin:

composite can ends. The technology is environmentally friendly and energy-efficient.

Coil coatings have been gaining in the appliance market. More OEMs have turned to precoated metal coils as an efficient manufacturing alternative to produce appliance panels, eliminating the needs for postformed coating processes. PVC organosol (copolymers of vinyl chloride and vinyl acetate) coatings for coil-coated can ends and bodies have been under environmental pressures and epoxy has been gaining as PVC coatings are replaced [230].

The growth of can coatings has been steady globally because of the expansion of new can plants in Asia-Pacific and South America in the 1980s and early 1990s, which made up for the stagnant growth of the U.S. market. However, growth of plastic bottles based on PET [poly (ethylene terphthalate)] has recently eroded the metal can position in beverage packaging, affecting epoxy can coatings growth. In addi-

The polymeric product is a mixture of methacrylic acid–styrene copolymer, SER, and graft copolymer of the unsaturated monomers onto the epoxy resin backbone. It is dispersible in water upon neutralization with an amine, and cured with an amino–formaldehyde resin. The technology revolutionized the can coatings industry in the 1970s which was primarily based on low solids, solventborne coatings. This waterborne epoxy coating system and its variations continue to be the dominant choices for interior beer and beverage can coatings globally today. They are also formulated with phenol–formaldehyde resole and used as interior coatings in the new two-piece food can plants in the United States.

UV-curable coatings based on cycloaliphatic epoxies are used on the exterior of some beer, beverage, and food cans, as well as food and

tion, new can fabrication technologies utilizing other polymers are being developed which may challenge the dominant position of epoxy coatings in metal cans.

In the 1980s, Toyo Seikan Co. of Japan successfully developed and commercialized TULC (Toyo Ultimate Laminate Cans), a revolutionary technology in which cans are fabricated using a deep draw process from metal coils laminated with thermoplastic polyester films [231, 232]. No epoxy coating is used in this technology. Special polyester film combinations were used (in a much higher thickness than typical epoxy coatings) to facilitate the demanding deep draw process while maintaining all of the other requirements of can coatings. This technology is a significant breakthrough with claimed benefits such as no solvent emission, lower energy and water usage, and excellent quality cans. The costs however are significantly higher than those of conventional cans, and the technology has found widespread application only in Japan where higher packaging costs are acceptable. Other companies such as British Steel have been actively promoting laminated cans as a way to produce differentiable packaging like shaped cans with very limited success. Higher cost is the biggest barrier to their broad commercialization. Recent developments include attempts to fabricate can ends and bodies from extrusion-coated metals by companies such as Alcoa. Thermoplastics like modified polyesters are providing challenges to epoxies in these new technologies due to their excellent formability. However, their resistance against aggressive drinks, foods, and retort are inferior to those of epoxies.

More recently in the United States, Campbell Soup Co. has successfully launched a new line of microwaveable, ready-to-eat plastic cans. These cans are constructed from a molded thermoplastic can body (polypropylene, high density polyethylene) and an easy-open-end (EOE) of coated metal. In addition, flexible pouches have made inroad as an alternative for metal cans in certain markets such as packaged tuna fish.

Recently, there have been debates in the can coatings industry concerning the potential health effects of residual bisphenol A and DGEBA in epoxy can coatings. The resin suppliers, can coatings producers, and can makers have jointly formed an industry group to coordinate a number of studies on this issue. Results indicated that epoxy can coatings, when properly formulated and cured, are safe and in compliance with global food contact regulations. Current regulatory guidelines such as the Specific Migration Levels for Europe set extractable limits of 1 mg/kg for DGEBA and 3 mg/kg for bisphenol A. Additional information is available in the references [233, 234]. Some polyester coatings have been developed as epoxy coating alternatives, but high costs and inferior pasteurization-resistance limit their uses [235].

### 20.2.3. Automotive Coatings

Automotive coatings are another major application for epoxy resins. The excellent adhesion and corrosion resistance properties of epoxies make them the overwhelming choice for automotive primers. One new, growing application is the use of epoxy–polyester or acrylic–GMA powders in primer-surfacer coatings. In addition, glycidyl methacrylate (GMA) is used as a comonomer in etch-resistant liquid top coats containing acrylic acid/anhydride [236] and in GMA-acrylic powder coatings for clear coats and automotive parts [220]. Epoxy powder coatings for automobiles are expected to grow significantly in the near future.

Electrodeposition processes using epoxy-based automotive primers were developed for anodic and cathodic systems. Anodic systems (AED) employ carboxylated epoxy resins neutralized with an amine. A typical binder is prepared by the esterification of the terminal epoxy groups of a solid resin (EEW = 500) with stoichiometric quantities of dimethylolpropionic acid to form a hydroxyl-rich resin. This intermediate is subsequently treated with a cyclic anhydride to form an acid functionalised polymer, which is then neutralized with the amine.

Significant advances in waterborne automotive coatings have been made by PPG Industries and others utilizing epoxies as co-resins in the 1970s. These coatings are used in cathodic electrodeposited (CED) systems, which are widely accepted for automobile primers. Many patents have been issued for this important technology [214]. Cathodic systems, which have superior corrosion resistance, have replaced anodic systems. A typical epoxy binder for cathodic electrodeposition is prepared by first forming a tertiary amine adduct from an epoxy resin and a secondary amine, followed by neutralization with an acid to form a water-soluble salt:

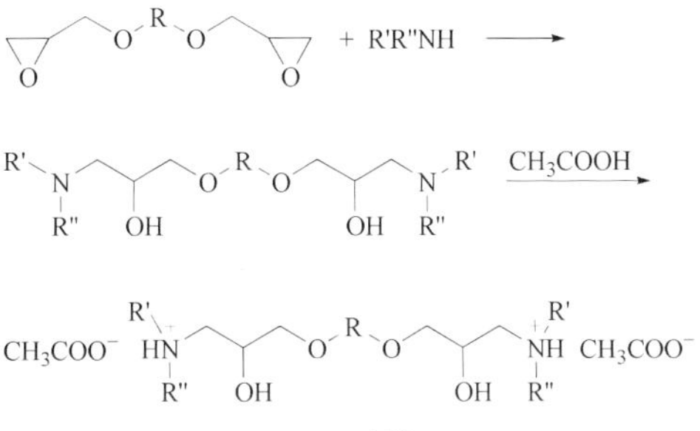

Cross-linking is achieved by reaction of the hydroxyl groups with a blocked isocyanate, which is stable at ambient temperature.

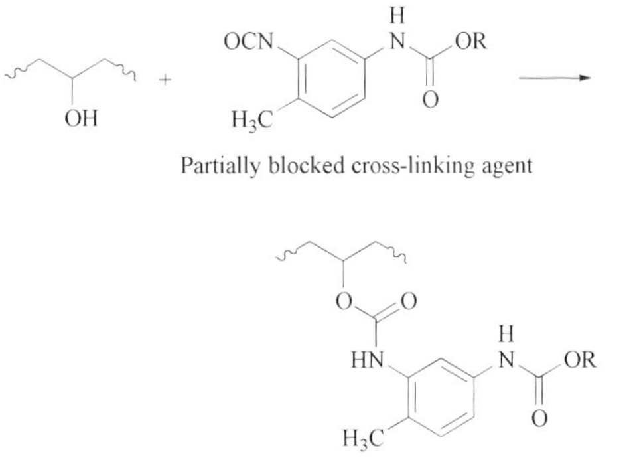

where R = 2-Ethylhexyl

The ability of the CED coating system to thoroughly coat all metal surfaces of the car and the resultant superior corrosion resistance was a significant breakthrough, enabling its dominant position in the global automotive industry.

PPG has continued to develop new generations of improved CED epoxy coatings [237]. Dupont, BASF, and a number of Japanese coating companies such as Nippon Paint and Kansai Paint have contributed to the epoxy primer coating technology by developing advanced coating systems to meet higher performance and regulatory requirements of the automotive industry [238–240]. The popular pigment systems based on heavy metals such as lead and chromium in primer coatings have been recently banned in certain countries, leading to efforts to develop new formulations with improved corrosion resistance. Nippon Paint has proposed pigment-free CED systems [241].

Epoxy–polyester and acrylic–GMA powder coatings have made significant advances recently in the area of primer-surfacer coatings. They offer better adhesion to topcoats and significantly improve chip resistance compared to the traditional liquid polyester and epoxy ester coatings. This translates to warranty cost reductions, leading many car manufacturers to convert to the powder coating technologies.

While epoxy coatings based on DGEBA and other aromatic epoxies are limited to undercoats and under-the-hood applications because of their poor UV resistance, GMA-based coatings have been developed for improved acid-etch performance automotive top coats. They compete with traditional acrylic polyol–melamine topcoats that are highly susceptible to acid rain-induced hydrolysis, and offer better mar resistance and less worker exposures than isocyanate-based topcoats [242, 243]. BMW has coverted to a GMA–acrylic powder clear coat developed by PPG.

### 20.3. Inks and Resists

Inks and resists comprise a relatively small but high value and growing market for epoxies and epoxy derivatives. In 2001, there were an estimated of 6800 MT of epoxies and epoxy derivatives used in this market to produce ink and resist formulations worth almost $400 million in the U.S. market. Epoxies are often used with other resins such as polyester acrylates and urethane acrylates in these formulations. The largest applications are lithographic and flexographic inks followed by electronic inks and resists.

Resist technology is widely used in the electronics industry to manufacture printed circuits. The resist (a coating or ink) is applied over a conducting substrate such as copper in a pattern to protect its surface during etching, plating, or soldering. Cure is either by radiation or heat. The uncured coating (or ink) is removed later by solvents. Solder masks perform similar functions in the manufacturing of printed circuit boards. The growth of the computer and electronics industries has fueled growth of epoxy-based inks and resists. The market is projected to grow at 10 % annually.

The primary resins used in this market are the radiation-curable epoxy acrylates, accounting for 60 % of the resins used. A small amount of cycloaliphatic epoxies are also used in UV-curable inks and resists. Phenol and cresol epoxy novolacs, and bisphenol A based epoxies are used in thermally cured formulations. The epoxy novolacs are used where higher heat resistance is needed such as in solder masks. Both free-radical and cationic-curable UV inks and colored base coats have grown rapidly because of the needs for higher line speeds, faster cleanup or line turnaround, less energy consumption, less capital for a new line, and fewer emissions.

A unique epoxy (epoxy chalcone) produced by Huntsman can be used for dual cure [244]:

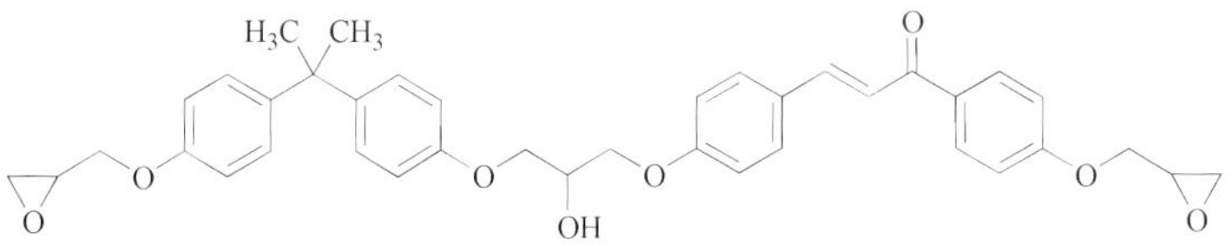

structural applications can be divided into three major areas: fiber-reinforced composites and electrical laminates; casting, encapsulation, and tooling; and adhesives. Within this segment, the largest applications are electrical laminates for PCB and composites made of epoxy and epoxy vinyl ester for structural applications.

## 21.1. Structural Composites

Epoxy resins and epoxy vinyl ester resins are well suited as fiber-reinforcing materials because they exhibit excellent adhesion to reinforcement, cure with low shrinkage, provide good dimensional stability, and possess good mechanical, electrical, thermal, chemical, fatigue, and moisture-resistance properties. Epoxy composites are formed by aligning strong, continuous fibers in an epoxy resin-curing agent matrix. Processes currently used to fabricate epoxy composites include hand lay-up, spray-up, compression molding, vacuum bag compression molding, filament winding, resin transfer molding reaction, injection molding, and pultrusion (see → Composite Materials, Section 4.2.).

Important fiber materials are surface-treated glass, boron, graphite (carbon), and aromatic polyamides (e.g. DuPont's Kevlar). In most composites the reinforcement constitutes ca. 65 % of the final mass. Orientation of the fibers is important in establishing the properties of the laminate. Unidirectional, bidirectional, and

Radiation-initiated free-radical cure is possible via the double bonds, while the epoxy groups are available for thermal cure. Epoxy chalcone is used as a photopolymerizable solder mask and in photoresists.

# 21. Structural Applications

Next to coatings, structural applications account for the second largest share of epoxy resin consumption (∼ 40 %). Epoxy resins in

random orientations are possible. The characteristics of the cured resin system are extremely important since it must transmit the applied stresses to each fiber. A critical region in a composite is the resin–fiber interface. The adhesive properties of epoxy resins make them especially suited for composite applications.

The most important market for epoxy composites is for corrosion-resistant equipment where epoxy vinyl esters is the dominant material of choice. Other smaller markets are automotive, aerospace, sports/recreation, construction, and marine. Because of their higher costs, epoxy and epoxy vinyl esters composites found applications where their higher mechanical strength and chemical and corrosion resistance properties are advantageous.

### 21.1.1. Epoxy Composites

Composites made with glass fibers usually have a bisphenol A based epoxy resin–diamine matrix and are used in a variety of applications including automotive leaf springs and drive shafts, where mechanical strength is a key requirement. A large and important application is for filament-wound glass-reinforced pipes used in oil fields, chemical plants, water distribution, and as electrical conduits. Low viscosity liquid systems having good mechanical properties when cured are preferred. These are usually cured with liquid anhydride or aromatic–amine hardeners. Similar systems are used for filament-winding pressure bottles and rocket motor casings. Other applications that use fiber-reinforced epoxy composites include sporting equipment, such as tennis racquet frames, fishing rods, and golf clubs, as well as industrial equipment. The wind energy field is emerging as a potential high growth area for epoxy composites, particularly in Europe where a number of new wind energy farms are planned. With windmill blades increasing in lengths (up to 50 m), the strength and fatigue properties of epoxy composites provide benefits over competitive chemistries.

In the aerospace industry, particularly in military aircraft construction, the use of graphite fiber-reinforced composites has been growing because of high strength-to-weight ratios. Some newer commercial airliners now contain up to 10 % by weight of composite materials. High performance polyfunctional resins, such as the tetraglycidyl derivative of methylenedianiline in combination with diaminodiphenylsulfone or nadic methyl anhydride, are used to provide good elevated temperature properties and humidity resistance. Handling characteristics are well suited to the autoclave molding technique primarily used in the manufacture of such components. The low viscosity and high $T_g$ of cycloaliphatic epoxies has led to their use in certain aerospace applications. Newer resins such as diglycidyl ether of 9,9′-bis(4-hydroxyphenylfluorene) have been developed.

While the overall growth of composites in the aerospace industry is continuing, epoxy has been facing stiff competition from other materials and the growth rate has been relatively small (2 % annually). While epoxies are still used in many exterior aircraft parts, carbon fiber composites based on bismaleimide and cyanate esters have shown better temperature and moisture resistance than epoxies in military aircraft applications. In the commercial aircraft arena, phenolic composites are now preferred for interior applications because of their lower heat release and smoke generation properties during fires. High performance thermoplastics, such as polysulfone, polyimides, and polyetherether ketone (PEEK), have also found some uses in aerospace composites.

### 21.1.2. Epoxy Vinyl Ester Composites

Epoxy vinyl ester composites are widely used to produce chemically resistant glass-reinforced pipes, stacks, and tanks by contact molding and filament-winding processes. Epoxy vinyl ester resins provide outstanding chemical resistance against aggressive chemicals such as aqueous acids and bases and are materials of choice for demanding applications in petrochemical plants, oil refineries, and paper mills. Epoxy vinyl ester composites are also used in demanding automotive applications such as engine and oil pan covers where high temperature performance is required. Exterior panels and truck boxes are also growth automotive applications for vinyl esters. However, in less demanding automotive applications, cheaper thermoplastics and thermosets such as unsaturated polyesters or furan resins are often used. In general, epoxy vinyl ester is considered to be a premium polyester resin with higher temperature and corrosion resistance properties at higher costs. It is used where the cheaper unsaturated polyesters cannot meet performance requirements. For the same reason, epoxy vinyl ester has not grown significantly in less demanding civil engineering applications. Other uses of epoxy vinyl ester composites include boat hulls, swimming pools, saunas, and hot tubs.

Improved versions of the high performance resin systems continue to be developed [245, 246]. Toughening of epoxies and epoxy vinyl esters has emerged as an area for investigation

[247]. Lower styrene content vinyl esters have been developed to reduce worker exposure. Performance enhancements with epoxy and vinyl ester nanocomposites have been reported in the literature, but commercialization has not been yet realized.

### 21.1.3. Mineral-Filled Composites

Epoxy mineral-filled composites are widely used to manufacture laboratory equipment such as lab bench tops, sinks, hoods, and other laboratory accessories. The excellent chemical and thermal resistance properties of epoxy thermosets make them ideal choices for this application. Typically, liquid epoxy resins of bisphenol A are cured with anhydrides such as phthalic anhydride, which provide good exotherm management and excellent thermal performance. The systems are highly filled with fillers such as silica or sand (up to 70 wt. %). Multifunctional epoxy novolacs can be added when higher chemical and thermal performance is needed.

## 21.2. Civil Engineering, Flooring, and Construction

Civil engineering is another large application for epoxies, accounting for up to 13 % of total global epoxy consumption. This application includes flooring, decorative aggregate, paving, and construction [248]. Key attributes of epoxies such as ease of installation, fast ambient cure, good adhesion to many substrates, excellent chemical resistance, low shrinkage, good mechanical strength, and durability make them suitable for this market. In the United States an estimated 20,000 MT of resins were used for flooring applications in 2000. The building boom in China has provided significant growth for this market during the past decade. Epoxy flooring compounds are expected to grow well as the construction industry becomes more aware of their benefits.

Epoxy resins are used for both functional and decorative purposes in monolithic flooring and in factory-produced building panel applications. Products include floor paints, self-leveling floors, trowelable floors, and pebble-finished floors. Epoxy floorings provide wear-resistant and chemical-resistant surfaces for dairies and food processing and chemical plants where acids normally attack concrete. Epoxies are also used in flooring for walk-in freezers, coolers, kitchens, and restaurants because of good thermal properties, slip resistance, and ease of cleanup. In commercial building applications, such as offices and lobbies, terrazzo-like surfaces can be applied in thin layers. Continuous seamless epoxy floors are competitive with ceramic tiles. They are usually applied by trowel over a prepared subfloor. Semiconductive epoxy/carbon black floorings are used in electronics manufacturing plants because of their ability to dissipate electrical charges. Decorative slip-resistant coatings are available for outdoor stair treads, balconies, patios, walkways, and swimming-pool decks. Epoxy aggregates are highly filled systems, containing up to 90 % of stones or minerals. They are used for decorative walls, floors, and decks.

Usually, two-component systems consisting of liquid epoxy resin, diluents, fillers (e.g. sands, stones, aggregates), pigments, thickening agents, and polyamine or polyamide curing agents are employed. Cycloaliphatic amines and their adducts are used when either better low temperature cure or adhesion to wet concrete is desired. The other components of the flooring formulation are as critical as the resin and hardener. Typical filler and pigment levels are 10 % for paving, 30 % for flooring, and 40 % or higher for decorative aggregates. Self-leveling floors consist of resin-hardener mixtures with low filler content or unfilled compositions with high gloss. In epoxy terrazzo floors, an epoxy binder replaces the cement matrix in a marble aggregate flooring, providing impact resistance, mechanical strength, and adhesion.

Epoxy systems for roads, tunnels and bridges are effective barriers to moisture, chemicals, oils, and grease. They are used in new construction as well as in repair and maintenance applications. Typical formulations consist of liquid epoxy resins extended with coal tar and diethylenetriamine curing agent. Epoxy resins are widely used in bridge expansion joints and to repair concrete cracks in adhesive and grouting (injectable mortar) systems. Epoxy pavings are used to cover concrete bridge decks and parking structures. Formulations of epoxy resins and polysulfide polymers in conjunction with polyamine curing agents are used for bonding concrete to concrete. After cleaning the old surface, the epoxy adhesive is applied and good adhesion between the old and the new concrete is obtained.

Recent developments in the construction and civil engineering industry include the development of "intelligent concrete" with self-healing capability in Japan [249]. Some of the systems are based on epoxy resins encapsulated in concrete which when triggered by cracks open and cure to repair the concrete.

## 21.3. Electrical Laminates

Printed wiring boards (PWB) or printed circuit boards (PCB) are used in all types of electronic equipment. In noncritical applications such as inexpensive consumer electronics, these components are made from paper-reinforced phenolic, melamine, or polyester resins. For more critical applications such as high end consumer electronics, computers, complex telecommunication equipment, etc., higher performance materials are required and epoxy resin-based glass fiber laminates fulfill the requirements at reasonable costs. This application constitutes the single largest volume of epoxies used in structural composites. In 2000, an estimated 200,000 MT of epoxy resins were used globally to manufacture PCB laminates.

Systems are available that meet the National Electrical Manufacturers Association (NEMA) G10, G11, FR3, FR4, FR5, CEM-1, and CEM-3 specifications. Both low viscosity liquid (EEW = 180–200) and high melting solid (EEW = 450–500) epoxy resins are used in printed circuit prepreg manufacture. Currently, the most widely used boards (> 85 %) are manufactured to the flame-retardant FR4 specification using epoxy thermosets. Flame retardance is achieved by advancing the liquid DGEBA epoxy resin with tetrabromobisphenol A (TBBA). This relatively low cost resin which contains about 20 wt. % bromine is the workhorse of the PCB industry. Epoxy resins based on diglycidyl ether of TBBA are also available, which allow the preparation of resins with even higher bromine content, up to 50 wt. %. Multifunctional epoxy resins such as epoxy novolacs based on phenol, bisphenol A, and cresol novolacs or the tetraglycidyl ether of tetrakis (4-hydroxyphenyl)ethane are used as modifiers to increase the glass-transition temperature ($T_g$ > 150 °C), thermal decomposition temperature ($T_d$), and chemical resistance.

The most commonly used curing agent for PWBs is dicyandiamide (DICY) catalyzed with imidazoles such as 2-methylimidazole (2-MI), followed by phenolic novolacs and anhydrides.

The epoxy–DICY systems offer the following advantages:

1. Cost effectiveness (DICY is a low equivalent weight, multifunctional curing agent)
2. Stable formulations
3. Excellent adhesion to copper and glass
4. Good moisture and solder resistance
5. Good processability

The primary disadvantage of the standard epoxy–DICY systems is their relatively low thermal performance ($T_g$ < 140 °C, $T_d$ = 300 °C), which limits their uses in more demanding applications such as the FR-5 boards and other high density circuit boards. Specialty epoxy–DICY systems are available with $T_g$ approaching 190 °C but at higher costs. Alternatively, high temperature epoxy systems are obtained using diaminodiphenyl sulfone (DDS) as curing agent and boron trifluoride monoethylamine ($BF_3$/MEA) complex, benzyldimethylamine (BDMA), or various imidazoles as catalysts. However, concerns over the toxicity of DDS have led to significant decrease of its use. More recently, higher thermally resistant laminates using novolac curing agents, including bisphenol A based novolacs, have become popular in the industry. However, brittleness is a significant disadvantage of these systems.

Prepreg is commonly prepared by passing the glass cloth through a formulated resin bath followed by heat treatment in a tower to evaporate the solvent and partially cure the resin to an intermediate or B stage. Prepreg sheets are stacked with outer layers of copper foil followed by exposure to heat and high pressure in a laminating press. This structure is cured (C-staged) at high temperature (150–180 °C) and pressure for 30–90 min. Attempts to develop continuous prepreg and laminating processes have only achieved limited commercialization. Laminate boards may be single-sided (circuitry printed on only one side), double-sided, or multilayered (3 to 50 layers) for high density circuitry boards. Electrical connections for mounted components are obtained via drilled holes which are plated with copper.

The 1990s witnessed the explosive growth of the personal computer, consumer electronics, and wireless telecommunication industries, resulting in significant demands for PWB based on epoxy resins. The PWB industry trends toward device miniaturization, multilayer laminates, high density circuitries, lead-free solder, and faster signal transmission speeds have resulted in increased performance requirements. For example, lead-free legislation which bans electronics containing lead in the European Union became law in 2003 with an implementation date of 2006. This legislation is expected to speed up the phase-out of lead-based solders globally, forcing the industry to use alternatives such as tin alloys which have much higher soldering temperatures, and thereby drives the need for epoxy systems with higher thermal performance.

The end-use industries' demands for PCB boards with better heat resistance [250, 251], higher glass-transition temperature ($T_g$), higher thermal decomposition temperatures ($T_d$), lower water absorption, lower coefficient of thermal expansion (CTE), and better electrical properties (dielectric constant $D_k$ and dissipation factor $D_f$) have led to the development of new, high performance epoxies and cross-linker systems [252]. Toughness is also becoming an issue as electrical connection holes are drilled in the highly cross-linked, high $T_g$ laminates.

Since reinforcing materials make up from 40 to 60 wt.% of the PCB laminates, their contributions to the laminate dielectric properties are significant. The standard reinforcing glass–cloth compositions in electrical laminates are designated E (electrical) glass. Woven E glass is most commonly used, but other reinforcing materials such as nonwoven glass mat, aramid fiber, S-2 glass, and quartz are available. In recent years, the PCB industry has been evaluating materials with better dielectric properties, but they are much more expensive than standard E glass (Table 25).

In recent years, environmental concerns over toxic smoke generation during fire and end-of-

Table 25. Reinforcing material comparison$^a$

| Reinforcing material | $D_k$ (at 1MHz) | $D_f$ (at 1MHz) | Relative cost |
|---|---|---|---|
| E glass | 6.5 | 0.003 | 1 |
| S-2 glass | 5.3 | 0.002 | 4 |
| D glass | 3.8 | 0.0005 | 10 |
| Quartz | 3.8 | 0.0002 | 30 |
| Aramid | 3.8 | 0.012 | 10 |

$^a$From Ref. [253].

life incineration of electronic equipment containing brominated products, particularly in Europe and Japan, have driven development efforts on halogen-free resins. This has resulted in a number of alternative products such as phosphorous additives and phosphor-containing epoxies [254–256]. Some examples of these phosphorous compounds are as follows:

9,10-Dihydro-9-oxa-10-phosphathenanthrene-10-oxide

6H-Dibenz(C,H)(1,2)oxaphosphorin-6-oxide

However, commercialization of phosphor-containing epoxies has been limited because of higher costs and other disadvantages such as poorer moisture resistance and lower thermal performance. In addition, concerns over phosphine gas emission during fires and potential leakage of phosphorous compounds in landfills have raised questions about their long-term viability. Alternatively, the industry has been researching new epoxy resins based on nitrogen, silicon, sulfur-containing compounds, and new phenolic resins as potential halogen-free, phosphor-free replacements. Inorganic fillers such as alumina trihydrate, magnesium hydroxide, and zinc borate have also been evaluated as flame-retardant alternatives in epoxy systems.

While brominated epoxy resin remains the workhorse of the PCB industry (FR-4 boards) because of its good combination of properties and cost, it is facing competition from other thermoset and thermoplastic materials as industry performance requirements increase. Thermosets with higher temperature performance ($> 180\,°C$ $T_g$) and lower dielectric properties include polyimides, cyanate esters, and bismaleimide–triazine (BT) resins. They are used alone or as blends with epoxies in high performance chip-packaging boards and military applications. GE's GETEK system is an interpenetrating network of polyphenylene oxide (PPO) in epoxy and has lower dielectric constant than standard epoxies. Polytetrafluoroethylene (PTFE) has a very low dielectric constant (Table 26) and is used primarily in high performance PCBs for military and high frequency (e.g. radars) applications. While these alternative materials offer certain performance advantages over standard epoxies, they are generally more expensive and more difficult to process. Thermosets such as polyimides, cyanate ester, and BT resins are very brittle and have higher water absorption than epoxies. PTFE has very poor adhesion to substrates, requiring special treatments. Consequently, they are limited to niche, high performance applications [250].

In flexible printed circuits, polyimide and polyester films are the preferred choices over epoxies. Molded interconnects based on heat-resistant thermoplastics such as polyether sulfone, polyether imide, and polyarylate have been developed to replace epoxy-based PCBs in certain applications. However, their uses are limited to special applications.

There has been a significant migration of the PCB laminate manufacturing capacity to Asia (mainly Taiwan and China) in the late 1990s. In 2001, 70% of epoxy resins used in PCB laminates was consumed in the region and the trend is expected to continue in the near future.

## 21.4. Other Electrical and Electronic Applications

### 21.4.1. Casting, Potting, and Encapsulation

Since the mid-1950s, electrical-equipment manufacturers have taken advantage of the good electrical properties of epoxy and the design freedom afforded by casting techniques to produce switchgear components, transformers, insulators, high voltage cable accessories, and similar devices.

In casting, a resin-curing agent system is charged into a specially designed mold containing the electrical component to be insulated. After cure, the insulated part retains the shape of the mold. In encapsulation, a mounted electronic component such as a transistor or semiconductor in a mold is encased in an epoxy resin based system. Coil windings, laminates, lead wires, etc, are impregnated with the epoxy system. Potting is the same procedure as encapsulation except that the mold is a part of the finished unit. When a component is simply dropped into a resin-curing agent system and cured without a mold, the process is referred to as dipping. It provides little or no impregnation and is used mainly for protective coatings.

The choice of epoxy resin, curing agent, fillers, and other ancillary materials depends on factors such as cost, processing conditions, and the environment to which the insulated electrical or electronic component will be exposed.

**Table 26.** Base resin systems used in PCB laminates

| Resin system | $T_g$, °C | $D_k$ (at 1MHz) | $D_f$ (at 1MHz) | Estimated relative resin cost | Laminate cost |
|---|---|---|---|---|---|
| Standard Epoxies | 135–140 | 4.6–4.8 | 0.015–0.020 | 1 | 1 |
| High performance Epoxy | 170–180 | 4.6–4.8 | 0.015–0.020 | 1.5–2 | 1.5 |
| PPO/Epoxy | 175–185 | 3.6–4.2 | 0.009–0.015 | 4–6 | 2–3 |
| BT/Epoxy | 170–220 | 3.9–4.2 | 0.008–0.013 | 8–15 | 2–5 |
| Polyimide | 260 | 3.9–4.4 | 0.012–0.014 | 5–16 | 3–6 |
| Cyanate ester | 230–260 | $3.5–3.7^a$ | $0.005–0.011^a$ | 5–16 | 4–8 |
| Polyester | 135–140 | $3.1–3.2^b$ | $0.004–0.014^b$ | | 7–10 |
| PTFE | NA | $2.1–2.5^b$ | $0.0006–0.0022^b$ | 40 | 15–50 |

$^a$Measured at 1GHz.
$^b$Measured at 10 GHz.

The type and amount of filler that can be incorporated into the system are very important and depend on the viscosity of the resin at the processing temperature. Filler loading reduces costs, increases pot life, improves heat dissipation, lowers exotherms, increases thermal shock resistance, reduces shrinkage, and improves dimensional stability.

The exotherm generated during the resin cure must be controlled to prevent damage to the electrical or electronic component. The exotherm is easily controlled during the production of small castings, pottings, and encapsulations. In the production of large castings, the excess heat of reaction must be dissipated in order to prevent locked-in thermal stresses. During the 1970s, the pressure gelation casting process was developed [257]; this method provides better temperature control and reduces cycle times. The heat generated by polymerization is used to heat the resin mass and is not dissipated in the mold.

Both DGEBA and cycloaliphatic epoxy resins are used in casting systems. Most systems are based on DGEBA resins cured with anhydride hardeners and contain 60–65 wt. % inert fillers. The cycloaliphatic resin systems exhibit good tracking properties and better UV resistance than DGEBA resins, the latter of which causes crazing and surface breakdown. An electrical current is more likely to form a carbonized track in aromatic-based resins than in nonaromatic ones. Their lower viscosity also facilitates device impregnation. The cycloaliphatic epoxies are often used as modifiers for DGEBA resin systems. This application represents a significant outlet for cycloaliphatic epoxies.

Amine curing agents are used in small castings, and anhydrides are used in large castings. Anhydrides are less reactive and have lower exotherms than amines. In addition, their viscosity and shrinkage are low and pot lives are longer.

### 21.4.2. Transfer Molding

Epoxy molding compounds (EMC) are solid mixtures of epoxy resin, curing agent(s) and catalyst, mold-release compounds, fillers, and other additives. These systems can be formulated by dry mixing or by melt mixing and are relatively stable when stored below room temperature. Molding compounds become fluid at relatively low temperatures (150–200 °C) and can be molded at relatively low pressures (3.5–7.0 MPa) by compression, transfer, or injection molding. Advantages of molding over casting are elimination of the mixing step immediately before use, improved handling and measuring procedures, and suitability for high production quantities. A typical standard EMC formulation contains approximately 30 % epoxies, 60 % filler, and 10 % of curing agents and other additives such as release agent.

An important application of epoxy molding compounds is the encapsulation of electronic components such as semiconductor chips, passive devices, and integrated circuits by transfer molding. Transfer molding is a highly automated, efficient method of encapsulation. High purity phenol and cresol epoxy novolacs and phenol and cresol novolacs and/or anhydride curing agents are used most often in semiconductor applications. For passive device encapsulation, standard epoxy novolacs can be used as blends with bisphenol A based solid resins. The ECN or EPN molding powders can be processed at relatively low pressures and provide insulation for the electronic components. Ionic impurities, i.e., NaCl or KCl, must be kept to a minimum, since trace quantities can cause corrosion and device failure. In addition, residual stress and thermal and mechanical shock resistances are issues that must be managed properly [258].

Efforts have been made to improve the high temperature performance of these systems by replacing the epoxy novolacs with other multifunctional epoxy resins. Hydrocarbon based epoxy novolacs (HEN) were developed to improve the moisture resistance of molding compounds. Crystalline epoxy resins derived from biphenol and dihydroxy naphthalenes were developed for high end semiconductor encapsulants using Surface Mount Technology (SMT). The emergence of SMT as a key semiconductor manufacturing technology requires epoxy molding compounds with a high filler loading capacity (up to 90 wt. %) to enhance solder crack resistance. SMT uses new solder alloys to attach components to the PCB board at high temperatures (215–260 °C). Solder reflow, delamination, and package cracks are problems often encountered with conventional molding compounds based on cresol epoxy novolacs. The high filler content helps lower costs, reduces moisture absorption, and decreases the thermal expansion coefficient of the system. Crystalline products with very low melt viscosity such as biphenyl epoxies facilitate the processing of the high silica filler formulations while maintaining other critical requirements: moisture resistance and electrical, thermal, and mechanical properties [61]. The majority of high purity epoxies used in epoxy molding compounds (EMC) for semiconductor encapsulations are supplied by Japanese producers and a few Asian companies.

## 21.5. Adhesives

Epoxy-based adhesives provide powerful bonds between similar and dissimilar materials such as metals, glass, ceramics, wood, cloth, and many types of plastics. In addition, epoxies offer low shrinkage, low creep, high performance over a wide range of usage temperatures, and nobyproducts (such as water) release during cure. The epoxy adhesives were originally developed for use in metal bonding in the aircraft industry [259, 260]. In aircraft wing assemblies, high strength epoxy adhesives are used in place of metal fasteners to avoid corrosion problems inherent with metal fasteners, to reduce weight, and to eliminate "point" distribution by spreading the load over a large area. Today, epoxy is the most versatile engineering/structural adhesive, widely used in many industries including aerospace, electrical/electronic, automotive, construction, transportation, dental, and consumer. The market is of high value, consuming 25,000 MT of epoxies in North America in 2001 worth almost $500 million.

The broad range of epoxy resins and curing agents on the market allows a wide selection of system components to satisfy a particular application. Although the majority of epoxy adhesives are two-pack systems, heat activated one-pack adhesives are also available. Low molecular weight DGEBA liquid resins are the most commonly used. Higher molecular weight (EEW = 250–500) DGEBA epoxy resins improve adhesive strength because of the increased number of hydroxyl groups in the

**Table 27.** Epoxy adhesive lap-shear strengths

| Hardener | Lap-shear strength,[a] MPa[b] |
|---|---|
| Aliphatic polyamine | 19 |
| Polythiol cohardener | 18 |
| Aromatic diamine | 24 |

[a] Adhesive strength.
[b] To convert MPa to psi, multiply by 145.

resin backbone. For applications requiring high temperature or improved chemical performance, the multifunctional epoxy phenol novolac and triglycidyl-$p$-aminophenol resins are employed. More recent products include vinyl epoxies. Adhesive systems modified with reactive diluents facilitate wetting of the substrate, allowing more filler to be added and modifying handling characteristics; however, adhesive strength is reduced. Toughened epoxy adhesives are available.

Polyamines or polyamides are the curing agents for ambient or slightly elevated temperature cures, and aromatic polyamines or anhydride hardeners are used for hot cures. These systems provide exceptional bonding strength but slower cure time. Boron trifluoride amine complexes and dicyandiamide are used in one-component adhesives. Polythiols (polysulfides, polymercaptans) are the fast-curing hardeners in "5-min." consumer epoxy formulations. The lap-shear strengths of a DGEBA epoxy cured with different hardeners are given in Table 27.

Cationically cured UV laminating adhesives based on cycloaliphatic epoxies are emerging as an alternative to solvent-based adhesives. The "dark cure" of cationics allows UV exposure and post lamination in line. This process does not require UV exposure "through" the plastic barrier material.

Epoxy adhesives are expected to grow at GDP (3–4 %) over the next decade. Increased usage in the automotive and recreational markets, and replacement of mechanical fasteners help offset the slowdown in the aerospace industry (see also Adhesive Compositions).

## 21.6. Tooling

Tools made with epoxy are used for producing prototypes, master models, molds and other parts for aerospace, automotive, foundry, boat building, and various industrial molded items [261]. Epoxy tools are less expensive than metal ones and can be modified quickly and cheaply. Epoxy resins are preferred over unsaturated polyesters and other free-radical cured resins because of lower shrinkage, greater interlaminar bond strength and superior dimensional stability.

Most epoxy-based tooling formulations are based on liquid DGEBA resins. Aliphatic polyamines, amidoamines, or modified cycloaliphatic amines are used for ambient temperature cure, and modified aromatic diamines and anhydrides are used for high temperature cure. When high heat resistance is required (> 350 °F) epoxy novolac resins can be employed. Reactive diluents such as aliphatic glycidyl ethers are often employed to permit higher filler load or to reduce the system viscosity for proper application. Fillers, reinforcing fibers, toughening agents, thixotropic agents, and other additives are often used depending on the desired application and final properties.

Tooling production uses four major processing methods: lamination, surface cast, splining, and casting. Lamination is made by alternating layers of glass cloth or fabric and formulated resin, usually on a framework of metal or plastic. Surface cast utilizes a filled resin compound that is applied onto the surface of a mold, which is later filled with a core material that adheres to the casting compound. Splining employs heavily filled formulations that are directly applied to a surface and manually molded or leveled to the desired shape, before or after curing, with the help of proper tools. Lastly, casting compounds are filled formulations that are directly poured or compressed into a mold coated with a release agent.

## 22. Health and Safety Factors

There have been many investigations of the toxicity of various classes of epoxy-containing materials (glycidyloxy compounds). The use and interpretation of the vast amount of data available has been obscured by two factors: (*1*) proper identification of the epoxy systems in question and (*2*) lack of meaningful classification of the epoxy materials. In general, the toxicity of many of the glycidyloxy derivatives is low, but the diversity of compounds found within this group does not permit broad generalizations for the class. Information on toxicity and safe handling of epoxy compounds are summarized in References [262, 263] and [264].

*Diglycidyl ether of bisphenol A.* Bisphenol A based epoxies are the most commonly used resins. Although unmodified bisphenol A epoxy resins have a very low order of acute toxicity, they should be handled carefully and personal contact should be avoided. Prolonged or repeated skin contact with liquid epoxy resins may lead to skin irritation or sensitization. Susceptibility to skin irritation and sensitization varies from person to person. Skin sensitization decreases with an increase in MW, but the presence of low MW fractions in the advanced resins may present a hazard to skin sensitization. Inhalation toxicity does not present a hazard because of low vapor pressure. DGEBA-based resins have been reported to cause minimal eye irritation. Toxicological studies support the conclusion that bisphenol A based epoxy resins do not present a carcinogenic or mutagenic hazard.

Because of the solvents used, solution of epoxy resins are more hazardous to handle than solid resins alone. Depending on the solvents used, such solutions may cause irritation to the skin and eyes, are more likely to cause sensitization responses, and are hazardous if inhaled.

*Epoxy phenol novolac resins.* Acute oral studies indicate low toxicity for these resins. Eye studies indicate only minor irritation in animals. The EPN resins have shown weak skin sensitizing potential in humans.

*Low MW epoxy diluents,* particularly the aromatic monoepoxides such as phenyl glycidyl ether (PGE) are known to have high toxicity and should be handled with care. They are capable of causing skin and eye irritation and sensitization responses in people. They may also present a significant hazard from inhalation.

*Curing agents.* In general, amine curing agents are much more hazardous to handle than the epoxy resins, particularly at elevated temperatures. Aliphatic amines and anhydrides are capable of serious skin or eye irritation, sensitization, and even burns. Other curing agents possess consideration variation in the degree of health hazards because of the variety of their chemical structures and it is impossible to generalize.

All suppliers provide material safety data sheets (MSDS), which contain the most recent toxicity data. These are the best sources of information and should be consulted before handling the materials.

## 23. Acknowledgments

The authors would like to acknowledge the contributions of Robert F. Eaton of the Dow Chemical Co. in Bound Brook, N.J., who contributed to the sections on cycloaliphatic epoxies and epoxidized vegetable oils and the cationic curing mechanism. We also would like to thank Timothy Takas of Reichhold who kindly reviewed the article. We are indebted to many colleagues in the Epoxy Products and Intermediates business at Dow Chemical for their assistance in many ways to make this article possible.

## Bibliography

"Epoxy Resins" in *Encyclopedia of Polymer Science and Technology* 3rd ed., by Ha Q. Pham and Maurice J. Marks, Dow Chemical, Freeport, TX; 10.1002/0471440264.pst119

1. S. J. Hartman, *The Epoxy Resin Formulators Training Manual*, The Society of the Plastics Industry, Inc., New York 1984, p. 1.
2. H. Lee, K. Neville, *Handbook of Epoxy Resins*, McGraw-Hill, Inc., New York 1967, reprinted 1982.
3. I. G. Farbenindustrie, DE 676 117, 1938, US 2 136 928, 1938 (P. Schlack).
4. Ciba-Geigy Corp., US2 324 483, 1943 (P. Castan).
5. DeVoe & Raynolds, US2 456 408, 1948 (S. O. Greenlee).
6. D. Swern, *Chem. Rev.* **45** (1949) 1.
7. E. O. C. Greiner, F. Dubois, M. Yoneyama, *Epoxy Resins, Chemical Economics Handbook (CEH) Marketing Research Report*, Stanford Research Institute (SRI) International, Menlo Park, CA 2001.
8. Dow Chemical data, the Dow Chemical Co., Midland, Mich.
9. W. L. Faith, D. B. Keyes, R. L. Clark, *Industrial Chemistry*, 3rd ed., John Wiley & Sons, Inc., New York 1965, p. 404.
10. H. H. Szmant, *Organic Building Blocks of the Chemical Industry*, John Wiley & Sons, Inc., New York 1989, p. 281.
11. P. H. Williams, *Encyclopedia of Chemical Technology*, 1st ed., Vol. 3, John Wiley & Sons, Inc., New York 1965, pp. 857, 865.
12. W. F. Richey, *Encyclopedia of Chemical Technology*, 4th ed., Vol. 6, John Wiley & Sons, Inc., New York 1993, pp. 140, 155.
13. Shell Oil Co., US 2 714 123, 1955 (G. F. Johnson); Solvay & CIE, BE 517 463, 1959.
14. Showa Denko K. K., JP 88 290 835, 1988 (N. Nagato, H. Mori, R. Ishioka); Showa Denko K. K., US 4 634 784, 1987 (N. Nagato, H. Mori, K. Maki, R. Ishioka).
15. Bisphenol A, Chem Systems Report, Sept. 2002.
16. General Electric Co., US 4 400 555, 1983 (Mendiretta); General Electric Co., WO 00/35847, 2000 (G. M. Kissinger, R. Sato).
17. W. G. Potter, *Epoxide Resins*, Springer-Verlag, New York 1970.
18. Shell Oil Co., US 2 575 558, 1951 (H. A. Newley, E. C. Shokal).
19. Union Carbide Corp., US 2 943 095, 1960 N. H. Reinking).
20. Shell Development, US 2 848 435, 1958 L. H. Griffin, J. H. Long); Reichhold Chemicals, Inc., US 2 921 049, 1960 (H. L. Moroson); Union Carbide Corp., US 3 069 434, 1962 (S. P. Spence, A. R. Grover, F. P. Klosek, R. E. Nicolson).
21. Dow Chemical Co., US 4 499 255, 1985 (C. S. Wang, H. Q. Pham, J. L. Bertram); Tohto Kasei K. K., JP 61/195111A, 1986 (T. Ogata, H. Nakanishi, M. Aritomi).
22. N. S. Enikolopyan, M. A. Markevitch, L. S. Sakhonenko, S. Z. Rogovina, V. G. Oshmyan, *J. Polym. Sci., Chem. Ed.* **20** (1982) 1231–1245.
23. D. O. Bowen, R. C. Whiteside, in R. F. Gould (ed.): *Epoxy Resins*, Advances in Chemistry Series 92, American Chemical Society, Washington, D.C. 1970, p. 48.
24. Shell Oil Co., US 2 643 239, 1953 (E. C. Shokal, H. A. Newley, T. E. Bradley); Shell Development Co.), US 2 879 259, 1959 (R. W. H. Tess).
25. Devoe & Raynolds Co., Inc., US 2 456 408, 1943 (S. O. Greenlee).
26. H. Batzer, S. A. Zahir, *J. Appl. Polym. Sci.* **19** (1975) 601.
27. W. Burchard, S. Bantle, S. A. Zahir, *Makromol. Chem.* **182** (1981) 145.
28. U. Fuchslueger, H. Stephan, H.-J. Grether, M. Grasserbauer, *Polymer* **40** (1999) 661–673.
29. Ciba-Geigy Corp., US 3 634 323, 1972 (R. M. Moran).
30. Shell Oil Co., US 3 477 990, 1969 (M. F. Dante, H. L. Parry).
31. Dow Chemical Co., US 3 948 855, 1976 (W. O. Perry).
32. Dow Chemical Co., US 4 302 574, 1981 (G. A. Doorakian, J. L. Bertram).
33. Shell Oil Co., US 4 358 578, 1982 (T. F. Brownscombe).
34. Dow Chemical Co., US4 366 295, 1982 (M. C. Tyler Jr., A. L. McCrary).
35. Dow Chemical Co.), US 4 808 692, 1989 (H. Q. Pham, L. A. Ho).
36. W. A. Romanchick, J. F. Geibel, *Org. Coat. Appl. Polym. Sci. Proc.* **46** (1982) 410.
37. J. F. Geibel, *Org. Coat. Plast. Chem.* **43** (1980) 545.
38. H. Batzer, S. A. Zahir, *J. Appl. Polym. Sci.* **21** (1977) 1843.
39. D. P. Sheih, D. E. Benton, ASTM Special Technical Publication STP 1119, Analysis of Paints and Related Materials: Current Techniques for Solving Coatings Problems, 1992, pp. 41–56.
40. Shell International Research, WO 01/16204 A1, 2001 (S. MK. Li); Resolution Research Nederland B. V., WO

Appl. 01/46287 A1, 2001 (R. R. Dominquez, H. Frank, S. MK. Li).
41. Dow Chemical Co., US 4 612 156, 1986 (B. W. Heinemeyer, S. D. Tatum).
42. J. E. White, H. C. Silvis, M. S. Winkler, T. W. Glass, D. E. Kirkpatrick, *Adv. Mater.* **12** (2000) 1791, 1800.
43. Shell Oil Co., GB 980 509, 1965.
44. Dow Chemical Co., US 5 275 853, 1994 (H. C. Silvis, J. E. White).
45. H. C. Silvis, C. N. Brown, S. L. Kram, J. E. White, *Polym. Prepr.* **36** (1995) no. 2, 178–179.
46. T. Glass, H. Pham, M. Winkler, in *Proceedings of the 58th SPE Annual Technical Conference and Exhibits (ANTEC 2000)*, Lauderdale, FL 2000.
47. Dow Chemical Co., US 5 134 201, 1992 (M. N. Mang, J. E. White).
48. Nan Ya Plastics Corp. Taiwan, US 6 512 075, 2003 (M. J. Tzou).
49. Tohto Kasei K.K., JP 08198949 A, 1996 (K. Ishihara, T. Sato, K. Aida, T. Hoshono).
50. J. R. Griffith, *CHEMTECH* **12** (1982) 290–293.
51. the United States of America, US 3 879 430, 1975 (J. R. Griffith, J. G. O'Rear).
52. T. E. Twardowski, P. H. Geil, *J. Appl. Polym. Sci.* **42** (1991) 69.
53. Devoe & Raynolds, US 2 521 912, 1950 (S. O. Greenlee).
54. Dow Chemical Co., US 4 785 061, 1988 (C. S. Wang, Z. K. Liao); Yuka Shell Epoxy Co., JP 200239346, 2000 (Y. Murata, B. Shigeki).
55. Dow Chemical Co., US 4 394 497, 1983 (D. L. Nelson, B. A. Naderhoff).
56. Shell Development Co., US 2 806 016, 1957 (C. G. Schwartzer).
57. K. L. Hawthorne, F. C. Henson, in R. S. Bauer (ed.): *Epoxy Resin Chemistry II*, ACS Symposium Series 221, American Chemical Society, Washington, D.C. 1983, Chapt. 7, pp. 135–151.
58. Union Carbide Corp., US 2 951 825, 1960), (N. H. Reinking, B. P. Barth, F. J. Castner).
59. T. J. Galvin, M. A. Chaudhari, J. J. King, *Chem. Eng. Prog.* **81** (1985) no. 1, 45–48.
60. Shell Internationale Research B. V., JP 3 315 436, 2002 (Y. Murata, Y. Nakanishi, M. Yosumura).
61. I. Ogura, T. Imada, Dai Nippon Inks & Chemical (DIC) Technical Review No. 5, 1999.
62. Dow Chemical Co., US 5 463 091, 1995 (J. D. Earls, R. E. Hefner Jr., P. M. Puckett); Dow Chemical Co., US 5 266 660, 1993 (R. E. Hefner Jr., J. D. Earls, P. M. Puckett).
63. H.-J. Sue, J. D. Earls, R. E. Hefner Jr., M. I. Villarreal, E. I. Garcia-Meitin, P. C. Yang, C. H. Cheatham, C. J. G. Plummer, *Polymer* **39** (1998) 4707–4714.
64. R. S. Bauer, in G. D. Parfitt, A. V. Patsis (eds.): *Organic Coatings, Science and Technology*, vol. 5, Marcel Dekker, New York 1983, pp. 1–33.
65. Henkel & Cie. GmbH, US3 288 789, 1966 (M. Budnowski, M. Dohr).
66. Internationale Octrool Maatschappij Octropa, GB 1 381 262, 1975 and US 4 147 737, 1979 (A. J. Sein, J. Reitberg, J. M. Schouten).
67. E. H. Catsiff, R. E. Coulehan, J. F. Diprima, D. A. Gordan, R. Seltzer, in R. S. Bauer (ed.): *Epoxy Resin Chemistry*, ACS Symposium Series 114, American Chemical Society, Washington, D.C. 1979, Chapt. 10, pp. 115–156.
68. R. Y. Ting, in C. A. May, Y. Tanaka (eds.): *Epoxy Resins Chemistry and Technology*, 2nd ed., Marcel Dekker, Inc., New York 1988, p. 551, 601.
69. H. Jahn, *J. Polym. Sci., Part C* **16** (1967) 1829, 1841.
70. Dow Chemical Co., US 4 273 921, 1980 (J. L. Bertram, P. S. Sheih).
71. Union Carbide Corp., US 2 716 123, 1953), (B. Phillips, F. Frostick Jr.).
72. P. K. T. Ording: *Waterborne and Solvent Based Epoxies and Their End User Applications*, John Wiley & Sons, Inc., New York 1996, pp. 57, 100.
73. J. C. Kenny, T. Ueno, K. Tsutsui, *J. Coat. Technol.* **68**, (1996) 855; T. Agawa, E. D. Dumain, *Proc. Waterborne High-Solids Powder Coat. Symp.* **24** (1997) 342–353; B. V. Gregorovich, I. Hazan, *Prog. Org. Coat.* **24** (1994) 131.
74. G. Webster: *Chemistry & Technology of UV & EB Formulation for Coatings, Inks & Paints*, vol. 2, John Wiley & Sons, Inc., New York 1997, pp. 41, 73.
75. Dow Chemical Co., US 3 367 992, 1968 (C. Bearden); Dow Chemical Co., US 3 524 901, 1970 (D. J. Najvar).
76. T. F. Anderson, V. B. Messick, in G. Pritchard (ed.): *Developments in Reinforced Platics 1*, Allied Science Publishers Ltd., London 1980, pp. 29–58.
77. Dow Chemical Co., US 4 397 970, 1983 (K. D. Campbell, H. G. Langer, P. H. Martin).
78. H. Jahn, P. Goetzky, in C. A. May, Y. Tanaka (eds.): *Epoxy Resins Chemistry and Technology*, 2nd ed., Marcel Dekker, Inc., New York 1988, pp. 1049, 1087.
79. Annu. Book ASTM Stand. Section 8 (Plastics). Web site: http://www.astm.org .
80. H. Pasch, R. Unvericht, M. Resch, *Angew. Makromol. Chem.* **212** (1993) 191–200; H. Pasch, J. Adrian, D. Braun, *GIT Spezial Separation* **21** (2001) no. 2, 104–108.
81. D. Crozier, G. Morse, Y. Tajima, *SAMPE J.* **18** (1982) no. 5, 17–22.
82. A. Durbetaki, *Anal. Chem.* **28** (1956) 2000.
83. R. Jay, *Anal. Chem.* **36** (1964) 667.
84. B. Dobinson, W. Hofmann, B. Stark, *The Determination of Epoxide Groups*, Pergamon Press, Elmsford, NY 1969.
85. G. H. Schneer, W. van Gilder, V. E. Hauser, P. E. Schmidt, *IEEE Trans. Electron. Devices* (1969) ED–15.
86. H. Lee, K. Neville, *Handbook of Epoxy Resins*, McGraw-Hill, Inc., New York 1967, reprinted 1982.
87. M. C. Paputa Peck, R. O. Carter III, S. B. A. Qaderi, *J. Appl. Polym. Sci.* **33** (1987) no. 1 77–86; G. Lachenal, Y. Ozaki, *Macromol. Symp.* **141** (1999) 283–292. 13th European Symposium on Polymer Spectroscopy, 1998.
88. W. B. Moniz, C. F. Poranski Jr., *Org. Coat. Plast. Chem.* **39**, (1978) 99–102.
89. E. Mertzel, J. L. Koenig, in K. Dusek (ed.): *Epoxy Resins and Composites III, Advances in Polymer Science 75*, Springer-Verlag, Berlin 1986, pp. 73, 112.
90. W. A. Dark, E. C. Conrad, L. W. Crossman Jr., *J. Chromatogr.* **91** (1974) 247–60.
91. D. J. Crabtree, D. B. Hewitt, *Liq. Chromatogr. Polym. Relat. Mater.* **8** (1977) 63–77, Chromatographic Science Series.
92. G. Eppert, G. Liebscher, *J. Chromatogr.* **238** (1982) 399.
93. S. A. Zahir, S. Bantle, in R. S. Bauer (ed.): *Epoxy Resin Chemistry II*, ACS Symposium Series 221. American Chemical Society, Washington, D.C. 1983, pp. 245, 262.
94. D. K. Hadad, in C. A. May, Y. Tanaka (eds.): *Epoxy Resins Chemistry and Technology*, 2nd ed., Marcel Dekker, Inc., New York 1988, pp. 1089, 1172.

95. W. Lwowski, in A. R. Katritsky, C. W. Rees (eds.): *Comprehensive Heterocyclic Chemistry*, vol. 7, Pergamon Press, Oxford 1984, pp. 1, 16.
96. R. E. Parker, N. S. Isaacs, *Chem. Rev.* **59** (1959) 737, 799.
97. Y. Tanaka, R. S. Bauer, in C. A. May, Y. Tanaka (eds.): *Epoxy Resins Chemistry and Technology*, 2nd ed., Marcel Dekker, Inc., New York 1988, pp. 465, 550.
98. L. Shechter, J. Wynstra, *Ind. Eng. Chem.* **48** (1956) 86.
99. I. T. Smith, *Polymer* **2** (1961) 95.
100. N. B. Chapman, R. E. Parker, N. S. Issacs, *J. Chem. Soc.* **2** (1959) 1925.
101. N. G. Rondan, M. J. Marks, S. Hoyles, H. Pham, paper presented at the *225th ACS National Meeting*, New Orleans, LA, March 2003.
102. R. F. Eaton, *Paint Coat. Ind.* (1999) June, 76–80.
103. L. Shechter, J. Wynstra, R. P. Kurkjy, *Ind. Eng. Chem.* **48** (1956).
104. K. Horie, H. Hiura, M. Sawada, I. Mita, H. Kambe, *J. Polym. Sci., A–1* **8** (1970) 1357.
105. K. Dusek, M. Ilavsky, S. Lunak, *J. Polym. Sci., Polym. Symp.* **53** (1975) 29.
106. C. V. Hare, *Protective Coatings: Fundamentals of Chemistry and Composition*, Technology Publishing Co., Pittsburg, PA 1994, pp. 187, 238.
107. CIBA (A.R.L) Ltd., GB 886 767, 1962 (P. Halewood).
108. W. R. Ashcroft, in B. Ellis (ed.): *Chemistry and Technology of Epoxy Resins*, 1st ed., Blackie Academic & Professional, Glasgow, U.K. 1993, pp. 37, 71.
109. M. Fedtke, F. Domaratius, A. Pfitzmann, *Polym. Bull.* **23** (1990) 381, 388.
110. Minnesota Mining & Manufacturing Co., US 3 390 124, 1968 (J. B. Kittridge, A. L. Michelli).
111. M. D. Gilbert, N. S. Schneider, W. J. McKnight, *Macromolecules* **24** (1991) 360.
112. S. A. Zahir, in G. D. Parfitt, A. V. Patsis (eds.): *Advances in Organic Coatings Science and Technology, Vol. IV: Sixth International Conference in Organic Coatings Science and Technology*, Technomic Publishing Co., Inc., Lancaster, PA 1982, p. 83.
113. M. J. Husband, in P. Oldring, G. Haywood, eds., *Resins for Surface Coatings, SITA Technology*, John Wiley & Sons Ltd., London 1987, pp. 63, 167.
114. P. J. Madec, E. Marechal, *Makromol. Chem.* **184** (1983) 323.
115. L. Matejka, J. Lovy, S. Pokorny, K. Bouchal, K. Dusek, *J. Polym. Sci., Polym. Chem. Ed.* **21** (1983) 2873.
116. H. Kunitomo, in *Proceedings of the 26th Technical Conference of the Japan Society of Epoxy Resin Technology*, Tokyo, July 2002, pp. 19, 33.
117. A. Gardziella, L. A. Pilato, A. Knop, *Phenolic Resins Chemistry, Applications, Standardization, Safety and Ecology*, 2nd ed., Springer, Berlin 1999.
118. A. J. Kirsch, *50 Years of Amino Coatings*, American Cyanamid Co., Wayne, NJ 1986.
119. T. M. Rees, *J. Oil Color Chem. Assoc.* **71** (1988) no. 2, 39, 41.
120. Jefferson Chemical, US 3 020 262, 1962 (G. P. Speranza); Dow Chemical Co., US 4 658 007, 1987 (M. J. Marks, R. A. Plepys).
121. M. J. Marks, *Polym. Mater. Sci. Eng. (Am. Chem. Soc., Div. Polym. Sci. Eng.)* **58** (1988) 864.
122. D. A. Shimp, F. A. Hudock, S. J. Ising, paper presented at 33rd SAMPE, Anaheim, CA, March 7–10, 1988.
123. B. A. Rozenberg, in K. Dusek (ed.): *Epoxy Resins and Composites II*, Advances in Polymer Science 75, Springer-Verlag, Berlin 1986, pp. 146, 156.
124. F. Ricciardi, W. A. Romanchick, M. M. Joullie, *J. Polym. Sci., Polym. Chem. Ed.* **21** (1983) 1475.
125. M. S. Heise, G. C. Martin, *Macromolecules* **22** (1989) 99.
126. J. D. B. Smith, *Org. Coat. Plast. Chem.* **39** (1978) 42–46.
127. R. J. Arnold, *Mod. Plast.* **41** (1964) no. 4, 149.
128. J. J. Harris, S. C. Temin, *J. Appl. Polym. Sci.* **10** (1966) 523.
129. J. V. Crivello, J. H. W. Lam, *Macromolecules* **10** (1977) 1307; *ACS Symp. Ser.* **114**, 1(1979).
130. R. F. Eaton, K. T. Lamb, paper presented at the *23th International Waterborne, Higher Solids and Powder Coatings Symposium*, New Orleans, LA, Feb. 1996.
131. J. M. Land, A. Aubuchon, C. Pundmann, W. L. Dechent, J. O. Stoffer, *Book of Abstracts, 211th ACS National Meeting*, New Orleans, LA, March 24–28, 1996, PMSE-181, American Chemical Society, Washington, D.C. 1996.
132. D. W. Brooker, G. R. Edwards, A. McIntosh, *J. Oil Color Chem. Assoc.* **52** (1969) 989–1034; G. K. Noren, *J. Coat. Technol.* **72** (2000) no. 905, 53–59.
133. J. E. Mark, B. Erman, *Rubber Elasticity: A Molecular Primer*, John Wiley & Sons, New York 1988.
134. T. I. Smith, *J. Polym. Sci., Polym. Symp.* **46** (1974) 97.
135. F. Lohse, R. Schmid, paper presented at the Fifth International Conference in Organic Coatings Science and Technology, Athens, FATIPEC, Liége, Belgium 1979.
136. I. Ogura, Dainippon Ink & Chemicals (DIC) Technical Review No. 7, 2001. Paper written in Japanese.
137. T. Kamon, H. Furukawa, in K. Dusek (ed.): *Epoxy Resins and Composites IV*, Advances in Polymer Science 80, Springer-Verlag, Berlin 1986, pp. 173, 202.
138. E. F. Oleinik, in Ref. 137, pp. 50, 99.
139. W. R. Ascroft, *Eur. Coat. J.* **4** (1991) 229, 241.
140. R. S. Bauer, *CHEMTECH* **10** (1980) Nov., 692.
141. C. Rooney, *Mod. Paint Coat.* **81** (1991) no. 5, 44–50.
142. J. K. Gillham, C. A. Grandt, in S. S. Labana (ed.): *Chemistry and Properties of Cross-Linked Polymers*, Academic Press, Inc., Orlando, FL 1977, pp. 491, 520.
143. J. D. Keenan, *J. Appl. Polym. Sci.* **24** (1979) 2375, 2387.
144. C. A. May (ed.): *Chemorheology of Thermosetting Polymers*, ACS Symposium Series 227, American Chemical Society, Washington, D.C. 1983.
145. D. Adolf, J. E. Martin, *Macromolecules* **23** (1990) 3700, 3704.
146. *General Guide to Formulating with Dow Epoxy Resins*, the Dow Chemical Co., Midland, MI 1983.
147. B. Ellis, in B. Ellis (ed.): *Chemistry and Technology of Epoxy Resins*, 1st ed., Blackie Academic & Professional, Glasgow, U.K. 1993, pp. 72, 116.
148. J. P. Pascault, H. Sautereau, J. Verdu, R. J. J. Williams, *Thermosetting Polymers*, Marcel Dekker, Inc., New York 2001.
149. J. C. Seferis, L. Nicolais (eds.): *The Role of Polymer Matrix in Processing and Structural Properties of Composites*, Plenum Press, New York 1983, pp. 127–145.
150. J. K. Gillham, *Encyclopedia of Polymer Science and Engineering*, 2nd ed., John Wiley & Sons, Inc., New York 1986, pp. 519, 524.
151. X. Wang, J. K. Gilham, *J. Coat. Technol.* **64** (1992) 37, 45.
152. H. E. Adabbo, R. J. J. Williams, *J. Appl. Polym. Sci.* **27** (1982) 1327, 1334.
153. J. B. Enns, J. K. Gillham, *J. Appl. Polym. Sci.* **28** (1983) 2567.
154. C. M. Tung, J. P. Dynes, *J. Appl. Polym. Sci.* **27** (1982) 569, 574; M. E. Smith, H. Ishida, *J. Appl. Polym. Sci.* **73** (1999) 593, 600.
155. T. Glauser, M. Johansson, A. Hult, *Polymer* **40** (1999) 5297, 5302.

156. K. Dusek, in K. Dusek (ed.): *Epoxy Resins and Composites III*, Advances in Polymer Science 78, Springer-Verlag, Berlin 1986, pp. 1, 59.
157. S. G. Croll, in K. L. Mittal (ed.): *Adhesion Aspects of Polymeric Coatings*, Plenum Press, New York 1983.
158. D. B. Adolf, J. E. Martin, *J. Comp. Mater.* **30** (1996) no. 1, 13–34.
159. J. C. Arnold, *Polym. Eng. Sci.* **35** (1995) no. 2, 165–169.
160. E. S.-W. Kong, in K. Dusek (ed.): *Epoxy Resins and Composites IV*, Advances in Polymer Science 80, Springer-Verlag, Berlin 1986, pp. 125, 172.
161. R. S. Durran, *J. Non-Cryst. Solids* **131–133** (1990) 497–504; D. J. Plazek, Z. N. Frund, *J. Polym. Sci., Bert. Polym. Phys.* **28** (1990) 431–448.
162. S. L. Maddox, J. K. Gillham, *J. Appl. Polym. Sci.* **64** (1997) 55–67.
163. L. W. Hill, *J. Coat. Technol.* **64** (1992) no. 808, 29.
164. C. Billaud, M. Vandeuren, R. Legras, V. Carlier, *Appl. Spectr.* **56** (2002) 1413–1421.
165. G. Wisanrakkit, J. K. Gilham, *J. Appl. Polym. Sci.* **41** (1990) 2885, 2929.
166. C. W. Macosko, D. R. Miller, *Macromolecules* **9** (1976) 199, 206; J. Mijovic, C. H. Lee, *J. Appl. Polym. Sci.* **29** (1989) 2155, 2170.
167. C. A. May (ed.): *Chemorheology of Themosetting Polymers*, ACS Symposium Series 227, American Chemical. Society, Washington, D.C. 1983.
168. S. D. Senturia, N. F. Sheppard Jr., in K. Dusek (ed.): *Epoxy Resins and Composites IV*, Advances in Polymer Science 80, Springer-Verlag, Berlin 1986, pp. 1, 47.
169. G. P. Johari, D. A. Wasylyshyn, *J. Polym. Sci., Polym. Phys.* **38** (2000) 122–126.
170. J. H. Flynn, in H. F. Mark, N. M. Bikales, C. G. Overberger, G. Menges, eds. *Encyclopedia Polymer Science and Engineering*, 2nd ed., Suppl. Vol., John Wiley & Sons, New York 1989, pp. 715, 723.
171. Annu. Book ASTM Stand. Web site: http://www.astm.org
172. J. M. Charlesworth, *Polym. Eng. Sci.* **28** (1988) 230.
173. M. J. Marks, *Polym. Mater. Sci. Eng.* **66** (1992) 365.
174. A. J. Lesser, E. Crawford, *J. Appl. Polym. Sci.* **66** (1997) 387, 395; E. Crawford, A. J. Lesser, *J. Appl. Polym. Sci., Part B: Polym. Phys.* **36** (1998) 1371, 1382.
175. C. W. Macosko, D. R. Miller, *Macromolecules* **9** (1976) 206, 211; D. R. Miller, E. M. Valles, C. W. Macosko, *Polym. Eng. Sci.* **19** (1979) 272, 283.
176. E. D. Crawford, A. J. Lesser, *Polym. Eng. Sci.* **39** (1999) Feb., 385, 392.
177. J. G. Williams, *Fracture Mechanics of Polymers*, Ellis Horwood, Chichester, U.K. 1984.
178. A. J. Kinloch, in K. Dusek (ed.): *Epoxy Resins and Composites 1*, Advances in Polymer Science 72, Springer-Verlag, Berlin 1986, pp. 1, 59.
179. W. J. Cantwell, H. N. Kausch, in B. Ellis (ed.): *Chemistry and Technology of Epoxy Resins*, 1st ed., Blackie Academic & Professional, Glasgow, U.K. 1993, pp. 144, 174.
180. J. M. Charlesworth, *Polym. Eng. Sci.* **28** (1988) 221.
181. E. Urbaczweski-Espuche, J. Galy, J. Ferard, J. Pascault, H. Sautereau, *Polym. Eng. Sci.* **31** (1991) 1572.
182. J. Galy, J. Gerard, H. Sautereau, R. Frassine, A. Pavan, *Polym. Networks Blends* **4** (1994) 105.
183. R. A. Dubois, D. S. Wang, D. Sheih, in C. B. Arends (ed.): *Polymer Toughening*, Marcel Dekker, Inc., New York 1996, pp. 381, 409.
184. R. A. Dubois, D. S. Wang, *Prog. Org. Coat.* **22** (1993) 161.
185. J. P. Pascault, H. Sautereau, J. Verdu, R. J. J. Williams, *Thermosetting Polymers*, Marcel Dekker, Inc., New York 2001, p. 400.
186. L. T. Drzal, in K. Dusek (ed.): *Epoxy Resins and Composites II*, Advances in Polymer Science 75, Springer-Verlag, Berlin 1986, pp. 3, 32.
187. R. G. Schmidt, J. P. Bell, in K. Dusek (ed.): *Epoxy Resins and Composites II*, Advances in Polymer Science 75, Springer-Verlag, Berlin 1986, pp. 33, 71.
188. Z. W. Wicks Jr., F. N. Jones, S. P. Pappas, *Organic Coatings Science and Technology*, 2nd ed., Wiley-Interscience, New York 1999, pp. 77, 111.
189. W. Brostow, R. D. Corneliussen (eds.): *Failure of Plastics*, Hanser, Munich 1989.
190. Celanese Corp., US 4, 051, 195, 1977 (M. F. McWhorter).
191. R. F. Eaton, K. T. Lamb, in *SPE Proceedings*, Epoxy Resin Formulators Division of the Society of Plastic Industry, Inc., Aspen, CO, May 1996 (Paper 7).
192. M. DiBenedetto, *Mod. Paint. Coat.* (Jul. 1980) 39.
193. J. Melloan, paper presented at the *Epoxy Resins Formulators Division of SPI Meeting*, Atlanta, GA, Nov. 1983, The Society of Plastics Industry, New York.
194. Z. Wang, J. Massam, T. J. Pinnavaia, in T. J. Pinnavaia, G. W. Beall (eds.): *Polymer-Clay Nanocomposites*, John Wiley & Sons, New York 2000, pp. 127, 149.
195. J. H. Park, S. C. Jana, *Macromolecules* **36** (2003) 2758–2768.
196. *Proceedings of POSS (Hybrid Plastics, Inc.) Nanotechnology Conference*, Huntington Beach, CA, Sept. 2002.
197. M. J. Biercuk, M. C. Llaguno, M. Radosavljevic, J. K. Hyun, A. T. Johnson, J. E. Fischer, *J. Appl. Phys. Lett.* **80** (2002) 2767, 2769.
198. R. A. Dubois, P. S. Sheih, *J. Coat. Technol.* **64** (1992) 51.
199. B. L. Burton, J. L. Bertram, in C. B. Arends (ed.): *Polymer Toughening*, Marcel Dekker, Inc., New York 1996, pp. 339, 379.
200. H. S. Sue, E. I. Garcia-Meitin, D. M. Pickleman, in N. P. Cheremisinoff (ed.): *Rubber-Modified High Performance Epoxies*, CRC Press, Boca Raton FL 1993, Chapt. 18, pp. 661, 700.
201. A. F. Yee, J. Du, M. D. Thouless, in D. R. Paul, C. R. Bucknall (eds.): *Polymer Blends*, vol. 2, John Wiley & Sons, Inc., New York 2000, pp. 226, 267.
202. Y. Huang, D. L. Hunston, A. J. Kinloch, C. K. Riew, in C. K. Riew, A. J. Kinloch (eds.): *Toughened Plastics 1*, Advances in Chemistry Series 233, American Chemical Society, Washington, D.C. 1993, pp. 1, 35.
203. C. K. Riew, E. H. Rowe, A. R. Siefert, *ACS Adv. Chem. Ser.* **154** (1976) 326.
204. H. Chen, N. Verghese, H. Pham, N. Jivraj, paper presented at the 9th Annual International Conference on Composites Engineering, San Diego, CA, July 2002.
205. Dow Chemical Co., US 4 708 996, 1987 (D. K. Hoffman, C. B. Arends).
206. H. J. Sue, *Polym. Eng. Sci.* **31** (1991) 275.
207. Dow Chemical Co., US 4 594 291, 1986; Dow Chemical Co., US 4 725 652, 1988 (J. L. Bertram, L. L. Walker, V. I. W. Stuart).
208. J. M. Dean, P. M. Lipic, R. B. Grubbs, R. F. Cook, F. S. Bates, *J. Polym. Sci., Part B: Polym. Phys.* **39** (2001) 2996, 3010; S. Ritzenthaler, F. Court, L. David, *Macromolecules* **35** (2002) 6245, 6254.
209. S. R. White, N. K. Sottos, P. H. Geubelle, J. S. Moore, M. R. Kessler, S. R. Sriram, E. N. Brown, S. Viswanathan, *Nature* **409** (2001) 794, 797.

210. E. O. C. Greiner, F. Dubois, M. Yoneyama, *Epoxy Resins*, Chemical Economics Handbook (CEH) Marketing Research Report, Stanford Research Institute (SRI) International, Menlo Park, CA 2001.
211. E. Linak, F. Dubois, M. Yoneyama, *Epoxy Surface Coatings*, Chemical Economics Handbook (CEH) Marketing Research Report, Stanford Research Institute (SRI) International, Menlo Park, CA 2001.
212. E. W. Flick, *Contemporary Industrial Coatings, Enviromentally Safe Formulations*, Noyes Publishers, Park Ridge, NJ 1985.
213. R. F. Eaton, K. T. Lamb, *J. Coat. Technol.* **68** (1996) 49.
214. PPG Industries, Inc., US 3 984 299, 1976 (R. D. Jerabek); PPG Industries, Inc., US 4 009 133, 1977 (J. E. Jones).
215. SCM Corp., US 4 212 781, 1980 (J. M. Evans, V. W. Ting).
216. E. C. Galgoci, P. C. Komar, *Paints Coat. Ind.* (1994) Aug., 50.
217. D. S. Kincaid, P. Komar, J. R. Hite, paper presented at the 24th International Waterborne, High Solids and Powder Coatings Symposium, New Orleans, LA, 2002.
218. M. D. Soucek, G. Teng, S. Wu, *J. Coat. Technol.* **73** (2001) Oct., 117, 125.
219. E. Linak, A. Kishi, *Thermoset Powder Coatings*, Chemical Economics Handbook (CEH) Marketing Research Report, Stanford Research Institute (SRI) International, Menlo Park, CA, 2002.
220. R. Amey, R. Farabaugh, *Mod. Paint Coat.* (1997) June, 28, 30.
221. H. Nowack, *Proc. PCE 2000* (2000) 365, 375.
222. P. Horinka, *Powder Coating Magazine* (Aug. 2002) 33, 48.
223. K. Buysens, K. Jacques, *Eur. Coat. J.* (Sept. 2001) 22, 26.
224. D. L. Steele, *Surface Coat. Aus.* (Oct. 1992) 6, 12; W. Wood, *J. Protective Coat. Linings* (Apr. 1987) 32, 38.
225. US 5 275 6451994 L. R. Ternoir, R. E. Foscante, R. L. Gasmens); Ameron International Corp., US 5 618 860, 1997 (N. R. Mowrer, R. E. Foscante, J. L. Rojas).
226. Glidden Co., US 5 508 325, 1996 (G. P. Craun, D. J. Telford, H. J. DeGraaf).
227. Shell Oil Co., US 4 119 595, 1978 (R. S. Bauer, J. A. Lopez).
228. J. T. K. Woo, V. Ting, J. Evans, C. Ortiz, G. Carlson, R. Marcinko, in R. S. Bauer (ed.): *Epoxy Resin Chemistry 2*, American Chemical Society, Washington, D.C. 1983, pp. 283, 300.
229. Glidden Co., US 5 532 297, 1996 (J. T. K. Woo, G. C. Pompighano, D. E. Awarski, K. A. Packard).
230. P. Newman, *The Canmaker* (Apr. 2001) 57.
231. D. Hayes, *The Canmaker* (Nov. 1998) 37, 40.
232. Toyo Seikan Kaisha Ltd., Japan, EP Appl. 493133 A2, 1992 (N. Sato, K. Imazu).
233. Bisphenol A: Information Sheet from the Global Industry Group, Safety of Epoxy Can Coatings (Oct. 2002) at http://www.bisphenol-a.org; Statement from CEPE (European Confederation of Paints, Printing Inks and Artists Colours Manufacturers), Feb. 2000.
234. P. Hitchin, *The Canmaker* (Oct. 2001) 43, 46.
235. BP Corp. North America Inc., US 6 472 480, 2002 (R. L. Anderson); Valspar Corp., US 6 235 102 B1, 2002 (G. G. Parekh, L. P. Seibel).
236. PPG Industries, US 4 650 718, 1987 (D. A. Simpson, D. L. Singer, R. Dobenko, W. P. Blackburn, C. M. Kania).
237. L. Mauro, *25th FATIPEC Congr.* **4** (2000) 1, 10.
238. BASF, EU Appl. 581175 A2, 1994 (K. Huemke, C. Sinn).
239. Kansai Paint Co. Ltd; Nippon Packaging K.K., JP 08003483 A2, 1996 (H. Haishi, M. Kume, H. Ishiiand, K. Myawaki).
240. Nippon Paint, US Appl. 2002139673 A1, 2002 (Y. Kojima, M. Yamada).
241. Nippon Paint Co., Ltd., Japan, EU Appl. 0974623 A2, 2000 (H. Sakamoto, T. Kawanami, I. Kawakami, T. Kokubun, T. Saito, S. Yoshimatsu).
242. G. Mauer, D. Singer, *PPG Technol. J* (1999) 63, 72.
243. Y. Okude, S. Ishikura, *Prog. Org. Coat.* **26** (1995) 197, 205.
244. G. E. Green, B. P. Stark, S. A. Zahir, *J. Macromol. Sci. Revs. Macromol. Chem.* **21** (1981/1982) 187.
245. P. Kelly, G. Pritchard (eds.): *Reinforced Plastics Durability*, CRC Press, Boca Raton, FL 1999, pp. 282–321.
246. M. R. Thoseby, B. Dobinson, C. H. Bull, *Br. Polym. J.* **18** (1986) 286.
247. U. Helrold, J. Mason, N. Verghese, H. Chen, H. Reddy, in *Proceedings of NACE Corossion 2003 Conference*, Houston, TX., March 16–20, 2003.
248. D. Kriegh (ed.): *Epoxies with Concrete*, American Concrete Institute, Detroit, MI1968; R. L. McGown, in *Proceedings of the 1991 Steel, Structures, Painting Council (SSPC) National Conference and Exhibition*, Long Beach, CA.
249. H. Mihashi, Y. Kaneko, *Transactions Mater. Res. Soc. Japan* **25** (2000) 557, 560.
250. S. Ehrler, *PC Fab* (Apr. 2002) 32, 38; Part 2, PC Fab (May 2002) 32, 36.
251. W. Christiansen, D. Shirrell, B. Aguirre, J. Wilkins, in *Proceedings of the Technical Conference*, IPC Printed Circuits Expo 2001, Anaheim, CA, Apr. 2001, pp. SO3-1-1, SO3-1-7.
252. J. Sharma, M. Choate, S. Peters, in *Proceedings of the Technical Conference*, IPC Printed Circuits Expo 2002, Long Beach, CA, Mar. 2002, pp. SO5-1-1, SO5-1-8.
253. D. Sober, in Base Material Basics, *IPC Printed Circuits Expo 2002*, Long Beach, Ca.
254. Ciba-Geigy Corp., US 5 506 313, 1996 (P. Flury, C. W. Mayer, W. Scharf, E. Vanoli).
255. C. S. Cho, L. W. Chen, Y. S. Chiu, *Polym. Bull.* **41** (1998) no. 1, 45, 52.
256. D. K. Luttrull, F. E. Hickman III, Future Circuits International, Mar. 2001.
257. Ciba-Geigy AG, US 3 754 071, 1973 (O. Ernst, E. Kusenberg, E. Hubler, H. R. Aus Der Au).
258. K. Ito, Y. Nakamura, *IEEE Electrical Insulation Mag.* **6** (1990) no. 4, 25, 32.
259. T. M. Goulding, in A. Pizzi, K. L. Mittal (eds.): *Handbook of Adhesive Technology*, 2nd ed., Marcel Dekker, Inc., New York 2003, pp. 823, 838.
260. A. F. Lewis, in C. A. May, Y. Tanaka (eds.): *Epoxy Resins Chemistry and Technology*, 2nd ed., Marcel Dekker, Inc., New York 1988, pp. 653, 718.
261. J. Sheehan, *the Epoxy Resin Formulators Training Manual*, The Society of the Plastics Industry, Inc., New York 1984, Chapt. XV, p. 175.
262. J. Waechter, *Patty's Industrial Hygiene and Toxicology*, 5th ed., Vol. 6, John Wiley & Sons, Inc., New York 2001, Chapts. 82 and 83, pp. 993, 1145.
263. *Epoxy Resin Systems Safe Handling Guide*, the Society of Plastics Industry (SPI), Inc., New York, Sept. 1997. Publication No. AE-155. Web site: http://www.plasticsindustry.org/about/epoxy/epoxy_guide.htm .
264. Epoxy Resins and Curing Agents, prepared by the Epoxy Resins Committee of the Association of Plastics Manufacturers in Europe (APME), Jan. 1996. Web site: http://www.apme.org/dashboard/presentation_layer_htm/dashboard.asp.

## General References

B. Ellis (ed.): *Chemistry and Technology of Epoxy Resins*, 1st ed., Blackie Academic & Professional, Glasgow, U.K. 1993.

H. Lee, K. Neville, *Handbook of Epoxy Resins*, McGraw-Hill, Inc., New York 1967. Reprinted 1982.

C. A. May, Y. Tanaka (eds.): *Epoxy Resins Chemistry and Technology*, 2nd ed., Marcel Dekker, Inc., New York 1988.

B. Sedlacek, J. Kahovec (eds.): *Crosslinked Epoxies*, Walter de Gruyter, Berlin, 1987.

K. Dusek (ed.): *Epoxy Resins and Composites I–IV*, Advances in Polymer Science 72, 75, 78, 80, Springer-Verlag, Berlin 1986.

*Epoxy Resins*, Advances in Chemistry Series 92, American Chemical Society, Washington, D.C. 1970.

*Epoxy Resin Chemistry*, ACS Symposium Series 114, American Chemical Society, Washington, D.C. 1979.

*Epoxy Resin Chemistry II*, ACS Symposium Series 221, American Chemical Society, Washington, D.C. 1983.

*The Epoxy Resin Formulators Training Manual*, the Society of Plastics Industry, Inc., New York 1984.

E. O. C. Greiner, F. Dubois, M. Yoneyama, *Epoxy Resins*, Chemical Economics Handbook (CEH) Marketing Research Report, Stanford Research Institute (SRI) International, Menlo Park, CA 2001.

J. W. Muskopf, S. B. McCollister, *Ullman's Encyclopedia of Industrial Chemistry*, 5th ed., Vol. A9, 1987, pp. 547–563.

## Further Reading

H. Q. Pham, M. J. Mark: *Epoxy Resins*, Kirk Othmer Encyclopedia of Chemical Technology, John Wiley & Sons, Hoboken, NJ, online DOI: 10.1002/0471238961.0516152407011414.a01.pub2.

# Phenolic Resins

WOLFGANG HESSE, Hoechst AG, Werk Kalle–Albert, Wiesbaden, Germany

JÜRGEN LANG, Dynea Erkner GmbH, Erkner, Germany

| | | | | | |
|---|---|---|---|---|---|
| 1. | Introduction | 1733 | 7.1.1. | Cross-Linked Novolacs | 1744 |
| 2. | Physical Properties | 1736 | 7.1.2. | Novolacs without Cross-Linking | 1745 |
| 3. | Raw Materials | 1737 | **7.2.** | **Resols** | **1746** |
| 3.1. | Phenols | 1737 | 7.2.1. | Water-Soluble Resols | 1746 |
| 3.2. | Aldehydes | 1737 | 7.2.2. | Resols in Organic Solvents | 1747 |
| 3.3. | Catalysts | 1738 | 7.2.3. | Alkylphenol Resols | 1748 |
| 4. | Production | 1738 | 7.3. | Phenolic Resins Modified by | |
| 4.1. | Novolacs | 1739 | | Natural Resins | 1748 |
| 4.2. | Resols | 1740 | 7.4. | Waterborne Paints, Phenol Ether | |
| 4.3. | Phenolic Resins Modified by | | | Resins | 1748 |
| | Natural Resins | 1741 | 7.5. | High-Temperature Coking | 1749 |
| 4.4. | Phenolic Resins with Special | | 7.6. | Phenolic Resins in Composite | |
| | Properties | 1742 | | Materials | 1749 |
| 4.5. | Wastewater | 1742 | 7.7. | Development Trends | 1749 |
| 5. | Storage and Transportation | 1743 | 8. | Economic Aspects | 1749 |
| 6. | Testing and Analysis | 1743 | 9. | Toxicology and Occupational | |
| 7. | Uses | 1744 | | Health | 1749 |
| 7.1. | Novolacs | 1744 | | References | 1749 |

## 1. Introduction

**History.** The first synthetic resins and plastics were produced by polycondensation of phenol with aldehydes. In 1872 VON BAYER first reported the reaction between phenol and aldehydes. The resins formed were, however, not of industrial, and certainly not of scientific interest.

The first plastic used on and industrial scale was phenolic resin, which is a polycondensation product of phenol and formaldehyde normally cured under heat and pressure The birth of phenolic resins and the start of a century of plastics usage are inseparable and are linked to a Belgian-born chemist, who worked in the United States, LEO HENDRIK BAEKELAND. From his work in electrochemistry and electrical engineering he recognized the lack of suitable insulating materials, so in 1904 he started to develop a hard moldable plastic that would be suitable for this type of application. In 1907 he was finally successful and his "Pressure and Heat" patent provided the basics for the utilization of phenolic resins on an industrial scale. The first production of phenolic resins started in 1909, mainly to provide molding compounds to be used as electrically insulating materials for plugs, switches, etc. The rapid development of the electrical industry at that time was one of the success factors of this new thermoset material. In 1910 subsidiary of the Rütgers company was established in Erkner to produce the new material, which became world famous under the trade name "Bakelite".

In addition to being used in the production of plastics, phenolic resins were also sought as replacements for natural resins, which were then used on a large scale for oil varnishes. In 1910 oil-soluble modified phenolic resins were produced by BEHRENDS by polycondensation of a suitable phenol, formaldehyde and rosin.

Between 1928 and 1931 phenolic resins gained increased importance through the treatment of resols with fatty oils to give air-drying varnishes. The main problem, an inadequate compatibility of phenolic resins with other varnish raw materials, was solved by using either alkylphenols or by the etherification of the hydroxymethyl groups of resols with monohydric alcohols.

These varnish applications and the use of phenolic resins as thermosets and electrical insulating materials were originally the main areas of interest. However, the availability of other polycondensates and above all polymers increasingly limited the market for phenolic resins from the mid 1930s onwards. Theoretical work on the composition and mechanism of the formation of phenolic resins was being carried out at that time by VON EULER, HULTZSCH, MEGSON, ZIEGLER, and others, which led to the development of new application areas for phenolic resins (i.e., as adhesives, printing-ink binders, waterborne paints, temperature-resistant binders, and laminated plastics).

The industrial development of phenolic resins is still continuing despite its long history, and such resins are likely to remain of considerable importance because the raw materials required to manufacture them can be obtained at reasonable cost from both petroleum and coal. In turn phenolic resins can be used as raw materials for synthetic fibers and in photoresists for the production of microchips which characterizes the continuing relevance of this group of resins [7]. Phenolic resins themselves are still used today and even find new application areas due to combining superior properties like heat resistance, chemical resistance, insulating properties and nontoxic combustion gases with reasonable cost. New developments have also been made regarding high performance materials in the aerospace and aircraft industry, and for light weight construction materials.

**Classification.** Phenolic resins are polycondensation products of phenols and aldehydes, in particular phenol and formaldehyde. (DIN EN ISO 10082 attempts to define the relevant terms and properties.)

The hydrogen atoms in the *para-* and both *ortho-*positions of the ring, relative to the hydroxyl group, may react with formaldehyde and thus cross-link to form a three-dimensional network. If at least one of these three positions bears a substituent other than hydrogen, cross-linking is no longer possible and comparatively low molecular mass compounds are formed as the end products of the polycondensation.

Phenolic resins are classified as novolacs and resols. In *resols* the polycondensation is base-catalyzed and the condensation degree is adjusted to the intended application. The molar ratio phenol/formaldehyde is normally < 1. Characteristic functional groups of this class of resins are the hydroxymethyl group and the dimethylene ether bridge. Both are reactive groups. During processing the polycondensation can be restarted by heating and/or addition of catalysts (i.e., resols are self-cross-linking). In the case of *novolacs* the polycondensation is brought to completion during the production process. Novolacs are phenols that are linked by alkylidene (usually methylene) bridges, without functional groups (apart from the phenolic hydroxyl groups), and cannot cure on their own. However, novolacs can be cross-linked by the addition of curing agents, such as formaldehyde, hexamethylenetetramine or resols, and give end products similar to resols.

The classification of phenolic resins as novolacs and resols is only strictly valid if phenols that are trifunctional towards formaldehyde are used as starting materials; this is because resols from bifunctional phenols cannot cross-link by themselves. Nevertheless, the polycondensates from substituted phenols are differentiated according to their characteristic groups as *alkylphenol novolacs* (alkylidene bridge) or *alkylphenol resols* (hydroxymethyl group, dimethylene ether bridge).

The third large group comprises *phenolic resins modified by natural resins*. Besides phenolic hydroxyl groups, they contain double bonds, ester links, and carboxyl groups.

**Novolacs.** Novolacs are obtained by a polycondensation reaction with an acidic catalyst (most common is oxalic acid) and a 1:< 1 molar ratio of phenol/formaldehyde. Normally novolacs are solids with melting points between 50°C and 130°C, which can be used in the form of flakes, pills, powder resins or solutions in organic solvents. Novolacs themselves are thermoplastic materials, but their main use is as thermosets using curing agents.

The first step in phenolic resin polycondensation is always the electrophilic attack of a carbonyl compound (generally formaldehyde) on the *para-* and/or *ortho-*positions of a phenol molecule (acid catalysis, Eq. 1) or a phenolate anion (base catalysis, Eq. 2).

Since hydroxymethyl-substituted phenols are more reactive than phenol itself, the hydroxymethylation continues. The hydroxymethyl compounds formed are unstable in acidic medium and are rapidly converted into compounds linked by methylene bridges (Eq. 3). This reaction also occurs in both the *ortho-* and *para-*positions. In basic media hydroxymethyl groups can be stable, but at elevated temperatures they react with the formation of methylene bridges according to Equation (3). To hinder polyalkylation by cross-linking, which would make further processing more difficult or impossible, less than one molecular equivalent of formaldehyde must be added per mol of phenol.

Novolacs are sometimes used as chemically unmodified synthetic resins. Their main application is based, however, on their capability to undergo cross-linking with hexamethylenetetramine. The reaction occurs at ca. 150°C according to Equation (4) [8].

**Resols.** Resols are obtained by polycondensations with basic catalysts and a molar ratio phenol/formaldehyde 1:> 1. Normally, they are used in the liquid form either dissolved in water or in organic solutions (alcohols and ketones). For special applications solid resols are in use.

In a strongly acidic medium, hydroxymethyl groups are rapidly converted into methylene bridges, therefore, the synthesis of resols can only be catalyzed by bases or the salts of weak acids or bases. By analogy with the novolacs the hydroxymethyl groups are formed in the *ortho*- or *para*-positions (Eq. 2). At temperatures above ca. 40°C the hydroxymethyl groups can react to form dimethylene ether bridges with the elimination of water, according to Equation (5). No catalyst is needed for this type of reaction. The dimethylene ether bridges formed are more stable in the *ortho-* than in the *para-*position. They can be converted into methylene bridges with the elimination of formaldehyde (Eq. 6); the formaldehyde liberated is then available for the formation of new hydroxymethyl groups if a suitable catalyst is present.

The hydroxymethyl groups of resols can also condense directly with other phenol molecules according to Equation (3). In resols therefore, three different types of formaldehyde-derived units occur: (1) relatively stable methylene bridges (the only type of linkage in novolacs), (2) hydroxymethyl groups, which are both capable of condensation reactions, and (3) dimethylene ether bridges.

The structures of resols depend not only on the choice of raw materials and their molar ratios, but also on the temperature of formation, concentration of raw materials, presence or absence of solvents, type of catalyst, and catalyst concentration. These parameters determine the structure to a much greater extent than in the case of novolacs.

The catalyst-containing synthetic resins formed are ready to be used industrially. However, the catalyst can also be removed, either before application or during production, as no catalyst is required for the conversion of hydroxymethyl groups into dimethylene ether bridges, a rise in temperature is sufficient.

To affect compatibility (i.e., miscibility) properties the hydrophilic character of resols can be lowered by etherification of the hydroxymethyl group with alcohols according to Equation (7).

$$\text{Ph}-CH_2OH + ROH \longrightarrow \text{Ph}-CH_2OR + H_2O \quad (7)$$

**Modified Phenolic Resins.** Rosin, a natural resin (→ Resins, Natural), contains abietic acid and its double bond isomers as main components. Resols react with the unsaturated centers of these resin acids to form polycarboxylic acids with methylene bridges (Eq. 8) [9].

(8)

X = H or –CH$_2$–

These condensation products from rosin and phenol–formaldehyde resin are known as albertol acids. They can be converted by esterification with polyols, or by salt formation, into higher molecular mass products which are readily soluble in nonpolar solvents but can release the solvent rapidly.

Albertol acids are also obtained by direct condensation of rosin, phenols, and formaldehyde.

Resols also undergo analogous reactions with other natural or synthetic unsaturated compounds such as fatty oils, rubbers, and polymer oils. A limited increase in the molecular mass or cross-linking can thus be achieved. Whether a particular reaction can be carried out successfully depends upon the ratio of the rates of autocondensation of the starting materials to cocondensation with the other reaction partners, but particularly on their mutual compatibility. Phenolic resins are therefore often classified as "water-soluble", "alcohol-soluble", "oil-soluble" etc.

Compatibility of resols with other components can be influenced in many ways, e.g., (1) by using ring-alkylated phenols as the raw materials; (2) by etherification of the hydroxymethyl groups with alcohols; (3) by cocondensations of the resol with natural resins; or by a combination of these measures.

## 2. Physical Properties

Phenolic resins are yellow to brown in color and the coloration can be very intense. Pale phenolic resins become colored immediately after production, during storage or processing. The coloration is enhanced in the presence of oxygen and alkali. The coloration is less intense only in the case of phenolic resins from *para*-alkylated phenols. Characteristic UV absorption maxima lie at 254 nm and 280 nm. The IR spectra of phenolic resins are described in [10].

Phenolic resins which are not cross-linked are commercially available as solids or solutions. Due to the self-curing properties of resols they have a limited shelf life. For particular applications, e.g., in thermosets, the polycondensation can be driven so far that the resins are no longer soluble but can only be swelled by organic solvents. The *softening point* of solid resins can be determined by the capillary melting point according to DIN 53 244, by the ring and ball method, or similar procedures. These temperatures are not melting points in the thermodynamic sense. They characterize a lowering of viscosity caused by a rise in temperature, as a result of which previously crushed resin particles can be observed to coalesce or undergo another change in form.

The *miscibility* of resins with solvents, usually described as "solubility," depends upon the structure of the resin, and ranges from solubility in water to that in naphtha. Resins often have limited miscibility with certain solvents. Typical solvents are alcohols and ketones.

The *viscosity* of phenolic resins or their solutions is measured at high concentrations, e.g., in 30–80% solution. Estimation of the degree of condensation from the viscosity is only possible for the same type of resin, because the molecular structure, in particular the presence of hydroxymethyl groups and dimethylene ether bridges, has a great effect on the viscosity. Usually, the results obtained do not depend on the type of viscometer used. Soluble phenolic resins have a broad *molecular mass distribution*; determination of which (usually by gel permeation chromatography) gives values of more than 50 000, depending upon the type of phenol monomer.

Cross-linked phenolic resins are hard substances which only have a small fracture strain and cannot be melted. Decomposition reactions begin at 120–250°C, depending on the molecular structure. There are, however, also types of phenolic resin (phenolic ether resins) which are stable for some time up to 300°C.

Phenolic resins can be plasticized. Their compatibility with plasticizers can be adjusted by introduction of hydrophilic or hydrophobic groups. Plasticizers increase the flexibility of the phenolic resin network by introducing compact spherical molecules such as tricresyl phosphate or molecules with aliphatic chains like polyglycols.

## 3. Raw Materials

### 3.1. Phenols

*Phenol* (→ Phenol). Unsubstituted phenol is by far the most important of this group.

*Cresols* (→ Cresols and Xylenols). Resins produced from cresols are more compatible with hydrophobic substances than those produced from phenol. Cresols are therefore important principally for the production of varnish resins. The cresol fractions obtained in tar distillation are mixtures of isomers comprising all three cresol isomers and usually also 2,6-dimethylphenol. For the application of these fractions their *meta*-cresol content (which is the most reactive component) is particularly important; thus, as only the *meta*-position is substituted, *meta*-cresol is trifunctional towards formaldehyde. Of the pure isomers *ortho*- and *para*-cresols are the usual raw materials for special applications.

*para-Alkylphenols* (→ Phenol Derivatives). *para*-Alkylphenols are bifunctional towards formaldehyde, i.e., resols produced from *para*-alkylphenols cannot cross-link directly on heating. If a mixed condensation of different phenols with formaldehyde is carried out, their different reaction rates towards formaldehyde must be taken into account. 4-*tert*-Butylphenol (*mp* 99°C), 4-isooctylphenol (*mp* 72–74°C), 4-hydroxybiphenyl (*mp* 167°C), and 4-nonylphenol (a liquid at room temperature) are mainly used. Other alkylphenols such as isopentylphenol, cyclohexylphenol, dodecylphenol, and cashew oil are also sometimes used. The latter mainly contains phenols with $C_{14}$-alkenyl substituents bonded to the *meta*-position.

*2,2-bis(4-Hydroxyphenyl)propane*, (diphenylolpropane, bisphenol A, Dian, *mp* 155–156°C) is tetrafunctional towards formaldehyde. Polycondensation of bisphenol A and formaldehyde gives relatively pale resins.

*Resorcinol* (1,3-dihydroxybenzene, *mp* 117°C) reacts particularly rapidly with formaldehyde and gives resins, which can be cross-linked even in an alkaline medium at room temperature.

Other phenol derivatives that can be used as starting materials are, for example, phenol ethers, phenols containing carboxyl groups, and phenoxyacetic acid.

### 3.2. Aldehydes

*Formaldehyde*. The most important aldehyde for the production of phenolic resins is formaldehyde (→ Formaldehyde), which is used as an aqueous solution at concentrations between 30 and 60%. Besides formaldehyde, the aqueous solutions contain formic acid (oxidation product of formaldehyde) and methanol. Methanol is added to inhibit formation of higher molecular mass polyoxymethylenes, which would precipitate from the aqueous solution. Methanol can intervene in the polycondensation reaction between phenol and formaldehyde by the formation of ethers, acetals, and hemiacetals, and the promotion of side reactions. Thus, knowledge of the methanol concentration (which should be as low as possible) is important for carrying out polycondensation reactions. Formaldehyde solutions can also be stabilized by certain amines. The methanol concentration can then be kept low.

*Paraformaldehyde* is a polyoxymethylene with a relatively low molecular mass. Paraformaldehyde depolymerizes in the reaction mixture. The quantity of formaldehyde liberated per unit time increases with decreasing molecular mass of the paraformaldehyde used as starting material. Since phenolic resin condensations can often take place very vigorously, it is important to assess the reactivity of the paraformaldehyde correctly.

*Higher aldehydes* are less important. In the presence of basic catalysts they frequently undergo side reactions and can therefore generally only be reacted with phenol in an acidic medium. Besides butyraldehyde, benzaldehyde, salicylaldehyde, acrolein, and crotonaldehyde, acetaldehyde, and glyoxal are also important. Glyoxal is tetrafunctional towards phenol and gives novolacs with a low content of low molecular mass condensation products.

*Hexamethylenetetramine* forms adducts with phenols and novolacs very easily, which cross-link on heating. Hexamethylenetetramine is particularly important in the production of thermosetting materials.

### 3.3. Catalysts

The catalytic action of *acids* on the condensation to produce novolacs is essentially a function of the hydrogen ion concentration. The nature of the anion is less important but must be taken into account because of possible side reactions (sulfonation, ester formation). Besides mineral acids, oxalic acid is often used. It decomposes on heating above 180°C and thus allows the production of catalyst-free novolacs.

*Bases* or *salts of weak acids and bases* are suitable as catalysts for resols and novolacs. The cations of catalysts are particularly important; whereas alkali hydroxides catalyze the reaction of phenol with formaldehyde in both the *ortho*- and *para*-positions, alkaline-earth metal hydroxides particularly favor the formation of *ortho*-products.

Ammonia, amines and hexamethylenetetramine have a catalytic effect. They can be incorporated into the resin by the Mannich reaction.

## 4. Production

The production of phenolic resins can involve several separate stages. The substitution and condensation reactions between phenols and aldehydes can be carried out separately in terms of both time and location if required, and also with various catalysts. In addition, hydroxymethyl groups or phenolic hydroxyl groups can be etherified and reactions with unsaturated compounds can be carried out. Resins are frequently distilled to give concentrated solutions or solid resins. Examples relating to production can be found in [11] and [12].

Resols are normally produced in a batch process. Continuous processes mainly using a kind of cascade are described in the literature but are not very common.

Principle steps of the batch process:

- Dosing of raw materials
- Heating to reach condensation temperature
- Condensation until desired parameters are reached
- Distillation of excess water (only if needed)
- Cooling
- Adjustment of final parameters
- Unloading and filtration of the resin

Resin production is normally done in vessels which can be heated and cooled by jackets or coils. They vessels are equipped with condensers for reflux or distillation and often vacuum can be applied.

In case of solvent based resins explosion proof devices are necessary. The vessel size also depends on the application, so for mineral wool resins or binders for wood materials vessels of 50 m$^3$ are in use. Due to the corrosive raw materials the vessel and the auxiliary facilities like tubes and storage tanks are made from stainless steel. Modern plants utilize higher concentrated formalin (about 50%) and molten (100%) phenol to save energy cost for heating and distillation and to minimize the amount of wastewater. In this case the storage facilities and all the tubes and dosing systems have to be heated. The phenolic resin plant shown in Figure 1 can be used for all steps in batchwise

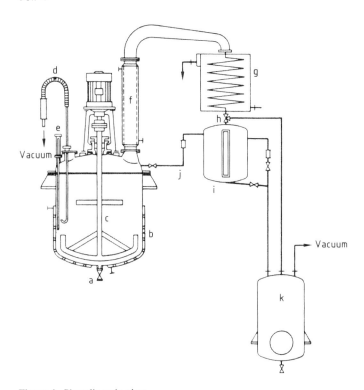

**Figure 1.** Phenolic resin plant
a) Outlet tap; b) Heater; c) Stirrer; d) Evacuation system; e) Thermometer; f) Reflux condenser; g) Main condenser; h) Three-way tap; i) Reflux and separation vessel; j) reflux return with siphon; k) Receiver

production. Substitution and condensation reactions between phenol and formaldehyde are strongly exothermic (the heat of reaction for the formation of methylene bridges in novolacs is −98.7 kJ/mol [11]) and can proceed very fast and vigorously. Therefore, the access to the condensers and overhead and outlet pipes must be kept very wide and the cooling capacity must be designed for rapid condensation and removal of the condensate. To achieve safe industrial production, phenol is always charged to the reactor first and formaldehyde is then added continuously to avoid a run-away reaction. Due to the low MAK and TLV values of the monomers (phenol MAK and TWA = 5 ppm, formaldehyde MAK = 0.5 ppm, TWA = 1 ppm, STEL = 2 ppm) working areas must be well ventilated.

Emptying the reactor vessel is not problematic for dissolved, low viscosity resins, but for solid phenolic resins—particularly resols with a high melt viscosity—rapid discharging and cooling of the resin is very important to avoid further condensation. Besides discharging into pans, removal by cooling conveyors or pelletizing machines has come into use. Phenolic resins which cannot cross-link, such as novolacs or alkylphenol resols, are processed in thin-layer evaporators or other continuous evaporation plants.

In general, phenolic resins are produced in small quantities of a very great variety of types, so that batchwise processes are more economic. Some types (e.g., novolacs for thermosets or resols for the production of wood glues) are, however, produced in large quantities, so that a continuous process is more favorable from the point of view of both obtainable throughput and properties of the substance. Processes described in the literature suggest the use of tubular reactors and series of tank reactors [13, 14].

### 4.1. Novolacs

In the formation of novolacs, substitution and condensation reactions occur simultaneously. In large reaction vessels formaldehyde is continuously metered to a phenol–catalyst mixture and the rate of addition is controlled depending on the heat evolved for safety reasons.

Under industrial conditions, analytical control of the course of the reaction is not necessary, because the degree of polycondensation and molecular structure of the resin are determined by the reaction conditions and the concentration of the starting materials. A temperature/time program is usually used.

**Example** [11]: phenol (1880 parts) is melted in a reaction vessel to which oxalic acid (50 parts) is added and heated to 100°C; an 30% aqueous formaldehyde solution (1560 parts) is then gradually metered in over ca. 2 h. The reaction mixture is held at 98°C for 5 h, and the volatile components are then distilled off under reduced pressure until a temperature of 160°C is attained at a pressure of 2.0 kPa, A novolac (1968 parts), with a softening point of 124°C, remains in the vessel.

## 4.2. Resols

***Resols from Trifunctional Phenols.*** The production of resols(see Fig. 2) differs from that of novolacs in that the reactions between phenol and formaldehyde are not allowed to go to completion, but are stopped at the stage where the self-curing resols are still liquid or soluble. The continuation of the condensation reaction beyond the resol stage (*A-stage*) leads to resins, which are no longer soluble, but can only be swelled and which are known as *resitols* (*B-stage*). The final cross-linking to form *resites* (*C-stage*) gives completely cross-linked plastics.

In resol production the degree of condensation must be controlled during the reaction. The degree of condensation is controlled by measurement of the viscosity, if necessary after dilution with solvents. Other possibilities include the determination of solubility, melting point, and B-time (see Chap. 6). The reaction temperatures for the production of resols are between room temperature and ca. 100°C. The molar ratio of phenol to formaldehyde can vary widely. The upper limit is determined by the functionality of the phenol towards formaldehyde (i.e., a molar formaldehyde/phenol ratio of 3:0). Resols are also formed, however, if an excess of phenol is used. In this case a major portion of the phenol does not react.

The reaction is controlled by the type and quantity of catalyst used and the temperature. High alkali concentrations (up to 1 mol per phenolic hydroxyl group) in combination with low temperatures promote the formation of hydroxymethyl groups and stabilize them. Low catalyst concentrations and particularly high temperatures promote the condensation reactions.

The catalyst can remain in the resin and, in the case of alkali hydroxides, render it water-soluble. These catalyst-containing resols can be used industrially. The catalysts can also be neutralized and, if necessary, removed by washing out or by filtration. The resols thus lose their water solubility partially or completely and must then be employed in polar organic solvents.

The heat of formation of hydroxymethyl groups is only −20.3 kJ/mol, as compared with −98.7 kJ/mol for the formation of

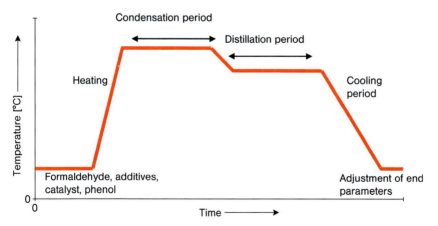

**Figure 2.** Production scheme of a resol

methylene bridges [11]. Thus, the heat of formation of resols is theoretically lower than that of novolacs. However, hydroxymethyl groups are only stable at low temperatures and can be converted into methylene bridges at higher temperatures. Therefore, in the production of resols a great quantity of heat is evolved if the reaction gets out of control. As a result of the high formaldehyde/phenol ratio resol production demands high safety requirements.

To *etherify* the hydroxymethyl groups, a solution of the resol is heated in, for example, butanol. The water formed is distilled off azeotropically with the organic solvent as entrainer, which is recycled to the reactor. Condensation reactions between the hydroxymethyl groups to form ether bridges occur simultaneously. The reaction is controlled by measurement of the viscosity and the solubility. Etherification of alkyd- and oil-modified phenolic resins should be carried out in the presence of a plasticizer, because the compatibility between the reaction components is thus established more quickly. The removal of solvents or water by distillation may be necessary during the various reaction stages or as a final step. During the distillation the condensation reactions continue. Distillation must therefore be carried out under mild conditions.

**Example** [11]. *Condensation*: phenol (846 g, 9 mol) and 30% formaldehyde solution (1600 g, 16 mol) are mixed at 20°C and the pH is adjusted to 9.5–10 by the addition of 38% aqueous sodium hydroxide solution (ca. 110 mL) with stirring. The mixture is heated to 60°C and kept for 4 h at this temperature. The decrease in formaldehyde concentration is monitored by titration with hydroxylamine hydrochloride. At the end of the reaction only ca. 1% formaldehyde is present. The mixture is cooled, and carefully neutralized almost completely with ca. 5% hydrochloric acid with stirring. The reaction solution must not become acidic and should have a pH of ca. 7.2–7.5. Water and excess phenol are subsequently distilled off under vacuum, such that the temperature may reach a maximum of 60°C (at 1.5 kPa). A viscous resin (1360 g) is obtained as residue.

*Etherification*: butanol (1360 g) and toluene (125 g) are added to the residue and the mixture is stirred at 40°C until the components have completely dissolved. After cooling, precipitated sodium chloride is filtered off. Butanol (900 g) is then added to the filtrate and the pH is adjusted to 6 with a concentrated solution of phosphoric acid in ethanol with vigorous stirring. The reaction mixture is then heated to boiling with stirring in a three-necked flask fitted with a Dean–Stark head which is filled with a 1:1 butanol/toluene mixture. In the Dean–Stark head, the water which has been distilled off is separated and the solvent is recycled into the reaction vessel. After ca. water (72 mL) has been distilled off, a sample of the reaction solution can be diluted at 20°C with xylene (4 parts) or petroleum ether (1.4 parts) without the resulting solution becoming cloudy (before the etherification a sample could only be diluted by 0.7 or 0.4 parts, respectively). The solution is concentrated (to 2620 g) under reduced pressure and traces of salt are filtered off at 20°C, after 48 h.

***Alkylphenol Resols.*** *para*-Alkylphenols can only react with formaldehyde in the *ortho*-positions. Hydroxymethyl groups or dimethylene ether bridges in the *ortho*-position differ from those in the *para*-position by their greater stability. In the production of alkylphenol resols the hydroxymethyl groups therefore remain relatively stable in alkaline medium even at elevated temperatures of between 60°C and 70°C. To prevent the conversion of dimethylene ether bridges to methylene bridges according to Equation (6), the production of alkylphenol resols is often performed in two steps. In the first step a resin, rich in hydroxymethyl groups, is produced in the presence of alkali. The catalyst is subsequently removed by neutralization and washing out. The substance is then heated to >100°C to carry out the actual polycondensation, the etherification of hydroxymethyl groups to form dimethylene ether bridges. Catalysts are not required in this step.

## 4.3. Phenolic Resins Modified by Natural Resins

The reactions of resols with rosin to give albertol acids are carried out in the melt at 100–200°C. The albertol acids are subsequently reacted with polyols to give polyesters or with metal oxides to give resinates (both procedures can also be combined). These reactions must be carried out at temperatures of up to 280°C because the tertiary carboxyl groups of the abietic acids are only esterified slowly and the melt viscosities and softening points of the rosin-modified phenolic resins can be very high. Suitable resol components are the reaction products of formaldehyde with phenol, cresols, alkylphenols, or bisphenol A, or mixtures of these. The properties of the resins are determined by the composition and structure of the resol, the resol/rosin ratio, and the nature and quantity of the polyols or metal oxides used for the subsequent reaction. The modified phenolic resins vary greatly with respect to viscosity, softening point, miscibility with solvents, and compatibility with varnish or printing ink raw materials.

The reaction sequence between rosin, resol, and esterification components can also be reversed. In this case the natural resin acids are first esterified and the esters are then reacted with resols.

## 4.4. Phenolic Resins with Special Properties

***Resols for Water-Soluble Varnishes.*** Water-soluble resols exist as phenolates that are unsuitable for varnishes because they do not form adhesive varnish films. For the production of self-curing binders for water-soluble varnishes, carboxyl groups are introduced into the resols. These carboxyl groups impart water solubility when they are converted into salts with amines or ammonia. The phenolic hydroxyl groups are reacted with chloroacetic acid according to the Williamson ether synthesis usually at the temperature of formation of the resol. Alkali is used as the condensation agent. To achieve water solubility it is not necessary to introduce carboxyl groups into all resol molecules. The reactions are generally carried out in the presence of auxiliary solvents, mostly glycol ethers, which then also remain in the varnish formulation [15].

Water-soluble phenolic resins for varnishes are mostly applied in combination with plasticizers; for example, epoxy resins or fatty oils. It is thus often sufficient to incorporate the groups that impart water solubility into one of the components. The second component is then carried into the aqueous phase by the first.

***Ether Resins from Resols and Novolacs.*** The phenolic hydroxyl groups can be etherified with, e.g., benzyl chloride or allyl chloride. The etherification always lowers the rate of curing. On the other hand, it reduces disadvantages which are due to the presence of phenolic hydroxyl groups, e.g., oxidizability, tendency towards coloration, and sensitivity towards alkalis. A strong decrease in viscosity is often associated with etherification. This is particularly the case with allyl ethers. The phenolic hydroxyl groups of resols and novolacs can also be etherified with epoxy compounds such as ethylene oxide, propylene oxide, or styrene oxide in the presence of catalytic quantities of alkali. The nucleophilic opening of the oxacyclopropane ring by the phenolic hydroxyl groups is very fast. It is thus possible to work in aqueous solution without significant nucleophilic attack of water at the epoxy group. Reaction of the epoxy compounds with hydroxymethyl groups only occurs if the epoxy compound is added in a nearly stoichiometric ratio to the phenolic hydroxyl group.

***Condensation Reactions between Phenol Ethers or Hydrocarbons with Formaldehyde and Phenol.*** Phenol ethers can form polycondensates with formaldehyde. Known examples are diphenyl ether, phenoxyacetic acid, diphenylglycerol ethers, and phenoxyisopropanol. The condensation reactions can only be carried out with acid catalysts. Depending on the polarity of the ether, strong acids must be used, sometimes in very large quantities. Aromatic hydrocarbons, e.g., xylene, are even less reactive than phenol ethers. The hydroxymethyl groups and dimethylene ether bridges are fairly stable in the phenol ether resins in spite of the high acid concentration and remain as reactive groups in the resin.

Because of the low tendency towards self-curing, these resins are usually reacted with (excess) phenol to form methylene bridges between the phenol ether (or hydrocarbon) and the phenol.

## 4.5. Wastewater

In the production of phenolic resins phenol-containing wastewater is produced, which can also be contaminated by formaldehyde, salts, acids and alkali, and organic solvents. This must be rendered harmless in a way that will not pollute the environment, for example by incineration polluted wastewater. In addition, phenol can be degraded very readily by bacteria. The purification in biological sewage plants—after dilution with other waste water, if necessary—is therefore very effective. It can be economic to extract the phenol before the incineration of the waste water.

## 5. Storage and Transportation

*Solid phenolic resins* which are sold in broken pieces, flakes, pastilles, or granules, can be packed in sacks, drums, or large containers. If the resins (e.g., rosin-modified types), have a tendency towards autoxidation, they should be stored in the absence of oxygen. *Phenolic resin solutions* are not corrosive and can therefore be stored in iron containers, if a possible coloration of the resin is acceptable. Otherwise the containers must be coated on the inside. *Liquid phenolic resins* are transported in tank cars, containers, or drums. When stored at room temperature, aqueous alkaline resol solutions often lose their viability after a few days or weeks. These solutions must be kept cool during transportation and stored in refrigerated areas. The storage life of solid resols at room temperature is generally longer; it is nevertheless advisable to keep all resols cool during storage. Certain phenolic resins have relatively low softening temperatures and compact during storage which impairs their storage stability.

## 6. Testing and Analysis

A summary of the testing methods is contained in DIN EN ISO 10 082

*General Tests.* Color index, viscosity, flash point, density, melting point (softening point) are measured according to DIN 53 244.

Determination of *water* is carried out by the Karl Fischer method.

*Hydroxyl Groups.* The total quantity of hydroxyl groups is determined by acetylation according to DIN 53 240. The presence of unsaturated groups, for example in modified natural resins, can give rise to values which are too high.

*Reactivity Determination for Resols: B-Time.* The cross-linking rate of resols, the transition from the A- into the B-stage is determined on a hot plate at a predetermined temperature, usually 120°C or 150°C.

*Differentiation between the Various Types of Phenolic Resin.* Neutral resol solutions in polar solvents give a blue or purple coloration with iron(III) chloride. With novolacs this coloration only appears after the addition of a small amount of hexamethylenetetramine. Alkylphenol novolacs often give a black-violet coloration. Modified natural resins do not show the color reaction, but can be identified by the Storch–Morawski color reaction (→ Resins, Natural) [16, 17].

*Chromatographic Methods.* Thin layer chromatography is very suitable for the determination of phenolic resins (e.g., monomer content) and can also be used for the quantitative determination of components by comparison of the spot size. The sodium salt of diazotized sulfanilic acid in aqueous solution, or iodine vapor can be used to develop the spots.

Gel permeation chromatography on polystyrene columns is applied to determine the molecular mass distribution of phenolic resins. In the low molecular mass range it also gives information on the concentrations of phenols, hydroxymethylated phenols etc. The columns must be calibrated with phenolic compounds [18].

Gas chromatography is suitable for the determination of free phenols. Phenols up to and including butylphenol can be detected by direct injection. Higher phenols are determined after esterification (acetylation, silylation).

*Spectroscopic Methods.* IR spectroscopy of phenolic resins is described in detail in [10]. The groups typical of phenolic resins can be determined and analyzed semiquantitatively.

Phenolic resins show UV absorption bands at 254 and 280 nm. UV spectroscopy is sometimes used for product monitoring; however, the shift of the bands due to other resin components or because of substitution must be taken into account.

NMR spectroscopy enables detailed examination of the molecular structure of phenolic resins and at the same time kinetic studies of the formation reactions under different conditions [19].

The most common analytical methods are described in the following standards:

- **DIN EN ISO 3251.** Paints, varnishes and plastics - Determination of nonvolatile-matter content (ISO 3251:2008); German version EN ISO 3251:2008
- **DIN EN ISO 8974.** Plastics - Phenolic resins - Determination of residual phenol content by gas chromatography (ISO 8974:2002); German version EN ISO 8974:2002
  Plastics - Phenolic resins - Determination of reactivity on a B-transformation test plate (ISO 8987:2005); German version EN ISO 8987:2005
- **DIN EN ISO 8989.** Plastics - Phenolic resins - Determination of water miscibility (ISO 8989:1995); German version EN ISO 8989:1998
- **DIN EN ISO 11402.** Phenolic, amino and condensation resins - Determination of free-formaldehyde content (ISO 11402:2004); German version of EN ISO 11402:2005
- **DIN ISO 8975.** Plastics; phenolic resins; determination of pH; identical with ISO 8975:1989
- **DIN EN ISO 9396.** Plastics - Phenolic resins - Determination of the gel time of resols under specific conditions using automatic apparatus (ISO 9396:1997); German version EN ISO 9396:2000
- **DIN EN ISO 8619.** Plastics - Phenolic resin powder - Determination of flow distance on a heated glass plate (ISO 8619:2003); German version EN ISO 8619:2004
- **DIN EN ISO 8988.** Plastics - Phenolic resins - Determination of hexamethylenetetramine content - Kjeldahl method, perchloric acid method and hydrochloric acid method (ISO 8988:2006); German version EN ISO 8988:2006
- **DIN EN ISO 11401.** Plastics - Phenolic resins - Separation by liquid chromatography (ISO 11401:1993); German version EN ISO 11401:1998

# 7. Uses

The use of phenolic resins is determined by their superior properties, which include:

- FST (fire–smoke–toxicity) properties
- Heat resistance
- Chemical resistance
- Moisture resistance
- High carbon yield, after decomposition
- Electrical insulating properties

Thus, from an economic point of view, phenolic resins are quite cost efficient.

## 7.1. Novolacs

### 7.1.1. Cross-Linked Novolacs

For industrial use novolacs can be delivered in flakes, prills, powder resin, dispersions, solutions in organic solvents, or even in the molten state.

For cross-linking, novolacs are processed together with *curing agents*, mainly hexamethylenetetramine. Cross-linking occurs at a sufficient rate at 140–160°C and can be carried out within a few minutes, particularly if a part of the polycondensation reaction has already taken place.

Novolacs are sometimes cross-linked with *resols*. This reaction gives resins with high hardness, high stability, but a lower fracture strain.

Cross-linked phenolic resins are much less flammable than thermoplastics. Phenolic resins can be rendered virtually nonflammable by the addition of usual plastics additives such as phosphates, borates, red phosphorus, phosphoric acid esters, and other flame retardant agents, mainly based on phosphorus/nitrogen synergies.

*Thermosets/Molding Materials.* Phenolic resin molding materials are produced with heated rollers or extruders. Treatment of a mixture of novolac, hexamethylenetetramine, fillers, and additives leads to an intermediate, which is the actual molding material. The polycondensation is driven so far that the molding material in the final molding step can still flow, but so that the final cross-linking occurs very quickly. Phenolic resin thermosets are standardized according to DIN EN 14598 part 1-3. The application properties of the final products depend very much on the choice of filler material. The processing properties, in contrast, are affected primarily by the novolacs used. Especially important processing properties are good flow characteristics and rapid curing. Molding technology which predominated until the end of the 1960s has in the meantime been substituted by transfer molding and injection molding, particularly in the mass production of less complicated preforms. Thus the expression "thermosetting molding materials" has come into common use.

*Grinding Wheels.* In the production of phenolic resin-bound grinding wheels, the grinding material (usually corundum of various granularities) is impregnated with a liquid phenol resol and mixed with a ground, pulverized mixture of phenol novolac and hexamethylenetetramine. The binder cures to a three-dimensional cross-linked resite under a carefully controlled temperature and pressure program. No defects may occur in this process because grinding wheels are exposed to high thermal and mechanical stresses.

*Friction Linings.* Brake linings and clutch facings are made of reinforced phenolic resins. Novolac–hexamethylenetetramine mixtures can also be used here as binders. For better heat conduction, copper wire or copper gauze is often incorporated. Novolacs modified with alkylphenols or cashew oil are used to adjust hardness and lubricating properties, particularly in brake linings.

**Reinforcing Resin for Rubber** (→ Rubber, 9. Chemicals and Additives). The hardness of rubber is increased by the incorporation of novolac–hexamethylenetetramine mixtures. This hardness can be so great as to enable fabrication of solid, tough molded articles, for example for bodywork parts. Often, however, as in car tire mixtures, only slight improvement of the rubber hardness is desired. The reinforcing effect is thought to be caused by strong intermolecular interactions between the cured, thermoplastic phenolic resin and the rubber-elastic vulcanized product. Both types of macromolecule are not bonded by covalent bonds.

***Novolac Powder Resins for Felt Bonding.*** Felts → Felts are used as acoustic and thermal insulation material for the automotive industry and household appliances.

Felt is a combination of reclaimed fabric, mainly cotton, and powder resin, which is used as a binder. Phenolic resins offer the best compromise regarding performance and economy in use.

The final properties of the phenolic resin bonded textile felt, including mechanical strength, damping behavior, thermal resistance and environmental behavior are determined by the composition of the felt and the processing parameters. Normally novolac–hexamethylentetramine mixtures are in use.

***Novolac Solutions for Car Filter Media.*** Novolacs dissolved in alcohols are widely used for the impregnation of filter papers for car filter media. They are cured by hexamethylenetetramine and/or resols. Novolac solutions in combination with resols are mainly used for oil and fuel filters and airfilters are produced with solvent based resols.

### 7.1.2. Novolacs without Cross-Linking

The use of novolacs which are not cross-linked is less important and is based particularly on their solubility and compatibility properties. As polyalkylidenephenols they have strong inter- and intramolecular interactions and can be adapted to the most varying requirements by alkyl substitution.

***Varnishes*** (→ Paints and Coatings, 2. Types, Section 13.2). The use of uncured novolacs in varnishes is the oldest application of phenolic resins. Because of their high susceptibility towards oxidation and coloration, particularly in the case of unsubstituted phenol novolacs, this application area has hardly developed further. Noncross-linked novolacs are now only important in specialist areas such as model lacquers and bituminous paints.

***Printing Technology.*** Novolacs are often added to aniline printing inks because of their high affinity for these dyes. Because aniline printing inks are often used for printing on food packaging materials, the novolacs must be phenol-free. Novolacs are also used in the formulation of ballpoint pastes.

Novolacs are applied, together with diazo compounds, to zinc or aluminum plates in a thin layer for the production of positive offset printing plates (→ Imaging Technology). Similar systems are used as photoresists for the photolithographic production of integrated circuits.

***Raw Materials for Epoxy Resins.*** The reaction of epichlorohydrin with novolacs (instead of bisphenol A) leads to polyfunctional epoxy resins (→ Epoxy Resins, Chap. 5). Their curing products are characterized by their particularly high usage temperatures.

***Alkylphenol Novolacs,*** particularly those based on 4-hydroxybiphenyl, 4-isooctylphenol, and formaldehyde are used for the production of copying paper.

***Tackifiers for Rubber*** (→ Rubber, 9. Chemicals and Additives). Nonvulcanized rubber mixtures are often not tacky enough to adhere to each other. This is a considerable disadvantage in the production of laminated rubber articles, e.g., car tires. Tackifiers are used in rubber technology to improve this. Besides mineral oils and natural and synthetic resins, alkylphenol novolacs are particularly effective and when used with other tackifiers have a synergistic effect. In addition to alkylphenol–formaldehyde novolacs, *para*-alkylphenols–acetaldehyde resins, and *para*-alkylphenol–acetylene resins are often used as tackifiers.

***Crude Oil Separators.*** Ethoxylation (i.e., reaction with ethylene oxide) of alkylphenol novolacs gives water-soluble, surface-active substances. If the degree of ethoxylation is suitably adjusted, they have de-emulsifying properties. Addition of small quantities to the crude oil that is extracted as an oil–saltwater emulsion effects a faster separation into an oil and an aqueous phase.

## 7.2. Resols

### 7.2.1. Water-Soluble Resols

Water-soluble condensation products of phenol and formaldehyde are used mainly as adhesives for wood or wood chips, as binders for organic and inorganic fibers and for granular material such as corundum or sand. Curing is performed thermally at 100–180°C and, if possible, under pressure. If curing is carried out under atmospheric pressure the cross-linking process must be carefully controlled and performed slowly to avoid the formation of bubbles or cavities.

The waste-gas from the thermal curing process contains phenol, formaldehyde, and sometimes solvent vapors. When legally stipulated limiting values are exceeded, the waste-gases must be incinerated.

Cross-linked phenolic resins have a low flammability. They can be rendered virtually nonflammable by additives. In the event of fire the flue gases are relatively nontoxic and, compared to the flue gases from other burning plastics, are relatively transparent. For example, this makes them particularly suitable in the interiors of vehicles.

***Binder Systems for Wood Fibers.*** Aqueous, alkaline resol solutions are used in the production of fiber boards (HDF, MDF), particle board, oriented strand boards (OSB) and plywood. Wooden substrates bound by phenolic resin are moisture-resistant in contrast to those bound by urea resins, and are therefore suitable for the production of exterior walls in buildings and related uses.

***Construction Adhesives.*** Resorcinol–formaldehyde resins are used as wood glues. Resorcinol is treated with a substoichiometric amount of formaldehyde in an alkaline medium and the stable solution is treated with more formaldehyde (usually in the form of paraformaldehyde) immediately before use. The polycondensation with resorcinol, which is very fast at, and even below, room temperature, is thus restarted. The resorcinol–formaldehyde resin cross-links three-dimensionally. Glues containing resorcinol resins have high mechanical stability and are highly water-resistant.

***Binding Agents for Molding Sand*** (→ Foundry Technology, Section 2.3.4). Phenol resols or novolac–hexamethylenetetramine mixtures that can be sufficiently diluted with water, are used as binders for casting molds, which consist mainly of sand. Mixtures of both resins are sometimes used for full molds. For hollow and thin-walled molds (Croning process), alcoholic solutions of novolac–hexamethylenetetramine adducts are used. To achieve special properties or to make the process cheaper, phenolic resins are sometimes modified with furan or urea resins.

***Laminates.*** High quality laminates for mechanical and electrical equipment are obtained by impregnation of paper, cotton, glass fiber, carbon fiber or aramide fiber with phenolic resins and subsequent thermal cross-linking. They are used as insulation and construction materials. For the production of laminates with high electrical performance, e.g., for printed circuit boards, unmodified resols and tungoil-plasticized resols are used together for impregnation. Both resins have different functions. The unmodified resins — with as low a molecular mass as possible — penetrate into the fiber and the tungoil-plasticized resins remain on the fiber surfaces and bind them together.

In *decorative laminates*, water-soluble, alkaline resol solutions can only be used for the production of core resins because of their dark coloration. Melamine resins are used as decorative facings. By modifying the phenolic resins different properties can be achieved such as postforming laminates, flame-retardant laminates, compact boards or laminate flooring, produced by HPL (high pressure)-, CPL (continuous press)- or short-cycle press technologies.

***Fiber Bonding.*** Organic or inorganic fibers or rock wool can be bonded to plates or nonwoven fabrics by impregnation with an aqueous resol solution or this can be sprayed on them. Curing is carried out continuously with heated rollers or in presses. Matting or insulating plates bound with phenolic resin have moderate heat-insulation properties in comparison to thermoplastic foams but have the advantage of being virtually nonflammable, and at the same time soundproof.

Nowadays the greatest amount of phenolic resins is used to produce glass-wool and stone –wool insulating materials.

***Phenolic Foam.*** Phenolic foam is used in three applications:

- Flower foam with open cells to store the necessary water for the flowers
- Insulation foam with closed cells with high insulation effects
- Mining foam to stabilize the stone in mines (operating mines or old mines)

Water-soluble resols can be used for the production of foamed plastic. For foaming, the addition of surfactants and the use of blowing agents are necessary. Phenolic resin foams have the advantage of low flammability. However, the corrosive effect of the strong acids used as hardeners limits their general use.

***Coated Abrasives.*** To produce abrasives such as sandpaper, abrasive cloth, or flexible sanding disks on vulcanized fiber, abrasive grains are applied to a moving backing cloth on one side. These grains are fixed together and to the surface of the backing cloth with a binder. For heat-resistant abrasives, phenol resols are preferably used as binders.

Since the carrier material (usually cellulose) is sensitive to high temperatures, a prolonged application of temperatures $>130°C$ must be avoided. Therefore, cross-linking of the phenol resol to resite takes a long time. Phenolic resins that are suitable for the production of abrasives must therefore be very reactive, to limit the curing time to a few hours. This naturally limits their storage life. The cross-linking reactions are carried out under atmospheric pressure in festoon dryers and the carrier materials, coated with phenolic resin and abrasive grains are introduced into a dryer where the temperature and humidity must be carefully controlled.

### 7.2.2. Resols in Organic Solvents

The resols produced as aqueous alkaline solutions can be dissolved in polar solvents after neutralization of the catalyst. Basically such solutions do not differ from aqueous solutions. Catalysts can be added to accelerate cross-linking. Since, however, phenolic resins are often used in combination with other components, these resins must be compatible with the other substances, and for this purpose their solution properties must be altered. This can be done using alkylated phenols as raw materials or in the etherification of the methylol groups with alcohols.

***Epoxy Resin–Phenolic Resin Coatings.*** Resols based on cresols are used in cross-linking together with epoxy resin. The enamels baked at temperatures around 200°C stand out by virtue of their very good ductility and very high chemical stability. They are principally used for the inner coating of food and drink cans, and other containers.

***Oil-Plasticized Resols.*** Resols—mainly based on cresol or bisphenol A—are modified by fatty oils or alkyds and are used as baking enamels. These varnishes have good chemical stability and very high ductility and are used, in particular, as inner coatings for containers. Other uses of oil-plasticized resols lie in the area of cylinder-head gaskets and electric insulation materials.

### 7.2.3. Alkylphenol Resols

Resols from *para*-alkylphenols are also soluble in nonpolar solvents. The solubility increases with increasing size of the alkyl group. *para*-Alkylphenol resols differ from other resols in their bifunctionality towards formaldehyde. Thus, *para*-alkylphenol resols cannot cross-link by themselves, but are used as cross-linking agents for other, polyfunctional substances. Examples are fatty oils, rubbers, long-chain epoxy resins, alkyd resins, and phenol novolacs.

***Vulcanization.*** Rubbers can be vulcanized by alkylphenol resols. As phenolic resins are considerably more expensive than sulfur, this possibility is only exploited for special cases, e.g., butyl rubber.

***Adhesive Resins.*** The properties of polychloroprene adhesives can be improved by natural and synthetic resins. Alkylphenol resols not only increase the stickiness of the adhesive but also its binding and heat stability. Since the amount of resin needed to be added is about half the quantity of polychloroprene, adhesive resins have considerable economic importance.

The alkylphenol resols do not act as such in polychloroprene adhesives but in the form of their magnesium and zinc complexes. These are formed with the magnesium and zinc oxides present in the polychloroprene adhesive. The complexes described as chelate polymers have a very high molecular mass, but are nevertheless readily soluble in aromatic hydrocarbons and even in naphtha [20].

## 7.3. Phenolic Resins Modified by Natural Resins

***Printing Inks.*** Rosin-modified phenolic resins are used as binders for book or offset printing inks for the following reasons: (1) they are soluble in mineral oil and aromatics, (2) compatible with other printing-ink raw materials, and (3) have a high melting point, which allows a rapid transition from the liquid to the solid state on printing. Phenol resins, modified by natural resins, are very varied so that they can be adapted to the continually increasing rate of working in the production of printed articles.

## 7.4. Waterborne Paints, Phenol Ether Resins

***Waterborne Paints.*** The amine and ammonium salts of resols containing carboxyl groups are water-soluble. They are used as baking enamels and give coatings which are resistant to chemicals but are brittle. To plasticize them they are mixed with fatty oils, alkyds, or unsaturated hydrocarbons (which have been rendered water-soluble by the addition of maleic anhydride, and subsequent salt formation). Their main use lies in the electrodeposition coating of car bodies and vehicle parts. See also, → Paints and Coatings, 2. Types, Chap. 13.

***Allyl Ether Resols.*** Resols etherified with allyl groups at the phenolic hydroxyl group have a very low viscosity and can be used as solvent-free or low-solvent binders in varnishes. Higher baking temperatures must be used for cross-linking than in the case of simple phenolic resins.

***Phenol Ether and Hydrocarbon–Formaldehyde Polycondensates.*** Mixed condensates from (1) phenol ethers–formaldehyde–phenol, or (2) hydrocarbons–formaldehyde–phenol are used in thermosets which have an increased stability towards chemicals and oxygen.

## 7.5. High-Temperature Coking

Phenolic resins, whether resols or novolacs, coke upon heating in the absence of oxygen. They can therefore be used as binders for refractories based, for example, on magnesite or dolomite, which are used for the lining of blast furnaces and similar constructions. Phenolic resins can also be used as high-temperature materials after coking and in some cases graphitization. These materials can be reinforced with temperature-resistant fibers.

## 7.6. Phenolic Resins in Composite Materials

Although phenolic resins are quite old materials there are new applications in the field of advanced composite materials, used, e.g. in the aerospace or aircraft industry. In particular, the superior FST properties of phenolic resin bonded composite materials are valued in these sophisticated industries. Here limiting aspects for the use of phenolic resins are associated with their mechanical properties such as low impact strength and toughness due to the rigid polymer network. To overcome such problems highly modified and plasticized resins are in use.

Some examples for the usage of phenolic resins in high performance applications are:

- Light weight honeycomb composites used for interior parts of airplanes
- Carbon–carbon composites (using the carbon network) for aircraft brakes, heat pipes, reentry vehicles, rocket motor nozzles, biomedical implants.
- Prepregs for ballistics like helmets, bullet proof jackets, armoured cars.

## 7.7. Development Trends

- Broader usage of modified phenolic resins in composite materials due to their superior FST properties and cost efficiency
- Combinations with other polymers
  - Modification of thermoplastics
  - Phenolic–melamine–urea resin combinations
  - Phenolic–epoxy resin combinations
  - Phenolic resins–acrylates combinations
- Usage of nano structures
- Usage of renewable materials in combination with synthetic raw materials for new binder systems

## 8. Economic Aspects

In 1978, 728 000 t phenolic resin were produced in the United States [21] and in 1994, 730 000 t [22].

In Europe in the years 2003–2009 about 650000–850000 t/a of phenolic resins were produced.

No conclusions about the turnover value can be drawn from the quantity data, for although phenolic resins for mass applications are cheap, very high prices are often demanded for specialist applications.

## 9. Toxicology and Occupational Health

Cured phenol–aldehyde resins are classified as nontoxic. The toxicity of phenolic resins is therefore assessed according to their content of unreacted monomers [23]. In Germany phenolic resin formulations with a content of more than 5% free phenol must be labeled with a skull and crossbones and those with more than 1% with a cross. Substances liberated in the thermal cross-linking of phenolic resins, particularly formaldehyde, are toxic and lead at least to irritation. Contact with phenolic resins can lead to allergies in sensitized people.

*Occupational Health.* No TLV or MAK values have been established for phenolic resins. The corresponding values of the most important monomers are as follows: formaldehyde TLV-TWA 1 ppm (1.2 mg/m$^3$), TLV-STEL 2 ppm (2.5 mg/m$^3$), MAK 0.5 ppm; phenol TLV-TWA 5 ppm (19 mg/m$^3$), MAK 5 ppm.

# References

## General References

1. A. Bachmann: *Phenoplaste*, Dtsch. Vlg. Kunststoffind., Leipzig 1973.
2. K. Hultzsch: *Chemie der Phenolharze*, Springer Verlag, Berlin 1950.

3 A. Knop, L. Pilato: *Phenolic Resins*, Springer Verlag, Berlin 1985.
4 R. Martin: *The Chemistry of Phenolic Resins*, Wiley, New York 1956.
5 N. Megson: *Phenolic Resin Chemistry*, Butterworths, London 1958.
6 H. Wagner, H.F. Sarx: *Lackkunstharze*, Hanser, München 1971.

## Specific References

7 N. Megson, *Chem. Ztg.* **96** (1972) no. 1/2, 14–19.
8 K. Hultzsch, *Kunststoffe* **42** (1952) no. 11, 385.
9 K. Hultzsch, *Kunststoffhandbuch*, vol. X: Duroplaste, Hanser, München 1968.
10 D. Hummel: *Atlas der Kunststoff-Analyse*, 2nd ed., Hanser, München 1984.
11 *Houben-Weyl, Methoden der organischen Chemie*, 4th ed., vol. **XIV/2**, Georg Thieme Verlag, Stuttgart, pp. 272–291.
12 *Ullmanns Encyklopädie der Technischen Chemie*, 3rd ed., vol. **13**, Urban und Schwarzenberg, München 1951–1970, pp. 453–479.
13 K.H. Jung, DE-OS 2 620 488, 1977.
14 F. Wilhelm, DE 1 570 349, 1972.
15 W. Brushwell, *Farbe + Lack* **82** (1976) no. 10, 917.
16 *Dtsch. Farben-Z.* **20** (1966) 363.
17 D. Braun, *Kunststoffe* **62** (1972) no. 1, 41.
18 E. Wagner, R. Greff, *J. Polym. Sci. Polym. Chem. Ed.* **9** (1971) 2193–2207.
19 J. Woodbrey et al., *J. Polym. Sci. Polym. Chem. Ed.* **3** (1965) 1079–1106.
20 K. Hultzsch, *Farbe + Lack* **77** (1971) 1165–1172.
21 *Chem. Eng. News* **67** (1989) June 19, 45.
22 K. Weissermel, H.-J. Arpe: *Industrielle organische Chemie*, Wiley-VCH, Weinheim 1998, p. 394.
23 H. Babisch et al., *Regul. Toxicol. Pharmacol.* **1** (1981) 90–109.

## Further Reading

P. W. Kopf: *Phenolic Resins*, Kirk Othmer Encyclopedia of Chemical Technology, 5th edition, vol. 18, p. 756–802, John Wiley & Sons, Hoboken, NJ, 2006, online: DOI: 10.1002/0471238961. 1608051411151606.a01.pub2.

L. Pilato (ed.): *Phenolic Resins: A Century of Progress*, 1. ed., Springer, Heidelberg 2010.

# Resins, Synthetic

**ULLMANN'S ENCYCLOPEDIA OF INDUSTRIAL CHEMISTRY**

GERD COLLIN, DECHEMA e. V., Frankfurt a. M., Federal Republic of Germany

ROLF MILDENBERG, formerly Rütgerswerke AG, Duisburg, Federal Republic of Germany

MECHTHILD ZANDER, formerly CdF, Forbach, France

HARTMUT HÖKE, Weinheim, Federal Republic of Germany

WILLIAM MCKILLIP, Advanced Resin Systems, Des Plaines, Illinois 60016, United States

WERNER FREITAG, Hüls Aktiengesellschaft, Marl, Federal Republic of Germany

WOLFGANG IMÖHL, Schering AG, Berlin, Federal Republic of Germany

| | | |
|---|---|---|
| 1. | Definition | 1751 |
| 2. | Hydrocarbon Resins | 1752 |
| 2.1. | Classification | 1752 |
| 2.2. | Raw Materials | 1752 |
| 2.2.1. | Indene – Coumarone Feedstocks | 1753 |
| 2.2.2. | Petroleum Resin Feedstocks | 1753 |
| 2.3. | Production | 1754 |
| 2.4. | General Properties | 1755 |
| 2.5. | Characterization and Analysis | 1755 |
| 2.6. | Resin Types | 1756 |
| 2.6.1. | Aliphatic Hydrocarbon Resins | 1756 |
| 2.6.2. | Aromatic Hydrocarbon Resins | 1757 |
| 2.6.3. | Terpene Resins | 1757 |
| 2.6.4. | Resins from Pure Monomers | 1757 |
| 2.6.5. | Modified Resins | 1758 |
| 2.7. | Uses | 1758 |
| 2.8. | Economic Aspects | 1761 |
| 3. | Resins by Condensation | 1762 |
| 3.1. | Furan Resins | 1762 |
| 3.1.1. | Monomers | 1762 |
| 3.1.2. | Production – Characteristics | 1762 |
| 3.1.3. | Uses | 1763 |
| 3.2. | Ketone and Aldehyde Resins | 1763 |
| 3.2.1. | Resins from Aliphatic – Aromatic Ketones | 1764 |
| 3.2.1.1. | Acetophenone – Formaldehyde Resins | 1764 |
| 3.2.1.2. | Modified Acetophenone Resins | 1766 |
| 3.2.2. | Resins from Aliphatic Ketones | 1766 |
| 3.2.2.1. | Methyl Ethyl Ketone – Formaldehyde Resins | 1766 |
| 3.2.2.2. | Acetone – Formaldehyde Resins | 1767 |
| 3.2.2.3. | Cyclohexanone Resins | 1767 |
| 3.2.2.4. | Cyclohexanone – Formaldehyde Resins | 1768 |
| 3.2.3. | Aldehyde Resins | 1768 |
| 3.3. | Polyamide Resins | 1769 |
| 3.3.1. | Physical and Chemical Properties | 1770 |
| 3.3.2. | Raw Materials | 1770 |
| 3.3.3. | Production | 1770 |
| 3.3.4. | Quality and Analysis | 1770 |
| 3.3.5. | Storage | 1770 |
| 3.3.6. | Uses | 1770 |
| 3.3.7. | Producers, Trade Names | 1771 |
| 4. | Natural and Synthetic Resins— Toxicology and Legislation | 1771 |
| 4.1. | Toxicology | 1771 |
| 4.1.1. | Rosins and Their Derivatives | 1771 |
| 4.1.2. | Hydrocarbon Resins | 1772 |
| 4.1.3. | Polyamide Resins | 1772 |
| 4.2. | Legal Aspects | 1772 |
| | References | 1773 |

## 1. Definition

The term "resin" is a versatile expression that is not sharply defined. Originally, the term was applied to low molecular mass natural products, which were usually defined as yellowish to brown- colored, transparent to opaque, soft to brittle, easily fusible, tacky and noncrystallizable materials that are soluble in most common organic solvents and virtually insoluble in water. With the development of the chemical industry, the term resin has also been applied to synthetic

Ullmann's Polymers and Plastics: Products and Processes
© 2016 Wiley-VCH Verlag GmbH & Co. KGaA, Weinheim
ISBN: 978-3-527-33823-8 / DOI: 10.1002/14356007.a23_089.pub2

materials that are used as substitutes for natural resins, or to materials with similar physical properties. In the international standards ISO 472 (Plastics — Vocabulary) and ISO 4618/3 (Paints and Varnishes — Vocabulary — Part 3: Terminology of Resins), resins are defined similarly, i. e., as solid, semisolid, or pseudosolid organic materials that have an indefinite and often high relative molecular mass, and generally soften or melt over a range of temperatures on heating. Very often, the term resin is also used to designate any polymer that is a basic material for plastics (e. g., polyethylene resin, → Plastics, General Survey). In the German standard DIN 55 958, resin ("Harz") is defined as a collective term for solid to liquid organic products. Their molecular mass distribution can vary within wide limits. Resins include a large variety of organic substances of different chemical composition, but with many similar physical properties.

Generally, resins are products with medium molecular mass. Their average molecular mass does not exceed 10 000 and, in most cases, is even lower, Resins have an amorphous and often complex structure. They do not exhibit a sharp melting point, but like every amorphous polymeric system, resins have a glass transition temperature and are characterized by their softening point, which denotes the transition temperature from a pseudosolid to a plastic state on heating.

Resins can be subdivided into natural and synthetic resins.

***Natural Resins.*** In ISO 4618/3, natural resins are defined as resins of vegetable or animal origin. The term includes rosins (gum, wood, or tall-oil rosins from tree and plant exudates; wood extracts; or byproducts from paper manufacturing), fossil resins such as amber; mined resins such as asphaltite; shellac as secretion product from an insect; and their main derivatives.

***Synthetic Resins.*** In ISO 4618/3, synthetic resins are defined as resins resulting from controlled chemical reactions such as polyaddition or polycondensation between well-defined reactants that do not themselves have the characteristics of resins. Synthetic resins are also obtained by polymerization (ISO 472) of unsaturated monomers. This term includes two subgroups:

1. Hydrocarbon resins, i.e., synthetic resins from coal tar, petroleum, and turpentine streams, produced by polymerization. These resins are used like natural ones, e.g., in combination with other polymers to impart special properties such as tack, flow, and hardness to a material.
2. Synthetic resins obtained mainly by addition polymerization and polycondensation, which are intermediates in the synthesis of higher molecular mass plastics.

## 2. Hydrocarbon Resins

### 2.1. Classification

The term hydrocarbon resin is generally applied to low molecular mass thermoplastic polymers that are composed mainly of carbon and hydrogen. With respect to their origin, hydrocarbon resins can be classified in four main groups:

1. Petroleum hydrocarbon resins, derived from cracked petroleum distillates
2. Indene – coumarone resins, based on coal-tar oils
3. Terpene resins from turpentine fractions
4. Resins from pure monomers

The oldest family of hydrocarbon resins is the indene – coumarone resins, discovered by KRÄMER and SPILKER in 1890 and produced on an industrial scale since about 1910. In the mid-1930s the terpene resins were developed, which are based on byproducts of rosin extraction. Only in the 1940s did the petroleum resins become known, initiated by the rapid development of the petrochemical industry. These resins established their position as the most important group of these products, Resins from pure monomers came onto the market in the mid-1970s.

### 2.2. Raw Materials

The main sources of raw materials for the chemical industry are pyrolysis processes [e. g., of coal in coke ovens and, to a larger extent, of petroleum fractions in steam crackers (→ Ethylene, Section 5.1.)].

Pyrolysis of coal gives high-temperature coal tar as byproduct of coke production. The most important products are:

| | |
|---|---|
| Coke | ca. 75 wt % |
| Coke-oven gas | 13 wt % |
| Crude tar | 3 wt % |
| Crude benzole | 1 wt % |

The first stage of coal-tar distillation gives water and the coal-tar light oil. The latter and the following carbolic oil fraction contain the aromatic unsaturated compounds that are useful in indene – coumarone resin production [4].

The pyrolysis of naphtha fractions can generate the following products and quantities, depending on the origin of the naphtha and the cracking conditions:

| | |
|---|---|
| Hydrogen | 1.0 wt % |
| Methane-rich gas | 15.0 wt % |
| Ethylene | 28.0 wt % |
| Ethane | 3.5 wt % |
| Propene | 14.0 wt % |
| $C_4$ fraction | 9.0 wt % |
| $C_5$ fraction | 3.0 wt % |
| Benzene | 7.5 wt % |
| $C_6$ fraction | 1.0 wt % |
| Toluene | 5.0 wt % |
| $C_8$ aromatics | 3.0 wt % |
| $C_9$ aromatics | 3.5 wt % |
| Residual oil | 5.5 wt % |

The following crude feeds are used as starting materials in the production of hydrocarbon resins:

1. $C_5$ streams that contain linear and cyclic olefins and diolefins as resin formers are used for the production of aliphatic hydrocarbon resins
2. $C_8 - C_{10}$ streams containing styrenes, indenes, and dicyclopentadiene (DCPD) are used for the production of aromatic hydrocarbon resins

The composition of the feedstock is very important for the properties of the final resin. It affects color, degree of unsaturation, softening point, molecular mass distribution, solubility, and compatibility of the product. The crude cuts can be processed in various ways to give suitable raw materials for the manufacture of different resin types.

### 2.2.1. Indene – Coumarone Feedstocks

Coal-tar light oil and carbolic oil, containing aromatic resin formers such as styrene, α-methylstyrene, vinyltoluenes, indene, methylindenes, and coumarone, are treated to remove phenols and pyridine bases by extraction.

Further treatment steps to improve resin color and properties are generally applied [5]. A typical composition of an indene – coumarone resin feedstock is as follows:

| | |
|---|---|
| Styrene | 2.0 wt % |
| α-Methylstyrene | 1.0 wt % |
| Alkylbenzenes | 30.0 wt % |
| Vinyltoluenes | 4.0 wt % |
| Dicyclopentadiene | 5.0 wt % |
| Coumarone | 7.0 wt % |
| Indene | 48.0 wt % |
| Methylcoumarones and methylindenes | 3.0 wt % |

### 2.2.2. Petroleum Resin Feedstocks

*$C_5$ Streams.* $C_5$ streams from depentanizer effluents consist of olefins and diolefins such as isoamylene, isoprene, piperylene, and cyclopentadiene, together with $C_5$ paraffins. A straight depentanizer $C_5$ stream gives only poor resin quality. The resulting resin is highly susceptible to oxidation. Lowering the cyclopentadiene content by heat soaking and subsequent distillation improves product quality. Modern $C_5$ resins are made from distilled concentrated piperylene streams that are obtained in isoprene recovery. Special qualities are produced from extractively distilled piperylene concentrates. Some typical compositions are given in Table 1.

The diolefin/monoolefin ratio and the nature of monoolefins present in the feedstock affect the overall course of the polymerization process and the molecular mass distribution of the final resin and are therefore carefully controlled.

*$C_8/C_9$ Streams* are obtained from a special separate downstream column. The BTX (ben-

**Table 1.** Typical composition (weight percent) of $C_5$ resin feeds from steam cracking

| Raw materials | Straight | Distilled concentrate | Extractive distillation concentrate |
|---|---|---|---|
| Cyclo- and dicyclopentadiene | 19.0 | 7.5 | 2.0 |
| Isoprene | 30.0 | 8.0 | 0.5 |
| Piperylene | 14.0 | 38.0 | 69.0 |
| Pentenes | 15.0 | 12.5 | 0.5 |
| Pentanes | 11.0 | 15.0 | 2.0 |
| Cyclopentene | 2.0 | 8.5 | 18.0 |
| Other hydrocarbons | 9.0 | 10.5 | 8.0 |

zene, toluene, xylenes) fraction is drawn off as the top product, and a $C_8/C_9$ material is obtained as a sidestream or residue, depending on the process. This material contains high-boiling unsaturated benzene homologues, such as styrene, α-methylstyrene, vinyltoluenes, indene, and methylindenes, together with some dicyclopentadiene that arises from the depentanizer bottoms by dimerization of cyclopentadiene. Suitable $C_8/C_9$ streams can be pretreated to improve resin color and properties. A typical composition is given below:

| | |
|---|---|
| Styrene | 2.0 wt % |
| α-Methylstyrene | 4.0 wt % |
| Vinyltoluenes | 20.0 wt % |
| DCPD and codimers | 6.0 wt % |
| Indene | 20.0 wt % |
| Methylindenes | 5.0 wt % |
| Naphthalene | 5.0 wt % |
| Other nonreactive aromatics | 38.0 wt % |

***Dicyclopentadiene concentrates*** are generated in the heat-soaking of depentanizer effluents and subsequent distillation of $C_5$ fractions. Besides dicyclopentadiene, such streams may contain codimers of cyclopentadiene with other dienes, such as butadiene or isoprene, and methyl derivatives of dicyclopentadiene. Dicyclopentadiene concentration may vary from high (90 – 95 %) through medium (70 – 80 %) to relatively low. A typical composition is given below:

| | |
|---|---|
| Dicyclopentadiene | 65 – 98 wt % |
| Codimers | 2 – 35 wt % |
| Benzene | 1 wt % |
| Sulfur | 150 ppm |

***$C_4$ streams*** contain mainly isobutenes. After removal of butadiene, a suitable feedstock for resin production may have the following composition:

| | |
|---|---|
| 1-Butene | ca. 10 wt % |
| *cis*- and *trans*-2-butene | 10 – 15 wt % |
| Isobutene | 15 – 30 wt % or more |
| *n*-butane | ca. 30 wt % |
| Isobutane | ca. 30 wt % |
| $C_{5+}$ components | balance |

***Pure Commercial Monomers.*** The pure commercial monomers styrene, α-methylstyrene, and vinyltoluenes are used individually, as blends, or together with terpenes or other selected unsaturated aliphatic compounds in resin manufacture. Styrene is produced in large quantities by alkylation of benzene and subsequent dehydrogenation. Its main use is in the manufacture of polystyrene. α-Methylstyrene is a byproduct of the large-scale production of phenol. Vinyltoluenes are produced in smaller quantities by alkylation of toluene, followed by dehydrogenation, Commercial vinyltoluene is a mixture of 64.5 % *m*- and 35.5 % *p*-vinyltoluene.

***Terpene Monomers.*** The raw material for the production of terpene resins is based on natural terpene monomers that are obtained from various sources such as naval stores (gum and wood turpentine) and kraft paper production (tall-oil turpentine). Citrus terpenes, obtained by extraction from orange peel in the production of orange juice, are also used as raw materials. The most important terpenes for resin manufacture are α-pinene, β-pinene, and dipentene (*dl*-limonene). Typical compositions of terpene oils are listed in Table 2.

## 2.3. Production

Hydrocarbon resins are produced by polymerization of the unsaturated compounds contained in the fractions described in Section 2.2 by using mainly Friedel – Crafts catalysts such as aluminum chloride and its complexes, as well as boron trifluoride and its complexes. Brønsted acids (e.g., sulfuric acid) may be also used. The term "catalyst" is scientifically incorrect with respect to cationic polymerization and should be replaced by the term initiatior – coinitiatior (initiating system) [6]. The reaction can be performed batchwise or continuously and includes the following steps:

1. Feed preparation (distillation, pretreatment, blending)
2. Polymerization

**Table 2**. Typical compositions of terpene oils (in wt %)

| | Gum turpentine | Wood turpentine | Tall-oil turpentine |
|---|---|---|---|
| α-Pinene | 58 – 65 | ca. 80 | 60 – 70 |
| β-Pinene | 23 – 35 | 0 – 2 | 20 – 25 |
| Camphene | 4 – 8 | | |
| Other terpenes | 5 – 8 | 15 – 20 | 6 – 12 |

3. Deactivation of iniating system
4. Resin separation

The most important reaction parameters are feed composition, total and relative concentration of resin formers in the feed, type of catalyst and cocatalyst, catalyst concentration, and temperature. These parameters must be carefully controlled. They affect the reaction widely, in terms of yield, type (polymerization, alkylation), molecular mass, and molecular mass distribution, The particular resin feed is contacted with the initiating system by vigorous stirring. The reaction is exothermic and must be strictly temperature controlled, After reaction, the initiating system is inactivated (e.g., by addition of water, alkali, lime, or ammonia). Then the unreacted feed and the low molecular mass oligomers are removed by distillation, usually in combination with a steam-stripping process. The residual resin is conditioned in flakes, bulk, or other forms.

The proportion of catalyst in the reactant medium influences molecular mass and composition of the resin. High concentrations give resins with lower molecular mass, and lower concentrations give resins with higher molecular mass. A low concentration of catalyst may influence resin composition by selectivity (i.e., only the most reactive components are transformed). Reaction temperature affects product molecular mass, yield, and color. Molecular mass decreases and yield increases with increasing temperature.

Dicyclopentadiene resins are usually manufactured by thermal polymerization of various dicyclopentadiene feedstreams that may also contain styrene or indene.

## 2.4. General Properties

Depending on raw material and polymerization conditions hydrocarbon resins may range from viscous liquids to very hard solids with light to amber coloration. Special water-white types also exist. Solubility and compatibility characteristics are a function of the special resin type. All unmodified types have in common an apolar structure that gives them good water repellency and high resistance to acid, alkali, and salt solutions. They are soluble in aromatic and most common aliphatic solvents, ethers, esters, ketones, and chlorinated hydrocarbons. Unmodified resins are compatible with a wide range of other resins and polymers. The absence or presence of double bonds in the resin structure determines whether the resin is stable and/or reactive. Thus, both resins with excellent heat and weather stability and highly reactive grades are commercially available. The structure of hydrocarbon resins is very complex, The major principles involved are the following:

1. Mixtures of mono- and diolefin monomers result in paraffinic and naphthenic structures; diolefins (DCPD type) give unsaturated resins
2. Vinyl aromatic monomers (styrenes, indenes) generate chain structures with high aromatic substitution and a low degree of unsaturation
3. Bicyclic terpenes react mainly to produce cyclic structures by rearrangement

## 2.5. Characterization and Analysis

Hydrocarbon resins are usually characterized by their softening point, color, degree of unsaturation, saponification number, density, and ash content, These properties are determined by standardized methods (see Table 3).

Because of their complex structure, hydrocarbon resins do not exhibit a definite melting point but soften on heating within a temperature interval, The *softening point* is defined as the temperature at which a resin, poured in a ring of defined diameter and thickness, flows under the pressure of a given weight. The most widely applied method is the ring-and-ball (RB) method; the older Krämer – Sarnow (KS) method is used more rarely. A more recent method is the Mettler method.

*Color* is determined optically in solution by comparison with a color standard on the Gardner scale or less commonly on the Barrett scale. For light colors the Saybolt, Hazen, or APHA method is used. A comparison of Barrett and Gardner color numbers is given in what follows:

| Barrett | Gardner |
|---|---|
| B 1/2 | 1 – 6 |
| B 1 | 7 – 10 |
| B 1 1/2 | 11 – 12 |
| B 2 | 12 – 13 |
| B 2 1/2 | 14 |
| B 3 | 15 |
| B 4 | 16 – 17 |
| B 5 | 18 |

Table 3. Characterization of hydrocarbon resins

| Property | Method (unit) | Standard |
|---|---|---|
| Softening point | ring and ball RB (°C) | ASTM D 3461–76, ISO 4625 |
| | Krämer – Sarnow KS (°C) | DIN 53 180, NFT 67 001 |
| | Mettler | ASTM D 3461–76 |
| Color | Gardner (50 % solution in toluene) | ASTM D 1544–80(0–18), ISO 4630 |
| | Barrett (8 % solution in toluene) | |
| | Saybolt, Hazen, APHA (for very light colors) | |
| Unsaturation | bromine number (g $Br_2$/100 g) | ASTM D 1159–84, DIN 51 774, part 1 |
| | Wijs iodine number (g $I_2$/100 g) | ASTM D 1959–69 |
| Acid number | mg KOH/g | DIN 51 558, DIN 53 402, ASTM D 974–80 |
| Saponification number | mg KOH/g | DIN 51 559, DIN 53 401 |
| Density | at 20 °C (g/cm$^3$) | DIN 51 757 |
| Flash point | Cleveland open cup (°C) | ISO 2592 |

*Unsaturation* is indicated by the iodine or bromine number. Neither number gives reliable information on the true degree of unsaturation, because along with addition reactions to the double bonds, substitution reactions may occur.

Other properties determining the end use of resins are solubility, compatibility, melt and solution viscosity, molecular mass, and molecular mass distribution. *Solubility and compatibility* are measured by determining the cloud point. Although generally accepted, this method is not standardized for resins. *Molecular mass and its distribution* are important resin characteristics and are determined by size exclusion (gel permeation) chromatography. Some information on resin structure is obtained by spectroscopic methods such as IR, UV, and NMR. The relation of aromatic to aliphatic protons evaluated by NMR is a useful tool for determining the aromaticity of a resin.

## 2.6. Resin Types

### 2.6.1. Aliphatic Hydrocarbon Resins

Aliphatic hydrocarbon resins are based on $C_5$ feedstreams and can be divided into general-purpose or basic resins, high-quality special piperylene resins, and dicyclopentadiene resins. Typical data for aliphatic petroleum resins are listed in Table 4.

*$C_5$ Olefin – Diolefin Resins.* Basic resins are produced from more or less narrow $C_5$ cuts and are available in large quantities as multipurpose resins. They have satisfactory color and color stability and a moderate degree of unsaturation. They are highly soluble in aliphatic solvents and compatible with a wide range of other resins, polymers, and waxes. Their softening point is usually between 70 and 100 °C (RB).

*Piperylene resins* exhibit an improved quality and are produced from highly concentrated streams that contain up to 70 % piperylene. They have properties similar to those of terpene resins because of their cycloaliphatic structure. Thus they are often called "synthetic terpene resins." They have excellent light stability and light color. A wide range of piperylene resins is produced — from viscous, liquid resins to hard resins.

*Dicyclopentadiene Resins.* Dicyclopentadiene resins can be divided in two main groups:

1. Those that contain predominantly dicyclopentadiene (DCPD), with high degrees of unsaturation (bromine number ca. 65 – 100 or higher) and which are distinctly heat- and oleoreactive
2. Those containing substantial amounts of aromatic structures (formed by co-reaction with unsaturated aromatics), with moderate unsaturation (bromine number ca. 55 – 57)

Table 4. Typical data for aliphatic petroleum resins

| Property | $C_5$ Olefin – diolefin resins | Piperylene resins |
|---|---|---|
| Softening point RB, °C | 70 – 100 | liquid – 115 |
| Gardner color | 5 – 7 | 3 – 7 |
| Bromine number, g $Br_2$/100 g | 15 | 25 |
| Density at 20 °C, g/cm$^3$ | 0.93 – 0.97 | 0.88 – 0.97 |

Table 5. Typical data for dicyclopentadiene resins

| Property | Straight DCDP resins | Aromatic-modified DCPD resins |
|---|---|---|
| Softening point (RB), °C | liquid – 160 | 10 – 150 |
| Gardner color | 6 – 11 | 5 – 15 |
| Bromine number, g Br$_2$/100 g | 95 – 100 | ca. 55 |
| Density at 20 °C, g/cm$^3$ | ca. 1.1 | 1.05 – 1.12 |

Because of the presence of reactive unsaturation these DCPD resins may be modified with other unsaturated components such as maleic anhydride, drying oils, or rosin (see Table 5).

***Polybutenes,*** manufactured from butadiene-free C$_4$ fractions, are low molecular mass copolymers of the butene isomers. These polybutenes are normally low-colored, viscous liquids that are chemically stable, oxidation resistant, nondrying, and tacky (→ Polybutenes, Chap. 2.)

### 2.6.2. Aromatic Hydrocarbon Resins

Aromatic hydrocarbon resins are based on coal-tar-derived indene – coumarone streams or petroleum-derived C$_8$/C$_9$ streams. Typical properties of aromatic hydrocarbon resins are listed in Table 6.

***Indene – Coumarone Resins [4], [5].*** The oldest family of hydrocarbon resins is now available with an acceptably light color. Indene – coumarone (IC) resins are on the market as viscous liquids and as solids with high softening points, up to 170 °C (RB). The resins are soluble in most common solvents (except low molecular mass alcohols) and compatible with a wide range of polymeric materials. They exhibit high water repellency and resistance to acid and alkali. When not stabilized, they are subject to yellowing by autoxidation under the influence of UV light.

***Petroleum Resins.*** The C$_9$ aromatic hydrocarbon resins are obtained from streams of cracked petroleum fractions that are purified to a greater or lesser extent, Depending on its origin and pretreatment, the raw material contains a varying percentage of dicyclopentadiene. The degree of unsaturation of the resin can thus be controlled by the choice of raw material. The properties of petroleum resins are generally similar to those of indene – coumarone resins, but petroleum resins are often characterized by a lower degree of unsaturation, higher light and heat stability, and a wider range of solubility and compatibility. They are produced in many different grades from viscous liquids to solids of high softening point.

### 2.6.3. Terpene Resins

Terpene resins are based on naturally occurring monomers. The presence of tertiary carbon in their structure makes them very susceptible to oxidation; thus they must be protected by antioxidants.

Terpene resins are usually light colored and soluble in most common solvents. They are widely compatible with other resins and polymers, and have excellent tackifying properties. They are available as viscous liquids and in solid form. Typical properties are presented in Table 7.

### 2.6.4. Resins from Pure Monomers

Pure monomers such as styrenes or isobutene are excellent resin formers for the manufacture

Table 6. Typical properties of aromatic resins

| Property | Coal/tar based | Petroleum based |
|---|---|---|
| Softening point RB, °C | liquid – 170 | liquid – 150 |
| Gardner color | 6 – 18 | 5 – 12 |
| Iodine number, g I$_2$/100 g | 35 – 105 | 25 – 90 |
| Density at 20 °C, g/cm$^3$ | 1.08 – 1015 | 1.06 – 1.11 |

Table 7. Typical data for terpene resins

| Property | Derived from | | |
|---|---|---|---|
| | α-Pinene | β-Pinene | d-Limonene |
| Softening point RB, °C | 25 – 135 | 10 – 135 | 100 – 135 |
| Gardner color | 3 – 5 | 1 – 4 | 1 – 4 |
| Bromine number, g Br$_2$/100 g | 25 – 30 | 15 – 30 | 27 – 28 |
| Density at 20 °C, g/cm$^3$ | 0.90 – 0.98 | 0.93 – 0.99 | 0.99 |

Table 8. Typical properties of resins from pure aromatic monomers

| Property | Styrene | α-Methylstyrene | α-Methylstyrene – vinyltoluenes |
|---|---|---|---|
| Softening point RB, °C | liquid – 160 | 70 – 145 | 75 – 120 |
| Gardner color | 1 – 3 | < 1 | < 1 |
| Bromine number, g $Br_2$/100 g | 0 | 2 – 8 | 2 |
| Density at 20 °C, g/cm$^3$ | 0.98 – 1.08 | 1.06 – 1.07 | 1.05 |

of very stable, light-colored to colorless (water-white) resins. Industrially, resins based on styrene or α-methylstyrene, or blends of mainly α-methylstyrene and vinyltoluenes, are produced. Resins from the copolymerization of terpenes and styrene are also available. Typical data of resins from pure aromatic monomers are given in Table 8.

### 2.6.5. Modified Resins

In the previous sections, resins are described that are obtained from steam cracker fractions or monomer mixtures. They contain predominantly carbon and hydrogen, but no polar or functional groups. With the exception of dicyclopentadiene resins, they are chemically nonreactive. This section deals with resins that are either obtained from special, modified feedstreams or are modified after polymerization. Thus, properties of aliphatic- or aromatic-type resins may be adapted by blending initial feedstreams or by addition of selected monomers, Mixed resins of $C_5 - C_9$ types, as well as terpene – aromatic resins, are available. Typical properties are given in Table 9.

Addition of dicyclopentadiene to the initial feedstream increases the degree of unsaturation and thus the reactivity of the final resin. Also coreaction of dienes (e.g., dicyclopentadiene with maleic anhydride) may be applied, Resins made by coreaction of aromatic or terpene resin formers with phenol are well known. Their properties are listed in Table 10.

The high reactivity of terpenes makes them suitable for copolymerization reactions with other monomers, e.g., styrenes or maleic anhydride.

Resins may also be modified by further reaction of the final resin; for example, dicyclopentadiene resins can react with maleic anhydride, unsaturated fatty acids, or rosin upon heat treatment. A very important modification is the catalytic hydrogenation of resins, which gives high-quality water-white resins with a different solubility and compatibility profile than the original resin. Dicyclopentadiene resins are hydrogenated in large quantities. Hydrogenated aromatic resins derived from steam cracker streams or pure monomers and hydrogenated terpene resins are also available. Typical properties are listed in Table 11.

### 2.7. Uses

Hydrocarbon resins are used in numerous fields of application because of their versatile properties. They are seldom used alone but mainly in conjunction with other materials to adapt their properties to special requirements. Hydrocarbon resins are mostly used as tackifiers, binders, processing aids, film formers, reinforcing agents, and extenders.

With regard to their specific properties, the following classification is often used:

- Hard resins (binders, film-formers) in the paint, varnish, printing and coatings industries

Table 9. Typical properties of mixed resins

| Property | $C_5 - C_9$ Resins | Terpene – aromatic resins |
|---|---|---|
| Softening point RB, °C | liquid – 105 | 105 – 115 |
| Gardner color | 2 – 10 | 2 – 5 |
| Iodine number, g $I_2$/100 g | < 25 | not available |
| Density, g/cm$^3$ | 0.86 – 1.07 | 1.02 |

Table 10. Phenol-modified resins

| Property | Based on | |
|---|---|---|
| | $C_9$ and indene – coumarone | Terpene |
| Softening point RB, °C | liquid – 120 | 95 – 155 |
| Gardner color | 5 – 8 | 4 – 11 |
| Hydroxyl number, mg OH/g | 56 – 112 | 80 – 260 |
| Density at 20 °C, g/cm$^3$ | 1.08 – 1.1 | 1.03 – 1.09 |

Table 11. Typical properties of hydrogenated resins

| Property | Based on | | | |
|---|---|---|---|---|
| | DCPD | $C_9$ Fraction | Aromatic monomer | Terpene |
| Softening point (RB), °C | 85 – 140 | 70 – 140 | liquid – 125 | 85 – 115 |
| Gardner color | < 1 | > 1 | crystal clear | < 1 |
| Bromine number, g $Br_2$/100 g | 2 – 3 | not available | not available | 10 – 20 |
| Density at 20 °C, g/cm$^3$ | 1.10 | 0.98 – 0.99 | 0.92 – 1.04 | 0.94 – 1.01 |

- Tackifiers in the adhesives, sealants, and rubber industry
- Processing aids (plasticizers) in the plastics industry (rubber goods, tires, PVC)
- Fixing agents, e.g., in wood protection
- Water repellents in various applications (e.g., wood protection, coatings)

The most important applications are in adhesives (construction, solvent- or water-based, hot-melt and pressure-sensitive adhesives), coatings (paints, varnishes, roadmaking), the rubber and plastics industry (mainly mechanical rubber goods and tires, also PVC), printing inks, and, to a lesser extent, in various other industrial fields. A generalized view of the main application areas by resin type is given in Table 12. The large group of modified resins is used in applications where special properties are required.

*Aliphatic resins* are used extensively in hot-melt adhesives; in pressure-sensitive adhesives based on natural or thermoplastic rubber; in coatings, road-marking formulations, mastics, caulks, and sealants; and in rubber formulations for mechanical goods and tires. *Aromatic $C_9$ and indene – coumarone* resins are widely applied in hot-melt formulations based on ethylene – vinyl acetate or thermoplastic rubber; in solvent-based and waterborne adhesive formulations; in the rubber industry (tires and mechanical goods); and in paints, varnishes, coatings, printing inks, sealants, wood protection, and concrete curing. *Terpene resins* are used almost exclusively in the adhesive industry as excellent tackifiers for pressure-sensitive adhesives and hot-melt formulations. *Dicyclopentadiene resins* are generally applied in paints, varnishes, and coatings where they are introduced in the binder system, and in the printing ink industry. *Water-white resins* are used in adhesives and coating formulations where light color and color stability are important. The liquid polybutylenes are used in sealants, caulks, mastics, and pressure-sensitive formulations. The large group of modified resins is used in applications where special properties are required.

***Adhesives* [7].** Most synthetic adhesives are based on a "backbone" polymer, tackifying resins, and a carrier (solvent, water, or heat) that gives the formulation the necessary consistency for application. Tackifying resins regulate the final rheological properties of the system (tack, specific adhesion, and wetting).

*Solvent-borne adhesives*, mainly based on SBR and polychloroprene rubber, are frequently combined with hydrocarbon resins (aromatic $C_9$ and indene – coumarone, aliphatic and mixed $C_5/C_9$), as well as rosin, rosin esters, terpene – phenolic, and *p-tert*-butylphenolic resins.

*Waterborne adhesives* are increasingly being used for environmental reasons. Polymer dispersions based on poly(vinyl acetates), latexes, or acrylates may be modified by addition of tackifying resins such as petroleum and indene – coumarone resins, terpene resins, as well as rosin esters, terpene – phenolic resins, and ketone resins.

Table 12. Main application areas by resin type

| Resin type | Adhesives | Coatings | Paints | Printing inks | Rubber |
|---|---|---|---|---|---|
| Aliphatic$C_5$ * | x | x | x | | x |
| Aromatic $C_9$ ** | x | x | x | x | x |
| Mixed aliphatic/aromatic | x | x | x | x | |
| Dicyclopentadiene | | x | x | x | x |
| Modified DCPD | | | x | | |
| Butene | x | x | | | |
| Pure monomer | x | x | x | | |
| Hydrogenated | x | x | | | |
| Terpene | x | | | | |

\* Petroleum resins;
\*\* Petroleum and indene – coumarone resins.

*Hot-melt adhesives* are thermoplastic solids which are liquified by heating for application. Backbone polymers for these adhesives are ethylene – vinyl acetate (EVA), thermoplastic rubbers, such as styrene – isoprene – styrene block copolymers, polyethylene, and polypropylene, formulated together with tackifying resins. Polyesters and polyamides, which are also used in hot-melt adhesives, do not need additional tackification. Hot-melt adhesives formulated with the above-mentioned polymers contain paraffins or waxes, tackifying resins, and fillers. Tackifying resins of various types may be used, mainly hydrocarbon resins (aliphatic $C_5$, aromatic $C_9$, indene – coumarone, mixed aliphatic – aromatic, terpene and hydrogenated resins), rosin and rosin esters, and terpene – phenolic resins.

*Pressure-sensitive adhesives* are defined as adhesives which in solvent-free form remain permanently tacky and adhere instantaneously with application of a minimum of pressure. Their main use is in tape and label manufacturing. Formulations are based on backbone polymers such as rubbers (natural rubber, SBR), and block copolymers, together with tackifying resins, or on acrylic resins. Tackifying resins for this use are aliphatic $C_5$, aromatic $C_9$, mixed $C_5/C_9$, pure monomer, terpene, and hydrogenated resins, as well as rosin and rosin esters.

**Sealants.** Sealants are based on polymers such as butyl rubber, polyisobutene, thermoplastic elastomers, SBR, neoprene, and polysulfides. Addition of tackifying resins is necessary to improve instantaneous adhesion. Practically all hydrocarbon resin types as well as rosin and modified rosin are usable. Other basic raw materials for sealants, such as silicones or urethanes, do not need additional tackification.

**Coatings.** Coatings [5, 8] range from protective to decorative and can be intended for temporary or permanent use. Hydrocarbon resins are extensively used, as illustrated by the following examples:

- Hot-melt coatings are mainly used in the packaging industry. Formulations and tackifiers used are very similar to those for hot-melt adhesives.

- Paints and varnishes, based on a variety of binder systems, may use rosin derivatives as well as hydrocarbon resins of the aromatic, aliphatic, and DCPD type as hard resins. Aromatic hydrocarbon resins are especially important in the formulation of aluminum paints.

- Road-marking material of the paint or hot-melt type uses mainly aliphatic $C_5$ resins but also $C_9$ types, modified by straight or cyclic aliphatics, as well as hydrogenated and special $C_9$ resins.

- Anticorrosion paints may be based on $C_5$, $C_9$ and DCPD resins, mixed $C_5/C_9$ resins, and phenol-modified aromatic resins.

- Concrete curing membranes, applied to concrete to avoid too rapid drying, contain hydrocarbon resins (DCPD, $C_5$ and $C_9$) in solution or dispersion.

- Wood protection by impregnation or surface treatment is carried out with, e.g., solutions containing aromatic $C_9$ resins because of their excellent water-repellent and fixing properties.

**Printing Inks.** Use of hydrocarbon resins in ink formulation takes advantage of their good compatibility with many ink components, e.g., alkyds, oils, and ink solvents, and gives the ink gloss and good resistance to water and alkali. Hydrocarbon resins are usually formulated together with the classical ink binders such as alkyd resins and modified rosins. High-softening-point hydrocarbon resins of the aromatic $C_9$ and indene – coumarone type, unmodified and modified, as well as DCPD and functional modified DCPD resins are used. The main use is in offset and gravure inks.

**Tires and Mechanical Rubber Goods.** Rubber stocks are compounded with a multitude of different additives: processing aids, plasticizers, reinforcing agents, tackifiers, pigments, antioxidants, and curing agents. Resins widely used in this application are generally referred to as processing aids, but in fact they have multiple functions. They promote processing by lowering viscosity, improve the building tack of unvulcanized mixtures, and may improve elongation, tensile strength and flex resistance of the vulcanized materials. Besides alkyl-phenolic resins, the largest volume resins for this application, considerable quantities of hydrocarbon resins, aromatic $C_9$ and indene – coumarone resins, $C_5$ aliphatic, mixed aromatic/aliphatic, and DCPD resins, as well as some unmodified rosin are used. The main criteria for choice are availability, consistency, and price.

***Other Uses.*** Hydrocarbon resins are used in many other applications, such as in compounding floor tiles, in the foundry industry as binder and lustrous carbon former, as oil additives, in investment casting, the ceramic industry, in textile treatment and waterproofing, as well as in gypsum waterproofing.

## 2.8. Economic Aspects [9]

The total capacity for synthetic hydrocarbon resin production is estimated to be about $1 \times 10^6$ t/a worldwide in 1991. Main capacities are located in the United States, Canada, and Brazil (ca. $0.45 \times 10^6$ t/a); Europe (roughly $0.25 \times 10^6$ t/a); and the Asia/Pacific region (about $0.3 \times 10^6$ t/a).

Aromatic resins represent a volume of ca. 44 %, $C_5$ and DCPD resins ca. 37 %, and hydrogenated resins ca. 19 %. Producers and trade names of hydrocarbon resins are listed in Table 13.

Table 13. Resin manufacturers and trade names

| | Trade names | Resin types[a] |
|---|---|---|
| **Europe** | | |
| Hexion (Netherlands) | Setalin | hydrocarbon-modified phenolic resins |
| Cray Valley (France) | Norsolene | $C_9$ and $C_5 – C_9$ resins |
| DRT (France) | Dercolyte, Dertophene | terpene and modified terpene resins |
| Exxon (France) | Escorez | $C_5$ and $C_5 – C_9$ resins, DCPD and hydrogenated DCPD resins |
| Eastman (Netherlands) | Hercures, Regalite Kristalex, Hercoflex | $C_5$ and $C_9$ resins $C_5 – C_9$ resins, resins from pure monomers, hydrogenated resins |
| Neville Chemical Europe (Netherlands) | Necires, Nevchem, Nevex, Nevbrite, Nevroz, Nevillac | $C_9$ and $C_5 – C_9$ resins, DCPD and modified DCPD resins, phenol-modified resins |
| Rütgers–VFT (Germany) | Novares | $C_9$ and modified $C_9$ resins, IC and phenol-modified IC resins |
| **United States** | | |
| Arizona | Zonatac, Sylvagum, Sylvares | terpene and modified terpene resins, $C_5$ and |
| Eastman | Eastotac, Tacolyn, Piccotac, Regalrez, Piccolastic, Kristalex, Piccotex, Endex | $C_5 – C_9$ resins, hydrogenated resins, DCPD resins, terpene and modified terpene resins, resins from pure monomers |
| Exxon | Escorez | $C_5$, $C_5 – C_9$, and $C_9$ resins, hydrogenated resins |
| Cray Valley | Wingtack, Norsolene | $C_5$ and $C_5 – C_9$ resins |
| Hexion | Setalin, Petro-Rez, Alpha-Rez | $C_9$ and modified $C_9$ resins, DCPD and modified DCPD resins |
| Neville | Nevillac, Nevchem, Nevoxy, Cumar, Nevex, Nevpene, Nevroz, Nebony, Nevtac | IC resins, $C_5$, $C_9$, $C_5 – C_9$, and DCPD resins and modified resins |
| Resinall | Resinall | $C_9$, modified $C_9$, and modified DCPD resins |
| SI Group | SP | terpene phenolic resins |
| **Japan** | | |
| Arakawa | Arkon | hydrogenated resin |
| Maruzen | Marukarez | $C_5$ and DCPD resins, hydrogenated $C_5$ and DCPD resins |
| Mitsui | Hi-rez, FTR | $C_5$ and $C_9$ resins, resins from pure monomers |
| Nippon Petrochemical | Nisseki Neopolymer | $C_9$ resins, liquid resins |
| Nippon Zeon | Quintone | $C_5$ and modified $C_5$ resins, DCPD and modified DCPD resins |
| Exxon | Escorez | $C_5$, $C_9$, $C_5 – C_9$, and DCPD resins, hydrogenated DCPD resins |
| Tosoh | Petcoal | $C_9$ and $C_5 – C_9$ resins |
| Yasuhara | YS, Polyster, Clearon | terpene, modified and hydrogenated terpene resins |
| Other countries | | |
| Korea | Kolon | $C_5$, $C_9$, and IC resins |
| Kolon Korea | | |
| P.R. China | | |
| Sunresin | Sunresin | $C_5$, $C_9$, and modified $C_9$ resins, modified DCPD resins |
| Taiwan | | |
| Yung Tung | Sunrez | $C_9$ resins |
| India | | |
| Indian Petrochemical Corp. | Petrez | $C_9$ resins |

[a] IC = indene – coumarone

## 3. Resins by Condensation

### 3.1. Furan Resins

The term "furan resins" describes generic polymers or resins that are derived from the condensation of furfural and/or furfuryl alcohol alone, and with a variety of coreactants. These include formaldehyde or a formaldehyde precursor such as paraformaldehyde, urea – formaldehyde concentrates or syrups, phenol, phenol – formaldehyde resins, aldehydes, ketones, and polyhydroxyl-substituted aromatics. Polymers or resins derived from 2,5-bis(hydroxymethyl) furan (BHMF) have been reported, With 2,5-BHMF, both condensation and addition polymers can be synthesized.

#### 3.1.1. Monomers

Although many chemicals possess a furan nucleus in their molecular structure, furfural and furfuryl alcohol are the only monomers of major industrial significance. *Furfural* [98-01-1] is produced commercially from agricultural byproducts such as corncobs, sugarcane bagasse, oat hulls, and wood residues [10] (→ Furfural and Derivatives, Chap. 2.). *Furfuryl alcohol* [98-00-0] is produced by hydrogenating furfural in a vapor- or liquid-phase process with a variety of catalysts based primarily on copper [11]. *2,5-Bis(hydroxymethyl)furan* [1883-75-6] is synthesized by selective hydroxymethylation of furfuryl alcohol [12] or by hydrogenation of the less readily available 5-hydroxymethylfurfural [13].

#### 3.1.2. Production – Characteristics

Furfural and furfuryl alcohol are converted to resins by *condensation polymerization*.

***Furfural-Based Resins.*** Homopolymerization of furfural, although studied extensively [14], has no industrial importance. Copolymers of furfural with phenol, phenol – formaldehyde resins, and other hydroxyl-substituted aromatics have been available commercially for many years. Since the acid-catalyzed reaction of furfural and phenol is difficult to control, most industrial processes involve the use of alkaline catalysts such as alkali hydroxides or carbonates. Furfural and ketones (usually acetone) condense with alkaline catalysts to form furfurylidene acetone monomer, which polymerizes to a thermoset resin in the presence of strong acid.

***Furfuryl Alcohol-Based Resins.*** The mechanism proposed up to the early 1950s for the polymerization of furfuryl alcohol was reported by DUNLOP and PETERS in their treatise on furans [10]. As a result of the growing importance of furfuryl alcohol resins since the 1950s, the chemistry of furfuryl alcohol polymerization initiated by heat, acid, and alumina has been studied extensively and has proved to be complex, Studies dealing with the identification of intermediates and byproducts by spectroscopic and chromatographic techniques [15–18], the reaction kinetics involved in polymerization [19–21], the structure of the resulting polymers [22], and the nature of the oxygen- or acid-catalyzed cross-linking of the initial resin have all led to a better understanding of this complex chemistry [23]. A comprehensive review examines the literature up to 1977 and interprets the resinification mechanisms [14]. A more recent review further clarifies the two-stage polymerization of furfuryl alcohol [12]. Based on accumulated data, furfuryl alcohol must be considered a bifunctional monomer in the initial stages of weak acid catalysis, and its "normal" reactions give linear chains or oligomers containing essentially two repeating units (1, 2), with 2 predominating:

$$\underset{1}{\text{―}\underset{O}{\bigcirc}\text{―}CH_2OCH_2\text{―}\underset{O}{\bigcirc}\text{―}}$$

$$\underset{2}{\text{―}\underset{O}{\bigcirc}\text{―}CH_2\text{―}\underset{O}{\bigcirc}\text{―}}$$

The polymerization or resinification of furfuryl alcohol is highly exothermic and, depending on the activity and concentration of the catalyst used, requires careful control of reaction temperature. Generally, weak acids such as carboxylic acids or metallic salts are used to produce oligomeric prepolymers with repeating moieties 1 or 2. Further conversion of these prepolymers to cross-linked thermoset resins is accomplished by the addition of strong inorganic or organic acids. Aromatic sulfonic acids are the preferred catalyst. Earlier investigators postulated cross-linking through the C-3 and C-4 positions of the furan ring. However, more recent studies [24], suggested, on the basis of analytical data, that the main cause of branching and cross-linking is a condensation reaction between methylene groups within a chain and a hydroxymethyl group at the end of another chain to yield structures such as 3.

[Chemical structure diagram showing furan rings with CH₂ groups combining with furfuryl alcohol, losing H₂O, to form structure 3]

***2,5-Bis(hydroxymethyl)furan.*** The 2,5-BHMF monomer behaves as a diol and undergoes esterification (base- catalyzed transesterification), alkoxylation, glycidyl ether formation, cyanoethylation, and urethane formation. In the presence of an acid catalyst, resinification of 2,5-BHMF occurs, similar to that of furfuryl alcohol, although the end groups of the resulting resin are hydroxymethyl, These resins are characterized by enhanced reactivity compared to acid- catalyzed furfuryl alcohol resins [25]. 2,5-Bis(hydromethyl)furan resins convert to highly cross-linked thermosetting polymers that demonstrate unique characteristics in composites, including a desirable high char yield on carbonization, low flame propagation, corrosion resistance, and superior high-temperature stability.

### 3.1.3. Uses

*Furan resins* are used in a variety of applications including sand cores and molds for metal casting; corrosion-resistant fiberglass-reinforced plastics; low flammability and low smoke generating composites and foams; carbonaceous products; polymer concretes; wood adhesives; and proppants in oil-well sand consolidation.

The use of *furfuryl alcohol resins* in the worldwide foundry industry remains the only important application. Total world production of furan resins for use as binders to produce sand cores and molds for metal castings exceeds 75 000 t/a.

## 3.2. Ketone and Aldehyde Resins

Ketone and aldehyde resins are obtained from aliphatic aldehydes and aliphatic or aliphatic – aromatic ketones by self- condensation, or more frequently by cocondensation with formaldehyde [26, 27]. Other monomers such as phenols or urea are also sometimes incorporated. Although these resins have been known for a long time, new types of resin and new areas of application are always being found. In paints and coatings, as in many other areas, these resins are used in combination with other binding agents, plasticizers, pigments, and process materials [28–40].

They are also used in printer and copier technology [41–43], the production of printed circuit boards [44], adhesives [45–48], binding agents for corrugated cardboard [49–51], molding sands [52, 53], and laminates [54–56]. The production of aqueous dispersions has also been described [57].

***Reaction Mechanism.*** The condensation of aliphatic or aliphatic – aromatic ketones with formaldehyde is base catalyzed [58]. Formaldehyde adds first to the ketone, with formation of the aldol:

$$R^1-\underset{\underset{O}{\|}}{C}-CH_2R^2 + \ddot{B} \rightleftharpoons R^1-\underset{\underset{O}{\|}}{C}-\bar{C}HR^2 + BH^+$$

$$+ HCHO \downarrow$$

$$R^1-\underset{\underset{O}{\|}}{C}-CHR^2-CH_2 + \ddot{B} \xleftarrow{+ BH^+} R^1-\underset{\underset{O}{\|}}{C}-CHR^2-CH_2$$
$$\phantom{R^1-\underset{\underset{O}{\|}}{C}-CHR^2-}\,OH \phantom{xxxxxxxxxxxxxxxxxx} O^-$$
$$\text{Aldol} \phantom{xxxxxxxxxxxxx} R^1 = CH_3, C_6H_5; R^2 = CH_3, H$$

The aldol is then either dehydrated to give vinyl ketones, which can polymerize or reacted with excess formaldehyde to form highly reactive methylol compounds. The reaction pathway followed depends on the molar ratio of ketone to formaldehyde and on reaction conditions (pH, temperature, concentration).

$$R^1-\underset{\underset{O}{\|}}{C}-CHR^2-CH_2 + \ddot{B} \rightleftharpoons R^1-\underset{\underset{O}{\|}}{C}-\bar{C}R^2-CH_2 + BH^+$$
$$\downarrow$$
$$R^1-\underset{\underset{O}{\|}}{C}-CR^2=CH_2 + OH^-$$

$$R^1-\underset{\underset{O}{\|}}{C}-CHR^2-CH_2 + HCHO \longrightarrow R^1-\underset{\underset{O}{\|}}{C}-CR^2\underset{CH_2OH}{\overset{CH_2OH}{<}}$$
$$\downarrow +HCHO$$
$$\text{(when } R^2 = H)$$
$$R^1-\underset{\underset{O}{\|}}{C}-\underset{CH_2OH}{\overset{CH_2OH}{C}}-CH_2OH$$

The resins are formed by polymerization of the vinyl ketones to give polymers with an alternating structure (see Fig. 1) or by complex condensation of the methylol compounds with one another or with other ketone molecules to give partially branched products. In industrial processes, both mechanisms often operate in parallel. An excess of formaldehyde can lead to reduction of carbonyl groups to hydroxyl

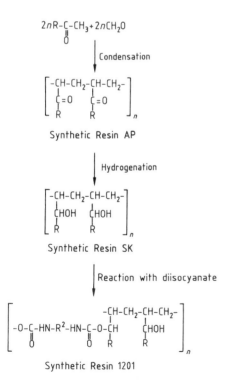

**Figure 1.** Resins from acetophenone and formaldehyde
R = Phenyl, $R^2$ = Aliphatic hydrocarbon.

groups [59]. The tendency of the vinyl ketone intermediate to polymerize is rather independent of the nature of the ketone used [60, 61], provided the vinyl group is not substituted.

### 3.2.1. Resins from Aliphatic – Aromatic Ketones

Industrially mainly acetophenone is reacted with formaldehyde. Derivatives of acetophenone with alkylated benzene rings, such as methyl-, ethyl-, *tert*-butyl-, or cyclohexylacetophenone can also be employed [62]. Besides the unmodified resins, their hydrogenation products and the resins derived from them are available commercially (see Fig. 1).

#### 3.2.1.1. Acetophenone – Formaldehyde Resins

Acetophenone – formaldehyde resins are light-colored nonhydrolyzable products, with sufficient lightfastness and softening points of 75 – 85 °C. They are readily soluble in aromatic hydrocarbons, esters, ketones, and glycol ethers, but insoluble in alcohols, aliphatic hydrocarbons, and mineral oils. Other physical and technical properties are listed in Table 14. The resins are compatible with nitrocellulose; copolymers of vinyl chloride; chlorinated rubber; maleic, urea, melamine, and ketone resins; plasticizers; and a few alkyd resins. They are incompatible with the majority of alkyd resins, drying oils, polyesters, and polyacrylates. Their relatively low price, gloss-improving properties, and capacity to increase the solid material content of coatings (i.e., the coating can be laden with a higher solids content and the viscosity remains constant when ketone resins are added) should be mentioned in particular.

*Production.* Light, soluble, hard resins are obtained by heating an aqueous mixture of acetophenone, formaldehyde, and potassium or sodium hydroxide to the boiling point. Solvents such as alcohols can also be used. Strong acids, such as sulfuric acid, or dehydrating salts, such as zinc chloride, are also suitable condensing agents but are less common. Quaternary nitrogen or phosphorus bases [63] are used as condensing agent to obtain particularly high-melting resins. Under suitable conditions, sodium sulfite leads to low molecular mass,

Table 14. Physical and technical properties of ketone and aldehyde resins

| | Synthetic Resin (Hüls) | | | | Laropal (BASF) | | | | |
|---|---|---|---|---|---|---|---|---|---|
| | AP | SK | 1201 | CA | K 80 | A 81 | A 101 | A LR 8768 | A LR 8825 |
| Resin type | ketone | aromatic | ketone | | | | | | |
| hydrogenated | ketone | modified | ketone | | | | | | |
| aliphatic | ketone | aliphatic | aldehyde | aldehyde | aldehyde | aldehyde | | | |
| Softening point, °C, DIN 53 180 | 76 – 82 | | | | 75 – 85 | 80 – 90 | 95 – 110 | ca. 80 | 90 – 100 |
| Softening point, °C, DIN 53 181 | | 110 – 120 | 155 – 170 | 92 – 108 | | | | | |
| Iodine color index (50 % solution) DIN 6162 | 1 – 2.5 | 0 – 1 | 0 – 1 | max. 2 | max. 2 | max. 3 | max. 5 | max. 5 | max. 4 |
| Acid value, mg KOH/g, DIN 53 402 | max. 0.1 | max. 0.1 | max. 0.1 | max. 0.3 | max. 1 | max. 3 | max. 3 | max. 3 | max. 3 |
| OHvalue, DIN 53 240 | | ca. 325 | ca. 210 | ca. 325 | | 40 – 60 | | | |
| Solubility in | | | | | | | | | |
| Alcohols | ± | + | + | + | ± | + | + | + | + |
| Esters | + | + | + | + | ± | + | + | + | + |
| Ketones | + | + | + | + | ± | + | + | + | + |
| Aromatics | + | – | ± | + | + | + | + | + | + |
| Aliphatics | – | – | – | – | + | ± | – | + | ± |
| Compatibility with nitrocellulose | + | + | + | + | + | + | + | + | + |
| Chlorinated rubber | + | – | – | + | + | + | + | + | + |
| Vinyl chloride polymers | + | – | – | + | ± | + | + | + | + |
| Polyacrylates | – | ± | ± | ± | ± | ± | ± | ± | ± |
| Urea resins | + | + | + | + | ± | + | + | + | + |
| Melamine resins | + | + | + | + | ± | + | + | + | + |
| Alkyd resins | ± | ± | ± | ± | + | + | + | + | + |

<sup>a</sup> + = compatible or soluble; ± = some products compatible or only miscible in certain ratios; – = incompatible or insoluble.

water-soluble condensates [64]. Formaldehyde is added mostly as a 30 % aqueous solution or as paraformaldehyde.

Larger quantities of alkali hydroxide and higher temperature result in resins that are infusible and not readily soluble, because the methylol intermediates cross-link with one another. Readily soluble, fusible condensates are obtained with ca. 0.2 mol of hydroxide per mole ketone and a sufficiently short reaction time, so that the molecular mass cannot rise above 400 – 800. Generally, increasing the reaction temperature and time results in a higher softening point, while the solubility and compatibility deteriorate. This is probably due to competing reaction mechanisms during condensation. If conditions are such that vinyl phenyl ketone is produced mainly as the intermediate, resins with low softening point and solubility in a broad range of solvents are formed. For example, reaction of 1 mol of acetophenone with 0.05 – 0.5 mol of formaldehyde gives an oil-compatible resin. If methylol intermediates are predominantly formed, the resins have higher melting points and are limited in their solubility and compatibility. Resins with these properties are obtained from acetophenone and formaldehyde in the ratio 1: 1.

*Uses.* Acetophenone – formaldehyde resins are used as favorably priced binding agents in coatings and printing inks, in combination with other coating raw materials such as nitrocellulose. They improve gloss, intensity, covering capacity, solid material content, adhesion, and drying. In nitrocellulose lacquers the ease of pigmenting and polishing on wood is increased. Acetophenone – formaldehyde resins are also used in vinyl chloride copolymer coatings for metal coating and road-marking paints. Adhesives and hot melts are also formulated with these resins.

***Trade Name.*** Synthetic Resin AP (Evonik, Germany).

### 3.2.1.2. Modified Acetophenone Resins

Hydrogenation of acetophenone – formaldehyde resins, in particular, gives products that differ markedly from the usual ketone and aldehyde resins and therefore open up new areas of application. The reaction of hydroxyl groups, which are formed from the keto groups on hydrogenation, with aliphatic diisocyanates gives high-quality special binding agents (see Fig. 1 and Table 14).

***Properties.*** The hydrogenated resins are very light-colored and particularly stable. They are completely free of formaldehyde, which can no longer be formed by retroaldol-type reactions. Depending on hydrogenation conditions, resins are obtained in which only the keto groups are converted into hydroxyl groups or in which the aromatic ring is also hydrogenated [65]. Only the former are available on the market. Their softening points lie in the range 110 – 120 °C and they are readily soluble in alcohols. The resins can be diluted with aromatics but are insoluble in aliphatics and mineral oil. Their compatibility with alkyd resins is better than that of nonhydrogenated products.

The hydroxyl values of these resins are > 300 mg KOH/g, The OH groups can be reacted with diisocyanates to give extremely high-melting resins (*mp* ca. 160 °C), which are also alcohol soluble. Diisocyanate-modified resins are even more stable and exhibit rapid solvent release.

***Production.*** Under mild hydrogenation conditions only the keto groups are hydrogenated (Fig. 1). The reaction is controlled by determination of the oxygen content and the melting point [66]. Hydrogenation can be carried out catalytically in solution or in the melt [67, 68]. Reactions with diisocyanate take place at slightly elevated temperature with the usual catalysts.

***Uses.*** Modified acetophenone – formaldehyde resins are high melting and stable. They are therefore used as hard resin components together with other binding agents in paints and coatings, printing inks, and adhesives, if high-quality coatings are required. The *hydrogenated resins* are used for lightfast nitrocellulose-based overprinting inks for paper. They are also used in ballpoint pen pastes, inks, and toners; as flow-control agents for powder coatings; as hard resin components in isocyanate reactive systems; and for modification of alkyd resins.

The *isocyanate-modified resins* are used in high-quality printing inks, in paper and wood varnishes, and in the formulation of coatings with good adhesion to plastic surfaces. They are particularly suitable for wetting and stabilizing aluminum pigments in paints and printing inks.

***Trade Names.*** Synthetic Resin SK (hydrogenated resin) and Synthetic Resin 1201 (diisocyanate-modified, Evonik, Germany).

### 3.2.2. Resins from Aliphatic Ketones

Resins are produced from linear aliphatic ketones — principally methyl ethyl ketone and acetone — and formaldehyde. Methyl isobutyl ketone is also used, mainly for the production of adhesives [69, 70]. Cyclohexanone and methylcyclohexanone are most important in the synthesis of alicyclic resins, although reactions of cyclopentanone, cycloheptanone, and cyclic ketones with longer side chains have also been described [71].

### 3.2.2.1. Methyl Ethyl Ketone – Formaldehyde Resins

Methyl ethyl ketone is still important for the production of binding agents for paints, coatings, and adhesives. These resins differ from alicyclic and aromatic ketone-based products in certain application-relevant properties, such as solubility in various solvents and compatibility with other lacquer raw materials. These properties may be explained by the high polarity of the ketone and its special behavior in the alkali-catalyzed condensation with formaldehyde [59].

***Physical Properties.*** Methyl ethyl ketone – formaldehyde resins are lightly colored and are soluble in polar organic solvents such as alcohols, esters, ketones, and glycol ethers. The hydroxyl values are in the range 80 – 190 mg KOH/g. The strongly polar resins have a total oxygen content of 21 – 29 wt % and are hygroscopic. Depending on production conditions, the molecular mass lies between 3000 and 5000, and the melting range between 80 and

120 °C. Other physical data and compatibilities are described in [27].

***Chemical Properties.*** Methyl ethyl ketone – formaldehyde resins are unhydrolyzable, The keto and hydroxyl groups and the carbon double bonds give the compounds a certain degree of chemical reactivity. Hydrogenation leads to clear, colorless resins [72]. The hydroxyl groups can be esterified with carboxylic acids or acid anhydrides [73] and can react with isocyanates [74].

***Production.*** Industrially, ketone resins are obtained by alkali-catalyzed condensation of methyl ethyl ketone and formaldehyde (molar ratio 1: 2 – 1: 2.5). Condensation is performed mainly batchwise [59, 75, 76]. Continuous production has also been described [77]. The unpurified ketone is treated with formaldehyde in the presence of water and sodium or potassium hydroxide as catalyst. By increasing the formaldehyde content, the melting range can be raised to 80 – 120 °C.

Special purification processes give light, high-melting products [78]. High softening points are also achieved by using phase-transfer catalysts [72]. The reaction mechanism is not completely understood and involves many concurrent reactions [27].

***Uses.*** Methyl ethyl ketone – formaldehyde resins are used in combination with film-forming agents such as nitrocellulose, acetylcellulose, cellulose ethers, and natural resins. They impart hardness, drying, polishing ability, and good lightfastness. The capability to gelatinize nitrocellulose is exploited in the production of high-gloss nitro lacquers. The free hydroxyl groups are used for cross-linking in isocyanate-reactive coatings, adhesives [79], and molding sands [74].

**3.2.2.2. Acetone – Formaldehyde Resins**
The alkali-catalyzed condensation of acetone and formaldehyde does not lead to useful, solid synthetic resins. The considerably greater hardening tendency of the unstable methylol intermediates from acetone in comparison to those of other carbonyl compounds [80, 81] leads to a cross-linked, insoluble final product. *Self-hardening precondensates* can, however, be produced, which are used alone or in combination with other hardenable precondensates, such as phenol resols.

***Uses.*** Hard foams [82] are obtained by foaming the methylol compounds in the presence of alkali hydroxide alone or of alkali hydroxide and an elastomer latex [83]. Sands can be hardened with acetone – formaldehyde precondensates for the production of moldings [84–86]. Rapidly hardening moldings made from cement have also been described [87, 88]. Acetone – formaldehyde condensates are used widely in paper [89] and wood adhesives [90]. Phenol resols are often cocondensed to obtain particularly weather-resistant wood materials and chipboard [91–93] or for the waterresistant adhesion of corrugated paperboard [94]. Other uses of acetone precondensates are in photoreceptors in electrophotography [95, 96] and in production of thickeners and emulsifiers [97, 98].

**3.2.2.3. Cyclohexanone Resins**
Cyclohexanone and methylcyclohexanone readily undergo self-condensation, in addition to mixed condensation with formaldehyde [99]. The former involves aldol condensation between the carbonyl group of one molecule and the activated methylene group of another. The carbonyl group of the intermediate then reacts with another cyclohexanone molecule, etc. The reaction occurs at elevated temperature and can be catalyzed by basic, acidic, or neutral agents.

***Properties.*** The softening points of the light, neutral resins are between 80 and 120 °C. Under normal conditions the resins are resistant to acids and bases. However, they eliminate water in the presence of acids at elevated temperature (> 80 °C), thus altering their properties significantly, Cyclohexanone resins are lightfast, soluble in many solvents, and compatible with most coating raw materials (see Table 14). They are more expensive than cyclohexanone – formaldehyde resins [99]. An example of the production process (according to [100, 101]) is given in [27].

***Uses.*** Cyclohexanone resins are used in coatings, mainly to improve intensity, gloss, and hardness. They can also improve light and weather resistance, as well as adhesion. In quantities of 5 – 50 % (based on the film-forming agent) these resins are added to coatings based on alkyd resins, vinyl chloride copolymers, chlorinated rubber, nitrocellulose, or oils.

Their use as carrier resins for pigment preparations has also been described [102].

*Trade Name.* Laropal K 80 (BASF).

### 3.2.2.4. Cyclohexanone – Formaldehyde Resins

Cyclohexanone can react with aldehydes, in particular formaldehyde, to give defined methylol compounds or resinous products. The molar ratio and the reaction conditions determine the final products. A large formaldehyde excess promotes the formation of methylol compounds, whereas basic catalysis leads to resin formation [103, 104]. Higher aldehydes can also be used instead of formaldehyde, but they have not gained any industrial importance. In addition to cyclohexanone, methylcyclohexanone or mixtures of cyclohexanone with aliphatic ketones are also used as the component with the activated methylene group [105]. The resins can be modified with phenols [100, 106], epoxides [107], polyesters [108], and sulfonamides [109]. Gritty products are obtained by the addition of dispersion agents [110]. Continuous production has been described in [111] and [112]. The lightfastness can be increased by hydrogenation [113] and treatment with reducing agents [114].

*Properties.* Cyclohexanone – formaldehyde resins do not exhibit the broad compatibility and solubility of cyclohexanone resins. However, they are cheaper and nevertheless sufficiently lightfast. They can be combined with a range of binders for coatings, but not with oils (see Table 14). The use of methylcyclohexanone as a raw material generally leads to better solubility and compatibility.

*Production.* Industrially, cyclohexanone is condensed with formaldehyde in the presence of alkali. An example of the production procedure is given in [27], where the modification with phenol is also described.

*Uses.* In many cases, cyclohexanone – formaldehyde resins are used to improve drying, hardness, intensity, gloss, and solid material content, In paints and coatings they are always used as additives in addition to other binding agents, for example, in alkyd – acrylate varnishes [115], cement paints [116], epoxy resin systems [117], and marine paints [118]. In addition to their use in conventional printing inks [119, 120], they are increasingly being used in UV-hardening printing inks [121–123]. Another important area of application is in adhesives [124–126] and sealants [127]. The use of these resins in optical recording media has also been described [128, 129].

*Trade Names.* Synthetic Resin AFS (Bayer), Synthetic Resin CA (Hüls), L2 Resin (Leuna), Krumbhaar type resins (Lawter).

### 3.2.3. Aldehyde Resins

In the 1970s, aldehyde resins synthesized from isobutyraldehyde, formaldehyde, and urea were developed to an industrial level. α-Ureidoalkylation formed the basis of the synthesis and is described in detail in [130]. Aldehyde resins from isobutyraldehyde and phenol are used in melt adhesives [131, 132].

*Isobutyraldehyde – Formaldehyde – Urea Resins.* Both the reaction of urea with enolizable aldehydes [133] and the reaction of urea, formaldehyde, and enolizable aldehydes [134] in an acidic medium lead, on subsequent treatment with alkali alkoxides, to resinous products. The process can be described as follows:

***Properties.*** The isobutyraldehyde – formaldehyde – urea resins are slightly colored and fairly lightfast. Their melting points lie in the range 80 – 110 °C. The solubility and compatibility with other coating binders can be varied, depending on the production process, Thus products soluble in aliphatics and all other common solvents are on the market. Their compatibility is, however, limited to acrylate resins and some cellulose derivatives (see Table 14). The available functional groups can be seen in the above formula scheme.

***Production*** [134]. Formaldehyde, isobutyraldehyde, and urea are reacted in the presence of sulfuric acid. The condensation is carried out for 3 h at 80 °C. After the addition of xylene the mixture is stirred for another 20 min at 75 °C, and the aqueous phase is separated. The organic phase is treated first with sodium methoxide and then neutralized with 75 % sulfuric acid. The resin resulting after work-up has a pale-yellow color and a softening point of 80 – 82 °C.

***Uses.*** Like ketone resins, aldehyde resins can be added to other coating binders, Improvements in hardness, gloss, intensity, flow, resistance to yellowing, and solid material content are achieved in coatings based on alkyd resins, nitrocellulose, or chlorinated polymers. By the addition of aldehyde resins, the overall formulation can often be made more economical (e.g., in powder coatings, where the flow is improved at the same time). Another area of use is in hot meltings and spray plastics for road markings. Some aldehyde resins are suitable as mixing resins for pigment pastes with broad solubility and compatibility. A high pigment-binding capacity and low solution viscosity have thus far not been achieved with ketone resins. For aqueous dispersions see [135].

***Trade Names.*** Laropal A types (BASF).

### 3.3. Polyamide Resins

Polyamides are thermoplastic polymers with repeating amide units (–CO–NH–) in the main chain (→ Polyamides).

Polyamide resins (PA resins) also belong to the polyamide group. However, PA resins are based on dimerized fatty acids, and their molecular mass is generally less than 15 000; thus they differ considerably from high molecular mass polyamides in their chemical structure and properties [136]. Whereas high molecular mass polyamides are important synthetic fibers and engineering plastics, polyamide resins are used as hotmelt adhesives, casting compounds, and binders for printing inks.

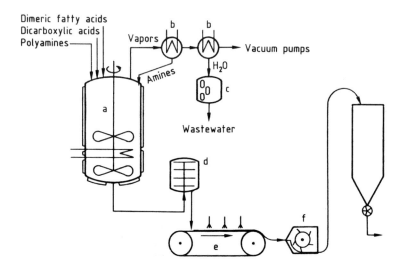

**Figure 2.** Polyamide resin production
a) Reaction vessel; b) Condenser; c) Separator; d) Filter; e) Cooling band; f) Granulator

### 3.3.1. Physical and Chemical Properties

The physical and chemical properties of PA resins can be varied within wide limits through the choice and ratio of raw materials used. Brittle, tough-hard, or flexible PA resins can be produced, which have very different tensile strengths and elongation values [137].

The density of PA resins is ca. 0.95 g/cm$^3$. The *softening point* is in the range of 80 – 250 °C, preferably between 90 and 190 °C. Melt viscosities are between 0.1 and 100 Pa · s at 220 °C. The water absorption of PA resins is between 0.5 and 3 % after 28-d storage in water and is thus much lower than the water absorption of industrially relevant polyamides, All PA resins show good resistance to chemicals. They are generally resistant to water, dilute alkali, weak acid, oils, and fats. Most PA resins are moderately stable to unstable toward esters, ketones, alcohols, and aliphatic hydrocarbons, Some are resistant to chlorinated hydrocarbons.

### 3.3.2. Raw Materials

The dimerized fatty acids required for the production of PA resins are obtained by heating mono- or polyunsaturated fatty acids, such as oleic, linoleic, or linolenic acids or their esters with or without a catalyst. The dimerized fatty acids formed have an open-chained, cyclic-unsaturated, or aromatic structure [138].

In addition to dimerized fatty acids, dicarboxylic acids are frequently used as comonomer. Examples are adipic acid, azelaic acid, sebacic acid, decamethylenedicarboxylic acid, terephthalic acid or its ester, or isophthalic acid. Ethylenediamine, hexamethylenediamine, 1,12-diamino-4,9-dioxadodecane, piperazine, dipiperidylpropane, and polyoxypropylenediamine are used as the polyamine component [139]. 6-Aminocaproic acid in the form of its lactam and 11-aminoundecanoic acid are the preferred aminocarboxylic acids.

### 3.3.3. Production

Polycondensation of the aforementioned monomers is carried out in a reactor equipped with a stirrer and condenser at 150 to 250 °C with exclusion of air (Fig. 2). The reaction is performed initially under normal pressure and then under reduced pressure, The water formed in the reaction is distilled off. When the condensation reaction has gone to completion the product is removed as a melt, cooled, and ground or poured into special molds.

### 3.3.4. Quality and Analysis

For quality testing and characterization of the products the following procedures are used:

The *softening range* is measured by the ring-and-ball method (DIN 52 011 or ASTM E-28). Melting points determined by differential scanning calorimetry or thermomechanical analysis, or microscopic melting points are sometimes given.

*Melt viscosities* are measured with rotation viscometers, The values obtained can vary considerably, depending on the type of viscometer used. For highly viscous products the *melt index* according to DIN 53 735 or ASTM D-1238 is sometimes determined. The *tensile strength* and *elongation* are determined according to DIN 53 455 or ASTM D-1708. The *water absorption* is determined according to DIN 53 495 or ASTM D-570. The *acid value* is determined according to DIN 53 402 and the *amine value* according to DIN 16 945.

### 3.3.5. Storage

Polyamide resins are usually supplied as pellets. Because many machines for further processing can only handle rods, cylinders, or rectangular blocks, PA resins are also supplied in these forms.

Since problems can arise in processing if the product is damp, moisture-proof packaging is recommended. With proper storage, the shelf-life of PA resins is virtually unlimited.

### 3.3.6. Uses

Thermoplastic PA resins are used as high-performance hotmelt adhesives [140], casting compounds [141], binders for printing inks, overprinting lacquers [142], and thixotropic agents for alkyd resin coatings [143].

In the shoe industry, PA hotmelt adhesives have been used for many years as an aid in assembly for bonding the upper leather to the insole (lasting). The furniture industry utilizes PA hotmelt adhesives for heat-resistant edge banding of chip or particle board.

In the automobile industry, PA hotmelt adhesives are used for interior trim. The bonding of air and oil filters by using PA hotmelt adhesives is particularly important. A relatively new area of application is the sealing of wires and cables into car-connectors by potting with PA resins, Connectors are being incorporated more frequently in the car because of the increased use of electronic control systems. In the electrical and electronics industries, PA resins are used because of their relatively high heat resistance and good insulating properties. In addition to bonding radio and television housings and loudspeakers, in particular coil winding, coil ends, and other electronic components are fixed and bonded with PA hotmelt adhesives.

In the telecommunications sector, PA hotmelt adhesives are used to bond and seal cables and heat-shrink components, and to protect telephone cables from moisture and corrosion at the ends or connection points. In the textile industry, PA resins are used for "cleaning-resistant" bonding of interlining to textiles [144].

PA resins are also used as binders for printing inks. Printing inks based on PA resins are particularly suitable for printing on nonporous materials, such as plastic films, because of their good adhesion, flexibility, and gloss, Since the printed films show good resistance to chemicals and abrasion, they are also used as overprint lacquers.

In the area of transparent lacquers, which are applied by gravure or flexographic printing, PA resins are used as cold-seal release lacquers, antislip varnishes, or heat-seal varnishes.

### 3.3.7. Producers, Trade Names

The PA resins are sold under the following trade names: Euremelt, Eurelon (Huntsman Advanced Materials, Switzerland); Versamid, Macromelt (Cognis, Germany); Reamide (Chemplast, Italy); Unirez (Arizona, United States); Flex-Ref (United States); Casamid (Swan, United Kingdom); Polymide (Sanyo, Japan).

## 4. Natural and Synthetic Resins — Toxicology and Legislation

### 4.1. Toxicology

#### 4.1.1. Rosins and Their Derivatives

For review, see [145].

*Crude Rosins* ($\rightarrow$ Resins, Natural, $\rightarrow$ Resins, Natural, Section 5.3., $\rightarrow$ Resins, Natural, Section 5.4., $\rightarrow$ Tall Oil). In contrast to refined rosins, crude rosins contain volatile, neutral components ($\rightarrow$ Terpenes), which may impose properties on them that resemble those of organic solvents.

*Resin Acids.* Abietic acid, the major component of *refined rosins*, has an acute oral $LD_{50}$ in mice of about 2500 mg/kg. The acute oral $LD_{50}$ of its *methyl ester* is > 5000 mg/kg in rats and rabbits. Resin acids were not mutagenic in the Ames test.

*Refined rosins,* original or chemically modified, have a negligible acute oral toxicity in rodents and guinea pigs, the $LD_{50}$ values ranging from 4000 to 12 000 mg/kg; no acute dermal toxicity was found in rabbits for pale rosins.

On oral administration, *dark rosins* with a high phenolic fraction and *rosin esters* showed no acute toxic symptoms.

In longterm feeding studies with pale and esterified rosins on rats and dogs (90 days or two years), no toxic symptoms were observed at the 0.2 % to 1 % dietary level. Therefore rosins and their derivatives may be regarded as safe.

Generally, rosin acids, rosins, and their derivatives are not considered skin irritants. However, a small portion of the population shows an allergic response on patch tests. A high sensitization level was reported among printers exposed to gum arabic, a constituent of printing ink [146].

*Fumes from rosins* may cause ocular and bronchial irritation: ca. 80 % of humans exposed to products of wood rosin at a level > 0.12 mg/m$^3$ experienced severe irritation of the eye, nose, or throat. The TLV for rosin core solder pyrolysis products is 0.1 mg/m$^3$ aliphatic aldehydes measured as formaldehyde.

### 4.1.2. Hydrocarbon Resins

The acute oral toxicity ($LD_{50}$) of hydrocarbon resins in rats is generally greater than 5 g/kg body weight, and they were found to be nonirritating to the skin and eye under experimental conditions [147]. In the EEC directives, most hydrocarbon resins are not classified as harmful, toxic, or irritant to the skin and eyes. Handling of resins requires the usual precautions for working with chemicals. Inhalation of dust or fumes (the latter generated on heating or combustion) must be avoided, More specific information is given in the manufacturers safety data sheets.

Food legislation concerning hydrocarbon resins is regulated by many national authorities but will likely be harmonized with the requirements of the FDA and the EU. The main attention is directed to migration of residual monomers, oligomers, and additives that might contaminate contiguous media by diffusion, and to the stability of resin formulations under application conditions.

In the United States, numerous resins are CFR approved substances for the following:

- 175.105: Adhesives
- 175.300: Resinous and polymeric coatings
- 176.170: Components of paper and paperboard in contact with aqueous and fatty food
- 176.180: Components of paper and paperboard in contact with dry food
- 177.1210: Closures with sealing gaskets for food containers
- 177.2600: Rubber articles intended for repeated use
- 178.3800: Preservatives for wood.

In the EU, the use of materials intended — in their final state — to come in contact with products for human consumption is regulated by the framework directive 89/109/EEC, followed by several subordinate directives, one of which concerns the use of plastics materials, e.g., in food packaging (90/128/EEC). Implementation by the EU member states was required by January 1, 1993.

Monomers and additives are not allowed to be used for food packaging unless listed in the annexes of the latter directive (positive-list principle). Not all hydrocarbon resins have yet been covered there. Therefore, an industrial motion to enclose those resins in the additive list, among others of the IC type, was directed to the European Scientific Committee for Food, but has not yet been settled.

### 4.1.3. Polyamide Resins

Polyamide resins do not need to be labeled [EC guideline 79/831/EEC (classification, packaging, and labeling of hazardous substances), GefStoffV]. Some PA resins are licensed by the FDA for contact with foods.

If PA resins are used in the molten state, vapors can sometimes be formed, which do not impair general health or have any toxic effects, according to the current state of knowledge. Nevertheless, extraction of the vapors above processing equipment and adequate ventilation of the workplace are recommended.

### 4.2. Legal Aspects

Food legislation concerning resins is regulated by many national authorities, but will likely be harmonized with the requirements of the FDA (United States) and the European Community. The main questions are directed to the migration of residual monomers, oligomers, and additives that might contaminate contiguous media by diffusion, and to the stability of resinous formulations under the conditions applied (→ Foods, 4. Food Packaging, Section 8.1., → Foods, 4. Food Packaging, Section 8.2.).

In the *European Community,* the application of resins in the food sector will be regulated in connection with materials intended to come in contact with foods [148, 149] and with use of food additives other than colors and sweeteners [150]. Several directives that are especially relevant [151–153] have already been adopted.

In the United States, the use of resinous materials in contact with food is regulated in the Code of Federal Regulations (CFR) under Title 21 (Food and Drugs): Besides the use of resins, other additives such as stabilizers, fillers, and linker reagents are also addressed (e.g., Parts 172, 175, 176).

Natural gums, synthetic as well as natural terpene resins, indene coumarone, and petroleum hydrocarbon resins may be used as chewing gum bases with limitations (121.1059, 121.90). Terpene resins are permitted for coatings in direct use (e.g., coating of fruits) when the precursors originate from wood (172.280). In indirect applications, most resins (synthetically or naturally derived) may be used.

Petroleum hydrocarbons are authorized for indirect use as components of paper and paperboard (Part 176).

# References

## General References

1. W. Vredenburgh, K. F. Foley, A. N. Scarlatti: Hydrocarbon Resins, in *Encycl. Polym. Sci. Eng.*, vol. 7, John Wiley, New York 1987, pp. 758 – 783.
2. E. W. Flick: *Industrial Synthetic Resins Handbook*, Noyes Publication, Park Ridge, N.J., 1989, pp. 214 –384.
3. R. Mildenberg, M. Zander, G. Collin: *Hydrocarbon Resins – Manufacture, Properties and Applications*, VCH, Weinheim, Germany 1997.
4. H.-G. Franck, G. Collin: *Steinkohlenteer*, Springer Verlag, Heidelberg 1968, pp. 92 – 95, 162 – 169.
5. G. Collin, W. Lücke: *Erdöl, Kohle, Erdgas, Petrochem.* **38** (1985) 120 – 126.
6. J. P. Kennedy, E. Marechal: *Carbocationic Polymerization*, John Wiley, New York 1982, pp. 10 –12.
7. E. W. Flick: *Handbook of Adhesive Raw Materials*, 2nd ed., Noyes Publication, Park Ridge, N.J., 1985, pp. 91 – 101.
8. R. Jordan: *Coating* **15** (1982) 335 – 337; **16** (1983) 12 – 14, 143 – 147.
9. DeWitt Comp.Inc., Inter-Tech Ltd., Directory for the Hydrocarbon Resin Industry, Houston 1991.
10. A. P. Dunlop, F. N. Peters: The Furans, *ACS Monogr. 119*, Rheinhold Publishing Corporation, New York 1953.
11. Süd-Chemie AG: "Furfural Hydrogenation, Catalysts and Chemistry," Technical Bulletin, München.
12. W. J. McKillip: "Chemistry of Furan Polymers," *ACS Symp. Ser.* **385** (1989) chap. 29.
13. Merck & Co., US 3 083 236, 1963 (V. Torleif, J. D. Garber, R. E. Jones).
14. A. Gandini, *Adv. Polym. Sci.* **25** (1977) 47 – 58.
15. J. B. Barr, S. B. Wallon, *J. Appl. Polym. Sci.* **15** (1971) 1079.
16. Y. Hachihama, T. Shono, *Technol. Rep. Osaka Univ.* **7** (1957) 479.
17. E. Wewerka, E. Loughran, K. J. Walters, *J. Appl. Polym. Sci.* **15** (1971) 1437.
18. E. Wewerka, *J. Polym. Sci.* **A19** (1971) 2703.
19. Y. Hachihama, T. Shono, *Technol. Rep. Osaka Univ.* **4** (1954) 413.
20. A. K. A. Rathi, M. J. Chanda, *J. Appl. Polym. Sci.* **18** (1974) 1541.
21. T. A. Krishnan, M. Chanda, *Angew. Makromol. Chem.* **43** (1975) 145.
22. R. T. Conley, I. J. Metil, *J. Appl. Polym. Sci.* **7** (1963) 37.
23. R. T. Conley, I. J. Metil, *J. Appl. Polym. Sci.* **7** (1963) 1083.
24. G. E. Maciel, I. S. Chuang, G. E. Myers, *Macromolecules* **17** (1984) 1087.
25. QO Chemicals Technical Data Bulletin no. 194, 1979.
26. H. Kittel: *Lehrbuch der Lacke und Beschichtungen*, vol. **I/1**, Verlag W. A. Colomb, Stuttgart-Berlin 1971, pp. 390 ff.
27. *Ullmann*, 4th ed., **15**, 547 – 552.
28. Chugoku Marine Paint, JP 49 092–137, 1972.
29. Kubo Takashi Paint, JP 60 124–666, 1983.
30. Nippon Seika KK, JP 60 018–551, 1983.
31. Dainippon Paint KK, JP 50 109–234, 1974.
32. Sakai Kagaku Koggo Co, JP 74 001–451, 1967.
33. G. Bassler, DD 140 053, 1978.
34. E. W. Flick: *Printing Ink Formulations*, Noyes Data Corp., Park Ridge, N.J. 1985.
35. W. Ritzerfeld, DE 3 219 893, 1982.
36. Toka Shikiso Chem. KK, JP 52 004–310, 1975.
37. Millmaster Onyx Corp., US 4 070 500, 1976.
38. Canon KK, JP 62 160–273, 1986.
39. Carre SA, FR 2 058 553, 1969.
40. Pentel KK, JP 56 065–062, 1979.
41. Fuji Photo Film KK, DE 2 648 498, 1975.
42. Hitachi KK, JP 61 002–161, 1984.
43. Mitsubishi Paper Mill, JP 53 017–346, 1976.
44. Fuji Photo Film Co, DT 2 232 185, 1971.
45. Annheuser Busch Inc., US 4 033 914, 1976.
46. Hitachi Chem. Co, JP 49 004–731, 1972.
47. H. Kempin, EP 239 658, 1986.
48. Sugiyama Sangyo Kag., JP 5 6024–474, 1979.
49. Aika Kogyo KK, JP 57 056–242, 1980.
50. Hohnen Oil KK, JP 6 0 34–775, 1979.
51. Oji Cornstarch KK, JP 59 152–969, 1983.
52. Aishin Kako KK, JP 55 149–742, 1979.
53. Kao Corp.JP 62 179–847, 1986.
54. Ciba Geigy AG, EP 176 484, 1984.
55. Fa. Munchen, DT 2 438 724, 1974.
56. Nippon Crown Cork KK, JP 57 189–035, 1981.
57. BASF AG, DE 3 406 474, 1984 (K. Fischer, H. Petersen, H. Kasch, E. Wistuba).
58. S. Patai: *The Chemistry of the Carbonyl Group*, Interscience, New York 1966, pp. 589 – 590.
59. F. Josten, *Fette Seifen Anstrichm.* **54** (1952) 673.
60. J. Scheiber: *Chemie und Technologie der künstlichen Harze*, vol. **1**, Wissenschaftliche Verlagsgesellschaft, Stuttgart 1961, pp. 164 – 165.
61. H. Wagner, H. F. Sarx: *Lackkunstharze*, Hanser Verlag, München 1971, pp. 84 – 85.
62. Chem. Werke Hüls, DT 892 975, 1940.
63. Chem. Werke Hüls, DE 3 324 287, 1983.
64. Farbenfab. Bayer AG, FR 1 595 632, 1967.
65. Chem. Werke Hüls, DE 3 334 631, 1983.
66. Chem. Werke Hüls, DT 870 022, 1944.
67. Farbenfab. Bayer AG, DT 907 348, 1940.
68. Chem. Werke Hüls, DT 826 974, 1949.
69. Arakawa Kagaku Kogyo, JP 01 069–682, 1987.
70. Weyerhauser Co, US 3 947 425, 1972.
71. BASF AG, EP 7 106, 1978.
72. Farbenfab. Bayer AG, DT 907 348, 1940.
73. Rheinpreussen, DE-AS 1 066 020, 1957.
74. Deutsche Texaco AG, DT 2 039 330, 1970.
75. Rheinpreussen, DT 890 866, 1941.
76. Dainichi Seika Co, JP 7 022 218, 1965.
77. Rheinpreussen, DT 1 155 909, 1957.
78. Rheinpreussen, DT 1 300 256, 1960.
79. C. Lüttgen: *Die Technologie der Klebstoffe*, 2nd ed., part 1, W. Panzegrau Verlag, Berlin 1959, p. 212.
80. F. Engelhardt, J. Wöllner, *Brennst. Chem.* **44** (1963) 52.

81. *Houben-Weyl*, 4th ed., **XIV/2**, p. 416.
82. VEB Farbenfab. Wolfen, DD 1 171 156, 1962.
83. Asahi Chem. Ind. Co, JP 7 0 38 425, 1966.
84. Deutsche Texaco AG, DT-AS 1 767 904, 1968.
85. Deutsche Texaco AG, DT 1 922 015, 1969.
86. Deutsche Texaco AG, DT 2 439 828, 1974.
87. Deutsche Texaco AG, DT-OS 2 353 490, 1973.
88. Kaluga Halurgy Res., SU 925 903, 1980.
89. D. C. Sistrunk, CA 2 007 361, 1989.
90. Casco Lab. Inc., EP 66 560, 1981.
91. Deutsche Texaco AG, DT 2 363 797, 1973.
92. Deutsche Texaco AG, DT 1 247 017, 1963.
93. Deutsche Texaco AG, DT 2 264 288, 1972.
94. Owens-Illinois Inc., US 3 591 534, 1968.
95. Canon KK, JP 56 051–748, 1979.
96. Canon KK, JP 57 078–045, 1980.
97. SKW Frostberg AG, DE 3 315 152, 1983.
98. SKW Frostberg AG, DE 3 429 068, 1984.
99. I.G. Farbenind., DT 511 092, 1925.
100. BIOS-Rep. no. 629, no. 743, no. 1243.
101. *Mod. Plast.* **25** (1948)no. 12, 119.
102. BASF, DT-OS 2 400 194, 1974 (K. Heinle, E. Herrmann, J. D. Stetten).
103. M. N. Tilitschenka, *Zh. Obshch. Khim.* **25** (1952) 64.
104. R. Bittner, P. Franke, H. Patzel,DL 83 412, 1970.
105. H. G. Rosenkranz et al., DT 1 262 600, 1965.
106. Hitachi Chem. Co., JP 7 2 41 648, 1969.
107. Howards of Ilford, GB 864 542, 1957.
108. VEB Leuna, DT 1 042 229, 1955.
109. M. Ikuta, K. Toyoshima, A. Shimizu,JP 7 1 40 875, 1966.
110. VEB Leuna, DL 12 433, 1953.
111. Rheinpreussen, DT 1 155 909, 1957.
112. Hitachi-Kasei, JP 15 098, 1964.
113. Hitachi Chemical KK, JP 50 013–415, 1973.
114. Howards of Ilford, GB 864 541, 1958.
115. Budalakk Festek, HU 029 441, 1982.
116. VEB Chem. Bitterfeld, DD 234 161, 1983.
117. Lechler Chemie GmbH, EP 41 200, 1980.
118. Toyo Soda Mfg. KK, JP 62 236–804, 1986.
119. H. G. Rosenkranz, P. Jodl, M. Schloffel,DL 75 746, 1969.
120. Toyo Ink Mfg. KK, JP 5 2 043–501, 1975.
121. Dainippon Ink Chem. KK, JP 56 093–776, 1979.
122. Suzuka Paint Mfg. KK, JP 50 150–794, 1974.
123. Suzuka Paint Mfg. KK, JP 50 156–594, 1974.
124. Dunlop Ltd., FR 2 221 507, 1973.
125. W. Kunzel, DL 136 149, 1977.
126. Vibac Spa., GB 2 140 439, 1983.
127. Union Carbide Cop., US 3 772 237, 1971.
128. TDK Corp., JP 59 124–896, 1982.
129. TDK Corp., JP 59 124–894, 1982.
130. H. Petersen, *Synthesis* 1973, no. 5, 243 – 292.
131. BASF, DE-OS 2 847 030, 1978 (H. Petersen *et al.*).
132. BASF, DE-OS 2 941 635, 1979 (H. Petersen, K. Fischer, H. Zaunbrecher).
133. BASF, DT-OS 2 757 176, 1977 (H. Petersen *et al.*).
134. BASF, DT 2 757 220, 1977 (H. Petersen *et al.*).
135. BASF, DE 3 406 473, 1984 (K. Fischer *et al.*).
136. D. E. Floyd: *Polyamide Resins*, Reinhold Publishing Corp., New York 1966.
137. W. Imöhl, *Adhäsion* **18** (1974)no. 1, 7 – 11.
138. W. Link, G. Spiteller, *Fat Sci. Technol.* **92** (1990) 135 – 138.
139. R. Jordan: *Schmelzklebstoffe, Rohstoffe-Herstellung*, Hinterwaldner Verlag, München 1985, p. 93.
140. W. Imöhl, *European Adhesives & Sealants* (1991)June, 17 – 18.
141. W. Imöhl in R. Hinterwaldner (ed.): *Verbund- und Vergußtechnik in der Elektrotechnik, Elektronik und Optik*, Hinterwaldner Verlag, München 1989, pp.115 – 122.
142. Schering AG, *Binders for Printing Inks*, company brochure, Bergkamen.
143. Washburn Co., US 2 663 649, 1952 (W. B. Winkler).
144. Schering AG, DE 2 534 121, 1975 (M. Drawert, E. Griebsch, W. Imöhl).
145. J. J. Domanski in D. F. Zinkel,J. Russell (eds.): *Naval Stores*, Pulp Chem. Assoc., New York 1989, p. 896.
146. W. Gronemeyer, E. Fuchs (eds.): *Karenz und Hyposensibilisierung bei Inhalations- und Insektengift-Allergie*, Dustrie, München 1983.
147. Nevcin Polymers BV: unpublished data, 1986; Reprotox Huntingdon, Germany: unpublished data, 1979.
148. Council Directive 89/109/EEC, 21 Dec. 1988.
149. H. J. Palmen, *Kunststoffe* **81** (1991) 736.
150. EEC Commission: III/3624/91/EN-Rev. 1.
151. Commission Directive 90/128/EEC, 23. Feb. 1990.
152. Council Directive 82/711/EEC, Oct. 18, 1982.
153. Council Directive 85/572/EEC, Dec. 19, 1985.

# Further Reading

R. D. Lowery: Hydrocarbon Resins, *Kirk Othmer Encyclopedia of Chemical Technology*, 5th edition, John Wiley & Sons, Hoboken, NJ,online DOI: 10.1002/0471238961.0825041812152305.a01.

H. Q. Pham, M. J. Marks: Epoxy Resins, *Kirk Othmer Encyclopedia of Chemical Technology*, 5th edition, John Wiley & Sons, Hoboken, NJ,online DOI: 10.1002/0471238961.0516152407011414.a01.pub2.

L. Pilato: *Phenolic Resins: a Century of Progress*, Springer, Berlin, Heidelberg 2010.

J. Tulla-Puche: *The Power of Functional Resins in Organic Synthesis*, Wiley-VCH, Weinheim 2008.

# Part 5

# Inorganic Polymers

# Inorganic Polymers

CHARLES E. CARRAHER JR., Florida Atlantic University, College of Science Boca Raton, Florida, USA
Florida Center for Environmental Studies, Palm Beach Gardens, Florida, USA

CHARLES U. PITTMAN JR., Mississippi State University, Department of Chemistry Mississippi, USA

| | | | | |
|---|---|---|---|---|
| 1. | Introduction | 1777 | 11. | Biomedical and Antimicrobial |
| 2. | Polysilanes, Polygermanes, and | | | Polymers . . . . . . . . . . . . . . . . . . . . . 1797 |
| | Polystannanes . . . . . . . . . . . . . . . . . | 1778 | 12. | Boron Polymers. . . . . . . . . . . . . . . 1798 |
| 3. | Poly(Carborane – Siloxanes), | | 13. | Aluminum Polymers . . . . . . . . . . 1801 |
| | Polycarbosilanes and Polysilazanes . | 1781 | 14. | Polymers with —Si—O—M—O— |
| 4. | Polyphosphazenes . . . . . . . . . . . . . | 1782 | | Backbones. . . . . . . . . . . . . . . . . . . . 1802 |
| 5. | Poly(Boron Nitride) . . . . . . . . . . . . | 1785 | 15. | Poly(Carbon Disulfide), Poly(Carbon |
| 6. | Poly(Sulfur Nitride) . . . . . . . . . . . . | 1785 | | Diselenide), and Polythiocyanogen . 1803 |
| 7. | Polysulfur . . . . . . . . . . . . . . . . . . . . | 1787 | 16. | Phosphorus-Containing Polymers . . 1804 |
| 8. | Organometallic Polymers . . . . . . . . | 1787 | 17. | Traditional and Metal-Matrix |
| 9. | Sol – Gel Inorganic Polymers . . . . . | 1791 | | Composites . . . . . . . . . . . . . . . . . . . 1806 |
| 10. | Inorganic Fibers and Whiskers . . . | 1795 | | References. . . . . . . . . . . . . . . . . . . . 1808 |

## 1. Introduction

Inorganic polymers are macromolecules linked by covalent bonds, whereby there is an absence or near-absence of organic units within the backbone. Inorganic polymers include many of the most important "natural" materials as well as many synthetic materials. Natural inorganic polymers include many of the rocks about us, diamond, graphite, sulfur, boric oxide, silica, polyphosphates, quartz, and glass. Sulfur, selenium, and tellurium form macromolecules. Chalcogenide glasses are three-dimensional polymers, as are the asbestoses and a number of silicates. Most of these are industrially important, and some exhibit properties that are utilized in high-technology applications. Thus, arsenic sulfide, a chalcogenide glass, is used as an infrared-transparent window, and modified zeolites as selective chelating agents and catalysts.

Many scientific societies, including the National Research Council [1], have emphasized that the major hindrance to progress in most technological areas is the lack of suitable materials. The search for new materials in the 1970s and 1980s generally focused on rearranging, mixing, blending, and copolymerizing well-known organic polymers. Although great success has been achieved, many new applications require materials whose performance greatly exceeds that of currently available materials. Such performance requirements can be achieved only by materials containing other elements in addition to those found in the usual carbon-based polymers. For example, although the first polysiloxane, polydimethylsiloxane, was prepared in 1924 by KIPPING [2], the initial products were infusible and insoluble. In fact, these properties were so unfavorable that polysiloxanes were not developed until more than five decades later. Today, polysiloxanes offer combinations of chemical, mechanical, and electrical properties not common to any other commercially available class of materials.

Major applications of inorganic polymers include their use as bulk and specialty building materials, biologically inert and biologically active materials, high-strength materials, catalysts, and speciality coatings. The breadth of the field in relation to more traditional polymer chemistry is exemplified by comparing the number and variety of elements typically found in commercial organic polymers (C, H, N, Cl, F, O) with the number and variety of readily available inorganic materials. The relative abundance of elements in the earth's crust also attests to the importance of noncarbon-based materials. In fact, the most abundant natural building materials are based on silicon – oxygen polymers, including the majority of rocks, soil, and sand.

The alternative of utilizing heteroatom backbones or noncarbon-based polymers is attractive because of the great variety of available reactants. Thus, a wide range of compounds can be exploited to obtain materials with specific properties. Furthermore, these materials are of interest because the resulting bonds may have bond energies greater than that of the C–C single bond in alkanes (ca. 343 kJ/mol). Approximate bond energies for other elements, in kilojoules per mole, are B–N 389, Si–N 439, Si–O 795, P–N 615, and Be–O 448. Bonding in these materials is a combination of ionic (nondirectional) and covalent (directional) components; the ionic character increases as the electronegativity difference between adjacent atoms increases. Furthermore, the tendency to form multiple bonds is high for many of the elements involved. The thermal stability of many inorganic and organometallic materials has been described [3].

As in inorganic chemistry there is a shift from purely inorganic compounds such as sodium chloride to metal- and nonmetal-containing compounds that contain organic moieties. This shift continues. Another trend is the movement to more diversified areas including ceramics, optical and thermal applications, composites, communications, and high-strength materials. This movement is fuelled by the unique properties offered by the presence of metal and inorganic atoms. This shift will continue. Probably the most rapidly growing area in materials is the emergence of carbon nanotubes.

Inorganic and organometallic polymers represent a rapidly growing field of chemical research and already have many applications [4–27]. Any division between inorganic and organometallic polymers is necessarily somewhat arbitrary. The classification used here is convenient but far from perfect. The reader is therefore referred to several other articles which review the scope of organometallic polymers and discuss many other systems which could also be regarded as inorganic polymers [8, 10–14, 42].

The variety of structures continues to increase and includes, along with hybrids, geometrically differing structures including ladders, cages, dendrimers, and hyperbranched and network structures.

## 2. Polysilanes, Polygermanes, and Polystannanes

Carbon has an unique ability to form strong single, double, and triple bonds to itself. Because of the relatively high bond energies (C–C single bonds are of the order of 340 kJ/mol) compounds containing carbon in their backbone do not readily cleave. By comparison, fellow members of group 14 have only recently been synthesized because of the low bond energies (generally on the order of 85 – 125 kJ/mol). Polysilanes may have been prepared by KIPPING in 1924 [2]. Only more recently have soluble, well-characterized polysilanes been synthesized [30–32]. (Topics related to polysilanes have been reviewed [33–41].)

***Production.*** Polysilanes have been synthesized by four methods:

1. Reductive coupling
2. Dehydrogenative coupling
3. "Masked" dienes
4. Anionic ring opening

Attempts to prepare polysilanes in the late 1940s and 1950s gave insoluble, intractable materials; the major route employed was the reaction of dichlorosilanes with sodium (Fig. 1). For example, polydimethylsilane [28883-63-8] was investigated at Union Carbide but found to be insoluble [43]. The key developments leading to soluble polysilanes were the introduction of large substituent groups to decrease order, the use of two different substituents, and the copolymerization of different monomers (Fig. 1) [44, 45]. This resulted in high molecular mass (>400 000), tractable, soluble polymers, which could be purified easily, cast into films, and molded.

Improved methods for preparing high molecular mass polysilanes are being sought,

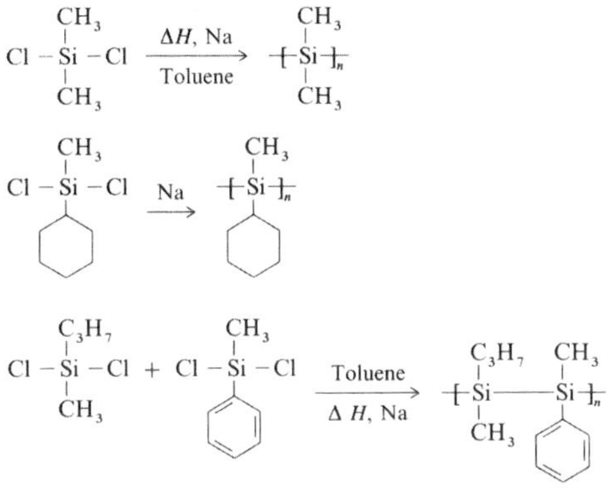

**Figure 1.** Synthesis of polysilanes

because the reaction of dichlorosilanes with sodium frequently gives low yields of high molecular mass polymers and because bimodal molecular mass distributions are obtained [43, 44, 46–50]. Carrying out the coupling reaction at low temperature by using ultrasonication gives monomodal polymers with low polydispersity [50].

Alternating copolymers are formed from the coupling of dichlorodisilanes with dilithium salts of 1,2-diethynyldisilanes [51] or dilithio polythiophenes [51, 52]. These products exhibit σ – π conjugation. Some of these materials also undergo solid-state transitions to liquid-crystalline mesophases [51].

Water-soluble products are formed by coupling silanes with oligomeric poly(ethylene glycol) side chains [53].

Dehydrogenative coupling of dialkyl and monoalkyl silanes in the presence of transition metal catalysts such as titanocene and zirconocene dialkyls has given polysilylenes [51–63]. Functionalized polysilanes have also been produced. When the side chains contain amine functional groups, the products can be made water-soluble by formation of an amine salt network [64–66]. Dendrimeric [67–69] products can also be formed by dehydrogenative coupling. These products exhibit σ conjugation in many directions and show good conductivity. They have been used as negative-type photoresists because exposure to the air or light gives insoluble products.

Polygermanes and polystannanes have also been prepared by similar routes [70–73]. Many of these materials show better σ conductivity than the corresponding polysilanes and have been used as light-emitting diodes [74]. Reviews are available on polysilane synthesis by Wurtz-type reductive coupling [75], dehydrocoupling of hydrosilanes to form polysilanes [76], and formation of silicon-containing dendrimers [77] and ladders [78].

*Properties and Uses.* Much of the interest in the polysilanes, polygermanes, and polystannanes involves their σ delocalization and their σ – π delocalization when coupled with arenes or acetylenes. This is not unexpected since silicon exists as a covalent network similar to diamond. In exhibiting electrical conductivity, germanium and tin show more typical "metallic" bonding. Some polystannanes have been referred to as "molecular metals" [79]. Conductivity is increased by doping [80], illumination [81], and application of an electric field. (reviews: [82–85]). The use of polysilane–metal complexes for organic semiconductors has been reviewed [86], as have organogermanium polymers [87] and organotin polymers [88].

Because of a number of interesting electronic and physical properties exhibited by polysilanes, a number of potential uses have been suggested and/or shown to exist, including precursors of β-SiC fibers [89–97], impregnation of ceramics [97, 98], polymerization initiators [99, 100], photoconductors for electrophotography [101–104], nonlinear optical materials [105–113], mid-UV photolithography bilayer materials [114–117], contrast enhancement layers in photolithography [118], deep UV-sensitive photoresists [116, 119–126], and self-developing by excimer laser or deep UV exposure [125–127]. The unusual absorption spectra of polysilanes have indicated potential uses in a number of conducting areas. The copolymer of polysilastyrene, when doped with arsenic pentafluoride, becomes a semiconductor. Conducting films have been reported [128].

Polysilanes display UV absorption due to silicon – silicon bonds of the backbone. For example, poly(cyclohexylmethylsilane) exhibits an absorption band at 326 nm [45]. This absorption is not found in saturated carbon chains. Polysilanes with phenyl side chains show strong absorption near 330 nm, resulting from interaction between the phenyl groups and the silicon backbone, which acts as a σ → σ* or σ → π* chromophore.

Polysilanes have a high sensitivity to degradation by UV light. After light is absorbed, chain cleavage occurs with measured quantum yields of 0.20 – 0.97. This has led IBM to examine polysilanes as resists for bilayer UV lithography [129–131]. They serve as excellent reactive-ion etching barriers for bilevel resist applications because a protective layer of silicon dioxide is formed during exposure of the polymers to oxygen plasmas [130, 131].

The wavelength at which maximum absorption occurs ($\lambda_{max}$) depends on the polysilane chain length. During irradiation nonlinear bleaching (decrease in UV absorption) occurs, because the polymer undergoes chain cleavage to give shorter fragments of lower absorptivity. As the irradiation continues, the polysilanes become increasingly transparent. This unique bleaching latency is useful in contrast-enhanced lithography [130, 131].

The $\lambda_{max}$ for UV absorption also depends on the conformation of the polysilane backbone [132, 133]. Polysilanes having a planar zigzag backbone conformation absorb at much longer wavelengths than those in which the backbone is disordered or helical. The solid materials are strongly thermochromic [132]. For example, the $\lambda_{max}$ of poly(di-n-hexylsilane) in solution or immediately after baking at 100 °C is 317 nm, but after 3 h at 21 °C films exhibit $\lambda_{max}$ = 371 nm.

Polysilanes are excellent photocatalysts for making high molecular mass vinyl polymers. On irradiation, chain cleavage occurs to give free radicals [134], which initiate vinyl polymerization and curing. The 3M Company has commercialized this process. Polysilane initiators are relatively insensitive to termination of polymerization by oxygen.

$$\text{+SiR}_2\text{+}_n \xrightarrow{h\nu} 2 \text{ }\sim\sim\text{SiR}_2\cdot + \text{R}_2\text{Si:}$$

$$\sim\sim\text{SiR}_2\cdot + \text{CH}_2\text{=CHX} \longrightarrow \sim\sim\text{SiR}_2\text{CH}_2\text{-}\overset{\cdot}{\text{C}}\text{HX}$$

R = alkyl, aryl, vinyl

Pioneering work by YAJIMA, [135] showed that poly(dimethylsilane) can be converted into β-SiC fibers with very high tensile strength (350 kg/mm$^2$) by a series of pyrolysis steps. Thermolysis of poly(dimethylsilane) at 470 °C gives a poly(carbosilane) which is fractionated and then melt-spun into fibers at 350 °C. The resulting fibers are oxidized at the surface in air at 190 °C (cross-linking occurs) to provide rigidity, then pyrolyzed at 1200 °C under nitrogen to give crystalline β-SiC (see also → Fibers, 13. Refractory Fibers).

$$\text{+Si(CH}_3\text{)}_2\text{+}_n \xrightarrow{470°C} \text{+SiCH}_2\text{+}_n(\text{H, CH}_3) \xrightarrow{\text{Melt-spun to fibers}}$$

$$\xrightarrow[\text{2) N}_2\text{, 1200°C}]{\text{1) Air, 190°C}} \beta\text{-SiC} + \text{H}_2 + \text{CH}_4$$

Objects have been manufactured from silicon carbide by Shin Nisso Kako using poly(silastyrene):

$$\text{+Si(H)(C}_6\text{H}_5\text{)-CH}_2\text{+}_n$$

These materials contain 20 – 30 vol% of microvoids, which lower the ultimate strength slightly but also inhibit crack growth, so that they can be machined to exact dimensions.

Delocalization in the Si–Si σ-bond framework makes polysilanes potential electrical conductors. Although neutral polysilanes are insulators, they become semiconductors when doped with AsF$_6$ or SbF$_5$ [46]. Conductivities up to 0.5 $\Omega^{-1}$ cm$^{-1}$ have been measured. Polysilanes are excellent photoconductors operating through hole migration with high hole mobilities ($10^{14}$ cm$^2$ V$^{-1}$ s$^{-1}$ at ambient temperature) [136]. This has possible applications in electrophotography and communications technology. The discovery that poly(methylphenylsilane) has nonlinear optical properties suggests eventual applications in laser technology [137].

A variety of linear, two- and three-dimensional polysilanes are under investigation for use as specialty caulks, adhesives, oils, sealants, viscosity regulators [46, 138, 139], biologically active materials [139, 140], as well as in electrophotography [141] and reprography [142]. Although these applications are largely in the research stage, the unique all-silicon backbone provides an alternative to carbon and siloxane polymers.

There is continued interest in the electronic properties and applications of polysilanes. Another aim is the synthesis of and studies on optically active polysilanes [143–145]. Thus, a series of enantiopure polysilanes were synthesized. It is believed that each of the polymers adopts helical conformations in solution with the same helical screwlike geometry. Thin films of **1** showed a second-order transition at about $-47\,°C$ ($T_g$) and a first-order transition at about $-8\,°C$ attributed to a helical-coil-associated transition. This helical-coil solid-state transition is reversible and may allow their use as chiroptical switches.

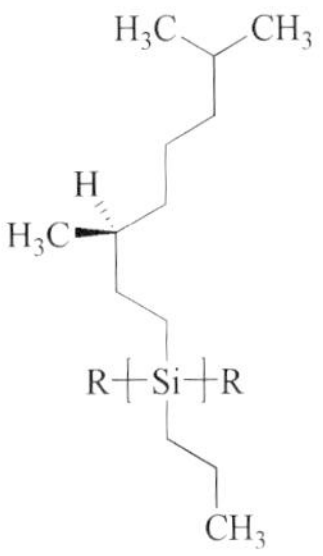

**1**

Similar behaviors were found for other optically pure polysilanes. This may allow their use as switches and memory devices with rewritable and write-once read-many (WORM) CDs. The temperature at which the helical structure switches is sensitive to molecular weight and may allow the introduction of mixtures of polysilanes with multiple structural "switches" available to them. Spin-coated and chemically grafted films of optically active polysilanes, such as **2**, on quartz showed helical coil shifting from optically active coils to optically inactive rigid-rod structures.

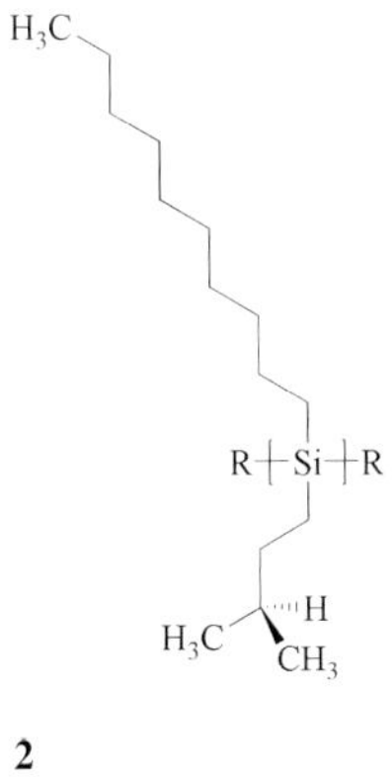

**2**

Similar results were found for optically active polygermanes [147]. The topic of photoprogrammable molecular hybrid materials for write-as-needed optical materials has been recently reviewed [148].

A polysilane copolymer containing a benzo-15-crown-5 ether, **3**, was synthesized. It showed high sensitivity and selectivity for the fluoride ion in nanomolar concentrations [149–152].

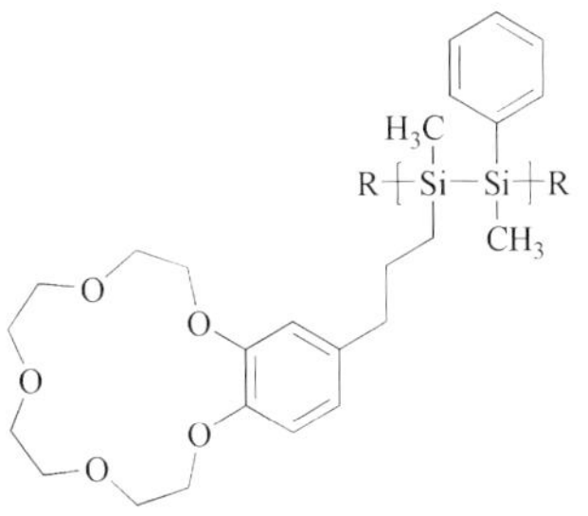

**3**

## 3. Poly(Carborane – Siloxanes), Polycarbosilanes and Polysilazanes

Silicon forms bonds with itself to give polysilanes, but it can also bond to other elements. Here several of these families are considered. (For polysiloxanes, see → Silicones).

*Poly(Carborane – Siloxanes).* The poly(carborane – siloxanes) have a linear structure **4** in which $R^1$ and $R^2$ can be alkyl, fluoroalkyl, or aryl groups. The main chain contains carborane polyhedra. Although the polyhedral carborane $C_2B_{10}H_{10}$ is most commonly employed in the preparation of poly(carborane –siloxanes), the *closo*-carboranes $C_2B_5H_7$ and $C_2B_{10}H_{12}$ have also been used as precursors [153, 154]. Major interest in this class of polymers results from the need for enhanced flame resistance and their high thermal and oxidative stability.

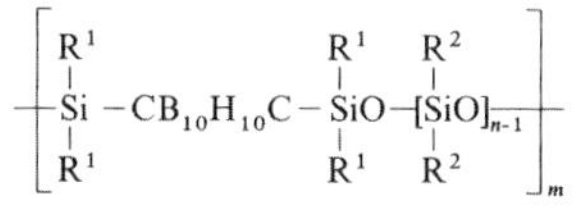

**4**

*Production.* The hydrolysis – condensation of carborane siloxane monomers containing terminal Si—Cl bonds is one synthetic route to the poly(carborane – siloxanes) [155]. Cohydrolysis – condensation leads to other compositions of **4**

$$Cl+SiO\underset{R^2}{\overset{R^2}{|}}\underset{n-1}{\overset{R^1}{|}}\underset{R^1}{\overset{R^1}{|}}SiCB_{10}H_{10}CSi+OSi\underset{R^2}{\overset{R^2}{|}}\underset{n-1}{\overset{}{}}Cl \xrightarrow{H_2O}$$

$$\left[\begin{array}{c}\underset{R^1}{\overset{R^1}{|}}\\+Si-CB_{10}H_{10}C-\underset{R^1}{\overset{R^1}{|}}SiO-[SiO]_{n-1}\\\underset{R^1}{\overset{}{}}\underset{R^2}{\overset{R^2}{}}\end{array}\right]_m$$

$$Cl-\underset{R^2}{\overset{R^2}{|}}Si-O-\underset{R^1}{\overset{R^1}{|}}SiCB_{10}H_{10}C\underset{R^1}{\overset{R^1}{|}}Si-O-\underset{R^2}{\overset{R^2}{|}}Si-Cl$$

$$+Cl-\underset{R^2}{\overset{R^2}{|}}Si+OSi\underset{R^2}{\overset{R^2}{|}}_{n-4}Cl \xrightarrow{H_2O}$$

$$\left[\begin{array}{c}\underset{R^1}{\overset{R^1}{|}}\\+Si-CB_{10}H_{10}C-\underset{R^1}{\overset{R^1}{|}}SiO-[SiO]_{n-1}\\\underset{R^1}{\overset{}{}}\underset{R^2}{\overset{R^2}{}}\end{array}\right]_m$$

Another route involves reacting a dihalosilane with a carborane-containing dihydroxydisilane [156]. Synthetic routes to the required monomers have been reported [157, 158]. More recently, poly(carborane–siloxane) alternating and block copolymers have been synthesized by employing a two-step polycondensation process. Initially, a trisiloxane is condensed with a dilithiocarborane [159, 160]. The product is then treated with dilithiodiacetylene to produce low molecular weight (up to 9000 Da) products. Hydrosilylation of the Si-H end group allowed the synthesis of carbohydrate-functionalized silicones [161].

***Properties and Uses.*** For polymers of series **4**, where R' = R = CH$_3$ and $n = 1 - 5$, the glass transition temperature ($T_g$) decreases with increasing siloxane content from $-42$ to $-88$ °C as $n$ increases from 1 to 5 [162]. Replacing R$^2$ by phenyl groups gives a completely amorphous polymer, but $T_g$ increases to $-12$ °C. When R$^2$ is a mixture of methyl and phenyl groups, $T_g$ is lower. Trifluoropropyl-modified polymers are amorphous and exhibit $T_g$ values that decrease with increasing $n$.

The thermal stability of poly(carborane – siloxanes) is excellent. In an inert atmosphere, rapid weight loss occurs only upon heating above 400 °C [162]. Phenyl-substituted polymers are more stable than their methyl analogues [163]. The amorphous polymers are elastomers and are formulated with fillers and vulcanized by using standard silicon technology. The high-temperature capabilities have resulted in the application of poly(carborane – siloxanes) as stationary phases in gas chromatography and the fabrication of these polymers into O-rings, gaskets, and wire coatings for operating temperatures >300 °C [153, 156].

***Polycarbosilanes*** have generally been made as precursors for silicon carbide. Recently, polysilaethylene and related functionalized products have been prepared [164–166]. Polysilaethylene has a low $T_g$ (about $- 140$ °C) and melts at about room temperature (ca. 25 °C). Ease of substitution of the chloro derivative by alkoxide, amine, and other functional groups may lead to a number of interesting products with derivative chemistry like that of polyphosphazenes. Fluorinated polysilaethylenes have also been made [167, 168]. It is suggested that these materials might exhibit piezo and pyroelectric properties similar to those of poly(vinyl fluoride).

***Polysilazenes and Poly(N-methylsilazenes).*** Most research focuses on the use of polysilazenes as precursors to ceramics [169–172]. These products have not proved to be very useful since the ceramic yield is low (about 50 % at 1000 °C) and the free carbon content is high (about 30 % at 1000 °C).

They are mainly synthesized by aminolysis and ammonolysis of dichlorosilanes [173, 174], deamination or redistribution reactions of aminosilazanes [175, 176], and more recently by the anionic and cationic ring-opening polymerization of cyclosilazanes. The latter products are soluble in typical organic liquids. The Si–N bond energy is greater than the Si–O bond energy. The temperature of initial weight loss for the products formed from the ring-opening reaction is about 50 °C higher than that of polydimethylsiloxane. Future applications will focus on areas similar to those of polysiloxanes themselves.

## 4. Polyphosphazenes

Polyphosphazenes are a well developed versatile class of inorganic polymers pioneered by Allcock and co-workers with over 2000 different polymers formed in the past four decades [177–186].

Polyphosphazenes (**5**) are polymers containing alternating nitrogen and phosphorus atoms in the backbone and alternating single and double bonds [187]. Phosphorus is pentavalent and has two pendant substituents (X and Y) which can be alkyl, aryl, alkoxy, aryloxy, arylamino, alkylamino, halogen, or pseudohalogen. Variation of X and Y leads to a wide variety of materials with highly diverse properties and applications [187–191].

$$\begin{array}{c} Y \\ | \\ \text{\textemdash}\!\!\text{\textemdash}\text{P}\!=\!\text{N}\text{\textemdash}\!\!\text{\textemdash}_n \\ | \\ X \end{array}$$

**5**

***Production.*** Polyphosphazenes can be prepared in several ways [187–191]. Reaction of ammonium chloride with phosphorus pentachloride yields a mixture of cyclic and linear oligomeric phosphazenes with a degree of polymerization (DP) of up to 20:

$$PCl_5 + NH_4Cl \longrightarrow \begin{array}{c} Cl \\ | \\ \text{\textemdash}\!\!\text{\textemdash}P\!=\!N\text{\textemdash}\!\!\text{\textemdash}_n \\ | \\ Cl \end{array} + HCl + \text{Cyclic products}$$

The ring opening polymerization (ROP) of hexachlorocyclotriphosphazene[*940-71-6*], [*16422-79-0*] (**6**) leads to a rubbery material, but when a carefully purified starting material is polymerized under controlled conditions, a degree of polymerization of up to 15 000 can be obtained and the polymer can be purified readily. The formation of cross-linked materials is a problem in this synthetic approach. A mechanistic study of the ROP of hexachlorocyclotriphosphazene in a titanium reactor using diglyme as solvent allowed the synthesis of linear polymer with little or no cross-linking [192].

The substitution of the chloro groups continues to serve as a much-used approach to obtaining other polyphosphazenes.

**6**

Poly(dichlorophosphazene) [*26085-02-9*] degrades slowly in a moist environment. However, the chlorine atoms are readily (and completely) displaced by many different nucleophiles. This reaction leads to a variety of highly stable polymers [187, 190, 191, 193–195].

$$\begin{array}{c} Cl \\ | \\ \text{\textemdash}\!\text{P}\!=\!\text{N}\!\text{\textemdash}_n \\ | \\ Cl \end{array} \xrightarrow{\text{NaOR}} \begin{array}{c} OR \\ | \\ \text{\textemdash}\!\text{P}\!=\!\text{N}\!\text{\textemdash}_n \\ | \\ OR \end{array} + NaCl$$

$$\xrightarrow{R_2NH} \begin{array}{c} NR_2 \\ | \\ \text{\textemdash}\!\text{P}\!=\!\text{N}\!\text{\textemdash}_n \\ | \\ NR_2 \end{array}$$

The condensation polymerization of appropriate *N*-silylphosphoramines has resulted in the successful preparation of the first polyphosphazene to have only alkyl substituents on phosphorus [196]:

$$Me_3SiN\!=\!\begin{array}{c} CH_3 \\ | \\ P\!-\!X \\ | \\ CH_3 \end{array} \longrightarrow Me_3SiX + \begin{array}{c} CH_3 \\ | \\ \text{\textemdash}\!\text{P}\!=\!\text{N}\!\text{\textemdash}_n \\ | \\ CH_3 \end{array}$$

**4**

Recently, polyphosphazenes have been prepared by room-temperature cationic (living) polymerization of phosphoranimines [197–199]. Certain initiators allow the formation of multiple initiating sites and the formation of star-branched polyphosphazenes with a narrow polydispersity. Block structures with polymers such as poly(ethylene oxide) have been made that contain reactive amine end groups [200]. Copolymers with polyurethanes have also been produced [201].

A number of polythionylphosphazenes have been produced with a variety of aryloxy, alkoxy, and amino appendages [202].

$$\left(\begin{array}{ccc} O & OR & OR \\ \| & | & | \\ S\!=\!N\!-\!P\!=\!N\!-\!P\!=\!N \\ | & | & | \\ OR & OR & OR \end{array}\right)_n$$

Polyphosphazenes have also been coupled with organosilicon compounds [203].

***Properties.*** The poly(organophosphazenes), unlike their precursors the poly(dichlorophosphazenes), are resistant to hydrolysis and can be

Table 1. Glass transition temperatures ($T_g$) and crystalline melting points ($T_m$) for some polyphosphazenes, $[P(Y)_2 = N]_n$

| Substituent (Y) | $T_g$, °C | $T_m$, °C |
|---|---|---|
| Cl | −63 | 30 |
| $OCH_3$ | −76 | |
| $OCH_2CF_3$ | −66 | 242 |
| $OC_6H_5$ | 6 | 390 |
| $N(H)C_6H_5$ | 105 | |

vulcanized to form elastomers. Copolymers with trifluoroethoxy and heptafluorobutoxy (1:1) groups are elastomeric with $T_g = -77\,°C$ [193], whereas $T_g$ values for poly(arylaminophosphazenes) are much higher (see Table 1). Poly-(aryloxyphosphazenes) tend to be intermediate between these two classes.

Polyphosphazenes exhibit some of the lowest known rotational barriers for skeletal backbone bonds in polymers (as low as 420 J/mol of repeating units), which is consistent with their low $T_g$ values. Two first-order transitions are usually found, with a temperature interval of ca. 150 – 200 °C [187, 190, 191]. The nature of the lower temperature transition resembles the change into a mesomorphic state similar to that observed in nematic liquid crystals. In practice, polyphosphazenes are usually soft just above this lower transition temperature, which allows compression molding of films to be carried out. Table 1 lists representative data.

The thermal stability of some polyphosphazenes is outstanding. Several classes exhibit onset of weight loss above 300 °C, as shown by thermogravimetric analysis (TGA), but these studies overestimate the stability. Poly(diphenoxyphosphazene) [28212-48-8], for example, undergoes rapid depolymerization above 150 °C. Fluoroalkoxyphosphazene vulcanizates have excellent resistance to solvents (e.g., lubricants and fuels) and display low volume swells after long immersion in lubricants [204]. Polyphosphazenes also display good flame-retardant properties, with oxygen index values in the range 20 – 70, and they give off only moderate amounts of smoke [205, 206].

*Uses.* The favorable properties of polyphosphazenes, as well as the high elastomer elongations, which are greater than those of polysiloxanes from −60 to 200 °C, have resulted in the use of various polyphosphazenes as gaskets, specialty damping materials, and petroleum piping for arctic applications. They have also been used as specialty membrane materials [195]. Polyphosphazene – salt complexes, which have the highest ionic conductivities at ambient temperature of all polymer – salt electrolytes, are reported to be semiconducting materials [195]. Other uses include catalytic templates [195], biomedical materials [205], and their most prominent use as specialty elastomers in O-rings, gaskets, and fuel hoses [187, 206–210].

The first commercial polyphosphazene elastomers were synthesized and subsequently developed by the Firestone Tire and Rubber Co. [211] and represent the first new class of semi-inorganic elastomers to be developed commercially since silicone rubber. Ethyl Corporation holds the rights to many commercial uses and produces both commercial polymers and finished products from this interesting class of materials [212].

Thin films of amino-substituted polythionyl phosphazenes exhibit an unusual permeability to molecular oxygen [213]. When a phosphorescent dye that is quenched by oxygen is dispersed in the polymer film, the product can be used as a visually interpretable oxygen sensor. One potential use is to follow differences in pressure across a surface area such as a wing of a plane.

Polyphosphazene materials containing a sulfonyl-based chromophore have been synthesized that show second-order nonlinear optical properties [214]. Polymer electrolytes based on polymethoxyethoxyethoxyphosphazenes (MEEP) show high room-temperature ionic conductivities with charge transport occurring in the presence of certain salts [215].

There have been several recent reports related to biological uses of polyphosphazenes [184]. Nanofibers of polyphosphazenes have been produced by electrospinning. These nanofibers promote the adhesion of bovine coronary artery endothelical cells and osteoblast-like MC3T3-E1 cells [216]. Polyphosphazenes containing carboxyl substituents have been investigated for their biological activity with possible use as microencapsulating materials [217]. Hydroxyapatite–polyphosphazene composites have been investigated as bone substitutes [218].

Polyphosphazenes have also been used as drugs themselves. Allcock and co-workers synthesized a variety of polyphosphazenes that acted as carriers for a cisplatin-like moiety. These products showed good inhibition in the ascites test and 5/7 survival after the eighth day for the P388 mouse test. Thus, these materials exhibited good inhibition of some cancer cell lines and lower toxicity. This and related work has been reviewed [218, 219].

Water-soluble polyphosphazenes are emerging as an important class of synthetic polymers with much of the effort aimed at biomedical applications [221]. Another area of continued emphasis is the construction of fuel cells. New photon-conductive fuel-cell membranes have been produced through the linkage of acidic functional groups to aryloxy side units attached to the polyphosphazene backbone [222].

A wide range of hybrid metallocene–phosphazene polymers have been synthesized, including ferrocene and ruthenocene [223].

Polyphosphazenes are considered as additive and inherently flame-retardant materials and have the potential to be more widely employed as fire-resistant materials and fireproofing agents [224].

## 5. Poly(Boron Nitride)

Boron nitride [10043-11-5] occurs in three crystalline forms — α, β, and γ (→ Boron Carbide, Boron Nitride, and Metal Borides, Chap. 2.). The α-form is a layered, hexagonal polymer much like graphite. A crystalline, tetrahedral network polymer similar to diamond also exists [225–228]. Although boron nitride is isoelectronic with the diamond and graphite forms of carbon, it exhibits different properties. For example, layered boron nitride is white and an electrical insulator. Thus its electrons appear to be highly localized in contrast to those of graphite. Like graphite, layered boron nitride is soft and easily machined. Articles are fabricated by sintering or machining. These articles can be used in air up to 800 °C or in an inert atmosphere up to 1600 °C. Boron nitride has a Young's modulus : density ratio of 20 (versus 4 for iron) and a tensile strength : density ratio of 8 (versus 1.7 for glass and 0.3 for iron). The tetrahedral form of boron nitride is almost as hard as diamond and is therefore employed as a substitute for diamond in both cutting and grinding and in jewelry. Industrial applications include its use as an abrasive, insulator, and refractory material.

The hexagonal modification of boron nitride has been prepared from combinations of cheap boron and nitrogen compounds, such as B(OH)$_3$, (NH$_2$)$_2$CO, and N$_2$, or KBH$_4$ and NH$_4$Cl [207–212, 225–231]. More recently, chemical vapor deposition (CVD) methods have used mixtures of BCl$_3$ and NH$_3$, BF$_3$ and NH$_3$, or B$_2$H$_6$ and NH$_3$ to produce hexagonal boron nitride [231–234].

Polymeric precursors to boron nitride are beginning to attract attention. The reaction of difunctionalized borazines with bis(trimethylsilyl)-amines gives soluble oils that form films which can be pyrolyzed at 1200 °C to yield a boron carbide – boron nitride ceramic, but removal of all of the carbon has not yet been achieved [229]. The use of trifunctional borazines (shown below) leads to hexagonal boron nitride in good yields, with good crystallinity via intermediates such as **7**:

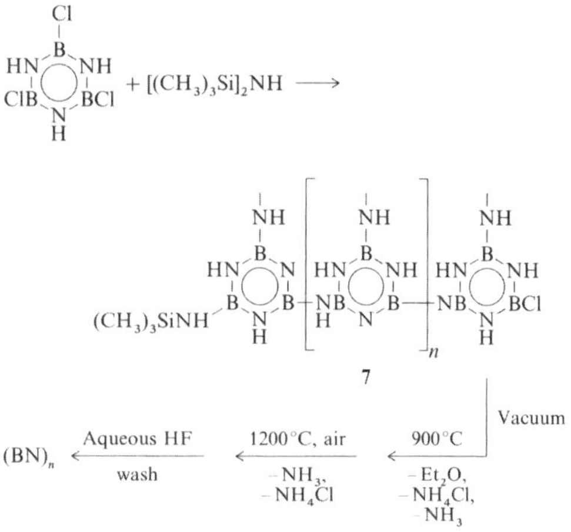

The formation of boron nitride nanocomposites has been reviewed [235].

## 6. Poly(Sulfur Nitride)

The topic of poly(sulfur nitride) has been reviewed [236].

Poly(sulfur nitride) [56422-03-8] was first prepared in 1910 by BURT, who passed vapors of cyclic tetrasulfur tetranitride [28950-34-7], S$_4$N$_4$, over heated silver gauze or silica wool

[237]. However, it was not studied extensively until the 1970s, when purer products were obtained [238]. Poly(sulfur nitride) films have a lustrous golden color and are very striking in appearance. Polymeric sulfur nitride is a crystalline, fibrous material that is malleable under mild pressure. Electron micrographs reveal that the crystals are composed of fiber layers which are stacked parallel to one another along the crystal axis; X-ray studies show that the polymer consists of almost planar chains of alternating sulfur and nitrogen atoms. All the S−N bond lengths are similar and correspond to a sulfur – nitrogen bond order intermediate between a single and a double bond. The polymer decomposes on heating at ca. 140 °C.

$$\begin{array}{c} S=N \\ \diagdown \\ S=N \end{array} \quad \begin{array}{c} S=N \\ \diagdown \\ S=N \end{array} \quad \text{or} \quad +N=S\frac{1}{1_n} \quad \longleftrightarrow \quad +N-S\frac{1}{1_n}$$

The polymer exhibits highly anisotropic electrical conductivity. Room-temperature conductivities as high as 3700 $\Omega^{-1}$ cm$^{-1}$ along the crystal axis have been measured [240, 241]. Conductivity increases markedly upon cooling to 4.2 K, and the polymer becomes superconducting at 0.26 K [242]. Poly(sulfur nitride) was the first example of a polymeric superconductor. The conductivity increases on doping with bromine, and $(SNBr_{0.4})_n$ at room temperature has a conductivity of 3.8×10$^4$ $\Omega^{-1}$ cm$^{-1}$ parallel to the fibers, but only 8 $\Omega^{-1}$ cm$^{-1}$ perpendicular to the fiber axis. This strong directionality is related to efficient conductivity along the molecular chain axis, whereas a lower degree of order in chain packing along the perpendicular axis results in lower conductivity.

The large-scale production of poly(sulfur nitride) does not appear to have attracted much attention because of its rather low thermal stability and because $S_4N_4$ is explosive.

Further, polymerization is slow (6 – 8 weeks). Efforts have aimed at developing alternate synthetic routes.

Poly(sulfur nitride) has been prepared from solution by addition of trimethylsilyl azide to $(NSCl)_3$ and other similar routes [243–245]. In these processes, a powdered product is obtained rather than the larger crystalline grains. Microcrystalline layers of poly(sulfur nitride) have been prepared on platinum electrodes by electrolysis of $[S_5N_5]^+$ [246, 247]. Additional synthetic routes have also been found [236]. Even so, the classical route appears to be the method of choice for the synthesis of pure poly(sulfur nitride).

The fine structure of poly(sulfur nitride) continues to be elucidated [236]. Currently, it is believed the polymer chains are planar and lie along the (crystallographic) $b$ axis that develops along the $a$ axis of $S_2N_2$. There are two chains in each unit cell with two SN units per chain. Disordered chains are present between the "ordered" chains. Secondary interaction between the chains is significant.

Polymer crystals are stable in air and water for several days. Stress and strain properties are like those of an anisotropic metal. The Young's modulus parallel to the axis of the chain is on the order of $10^{10}$ N/m$^{-2}$.

Poly(sulfur nitride) becomes superconductive at 0.26 K [248]. This property has been extensively studied [236]. The onset of superconductivity $T_c$ increases with increasing pressure up to about 9000 bar, whereupon it decreases, but at higher pressures $T_c$ again increases to about 3 K. It is possible that at higher pressures small amounts are converted to a −SNNS− polymer [249]. Poly(sulfur nitride) exhibits the Meissner – Ochsenfeld effect (→ Superconductors, Section 2.2.) below about 20 K. Below 4.2 K negative transverse magnetoresistance occurs in the presence of weak magnetic fields (<3 T), though a normal positive magnetoresistance is found above 77 K or in stronger fields [236].

Thin films have been produced with properties that at times are similar to and under certain conditions appears to differ from those of the crystals themselves. The reason(s) for these differences are not currently fully understood. The ability to make large crystals has been reported [250, 251].

Recent interest has focused on electrical applications because of its unusual magnetic and electrical properties. Suggested uses include light-emitting diodes, solar cells, and transistors [252, 253]. A poly(sulfur nitride) – Ga – As solar cell with an efficiency of 6.2 % has been described [254]. Blue light-emitting diodes have been made on ZnS with either gold or poly(sulfur nitride) barrier electrodes [255]. The polymer gives barriers about 0.75 eV higher with a hundredfold increase in quantum efficiency in comparison to the gold-based electrode. A power output of 440 W h/kg at 200 mA was found for a lithium cell with a poly(sulfur nitride) electrode [248]. Patents for the use of poly(sulfur nitride) have been issued for a variety of uses including the cathode in batteries [256, 257], coatings on an image recording sheet [258], etching mask [259–261], explosive initiator [262], electroconductive resins [263], multilayered wiring patterns [264], sublimable polishing grains for semiconductor devices [265], and in light-emitting diodes [266].

Extensive work has been done on poly(sulfur nitride) electrodes [236]. A ruthenium(II)-modified poly(sulfur nitride) electrode generated hydrogen continuously in aqueous sulfuric acid solution at $-0.1$ V under visible light [267]. Molybdenum(VI) electrodes using film and crystalline poly(sulfur nitride) can convert ethyne to ethene at rate $10^6$ times that achieved by chemical systems [268]. Paste electrodes of poly(sulfur nitride) exhibit metallic electrochemical behavior in nonaqueous liquids [269].

## 7. Polysulfur

The topic of sulfur-based polymers had been reviewed [270–272].

Because of its removal from industrial waste to meet environmental standards, sulfur has been stockpiled and is available in large quantities at a modest price. Although the stable form at room temperature is cyclooctasulfur, $S_8$, linear polysulfur is formed on heating. Unfortunately, the thermodynamically stable form of sulfur is the $S_8$ monomer and the polymer undergoes depolymerization after some time. Various methods have been used to retard depolymerization to the $S_8$ monomer. Removal of minute amounts of the monomer through judicious extraction with carbon disulfide retards depolymerization to some extent. Probably the most effective method involves the addition of olefins, such as limonene, cyclopentadiene, and myrcene [273]. Some of these result in polysulfur that is stable for more than five years. Such stabilized polysulfur has been incorporated into concrete and asphalt mixes to strengthen them. Concrete blocks, posts, highway pavement, and parking tire restrainers that contain stabilized polysulfur are produced.

## 8. Organometallic Polymers

The topic of organometallic polymers has been reviewed [15–28].

A variety of organometallic polymers have been synthesized for a number of applications. Examples are shown in Figure 2. The metals may be an integral part of the polymer chain (**8**) or side chain (**9**); they can be connected through carbon (**10**) or other atoms (**11**); they may be bonded by coordination (**11**), covalent (**8–10**), or p–metallocene-like bonding (**12**); or they can be part of a linear (**10–12**) or cross-linked (**9**) product. A classical coordination polymer is described by **13**, and stacked metal-chain complexes are depicted by **14**, where the circles represent organic macrocycles such as phthalocyanine.

Polymeric metal phosphinates may have single-, double-, or triple-bridged structures, such

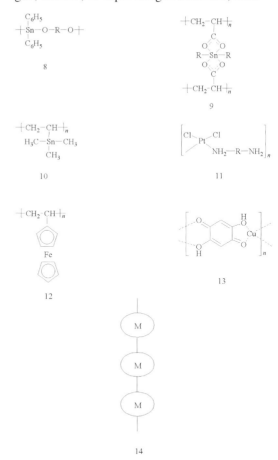

**Figure 2.** Some organometallic polymers

as **15**, **16**, and **17**, respectively. Such structures are known for aluminum, beryllium, cobalt, chromium, nickel, titanium, and tin [274]. They form films, with thermal stability up to 450 °C claimed in some instances. The chromium(III) polyphosphinates have been used as thickening agents for silicone grease to improve its high-pressure physical properties.

$$\left[ M\underset{}{\overset{O-PR_2-O}{\phantom{XXXX}}} \right]_n$$

**15**

$$\left[ \underset{O-PR_2-O}{\overset{O-PR_2-O}{M}} M \underset{O}{\overset{O}{\phantom{X}}} \right]_n$$

**16**

$$\left[ M \underset{PR_2}{\overset{PR_2}{\underset{O}{\overset{O}{\phantom{X}}}}} M \right]_n$$

**17**

The metal moiety of an organometallic polymer imparts particular properties. First, the metal atom is typically a site for reaction with Lewis bases and redox reactants. These sites have been exploited for catalysis and may be responsible for internal rearrangement and degradation. Second, the metal generally contributes to polymer stiffness through added volume occupation and π-bond formation. Thus, flexibility must be derived from the organic moiety. Third, the metal-containing portion typically contributes to the dipolar, ionic character of the products. The high chemical reactivity of the metal site results in polymers with poor thermal stability. For example, products derived from bis(cyclopentadienyl)titanium dichloride [*1271-19-8*] typically begin to undergo small weight losses between 60 and 150 °C due to the evolution of cyclopentadiene and formation of a more thermally stable material. Because of the tendency of the metal site to undergo reaction, care is needed in selecting solvents so that "permanent" bonding does not occur between the metal-containing moiety and the solvent.

The variety of structures and properties offered by organometallic polymers is enormous and cannot be summarized in a representative fashion in this short chapter. The reader is referred to reviews on organometallic polymers [11–14, 42, 275–288]. Important uses of organometallic polymers are as follows:

1. Catalysis [286, 273, 289]
2. Biological uses, general [285, 290–301]
3. Additives: coatings, textiles, plastics, and paper [285, 302, 289, 303–306]
4. Electrical conductors, semiconductors, and piezoelectric devices [307–315]
5. Electrocatalysis, photoelectrocatalysis, photovoltaic cells, specialized electrodes, sensor devices, electrochromism, electroluminescence, and multicolor displays [285, 286, 310, 316–322]
6. Analytical reagents [323–327]
7. Photoprotective agents [302, 289, 303–306]
8. Thickening agents [274]
9. Energy-transfer agents [329–331, 311, 312]
10. Permanent coloring agents [302–306]
11. Composites [332, 333]
12. Metal deposition [334–338]
13. Solar energy conversion [339, 340]
14. Ceramic precursors [341–346]
15. Superconductivity [347]
16. Functionalized electrodes (electrocatalysis, photoelectrocatalysis, photovoltaic cells) [348–351]
17. Ferromagnetic materials [352]
18. Photoconductive materials [326, 327, 351]
19. Poly(vinyl alcohol) heat stabilizers [285]
20. Electrochemical sensors [282]
21. Nonlinear optics [282]
22. One-dimensional conductivity [282]
23. Nanostructures and nanotechnology [286].

While organometallic and metal-containing polymers can be synthesized in a number of ways, the three most common routes are condensation, coordination, and addition.

***Condensation Polymers.*** Carraher and co-workers have produced a wide variety of organometallic condensation polymers based on the Lewis acid–base concept. These have been reviewed [15, 17–21, 29]. Polymers have been produced from Lewis bases containing amine, alcohol, acid, thiol, and related units including a number of drugs such as ciprofloxacin and acyclovir. Lewis acids containing such metals and metalloids as Ti, Zr, Hf, V, Nb, Si, Ge, Sn, Pb, As, Sb, Bi, Mn, Ru, P, Co, and Fe have been employed. These compounds have potential uses in the biomedical arena as antifungal, antibacterial, anticancer, anti-Parkinson's, and antiviral drugs. These polymers show promise in a wide variety of other areas including electrical, catalytic, and solar-energy conversion [17, 18]. Polymers referred to as polydyes, because of the presence of dye moieties in the polymer backbones, have been used to impregnate paper products, plastics, rubber, fibers, coatings, and caulks, giving the impregnated material color, (often) added biological resistance, and special photo properties [21]. Some of the polymers, including the metallocene-containing products, show the ability to control laser radiation [21]. Depending on the range of radiation, laser energy can be focused, allowing the material to be cut readily, or it can be dispersed, imparting to the material containing the polymer added stability towards the radiation.

A number of these polymers exhibit a phenomenon called "anomalous fiber formation," reminiscent of metallic whiskers [353].

There are more organometallic compounds containing tin than any other metal. These materials are employed to improve the heat stability of PVC when it is formed into pipe. Today, because polymers leach much slower than small molecules, organotin polymers are being used as heat stabilizers in PVC pipe [18].

The industrial emphasis on tin-containing polymers is the result of Federal laws that prohibit the use of leachable, monomeric organotin compounds in a variety of coatings and protective applications. This resulted in a move to polymeric materials that did not suffer the same leachability and which were allowed by law.

Much of the recent activity with organotin compounds, including polymers, involves their use to inhibit a wide variety of microorganisms at low concentrations. These microorganisms include a variety of cancers, bacteria, yeasts, and viruses. Some simple organotin polyethers **18** based on hydroxyl-terminated poly(ethylene glycols) inhibit a wide variety of cancers including ones associated with bone, lung, prostate, breast, and colon [354, 355].

**18**

Some of these organotin polymers are water-soluble, and this allows for medical applications utilizing the material in simple pills.

A wide variety of organotin compounds developed on the basis of known antiviral drugs such as acyclovir and known antibacterial agents such as ciprofloxacin, norfloxacin, cephalexin (**19**), and ampicillin inhibit a wide variety of viruses including those responsible for many of the common colds, chicken pox, small pox, shingles, and herpes simplex [355].

**19**

Some of the organotin-containing polymers inhibit *Candida albicans*, the yeast responsible for yeast infections in humans, better than commercially available applications while leaving the normal flora unharmed [353, 358]. Others, such as **20** inhibit methicillin-resistant *Staphylococcus aureus* (MRSA) preferentially [359, 360].

**20**

Organotin polyamines containing the plant growth hormone kinetin (**21**) increase the germination of damaged seeds and thus is may help in providing food in third-world countries [361]. It also increases the germination rate of sawgrass seed from about 0 to over 50 % and is an important agent in replacing the "sea of grass" (actually sawgrass) in the Everglades.

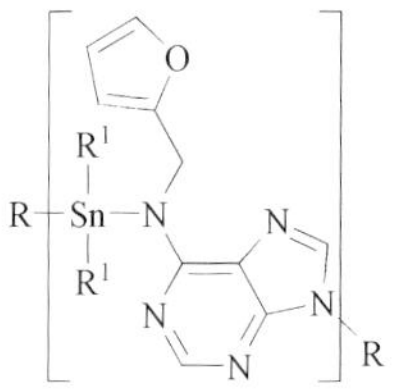

**21**

***Coordination Polymers.*** The drive for the synthesis and characterization of synthetic coordination polymers was catalyzed by work supported and conducted by the U.S. Air Force in a search for materials which exhibit high thermal stability [362]. Attempts to prepare highly stable, tractable coordination polymers were disappointing. Typically, only oligomeric products were formed and the monomeric versions were often more stable than their polymeric counterparts.

Coordination polymers can be prepared by a number of routes, with the three most common being

1. Preformed metal coordination complexes polymerized through functional groups where the actual polymer-forming step may be a condensation or addition reaction.

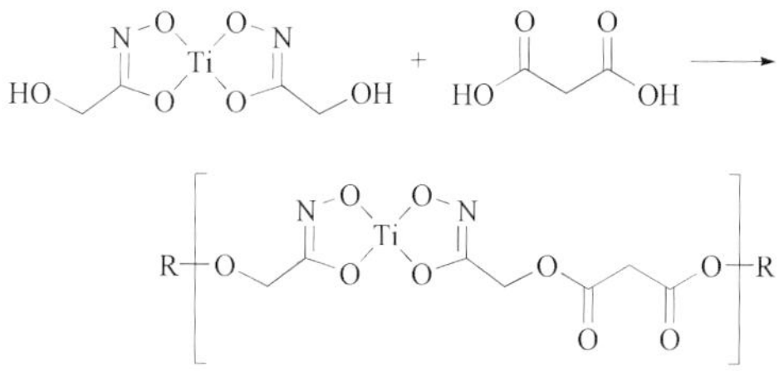

2. Reaction with polymer-containing ligands.

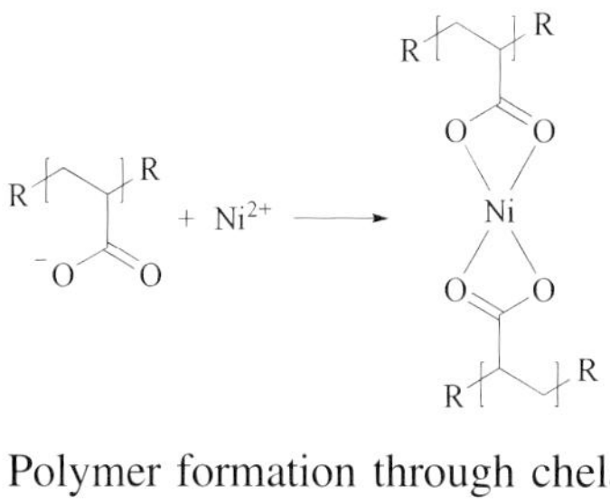

3. Polymer formation through chelation.

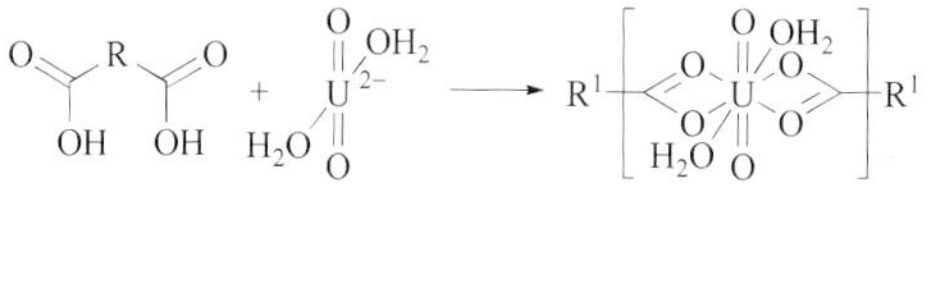

The last two processes were employed to recover the uranyl ion [19], the natural water-soluble form of uranium oxide, which acts as a heavy-metal toxin. By using salts of dicarboxylic acids and poly(acrylic acid) the uranyl ion was removed to $10^{-5}$ M and the resulting product was much less toxic and convertible to uranium oxide by heating.

Many of the organometallic polymers are semiconductors with bulk resistivities in the range of $10^3$ to $10^{10}\Omega$ cm suitable for specific semiconductor use [18]. Further, some exhibit interesting photoproperties.

In 1964, Rosenberg and co-workers found that bacteria failed to divide but continued to grow. After much effort they found that the cause of this anomalous growth was a broken electrode and eventually identified the chemical as *cis*-dichlorodiammineplatinum(II). This compound is now licensed under the name Platinol and is also known as cisplatin. Cisplatin is the most widely used anticancer drug. The toxic effects of cisplatin have been reduced though placement of the platinum moiety into various platinum coordination polymers [17]. Some of these polymers inhibit various cancer growths with much less toxic effects. Many of them are also very active antiviral agents and some are able to prevent the onset of virally related juvenile diabetes in test animals [17]. Cisplatin derivatives formed from reaction of tetrachloroplatinum(II) and methotrexate [27, 363] inhibited a wide range of viruses in the nanograms per milliliter range. Similar results were found for the analogous product formed from reaction with tilorone (**22**) [27, 364].

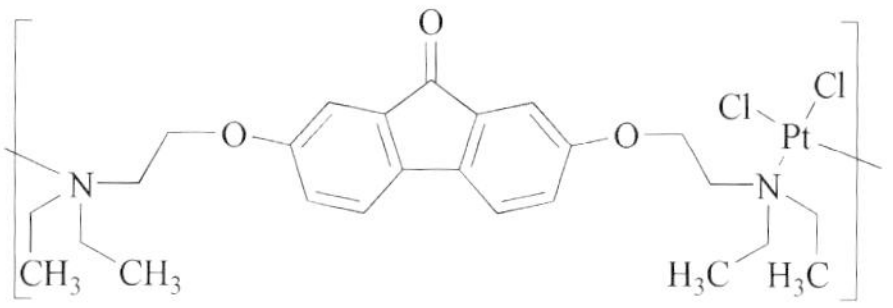

**22**

***Addition Polymers.*** Many of the addition polymers are discussed elsewhere in this article. These include sulfur nitride polymers, polyphosphanes, and polysilanes.

One area of active interest in ceramics is the formation of ceramics that may contain some fiber structure. Currently, ceramics, while very strong, are very brittle. Introduction of thermally

**Table 2.** Nonoxide ceramics produced from the pyrolysis of polymeric materials

| Polymer(s) | Resultant ceramic |
|---|---|
| Poly(phosphonitric chlorides) | PN |
| Polysilanes, polycarbosilanes | SiC |
| Polyphenylborazole | BN |
| Polytitanocarbosilanes | Si – Ti – C |
| Polysilazanes | $Si_3N_4$, Si – C – N |

stable fiberlike materials may impart the ceramics with some flexibility before cleavage. Such materials can be considered as ceramic composites in which the matrix is the ceramic portion and the fibers are the thermally stable fibers. Introduction of the fibers during the ceramic-forming step is a major obstacle that must be overcome. Carbon fibers have been investigated, as have other high-temperature materials such as polysilanes. Polysilanes are formed from a six-membered ring by extended heating at 400 °C.

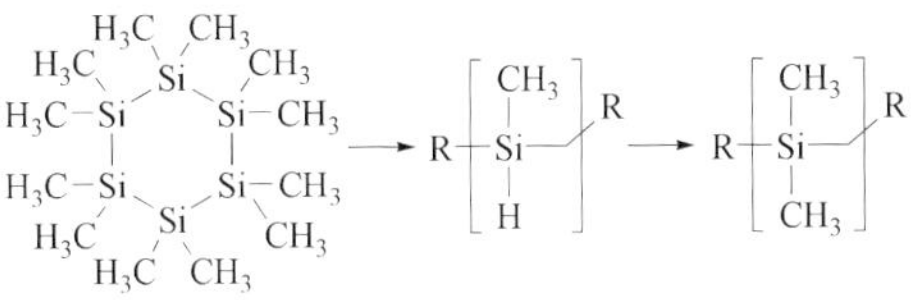

Further heating gives silicon carbide.

Table 2 lists a number of nonoxide ceramics that have been produced from the pyrolysis of polymers.

The landmark discovery of ferrocene by Kealy and Paulson in 1951 marked the beginning of modern organometallic chemistry [365]. The first organometallic addition polymer was polyvinylferrocene, synthesized by Arimoto and Haven in 1955 [366, 367], but it was about another decade until the work of Pittman [368], Hayes and George [368], and Baldwin and Johnson [370] allowed the launch of ferrocene-containing polymers.

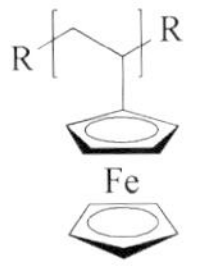

A large number of vinyl organometallic monomers have been prepared, homopolymerized, and copolymerized with classic vinyl monomers [368]. These include polymers containing Mo, W, Fe, Cr (**23**), Ir, Ru, Ti (**24**), Rh (**25**), and Co.

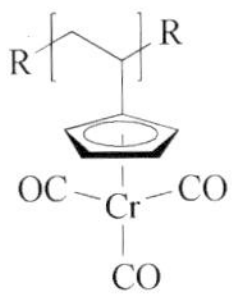

**23**

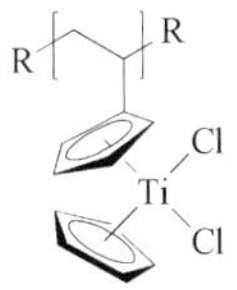

**24**

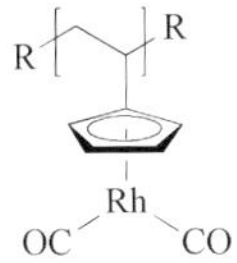

**25**

Neuse acted as an early catalyst in the development of metal-containing polymers including the use of ferrocene-containing polymers to fight cancer [371]. One key feature of ferrocene is its ability to donate an electron from a nonbonding high-energy MO resulting in the transformation of a neutral, diamagnetic site into the paramagnetic ferricenium radical cation. This occurs within a typical chemical environment and within selected biological environments.

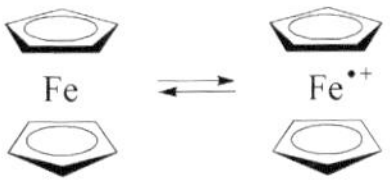

This ferricenium radical cation, as do other free radicals, readily recombines with other free radicals. Neuse and others have successfully inhibited a wide variety of cancers using ferrocene-containing polymers [372]. Often the employed compounds are elaborately designed with special backbones and a ferrocene-containing unit as a tether dangling from the polymer backbone.

## 9. Sol – Gel Inorganic Polymers

In the 1980s, several symposia were devoted to sol – gel inorganic polymer preparation, indicating the explosive growth in this field [373–375]. This field has continued explosive growth under additional "titles" such as organic/inorganic composite materials. Reviews are given in [376–384].

In sol – gel processes, ceramic polymer precursors are formed in solution at ambient temperature; shaped by casting, film formation, or fiber drawing; and then consolidated to furnish dense glasses or polycrystalline ceramics [385]. The most common sol – gel processes employ alkoxides of elements such as silicon, boron, titanium, and aluminum. In alcohol – water solution, the alkoxide groups are removed stepwise by hydrolysis under acidic or basic catalysis and replaced by hydroxyl groups, which then form $-M-O-M-$ linkages. Thus, branched polymeric chains grow and interconnect, as illustrated below for a silicate sol. Gelation eventually occurs as the growing polymers link together to form a network that spans the entire solution volume. At this point (the gel point), both the viscosity and the elastic modulus increase rapidly.

$$Si(OR)_4 + H_2O \rightleftharpoons (RO)_3Si-OH + ROH$$

$$2\,(RO)_3Si-OH \rightleftharpoons (RO)_3Si-O-Si(OR)_3 + H_2O$$

$$(RO)_3Si-O-Si(OR)_2OH \underset{-H_2O}{\overset{H_2O}{\rightleftharpoons}} (RO)_3SiOSiOSiOSiOR \text{ (with OR groups)}$$

$$\downarrow -ROH$$

$$SiO_2\ (glass\ or\ ceramic) \overset{\Delta H}{\leftarrow} \text{Chain extension and cross-linking}$$

The gel can be viewed as a viscoelastic material composed of interpenetrating liquid and solid phases. The solid network retards the escape of the liquid and prevents structural collapse. Gelation is advantageous in processing because the gel freezes in shapes. Thus, the sudden increase in viscosity locks in place the shapes formed by casting, drawing of fibers, or film formation. Some sols and gels may be oriented or modified by drawing or shearing. The gel can then be dried by evaporation to form a *xerogel* or by supercritical fluid extraction to give an *aerogel*. Consolidation to dense glasses or ceramics is finally carried out by thermal treatment and sintering. Figure 3 gives some generalized sol –gel processes for ceramic or glass preparation.

Both aerogels and xerogels have high surface area (>500 m$^2$/g) and small pore diameter (<20 nm). They have been used as ultrafiltration media, antireflective coatings, and catalyst supports. Final densification is accomplished by viscous sintering in which the viscosity exceeds $10^{12}$ Pa · s.

The rate of silicate sol and gel formation is exceptionally sensitive to pH and to the water – alcohol molar ratio in the reaction medium, as is the solubility of the amorphous silica that is formed. Silica networks are based on $[SiO_4]^{4-}$ tetrahedra modified by $[O_3Si-O^-M^+]$ units. Addition of $B_2O_3$, $Al_2O_3$, $TiO_2$, $ZrO_2$, and related network-forming units can result in the formation of mixed glasses and ceramics. The sol – gel approach uses the corresponding metal alkoxides, as illustrated below, for the formation of borosilicate glasses:

$$NaOR + B(OR)_3 + Si(OR)_4 \xrightarrow[-HOR]{H_2O}$$

$$NaOH + B(OH)_3 + Si(OH)_4 \xrightarrow[\text{via sol and gel}]{-H_2O}$$

$$(Na_2O \cdot B_2O_3 \cdot SiO_2)\cdot H_2O \xrightarrow[\Delta]{-H_2O} Na_2O \cdot B_2O_3 \cdot SiO_2$$

The chemistry of the sol – gel process can thus be tailored to design new polymeric glasses or ceramics. For example, two or more metal alkoxides can be mixed in varying ratios to form mixed sol – gel polymers. Mixing $Zr(OEt)_4$ with $Si(OEt)_4$ could lead to a gel with structure **26**. Similarly, a zirconia – alumina – silicate gel polymer of structure **27** could be obtained from $Zr(OR)_4$, $Al(OR)_3$, and $Si(OR)_4$.

$$C_2H_5O-Si(OC_2H_5)-O-Si(O)(OC_2H_5)-O-Zr(OC_2H_5)(O)-O-Si(OC_2H_5)(O)-O-Zr(OC_2H_5)_3$$
with additional $C_2H_5O-Si(OC_2H_5)-O$ branches

**26**

$$\sim Si(OR)(OR)-Zr(OR)(OR)-Al(O)(OR)-Si(OR)(OR)-Si(O)(OR)-Al(O\sim)$$
with $\sim O-Si(OR)_2$ branch

**27**

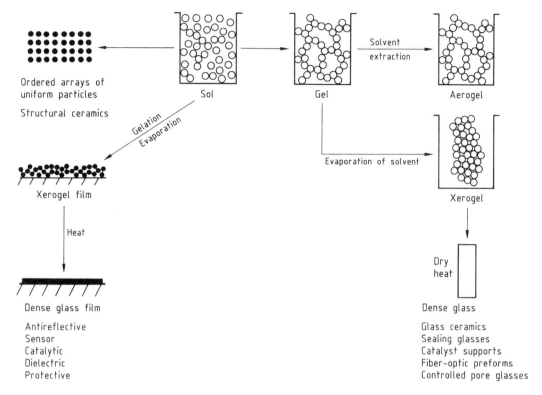

**Figure 3.** Schematic of the sol – gel process

The development of organically modified silicates (ceramers) has resulted in a variety of new materials [386]. Thus, vinyl groups, epoxides, and methacrylate functions have been polymerized to give interpenetrating networks with a wide variety of structures and properties [387–389]:

Applications of ceramers include adhesives for glass surfaces [387], protective coatings for medieval stained glass [390], and scratch-resistant coatings for plastic eyeglass lenses.

Sol – gel preparations of tetraethoxysilane – water – alcohol – HCl can be spun into fibers once the appropriate viscosity has been reached. These fibers are only slightly weaker than silica-glass fibers; $ZrO_2$–$SiO_2$ fibers have also been made in this way. Sol – gel processing of Ti($i$-$C_3H_7O)_4$ – $H_2O$ – ethanol – HCl gives thin film coatings of controlled thickness which exhibit interference colors that vary with film thickness and they are n-type semiconductors [391].

Hybrid materials have been made by incorporating end-capped poly(tetramethylene oxide) (PTMO) blocks of molecular mass 650 – 2000 into tetraethoxysilane sol – gel glasses [392].

$$Si(OC_2H_5)_4 + HO+CH_2CH_2CH_2CH_2O+_n H \longrightarrow$$
$$PTMO$$
$$(C_2H_5O)_3Si-O+PTMO+Si(OC_2H_5)_3 \xrightarrow{H^+, H_2O}$$
$$(HO)_3Si-O+PTMO+Si(OH)_3 \xrightarrow{Si(OH)_4}$$

These materials exhibit high extensibility. They contain interdispersed organic polymer and inorganic polymer regions. Although the PTMO blocks are well dispersed, some degree of local phase separation occurs. Titanium isopropoxide has been used to incorporate $TiO_2$ into these block systems [392].

Thus, although sol–gel chemistry has been extensively investigated since the early 1960s [394, 395], only now is it rapidly becoming important in material synthesis.

*Uses.* Sol–gels offer properties between those of organic polymers and inorganic materials. They are particularly useful in areas where they can replace materials to give better properties, better processability, and/or lower cost. Thin-film applications are promising, and the production of numerous nonlinear optical materials has been reported [376]. Thin-film protective coatings are being considered in a number of areas. Hot-melt adhesives for glass containers have been developed [377]. Hard contact lens materials have also been developed [377]. Another area of practical application is the reinforcement of plastics and elastomers [377]. Recent production of sol–gels with nanoscale pores allows their potential use as catalysts, porous supports, and as selective absorbents [377].

*Aerogels.* The topic of aerogels has reviewed [396, 397] (see also → Aerogels). Aerogels are highly porous materials in which the pore sizes are truly on a molecular level, less than 50 nm in diameter. This gives a material with the highest known internal surface area per unit weight, over 1000 $m^2/g$.

Porous materials can be either open-pored such as a common sponge, or closed-pored such as the bubble-wrap packaging. Aerogels are open-pored materials such that unbonded material can move from one pore to another.

While in the gel state the pre-aerogel has some flexibility, but solid aerogels behave as a fragile glass. It may be very strong in comparison to its weight, but it is very light. Aerogels are more durable when under compression. Compression can be simple such as sealing the aerogel sample in a typical food sealer packing. Aerogels are best cut using a diamond-coated saw similar to that used to slice rocks.

When handled, aerogel samples initially appear to exhibit some flexibility but then burst into millions of pieces. As force is applied, there is little bond flexing, so that the applied kinetic energy results in the collapse of the network with the force of impact spread over a large part of the aerogel and over some time because of the time required to transfer this energy from one cell to another within the aerogel matrix. Because the aerogel is open-pored, gas contained within the solid is forced outwards as collapse occurs. The frictional forces caused by the gas passing through a restricted opening are indirectly proportional to the square of the pore diameter. Because the pore sizes are so small, the rapidly moving gas also absorbs a lot of the energy. Thus, energy is absorbed by the aerogel through both collapse of the solid network structure and release of the gas within the aerogel.

Aerogels that are about 2–5 nm in diameter have large surface to volume ratios on the order of $10^9$ $m^{-1}$ and high specific surface areas approaching 1000 $m^2/g$. Such large surface to volume ratios makes the surface particularly active with potential in catalysts, absorbents, and catalyst substrates.

The precise chemical makeup of the surface depends on the materials used to make the aerogel and method of processing. Typical aerogel sequences produce products whose surfaces are rich in hydroxyl groups. Because of the high surface area, SiOH groups act as weak acids and are reactive in typical Lewis acid–base reactions. The many hydrogen-bonding hydroxyl groups at the surface make aerogels extremely hygroscopic. Dry aerogel materials will increase their weight by 20 % through uptake of moisture from the air. This absorption is reversible and appears to have little or no effect on the aerogel. Water is removed by heating to 100–120 °C.

While adsorption of water vapor has little effect on aerogels, contact with liquid water has devastating effects. When water enters the nanometer-sized pores, the surface tension of the water exerts capillary forces sufficient to fracture the silica backbone, resulting is a collapse of the complex matrix structure. This tendency to be attacked by water is overcome through conversion of the surface polar OH groups to nonpolar OR groups, where R is typically a trimethylsilyl group, though any aliphatic group would work. Conversion can be accomplished within the wet stage (pre-aerogel) or after supercritical drying. These treatments result a "hydrophobic" aerogel which is stable in water.

The pore size of aerogels varies. IUPAC classifies materials with pore sizes of less that 2 nm as "microporous", 2 – 50 nm as "mesoporous", and greater than 50 nm in diameter as "macroporous". While aerogels have some pores that fall within the micropore range, the majority of pores are in the mesopore region.

The good visible-light transmission and good insulting power make aerogel materials of interest in window manufacturing. The visible transmission spectra of light shows little absorption in the range of about 300–2700 nm, and the good visible-light "window" makes aerogels attractive for day-lighting applications. Aerogels provide about 40 times more insulation than fiberglass. While aerogels may eventually be used as the entire window component, for the present time they act as the material sandwiched between two pains of clear plastic or glass. Thermoglass is generally simply glass sheets that are separated by a vacuum. The seals on such thermoglass often spring small leaks causing diminished insulation properties. Aerogel inner cores will not suffer from this problem. Currently, about 40 – 50 % of a house's heating bill literally goes out the window because of heat lost through windows. A single one-inch thick glass pane of aerogel offers the insulation equivalent to over 30 windowpanes of R-20 insulation rated glass.

Another commercial area that is being considered is the use of aerogels as nanocomposite materials. In one approach, material is added to the silica sol before gelation. The material can be inorganic, organic, polymeric, bulk fibers, woven cloths, etc. The additional material must be able to withstand the subsequent processing steps including carbon dioxide drying. The added material must be present in a somewhat homogeneous manner throughout the system. Gentle agitation appears to be sufficient to give a product with decent homogeneity. Aerogels may be good materials for optical sensors. They have good visible transparency, high surface area, good temperature and chemical stabilities, and facile transport of gases through their pores.

## 10. Inorganic Fibers and Whiskers

Inorganic fibers have been important materials for over 50 years. Asbestos, glass, and carbon fibers are particularly well known. Many of the materials described herein are discussed in more detail under → Fibers, 11. Inorganic Fibers, Survey.

Many inorganic fibers exhibit very high tensile strength and thermal stability (>1000 °C) but suffer from high density and, high cost. This chapter discusses the preparation and properties of some inorganic fibers which are of interest in the production of high-performance materials, particularly advanced resin-matrix, ceramic-matrix, and metallic-matrix composites. The patent literature on inorganic fibers and composites is extensive and growing rapidly [392].

*Asbestos,* once widely used as insulation, now has more limited applicability because of its health hazard. Small fibers lodge in the lungs and can cause lung cancer and other lung damage. Asbestos-containing shingles, paper, welding rods, floor tiles, and brake linings are, however, still manufactured. These products avoid fibers of length 5 – 20 µm, which are the most dangerous. *Glass fibers* are used widely as insulation and in fiber-reinforced plastics. *Carbon fibers* are of increasing importance in high-performance composites. Several metallocene-type condensation polymers (e.g., **28**) exhibit fiber formation directly upon polymerization or upon application of stress [399].

$$\left[\begin{array}{c}\bigcirc\\\mathrm{M}-\mathrm{O}-\mathrm{R}\\\bigcirc\end{array}\right]_n$$

**28**

Inorganic fibers composed of alumina, silica, titania, zirconia, aluminum silicate, boron, boron carbide, boron nitride, silicon boride, silicon carbide, silicon nitride and others are known. Polymeric *alumina, titania,* and *zirconia fibers* are produced industrially [400–403]. Zirconia textiles are produced by Union Carbide under the trade name Zircar.

*Alumina fibers* with a high $Al_2O_3$ content have high temperature stabilities and high moduli.

They have been made by dispersing aluminum salts on rayon fiber, heating to burn off the organic material, and sintering. Alternatively, slurries of hydrated alumina can be extruded, dried, and heated to give polycrystalline filaments [404, 405]. Fibers may also be extruded and drawn from a viscous organic solution, followed by heating to high temperature [406]. Polyaluminoxanes are spun into fibers and sintered to give $Al_2O_3$ fibers, or mixed with silicon compounds, spun, and calcined at 1000 – 1200 °C to give aluminosilicate fibers.

High-purity *silicon dioxide fibers* can be used up to 1100 °C and are of interest for advanced composites [407]. They are made by melting a glass composed of 75 % $SiO_2$ and 25 % $Na_2O$, and then spinning and stretching the fiber simultaneously at 1100 °C. The fibers are then acid-leached to remove $Na_2O$ and dried above 300 °C [408]. In another process, silicon alkoxide polymers are spun by a sol – gel process to make high-purity $SiO_2$ fibers. The fibers are heated to 1000 °C to give a 99.999 % pure quartz fiber which is stable at 1000 °C [409].

A variety of *metal oxide fibers* are now available. Du Pont has developed an alumina – zirconia fiber with a tensile strength >2100 MPa and a modulus of 380 GPa for use in high-performance composites [410]. Even after exposure to 1400 °C for 100 h, its tensile strength was 1400 MPa and its modulus remained unchanged. Stabilized zirconia and alumina – silica – boria fibers have also been produced [408]. The patent literature reports fibers of such materials as alumina – chromia, thoria (thorium oxide), titania, titania – silica, thoria – silicate, zirconia – alumina, zirconium silicate, and zirconia – yttria, as well as whiskers of these and many other metal oxides [411].

*Boron – tungsten fibers* have densities of 2.4 – 2.6 g/cm$^3$ and elastic moduli as high as 390 GPa. However, they lose both tensile strength and modulus at high temperature. When coated with silicon carbide and boron carbide, they retain about half their initial strength in air up to 800 – 1000 °C. *Boron fibers* are much less dense and are, therefore, useful in the construction of lightweight composites [411] found in tennis racket frames and aircraft. However, their cost is high because of production complexities. Processes involve chemical vapor deposition of boron onto carbon fiber and other cores, followed by a variety of treatments [412]. *Boron carbide fibers* are made from mixtures of methane and boron trichloride by using CVD to create a boron carbide layer on tungsten core filaments [392]. *Boron nitride fibers* are used with boron nitride matrices to give composites that are employed as electrical insulators which are also good thermal conductors. Boron nitride fiber mats are useful as electric cell separators in lithium sulfide batteries [392]. The fibers are made from boric oxide fibers that are nitrided by ammonia at ca. 800 °C followed by stretching at 2000 °C. They have a tensile strength of 2100 MPa and a modulus of 345 GPa.

*Silicon carbide* can withstand 1800 °C under oxidizing conditions. Two methods of production are employed. In the first, CVD is used to deposit silicon carbide on tungsten filaments. In the second method, polysilanes are melt-spun to give a precursor fiber, which on heat treatment yields fibers 10 – 30 μm in diameter (see Chap. 2), [135, 413, 414]. Silicon carbide fibers are more resistant to oxidation at high temperature than carbon fibers and have greater compressive strength and electrical resistance. *Silicon carbide – carbon* fibers, can be spun from mixtures of pitch and a poly(dimethylsilane), followed by vacuum heating at 1400 °C [415]. In a similar fashion, melt-spinning of a polycarbosilazane followed by high-temperature treatment gives *silicon carbide – nitride* fibers with a tensile strength of 1.2 MPa and a modulus of 187 MPa, which resist oxidation up to 1200 °C [416].

*Silicon nitride fibers* are used in advanced composites for aircraft, radomes, and electrical components. They are prepared either by reduction of silica in the presence of nitrogen in an electric furnace [408] or by nitriding polycarbosilane precursor fibers with ammonia [418].

Sol – gel *polytitanocarbosilanes* have been prepared and used to melt-spin a precursor fiber which, on pyrolysis in nitrogen at 800 – 1500 °C, gives Si – C – Ti – O fibers. At 1300 °C, a tensile strength of 3.0 GPa and a modulus of 220 GPa were achieved [419]. Polycarbosilanes can be treated with titanium tetrabutoxide to give a polymer from which fibers may be spun. After heat treatment, Si –

C – Ti – C fibers with a Young's modulus of 75 – 150 GPa and tensile strength of 0.1 – 11 GPa were obtained [420].

More recently, a number of inorganic materials have also been produced as whiskers (see also → Whiskers). In general, the tensile strengths of whiskers are greater than those of fibers. For instance the approximate tensile strength of graphite in the bulk form is 1000 MPa, for fibers it is 2800 MPa, and for whiskers it is 15 000 MPa. Carbon whiskers are produced by depositing a layer of pyrocarbon from the vapor phase on a catalytically grown carbon filament [421].

Inorganic fibers and whiskers exhibit some of the highest tensile strengths recorded. They are useful when light weight and high strength are needed. Uses include turbine blades, heat-resistant re-entry vessels, golf club shafts, automotive radio aerials, fishing poles, dental fillings, and parts of aircraft, including stealth aircraft [422]. Many of these inorganic fibers and whiskers are used as reinforcing agents in composites [422].

## 11. Biomedical and Antimicrobial Polymers

Parts of this topic are covered in Chapter 8.

***Silicones.*** Silicones (polysiloxanes) are polymers having a silicon – oxygen backbone. Among the numerous materials having biomedical applications, silicon polymers play an important and versatile role. Medical devices made from polysiloxanes include artificial wrist, toe, and finger joints; hip implants; oviductal plugs [423, 424]; brain membranes; mammary implants; tracheotomy vents; and artificial hearts. For a detailed description of polysiloxanes, see → Silicones [412, 426–428]. This chapter deals with their biomedical applications [416]. Polysiloxanes are widely applied in medicine for numerous reasons including their generally extremely low toxicity, high gas permeability, good mechanical properties, and the wide range of fabrication processes that can be used (e.g., coating, encapsulation, casting, molding, sealing, and extrusion). Furthermore, they are generally inert to the body and to the immune system. The biomedical applications of polysiloxanes are reviewed in [290]. Monomeric and oligomeric siloxanes are generally not sold as medical-grade materials and may contain toxic trace contaminants that are not important in industrial applications but may be critical in biomedical applications. Proper procedures must be adopted to ensure quality control and absence of contaminants that may be toxic and may adversely affect long-term, repeated use of the end product.

Polysiloxanes are extensively employed in biomedical devices as specialty surface treatments [416, 429, 430]. For example, octadecylpolysiloxane (Siliclad, Glassclad 18) is applied to hydrophilic surfaces to reduce blood – surface interactions and protein adsorption [430, 420]. Other polysilane-derived surface treatments are sold under tradenames such as Glassclad HP, D.C. 5700, Glassclad 6C, Dri-Film, and Glassclad IM [290, 297]. Silicone membranes play critical roles in oxygenation and dialysis. Polysiloxanes are about ten times more permeable to oxygen than low-density polyethylene and a hundred times more permeable than nylon or butyl rubber [431]. Therefore, poly(dimethylsiloxane) [1646-73-7] soft lenses and poly(dimethylsiloxane-*co*-methacrylate) hard lenses constitute a major portion of the contact lens market. Silicone-based interpenetrating polymer networks give clean, resilient gels with cohesive properties, which are used in breast implants [432–434, 290]. Bilayer membranes employing a top layer of cured silicone polymer are used as artificial skin [442, 443].

Low-consistency, room-temperature vulcanizing silicones cure to give low modulus mechanical plugs when injected into the fallopian tubes [423, 424]. This blocks the ovum from reaching the uterus and serves to prevent conception. A retrieval ring at the uteral end makes the procedure reversible. Silicone polymers are used to make dental impression materials [436–439]. Vinyl addition cured systems are used in biomedical encapsulation and mold making [440] and in maxillofacial prostheses. High-consistency elastomers are used in compression molding of shunts, flexing joints, and catheter tubing [441]. Silicone – urethanes (e.g., cardiothane 51) are used in blood pumps, intraaortic balloons, and artificial hearts because of their good fatigue strength, flexibility, toughness, and retarded interaction with plasma proteins [442].

Silicone – polycarbonate block copolymers can be injection-molded, extruded, and cast into films. Their synthesis is described in [443, 423, 424]. They have high tear strength as well as high oxygen and water permeability. Applications include blood oxygenation, dialysis, and microelectrode materials [444–446].

***Metal-Containing Polymers.*** A few metal-containing polymers are used as therapeutic agents in the treatment of cancer [447–451], juvenile diabetes [452–454], Cooley's anemia [455, 456], and in the control of bacteria [451, 457–459] and viruses [451–454]. The polymer derived from the reaction of tetrachloroplatinate and methotrexate can inhibit the virus that causes juvenile diabetes-like symptoms [452, 453].

Tin-containing compounds are inhibitory to certain bacteria. A variety of methyl-, ethyl-, and propyl-substituted organotin polymers effectively inhibit *Pseudomonas aeruginosa* strains which are responsible for about one-half of the deaths of burn patients. Some suggested uses are as additives to paints, insulation, bandages, and clothing [461]. Degradation resistance has been obtained by in situ polymerization of tin-containing esters in wood [458, 459]. Impregnated poles and planks show total stability when exposed to seaside environment or seawater for more than five years [462, 463]; untreated wood is destroyed within a year. Similar results have been obtained for tin-containing cross-linked coatings in the control of ship hull fouling. Nevertheless, plants and sea creatures survive within centimeters of test objects, suggesting that the use of tin-containing polymers might be environmentally acceptable. Tin-containing polymers have been examined in the United States and Japan as antifouling coatings [464–467, 462, 463] and mildew-resistant paints [298]. Commercial products are available; however, environmental concerns in harbor areas remain.

## 12. Boron Polymers

The topic of boron polymers has been reviewed [299, 22, 468].

Organoboron has been incorporated into polymers by a variety of techniques. In the 1920s STOCK first created a boron hydride polymer during his work on boron hydrides. Much of the current interest in boron-containing polymers is a consequence of three factors. First, the presence of a low-lying (low-energy) vacant p orbital allows its use in moving electrons in a conjugated system. This is being taken advantage of through the synthesis of various π-conjugated systems and their use in optical and sensing applications. This includes use in light-emitting diodes (LEDs), nonlinear optical systems, energy storage in batteries, and the construction of sensing devices. One such polymer structure is **29**.

**29**

These polymers are mainly synthesized employing the hydroboration reaction, that is, the addition of a boron hydride to a double or triple bond.

$$HC\equiv CH + \underset{R}{\overset{R}{\diagdown}}BH \longrightarrow \underset{R'}{\overset{H_2C=}{\diagdown}}B-R$$

This low-lying vacant p orbital also allows boron-containing polymers to be luminescent with potential optical applications. Many of these luminescent materials are NLO materials.

The second reason for interest in boron polymers involves their use as catalysts. While the boron atom can be used as the site of catalytic activity, more effort has involved the use of boron-containing materials as blocking and protecting agents and as cocatalysts. They are increasingly being used in catalytic asymmetric syntheses.

The third reason for interest in boron polymers involves the ability of many boron compounds to form cocoons about objects, which provides a ready method for coating wires and fibers [469]. Thus boron-containing polymers and monomers have been employed to form a surface layer of intumescent protective char that acts as a barrier to oxygen, protecting the wires and fibers from ready oxidation. These coatings also provide flame retardancy to the coated materials. Boron-containing units have been incorporated into polymers to give materials that have added flame resistance through char formation. Somewhat related to this is the use of boron-containing polymers in forming high-strength fibers and whiskers for use in composites. Finally, boron has a high capture cross section for neutrons so effort has gone into using this nuclear characteristic.

Today there exist a wide variety of boron-containing polymers including ring systems such as borazines, boroxines, and triphosphatoborins

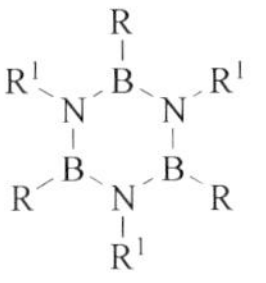

**Borazine**

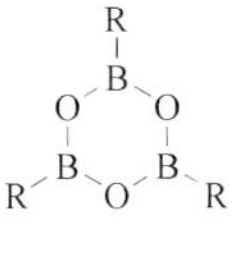

**Boroxine**

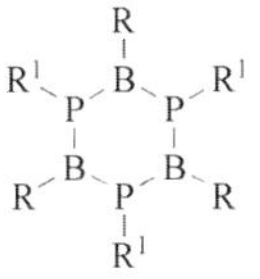

**Triphosphatoborin**

as well as metal-, metalloid-, and ferrocene-containing polymers, each offering their own potential for exhibiting desired properties.

A variety of boron-containing polymers have been made by heating trialkyl phosphates with boron trichloride at 300 °C [300, 470]. At 900 °C this reaction produces boron phosphate glass. The boron polymers form networks prior to glass formation.

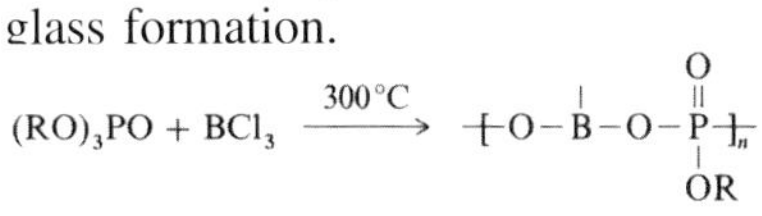

$$(RO)_3PO + BCl_3 \xrightarrow{300\,°C} {+}O-B-O-P{+}_n$$

Reaction of benzene boronic acid with diphenylsilane gives transparent, film-forming, polymeric borosiloxanes which can be drawn into fibers [471]. Similar materials resulting from boronic esters and chlorosilanes form electrically insulating films when baked on copper plates [472]. A variety of polymers containing Si−O−B units have been reported [473]:

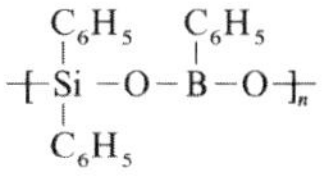

Boric acid forms condensation polymers with polyhydroxy compounds. With ethylene glycol and phenol, a tacky polymer is formed that is hydrolyzed readily by water [474]. Many such compositions have been studied as adhesives, binders, coatings, and resins.

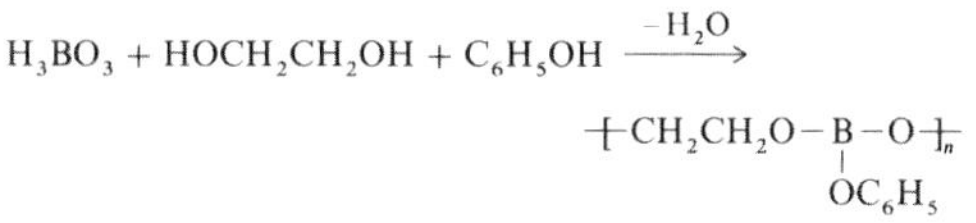

$$H_3BO_3 + HOCH_2CH_2OH + C_6H_5OH \xrightarrow{-H_2O}$$

Polymers with all-boron backbones [475] or alternating boron – phosphorous backbones [301] have been reported. Pyrolysis of dimethylphosphineborane in the presence of a base gives linear high molecular mass polymers [476]. In many cases, ring systems form instead of linear polymers, with cyclic trimers and tetramers predominating. Use of triethylamine as a chain end blocking group favors the formation of linear polymers.

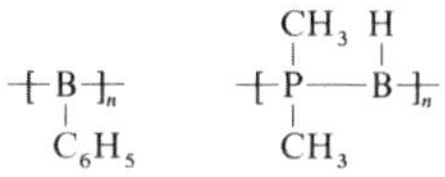

The action of water on diboron tetrachloride gives subboric acid, $B_2(OH)_4$, which when heated at 220 °C loses water to yield a white boron monoxide polymer [477]. This remains unchanged up to 500 °C, but at 650 °C is transformed into brown boron monoxide which has lower solubility in water.

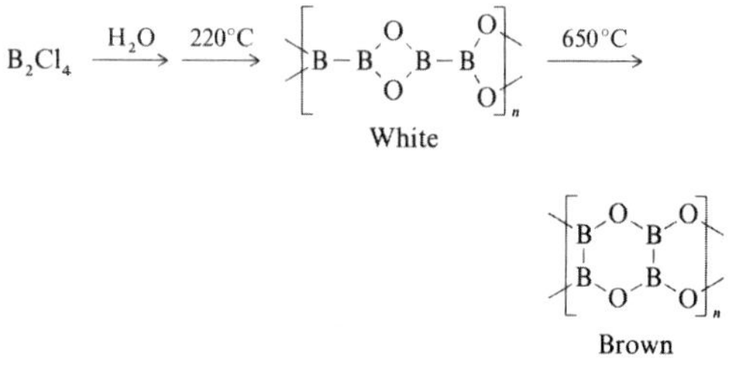

Low molecular mass polymers result from the reaction of 1,1′-dilithioferrocene with phenylboron dichloride, followed by treatment with water [478]:

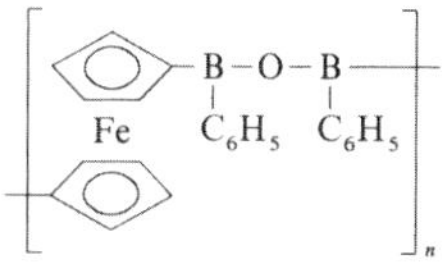

$$\left[ \text{Fe}(C_5H_4)_2 - B(C_6H_5) - O - B(C_6H_5) - \right]_n$$

A large number of polyborazines have been prepared, and many are reviewed in [479]. The condensation of bis(alkylthio)borazines with diphenylsilicondiol gives borazine – siloxane polymers [480]. Cross-linked materials result from the corresponding reactions with trialkylthioborazines:

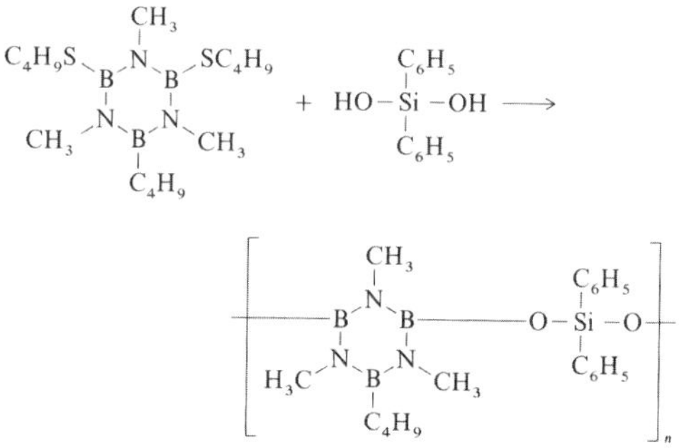

$$C_4H_9S-B(CH_3)-N(CH_3)-B(N)-B(SC_4H_9) + HO-Si(C_6H_5)_2-OH \longrightarrow$$

$$\left[ -B-N(CH_3)-B(C_4H_9)-N-B(CH_3)- O-Si(C_6H_5)_2-O- \right]_n$$

Reaction of β-chloroborazines with sodium sulfide, sodium disulfide ($Na_2S_2$), or sodium tetrasulfide ($Na_2S_4$) leads to polyborazines with sulfur linkages between the rings and molecular masses up to 53 000 [481]. Other polyborazines with oxygen [482], nitrogen [483], and phosphite bridges [484] between rings have also been prepared.

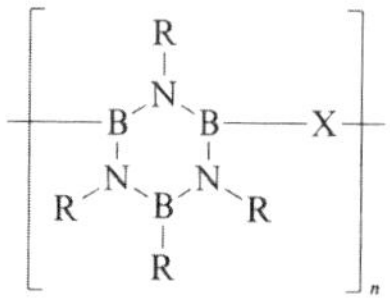

$$\left[ \begin{array}{c} R \\ N \\ B \diagdown B - X \\ R-N \diagup N-R \\ B \\ R \end{array} \right]_n$$

$$X = -(S)_x-, -O-, -RN-, -O-\underset{CH_3}{\overset{O}{\underset{\|}{P}}}-O-$$

Carborane polymers with a variety of structures have been made from the $C_2B_{10}H_{10}$ icosahedral cage systems, in which the carbon atoms are located ortho, meta, or para to one another. The best known examples are the Dexsil polymers of the Olin Corporation [485, 486]. This series has siloxane groups attached to the carbon atoms of *ortho*-carborane (1,2-$C_2B_{10}H_{12}$) and varying numbers of dimethylsiloxane groups between adjacent cages:

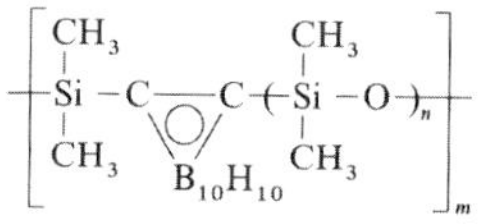

$$\left[ -Si(CH_3)_2-C(B_{10}H_{10})C-Si(CH_3)_2-O-\right]_m$$

$n = 2$, Dexsil 200
$n = 3$, Dexsil 300
$n = 4$, Dexsil 500

Many other groups have been used to connect carborane cages. Treatment of dilithiocarborane with phosphorus trichloride and sodium azide leads to the very unusual structure [487]:

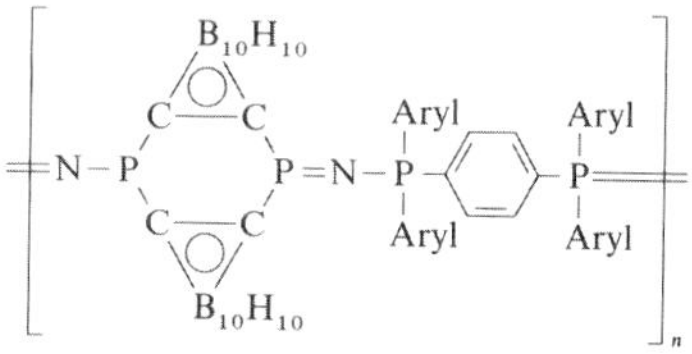

$$\left[ =N-P\underset{C(B_{10}H_{10})C}{\overset{C(B_{10}H_{10})C}{\diagup\diagdown}}P=N-P(Aryl)_2-Aryl-P(Aryl)_2= \right]_n$$

Dilithiocarboranes are reacted with dialkyl metal dihalides to introduce metal functions between the cages. For example, reaction of *ortho*- or *para*-dilithiocarborane with dimethyldichlorostannane introduces dimethyltin bridges between the cages. Dimethylgermanium bridges are introduced in a similar manner [488, 489].

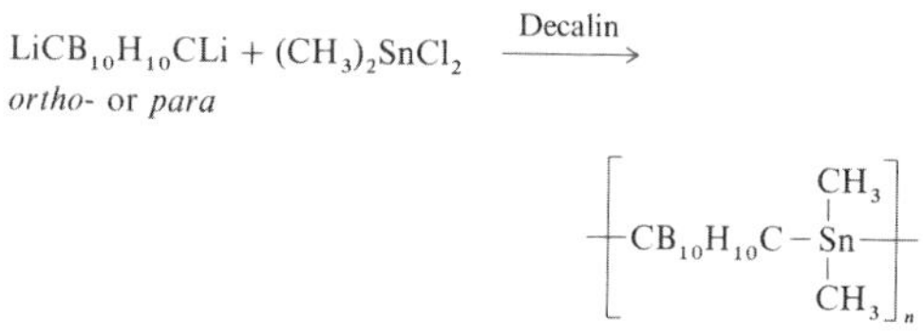

$$\text{LiCB}_{10}\text{H}_{10}\text{CLi} + (\text{CH}_3)_2\text{SnCl}_2 \xrightarrow{\text{Decalin}}$$
*ortho*- or *para*

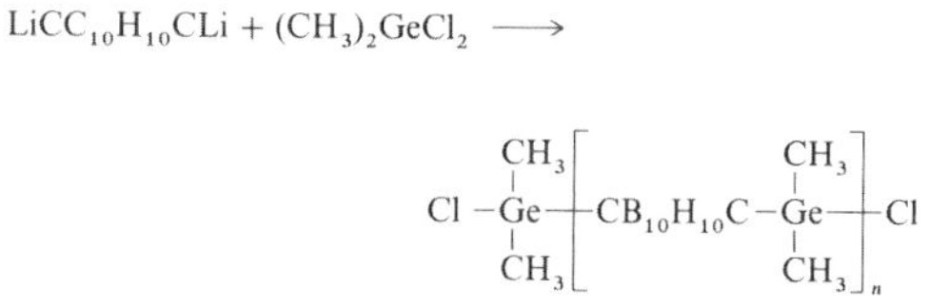

$$\left[ -CB_{10}H_{10}C-Sn(CH_3)_2- \right]_n$$

$$\text{LiCC}_{10}\text{H}_{10}\text{CLi} + (\text{CH}_3)_2\text{GeCl}_2 \longrightarrow$$

$$Cl-Ge(CH_3)_2-\left[ -CB_{10}H_{10}C-Ge(CH_3)_2- \right]_n-Cl$$

The dimethyltin-bridged *para*-carborane polymer softens only above 420 °C and decomposes at 425 °C, whereas its ortho analogue melts at 250 – 255 °C. Random mixtures of *meta*-carborane (1,7-$C_2B_{10}H_{12}$) and *para*-carborane (1,12-$C_2B_{10}H_{12}$) have been used to prepare dialkyltin-bridged polymers.

A polymer bridged by $-Ge(CH_3)_2NH-$ units was prepared from dilithiocarborane by using excess dimethyldichlorogermane and then ammonia:

$$LiCC_{10}H_{10}CLi + (CH_3)_2GeCl_2 \longrightarrow$$

$$Cl-\underset{\underset{CH_3}{|}}{\overset{\overset{CH_3}{|}}{Ge}}-CB_{10}H_{10}C-\underset{\underset{CH_3}{|}}{\overset{\overset{CH_3}{|}}{Ge}}-Cl \xrightarrow{NH_3}$$

$$\left[\underset{\underset{CH_3}{|}}{\overset{\overset{CH_3}{|}}{Ge}}-CB_{10}H_{10}C-\underset{\underset{CH_3}{|}}{\overset{\overset{CH_3}{|}}{Ge}}-\overset{H}{N}\right]_n$$

Borazine bridges have also been introduced [490]:

$$\left[\begin{array}{c}\text{Aryl}\\|\\\text{N}\\B\diagup\quad\diagdown B-CB_{10}H_{10}C-\\\text{Aryl}\diagup N\diagdown B\diagup N\diagdown\text{Aryl}\\|\\CH_3\end{array}\right]_n$$

Refluxing the bis(sulfinyl chloride) of *meta*-carborane in water gives a poly(carboranethio-sulfinate), whereas treatment with dilithiocarborane gives the corresponding poly(carboranesulfide) (*mp* 219 – 222 °C) [491]:

$$ClS-CB_{10}H_{10}C-SCl \xrightarrow[\text{Reflux}]{H_2O} \left[S-CB_{10}H_{10}C-\overset{\overset{O}{\|}}{S}\right]_n$$

$$LiCB_{10}H_{10}CLi \longrightarrow \left[CB_{10}H_{10}C-S\right]_n$$

The functionalization of polymers with Lewis acidic boron sites has allowed the synthesis of many materials, some of which are being studied as catalysts, luminescent materials, and for the design of sensor systems for nucleophiles [492].

## 13. Aluminum Polymers

Polymers containing only aluminum atoms in the backbone are unknown. Aluminum – oxygen and aluminum – nitrogen backbones are known, but they are hydrolyzed readily by aqueous acid and alkali. Polymers with an aluminum – oxygen backbone are usually obtained by reaction of aluminum alkoxides with water, diols, organic acids, or amides. They occur as gums to brittle solids; applications include use as cross-linking agents, drying agents, gelling agents, fuel additives, water repellents, paints, varnishes, linoleum, and inks [493–497]. An idealized linear structure is

$$\left[\begin{array}{c}Al-O\\|\\X\end{array}\right]_n$$

where X can be RO– [498], RCOO– [499], $R_3SiO-$ [498, 500], or R(RO)P(O)O– [501]. These polymers are highly cross-linked, and the aluminum is actually four-coordinate; cyclic structures often predominate.

To increase the hydrolytic stability of Al–O bonds and permit the synthesis of linear polymers, four-coordinate aluminum ethyl acetoacetate complexes have been made as chelate rings along the chain [502]. The use of diols gives hydrolytically unstable polymers [503].

$$\left[\overset{O\diagup Al\diagdown O}{\underset{H_3C\diagup\diagdown OC_2H_5}{}}\right]_n \quad \left[\overset{O\diagup Al-O-(CH_2)_4-O}{\underset{R\diagup\diagdown R}{}}\right]_n$$

A variety of coordination polymers involving aluminum chelates have been made, but these are considered true organometallic polymers and are not discussed here. ANDRIANOV pioneered the use of $R_3SiO-$ blocking groups to prevent formation of three-dimensional networks. Although oligomers such as

$$R_3SiO\left[\begin{array}{c}Al-O-\overset{\overset{O}{\|}}{P}-O\\|\quad\quad|\\OSiR_3\quad OSiR_3\end{array}\right]_n SiR_3$$

have been prepared [500], useful higher polymers remain elusive.

Poly(aluminosiloxanes) are polymers containing an Si–O–Al–O backbone. A typical example results from the reaction of the sodium salt of a poly(dimethylsiloxane) with aluminum chloride.

$$\text{NaO}\left[\begin{array}{c}\text{CH}_3\\|\\\text{Si}-\text{O}\\|\\\text{CH}_3\end{array}\right]_n\text{Na} + \text{AlCl}_3 \longrightarrow$$

$$\left[\begin{array}{c}\text{CH}_3\\|\\\text{Si}-\text{O}\\|\\\text{CH}_3\end{array}\right]_n\left[\begin{array}{c}\text{O}\\|\\\text{Al}-\text{O}\end{array}\right]_m + \text{NaCl}$$

Polymers with silicon – aluminum ratios of 0.8 – 23 have been made [504, 505]. Polymers with low silicon – aluminum ratios are brittle and insoluble, indicating a three-dimensional structure; those with silicon – aluminum ratios of 7 – 23 are soluble and stable in both polar and nonpolar solvents even at 150 °C.

The poly(aluminophenylsiloxane) having a silicon – aluminum ratio of 4 is completely infusible but very soluble in organic solvents; it can be plasticized. A ladder structure has been assigned:

$$\left[\begin{array}{c}\text{C}_6\text{H}_5\quad\text{C}_6\text{H}_5\\|\quad\quad|\\\text{Si}-\text{O}-\text{Si}\\|\quad\quad|\\\text{O}\quad\quad\text{O}\\|\quad\quad|\\\text{Si}-\text{O}-\text{Al}\\|\\\text{C}_6\text{H}_5\end{array}\right]_n$$

Regular poly(aluminosiloxane) structures are prepared by polycondensation of dialkyldiacetoxysilanes with aluminum alkoxides.

$$(\text{CH}_3)_2\text{Si}(\text{OOCCH}_3)_2 + \text{Al}(\text{OR})_3 \longrightarrow$$

$$\left[\begin{array}{c}\text{CH}_3\\|\\\text{Si}-\text{O}-\text{Al}-\text{O}\\|\quad\quad|\\\text{CH}_3\quad\text{OR}\end{array}\right]_n + 2\,\text{CH}_3\text{CO}_2\text{R}$$

## 14. Polymers with $-\text{Si}-\text{O}-\text{M}-\text{O}-$ Backbones

The early driving force in preparing poly-(metallosiloxanes) was to enhance thermal stability. The group 13 elements boron and aluminum have been incorporated into siloxane chains (see Chaps. 12 and 13). The metal can also be from group 14 (Ge, Sn, Pb), 15 (P, As, Sb), or 16 (S), or a transition metal (Ti, V, Cr, Fe, Ni, Zr, Hf). The thermal and hydrolytic stabilities of these polymers are usually inferior to those of the parent poly(organosiloxanes), but a variety of property modifications have been achieved. The hydrolytic stability (to HCl) of poly(titanoxane–phenylsiloxane) was shown to be greater than that of its aluminum analogue poly(aluminoxane – phenylsiloxane), but less than that of poly-(diphenylsiloxane). Thus, the order of stability with respect to cleavage by hydrochloric acid at 90 °C is $-\text{Si}-\text{O}-\text{Si}-\text{O}-> -\text{Si}-\text{O}-\text{Ti}-\text{O}-> -\text{Si}-\text{O}-\text{Al}-\text{O}-$.

Random $-\text{Si}-\text{O}-\text{Ge}-\text{O}-$ and $-\text{Si}-\text{O}-\text{Sn}-\text{O}-$ polymers can be made by the cohydrolysis of dialkyldichlorosilanes with dialkyldihalogermanes or dialkyldihalostannanes [506]. Low molecular mass components are treated further with sulfuric acid to give rubber-like materials.

$$(\text{CH}_3)_2\text{SiCl}_2 + (\text{CH}_3)_2\text{GeBr}_2 \xrightarrow{\text{H}_2\text{O}}$$

$$\left[\begin{array}{c}\text{CH}_3\\|\\\text{Si}-\text{O}\\|\\\text{CH}_3\end{array}\right]_n\left[\begin{array}{c}\text{CH}_3\\|\\\text{Ge}-\text{O}\\|\\\text{CH}_3\end{array}\right]_m + \text{HBr} + \text{HCl}$$

The random incorporation of tin is achieved by cohydrolysis of dihalostannanes with dihalosilanes [507] or by reaction of dialkyltin oxides with a silanediol [508].

$$\text{R}_2^1\text{SnO} + \text{R}_2^2\text{Si}(\text{OH})_2 \longrightarrow \left[\begin{array}{c}\text{R}^2\\|\\\text{Si}-\text{O}\\|\\\text{R}^2\end{array}\right]_n\left[\begin{array}{c}\text{R}^1\\|\\\text{Sn}-\text{O}\\|\\\text{R}^1\end{array}\right]_m + \text{H}_2\text{O}$$

The most stable polymers in this series have high silicon: tin ratios. Regular alternating structures are achieved by reaction of dialkyldiacetoxystannanes with dialkyldialkoxysilanes [509]:

$$\text{R}_2^1\text{Si}(\text{OR}^2)_2 + \text{R}_2^3\text{Sn}(\text{OOCCH}_3)_2 \longrightarrow \left[\begin{array}{c}\text{R}^1\quad\text{R}^3\\|\quad\quad|\\\text{Si}-\text{O}-\text{Sn}-\text{O}\\|\quad\quad|\\\text{R}^1\quad\text{R}^3\end{array}\right]_n$$

Poly(titanosiloxanes) can be prepared by cohydrolysis of dichlorosilanes with titanium alkoxides to give random incorporation of $-\text{Si}(\text{R})_2\text{O}-$ and $-\text{Ti}(\text{R})_2\text{O}-$ units. Alternatively, diphenylsilanediol can react with titanium alkoxides [510]. The resulting polymers are hard coatings, which are opaque to UV light and suitable for use as refrigeration enamels [510].

$$Ph_2Si(OH)_2 + Ti(OR)_4 \longrightarrow \left[ \begin{array}{c} OR \\ | \\ Ti-O-Si-O \\ | \\ OR \quad Ph \end{array} \right]_n + 2\ ROH$$

Partial hydrolysis of alkylalkoxysilanes with titanium alkoxides, acylates, or alkylamides gives copolymers that can be used to help catalyze the drying of printing inks [511, 512]. Similar silicon – zirconium ($-Si-O-Zr-O-$) preparations have also been reported [513]. A different type of titanium unit is represented by the titanocene fragment. Condensation of dimethylsilanediol [514] or its disodium salt [515] with titanocene dichloride results in the incorporation of titanocene units into the polymers.

$$(CH_3)_2Si(OH)_2 \text{ or } (CH_3)_2Si(ONa)_2 + Cl-Ti(Cp)_2-Cl \longrightarrow$$

$$\left[ \left( \begin{array}{c} CH_3 \\ | \\ Si-O \\ | \\ CH_3 \end{array} \right)_m \left( Ti(Cp)_2-O \right) \right]_n$$

A related polymer containing $-O-Si-S-Ti-S-Si-$ units in the backbone is made by polycondensation of dimercaptotitanocene [516]:

$$Cl-Si(CH_3)_2-O-Si(CH_3)_2-Cl + HS-Ti(Cp)_2-SH \longrightarrow$$

$$\left[ S-Ti(Cp)_2-S-\left( \begin{array}{c} CH_3 \\ | \\ Si-O \\ | \\ CH_3 \end{array} \right)_m \begin{array}{c} CH_3 \\ | \\ Si \\ | \\ CH_3 \end{array} \right]_n$$

Reactive silicones and titanium alkoxides give polymers with $-Si-O-Ti-O-$ backbones when heated in petroleum solvents. These are said to be suitable for making waterproof leather and textiles [517, 518].

Polymers with titanium–oxygen backbones, related to siloxanes, are prepared by heating trialkoxytitanium acylates and dialkoxytitanium diacylates at 75 °C. These materials have low molecular mass, and whether they are linear polymers or cyclic oligomers is as yet unresolved. Titanium alkoxides polymerize on treatment with acetic acid, but problems with cross-linking and three-dimensional network formation complicate this method.

$$Ti(OR)_4 + CH_3COOH \longrightarrow$$

$$\left[ \begin{array}{c} OR \\ | \\ Ti-O \\ | \\ OR \end{array} \right]_n + ROCOCH_3 + ROH$$

Incorporation of a group 15 element is illustrated by the preparation of $-Si-O-As-O-$ chain polymers. Condensation of dimethyldichlorosilane with $CH_3AsO(OH)_2$ gives a transparent, rubbery polymer that is used as an impregnation rubber additive [519]. The arsenic content inhibits the growth of molds.

$$2\ (CH_3)_2SiCl_2 + HO-\underset{\underset{CH_3}{|}}{\overset{\overset{O}{\|}}{As}}-OH \longrightarrow$$

$$Cl-\underset{\underset{CH_3}{|}}{\overset{\overset{CH_3}{|}}{Si}}-O-\underset{\underset{CH_3}{|}}{\overset{\overset{O}{\|}}{As}}-O-\underset{\underset{CH_3}{|}}{\overset{\overset{CH_3}{|}}{Si}}-Cl \longrightarrow$$

$$\left[ \left( \begin{array}{c} CH_3 \\ | \\ Si-O \\ | \\ CH_3 \end{array} \right)_{12} \left( \begin{array}{c} O \\ \| \\ As-O \\ | \\ CH_3 \end{array} \right) \right]_n$$

## 15. Poly(Carbon Disulfide), Poly(Carbon Diselenide), and Polythiocyanogen

The delocalized chalcogen-based polymers are of special interest due to the observation of superconductivity in $[SN]_n$ [520] and $[Se]_n$ [521]. BRIDGMAN first showed that carbon disulfide can be polymerized under high pressure (>4.5 GPa) at 175 °C to give a black semiconducting solid [522]. The $(CS_2)_n$ prepared under high pressure was later assigned a head-to-tail linear structure [523]. The head-to-head structure is 16.3 kJ/mol lower in energy and, by analogy to $(CSe_2)_n$, might be the more likely structure [524].

$$\begin{bmatrix} & \overset{S}{\underset{\|}{}} & & \overset{S}{\underset{\|}{}} & \\ -S- & C & -S- & C & - \\ \end{bmatrix}_n$$
Head-to-tail

$$\begin{bmatrix} & \overset{S}{\underset{\|}{}} & & & \\ -S- & C & & -S- & \\ & & C & & \\ & & \overset{\|}{S} & & \\ \end{bmatrix}_n$$
Head-to-head

Reaction products vary with experimental conditions. CHAN and JONSHER isolated a low-conductivity material ($<10^{-13}$ S/cm) and a semiconducting material ($10^{-3}$ S/cm) which consisted of free sulfur in a $(CS_{2-x})$ matrix [525]. They also proposed a cross-linked structure.

Anionic polymerization of carbon disulfide, initiated by sodium dispersed in dipolar aprotic solvents, results in polymers with less than two sulfur atoms per carbon [524, 526]. Poly(carbon disulfide) has also been prepared by photolysis [527], irradiation [528], and plasma deposition [529] to give structures described as $[C_3S_2]_n$, $[CS(S)_m]_n$, or mixtures [530].

Unlike carbon disulfide, carbon diselenide can be polymerized both at high pressure and under ambient conditions. The products obtained range from insulators to metals and superconductors. Carbon diselenide polymerizes at a rate of about 1 wt% per month at ambient temperature and pressure [531, 532]. When dissolved in solvents such as methylene chloride and dioxane, it polymerizes readily under high pressure (ca. 0.5 GPa) at 100 °C [533]. This polymer is an amorphous insulator. When the polymer is heated to 130 – 160 °C, a black semiconducting ($10^{-3}$ S/cm) material results. After high-pressure polymerization and annealing at 130 °C, a crystalline product is obtained with a room-temperature conductivity of 50 S/cm; it has been assigned a two-dimensional sheet structure [534–536]. The polymer is thought to have a head-to-head structure [531, 532], which is significantly more stable than the head-to-tail structure [524]:

$$\begin{bmatrix} & \overset{Se}{\underset{\|}{}} & & & \\ -Se- & C & & -Se- & \\ & & C & & \\ & & \overset{\|}{Se} & & \\ \end{bmatrix}_n$$

It is typically contaminated with trigonal selenium domains, depending on annealing and polymerization conditions. The observed crystallinity is due to both small and large trigonal selenium crystallites within the polymer matrix, and this selenium is the reason for the high conductivity [524]. As selenium crystallites form, the polymer becomes highly disordered, with the formation of $-C-C-$, $-C=Se-$, and $-C-Se-Se-C-$ units.

*Carbon selenosulfide* (CSeS) does not polymerize in solution under conditions similar to those used to polymerize carbon diselenide [524]. Polymerization of carbon diselenide in the presence of excess carbon disulfide gives only the selenium-containing product. The slow polymerization of carbon diselenide at atmospheric pressure and 25 °C gives a black product with an electrical conductivity $<10^{-6}$ S/cm, consistent with a band gap of ca. 2 eV [524].

*Polythiocyanogen*. Polythiocyanogen or parathiocyanogen is formed by the spontaneous polymerization of thiocyanogen $(SCN)_2$ [537]. While a number of structures are possible, a linear structure with the SCN atoms connected with a bond order of 1.5 is currently preferred [538]. The polymer is a semiconductor that can be doped with halogens [538].

A more recent synthesis and study of thiocyanogen polymers were reported [539]. These materials have mainly a sulfur – carbon – nitrogen repeat unit. Related compounds except including materials containing selenium – carbon – nitrogen repeat units were prepared from reaction of AgNCS with bromine and AgNCSe with iodine, which led to spontaneous formation of polymer [540]. Another new class of sulfur-containing polymers was synthesized from sulfur dicyanide. These polythiocyanogens are easily formed in the absence of water under moderate heating [541].

## 16. Phosphorus-Containing Polymers

Phosphorus-containing polymers are among the three most important basic blocks of life. They serve as the building blocks of the critical nucleic acids, DNA, RNA, phosphosaccharides, phosphoproteins, and related biomaterials.

Here, we will focus only on synthetic phosphorus-containing polymers. Reviews are given in [544–549]. Reviews deal with organic derivatives of pentavalent phosphorus [550], phosphorus-based flame retardants [551], organic phosphites as polymer stabilizers [552], and phosphorus-containing polymers for drug delivery [553]. The reviews cover the synthesis and uses of organic and inorganic phosphorus polymers having a wide variety of compositions and structures. Polymer chains containing phosphorus can be made by a variety of approaches, including those summarized below:

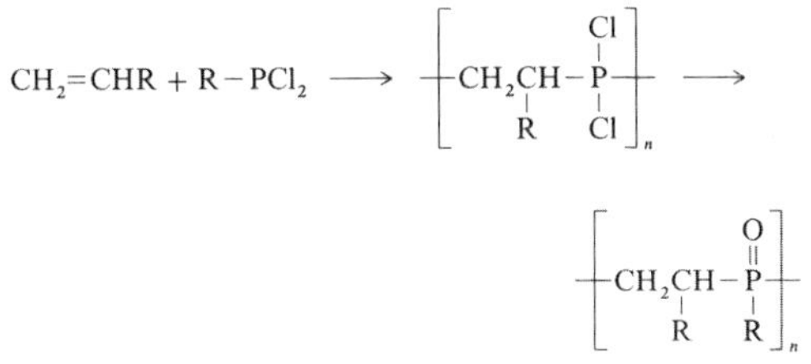

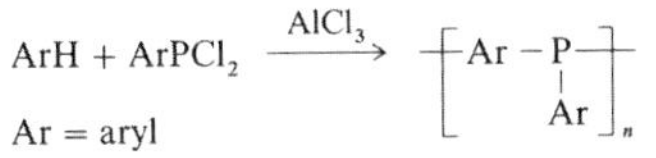

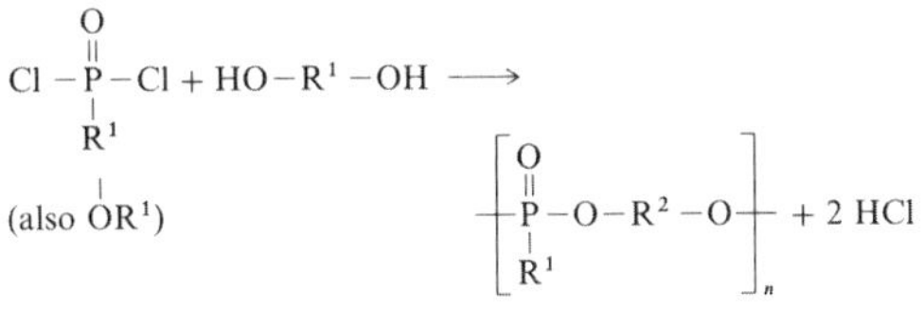

$$CH_2=CHR + R-PCl_2 \longrightarrow \left[ CH_2CH-\underset{\underset{Cl}{|}}{\overset{\overset{Cl}{|}}{P}}-\underset{R}{|} \right]_n \longrightarrow$$

$$\left[ CH_2CH-\underset{\underset{R}{|}}{\overset{\overset{O}{\|}}{P}}-\underset{R}{|} \right]_n$$

$$ArH + ArPCl_2 \xrightarrow{AlCl_3} \left[ Ar-\underset{Ar}{\overset{|}{P}} \right]_n$$

Ar = aryl

$$Cl-\underset{\underset{(also\ OR^1)}{R^1}}{\overset{\overset{O}{\|}}{P}}-Cl + HO-R^1-OH \longrightarrow \left[ \underset{\underset{R^1}{|}}{\overset{\overset{O}{\|}}{P}}-O-R^2-O \right]_n + 2\ HCl$$

R = alkyl, aryl, alkoxy, dialkylamido, diarylamido

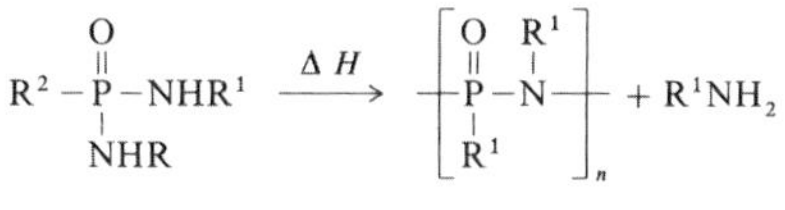

$$R^2-\underset{\underset{NHR}{|}}{\overset{\overset{O}{\|}}{P}}-NHR^1 \xrightarrow{\Delta H} \left[ \underset{\underset{R^1}{|}}{\overset{\overset{O\ \ R^1}{\|\ \ |}}{P}}-N \right]_n + R^1NH_2$$

Organophosphorus π-conjugated polymers have been reviewed [555]. While vinyl polymers such as polyacetylene, polypyrrole, and polythiophene are widely known as conductors after they are doped, a variety of phosphorus-containing structures are also capable of forming such π-conjugated structures. The focus is on phosphorus-containing moieties with an unbonded electron pair on the phosphorus atom. The unbonded electron pair is capable of bridging the electron gap between conjugated units, creating increased electrons that can promote conductivity when doped.

Structures of some of these central units are shown below.

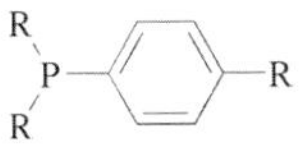

**Arylphosphane**

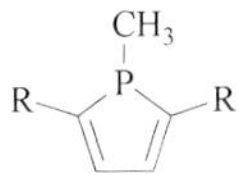

**Phosphole**

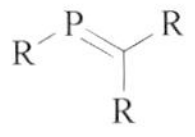

**Phosphalkene**

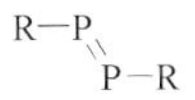

**Diphosphene**

Some of the arylphosphanes have exhibited nonlinear optical behavior. They show potential application in constructing organic light-emitting diodes (OLEDs). Efforts are underway to employ unbonded-electron phosphorus-containing units as end groups for organic conducting polymers such as polythiophenes (**30**) with the intent to used these unbonded electron sites as metal-chelating units to materials with possible conductivity, luminescence, and interesting optical and magnetic properties.

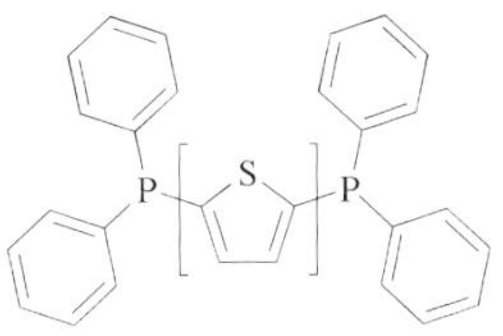

**30**

Complexes with a wide variety of metals and organometallics have been made and show promise in a variety of application areas including light-induced conductors. This is an area of great promise in a wide variety of areas including communications, electronics, solar energy conversion, and catalysis.

Polymers with a glasslike structure are formed by cocondensation of metal alkoxides or metal acetates with phosphorus oxychloride or phosphoric acid:

$$Si(OC_2H_5)_4 + POCl_3 \longrightarrow -O-\underset{\underset{O}{|}}{\overset{\overset{O}{\|}}{Si}}-O-\underset{\underset{OC_2H_5}{|}}{\overset{\overset{O}{\|}}{P}}-O-\underset{\underset{O}{|}}{\overset{\overset{O}{\|}}{Si}}-O-$$

$$Sn(OR)_4 + H_3PO_4 \longrightarrow -O-\underset{\underset{O}{|}}{\overset{\overset{O}{\|}}{Sn}}-O-\underset{\underset{O}{|}}{\overset{\overset{O}{\|}}{P}}-O-\underset{\underset{O}{|}}{\overset{\overset{O}{\|}}{Sn}}-O-$$

$$B(OCCH_3)_3 + PO(OC_2H_5)_3 \longrightarrow$$
$$\overset{\|}{O}$$

$$3\,CH_3CO_2C_2H_5 + {\left[-O-\underset{|}{B}-O-\underset{\underset{O}{|}}{\overset{\overset{O}{\|}}{P}}-\right]}_n$$

$$NaH_2PO_4 + NaH_2AsO_4 \longrightarrow {\left[-O-\underset{\underset{O}{|}}{\overset{\overset{O}{\|}}{P}}-O-\underset{\underset{O}{|}}{\overset{\overset{O}{\|}}{As}}-\right]}_n$$

Such reactions can be used to prepare artificial glass, paint and lubricant additives, plasticizers, and adhesives.

An interesting class of linear phosphorous – boron backbone polymers has been reported [543]. When the dimethylphosphine – borane complex is pyrolyzed, a linear polymer with $n = 80$ is obtained which is soluble in hot benzene:

$$(CH_3)_2PH \cdot BH_3 \longrightarrow {\left[\underset{\underset{CH_3}{|}}{\overset{\overset{CH_3}{|}}{P}}-\underset{\underset{H}{|}}{B}\right]}_n + H_2$$

A similar process yields borophane polymers from tris(trialkylsilyl)phosphines [543]:

$$[(CH_3)_3Si]_3P \cdot BC_6H_5Cl_2 \longrightarrow {\left[B-\underset{\underset{Si(CH_3)_3}{|}}{\overset{\overset{C_6H_5}{|}}{P}}\right]}_n$$

An interesting application of electrophilic substitution involves the incorporation of phosphorus bridges between ferrocene units [544, 556]. However, this route only gives low molecular mass materials.

$$Fe(C_5H_5)_2 + C_6H_5PCl_2 \xrightarrow[\text{sulfolane or melt}]{ZnCl_2,} {\left[Fe(C_5H_5)(C_5H_4)-\underset{\underset{C_6H_5}{|}}{\overset{\overset{O}{\|}}{P}}-Cl\right]}_n$$

$$\xrightarrow[H_2O]{H_2O_2,} {\left[Fe(C_5H_5)(C_5H_4)-\underset{\underset{C_6H_5}{|}}{\overset{\overset{O}{\|}}{P}}-OH\right]}_n$$

Polyphosphazenes are discussed in Chapter 4. However, a few carboranylphosphazene polymers have been reported [557]. Ring-opening polymerization of the carboranyl-substituted phosphazene ring proceeds without complications to yield a high molecular mass polymer:

[carboranylphosphazene ring-opening polymerization scheme with $B_{10}H_{10}$ carborane substituents]

## 17. Traditional and Metal-Matrix Composites

The topic of metal-matrix composites has been recently reviewed [558–562] (see also → Metal-Matrix Composites). Space age or advanced composites were front stage in the early 1960s with the development of high-modulus whiskers and filaments. These whiskers and filaments continue to be developed. Many fibers for both traditional and metal-matrix composites are inorganic polymers and are among the strongest materials made. They have tensile strengths on the order of 4 GPa and tensile strength/density ratios of about one thousand.

For polymer-intense composites, the matrix materials are organic polymers. For metal-matrix composites (MMCs), the matrix materials are typically a metal or less likely an alloy. Often employed metals include aluminum, copper, copper alloys, magnesium, titanium, and superalloys.

In polymer-matrix composites the non-continuous phase or reinforcement material is a fiber such as glass, carbon fibers (graphite), aromatic nylons, and a number of inorganic fibers including tungsten carbide (WC), titanium carbide (TiC), zirconia ($ZrO_2$), and alumina ($Al_2O_3$). For the metal-matrix composites, the discontinuous phase generally exists as fibers, wires, whiskers, and particulates. Some of the discontinuous-phase materials overlap with the listing given for polymer-intense composites including tungsten carbide, titanium carbide, alumina, and graphite, but there are a number of other materials whose fibers and whiskers are employed, including silicon carbide and boron carbide. By volume, the amount of whiskers, wires, and particulates is greater for the metal-matrix composites.

As in the case with polymer-intense composites, the matrix and fiber must be matched for decent properties. Below is a listing of typical matrix/fiber mixes.

In comparison to single-metal materials such as aluminum, copper, and iron, MMCs generally have

- Higher strength-to-density ratio
- Better fatigue and wear resistance
- Better high temperature strength
- Lower creep related to lower coefficients of thermal expansion
- Greater stiffness

In comparison to polymer-intense composites, MMCs offer

- No moisture absorption
- Greater fire resistance
- Higher use temperatures
- Greater radiation resistance
- Greater stiffness and strength
- Higher thermal and electrical conductivities

MMCs also have some disadvantages in comparison to polymer-matrix composites. These include

- Higher cost
- Newer and less developed technology and scientific understanding
- Generally more complex fabrication
- Greater weight

**Table 3.** Typical metal/composite matches

| Matrix material | Discontinuous phase | |
|---|---|---|
| | Form | Material |
| Aluminum | fibers | boron |
| | | alumina |
| | | graphite |
| | | alumina – silica |
| | | silicon carbide |
| | whiskers | silicon carbide |
| | particulates | silicon carbide |
| | | boron carbide |
| Titanium | fibers | boron (coated) |
| | particulates | silicon carbide |
| Titanium carbide | | |
| Copper | fibers | graphite |
| | | silicon carbide |
| | particulates | boron carbide |
| | | silicon carbide |
| | | titanium carbide |
| Magnesium | fibers | alumina |
| | | graphite |
| | particulates | boron carbide |
| | | silicon carbide |
| | whiskers | silicon carbide |

As noted above, the range of fibers employed does not precisely overlap with those employed for organic composites. Because the formation of the metal-matrix composites generally requires melting of the metal matrix the fibers must have some stability to relatively high temperatures. Such fibers include graphite, silicon carbide, boron, alumina – silica, and alumina fibers. Most of these are available as continuous and discontinuous fibers.

As with organic-matrix composites, the orientation of the reinforcing material determines whether the properties will be isotropic or oriented in a preferential direction, so that the strength and stiffness are greater in the direction of the fiber orientation.

There are a number of differences between traditional and metal-matrix composites. First, the metal matrix can also have considerable strength itself, so its contribution to the overall strength is more important than for organic-matrix composites. Second, the difference in the coefficient of expansion between the reinforcing material and metal-matrix is often greater. Because of the wider ranges of use temperature that are often required for metal-matrix composites, these differences become more important. They can result in large residual stresses in metal-matrix composites that may result in yielding. A third difference is related to the relative lower flexibility of metal-matrix composites. This leads to greater concerns in the marrying or joining of such composite parts. Many methods of joining these composite parts have been developed. A fourth difference is the possible greater reactivity between the matrix and fiber for metal-matrix composites. This limits combinations but has been overcome in many situations. One major approach is to place a barrier coating on the reinforcement. For example, application of boron carbide as a barrier coating on boron fibers allows their use to reinforce titanium. Because these coatings can be "rubbed" off by usage, composites made from these coated reinforcements should be monitored more closely and more often.

Metal-matrix composites are finding use in the military and in the aerospace industry as high-strength materials [558–562].

# References

1. NRC: *Polymer Science and Engineering*, National Academy Press, Washington, D.C. 1981.
2. F. S. Kipping, *J. Chem. Soc.* **125** (1924) 229.
3. C. Carraher, *J. Macromol. Sci. Chem.* **A17** (1982) no. 8, 1293.
4. F. G. A. Stone, W. A. G. Graham: *Inorganic Polymers*, Academic Press, New York 1962.
5. K. A. Andrianov: *Metalorganic Polymers*, Interscience, New York 1962.
6. S. N. Borisov, M. G. Voronkov, E. Lukevits: *Organosilicon Heteropolymers and Heterocompounds*, Plenum Press, New York 1970.
7. H. R. Allcock: *Phosphorus-Nitrogen Compounds*, Academic Press, New York 1972.
8. H. R. Allcock: *Heteroatom Ring Systems and Polymers*, Academic Press, New York 1967.
9. M. G. Voronkov, V. P. Mileshkevich, Yu. A. Yuzhelevskii: *The Siloxane Bond*, Consultants Bureau, New York 1978.
10. A. L. Rheingold: *Homoatomic Rings, Chains and Macromolecules of Main Group Elements*, Elsevier Scientific Publ. Co., New York 1977.
11. C. E. Carraher, J. E. Sheats, C. U. Pittman, Jr., (eds.): *Organometallic Polymers*, Academic Press, New York 1978.
12. C. E. Carraher, J. E. Sheats, C. U. Pittman, Jr., (eds.): *Advances in Organometallic and Inorganic Polymers*, Marcel Dekker, New York 1982.
13. M. Zeldin, K. J. Wynne, H. R. Allcock (eds.): "Inorganic and Organometallic Polymers," *ACS Symp. Ser.* **360** (1988).
14. C. U. Pittman, Jr., C. E. Carraher, Jr., J. R. Reynolds in H. Mank, N. Bikales, C. Overberger andG. Menges (eds.): *Encyclopedia of Polymer Science and Engineering*, "Organometallic Polymers," vol. **10**, 2nd ed., J. Wiley, 1987, p. 541.
15. A. Abd-El-Aziz, C. Carraher, C. Pittman, J. Sheats, M. Zeldin: Macromolecules Containing Metal and Metal-Like Elements, vol. 1, *A Half Century of Metal- and Metalloid-Containing Polymers*, Wiley, Hoboken, NJ, 2003.
16. A. Abd-El-Aziz, C. Carraher, C. Pittman, J. Sheats, M. Zeldin: Macromolecules Containing Metal and Metal-Like Elements,. vol. **2**. *Organoiron Polymers*, Wiley, Hoboken, NJ, 2004.
17. A. Abd-El-Aziz, C. Carraher, C. Pittman, J. Sheats, M. Zeldin, Macromolecules Containing Metal and Metal-Like Elements. vol. 3. *Biomedical Applications*, Wiley, Hoboken, NJ, 2004.
18. A. Abd-El-Aziz, C. Carraher, C. Pittman, M. Zeldin, Macromolecules Containing Metal and Metal-Like Elements. vol. **4**. *Group IVA Polymers*, Wiley, Hoboken, NJ, 2005.
19. A. Abd-El-Aziz, C. Carraher, C. Pittman, M. Zeldin, Macromolecules Containing Metal and Metal-Like Elements. vol. **5**. *Metal-Coordination Polymers*, Wiley, Hoboken, NJ, 2005.
20. A. Abd-El-Aziz, C. Carraher, C. Pittman, M. Zeldin: Macromolecules Containing Metal and Metal-Like Elements. vol. 6. *Transition Metal-Containing Polymers*, Wiley, Hoboken, NJ, 2006.
21. A. Abd-El-Aziz, C. Carraher, C. Pittman, M. Zeldin: Macromolecules Containing Metal and Metal-Like Elements. vol. 7. *Nanoscale Interactions of Metal-Containing Polymers*, Wiley, Hoboken, NJ, 2006.
22. A. Abd-El-Aziz, C. Carraher, C. Pittman, M. Zeldin: Macromolecules Containing Metal and Metal-Like Elements. vol. **8**. *Boron-Containing Polymers*, Wiley, Hoboken, NJ, 2008.
23. I. Manners: *Synthetic Metal-Containing Polymers*, Wiley-VCH, Hoboken, NJ, 2004.
24. A. Abd-El-Aziz, I. Manners: *Frontiers in Transition Metal-Containing Polymers*, Wiley, Hoboken, NJ, 2007.
25. R. Archer: *Inorganic and Organometallic Polymers*, Wiley, Hoboken, NJ, 2004.
26. V. Chandrasekhar, *Inorganic and Organometallic Polymers*, Springer, NY, 2005.
27. A. Abd-El-Aziz, C. Carraher, C. Pittman, M. Zeldin: *Recent Advances in Inorganic and Organometallic Polymers*, Springer, NY, 2008.
28. D. P. Gates, *Annu. Rep. Prog. Chem. Sect. A* **101** (2005) 452.
29. C. Carraher, *J. Inorg. Organomet. Polym. Mater.* **15** (2005) 121.
30. R. West *et al.*, *J. Amer. Chem. Soc.* **103** (1981) 7352.
31. J. Wesson, T. Williams, *J. Polymer Sci., Polym. Chem. Ed.* **17** (1979) 2833.
32. R. Trujillo, *J. Organometallic Chem.* **198** (1980) C27.
33. S. Sawan, S. Ekhorutomwen, in J. Salamone (ed.): *Polymeric Materials Encyclopedia*, vol. **9**, CRC Press, Boca Raton, FL 1996, p. 6722.
34. S. Hayase, in J. Salamone (ed.): *Polymeric Materials Encyclopedia*, vol. **9**, CRC Press, Boca Raton, FL 1996, p. 6734.
35. K. Matyjaszewski, in J. Salamone (ed.): *Polymeric Materials Encyclopedia*, vol. **9**, CRC Press, Boca Raton, FL 1996, p. 6741.
36. A. Soum *et al.*, in J. Salamone (ed.): *Polymeric Materials Encyclopedia*, vol. **9**, CRC Press, Boca Raton, FL 1996, p. 6747.
37. C. Tien, A. Savoca, in J. Salamone (ed.): *Polymeric Materials Encyclopedia*, vol. **9**, CRC Press, Boca Raton, FL 1996, p. 6758.
38. E. Hengge, *J. Inorg. Organomet. Polym.* **3** (1993) no. 4, 287.
39. S. Thames, K. Panjnani, *J. Inorg. Organometal. Polym.* **6** (1996) no. 2, 69.
40. A. Eckhardt, W. Schnabel, *J. Inorg. Organometal. Polym.* **6** (1996) no. 2, 95.
41. K. Matyjaszewski, *J. Inorg. Organometal. Polym.* **1** (1991) no. 4, 463.
42. M. D. Rausch *et al.*, New Vinyl Organometallic Monomers: Synthesis and Polymerization Behavior in B. M. Culbertson,

C. U. Pittmann, Jr. (eds.): *New Monomers and Polymers*, Plenum Press, New York 1984; F. R. Hartley: *Supported Metal Complexes*, Reidel Publishing Co., Boston 1985; J. E. Sheats, C. E. Carraher, Jr., C. U. Pittman, Jr. *Metal-Containing Polymeric Systems*, Plenum Press, New York 1985; C. U. Pittman, Jr., Vinyl Polymerization of Organic Monomers Containing Transition Metals in E. J. Becker, M. Tsutsui (eds.): *Organometallic Reactions*, vol. **6**, Wiley-Interscience, New York 1977, p. 1; C. U. Pittman, Jr. Polymer Supported Catalysts in G. Wilkinson and F. G. A. Stone (eds.): *Comprehensive Organometallic Chemistry*, vol. **8**, Chap. 55, Pergamon Press, Oxford 1982.
43 J. P. Wesson, T. C. Williams, *J. Polym. Sci. Polym. Chem. Ed.* **19** (1981) 65.
44 X. H. Zhang, R. West, *J. Polym. Sci. Polym. Chem. Ed.* **22** (1984) 159, 225.
45 R. West, L. D. David, P. I. Djurovich, H. Yu, R. Sinclair, *Am. Ceram. Soc. Bull.* **62** (1983) no. 8, 899.
46 R. West, *J. Organomet. Chem.* **300** (1986) 327.
47 R. West *et al.*, *J. Am. Chem. Soc.* **103** (1981) 7352.
48 L. A. Harrah, J. M. Zeigler, *J. Polym. Sci. Polym. Lett. Ed.* **23** (1985) 209.
49 J. M. Zeigler, *Polym. Prepr. Am. Chem. Soc. Div. Polym. Chem.* **28** (1987) no. 1, 424.
50 K. Matyjaszewski, Y. L. Chen, H. K. Kim: "Inorganic and Organometallic Polymers," *ACS Symp. Ser.* **360** (1988).
51 R. West, S. Hayase, T. Iwahara, *J. Inorg. Organometal. Polym.* **1** (1991) 545.
52 J. Wilderman, J. Herrema, G. Hadziioannou, E. Schomaker, *J. Inorg. Organometal. Polym.* **1** (1991) 567.
53 C. Van Wakreem, T. Cleiji, J. Zwikker, L. Jenneskens, *Macromolecules* **28** (1995) 8696.
54 J. Herrema *et al.*, *Macromolecules* **28** (1995) 8102.
55 C. Aitken, J. Harrod, E. Samuel, *Organomet. Chem.* **279** (1985) C11.
56 C. Aitken, J. Harrod, E. Samuel, *J. Amer. Chem. Soc.* **108** (1986) 4059.
57 J. Harrod, *ACS Symposium Ser.* **360** (1988) 89.
58 H. Woo, T. Tilley, *J. Amer. Chem. Soc.* **111** (1989) 3757.
59 T. Tilley, *Accounts Chemical Research* **26** (1993) 22.
60 J. Banovetz, K. Stein, R. Waymouth, *Organometallics* **10** (1991) 3430.
61 J. Corey, X. Zhu, T. Bedard, L. Lange, *Organometallics* **10** (1991) 924.
62 H. Li, F. Gauvin, J. Harrod, *Organometallics* **12** (1993) 575.
63 C. Forsyth, S. Nolan, T. Marks, *Organometallics* **10** (1991) 2543.
64 P. Bianconi, T. Weidman, *J. Amer. Chem. Soc.* **110** (1988) 2342.
65 W. Szymanski, G. Visscher, P. Bianconi, *Ultrasonics* **28** (1990) 310.
66 W. Szymanski, G. Visscher, P. Bianconi, *Macromolecules* **26** (1993) 869.
67 A. Sekignchi, M. Naugo, C. Kabuto, H. Sakurai, *J. Amer. Chem. Soc.* **117** (1995) 4195.
68 H. Suzuki, Y. Kimata, S. Satoh, A. Kuriyama, *Chem. Lett.* (1995) 293.
69 J. Lambert, J. Pfing, C. Stern, *Angew. Chem. Int. Ed. Engl.* **34** (1995) 98.
70 T. Imori, T. Tilley, *J. Chem. Soc. Commun.* (1993) 1607.
71 N. Devylder, M. Hill, K. Molloy, C. Price, *Chem. Commun.* (1996) 711.
72 V. Lu, T. Tilley, *Macromolecules* **29** (1996) 5763.
73 J. Reichi, C. Popoff, L. Gallagher, E. Remsen, D. Berry, *J. Amer. Chem. Soc.* **118** (1996) 9430.
74 M. Fujino, T. Hisaki, N. Matsumoto, *Macromolecules* **28** (1995) 5017.
75 R. G. Jones, S. J. Holder, *Polym. Int.* **55** (2006) 711.
76 J. Corey, *Adv. Organomet. Chem.* **51** (2004) 1.
77 C. Zhou, R. Guan, S. Feng, *Chem. J. on Internet* **5** (2003).
78 H. Matsumoto, *Organomet. News* **4** (2006) 128.
79 S. Adams, J. Drager, *Angew. Chem. Int. Ed. Engl.* **26** (1987) 1255.
80 J. Mark, H. Allcock, R. West: *Inorganic Polymers*, Prentice Hall, New York 1992.
81 H. Suzuki, H. Meyer, J. Simmerer, J. Yang, D. Naarer, *Adv. Mater.* **5** (1993) 743.
82 C. Marschner, J. Baumgartner, A. Wallner, *Dalton Trans.* 2006, 5667.
83 H. Sakurai, T. Sanji, *Kogyo Zairyo* **54** (2006) 56.
84 B. Xu, J. Zuo, Y. Ren, S. Huang, *Youjigui Cailiao* **19** (2005) 33.
85 S. Nespurek, G. Wang, K. Yoshino, *J. Optoelectron. Adv. Mater.* **7** (2005) 223.
86 L. Sacarescu, R. Ardeleanu, G. Sacarescu, M. Simionescu, *Trends Organomet. Chem. Res.* 2005, 85.
87 C. Carraher: Macromolecules Containing Metal and Metal-Like Elements. vol. 4. *Group IVA Polymers*, Wiley, Hoboken, NJ, 2005, chap. 9.
88 C. Carraher: Macromolecules Containing Metal and Metal-Like Elements. vol. 4, *Group IVA Polymers*, Wiley, Hoboken, NJ, 2005, chap. 10.
89 R. West, L. David, P. Djurovich, H. Yu, *Am. Ceram. Soc. Bull.* **62** (1983) 899.
90 M. Kumada, K. Tamao, *Adv. Organomet. Chem.* **6** (1968) 19.
91 R. Barney, G. Chandra: *Encyclopedia of Polymer Science and Engineering*, 2nd ed., vol. **13**, John Wiley & Sons, New York 1988, p. 312.
92 C. Schilling, *Br. Polymer J.* **18** (1986) 355.
93 J. Pilot *et al.*, in A. Bassindale, P. Gaspar (eds.): *Frontiers in Organosilicon Chemistry*, Royal Society of Chemistry, Cambridge 1991.
94 W. Schmidt *et al.*, *Chem. Mater.* **3** (1991) 257.
95 Y. Shieh, S. Sawan, *Polymer Preprints* **33** (1992) 1042.
96 D. Seyferth *et al.*, *Polymer Preprints* **34** (1992) no. 1, 223.
97 D. Seyferth, P. Czubarow, *Chem. Mater.* **6** (1994) no. 1, 10.
98 S. Yajima, Y. Hasegawa, L. Hayasi, M. Iimura, *J. Mater. Sci.* **13** (1978) 2569.
99 R. West, A. Wolff, D. Peterson, *J. Radiation Curing* **35** (1986) 40.
100 A. Wolf, R. West, *Applied Organomet. Chem.* **1** (1987) 7.
101 R. Miller, G. Wallraff, N. Clecak, R. Sooriyakumaran, *PMSE* **60** (1989) 40.
102 A. Eckhardt, N. Yars, T. Wollny, W. Schnabel, S. Nespurek, *Ber. Bunsen-Ges. Phys. Chem.* **98** (1994) 853.
103 G. VanderLaan *et al.*, *Macromolecules* **27** (1994) no. 7, 853.
104 I. Kminek *et al.*, *Collect. Czech. Chem. Commun.* **58** (1993) no. 10, 2337.
105 F. Kajar, J. Messier, C. Rosilio, *J. Appl. Phys.* **60** (1986) 3040.
106 C. Callender, L. Robitaille, M. Lecleric, *Opt. Eng. (Bellingham, Wash.)* **32** (1993) 2246.
107 S. Grigoras, T. Banton *et al.*, *Annu. Tech. Conf. Soc. Plast. Eng.* **50** (1992) no. 2, 2265.
108 S. Grigoras *et al.*, *Polymer Preprints* **33** (1992) 655.
109 K. Matyjaszewski, J. Chrusciel, H. Kim, J. Maxka, *Polymer Preprints* **34** (1993) 67.
110 R. Corriu, W. Douglas, Z. Yang, Y. Karakus, G. Cross, D. Bloor, *J. Organomet. Chem.* **69** (1993) 455.
111 S. Grigoras *et al.*, *Synth. Met.* **49** (1992) 293.
112 Y. Iwasa, T. Hasegawa, T. Koda, Y. Tokura, H. Tachibana, Y. Kawabata, *Synth. Met.* **50** (1992) 415.

113. F. Kajzar, J. Messier, C. Rossilio, *J. Appl. Phys.* **60** (1986) 3040.
114. D. Hofer, R. Miller, G. William, *APIE Advances in Resist Technology* **16** (1984).
115. R. Miller *et al.*, *Polym. Eng. Sci.* **29** (1989) 882.
116. T. Weidman, A. Joshi, *Proc. SPIE-Int. Soc, Opt. Eng.; Advances in Resist Technology and Processing* **1925** (1993) 145.
117. G. Wallraff *et al.*, *Polym. Mater. Sci. Eng.* **66** (1992) 105.
118. D. Hofer, R. Miller, C. Wilson, A. Neurath, *SPIE-Advances in Resist Technology* **469** (1984) 108.
119. C. Rosilio, A. Rosilio, B. Serre, *Microelectronic Eng.* **6** (1987) 399.
120. P. Trefonas, R. West, R. Miller, D. Hofer, *J. Polym. Sci., Polymer Lett. Ed.* **21** (1983) 823.
121. A. Gozdz, *Polym. Adv. Technol.* **5** (1994) 70.
122. A. Joslie, T. Weidman, A. Johnson, J. Miner, D. Ibbotson, *Proc. SPIE-Int. Soc. Opt. Eng., Adv. Resist Technology and Processing* **1925** (1993) 709.
123. M. Sakata, T. Ito, Y. Yamashita, *J. Photopolym. Sci. Technol.* **3** (1990).
124. R. Miller *et al.*, "Polymer Microlithography," *ACS Symp. Ser.* (1989).
125. J. Su, T. Hsu, S. Sawan, J. Lavine, *Polymer Preprints* **33** (1992) 1046.
126. C. G. Wilson, in L. Thompson, G. Wilson, M. Bowden (eds.): *Introduction to Microlithography*, American Chemical Society, Washington DC 1983.
127. J. Zeigler, L. Harrah, A. Johnson, *SPIE-Advances in Resist Technology and Processing* **539** (1985) 166.
128. *Chem. Abstr.* **107** (1987) 218592u.
129. I.B.M. Corp., US 4 464 460, 1984 (H. Hiraoka).
130. D. C. Hofer, R. D. Miller, C. G. Willson, *J. Soc. Photo Opt. Instrum. Eng.* **469** (1984) 16.
131. D. C. Hofer, R. D. Miller, C. G. Willson, A. R. Neureuther, *J. Soc. Photo Opt. Instrum. Eng.* **469** (1984) 108.
132. R. D. Miller *et al.*: "Inorganic and Organometallic Polymers," *ACS Symp. Ser.* **360** (1988).
133. J. Michl *et al.*: "Inorganic and Organometallic Polymers," *ACS Symp. Ser.* **360** (1988).
134. R. West, A. R. Wolff, D. J. Peterson, *J. Radiat. Curing* **13** (1986) 35.
135. Y. Hasegawa, M. Iimura, S. Yajima, *J. Mater. Sci.* **15** (1980) 1209.
136. R. G. Kepler, J. M. Zeigler, L. A. Harrah, S. R. Kurtz, *Phys. Rev. B* **35** (1987) 2818.
137. F. Kazar, J. Messier, C. Rosilio, *J. Appl. Phys.* **60** (1986) 3040.
138. R. West "Organopolysilanes" in G. Wilkinson, F. G. A. Stone, E. Abel (eds.): *Comprehensive Organometallic Chemistry*, vol. 9, Pergamon Press, Oxford, U.K., 1983.
139. R. Seymour, C. Carraher: *Polymer Chemistry*, 7th Ed., Taylor & Francis, New York, 2008.
140. C. Gebelein, C. Carraher: *Polymeric Materials in Medication*, Plenum Press, New York 1985.
141. R. West, J. Maxka, R. Sinclair, P. Cotts, *Polym. Prepr. Am. Chem. Soc. Div. Polym. Chem.* **28** (1987) no. 1, 387.
142. J. F. Harrod, *Polym. Prepr. Am. Chem. Soc. Div. Polym. Chem.* **28** (1987) no. 1, 403.
143. W. Peng, M. Motonaga, J. R. Koe, *J. Am. Chem. Soc.* **126** (2004) no. 13, 822.
144. S. Y. Kim, M. Fujiki, A. Ohira, G. Kwak, Y. Kawakami, *Macromolecules* **37** (2004) 4321.
145. A. Ohira, K. Okoshi, M. Fujiki, M. Kunitake, M. Naito, T. Hagihara, *Adv. Mater.* **16** (2004) 1645.
146. A. Saxena, G. Guo, M. Fujiki, Y. Yang, A. Ohira, K. Okoshi, M. Naito, *Macromolecules* **37** (2004) 3081.
147. M. Motonaga *et al.*, *J. Organomet Chem.* **685** (2003) 44.
148. B. G. Potter, K. Simmons-Potter, H. Chandra, G. M. Jamison, W. J. Thomes, *J. Non-Cryst. Solids* **352** (2006) 2618.
149. L. Sacarescu, R. Ardeleanu, G. Sacarescu, M. Simionescu, *Eur. Polym. J.* **40** (2004) 465.
150. R. Ardeleanu, I. Mangalagiu, G. Sacarescu, M. Simionescu, L. Sacarescu, *Macromol. Rapid Commun.* **25** (2004) 5873.
151. A. Saxena, M. Fujiki, M. Naito, K. Okoshi, G. Kwak, *Macromolecules* **37** (2004) 5873.
152. A. Saxena, M. Fujiki, R. Rai, S. Y. Kim, G. Kwak, *Macromol. Rapid Commun.* **25** (2004) 1771.
153. R. W. Williams, *Pure Appl. Chem.* **29** (1972) 569.
154. E. N. Peters, *J. Macromol. Sci. Rev. Macromol. Chem.* **C17** (1979) 173.
155. K. O. Knollmueller, R. N. Scott, H. Kawasnik, J. R. Sieckhaus, *J. Polym. Sci. Polym. Chem. Ed.* **9** (1971) 1071.
156. E. N. Peters *et al.*, *Rubber Chem. Technol.* **48** (1979) 14.
157. S. Papetti, B. B. Schaeffer, A. P. Gray, T. L. Heying, *J. Polym. Sci. Polym. Chem. Ed.* **4** (1966) 1623.
158. E. N. Peters, D. D. Stewart, *J. Polym. Sci. Polym. Lett. Ed.* **17** (1979) 405.
159. M. K. Kolel-Veetil, H. W. Beckham, T. M. Keller, *Chem. Mater.* **16** (2004) 3162.
160. G. Cai, W. P. Weber, *Polymer*, **45** (2004) 294.
161. D. Henkensmeier, B. Abele, A. Candussio, J. Thiem, *Macromol. Chem. Phys.* **205** (2004) 1851.
162. M. B. Roller, J. K. Gillham, *Polym. Eng. Sci.* **8** (1974) 567.
163. E. N. Peters *et al.*, *J. Polym. Sci. Polym. Phys. Ed.* **15** (1977) 723.
164. L. Interrante *et al.*, *J. Amer. Chem. Soc.* **116** (1994) 12025.
165. H. Wu L. Interrante, *Macromolecules* **25** (1992) 1840.
166. I. Rushkin, L. Interrante, *Macromolecules* **28** (1995) 5160.
167. M. Lienhard *et al.*, *J. Amer. Chem. Soc.* **116** (1994) 12020.
168. M. Lienhard, C. Weigand, T. Apple, B. Farmer, L. Interrante, *Polymer Preprints* **39** (1998) no. 2, 808.
169. S. Yamjima, Y. Hasegawa *et al.*, *J. Mater. Sci.* **12** (1978) 2569.
170. K. Wynne, R. Rice, *Ann. Rev. Mater. Sci.* **14** (1984) 297.
171. R. Laine, Y. Blum *et al.*, *ACS Symp. Ser.* **360** (1988) 142.
172. J. Mark, H. Allcock, R. West, *Inorganic Polymers*, Prentice Hall, Englewood Cliffs, NJ 1992, Chap. 6.
173. D. Seyferth, G. Wiseman, C. Prud'homme, *J. Am. Ceram. Soc.* **66** (1983) C13.
174. Rhône-Poulence, EP 197 863 B1, 1988 (J. Lebrun, H. Porte).
175. Bayer, US 3 853 567, 1974 (W. Verbeck).
176. Dow Corning, US 4 340 619, 1982 (J. Gaul, Jr.).
177. H. Allcock, *Adv. Mater.* **6** (1994) 106.
178. H. Allcock, S. Kuharcik, *J. Inorg. Organometal. Polym.* **6** (1996) no. 1, 1.
179. P. Bortolus, M. Gleria, *J. Inorg. Organometal. Polym.* **4** (1994) no. 2, 95 and 4 (1994) no. 3,205.
180. *J. Inorg. Organometal. polym.* **6** (1996) no. 3, and 6 (1996) no. 4.
181. G. Bosscher, A. Jekel, J. van der Grampel, *J. Inorg. Organometal. Polym.* **7** (1997) no. 1, 19.
182. L. Nair, Y. Khan, C. Laurencin: *Introduction to Biomaterials*, CRC Press, Boca Raton, FL, 2006.
183. H. R. Allcock,: *Chemistry and Applications of Polyphosphazenes*, Wiley, Hoboken, 2003.
184. L. Nair, D. Lee, C. Laurencin: *Handbook of Biodegradable Polymeric Materials and Their Applications*, American Scientific Publishers, Stevenson Ranch, CA, 2006.
185. M. Gleria, R. De Jaeger: *Topics in Current Chemistry. 250. New Aspects in Phosphorus Chemistry V*, Springer, NY, 2005.

186. H. R. Allcock: *Synthesis and Characterization of Poly(organophosphazenes)*, Nova Science Publishers, Hauppauge, NY, 2004.
187. H. R. Allcock: *Phosphorus-Nitrogen Compounds*, Academic Press, New York 1972.
188. L. Hagnauer, C. Carraher, J. Sheats, C. U. Pittman, Jr. (eds.): *Advances in Organometallic and Inorganic Polymer Science*, Chap. 17, Marcel Dekker, New York 1982.
189. R. Neilson, P. Wisian-Neilson in C. E. Carraher, Jr., J. E. Sheats, C. U. Pittman, Jr. (eds.): *Advances in Organometallic and Inorganic Polymer Sciences*, Chap. 19, Marcel Dekker, New York 1982, p. 425.
190. H. R. Allcock, *CHEMTECH* **5** (1975) 552.
191. H. R. Allcock, *Angew. Chem. Int. Ed. Engl.* **16** (1977) 147.
192. A. K. Andrianov, J. Chen, M. LeGolvan, *Macromolecules* **37** (2004) 414.
193. S. H. Rose, *J. Polym. Sci. Part. B* **6** (1968) 837.
194. C. Carraher, *J. Macromol. Sci. Chem.* **A17** (1982) no. 8, 1293.
195. H. R. Allcock, *Polymer Prepr. Am. Chem. Soc. Div. Polym. Chem.* **28** (1987) no. 1, 437.
196. R. H. Neilson et al., in M. Zeldin, K. J. Wynne and H. R. Allcock (eds.): "Inorganic and Organometallic Polymers," Chap. 22, *ACS Symp. Ser.* **360** (1988).
197. H. Allcock, J. Nelson, S. Reeves, C. Honeyman, I. Manners, *Macromolecules* **29** (1996) 7740.
198. J. Nelson, H. Allcock, *Macromolecules* **30** (1997) 1854.
199. H. Allcock, S. Reeves, J. Nelson, C. Crane, *Macromolecules* **30** (1997) 22B.
200. J. Nelson, H. Allcock, *Polymer Reprints* **39** (1998) no. 2, 631.
201. I. Dez, R. de Jaeger, *J. Inorg. Organometal. Polym.* **6** (1996) no. 2, 111.
202. D. Gates, I. Manners, *J. Chem. Soc., Dalton Trans.* (1997) 2525.
203. H. Allcock, S. Kuharcik, *J. Inorg. Organometal. Polym.* **5** (1995) no. 4, 307.
204. J. C. Vicic, K. A. Reynard, *J. Appl. Polym. Sci.* **21** (1977) 3185.
205. R. Singler, G. L. Hagnauer in C. Carraher, J. Sheats, C. U. Pittman, Jr. (eds.): *Organometallic Polymers*, Chap. 26, Academic Press, New York 1978.
206. R. E. Singler, N. S. Schneider, G. L. Hagnauer, *Polym. Eng. Sci.* **15** (1975) 321.
207. H. R. Penton, *Pol. Prepr. Am. Chem. Soc. Div. Polym. Chem.* **28** (1987) no. 1, 437.
208. C. Kim, H. R. Allcock, *Polym. Prepr. Am. Chem. Soc. Div. Polym. Chem.* **28** (1987) no. 1, 446.
209. R. Singler, R. Willingham, R. Lenz, A. Furukana, H. Finkelmann, *Polym. Prepr. Am. Chem. Soc. Div. Polym. Chem.* **28** (1987) no. 1, 448.
210. G. S. Kyker, T. A. Antkowiak, *Rubber Chem. Technol.* **47** (1974) 32.
211. D. F. Lohr, J. A. Beckman, *Rubber and Plastics News* **16** (1982).
212. H. R. Penton in M. Zeldin, K. J. Wynne and H. R. Allcock (eds.): "Inorganic and Organometallic Polymers," Chap. 21, *ACS Symp. Ser.* **360** (1988).
213. Z. Pang et al., *Adv. Mater.* **8** (1996) 768.
214. Z. Li, C. Huang, J. Hua, J. Qin, Z. Zang, C. Ye, *Macromolecules* **37** (2004) 371.
215. R. Frech, S. York, H. Allcock, C. Kellam, *Macromolecules* **37** (2004) 8699.
216. L. S. Nair, S. Bhattacharyya, J. D. Bender, Y. Greish, P. W. Brown, H. Allcock, C. Laurencin, *Biomolecules* **5** (2004) 2212.
217. A. K. Andrianov, Y. Svirkin, M. LeGolvan, *Biomacromolecules* **5** (2004) 1999.
218. Y. E. Greish, J. Blender, S. Lakahmi, P. W. Brown, H. R. Allcock, C. Laurencin, *Biomaterials* **26** (2005) 1.
219. D. Louda-Siegmann, C. Carraher: Macromolecules Containing Metal and Metal-Like Elements. Vol. 4. *Group IVA Polymers*, Wiley, Hoboken, NJ, 2005, Chap. 7.
220. H. R. Allcock, R. Allen, J. O'Brian, *J. Chem. Soc. Chem. Comm.* 1976, 717.
221. A. Andrianov, *J. Inorg. Organomet. Polym. Mater.* **16** (2006) 397.
222. H. R. Allcock, R. Wood, *J. Polym. Sci. B.* **44** (2006) 2358.
223. H. R. Allcock, *J. Inorg. Organomet. Polym. Mater.* **15** (2005) 57.
224. C. Allen, D. Hermandez-Rubio: *Application Aspects of Poly (organophosphazenes)*, Nova Science Publishers, Hauppauge, NY, 2004.
225. W. Balmain, *Phil. Mag.* **21** (1842) 10.
226. R. Wentorf, *J. Chem. Phys.* **26** (1957) 956.
227. T. O'Connor, *J. Amer. Chem. Soc.* **84** (1962) 1753.
228. F. Bundy, R. Wentorf, *J. Chem. Phys.* **38** (1963) 1144.
229. *Gmelin*, Boron Compounds, 2nd Suppl. 1(1983) 304.
230. N. J. Archer, *Spec. Publ. Chem. Soc.*No. 30(1977) 167.
231. C. S. Kalyoncu, *Ceram. Eng. Sci. Proc.* (1985) 1356.
232. T. Takahashi, H. Itoh, A. Takeuchi, *J. Cryst. Growth* **47** (1979) 245.
233. M. Sano, M. Aoki, *Thin Solid Films* **83** (1981) 247.
234. K. L. Chopra et al., *Thin Solid Films* **126** (1985) 307.
235. K. Gonslaves, X. Chen, in J. Salamone (ed.): *Polymeric Materials Encyclopedia*, vol. 5, CRC Press, Boca Raton, FL 1996, p. 3259.
236. A. J. Banister, I. B. Gorrell, *Adv. Mater.* **10** (1998) no. 17, 1415.
237. F. P. Burt, *J. Chem. Soc.* 1910, 1171.
238. C. Mikulski et al., *J. Am. Chem. Soc.* **97** (1975) 6358.
239. M. Akhtar et al., in C. Carraher, J. Sheats, C. U. Pittman, Jr. (eds.): *Organometallic Polymers*, Chap. 30, Academic Press, New York 1978.
240. R. H. Baughman, P. A. Apgar, R. R. Chance, A. G. MacDiarmid, A. F. Garito, *J. Chem. Phys.* **66** (1977) 401.
241. A. G. MacDiarmid in R. B. King (ed.): *Inorganic Compounds with Unusual Properties*,Chap. 6, American Chemical Society, Washington D.C. 1976.
242. R. L. Green et al., *Lect. Notes Phys.* **65** (1977) 603.
243. J. Passmore, M. Rao, *J. Chem. Soc., Chem. Commun*, 1980 1268.
244. F. Kennett, G. MacLean, J. Passmore, M. Rao, *J. Chem. Soc., Dalton Trans* **1982** 851.
245. A. Banister, Z. Hauptam, J. Passmore, C. Wong, P. White, *J. Chem. Soc., Dalton Trans* 1986 2371.
246. A. Banister, Z. Haupton, A. Kendrick, *J. Chem. Soc., Chem. Commun.* 1983 1016.
247. A. Banister, Z. Hauptman, A. Kendrick, R. Small, *J. Chem. Soc., Dalton Trans* 1987 915. H. Fritz, R. Bruchhaus, *Z. Naturforsch. B.* **38** (1983) 1375.
248. R. L. Greene, G. B. Street, L. J. Suter, *Phys. Rev. Lett.* **34** (1975) 577.
249. W. H. Jones, R. Bardo, *J. Phy. Chem.* **97** (1993) 4974.
250. I. Kakada, *J. Cryst. Growth* **55** (1981) 447.
251. H. Kahlert, B. Kundu, *Mater. Res. Bull.* **11** (1976) 967.
252. R. Scranton, *J. Appl. Phys.* **48** (1977) 3838.
253. R. Scranton, J. Best, J. McCaldin, *J. Vac. Sci. Technol.* **14** (1977) 930.
254. M. Cohen, J. Harris, *Appl. Phys. Lett.* **33** (1978) 812.
255. A. Thomas, J. Woods, Z. Hauptman, *J. Phys. D: Appl. Phys.* **16** (1983) 1123.
256. Asahi Chemical Industry Co. Ltd., JP 55 137 671, 1980; 58 93 164, 1983; 58 93 177, 1983; 59 49 155, 1984.

257 Electricité de France Service National, EP 97 078 B1, 1986 (J. Badoz et al.).
258 Sony Corp., JP 55 108 944 A, 1980 (T. Yoshioka, M. Ogawa, K. Hayashi).
259 Sony Corp., EP 517 165 A1, 1992; EP 517 165 B1, 1997; JP 07 230 984, 1955 (S. Kadomura).
260 Sony Corp., JP 07 263 420, 1995 (T. Nagayama).
261 Sony Corp., JP 07 263 426, 1995 (N. Akiba).
262 The United States of America as represented by the Secretary of the Army, US 4 206 705, 1980 (Z. Iqbal, H. Fair, D. Downs).
263 Nippon Denso Co. Ltd., JP 59 129 261, 1984 (H. Murametsu).
264 Sony Corp., JP 07 130 735, 1995 (T. Nagayama).
265 Sony Corp., JP 09 148 285, 1997 (J. Sato).
266 Toshiba Corp., JP 05 136 458, 1993 (T. Kamakura).
267 H. B. Mark, A. Voulgaropoulos, C. Meyer, *J. Chem. Soc., Chem. Commun.* (1981) 1021.
268 J. Rubinson, T. Behymer, H. Mark, *J. Amer. Chem. Soc.* **104** (1982) 1224.
269 R. Nowak, C. Joyal, D. Weber, *J. Electroanal. Chem.* **143** (1983) 413.
270 M. Akiba, T. Takada, *Purasuchikkusu* **55** (2004) 77.
271 T. Yasuda, *Ryusan Kogyo* **56** (2003) 113.
272 A. Kuznetsov, O. Kulikova, *Prog. Rubber Plastics Technol.* **16** (2000) 255.
273 B. R. Currell, A. J. Williams, A. J. Mooney and B. J. Nash, *Adv. Chem. Ser.* **1** (1974) 1.
274 B. P. Block, *Inorg. Macromol. Rev.* **1** (1970) no. 2, 115.
275 J. Sheats, C. Carraher, C. Pittman, M. Zeldin, B. Currell: *Inorganic and Metal-Containing Polymeric Materials*, Plenum, New York 1990.
276 C. Pittman, C. Carraher, M. Zeldin, B. Culbertson, J. Sheats: *Metal-Containing Polymeric Materials*, Plenum, New York 1996.
277 M. Zeldin, K. Wynne, H. Allcock: *Inorganic and Organometallic Polymers*, ACS, Washington, DC 1998.
278 Macromolecule-Metal Complexes III, Seclected Papers from seminar of July 1989; *J. Macromol. Sci. Chem.* **A27** (1990) nos. 9 – 11, 1109.
279 I. Manners, *Ann. Rep. Prog. Chem., Sect. A. Inorg. Chem.* **88** (1991) 77; **89** (1992) 93; **90** (1993) 103;91 (1994) 131; **92** (1995) 127; **93** (1996) 129.
280 J. Mark, H. Allcock, R. West, *Inorganic Polymers*, Prentice Hall, New York 1992.
281 J. Sheats, C. Carraher, C. Pittman, M. Zeldin, in C. Craver, C. Carraher (eds.): *Applied Polymer Science*, Elsevier, New York 1999.
282 S. Takahashi, K. Sonogashira, in J. Salamone (ed.): *Polymeric Materials Encyclopedia*, vol. **6**, CRC Press, Boca Raton, FL 1996, p. 4804.
283 A. D. Hunter, A. Guo, in J. Salamone (ed.): *Polymeric Materials Encyclopedia*, vol. **6**, CRC Press, Boca Raton, FL 1996, p. 4813.
284 T. Endo, I. Tomita, in J. Salamone (ed.): *Polymeric Materials Encyclopedia*, vol. **6**, CRC Press, Boca Raton, FL 1996, p. 4822.
285 R. Wei, L. Ya, W. Jinguo, X. Qifeng, in, J. Salamone (ed.): *Polymeric Materials Encyclopedia*, vol. **6**, CRC Press, Boca Raton, FL 1996, p. 4826.
286 W. L. Driessen (ed.): *Macromolecule-Metal Complexes*, Huthig & Wepf, Oxford, CT 1998.
287 F. Ciardelli, E. Tsuchida, D. Wohrle (eds.): *Macromolecule-Metal Complexes*, Springer-Verlag, Berlin 1996.
288 A. Pomogailo, V. Savost'yanov: *Synthesis and Polymerization of Metal-Containing Monomers*, CRC Press, Boca Raton, New York 1994.
289 E. Neuse, H. Rosenberg: *Metallocene Polymers*, Marcel Dekker, New York 1970.
290 Dow Corning, GB 1 582 081, 1980.
291 C. Carraher, F. Li, D. Siegman-Louda, C. Butler, J. Ross, *PMSE* **77** (1997) 499.
292 C. Carraher, V. Saurino, C. Butler, D. Sterling, *PMSE* **72** (1995) 192.
293 C. Carraher, D. Sterling, C. Butler, T. Ridgway, *Biotechnical Polymers*, Chap. 13, Technomic, Lancaster, PA 1995.
294 C. Carraher, M. Nagata, H. Stewart, S. Miao, S. Carraher, A. Gaonkar, C. Highland, F. Li, *PMSE* **79** (1998) 52.
295 C. Carraher, A. Gaonkar, H. Stewart, S. Miao, D. Mitchell, C. Barosy, M. Colbert, R. Duffield, *Polymer Reprints* **38** (1997) no. 2, 572.
296 C. Carraher, H. Stewart, W. Soldani, J. Dela Torre, B. Pandya, L. Reckleben, *Metal-Containing Polymeric Materials*, Plenum, NY 1996.
297 I. V. Yannas, J. F. Burke, *J. Biomed. Mater. Res.* **14** (1980) 65, 107.
298 R. W. Drisko, T. B. O'Neill, L. K. Schwab, Technical Note no. N-1480, Naval Facilities Engineering Command, Point Hueneme, CA, May 1977.
299 Y. Chujo, in J. Salamone (ed.): *Polymeric Materials Encyclopedia*, vol. 6, CRC Press, Boca Raton, FL 1996, p. 4795.
300 W. Gerrard, *J. Chem. Soc.* 1960, 3170.
301 A. B. Burg, *Angew. Chem.* **72** (1960) 183.
302 C. Carraher, R. Schwarz, J. Schroeder, M. Schwarz, *J. Macromol. Sci. Chem.* **A15** (1981) no. 5, 773.
303 C. Carraher et al., *Polym. Mater.* **55** (1986) 469.
304 C. Carraher, V. Foster, R. Linville, D. Stevison, *Polym. Mater.* **56** (1987) 401.
305 C. Carraher, R. Schwarz, J. Schroeder, M. Schwarz, H. M. Molloy, *Org. Coat. Plast. Chem.* **43** (1981) 798.
306 C. Carraher, R. Schwarz, J. Schroeder, M. Schwarz in C. Carraher, J. Preston (eds.): *Interfacial Synthesis*, vol. 3, Chap. 6, Marcel Dekker, New York 1982.
307 C. Carraher et al., *Org. Coat. Plast. Chem.* **44** (1982) 753.
308 J. Wrobleski, D. B. Brown, *Inorg. Chem.* **18** (1979) 498, 2738.
309 H. Matsuda, H. Nakaniski, M. Kato, *J. Polym. Sci. Polym. Lett. Ed.* **22** (1984) 107.
310 J. E. Sheats, C. Carraher, C. Pittman (eds.): *Metal-Containing Polymeric Systems*, Plenum Press, New York 1985.
311 X. Zhang, R. West, *J. Polym. Sci. Polym. Chem. Ed.* **22** (1984) 159.
312 R. Liepins, M. Timmons, N. Morosoff, J. Surles in J. Sheats, C. Carraher, C. Pittman (eds.): *Metal-Containing Polymeric Systems*, Plenum Publishing, New York 1985, pp. 225 – 236.
313 H. Yasuda et al., in J. Sheats, C. Carraher, C. U. Pittman (eds.): *Metal-Containing Polymeric Systems*, Plenum Publishing, New York 1985, pp. 275 – 290.
314 C. Carraher et al., *Organic Coatings Plastics Chemistry* **44** (1981) 753.
315 E. Savinova et al., *J. Mol. Catal.* **32** (1985) 149.
316 M. Kaneko, A. Yamada, *Adv. Polym. Sci.* **55** (1983).
317 M. S. Wrighton (ed.): *Interfacial Photoprocesses: Energy Conversion and Synthesis*, American Chemical Society, Washington 1980.
318 U. T. Müller-Westerhoff, A. I. Nazzal, US 4 379 740, 1983.
319 F. C. Anson et al., *J. Am. Chem. Soc.* **106** (1984) 59.
320 E. R. Savinova et al., *J. Mol. Catal.* **32** (1985) 149, 159.
321 in [11] pp. 137 – 147.
322 B. J. Spalding, *Chemical Week* **135** (1986) no. Oct. 8, 29.
323 C. Carraher, S. Tsuji, J. DiNunzio, W. Feld, *Polym. Mater.* **55** (1986) 875.

324 C. Carraher, J. Schroeder, *Polym. Prepr.* **16** (1975) 659, *J. Polym. Sci. Polym. Lett. Ed.* **13** (1975) 215.
325 L. Donaruma, *Polym. Prepr. Am. Chem. Soc. Div. Polym. Chem.* **22** (1981) 1.
326 B. Jiang et al., *Coord. Chem. Rev.* **171** (1998) 365.
327 W. E. Jones, L. Hermans, B. Jiang, in V. Ramamurthy, K. Schanze (eds.): *Molecular and Supramolecular Photochemistry*, vol. **2**, Marcel Dekker, New York 1999.
328 C. Pittman, *J. Paint Technol.* **39** (1967) 585.
329 N. E. Wolff, R. J. Pressley, *Appl. Phys. Lett.* **2** (1963) 152.
330 in [5] p. 425.
331 J. E. Sheats et al., in B. M. Culbertson, C. U. Pittman, Jr. (eds.): *New and Unusual Monomers and Polymers*, Plenum Press, New York 1983, pp. 83 – 98; *Polym. Mater. Sci. Eng.* **49** (1983) no. 2, 363.
332 N. Bilow, H. Rosenberg, *J. Polym. Sci. Part A-1* **7** (1969) 2689.
333 N. Bilow, H. Rosenberg, US 3 640 961, 1972.
334 C. U. Pittman, Jr., R. F. Felis, *J. Organomet. Chem.* **72** (1974) 389.
335 C. U. Pittman, R. F. Felis, *J. Organomet. Chem.* **72** (1974) 399.
336 R. Tannenbaum, E. P. Goldberg, C. L. Flenniken in J. Sheats, C. Carraher, C. Pittman (eds.): *Metal-Containing Polymeric Systems*, Plenum Publishing, New York 1985, pp. 303 – 340.
337 C. Carraher, X. Xu, *PMSE* **75** (1996).
338 C. Carraher, X. Xu: *Modification of Polymers*, Plenum, New York 1997.
339 C. Carraher, V. Forster, R. Linville, D. Steverson, R. Ventatachalam, *Adhesives, Sealants, and Coatings for Space and Harsh Environments, Chap. 19*, Plenum, NT 1988.
340 C. Carraher, A. Taylor-Murphy, *PMSE* **76** (1997) 409.
341 S. Yajima, H. Hayashi, M. Omori, *Chem. Letters.* (1975) 931 and 1209.
342 R. West, *Am. Ceram. Soc. Bull.* **62** (1983) 825.
343 D. Seyferth, G. H. Wiseman, *J. Amer. Ceram. Soc.* **67** (1984); Massachusetts Institute of Technology, US 4 482 669, 1984 (D. Seyferth, G. H. Wiseman).
344 R. Laine, Y. Blum, A. Chow, R. Hamlin, K. Schwartz, D. Rowecliff, *Polymer Reprints* **28** (1987) no. 1, 393.
345 M. Arai, S. Sakurada, T. Isoda, T. Tomizawa, *Polymer Reprints* **28** (1987) no. 1, 407.
346 C. Narula, R. Paine, R. Schaeffer, *Polymer Reprints* **28** (1987) no. 1, 454.
347 C. Carraher, H. Zhuang, F. Medina, D. Baird, R. Pennisi, B. Landreth, F. Nounou, *PMSE* **62** (1990) 633.
348 M. Kaneko, A. Yamada, *Adv. Polym. Sci.* (1983) 55.
349 US 4 379 740, 1983 (U. Muller-Westerhoff, A. Nazzal).
350 F. Anson, *J. Amer. Chem. Soc.* **106** (1984) 59.
351 B. J. Spaulding, *Chemical Week, Oct 8 (1986) 29 and C. M. Elliott, Abstracts, 193 ACS Meeting, Denver CO, April, 1987, INOR 232.*
352 M. MacLachlan, P. Aroca, N. Coombs, I. Manners, G. Ozin, *Adv. Mater.* **10** (1998) 144.
353 G. Barot, C. Carraher: *Recent Advances in Inorganic and Organometallic Polymers*, Springer, NY, 2008.
354 C. Carraher, M. Roner, G. Barot, *J. Inorg. Organomet. Polym. Mater.* **17** (2007) 595.
355 C. Carraher et al., *J. Inorg. Organomet. Polym. Mater.* **16** (2006) 249.
356 M. Roner, C. Carraher, J. Roehr, K. Bassett, *J. Polym. Mater.* **23** (2006) 153.
357 US 5043463, 1991 (C. Carraher, C. Butler).
358 C. Carraher, C. Butler, L. Reckleben, *Cosmetic and Pharmaceutical Applications of Polymers*, Plenum, NY, 1991.
359 US 584760, 1998 (C. Carraher, C. Butler).
360 C. Butler, C. Carraher, *Polym. Mater. Sci. Eng.* **80** (1999) 365.
361 C. Carraher, H. Stewart, S. Carraher, M. Nagata, S. Miao, *J. Polym. Mater.* **18** (2001) 111.
362 C. Pittman, C. Carraher: *Frontiers in Transition Metal-Containing Polymers*, Wiley, Hoboken, NJ, 2007.
363 M. Roner, C. Carraher, S. Dhanji, *Polym. Mater. Sci Eng.* **93** (2005) 410.
364 M. Roner, C. Carraher, S. Dhanji, *Polym. Mater. Sci. Eng.* **92** (2005) 499.
365 T. J. Kelly, P. L. Pauson, *Nature* **168** (1951) 1039.
366 F. S. Arimoto, A. C. Haven, *J. Am. Chem. Soc.* **77** (1955) 6295.
367 US 2821512, 1958 (A. C. Haven).
368 C. U. Pittman, *J. Inorg. Organomet. Polym. Mater.* **15** (2005) 33.
369 M. G. Baldwin, K. E. Johnson, *J. Polym. Sci.* **5** (1967) 2901.
370 M. George, G. Hayes, *Polymer* **15** (1974) 397.
371 E. W. Neuse, *J. Inorg. Organomet. Polym. Mater.* **15** (2005) 1.
372 E. W. Neuse: *Macromolecules Containing Metal and Metal-Like Elements. Vol. 3. Biomedical Applications*, Wiley, Hoboken, NJ, 2004, Chap. 6.
373 *Proceedings of the IIIrd Internat. Conference on Ultrastructure Processing of Glasses, Ceramics and Composites*, Feb. 24 – 27, 1987.
374 C. J. Brinker, D. Clark, D. R. Ulrich (eds.): *Better Ceramics Through Chemistry II*, Materials Res. Soc., Pittsburgh, Pa., 1986.
375 L. L. Hench, D. R. Ulrich (eds.): *Science of Ceramic Chemical Processing*, Wiley-Interscience, New York 1986.
376 J. O'Reilly, B. Coltrain, in J. Salamone (ed.): *Polymeric Materials Encyclopedia*, vol. **6**, CRC Press, Boca Raton, FL 1996, p. 4772.
377 G. Wilkes, J. Wen, in J. Salamone (ed.): *Polymeric Materials Encyclopedia*, vol. **6**, CRC Press, Boca Raton, FL 1996, p. 4782.
378 Y. Chujo, in J. Salamone (ed.): *Polymeric Materials Encyclopedia*, vol. **6**, CRC Press, Boca Raton, FL 1996, p. 4793.
379 H. Benoit, S. Lambert, N. Job, J. Pirad, *Catal. Prep.* 2007, 163.
380 G. Oye et al., *Adv. Colloid Interface Sci.* **123** (2006) 17.
381 J. Hajek, D. Murzin, *Nanocatalysis* (2006) 9.
382 B. Coltrain, L. Kelts, *Surfactant Sci. Ser.* **131** (2006) 637.
383 D. Avnir, T. Coradin, O. Lev, J. Livage, *J. Mater. Chem.* **16** (2006) 1013.
384 S. Sakka, *J. Sol-Gel Sci. Technol.* **37** (2006) 135.
385 C. J. Brinker et al., "Inorganic and Organometallic Polymers," *ACS Symp. Ser.* **360** (1988); D. R. Ulrich, *CHEMTECH* **18** (1988) 242.
386 H. K. Schmidt, "Inorganic and Organometallic Polymers," *ACS Symp. Ser.* **360** (1988).
387 H. Schmidt, H. Scholze, G. Tunker, *J. Non Cryst. Solids* **80** (1986) 557.
388 H. Schmidt, H. Scholze, *Springer Proceedings in Physics*, vol. **6**, Aerogels, Springer Publishing Co., 1986, p. 49.
389 G. Philipp, H. Schmidt, *J. Non Cryst. Solids* **63** (1984) 261.
390 G. Tunker, H. Patzelt, H. Schmidt, H. Scholze, *Glastechn. Ber.* **59** (1986) 272.
391 S. Sakka, *Polym. Prepr. Am. Chem. Soc. Div. Polym. Chem.* **28** (1987) no. 1, 430.
392 P. Brooke, H. Schurmans, J. Verhoest: *Inorganic Fibers and Composite Materials: A Survey of Recent Developments*, Pergamon Press, Oxford, UK, 1984.
393 H. Huang, R. H. Glaser, G. L. Wilkes, "Inorganic and Organometallic Polymers," *ACS Symp. Ser.* **360** (1988).
394 H. Dislich, *Angew. Chem. Int. Ed. Engl.* **106** (1970) 363.
395 B. E. Yoldas, *J. Mater. Sci.* **12** (1977) 203; **14** (1979) 1843.

396 G. Pajonk, J. Regalbuto, *Aerogel Synthesis*, CRC Press, Boca Raton, FL, 2007.
397 S. Jones, *J. Sol-Gel Sci. Technol.* **40** (2006) 351.
398 C. Carraher, *Polym. News* **30** (2005) 62,386.
399 C. Carraher, *CHEMTECH* 1972, 741.
400 V. C. Farmer, A. R. Fraser, J. M. Tait, *J. Chem. Soc. Chem. Commun.* 1977, 462.
401 K. Wada et al., *Clay Miner.* **8** (1970) 487.
402 G. Alberti, U. Costantino, *J. Chromatogr.* **102** (1974) 5.
403 N. H. Ray: *Inorganic Polymers*, Chap. 7, Academic Press, New York 1978.
404 J. D. Birchall, J. A. Bradbury, J. Dinwoodie in W. Watt, B. V. Perov (eds.): *Handbook of Composites*, North Holland, Amsterdam 1985, p. 115.
405 J. D. Birchall: "Inorganic Fibers" in M. B. Bever (ed.): *Encyclopedia of Material Sciences and Engineering*, Pergamon Press, Oxford, UK 1986, p. 2333.
406 Sumitomo Chemical Co., US 4 101 615, 1973 (S. Horikiri, K. Tsuji, Y. Abe); 4 152 149, 1977 (S. Horikiri, K. Tsuji, Y. Abe).
407 J. Tretzel et al., Production, Processing and Application of Enka Silica Fibers, *57th Annual TRI Conference*, Charlotte, N.C., April 21 – 22, 1987.
408 W. C. Miller in M. Grayson (ed.): *Encyclopedia of Textiles, Fibers and Nonwoven Fabrics*, J. Wiley and Sons, New York 1984, p. 443.
409 *Chem. Week* **13** (1987) no. April 22.
410 J. C. Romine, *Ceram. Eng. Sci. Proc.* **8** (1987) 755.
411 J. O. Carlsson: "Boron Fibers" in M. B. Bever (ed.): *Encyclopedia of Material Science and Engineering*, Pergamon Press, Oxford, UK, 1986, p. 402.
412 W. Noll: *Chemistry and Technology of Silicones*, Academic Press, New York 1968.
413 J. O. Carlsson: "Silicon Carbide Fibers" in M. B. Beuer (ed.): *Encyclopedia of Material Science and Engineering*, Pergamon Press, Oxford, UK, 1986, p. 4406.
414 S. Yajima, Y. Hasegawa, J. Hayashi, M. Imura, *J. Mater. Sci.* **13** (1978) no. 12, 2569.
415 Y. Hasegawa, K. Okamura, *Yogyo Kyokaishi* **95** (1987) 99, 103.
416 B. Arkles, *CHEMTECH* **13** (1983) 542.
417 W. Verbeck, DE-OS 2 218 960, 1973.
418 K. Okamura, M. Sato, Y. Hasegawa, *Ceram. Int.* **13** (1987) 55.
419 T. Yamamura et al., "Proc. World Congress on High Tech Ceramics, 6th Int. Meeting on Modern Ceramics Technologies," *Mater. Sci. Monogr.* **38** (1987).
420 E. Sako, O. Kawamura, US 3 746 196, 1973.
421 N. Tsubokawa, T. Yoshihara, H. Ueno, in J. Salamone (ed.): *Polymeric Materials Encyclopedia*, vol. **2**, CRC Press, Boca Raton, FL 1996, p. 973.
422 C. Carraher: *Polymer Chemistry: An Introduction*, 4th ed., Dekker, New York 1996, p. 122.
423 D. G. LeGrand, J. Magila, *Eng. Sci.* **10** (1970) 349.
424 J. Riffle, R. Freelin, A. Banthia, J. McGrath, *J. Macromol. Sci. Chem.* **A15** (1981) 967.
425 K. Okamura, *Composites* **18** (1987) 107.
426 W. Lynch: *Handbook of Silicon Rubber Fabrication*, Van Nostrand, New York 1978.
427 M. Ranney: *Silicones*, Noyes Data Corp., Park Ridge, N.J., 1977.
428 B. Arkles, W. Peterson, R. Anderson: *Organosilicon Compounds: Register and Review*, Petrarch Systems, Bristol, Pa., 1982.
429 A. Isquith, L. Abbot, P. Walters, *Appl. Microbiol.* **24** (1972) 859.
430 B. Arkles, W. Brinigar, A. Miller in D. Leyden, W. Collin (eds.): *Silylated Surfaces*, Gordon Breach, New York 1980.
431 W. Robb, *Ann. N.Y. Acad. Sci.* **146** (1968) 119.
432 M. Nelson, US 3 020 260, 1962.
433 E. Jeram, US 4 072 635, 1978.
434 T. Cronin, US 3 293 663, 1965.
435 N. Dagalakis et al., *J. Biomed. Mater. Res.* **14** (1980) 511.
436 R. A. Erb, US 4 245 623, 1981.
437 R. A. Erb, T. P. Reed, *J. Reprod. Med.* **23** (1979) 65.
438 A. Zaffaroni, US 3 905 360, 1975.
439 P. Hittmaier, W. Hechtel, H. Eckhart, US 4 035 453, 1977.
440 C. N. Raptis, R. Yu, J. G. Knapp, *J. Prosthet. Dent.* **4414** (1980) 447.
441 A. B. Swanson, US 3 875 594, 1975.
442 T. Honda et al., *J. Thorac. Cardiovasc. Surg.* **69** (1975) 92.
443 D. Laurin, US 3 994 988, 1976.
444 E. Pierce, N. Dibelus, *Trans. Am. Soc. Artif. Intern. Organs* **14** (1968) 220.
445 J. Montalvo, US 3 869 354, 1975.
446 J. Brown, G. Slusarczuk, O. LeBlanc, US 3 743 588, 1971.
447 H. Allcock in C. Carraher, J. E. Sheats, C. U. Pittman, Jr. (eds.): *Organometallic Polymers*, Academic Press, New York 1978, p. 283.
448 H. Allcock, R. Allen, J. O'Brien, *J. Chem. Soc. Chem. Commun.* 1976, 717.
449 C. Carraher, W. J. Scott, J. Schroeder, D. J. Giron, *J. Macromol. Sci. Chem.* **A15** (1981) no. 4, 625.
450 C. Carraher et al., in C. Gebelein, C. Carraher (eds.): *Polymeric Materials in Medication*, Chap. 15, Plenum Publishing, New York 1985.
451 C. Carraher in C. Gebelein, C. Carraher (eds.): *Bioactive Polymeric Systems*, Chap. 22, Plenum Publishing, New York 1985.
452 C. Carraher, I. Lopez, D. Giron, *Polym. Mater.* **53** (1985) 644.
453 C. Carraher, I. Lopez, D. Giron in C. Gebelein (ed.): *Advances in Biomedical Polymers*, Plenum Publishing, New York 1987, p. 311.
454 C. Carraher, N. Bigley, M. Trombley, D. Giron, *Polym. Mater.* **57** (1987) 177.
455 A. Winston, in C. Gebelein, C. Carraher (eds.): *Bioactive Polymeric Systems*, Chap. 21, Plenum Publishing, New York 1985.
456 A. Winston, *Polym. News* **10** (1984) 6.
457 C. Carraher et al., in C. Carraher, C. Gebelein (eds.): *Biological Activities of Polymers*, Chap. 2, American Chemical Society, Washington, D.C., 1982.
458 D. Anderson, J. Mendoza, B. Garg, R. V. Subramanian in C. Carraher, C. Gebelein (eds.): *Biological Activities of Polymers*, Chap. 3, American Chemical Society, Washington, D.C., 1982.
459 R. V. Subramanian, B. Garg, J. Corredor in C. Carraher, J. Sheats, C. U. Pittman, Jr. (eds.): *Organometallic Polymers*, Chap. 19, Academic Press, New York 1978.
460 C. Carraher et al., *Polym. Mater.* **57** (1987) 186.
461 C. Carraher, D. Giron, J. Schroeder, C. McNeely, US 4 312 981, 1982.
462 S. Matsuda, H. Kudara, US 4 174 339, 1979.
463 Shipcare International,10 (1978) no. 11, 23.
464 J. A. Montemarano, E. J. Dyckman, *J. Paint Technol.* **47** (1975) 59.
465 A. T. Phillip, *Prog. Org. Coat.* **2** (1973/74) 159.
466 J. Leebrick, US 3 167 473, 1965.
467 D. Atherton, J. Verborgt, M. A. Winkeler, *J. Coat. Technol.* **51** (1979) 88.
468 D. Gabel, *Science Synthesis* **6** (2004) 1277.

469 K. Shen, D. Ferm, *Recent Adv. Flame Retardancy Polym. Mater.* **11** (2000) 213.
470 R. Bedell, M. J. Frazer, W. Gerrard, *J. Chem. Soc.* 1960, 4037.
471 R. W. Upson, US 2 511 310, 1950.
472 K. Hizawa, E. Nojimoto, JP 4791, 1953.
473 J. M. Thompson: "Other Boron-Containing Polymers" in M. F. Lappert, G. J. Leigh (eds.): *Developments in Inorganic Polymer Chemistry*, Elsevier, Amsterdam 1962, p. 57.
474 R. M. Washburn, E. Levens, C. F. Albright, F. A. Billig: "Metal-Organic Compounds," *Adv. Chem. Ser.* **23** (1959) 139, 153.
475 W. Kuchen, R. Brinkmann, *Angew. Chem.* **72** (1960) 564.
476 R. I. Wagner, F. F. Caserio, *J. Inorg. Nucl. Chem.* **11** (1959) 259.
477 T. Wartik, E. F. Apple, *J. Am. Chem. Soc.* **77** (1955) 6400.
478 H. Rosenberg, F. L. Hedberg, Tech. Rept. AFML-TR-69–68, AD 863, 817 (1969).
479 H. Steinberg, R. J. Brotherton: *Organoboron Chemistry*, vol. **2**, J. Wiley and Sons, New York 1966, pp. 329 – 340, 363 – 364, 418.
480 B. M. Mikhailov, A. F. Galkin, *Bull. Acad. Sci. USSR, Div. Chem. Sci. (Engl. Transl.)* **12** (1963) 575.
481 R. H. Toeniskoetter, R. Didchenko, CA 685 580, 1964 to Union Carbide.
482 R. H. Toeniskoetter, K. A. Killip, *J. Am. Chem. Soc.* **86** (1964) 690.
483 V. Gutmann, A. Metter, R. Schlegel, Monatsh. Chem. **94** (1963) 1071.
484 G. Chainani et al., *J. Appl. Chem.* **15** (1965) 372.
485 K. O. Knollmuller, R. N. Scott, H. Kwasnik, J. F. Sieckhaus, *J. Polym. Sci. Polym. Chem. Ed. A1,* 9(1971) 1071.
486 S. Papetti et al., *J. Polym. Sci. Polym. Chem. Ed.* **4** (1966) 1623.
487 R. P. Alexander, H. Schroeder, *Angew. Chem.* **76** (1964) 278.
488 H. A. Schroeder et al., *Inorg. Chem.* **8** (1969) no. 11, 2444.
489 H. A. Schroeder, *Inorg. Macromol. Rev.* **1** (1970) 45.
490 I. B. Atkinson, D. B. Clapp, W. F. Flavell, *Polym. Prepr. Am. Chem. Soc. Div. Polym. Chem.* **13** (1972) 770.
491 N. Semenuk, S. Papetti, H. Schroeder, *Inorg. Chem.* **8** (1969) no. 11, 2441.
492 J. Frieder, *Coord. Chem. Rev.* **250** (2006) 1107.
493 J. Rinse, *Ind. Eng. Chem.* **56** (1964) no. 5, 42.
494 DuPont, US 2 744 071, 1956 (C. N. Theobald).
495 K. C. Pande, R. C. Mehrotra, *Z. Anorg. Allg. Chem.* **286** (1956) 291.
496 Peter Lunt and Co., GB 573 083, 1945 (E. J. Lush).
497 Hardman and Holden Ltd., GB 772 144, 1957 (J. Rinse).
498 A. L. McCloskey et al., WADC Tech. Rept. 59–761 (1959) pp. 117 – 200.
499 Koninklijke Industrieele Maatschappij voorheen Noury, GB 783 679, 1957 (N. V. Van der Lande).
500 K. A. Andrianov: *Metalorganic Polymers*, Chap. 5, Interscience Publishers, New York 1965.
501 K. A. Andrianov, V. M. Novikov, *Vysokomol. Soedin.* **1** (1959) 1390.
502 V. Kugler, *J. Polym. Sci.* **29** (1958) 637.
503 T. R. Patterson et al., *J. Am. Chem. Soc.* **81** (1959) 4213.
504 J. I. Jones: Polymetallosiloxanes in M. L. Lappert, G. J. Leigh (eds.): *Developments in Inorganic Polymer Chemistry*, Part I, Part II, Chap. 7, 8, Elsevier, New York 1962.
505 D. C. Bradley: Polymeric Metal Alkoxides, Organometalloxanes and Organometalloxanosiloxanes in F. G. A. Stone, W. A. G. Graham (eds.): *Inorganic Polymers*, Chap. 7, Academic Press, New York 1962.
506 I. K. Stavitskii, *Vysokomol. Soedin.* **1** (1959) 1502.
507 S. N. Borisov, N. G. Sviridova, *Vysokomol. Soedin.* **3** (1961) 50.
508 R. D. Crain, P. E. Koenig, *WADC Tech. Rept. 69–427*(1960) p. 15.
509 F. A. Henglein, R. Lang, L. Schmack, *Makromol. Chem.* **22** (1957) 103.
510 Du Pont, US 2 512 058, (H. C. Gulledge).
511 T. Boyd, GB 728 751, 1955.
512 T. Boyd, US 2 716 656, 1955.
513 A. Hancock, R. Sidlow, *J. Oil Colour Chem. Assoc.* **35** (1952) 28.
514 W. H. Post, W. T. Schwartz, Jr., AD 255 545, U.S. Dept. Commerce Office Tech. Serv., Washington D.C., 1961.
515 K. A. Andrianov, S. V. Pichkhadze, *Vysokomol. Soedin.* **3** (1961) 577.
516 C. Ungurenasu, I. Haidec, *Rev. Roumaine Chim*, **13** (1968) 957.
517 C. C. Currie, US 2 672 455, 1954.
518 Monsanto Inc., GB 723 989, 1955.
519 R. M. Kary, US 2 646 440, 1953.
520 W. P. Gill et al., *Phys. Rev. Lett.* **38** (1977) 1305.
521 F. B. Bundy, K. T. Dunn, *J. Chem. Phys.* **71** (1979) 1550.
522 P. W. Bridgman, *Proc. Am. Acad. Arts Sci.* **74** (1941) 399.
523 E. Whalley, *Can. J. Chem.* **38** (1960) 2105.
524 Y. Okamoto, Z. Iqbal, R. H. Baughman, *J. Macromol. Sci. Chem.* **A25** (1988) no. 5 – 7, 799.
525 W. S. Chan, A. K. Jonsher, *Phys. Status Solidi* **32** (1969) 749.
526 J. Tsukamoto, A. Takahashi, *Jpn. J. Appl. Phys.* **25** (1986) L338.
527 M. Berthelot, *Ann. Chim. Phys.* **11** (1936) 15.
528 Y. Asano, *Jpn. J. Appl. Phys.* **22** (1983) 1618.
529 L. A. Wall, D. W. Brown, *J. Polym. Sci. Part C* **4** (1964) 1151.
530 Z. Iqbal et al., *J. Chem. Phys.* **85** (1986) 4019.
531 G. Grimm, H. Metzger, *Ber. Dtsch. Chem. Ges.* **69** (1936) 1356.
532 A. J. Brown, E. Whalley, *Inorg. Chem.* **7** (1968) 1254.
533 Y. Okamoto, P. S. Wojciechowski, *J. Chem. Soc. Chem. Commun.* 1982, 386.
534 H. Kobayashi et al., *Chem. Lett.* 1983, 1407.
535 H. Kobayashi, A. Kobayashi, Y. Sasaki, *Mol. Cryst. Liq. Cryst.* **118** (1985) 427.
536 A. Kobayashi et al., *Bull. Chem. Soc. Jpn.* **59** (1986) 3821.
537 H. Emeleus, J. Anderson: *Modern Aspects of Inorganic Chemistry*, 3rd ed., Routledge & Kegan, London 1960.
538 F. Cataldo, *J. Inorg. Organomet. Polym.* **7** (1997) no. 1, 35.
539 W. R. Bowman,, J. Colin, P. Kilian, A. Slawin, P. Wormald, J. D. Woolins, *Chem. Eur J.* **12** (2006) 6366.
540 C. Burchell, S. Aucott, S. Robertson, A. Slawin, J. D. Woolins, *Phosphorus Sulfur Silicon Related Elements* **179** (2004) 865.
541 F. Cataldo, Y. Keheyan, *Polyhedron* **21** (2002) 1825.
542 Y. L. Gefter: *Organophosphorus Monomers and Polymers*, Pergamon Press, London 1962.
543 H. Noth, W. Schragle, *Z. Naturforsch. B* **166** (1961) 473.
544 C. U. Pittman, Jr., *J. Polym. Sci. Polym. Chem. Ed.* **5** (1967) 371.
545 C. Shimasaki, H. Kitano, in J. Salamone (ed.): *Polymeric Materials Encyclopedia*, vol. **7**, CRC Press, Boca Raton FL 1996.
546 *Phosphorus, Sulfur, Silicon, Relat. Elem.*, ongoing journal.
547 D. Corbridge: Studies in Inorganic Chemistry 10: Phosphorus, An Outline of Its Chemistry, *Biochemistry and Technology*, 4th ed., Elsevier, Amsterdam 1990.
548 G. Borissov, in J. Salamone (ed.): *Polymeric Materials Encyclopedia*, vol. **7**, CRC Press, Boca Raton, FL 1996, p. 5092.
549 D. Liaw, D. Wang, in J. Salamone (ed.): *Polymeric Materials Encyclopedia*, vol. **9**, CRC Press, Boca Raton, FL 1996, p. 6569.

550 A. Shanov, Kh. Kabard-Balkar, *Plast. Massy* **6** (2006) 21.
551 S. Levchik, E. Weil, *J. Fire Sci.* **24** (2006) 345.
552 W. Habicher, I. Bauer, J. Pospisil, *Macromolecular Symposia, Polymers in Novel Applications*, vol. **225**, Wiley, 2005, p. 147.
553 M. Chaubal, A. Sen Gupta, S. Lopina, D. Bruley, *Crit. Rev. Therapeutic Drug Carrier Syst.* **20** (2003) 295.
554 A. McWilliams, H. Dorn, I. Manners: *Topics in Current Chemistry 220. New Approaches in Phosphorus Chemistry*, Springer-Verlag, 2002, p. 141.
555 T. Baumgartner, R. Reau, *Chem. Rev.* **106** (2006) 4681.
556 E. W. Neuse, G. J. Chris, *J. Macromol. Sci.* A-1(1967) 371.
557 H. R. Allcock: "Rings, Clusters and Polymers of the Main Group Elements," *ACS Symp. Ser.* **232** (1983).
558 N. Chawla, K. Chawla, *Journal of Metals* **58** (2005) 67.
559 Metal and Ceramic Matrix Composites, Oxford-Kobe Materials Seminar, Kobe, Japan, Sept. 19 – 22, 2000, Institute of Physics Publishing, Bristol, UK, 2004.
560 T. Yamauchi, *Chuzo Kogaku* **76** (2004) 1014.
561 L. Lu, M. Gupta, *Recent Res. Dev. Mater. Sci. Eng.* **2** (2003) 125.
562 R. Mason, L. Gintert, M. Singleton, D. Skelton, *Adv. Mater. Processes* **162** (2004) 37.

# Further Reading

A. S. Abd-El-Aziz, C. E. Carraher, C. U. Pittman, M. Zeldin (eds.): *Inorganic and Organometallic Macromolecules*, Springer Science+Business Media LLC, New York, NY 2008.

V. Chandrasekhar: *Inorganic and Organometallic Polymers*, Springer, Berlin, Heidelberg 2010.

R. de Jaeger, M. Gleria (eds.): *Silicon-based Inorganic Polymers*, Nova Science Publishers, New York 2008.

A. K. Roy: *Inorganic High Polymers, Kirk Othmer Encyclopedia of Chemical Technology*, 5th edition, John Wiley & Sons, Hoboken, online: DOI: 10.1002/0471238961.09141518181525.a01.pub2(December2009)

J. E. Mark, H. R. Allcock, R. West: *Inorganic Polymers*, 2. ed., Oxford Univ. Press, New York, NY 2005.

# Author Index

## A

Abdou-Sabet, Sabet, Advanced Elastomer Systems Akron, United States, *Thermoplastic Elastomers*, 3
Abts, Georg, Bayer MaterialScience AG, Leverkusen, Germany, *Polycarbonates*, 2
Adam, Norbert, Bayer MaterialScience AG, Leverkusen, Federal Republic of Germany, *Polyurethanes*, 3
Albrecht, Klaus, Evonik Industries AG, Darmstadt, Germany, *Polymethacrylates*, 2
Allsopp, Michael W., Independent PVC Technology Consultant, Heswall, England, *Poly(Vinyl Chloride)*, 3
Avar, Geza, Bayer MaterialScience AG, Leverkusen, Federal Republic of Germany, *Polyurethanes*, 3

## B

Bailey, Frederick E., Union Carbide Technical Center, Charleston WV, United States, *Polyoxyalkylenes*, 2
Baumann, Franz Erich, Degussa AG - High Performance Polymers, Marl, Germany, *Thermoplastic Elastomers*, 3
Blankenheim, Herbert, Bayer MaterialScience AG, Leverkusen, Federal Republic of Germany, *Polyurethanes*, 3
Breulmann, Michael, BASF SE, Ludwigshafen, Germany, *Polymers, Biodegradable*, 3
Brügging, Wilhelm, Hüls Aktiengesellschaft, Marl, Federal Republic of Germany, *Polyesters*, 2
Bryant, Robert G., NASA Langley Research Center, Hampton VA, USA, *Polyimides*, 2
Burkhardt, Gert, Institut für Kunststoffverarbeitung, Aachen, Germany, Plastics Processing, *1. Processing of Thermoplastics*, 1; *Plastics, Processing, 2. Processing of Thermosets*, 1; *Plastics Processing, 3. Machining, Bonding, Surface Treatment*, 1
Bussink, Jan, GE Plastics BV, Bergen op Zoom, The Netherlands, *Poly(Phenylene Oxides)*, 2; *Polymer Blends*, 3; *Polymers, High-Temperature*, 3

## C

Cadogan, David F., ICI Chemicals and Polymers, Runcorn, Cheshire, United Kingdom, *Plasticizers*, 1
Carlowitz, Bodo, Königstein im Taunus, Federal Republic of Germany, *Plastics, Properties and Testing*, 1
Carlson, D. Peter, Du Pont de Nemours & Co., Polymer Products Dept., Experimental Station Laboratory, Wilmington, United States, *Fluoropolymers, Organic*, 2
Carraher, Charles E. Jr., Florida Atlantic University, College of Science, Boca Raton, Florida, USA, *Inorganic Polymers*, 6
Chiriac, Constantin I., Institute of Macromolecular Chemistry, "Petru Poni", Jassy, Romania, *Polyureas*, 3
Collin, Gerd, DECHEMA e. V. Frankfurt a. M., Federal Republic of Germany, *Resins, Synthetic*, 5

## D

Diem, Hans, BASF Aktiengesellschaft, Ludwigshafen, Germany, *Amino Resins*, 5
Dorf, Ernst-Ulrich, Bayer AG, Leverkusen, Germany, *Polymers, High-Temperature*, 3
Dufils, Pierre Emmanuel, Solvay, Tavaux, France, *Poly(Vinylidene Chloride)*, 3

## E

Eckel, Thomas, Bayer MaterialScience AG, Leverkusen, Germany, *Polycarbonates*, 2

Ehrenstein, Gottfried W., Universität Erlangen-Nürnberg Erlangen-Tennenlohe, Federal Republic of Germany, *Reinforced Plastics*, 3

Eichenauer, Herbert, Bayer AG, Dormagen, Germany, *Polystyrene and Styrene Copolymers*, 3

Elgert, Karl-Friedrich, Institut für Textilchemie, Denkendorf, Federal Republic of Germany, *Plastics, Analysis*, 1

Elias, Hans-Georg, Michigan Molecular Institute, Midland, United States, *Plastics, General Survey, 1. Definition, Molecular Structure and Properties*, 1; *Plastics, General Survey, 2. Production of Polymers and Plastics*, 2; *Plastics, General Survey, 3. Supermolecular Structures*, 2; *Specialty Plastics*, 3; *Plastics, General Survey, 4. Polymer Composites*, 1; *Plastics, General Survey, 5. Plastics and Sustainability*, 1

Estes, Leland L., E. I. Du Pont de Nemours & Co., Nashville, Tennessee, USA, *Fibers, 4. Polyamide Fibers*, 4

## F

Fischer, Ingo, VESTOLIT GmbH, Marl, Germany, *Poly(Vinyl Chloride)*, 3

Frankenburg, Peter E., E. I. Du Pont de Nemours Wilmington, Delaware, USA, *Fibers, 10. Polytetrafluoroethylene Fibers*, 4

Fredric, L. Buchholz, The Dow Chemical Company, Midland, Michigan, United States, *Polyacrylamides and Poly(Acrylic Acids)*, 2

Freitag, Werner, Hüls Aktiengesellschaft, Marl, Federal Republic of Germany, *Resins, Synthetic*, 5

Friederichs, Wolfgang, Bayer MaterialScience AG, Dormagen, Federal Republic of Germany, *Polyurethanes*, 3

Frushour, Bruce G., Monsanto Chemical Company Springfield, MA, United States, *Polystyrene and Styrene Copolymers*, 3

Fuss, Robert, Kuraray Europe GmbH, Frankfurt am Main, Germany, *Polyvinyl Compounds, Others*, 3

## G

Gabara, Vlodek, Retired from E.I. Du Pont de Nemours and Company, Richmond, Va., USA, *High-Performance Fibers*, 4

Gahleitner, Markus, Borealis Polyolefine GmbH, Linz, Austria, *Polypropylene*, 2

Giersig, Manfred, Bayer MaterialScience AG, Leverkusen, Federal Republic of Germany, *Polyurethanes*, 3

Graupner, René, Lanxess Deutschland GmbH, Leverkusen, Germany, *Polyaspartates and Polysuccinimide*, 2

van de Grampel, Hendrik T., GE Plastics BV, Bergen op Zoom, The Netherlands, *Poly(Phenylene Oxides)*, 2; *Polymers, High-Temperature*, 3; *Polymer Blends*, 3

De Grave, Isidoor, BASF Aktiengesellschaft, Ludwigshafen, Federal Republic of Germany, *Foamed Plastics*, 4

## H

Halfmann, Michael, Bergische Universität Wuppertal, Wuppertal, Federal Republic of Germany, *Polyurethanes*, 3

Hallensleben, Manfred L., Universität Hannover, Institut für Makromolekulare Chemie, Hannover, Germany, *Polyvinyl Compounds, Others*, 3

Hamielec, Archie E., Institute for Polymer Production Technology, McMaster University, Canada, *Polymerization Processes, 2. Modeling of Processes and Reactors*, 1

Haubs, Michael, Ticona GmbH, Frankfurt am Main, Germany, *Polyoxymethylenes*, 2

Herth, Gregor, BASF SE, Ludwigshafen, Germany, *Polyacrylamides and Poly(Acrylic Acids)*, 2

Herzog, Ben, INVISTA Intermediates, Wichita, Kansas, United States, *Polyamides*, 2

Hesse, Wolfgang, Hoechst AG, Werk Kalle–Albert, Wiesbaden, Germany, *Phenolic Resins*, 5

Hintzer, Klaus, 3M Dyneon, Burgkirchen, Germany, *Fluoropolymers, Organic*, 2

Höke, Hartmut, Weinheim, Federal Republic of Germany, *Resins, Synthetic*, 5

Howick, Christopher J., ICI Chemicals and Polymers, Runcorn, Cheshire, United Kingdom, *Plasticizers*, 1
Hüsgen, Ulrich, Institut für Kunststoffverarbeitung, Aachen, Federal Republic of Germany, *Plastics Processing, 3. Machining, Bonding, Surface Treatment*, 1; *Plastics Processing, 1. Processing of Thermoplastics*, 1; *Plastics, Processing, 2. Processing of Thermosets*, 1

## I

Imöhl, Wolfgang, Schering AG, Berlin, Federal Republic of Germany, *Resins, Synthetic*, 5
Immel, Wolfgang, BASF Aktiengesellschaft, Ludwigshafen, Federal Republic of Germany, *Polybutenes*, 1

## J

Janocha, Siegfried, Hoechst AG, Werk Kalle, Wiesbaden, Federal Republic of Germany, *Films*, 4
Jeremic, Dusan, Borealis Polyolefine GmbH, Linz, Austria, *Polyethylene*, 2
Jones, Frank N., Coatings Research Institute, Eastern Michigan University, Ypsilanti, MI, USA, *Alkyd Resins*, 5
Jünger, Oliver, Ticona GmbH, Frankfurt, Germany, *Polymers, High-Temperature*, 3

## K

Kabelka, Josef, Universität Erlangen-Nürnberg Erlangen-Tennenlohe, Federal Republic of Germany, *Reinforced Plastics*, 3
Kalwa, Matthias, Institut für Kunststoffverarbeitung, Aachen, Germany, *Plastics Processing, 3. Machining, Bonding, Surface Treatment*, 1; *Plastics Processing, 1. Processing of Thermoplastics*, 1; *Plastics, Processing, 2. Processing of Thermosets*, 1
Kaminsky, Walter, Universität Hamburg, Institut für technische und makromolekulare Chemie, Hamburg, Federal Republic of Germany, *Polyesters*, 2
Kaul, Bansi Lal, Sandoz Huningue S.A., Huningue, France, *Plastics, Additives*, 1
Kausch, Michael, Bayer AG, Dormagen, Germany, *Fibers, 6. Polyurethane Fibers*, 4
Klein, Thomas, Lanxess Deutschland GmbH, Leverkusen, Germany, *Polyaspartates and Polysuccinimide*, 2
Koch, Hartmut, Shell Chemical International, Shell Center, London, United Kingdom, *Polybutenes*, 1
Kohan, Melvin I., MIK Associates, Wilmington, Delaware, United States, *Polyamides*, 2
Koleske, Joseph V., Consultant, 1513 Brentwood Road, Charleston WV, United States, *Polyoxyalkylenes*, 2
Kontoff, Jeffrey R., Monsanto Chemical Company, Springfield MA, United States, *Polystyrene and Styrene Copolymers*, 3
Köpnick, Horst, Bayer AG, Dormagen, Federal Republic of Germany, *Polyesters*, 2
Künkel, Andreas, BASF SE, Ludwigshafen, Germany, *Polymers, Biodegradable*, 3
Krämer, Horst, Hüls Aktiengesellschaft, Marl, Federal Republic of Germany, *Polyester Resins, Unsaturated*, 2
Kurz, Klaus, Ticona GmbH, Frankfurt am Main, Germany, *Polyoxymethylenes*, 2

## L

Lang, Jürgen, Dynea Erkner GmbH, Erkner, *Phenolic Resins*, 5
Larimer, Donald-Richard, Bayer MaterialScience AG, Leverkusen, Federal Republic of Germany, *Polyurethanes*, 3
Lohmar, Jörg, Degussa AG - High Performance Polymers, Marl, Germany, *Thermoplastic Elastomers*, 3

## M

Maier, Udo, Bayer MaterialScience AG, Leverkusen, Federal Republic of Germany, *Polyurethanes*, 3
Malet, Frédéric L.G., Arkema, CERDATO Serquigny, France, *Thermoplastic Elastomers*, 3
Marks, Maurice J., Dow Chemical, Freeport, Texas, *Epoxy Resins*, 5
Matthias, Günther, BASF Aktiengesellschaft, Ludwigshafen, Germany, *Amino Resins*, 5
Maul, Jürgen, Hüls AG, Marl, Germany, *Polystyrene and Styrene Copolymers*, 3
Mawer, Ralph L., Shell Chemical International, Shell Center, London, United Kingdom, *Polybutenes*, 1
McKillip, William, Advanced Resin Systems Des Plaines, Illinois, United States, *Resins, Synthetic*, 5
Menault, Jacques, St-Foy-les-Lyons, France, *Fibers, 9. Polyvinyl Fibers*, 4

Mestemacher, Steve A., E. I. DuPont de Nemours & Co., Inc., Parkersburg, West Virginia, United States, *Polyamides*, **2**

Meyer-Ahrens, Sven, Bayer MaterialScience AG, Leverkusen, Federal Republic of Germany, *Polyurethanes*, **3**

Mildenberg, Rolf, formerly Rütgerswerke AG, Duisburg, Federal Republic of Germany, *Resins, Synthetic*, **5**

Morimoto, Osamu, Nippon Chemtec Consulting Inc., Osaka, Japan, *Fibers, 9. Polyvinyl Fibers*, **4**

Moritz, Ralf-Johann, Lanxess Deutschland GmbH, Leverkusen, Germany, *Polyaspartates and Polysuccinimide*, **2**

Mülhaupt, Rolf, Institute for Macromolecular Chemistry, Freiburg, Germany, *Plastics, General Survey, 2. Production of Polymers and Plastics*, **2**; *Plastics, General Survey, 3. Supermolecular Structures*, **2**; *Plastics, General Survey, 4. Polymer Composites*, **1**; *Plastics, General Survey, 5. Plastics and Sustainability*, **1**; *Plastics, General Survey, 1. Definition, Molecular Structure and Properties*, **1**

Mummy, Florian, Kuraray Europe GmbH, Frankfurt am Main, Germany, *Polyvinyl Compounds, Others*, **3**

# N

Naarmann, Herbert, BASF Aktiengesellschaft, Ludwigshafen, Federal Republic of Germany, *Polymers, Electrically Conducting*, **3**

Noble, Karl-Ludwig, Bayer MaterialScience AG, Leverkusen, Federal Republic of Germany, *Polyurethanes*, **3**

Nogaj, Alfred, Bayer AG, Dormagen, Germany, Fibers, 8. *Polyacrylonitrile* Fibers, **4**

# O

Olivieri, Paolo, Moplefan Milano, Italy, *Fibers, 7. Polyolefin Fibers*, **4**

Ostlinning, Edgar, Bayer AG, Leverkusen, Germany, *Polymers, High-Temperature*, **3**

Osugi, Tetsuro, Nippon Chemtec Consulting Inc., Osaka, Japan, *Fibers, 9. Polyvinyl Fibers*, **4**

Ott, Karl-Heinz, Bayer AG, Dormagen, Germany, *Polystyrene and Styrene Copolymers*, **3**

Ouhadi, Traz, Advanced Elastomer Systems, Brussels, Belgium, *Thermoplastic Elastomers*, **3**

# P

Pagilagan, Rolando U., E. I. DuPont de Nemours & Co., Inc., Parkersburg, West Virginia, United States, *Polyamides*, **2**

Parker, David, ICI Advanced Materials, Wilton, Middlesborough, United Kingdom, *Polymers, High-Temperature*, **3**

Paulik, Christian, Institute of Chemical Technology of Organic Materials, Johannes Kepler University Linz, Linz, Austria, *Polypropylene*, **2**

Pellegrini, Antonio, Moplefan Milano, Italy, Fibers, 7. *Polyolefin Fibers*, **4**

Penzel, Erich, BASF Aktiengesellschaft, Ludwigshafen, Federal Republic of Germany, *Polyacrylates*, **2**

Pham, Ha Q., Dow Chemical, Freeport, Texas, *Epoxy Resins*, **5**

Philipp, Sabine, BASF SE, Ludwigshafen, Germany, *Polymers, Biodegradable*, **3**

Pittman, Charles U. Jr., Mississippi State University, Department of Chemistry, Mississippi, USA, *Inorganic Polymers*, **6**

Plummer, Lawrence, E. I. Du Pont de Nemours and Co., Inc., Engineering Polymers Wilmington, United States, *Thermoplastic Elastomers*, **3**

Porth, Hans-Christoph, VESTOLIT GmbH, Marl, Germany, *Poly(Vinyl Chloride)*, **3**

Pötsch, Gerhard, Institut für Kunststoffverarbeitung, Aachen, Germany, *Plastics Processing, 3. Machining, Bonding, Surface Treatment*, **1**; *Plastics Processing, 1. Processing of Thermoplastics*, **1**; *Plastics, Processing, 2. Processing of Thermosets*, **1**

# R

Redmond, Kate, E. I. DuPont de Nemours & Co., Inc., Wilmington, Delaware, United States, *Polyamides*, **2**

Reimer, Valentine, BASF SE, Ludwigshafen, Germany, *Polymers, Biodegradable*, **3**

Reinking, Klaus, Bayer AG, Leverkusen, Germany, *Polymers, High-Temperature*, **3**

Rhein, Thomas, Röhm GmbH, Darmstadt, Germany, *Polymethacrylates*, **2**
Rinno, Helmut, Hoechst Aktiengesellschaft, Frankfurt/Main, Federal Republic of Germany, *Poly(Vinyl Esters)*, **3**
Röhrl, Eckhard, BASF Aktiengesellschaft, Ludwigshafen, Federal Republic of Germany, *Foamed Plastics*, **4**
Rüter, Jörn, Hüls Aktiengesellschaft, Marl, Federal Republic of Germany, *Polyesters*, **2**
Ryan, Larry M., E. I. Du Pont de Nemours and Co., Inc., Engineering Polymers Wilmington, United States, *Thermoplastic Elastomers*, **3**

## S

Sattler, Helmut, Königstein-Schneidhain, Germany, *Polyester Fibers*, **3**
Schade, Christian, BASF AG, Ludwigshafen, Germany, *Polystyrene and Styrene Copolymers*, **3**
Schmidt, Manfred, Bayer AG, Dormagen, Federal Republic of Germany, *Polyesters*, **2**
Schmiegel, Walter, Du Pont de Nemours & Co., Polymer Products Dept., Experimental Station Laboratory, Wilmington, United States, *Fluoropolymers, Organic*, **2**
Schmitt, Wilhelm Friedrich, VESTOLIT GmbH, Marl, Germany, *Poly(Vinyl Chloride)*, **3**
Schmitz, Peter, Hoechst AG, Werk Kalle, Wiesbaden, Federal Republic of Germany, *Films*, **4**
Schoene, Werner, BASF AG, Ludwigshafen, Germany, *Fibers, 7. Polyolefin Fibers*, **4**
Schornick, Gunnar, BASF SE, Ludwigshafen, Germany, *Polyacrylamides and Poly(Acrylic Acids)*, **2**
Schröder, Gerd, BASF Aktiengesellschaft, Ludwigshafen, Federal, Republic of Germany, *Poly(Vinyl Ethers)*, **3**
Schröer, Hans, Bayer AG, Dormagen, Germany, *Fibers, 6. Polyurethane Fibers*, **4**
Schubert, Frank, Evonik Industries AG, Marl, Germany, *Polymers, High-Temperature*, **3**
Schweizer, Michael, ITCF Denkendorf, Denkendorf, Germany, *Fibers, 6. Polyurethane Fibers*, **4**; *Fibers, 7. Polyolefin Fibers*, **4**; *Fibers, 8. Polyacrylonitrile Fibers*, **4**; *Fibers, 10. Polytetrafluoroethylene Fibers*, **4**; *Polyester Fibers*, **3**; *Fibers, 4. Polyamide Fibers*, **4**
Schwenzer, Claus, Institut für Kunststoffverarbeitung, Aachen, Germany, *Plastics Processing, 3. Machining, Bonding, Surface Treatment*, **1**; *Plastics Processing, 1. Processing of Thermoplastics*, **1**; *Processing, 2. Processing of Thermosets*, **1**
Sextro, Guenter, Hoechst AG, Frankfurt, Germany, *Polyoxymethylenes*, **2**
Siegenthaler, Kai O., BASF SE, Ludwigshafen, Germany, *Polymers, Biodegradable*, **3**
Skupin, Gabriel, BASF SE, Ludwigshafen, Germany, *Polymers, Biodegradable*, **3**
Stamm, Manfred, Max-Planck-Institut für Polymerforschung, Mainz, Federal Republic of Germany, *Plastics, Properties and Testing*, **1**
Stickler, Manfred, Röhm GmbH, Darmstadt, Germany, *Polymethacrylates*, **2**
Süling, Carlhans, Bayer AG, Leverkusen, Germany, *Fibers, 8. Polyacrylonitrile Fibers*, **4**

## T

Tanasă, Fulga, Institute of Macromolecular Chemistry "Petru Poni", Jassy, Romania, *Polyureas*, **3**
Tobita, Hidetaka, University of Fukui, Fukui, Japan, *Polymerization Processes, 1. Fundamentals*, **1**; *Polymerization Processes, 2. Modeling of Processes and Reactors*, **1**

## V

Vermeire, Hans F., Shell Research SA, Louvain-La-Neuve, Belgium, *Thermoplastic Elastomers*, **3**
Vianello, Giovanni, European Vinyls Corporation (IT), Porto Maguero, Italy, *Poly(Vinyl Chloride)*, **3**
Vinas, Jérôme, Solvay Brussels, Belgium, *Poly(Vinylidene Chloride)*, **3**

## W

Wagener, Reinhard, Fresenius University of Applied Sciences, Idstein, Germany, *Polymers, High-Temperature*, **3**
Wagner, Robert A., Dynea Austria GmbH, Krems/Donau, Austria, *Amino Resins*, **5**
Weber, Heinz, BASF Aktiengesellschaft, Ludwigshafen, Federal Republic of Germany, *Foamed Plastics*, **4**
Wehrmann, Rolf, Bayer MaterialScience AG, Leverkusen, Germany, *Polycarbonates*, **2**

Weigand, Eckehard, Bayer MaterialScience AG, Leverkusen, Federal Republic of Germany, *Polyurethanes*, 3
Wheatley, Gary W., Humberside Polytechnic, Hull, United Kingdom, *Polymers, High-Temperature*, 3
Wittbecker, Friedrich-Wilhelm, Bergische Universität Wuppertal, Wuppertal, Federal Republic of Germany, *Polyurethanes*, 3
Wolf, Karl-Heinz, Bayer AG, Dormagen, Germany, *Fibers, 6. Polyurethane Fibers*, 4
Wolf, Rainer, Sandoz Huningue S.A., Huningue, France, *Plastics, Additives*, 1
Wussow, Hans-Georg, Bayer AG, Dormagen, Germany, *Thermoplastic Elastomers*, 3
Wussow, Hans-Georg, Bayer MaterialScience AG, Leverkusen, Federal Republic of Germany, *Polyurethanes*, 3

# Y
Yamamoto, Motonori, BASF SE, Ludwigshafen, Germany, *Polymers, Biodegradable*, 3

# Z
Zander, Mechthild, formerly CdF, Forbach, France, *Resins, Synthetic*, 5
Zipplies, Tilman, 3M Dyneon, Burgkirchen, Germany, *Fluoropolymers, Organic*, 2

# Subject Index

## A

ABA block copolymers **2:**902
Abril Wax **1:**554
Aclon **2:**629
Aclyn **1:**572
AC Polyethylene **1:**554
Acramin **2:**692
Acronal **2:**690–692
Acrosol **2:**691
Acrylamide **2:**661
   kinetics of free-radical polymerization in **2:**661
2-Acrylamido-2-methylpropanesulfonic acid
Acrylate rubber (ACM) **3:**1021
Acrylates (esters)
Acrylic acid **2:**660
   as ethylene copolymer **2:**828
   kinetics of free-radical polymerization in **2:**661
   production via oxidation of propene **2:**660
Acrylite **2:**893, 895–896
Acrylite FF **2:**895
Acryloid **1:**575, **2:**690
Acrylonitrile–butadiene–styrene (ABS) blends **3:**999, **3:**1023
   overview of **3:**1023
Acrylonitrile–butadiene–styrene (ABS) graft polymers **3:**1001
Acrylonitrile–butadiene–styrene (ABS) graft rubber
   bimodal **3:**1006
   production of **3:**1008
   unimodal **3:**1006
Acrylonitrile–butadiene–styrene (ABS) matrix resins **3:**1002
Acrylonitrile–butadiene–styrene (ABS) polymers **3:**983, **3:**999, 1202
   additives for **3:**1015
   flame retardants for **3:**1018
   glass-fiber-reinforced **3:**1018
   heat-resistant **3:**1018
   production processes **3:**1004

   recycling **3:**1020
Acrylonitrile–styrene–acrylate (ASA) polymers **3:**998, **3:**1020
Acrylonitrile–styrene grafted onto acrylic rubber **3:**1212
Acrypet **2:**893
ACS polymers **3:**1020
Activex **1:**577
Additives
   biodegradable polyester films **3:**1249
   for ABS polymers **3:**1015
   for polyamide **2:**700–701, 728
   for polyesters **2:**796, 802
   for polyoxymethylenes **2:**915, 920–921, 923
   polypropylene **2:**957
   for poly(vinyl chloride) **3:**1115
   for UP resins **2:**789–790
Adflex **3:**1369
Adhesives
   SBC in **3:**1398
   surface tension **1:**453–454, 457
Adipic acid
   as polyamide monomer **2:**727
   raw material for thermoplastic polyesters **2:**796
   in resins **2:**782–784
Advanced fibers **4:**1541
Advastab **1:**550
Advawax **1:**554
AES polymers **3:**1020
Aflon **2:**629
Aflon COP **2:**624
AgeRite **1:**535
Airflex **2:**691, **3:**1172
Akulon **2:**728
Albigen A **3:**1158
Alcoholysis
   of polyesters **2:**792–793, 795–796, 813
Alcotex **3:**1144
Alcotex 78 **3:**1116

Alcotex 72.5/B72 **3:**1116
Alcotex 552P **3:**1116
Algoflon **2:**608
Alkanox **1:**535
Alkyd resins **4:**1597–1609, 1612–1614
Alpolit **2:**786
AL-PTBBA **1:**572
Altek **2:**786
Altuglas **2:**889, 893–896
Altulex **2:**889, 893–896
Aluminum trihydroxide
   filler in plastics **1:**566
   as flame retardant **1:**559, 568
Amgard **1:**555
Amgard CRP **1:**555
Amide interchange **2:**697, 707
   *see also Transamidation*
Amilan **2:**728
Amines, hindered
   light stabilizers in plastics **2:**920
Amino resin
   molding compounds **1:**407, 412, 419–420,
      423–426, 429–433, 435–438
Ammonium polyphosphate
   as flame retardant **1:**559, 568
Amoco process **4:**1454
Amodel **2:**728
Amylopectin
   for biodegradable polymers **3:**1234–1235, 1237,
      1251–1253, 1255, 1257
Amylose
   for biodegradable polymers **3:**1234–1235, 1237,
      1251–1253, 1255, 1257
*tert*-Amyl peroxyneodecanoate
   as initiator in poly(vinyl chloride) production
      **3:**1118
Anox **1:**535
Anoxyn **1:**535
Antiemetics **2:**601, **4:**1405, 1595, 1775
Antimony pentoxide
   for flame retardation **1:**556
Antimony trioxide
   for flame retardation **2:**784
   as nucleation agent **2:**920
Antioxidants
   classification **1:**562, 572, 574, 576
Antistatic agent
   for plastics **1:**532, 554, 568, 570–571,
      576
Apec **2:**778
Appretan **3:**1172
Apyeil **4:**1544
Apyral **1:**555

Aramide **4:**1543, 1544
Aramid fibers
   filler in plastics **1:**566
Aramids **1:**552, 554, 566, 568
   production **4:**1441
Aranox 1: **1:**535
Arlen 2: **1:**728
Armoslip **1:**554
Armostat **1:**574
Arnite **2:**810
Aropol **2:**786
Atmer **1:**554, 574
Atom-transfer radical polymerization (ATRP) **1:**291,
      **1:**356, **2:**888
Austrostab **1:**550
Autoclave reactor
   polyethylene production **2:**844
Avalon **3:**1380
Azelaic acid
   as polyamide monomer **2:**700
   raw material for thermoplastic polyesters **2:**792
Azeotropic method
   in polyester production **2:**794
Az-cup **1:**569
Azobis(isobutyronitrile) **1:**577
   as initiator in polyacrylate production **2:**684
   as initiator in poly(vinyl chloride) production
      **3:**1118
Azocel **1:**577
Azodicarbonamide
   as chemical blowing agent in plastics **4:**1570
Azubol **1:**577

**B**
Baco Superfine **1:**555
Bag molding
   for polymer composite fabrication **3:**1325–1326,
      1328, 1339
Bags
   organic waste **3:**1231–1232, 1234, 1242,
      1252–1253
   shopping **3:**1231, 1242, 1253, 1255
Barex 210 **3:**997
Barfilex process **4:**1507–1508
Barostab **1:**550
BASF belt process
   for high molecular mass polyisobutylenes **2:**754,
      757, 759–760
Basofil **4:**1550
Basoplast **2:**691
Bayblend **2:**778
Baycoll **2:**796
Bayflex **2:**796

Baygal **2**:796
Baymod **1**:575
Baymod A **3**:1022
Baypure **3**:1242
Baytec **2**:796
Baytherm **2**:796
Beetle **2**:786
Benzocyclobutene (BCB) **3**:1350–1351
Benzoin ether
  as photosensitizers **2**:785
*see also ω, ω-Dimethoxy-ω-phenylacetophenone*
Biocycle **3**:1240
Biodegradation
  of polymers **3**:1232–1234, 1240, 1247
Biogreen **3**:1240
Biomax **3**:1242
Bionolle **3**:1241, 1257, 1259
Biopol **3**:1232, 1240–1242, 1247, 1258–1259
Biorefinery
  biobased monomers from **1**:36, 152
Birefringence
  in plastics **1**:471, 480, 485, 492–493, 498–500, 502, 507, 509, 511, 519–520
  use in fiber analysis **4**:1459–1460, 1466–1468, 1474, 1479, 1482
Bis(4-aminocyclohexyl)methane-terephthalic acid copolymer **2**:699
2,2-Bis(bromomethyl)-1,3-propanediol **1**:564
Bis(4-*tert*-butyl cyclohexyl) peroxydicarbonate
  as initiator in poly(vinyl chloride) production **3**:1118
Bis(2-ethylhexyl) peroxydicarbonate
  as initiator in poly(vinyl chloride) production **3**:1118
Bis(2-ethylhexyl) peroxydicarbonate **3**:1118
1,4-Bis(2-hydroxyethoxy)benzene
  melt-spun spandex production **4**:1490
  raw material for thermoplastic elastomers **3**:1377
1,4-Bis(2-hydroxyethoxy)benzene
  melt-spun spandex production **4**:1490
1,4-Bis(hydroxymethyl)cyclohexane
  for polyester fibers **4**:1482
2,5-Bis(hydroxymethyl)furan **4**:1762-1763
Bis(hydroxymethyl)tricyclodecane
  in resins **2**:782-784
1,1-Bis(4-hydroxyphenyl)-3,3,5-trimethylcyclohexane [129188-99-4] (BP-TMC) **2**:764
Bisphenol A polycarbonate [24936-68-3] (BPA-PC; 2,2-bis(4-hydroxyphenyl)propane polycarbonate) **2**:763
  with additives **2**:767
  modified **2**:765
  properties **2**:764
Bisphenol A polysulfone **3**:1296
Bisphenol Z [843-55-0] (BPZ; 1,1-bis(4-hydroxyphenyl)cyclohexane) **2**:764
$N,N'$(Bis-stearoyl)ethylenediamine
  as release agent **2**:920
Bis(4-*tert*-butyl cyclohexyl) peroxydicarbonate **3**:1118
Bitumen
  SBC in **3**:1399
Blankophor **1**:577
Blendex **1**:575, **3**:1022
Blow molding **2**:853
  biodegradable polyesters **3**:1231, 1236–1237, 1243, 1245–1251, 1257
  polypropylene **2**:967
Blown-film extrusion **2**:851
  biodegradable polyesters **3**:1231, 1236–1237, 1243, 1245–1251, 1257
Boron nitride
  as nucleation agent **2**:920
Borstar polyolefin process **2**:954
Borstar process **2**:851
BP-UVA 1 **1**:546
BP-UVA 2 **1**:546
Bärolub **1**: **1**:554
Bärostab **1**:550
Bärostat **1**:574
BT-UVA 1 **1**:545
BT-UVA 2 **1**:545
Bulk molding compounds (BMC) **2**:785–786, 789
  for polymer composite fabrication **3**:1325–1326, 1328, 1339
Bulk polymerization **1**:334
Busan **1**:555
Butacite **3**:1150
Butadiene–styrene copolymer latex
  for production of ABS graft base **3**:1007
1,3-Butanediol
  in resins **2**:782–784
1,4-Butanediol
  for polyester fibers **4**:1482
1,4-Butanediol
  raw material for thermoplastic elastomers **3**:1381
Butenes
  as ethylene copolymer **2**:828
*n*-Butyl acrylate
  as ethylene copolymer **2**:828
*sec*-Butyl acrylate
*tert*-Butyl acrylate
Butyl benzyl phthalate (BBP)
  as plasticizer **3**:1129

*tert*-Butyl hydroperoxide (TBHP)
　as initiator for poly(vinyl ester) production **3:**1165, 1170, 1172–1173
4,4′-Butylidenebis(6-*tert*-butyl-3-methylphenol)
　as antioxidant **2:**917, 919
*tert*-Butyl peroxyneodecanoate
　as initiator in poly(vinyl chloride) production **3:**1118
*tert*-Butyl peroxyoctanoate
　as initiator for poly(vinyl ester) production **3:**1165, 1170, 1172–1173
　as initiator in poly(vinyl chloride) production **3:**1118

## C

Cadmium pigments
　in plastics **1:**534, 545, 551, 555, 560, 565, 569–571
Cadon **1:**575
Calcium carbonate
　as acid trapping agent **2:**920
Calcium carbonicum praecipitatum (ccp) **1:**566
Calcium stearate
　as lubricant **3:**1015
Calendering
　of thermoplastics **1:**367, 371
Calibre 2: **2:**778
CAMPUS data bank **2:**886
ε-Caprolactam-hexamethyleneterephthalamide copolyamide **2:**699
Carother's law **3:**1299
　fibers for, based on poly(ethylene terephthalate) **4:**1453–1455, 1459–1460, 1464, 1468, 1480, 1484
Carstab **1:**535
Casamid **4:**1771
Cast film extrusion **2:**852
Cast-film extrusion **3:**1231, 1248
Casting
　of acrylic glass **2:**885, 889–890, 892
　of polymers **1:**368, 372, 379, 407, 410
Castor oil, dehydrated
　typical fatty acid composition **4:**1601
Catafor **1:**574
Cataloy process
　in thermoplastic polyolefin production **3:**1367
Catalysis, free-radical
　polyethylene production **2:**829
Catalyst
　polypropylene production **2:**944
　for polymerization **2:**698, 701, 705, 707, 710
CA-UVA-1 **1:**535
Cavco **1:**569

CC **1:**528–529, 532, 534–537, 544–548, 550, 552–554, 556, 559–560, 564–569, 572, 574, 576
Celanex **2:**810
Cellcom **1:**577
Cellmic **1:**577
Cellophane
　PVDC coating **3:**1192–1193
Cellulose ester
　as protective colloids in suspension polymerization **2:**664
Cellulose esters **3:**1246
Cellulose ether
　as protective colloids in suspension polymerization **2:**664
Cellulose films **3:**1246
Celogen **1:**577
Centrex **3:**1021
Cevian **3:**996
Chain polymerization
　effect of reactor types **1:**329
　modeling of processes and reactors **1:**326
Chain polymerization **1:**271
Chain-transfer agent (CTA)
　in polymerization processes **1:**328
Charge-transfer complex
　electrically conducting **3:**1261–1263, 1265, 1268, 1274, 1277
Char Guard 329 **1:**555
Chemstat **1:**574
Chinfunex **4:**1544
Chlorinated paraffins
　flame retardants in plastics **1:**555
Chlorofiber
　properties **4:**1529–1534, 1536
Chlorotrifluoroethylene-ethylene copolymers [25101-45-5] **2:**630
　production **2:**630
　properties **2:**630
Chromatography **1:**231, 233–234, 238, 243, 252–253, 258, 260
　*see also under individual names*
　polymer fractionation by size exclusion chromatography (SEC) **1:**252
　*seealso under individual names*
Chromium(III) oxide pigments
　in plastics **1:**534, 545, 551, 555, 560, 565, 569–571
Cinnamate ester
　as light stabilizer **3:**1015
Civic **2:**786
Clarifex **1:**572
Clausius–Mosotti equation **1:**504
Clear polystyrene molding materials **3:**982
CN 1197 **1:**555

Coad **1**:554
Coating
  based on alkyd resins **4**:1604
  epoxy resin **1**:413
  from polyureas **3**:1040, 1046–1047
  gasplating **1**:466
  gel coating of glass-fiber-reinforced plastics **1**:411, 414
  low-solvent **4**:1597, 1609
  polyacrylates in **2**:675, 693
  of polymers **1**:439, 442, 447, 459, 462
  thermoplastic **1**:367–368, 370–373, 377–379, 382, 389, 393, 398, 400–403
Coconut oil
  typical fatty acid composition **4**:1601
Coldac **2**:691
Collacral **3**:1158
Colorimetry **1**:471, 509–510
Color triangle **1**:509–510
Comboloob **1**:554
Combustion
  of plastics **1**:471–472, 474, 477–478, 481, 483, 489, 491, 493, 498–499, 503–504, 506–507, 510, 512–513, 515–516, 518
Composting
  biodegradable polymers **3**:1231–1237, 1239, 1241, 1243–1244, 1247–1249, 1251–1259
Compression molding
  of polymers **1**:368, 372, 379
Compression molding compounds
  for UP resins **2**:789–790, 1751–1752, 1762–1767, 1769–1770
  resins by **4**:1751, 1762
Cone ribbon blade reactor **1**:320
Continuous stirred-tank reactor (CSTR)
  for polystyrene production **3**:985
Copolymerization **1**:283
  composition of copolymers **1**:284
  kinetics **1**:287
  long-chain approximation **1**:284
  Markov chain **1**:284
Copolymerization
  of acrylamide and acrylic acid **2**:662
  of polyamides **2**:698, 707–709, 717, 724, 726, 728
  reactivity ratios for **2**:662
  of TFE and PPVE **2**:620
Copolymerization, solid-state
Copolymers **1**:19
  for poly(vinylidene chloride) production **3**:1181, 1343, 1348, 1350, 1353–1361,
  in specialty plastics **3**:1343
Copoly($p$-phenylene/3,4-diphenyl ether terephthalamide) (POP) **4**:1543

Copolyureas **3**:1029–1031, 1034–1037, 1040–1042, 1047
Corialgrund **2**:675, 692
Corrosion inhibitors
Coupling agents
  for plastics **1**:532, 554, 568, 570–571, 576
Crastin **2**:810
Craston **3**:1303–1304, 1307
Crestapol **2**:796
Crilat **2**:690
Cristamid **2**:728
Crodacid **1**:554
Crodamid **1**:554
Crystic **2**:786
Cumene hydroperoxide
  as initiator for poly(vinyl ester) productions **3**:1165, 1170, 1172–1173
Cumyl peroxyneodecanoate **3**:1118
  as initiator in poly(vinyl chloride) production **3**:1118
Cunox **1**:535
Curing
  of UP resins **2**:781–783, 785–788, 790
Cyanox **1**:535
Cyastat **1**:574
1,4-Cyclohexanedimethanol
  for polyester fibers **4**:1482
Cycoloy **2**:778
Cyrolite G 20 **2**:894
Cyrolite XT polymer **2**:894
Cytop [101182-89-2] **2**:637

**D**

Dai-el **2**:641
Daltogel **2**:796
Daltolac **2**:796
Daltorez **2**:796
Dart drop index (DDI) **2**:824
Decabromodiphenyl oxide **1**:563
Dechlorane Plus **1**:564
Degalan **2**:690
Degenerative-transfer radical polymerization (DTRP) **1**:291
Dehydat **1**:574
Delpet **2**:893–894
Delrin **2**:911, 913, 922, 924
Denka B-24 **3**:1116
Denka HS-80 **3**:1116
Denka MP-10 **3**:1116
Denka Poval **3**:1144
Denka W–20N **3**:1116
Depro **3**:1369
Desmopan **3**:1380

Desmophen **2:**690, 796
Dexflex **3:**1369
Diakon **2:**893–894, 896
Diakon APA **2:**896
Dialac **3:**1021
Dibenzoyl peroxide
   as initiator for poly(vinyl ester) productions
      **3:**1165, 1170, 1172–1173
2,6-Di-*tert*-butyl-4-methylphenol (BHT)
2,6-Di-*tert*-butylphenol
   antioxidant **1:**527–535, 546, 568, 570
Dibutyl phthalate (DBP)
   as plasticizer **3:**1129
Dicetyl peroxydicarbonate **3:**1118
   as initiator in poly(vinyl chloride) production
      **3:**1118
1,1-Dichloroethylene *see Vinylidene chloride
   (1,1-dichloroethylene)* **3:**1181, 1194, 1196,
      1882, 1995
DIC-PPS **3:**1307
Dicyanodiamide
   as acid trapping agent **2:**920
4,4′-Dicyclohexylmethane diisocyanate
   raw material for thermoplastic elastomers
      **3:**1377
Dicyclohexyl peroxydicarbonate **3:**1118
   as initiator in poly(vinyl chloride) production
      **3:**1118
Didecanoyl peroxide **3:**1118
   as initiator in poly(vinyl chloride) production
      **3:**1118
1,1-Difluoroethylene elastomers
   production **2:**642
1,1-Difluoroethylene elastomers
   properties **2:**642
Diisocyanates
   for polyurethane fibers **4:**1487–1488
Diisodecyl adipate (DIDA)
   as plasticizer **3:**1129
Diisopropyl peroxydicarbonate
Dilauroyl peroxide **3:**1118
   as initiator in poly(vinyl chloride) production
      **3:**1118
Dilauryl peroxide
   as initiator for poly(vinyl ester) productions
      **3:**1165, 1170, 1172–1173
ω,ω-Dimethoxy-ω-phenylacetophenone **2:**785
   *see also Benzoin ether*
   raw material for thermoplastic elastomers
      **3:**1376
Dimethyl methylphosphonate **1:**556, 564
Dimethyl terephthalate
   poly(ethylene terephthalate) from **4:**1454–1455

Dimethyl terephthalate (DMT)
   for poly(ethylene terephthalate) **4:**1453–1455,
      1459–1460, 1464, 1468, 1480, 1484
Dimyristyl peroxydicarbonate **3:**1118
   as initiator in poly(vinyl chloride) production
      **3:**1118
Dioctadecanyl disulfide
   antioxidant **1:**527–535, 546, 568, 570
Dioctanoyl peroxide **3:**1118
   as initiator in poly(vinyl chloride) production
      **3:**1118
Dioctyl adipate (DOA)
   as plasticizer **3:**1129
Dioctyl azelate
   as plasticizer **3:**1129
Dioctyl sebacate
   as plasticizer **3:**1129
Diorez **2:**796
1,3,2-Dioxaphosphorinane-2,2′-oxybis-(5,5-dimethyl-
   2,2′-disulfide) **1:**531
Direct spinning
   of nylons **4:**1446
Dispersion polymerization **1:**337
Ditridecyl phthalate
   as plasticizer **3:**1129
Divergan **3:**1158
Dong II **3:**1369
   of polyphenylene **3:**1271
Dopral **1:**554
Dough molding compounds (DMC)
   for polymer composite fabrication **3:**1325–1326,
      1328, 1339
Doverphos **1:**535
   in emulsion polymerization **2:**664
Dry spinning
   polyacrylonitrile fibers **4:**1513–1514, 1518, 1521,
      1523–1524, 1526
   polyurethane fibers **4:**1487–1488, 1491
   poly(vinyl alcohol) fiber **4:**1529, 1533, 1535
   poly(vinyl chloride) fiber **4:**1529, 1532, 1536
Dry spinning, continuous
   polyacrylonitrile fibers **4:**1513–1514, 1518, 1521,
      1523–1524, 1526
Duracon **2:**912, 922
Durastrength **1:**575
Durethan **2:**728
Durolon **2:**778
Durolube **1:**554
Dyes
   in plastics **1:**534, 545, 551, 555, 560, 565, 569–571
Dynamar **1:**554
Dynapol L **2:**810
Dynapol P **2:**810

Dynapol S **2**:810
Dynasilan **1**:569
Dyneema **4**:1556
Dyneon **2**:608, **2**:617, **2**:620, **2**:632, **2**:641
Dyneon ET **2**:624
Dynyl **3**:1388
Dytron **3**:1376

# E
Eastobrite **1**:577
EC-1 **1**:572
Ecdel **3**:1386
Ecoflex **3**:1233–1235, 1242, 1244–1246, 1257–1259
Ecovio **3**:1242–1246, 1259
Edenol **1**:550
Ekonol **3**:1309–1310, 1321
Elastamax **3**:1369
Elastamine **3**:1387
Elasthane **4**:1487
Elastollan **3**:1380
Elastomer
   mechanical properties **1**:471–472, 475, 478, 480, 494–496, 510
   thermoplastic **3**:1365
Electroplating
   of plastics **1**:439, 453–454, 459, 461, 465, 467–469
Elix **1**:575
Elvacite **2**:690
Elvaloy **1**:575
Elvanol **3**:1144
Embilizer **1**:550
EMMA (Equatorial Mount with Mirrors for Acceleration) test **1**:545
   in production of poly(vinyl esters) **3**:1165, 1170, 1172–1173
Emulsion polymerization **1**:342
   ab initio **1**:342
   kinetics **1**:344
Emultex **2**:691, 1172
Energy elasticity **1**:480
Enmat **3**:1240, 1257
EnPol **3**:1241, 1257
Entropy elasticity **1**:480, 518
Envirostrand **1**:555
Enzymes
   biodegradation of polymers **3**:1233
Epolene **1**:554, 569
Epoxy resins
   brominated **1**:556, 559, 563–565
   processing **1**:407, 410–415, 417–420, 422–424, 426–427, 429–433, 436–438
   reinforcement with glass fibers **1**:417–418, 433, 436

Erdmenger principle
Esrel **3**:1386
Estabex **1**:550
Estane **3**:1380
   of dicarboxylic acids **2**:792, 797
Estol **1**:554, 575
Estolan **2**:796
Ethanox **1**:535
Ethanox 330 **3**:1381
Ethyl acrylate
   as ethylene copolymer **2**:828
Ethylene (ethene)
   polyethylene production **2**:841
   polymerization-grade **2**:841
Ethylene bis(tetrabromophthalimide) **1**:563
Ethylene copolymers **2**:827
   high-pressure **2**:845
   types of **2**:829
   types of **2**:828
   for poly(ethylene terephthalate) **4**:1453–1455, 1459–1460, 1464, 1468, 1480, 1484
   raw material for thermoplastic elastomers **3**:1381
   in resins **2**:782–784
Ethylene–propene–diene rubber (EPDM) **3**:987, 1211
Ethylene–propene rubber (EPR) **3**:1021
Ethylene terpolymers
   high-pressure **2**:845
Euderm **2**:692
Eudragit E-30 D **2**:896
Eudragit RS **2**:897
Eurelon **4**:1771
Euremelt **4**:1771
Everflex **3**:1172
Exolit **1**:555
Exolit 405 **1**:555
Expandex **1**:577
Extruded polystyrene (XPS) **3**:983
Extrusion
   biodegradable polymers **3**:1231–1237, 1239, 1241, 1243–1244, 1247–1249, 1251–1259
   of plastics **1**:404
   of PPS **3**:1282, 1301–1306
Extrusion blow molding
   biodegradable polyesters **3**:1231, 1236–1237, 1243, 1245–1251, 1257
Extrusion coating **2**:853
   biodegradable polyesters **3**:1231, 1236–1237, 1243, 1245–1251, 1257

Extrusion spinning
 of nylons **4:**1446
ExxonMobil PP process **2:**954
Exxon process
 for medium molecular mass polyisobutylenes **2:**754, 756–757, 759
Exxtral **3:**1369

**F**

Fatty acid salts
 as release agents **2:**911, 920
Fatty alcohols
 as release agents **2:**911, 920
Feast reaction (retro-cycloaddition)
 synthesis of electrically conducting polymers **3:**1262
Fenilon **4:**1544
Ferrocene (dicyclopentadienyliron)
 smoke inhibitor **1:**559, 563
Fibers
 determination of fine structuredegree of crystallinity, determination **4:**1458, 1462–1463
 gas-fading resistance, test method 1502,
 light resistance, test methods 1502–1503,
 oxidation resistance, test method **4:**1502
 high-modulus **4:**1550
 high-strength **4:**1550
 polyaryletherketone **3:**1281, 1291
 poly(phenylene sulfide) **3:**1281–1282, 1298, 1306, 1315
 polypropylene **2:**968
Fibers, absorbent **4:**1524
Fibers, bicomponent **4:**1525
Fibers, polyacrylonitrile fibers **4:**1513–1514, 1518, 1521, 1523–1524, 1526
Fibers, polyamide fibers **4:**1435
Fibers, polyester fibers **4:**1453, 1459–1460, 1466–1468, 1474, 1479, 1482
 types and properties **4:**1466, 1475
Fibers, polyolefin fibers **4:**1495, 1497, 1511
Fibers, polytetrafluoroethylene fibers **4:**1539–1540
Fibers, polyurethane fibers **4:**1487–1488
Fibers, polyvinyl fibers **4:**1529, 1536–1537
Fibers, shrinkable **4:**1525
Fibers, synthetic organic
 differential thermal analysis **4:**1462
 from polyureas **3:**1040, 1046–1047
 natural draw ratio **4:**1457, 1464–1465
 orientation measurement by optical birefringence **4:**1456–1457, 1463

 stress-strain tests **4:**1457–1458, 1463–1464, 1466–1467, 1469, 1475–1477
 thermal measurements **4:**1463
Ficel **1:**577
Filament winding
 for polymer composite fabrication **3:**1325–1326, 1328, 1339
Filler
 for plastics **1:**532, 554, 568, 570–571, 576
 in resins **2:**782–784
 from polyureas **3:**1040, 1046–1047
 polyaryletherketone **3:**1281, 1291
 poly(phenylene sulfide) **3:**1281–1282, 1298, 1306, 1315
 polypropylene **2:**970
Film yarn
 from polyolefines **4:**1508
Firebloc **1:**555
Firebrake ZB **1:**555
Fire Shield **1:**555
Flameguard **1:**555
Flame retardant
 brominated **1:**556, 559, 563–565
 chlorinated paraffin **1:**552, 556, 563–565
 organophosphorus **1:**555–556, 559–560, 565
 for plastics **1:**532, 554, 568, 570–571, 576
 toxicology **1:**527, 547, 565, 570
Flammability
 of polyoxymethylenes **2:**911–912, 914, 916, 921
 of PPE–HIPS blends **2:**927, 929–931, 933–935
 of PS and HIPS **3:**987
 of Ultem **3:**1285–1286
Flectol **1:**535
Flemion **3:**1356–1357, 1360–1362
Flexathene **3:**1369
Flexbond **2:**691
Flexcryl **2:**691
Flocculation
 polyacrylamides used in **2:**669
Flow rate ratio (FRR)
 polyethylene **2:**822
Fluidized-bed reactor
 polyethylene production **2:**848
Fluolite **1:**577
Fluon **2:**608
Fluorel **2:**641
Fluorescence spectroscopy
Fluorescent pigments **1:**571
Fluorinated ethylene-propene (FEP) resins [25067-11-2] **2:**617
 properties **2:**618
Fluoroelastomers **2:**640
 containing ethylene **2:**651

thermoplastic **2:**640
toxicology **2:**653
Fluoroethylene-vinylether (FE-VE) copolymer **2:**652
Fluoroplastics **2:**606
toxicology **2:**653
Fluoropolymers, organic **2:**603
composition **2:**605
monomers for **2:**604
Fluorothermoplastics **2:**606
Fomrez **2:**796
Fondocryl **2:**692
Food packaging
biodegradable polymers **3:**1231–1237, 1239, 1241, 1243–1244, 1247–1249, 1251–1259
Fortilene **3:**1369
Fortrell **2:**810
Fortron **3:**1303–1304, 1307
Free-radical polymerization **1:**272
chain transfer to small molecules **1:**277
dead polymer chain formation **1:**280
free radical generation **1:**273
kinetics **1:**277
types of reaction **1:**272
in poly(vinyl ether) production **3:**1175–1179
Friedel–Crafts polycondensation
of 4-phenoxybenzoyl chloride **3:**1292
Frye–Horst theory **1:**548
Ftoroplast **2:**608
Furfuryl alcohol (Furanol) **4:**1762–1763

## G
Garaflex O **3:**1369
Garbafix **1:**535
Gel dyeing
polyacrylonitrile fibers **4:**1513–1514, 1518, 1520–1521, 1523–1524, 1526
Gellal **1:**572
Geloy **3:**1021
Gel-permeation chromatography (GPC)
*see also Chromatography; Size exclusion chromatography (SEC)*
Gelvatol **3:**1144
Geniset MD **1:**572
Genitron **1:**577
Glass fiber-reinforced plastics (GRP) **1:**488
curable molding compounds, flow–cure behavior of **1:**423–424
processing **1:**407, 410–415, 417–420, 422–424, 426–427, 429–433, 436–438
of polyamides **2:**698, 707–709, 717, 724, 726, 728
Glycmonos **1:**554
Glycolube **1:**554, 574
Gohsenol **3:**1144

Gohsenol GH20 **3:**1116
Gohsenol KH17 **3:**1116
Gohsenol KH20 **3:**1116
Gohsenol KP08 **3:**1116
Gohsenol LL02 **3:**1116
Gohsenol LW-200 **3:**1116
Goodrite **1:**535
Gorham process
poly(*p*-xylylenes) from **3:**1343, 1345–1350
Green Tower **2:**778
Grid spinning
of nylons **4:**1446
Grignard coupling
synthesis of electrically conducting polymers **3:**1262
Grilamid **2:**728, **3:**1388
Grilon **2:**728, **3:**1394
Grilpet **2:**810
Grivory **2:**728
Grubb's method
synthesis of electrically conducting polymers **3:**1262

## H
Halar **2:**630
Halochromy **1:**548
Halon ET **2:**624
HALS 1 **1:**545–546
HALS 2 **1:**546
HALS 3 **1:**545
Halstab **1:**550
Harochem **1:**554
Harogel **1:**554
Harowax **1:**554
Heat distortion temperature
of PPE–HIPS blends **2:**927, 929–931, 933–935
Heat-resistant fibers **4:**1541, **4:**1542
fiber properties **4:**1547
Heat setting
of poly(ethylene terephthalate) **4:**1453, 1459
Hebron **1:**550, 577
Hetron **2:**786
Hexabromocyclododecane **1:**556, 563
Hexachloroendomethylenetetrahydrophthalic (HET) acid
in resins **2:**782–784
Hexafluoropropene (HFP) **2:**617
Hexamethyleneadipamide-
hexamethyleneterephthalamide copolyamide **2:**699
1,6-Hexamethylene bis[3-(3,5-di-*tert*-butyl-4-hydroxy-phenyl)propionate]
as antioxidant **2:**917, 919

Hexamethylene diisocyanate (HDI)
  raw material for thermoplastic elastomers
    **3**:1377
Hexamoll DINCH **3**:1153
Hexene
  as ethylene copolymer **2**:828
*n*-Hexyl acrylate
*n*-Hexyl methacrylate
Hifax **3**:1369
High-functional fibers **4**:1541
High-performance fibers **4**:1541
High-strength polyethylene (PE) **4**:1555
High-viscosity reactor **1**:321
HIPS **3**:983
  synthesis **3**:984
Homopolyureas **3**:1029, 1035, 1041
Hopelex **2**:778
Horizone process **2**:952
Horticulture
  biodegradable polymers **3**:1231–1237,
    1239, 1241, 1243–1244, 1247–1249,
    1251–1259
Hostaform **2**:912–914, 916, 922, 924, 926
Hostalub **1**:554, 572
Hostalux **1**:577
Hostamont **1**:554, 572
Hostanox **1**:535
Hostapren **1**:575
Hostastab **1**:550
Hostastat **1**:574
Hostatron **1**:577
Hycar **2**:692
Hydral **1**:555
Hydrocarbons, aromatic 1344
  in specialty plastics **3**:1343
Hydrocerol **1**:577
Hydrogen peroxide
  as initiator for poly(vinyl ester) productions
    **3**:1165, 1170, 1172–1173
Hydrolysis
  of polyesters **2**:792–793, 795–796, 813
1-Hydropentafluoropropene [32552-63-9]
  **2**:641
1,4-Hydroquinone
2-Hydroxybenzophenones
  as light stabilizer **3**:1015
2-Hydroxyphenylbenzotriazoles
  as light stabilizer **3**:1015
2-Hydroxyphenyltriazines
Hydroxypropyl acrylate
Hyflon AD **2**:636
Hyflon PFA **2**:620
Hytrel **2**:810, 902

I
Imhausen process **4**:1454
Imperon **2**:692
Impet **2**:810
Infino **2**:778
Initiator
  cationic, in poly(vinyl ether) production
    **3**:1175–1179
  for free-radical polymerization **2**:661
  radical polymerization **2**:832
  redox **2**:662
  thermal **2**:661
Injection molding **2**:853
  biodegradable polyesters **3**:1231, 1236–1237,
    1243, 1245–1251, 1257
  FEP **2**:619
  of PPS **3**:1282, 1301–1306
  processing of polyoxymethylenes **2**:914, 923
Injection stretch blow molding
  biodegradable polyesters **3**:1231, 1236–1237,
    1243, 1245–1251, 1257
Innovene G process **2**:847
Innovene process **2**:952
Inovol PA 4 **3**:1116
Inovol PA 6 **3**:1116
Inovol PA 7 **3**:1116
Interstab **1**:550, 554
Interstat **1**:574
Ionol **1**:535
Ionomers **1**:81, **1**:199
Ionox **1**:535
Irgafos **1**:535
Irganox **1**:535
Irganox 1098 **3**:1381
Irgastab **1**:550
Irgawax **1**:554
Iriodin **1**:571
Irogran **3**:1380
Iron oxide pigments
  in plastics **1**:534, 545, 551, 555, 560, 565,
    569–571
Isomerization
  of polymers **1**:238, 240, 242, 244, 251,
    253, 255
Isonox **1**:535
  in resins **2**:782–784
4-Isopropylaminodiphenylamine
  antioxidant **1**:527–535, 546, 568, 570
Iupilon **2**:778
Iupital **2**:912, 922
Ixef **2**:728
IXOL B 251 **1**:564
Izod impact strength **2**:715, 719–721

## J

Jeffamine **3:**1387
Joncryl **2:**690
J-Prene **3:**1369
JSR-AES **3:**1021

## K

Kallodoc **2:**896
Kalrez Perfluorocarbon Elastomer Parts **2:**648
Kamax **2:**893
Kane ACE **1:**575, **2:**896, **3:**1022
Katzschmann process **4:**1454
Kelburon **3:**1369
Keltan TP **3:**1369
Kemamide **1:**554
Kemester **1:**554
Kemfluid **1:**554
Kemistab **1:**550
Kempore **1:**577
Kenamine **1:**574
Ken-Stat **1:**574
Kepital **2:**912, 922
Kermel **4:**1548–1549
Kevlar **4:**1550, **4:**1551
KF Polymer **2:**632
Kinetics
 in VCM polymerization **3:**1118
KM-21 **4:**1544
Kneader
 manufacture of poly(vinyl alcohol) in **3:**1143
Koblend **3:**1021
Kodel **1:**577
Kollidon **3:**1158
Kopel **3:**1386
Kostil **3:**996
Kralastic **3:**1022
Kraton **1:**575
Kraton D-1107 **3:**1398
Kraton D-1320X **3:**1398
Krystalflex **3:**1380
Kubin distribution **1:**26
Kuhn expression **1:**249
Kuhn length **1:**33
Kynar **2:**632
Kynol [9003-55-8] **4:**1550

## L

LA *see Lactic acid*
Lacea **3:**1239
Lacstar **2:**691
Lactides
 isomers **3:**1238
Lamellon **2:**786
Laminates **1:**407, 416, 423, 428–429, 436
 *see also Composites*
Lankroplast **1:**554
Laripur **3:**1380
Larodur **2:**690
Laropal A types **4:**1769
Laropal K 80 **4:**1768
Larostat **1:**574
Larton **3:**1307
Lastab **1:**550
Lasumit **1:**535
Laxtar **3:**1321
Leguval **2:**786
Lemol **3:**1144
Leona **2:**728
Leucopur **1:**577
Levapren **1:**575
Levatherm F **2:**641
Lexan **2:**778
Life-cycle assessments
 biodegradable polymers **3:**1231–1237, 1239, 1241, 1243–1244, 1247–1249, 1251–1259
 polypropylene **2:**975
Light stabilizers **1:**527–528, 530, 532, 535–537, 544–546, 555
 classification **1:**562, 572, 574, 576
Limiting oxygen index (LOI)
 for plastics **1:**532, 554, 568, 570–571, 576
Linear low-density polyethylene (LLDPE) **2:**845
Linear polymerization **1:**350
 chain-transfer reaction **1:**350
 zero-one system **1:**350
Linseed oil
 typical fatty acid composition **4:**1601
Liquid crystal **1:**471, 473, 476–477
 types **1:**76, **1:**194
Liquid crystal, nematic
 Vectra **2:**803, 808–812
 Xydar **2:**803, 808–812
Liquid crystal polymers **3:**1281, 1307
 anisotropy **3:**1304, 1312
 aromatic groups **3:**1296, 1310
 bridging groups **3:**1310
 flame retardancy **3:**1291, 1311
 geometric arrangements **3:**1308
 impact strength **3:**1286, 1294, 1297–1298, 1302–1304, 1311–1312, 1314
 linear thermal expansion **3:**1308, 1317
 mechanical stress **3:**1304, 1312–1313, 1315
 mesophases **3:**1308
Listab **1:**554
Litex **2:**692

Living polymerization **1**:288
  ideal **1**:289, l21_l01
Lomod **2**:810, **3**:1386
Lower critical solution temperature (LCST) **2**:668
Lowinox **1**:535
Loxamid **1**:554
Loxiol **1**:554
L2 Resin **4**:1768
Lubrex **1**:554
Lubricants
  classification **1**:562, 572, 574, 576
  for plastics **1**:532, 554, 568, 570–571, 576
Lubriol **1**:554
Lubrox **1**:554
Lucel **2**:912
Lucite **2**:690, 889, 893–896
Luhydran **2**:690
Lumiflon **2**:652
Lumitol **2**:690, 796
Luperfoam **1**:577
Lupoy **2**:778
Luprenal **2**:690
Luran **3**:996
Luran S **3**:1021
Lutamer **3**:1268
Lutonal **2**:691, **3**:1177–1178
Luvican **3**:1160
Luviskol **3**:1158
Luwax V **3**:1178

**M**
MA **1**:527–537, 544–556, 559–560, 562–577
Macrodiols
  for synthetic fibers **4**:1491
Macromelt **4**:1771
MACR polymers **3**:1023
Macrynal **2**:690
  as flame retardant **1**:559, 568
Magnifin **1**:555
Makroblend **2**:778
Makrolon **2**:778
Manhaden oil
  typical fatty acid composition **4**:1601
Manton–Gaulin homogenizer **3**:1007
Marco **2**:786
Mark **1**:534–535, 546, 550, 553–554, 556, 565, 569, 571–572, 574
Mark–Houwink equation **2**:749
Mark–Houwink–Sakurada relationship **2**:666
Mark NA **1**:572
Markstat **1**:574
Martinal **1**:555
Matrix copolymers **3**:1009

Meister **1**:550
Melamine–formaldehyde resins
  as nucleation agents **2**:920
Melinite **2**:810
Melio Resin A **2**:692
Melt flow index (MFI) **3**:1300–1301, **4**:1497–1498
  polyethylene **2**:821
Melt flow rate **2**:821.
  see also Melt flow index (MFI)
Melt spinning
  polyacrylonitrile fibers **4**:1513–1514, 1518, 1521, 1523–1524, 1526
  polypropylene fibers by **4**:1495, 1497–1498, 1503–1504, 1511
  polytetrafluoroethylene fibers **4**:1539–1540
  polyurethane fibers **4**:1487–1488, 1492
Melt volume index (MVI) **3**:1301
Membrane osmometry **1**:231, 244–245
Membranes
  from polyureas **3**:1040, 1046–1047
Mesogens **3**:1308–1309, 1319
Metablen **1**:575, **2**:896, **3**:1022
Metal deactivator
  antioxidants **1**:527–535, 546, 568
Metallic state
  of polyaniline **3**:1261, 1273–1274
Metallization
  of plastics **1**:439, 453–454, 459, 461, 465, 467–469
Metallocenes
  polyethylene production **2**:838
  polypropylene production **2**:946
Metal oxide pigments **1**:570
Metastab **1**:550
Methacrylate–acrylonitrile–butadiene–styrene (MABS) **2**:894, **3**:1022
Methacrylate–acrylonitrile–butadiene–styrene resin matrix **3**:1016
Methacrylate–butadiene–styrene (MBS) **2**:894, **3**:1022
Methacrylic acid **2**:661
  as ethylene copolymer **2**:828
  kinetics of free-radical polymerization in **2**:661
4-Methoxyphenol (MEHQ)
  polymerization inhibitor **2**:660.
see also Methyl ether hydroquinone
Methyl acrylate
  as ethylene copolymer **2**:828
Methylalkoxy copolymer (MFA) **2**:620
  processing **2**:622
  properties **2**:622
Methyl aluminoxane (MAO) **2**:838
2-Methyl-2,5-dioxo-1,2-oxaphospholane **1**:564

2,2′-Methylenebis(4-methyl-6-*tert*-butylphenol)
   as antioxidant **2**:917, 919
Methylene diphenylene isocyanate (MDI)
   raw material for thermoplastic elastomers **3**:1376
Methyl ether hydroquinone **2**:660.
   *see also 4-Methoxyphenol (MEHQ)*
   as ethylene copolymer **2**:828
Methyl methacrylate [36426-74-1] (MMA) **2**:885
α-Methylstyrene–acrylonitrile (AMS–AN) copolymer **3**:998, **3**:1010
Methyl vinyl ether **3**:1175–1176.
   *see also Vinyl methyl ether*
α-Methylvinyl ether **3**:1175–1176
   *see also Methyl vinyl ether*
β-Methylvinyl ethers **3**:1175–1176
   *see also Methyl vinyl ether*
M5 fiber **4**:1554
Michael's reaction
   of acrylic acid **2**:661
Micral **1**:555
Micro-ken **1**:554
   in ABS systems **3**:1016
Millad **1**:572
Milling
   of plastics **1**:404
Miniemulsion polymerization **1**:349
Minlon **2**:728
Miractran **3**:1380
Mirel **3**:1240, 1257–1258
Mirrors
   chemical deposition **1**:439, 455, 460, 462
Mitsui Hypol bulk propylene process **2**:954
Mixed spinning
   poly(vinyl alcohol) fiber **4**:1529, 1533, 1536
Mold
   for plastics **1**:407, 410–411, 413–414, 420, 427, 437
Molding compounds **1**:407, 412, 419–420, 423–426, 429–433, 435–438
Molding machines **1**:424, 437
Mold Pro **1**:554
Mold Wiz **1**:554
   equilibrium distribution of **2**:702
Molecular mass determination
   of polymers **1**:238, 240, 242, 244, 251, 253, 255
Molecular mass distribution (MMD) **1**:319
   of polyesters **2**:792–793, 795–796, 813
Molecular mass, number-average
   determination **2**:804
Molecular mass, viscosity-average
   determination **2**:804
Molecular mass, weight-average
   determination **2**:804

Monomer scavenging **2**:665
Mowilith **2**:690, 1172, 1174
Mowiol **3**:1144
Mowiton **3**:1172
MuCell process **2**:972
Multiflex **3**:1369
Multilon **2**:778
Multron **2**:796
Myvaplex **1**:554, 574
Myverol **1**:554

N
Nafion **3**:1356–1357, 1360, 1362
Naftolube **1**:554
Naftovin **1**:550
Naftozin **1**:554
Nanocomposites
   polymeric **1**:128, **1**:211
2,6-Naphthalenedicarboxylic acid
   raw material for thermoplastic elastomers **3**:1381
1,5-Naphthalene diisocyanate (NDI)
   raw material for thermoplastic elastomers **3**:1377
Natural rubber–polyolefin thermoplastic vulcanizates **3**:1373
Naugard **1**:535
Naugard 445 **3**:1381
Neocellborn **1**:577
Neocryl **2**:690
Neoflon **2**:617, 620, 629, 632
Neoflon EP **2**:624
Neopentyl glycol
   in resins **2**:782–784
Neothane **3**:1380
Newstat **4**:1544
NIC-3 **1**:546
Nippolan **2**:796
*N,N*-Dimethyl acrylamide **2**:661
Nomex **4**:1542
Nonlinear polymerization **1**:308, **1**:353
Nonox **1**:535
Nonwoven fabrics
   poly(ethylene terephthalate) **4**:1453–1455, 1459–1460, 1464–1465, 1468, 1480, 1484
Nonwoven fabrics (nonwovens)
   spunbonded from poly(ethylene terephthalate) **4**:1453–1455, 1459–1460, 1464, 1468, 1480, 1484
Norpol **2**:786
Norrish reaction **1**:536–537, 544
Norsodyne **2**:786
Noryl **2**:927, 929–930, 932–935
Novadur **1**:575
Novalar **1**:575

Novalast **3**:1369
Novalene **1**:575
Novarex **2**:778
Novolen process **2**:953
Nuclear magnetic resonance spectroscopy (NMR)
    2D NMR **1**:257
    3D NMR **1**:257–258
Nucleating agent
    for plastics **1**:532, 554, 568, 570–571, 576
Nyacol **1**:555
Nylon 2
    high molecular mass derivatives **4**:1437
Nylon 3
    C-alkyl derivatives **4**:1438
Nylon 6
    as blend component **3**:1210, 1212
    synthesis **2**:697, 700, 709–710
    as blend component **3**:1210, 1212
    synthesis **2**:697, 700, 709–710
Nylon filament yarns
    coarse **4**:1447
    fine-denier **4**:1447
Nylon flakes **4**:1445
    spinning **4**:1445
Nylons. *See also* Fibers, polyamide fibers; Polyamide fibers; Polyamides
    basic parameters of commercial **2**:710
    black fibers **4**:1445
    properties **4**:1437
    semiaromatic **2**:698, 713, 717–718
    spin dyeing **4**:1444
    wires and bristles **4**:1447
Nylon staple fibers **4**:1448

## O

Octabromodiphenyl oxide **1**:563
Octene
    as ethylene copolymer **2**:828
Ohm's law **3**:1264
Okstan **1**:550
Oncor **1**:555
Ontex **3**:1369
Oppanol B100 **2**:759
Oppanol B200 **2**:759
Optiblanc **1**:577
Optical birefringence
    use in fiber analysis **4**:1459–1460, 1466–1468, 1474, 1479, 1482
Optical brighteners (fluorescent whitening agents)
    in plastics **1**:534, 545, 551, 555, 560, 565, 569–571
Optorez OZ 1000 **2**:893
Orgamid **2**:728

Orgater **2**:810
Oroglas **2**:893–894
Oroglas HT 121 **2**:893
Osmotic pressure **1**:243–245, 247
Oxford yarns **4**:1448
Oxi-chek **1**:535

## P

Packaging
    biodegradable polymers **3**:1231–1237, 1239, 1241, 1243–1244, 1247–1249, 1251–1259
Paints
    polyacrylates in **2**:675, 693
Paints and coatings
    alkyd coatings **4**:1605–1606, 1612
Palatal **2**:786
Pandex **3**:1380
Paralinx **2**:896
Paraloid **1**:575, 690
Paraloid K **2**:896
Particle size distribution **2**:688
    determination of **2**:688
    of emulsion polymers **2**:686
Parylenes **3**:1346, 1349–1350
PASP **2**:736–737, 739, 741, *see* Polyaspartates
PBAT *see* Poly[(butylene adipate)-co-(butylene terephtalate)]
PBT **2**:796, 802, 804–807, 809–812
Pearlthane **3**:1380
Pebax **3**:1394
PEG *see* Poly(ethylene glycol)
Pelesta **3**:1394
Pellethane **3**:1380
Pelletization
    of polymers **1**:368, 372, 379
P.E.R. **3**:1369
Perapret **2**:692
Perfluoroalkoxy resins **2**:623
Perfluoroalkoxy (PFA) resins [26655-00-5] **2**:620
    processing **2**:622
    properties **2**:622
Perfluoro(methyl vinyl ether) (PMVE) **2**:620
Perfluoropolymers, amorphous **2**:636
    production **2**:636
    properties **2**:637
Perfluoro(propyl vinyl ether) (PPVE) **2**:620
Perkanox **1**:535
Permeability
    of polyoxymethylenes **2**:911–912, 914, 916, 921
Permeation
    in plastics **1**:471, 480, 485, 492–493, 498–500, 502, 507, 509, 511, 519–520
Perspex **2**:889, 895–896

PHA *see Polyhydroxyalkanoates*
Phenolic resins
   molding compounds **1**:407, 412, 419–420, 423–426, 429–433, 435–438
Phillips catalysts
   polyethylene production **2**:835
Phos-Chek **1**:555
Phosphites
   as antioxidant **3**:1015
Phosphonites
   antioxidant **1**:527–535, 546, 568, 570
Phthalic acid (1,2-benzene dicarboxylic acid)
   in resins **2**:782–784
Pibiflex **3**:1386
Pibiter **2**:810
Pigment Orange 68 **1**:571
Pigments, inorganic
   in plastics **1**:534, 545, 551, 555, 560, 565, 569–571
Pigments, organic
   in plastics **1**:534, 545, 551, 555, 560, 565, 569–571
Pigment Yellow 180 **1**:571
Pigment Yellow 182 **1**:571
Pigment Yellow 183 **1**:571
Pigment Yellow 192 **1**:571
PLA *see Poly(lactic acid)*
Plasma polymerization **1**:455, 466–467
Plasticization
   external **1**:496–497, 504, 513, 520
   internal **1**:471, 480, 492, 495–498, 502, 513
Plastics
   amorphous **1**:471–478, 480, 482, 493, 496, 498–502, 510, 512
   antioxidants for **1**:531–532
   crack resistance **1**:492
   creep modulus **1**:483–484, 488, 521
   damaging deformation **1**:495
   damaging force **1**:495
   decomposition temperature range **1**:515
   self ignition temperature **1**:515
   deformation **1**:471, 475, 480–481, 483, 485, 487, 491–496, 506, 518–520, 522
   density **1**:233–234, 246, 249, 471–472, 476, 484–485, 496, 499–500, 502, 504, 510, 512, 514, 516–517, 521–522
   dielectric properties **1**:471, 503–505
   drawing **1**:367, 380, 392, 401–403
   electrical properties **1**:471, 502
   extruders for **1**:383, 385
   fatigue **1**:487, 489
   flammability **1**:234, 236
   gloss **1**:471, 506–508
   heat penetration coefficient **1**:521
   impact strength **1**:556, 565, 567, 575
   liquid absorption and swelling **1**:471, 510
   for noise suppression **1**:486, 497
   optical properties **1**:471, 478, 506
   particle-size distribution **1**:517
   pigments for **1**:571
   polarity and solubility **1**:453
   refractive index **1**:493, 504, 506–508, 521
   resistance **1**:439, 454, 457–459, 461–462, 466–467
   resistivity **1**:497, 502–503, 522
   saponification number **1**:237–238
   shear modulus **1**:474, 478–479, 483, 486–487, 521
   solubility **1**:231–233, 235, 238–239, 250
   specific heat **1**:471, 474, 498–500, 521
   static modulus of elasticity **1**:487–488
   surface tension **1**:453–454, 457
   thermal conductivity **1**:471, 489, 498–500, 522
   thermal degradation **1**:234
   thermal diffusivity **1**:500, 521
   thermal expansion **1**:471, 474, 497, 500–501
   yellowing **1**:509, 533, 535, 546
Plastics, additives **1**:527
   mixing **1**:411–413, 415, 419, 421
Plastics, analysis **1**:231
Plastics, fiber–reinforced **1**:411, 414
   aramid fibers **3**:1325, 1331–1332, 1335
   continuous fibers **3**:1325, 1330, 1332–1333, 1335–1336
   discontinuous fibers **3**:1325, 1332–1333
   glass fiber–reinforced, processing of **1**:444
   glass fibers **3**:1325, 1330–1331, 1335, 1337
   graphite (carbon) fibers **3**:1325, 1330
   molds for **1**:412, 433, 435
Plastics, general survey **1**:3, **1**:149, **1**:187, **1**:205, **1**:223
   manufacture **1**:62, **1**:178
   nomenclature **1**:6
   sustainability of plastics **1**:137
Plastics, processing **1**:367, 407, 443
*see also Polymers*
   properties and testing **1**:471, 477
   accelerated weathering test **1**:545
   ball indentation hardness **1**:489
   bending vibration test **1**:486
   fluctuating-load teststeady-load test **1**:487
   lubricant testing **1**:552
   Martens test **1**:490
   plastograph **1**:549, 553
   reaction to fire test **1**:516

Plastics, processing (*Continued*)
  Shore hardness value **1**:489
  test methods for antioxidants **1**:532
  torsional vibration test **1**:485–486
Plastics, surface treatment
  chemical metal deposition **1**:439, 455, 460, 462
  corona treatment **1**:457, 464–466
  flame treatment **1**:465
  flocking **1**:439, 455–456
  fluorination **1**:455, 465, 467, 469
  hot embossing **1**:439, 455
  painting and printing **1**:439, 455–456
  physical metal deposition **1**:439, 455, 463
  plasma treatment **1**:457, 465
Plastisol **3**:1129
  influence of plasticizers on **3**:1129
  influence of resins on **3**:1129, **3**:1130
Plastistrength **2**:896
Plastomers **2**:856, **3**:1369
Platamid **2**:728
Plexalloy-F-PAB **2**:894
Plexidon **2**:896
Plexiglas **2**:886, 889, 893–896
Plexiglas GS 233 **2**:886
Plexiglas hw 55 **2**:893
Plexiglas XT 7H **2**:886
Plexiglas zk **2**:894
Plexigum **2**:690
Pleximid **2**:893
Plextol **2**:690
Pocan **2**:810
Polivic S202 **3**:1116
Polivic S404W **3**:1116
Polivinol **3**:1144
Poly (tetramethylene ether glycol)
  for synthetic fibers **4**:1491
Poly (trimethylene terephthalate)
  fibers **4**:1453–1485
Poly(1,4-dimethylenecyclohexane terephthalate)
  fibers **4**:1453–1485
Polyacetylene
  cross-linked **3**:1262, 1264, 1267, 1270
  stretched **3**:1265, 1267, 1273
*cis*-Polyacetylene **3**:1261
Polyacrylamide **2**:659, 667
  cross-linked **2**:670
  hydrolized **2**:669
  toxicology **2**:671
Poly(2-acrylamido-2-methylpropanesulfonic acid) **2**:668
Polyacrylates **2**:675–676, 681, 685, 687–689, 691–693
  toxicology **2**:675, 685

Poly(acrylic acid) **2**:666
  toxicology **2**:671
Polyacrylonitrile fibers **4**:1513–1514, 1518, 1521, 1523–1524, 1526
  aftertreatment **4**:1514, 1518, 1520–1521, 1524
  dyeing **4**:1513–1515, 1520–1521, 1523–1525
  production **4**:1513–1514, 1516, 1518–1520, 1525–1527
  properties **4**:1513–1516, 1518, 1521, 1523, 1525
  spinning **4**:1513–1518, 1520–1521, 1523–1525
  types **4**:1513, 1522–1523, 1526
Poly(amic acid) **2**:864
Polyamidation **2**:697, 701, 706–707
Polyamide fibers **4**:1435
  delustering **4**:1444
  dyeing **4**:1443
  optical brighteners **4**:1445
Polyamide fibers, aromatic
  meta-oriented **4**:1542, **4**:1543
Polyamide fibers *see also* Fibers; Fibers, polyamide fibers
Polyamide-imide fiber **4**:1548, **4**:1549
Polyamide monomers
  Ecological aspects **2**:697, 725
Polyamide–poly(2,6-dimethyl-1,4-phenylene ether) blends **2**:927, 936
Polyamides (PA) **2**:697–702, 704, 707–709, 712–713, 715, 717–728,
  *see also Nylons; Polyamide fibers*
  as acid trapping agents **2**:920
  biodegradable polymer **3**:1231–1237, 1239, 1241–1244, 1247–1249, 1251–1259
  as blend component **3**:1210, 1212
  blends and alloys **2**:697, 720, 722
  conductive **2**:697, 719, 721
  filled and reinforced **2**:697, 719, 723
  as flame retardants **1**:559, 568
  modeling of processes and reactors **1**:323
  monomers for **2**:700
  recycling **2**:697, 711, 725–726
  special requirements **4**:1442
  toxicology **2**:697, 725
Poly(ammonium acrylate) **2**:666
Polyaryletherketones (PAEK)
  compounds and blends **3**:1295
  with high glass-transition temperatures **3**:1293
  high-temerature polymer **3**:1281
Polyaspartates (PASP) **2**:733–745
  by polymerization of protected aspartic acids **2**:734
  by solid-phase thermal polycondensation **2**:735
  dispersion effect **2**:739
  threshold inhibitor **2**:739

Poly(aspartic acids) **2**:734, 738
Polyazepine
Polyazole fibers **4**:1554
Polyazulene
Polybenzenes
   in specialty plastics **3**:1343
Polybenzimidazole fiber (PBI) **4**:1548
Polybenzocyclobutenes (PBCB)1343,1350
   in specialty plastics **3**:1343
Poly(benzyl methacrylate)
   physical properties of **2**:886–887
Polybond **1**:569
Poly(bornyl methacrylate)
   physical properties of **2**:886–887
Polybutadiene
   as additive in PS molding materials **3**:982
Poly(1-butene) **2**:747–754
   crystal modifications of **2**:749
   in high-performance pipes **2**:753
   in packaging **2**:747–748, 752–754, 761
Poly(1-butene), atactic **2**:748
Poly(1-butene), isotactic **2**:748–749
Poly[(butylene adipate)-*co*-(butylene terephtalate)] (PBAT)
   degradation **3**:1231–1235, 1239–1243, 1246–1247, 1256–1259
   ecotoxicological tests **3**:1234
   films **3**:1231–1232, 1237–1250, 1255–1256, 1259
Poly(butylene terephthalate) (PBT)
   *see also Polyesters*
   production of **2**:791–792, 796, 799–802, 805–806, 809, 812–813
   properties of **2**:795, 805–809
Poly(*n*-butyl methacrylate)
   physical properties of **2**:886–887
Poly(*sec*-butyl methacrylate)
   physical properties of **2**:886–887
Poly(*tert*-butyl methacrylate)
   physical properties of **2**:886–887
Polycarbonate–polyester blends **2**:772
Polycarbonates **2**:763
   aliphatic **2**:768
   as blend component **3**:1210, 1212
   blends **2**:768
   interfacial polycondensation **2**:773
   melt transesterification **2**:774
Poly[(chloro-1,4-phenylene)-1,2-ethanediyl] (PPX-C) **3**:1346
Polychlorotrifluoroethylene (PCTFE) [9002-83-9] **2**:629
   properties **2**:629
Polyclar **3**:1158
Polycondensation
   bifunctional **1**:56, **1**:172
   multifunctional **1**:59, **1**:174
Poly(cyclohexyl methacrylate)
   physical properties of **2**:886–887
Poly(*n*-decyl methacrylate)
   physical properties of **2**:886–887
Poly[(dichloro-1,4-phenylene)-1,2-ethanediyl] (PPX-D) **3**:1346
Poly(*N,N*-dimethylacrylamide) **2**:668
Poly(1,4-dimethylenecyclohexane terephthalate) **2**:809
   fibers **4**:1479
Poly(1,4-dimethylenecyclohexane terephthalate) (PDCT) **4**:1479
Poly(2,6-dimethyl-1,4-phenylene ether) (PPE)
   polymer blends **3**:1197–1199, 1203, 1216, 1223, 1226–1227
Poly($\beta,\beta$-dimethyl-$\beta$-propriolactam) **4**:1438
Polydis TR **1**:554
Poly(dodecyl methacrylate)
   physical properties of **2**:886–887
   molding compounds **1**:407, 412, 419–420, 423–426, 429–433, 435–438
Polyester resins, unsaturated **2**:781
Polyesters
   modeling of processes and reactors **1**:324
Polyesters **2**:805
   *see also Poly(ethylene terephthalate); Poly (butylene terephthalate)*
   anisotropy **2**:809
   as blend component **3**:1210, 1212
   heat-resistant **2**:809
   recycling **2**:791, 794, 812–813
   toxicology **2**:791, 811
Polyesters, aliphatic
   biodegradation **3**:1231–1235, 1239–1243, 1246, 1256, 1258–1259
   synthesis **3**:1231–1232, 1236, 1238–1242, 1257
Polyesters, aliphatic/aromatic
   biodegradation **3**:1231–1235, 1239–1243, 1246, 1256, 1258–1259
   synthesis **3**:1231–1232, 1236, 1238–1242, 1257
Polyesters, aromatic
   production of **2**:791–792, 796, 799–802, 805–806, 809, 812–813
Polyesters, biodegradable
   additives **3**:1231–1232, 1235, 1246–1247, 1249–1250
   films **3**:1231–1232, 1237–1250, 1255–1256, 1259
   multilayer films **3**:1231, 1250
   processing **3**:1231–1232, 1236–1237, 1241–1242, 1246–1247, 1249–1250, 1252, 1256, 1259
   properties of **3**:1238, 1240, 1242–1243, 1245–1247, 1250, 1259

Polyesters, crystalline
    production of **2:**791–792, 796, 799–802, 805–806, 809, 812–813
Polyesters, liquid crystalline (LCP) **4:**1554, **4:**1555
Polyetheramides **3:**1388
Polyether diols
    for polyurethane fibers **4:**1487–1488
Polyetherester amides **3:**1387
Polyetherether ketone (PEEK) [31694-16-3] **4:**1549, **4:**1549–1550
    polymers, high-temperature **3:**1281
    production **3:**1281–1283, 1290–1292, 1295–1296, 1299, 1301, 1305, 1307, 1319, 1321
Polyetherimides
    as blend component **3:**1210, 1212
    monomers for **3:**1292, 1319
    polymers, high-temperature **3:**1281
    production by melt polycondensation **3:**1283
Polyetherketone (PEK)
    polymers, high-temperature **3:**1281
    production **3:**1281–1283, 1290–1292, 1295–1296, 1299, 1301, 1305, 1307, 1319, 1321
Polyetherketone ether ketone ketone (PEKEKK)
    production **3:**1281–1283, 1290–1292, 1295–1296, 1299, 1301, 1305, 1307, 1319, 1321
Polyetherketone ketone (PEKK)
    production **3:**1281–1283, 1290–1292, 1295–1296, 1299, 1301, 1305, 1307, 1319, 1321
Polyethylene **2:**817.
    *see also Polyolefins; Polyethylene, etc.*
    block copolymers **2:**821
    catalysts **2:**834
    comonomers **2:**841
    crystal structure **1:**472, **2:**820
    dynamic coefficient of friction **1:**491, 497
    environmental aspects **2:**856
    molecular structure **2:**819
    pyrolysis **1:**231, 234, 237–238, 258
    recycling **2:**856
    relaxation spectra **1:**482–483, 521
    unimodal and multimodal **2:**825
    $p$–$V$–$T$ diagram **1:**484
Poly(ethylene glycol) (PEG) **3:**1242, 1258
    *see also Poly(ethylene oxide) (PEO)*
Poly(ethylene glycol) (PEO)
    raw material for thermoplastic elastomers **3:**1381
Polyethylene, high-density (HDPE)
    films **3:**1231–1232, 1237–1250, 1255–1256, 1259
    films and split yarns from 1495, 1505, 1509
    tapes **4:**1495, 1497, 1506–1510
Polyethylene, low-density (LDPE)
    as blend component **3:**1210, 1212
    dielectric properties **1:**471, 503–505

Poly(ethylene oxide) (PEO) **2:**899,
    *see also Poly(ethylene glycol) (PEG)*
Poly(ethylene terephthalate) (PETP)**2:**791, **4:**1454, 1482–1483, 1485
*see also Polyesters*
    enthalpy curve **1:**499
    granulate **4:**1455–1456, 1462
    properties **2:**791–792, 794–799, 803–811
Poly(ethylene terephthalate), dyed
    properties **4:**1453–1454, 1456, 1459–1464, 1466, 1468–1472, 1474–1477, 1479–1480, 1485
Poly(ethylene terephthalate) fibers **4:**1453, 1465
    blends with cotton, wool etc. **4:**1460, 1467
    continuous-filament yarns **4:**1453, 1468, 1471
    copolymerization **4:**1461
    crystallization **4:**1455, 1457, 1462–1464, 1476, 1478
    deformation **4:**1459, 1473, 1479–1480
    differential thermal measurements **4:**1463
    drawing **4:**1453, 1457–1459, 1463–1466, 1468–1472, 1476
    dyeing **4:**1453, 1456, 1459–1461, 1464, 1467–1468, 1471, 1479
    filament yarns **4:**1453, 1468, 1471, 1473
    flame-retardant **4:**1461, 1468
    fully oriented yarns (FOY) **4:**1470
    glass transition temperature **4:**1456–1460, 1462–1463, 1476, 1479–1480
    highly oriented yarn (HOY) **4:**1469
    high-tenacity filament **4:**1471
    high-tenacity filament **4:**1471
    homopolymer **4:**1461
    low orientation yarn (LOY) **4:**1468
    mechanical tests **4:**1463
    melt-spun **4:**1459
    melt-spun **4:**1455, 1459
    monofilaments **4:**1457, 1466, 1472, 1475–1477, 1480
    partially oriented yarn (POY) **4:**1469
    physical structure **4:**1462
    production **4:**1453–1454, 1458–1459, 1465, 1469–1470, 1474–1476, 1480–1482, 1484–1485
    spunbonds **4:**1453, 1474, 1480
    staple fibers and yarns from **4:**1465
    textured yarns **4:**1469–1470
    thermal aftertreatment **4:**1458
    thermal measurements **4:**1463
    tire-cord material **4:**1474
    tire-cord material **4:**1474
    very high molecular mass **4:**1455
Polyethylene, very low-density
    as blend component **3:**1210, 1212

Poly(2-ethylhexyl methacrylate)
  physical properties of **2**:886–887
Poly(ethyl methacrylate)
  physical properties of **2**:886–887
Poly(ethylthioethyl methacrylate)
  physical properties of **2**:886–887
Polyflon **2**:608
Poly(furfuryl methacrylate)
  physical properties of **2**:886–887
Poly(hexadecyl methacrylate)
  physical properties of **2**:886–887
Poly(*n*-hexyl methacrylate)
  physical properties of **2**:886–887
Polyhydantoins
  polyureas as starting materials for **3**:1029
Polyhydrouracils
  polyureas as starting materials for **3**:1029
Polyhydroxyalkanoates (PHA)
  biodegradation **3**:1231–1235, 1239–1243, 1246, 1256, 1258–1259
  derivatives **3**:1231, 1239, 1241, 1246
  synthesis **3**:1231–1232, 1236, 1238–1242, 1257
Poly(2-hydroxyethyl methacrylate)
  physical properties of **2**:886–887
Polyimide fiber P84 **4**:1549
Polyimides **2**:859
  by amic acid route **2**:863
  by imide-containing monomers **2**:870
  dianhydrides and diamines for **2**:865
  one-step formation **2**:870
  photoimageable **2**:877
  thermoplastic resins and powders **2**:860
  thermosetting **2**:872
  intramolecularly doped **3**:1277
Poly(isobornyl methacrylate)
  physical properties of **2**:886–887
Polyisobutylene, high molecular mass **2**:754, 757, 759–760
Polyisobutylene, low molecular mass **2**:754–756, 759
Polyisobutylene, medium molecular mass **2**:754, 756–757, 759
Polyisobutylenes **2**:747, 754–760
  copolymers **2**:747, 753, 760
  properties dependent on degree of polymerization **2**:759
Poly(isobutyl methacrylate)
  physical properties of **2**:886–887
Poly(*N*-isopropylacrylamide) **2**:668
Poly(isopropyl methacrylate)
  physical properties of **2**:886–887
Poly(lactic acid) (PLA) **3**:1231
  *see also Polylactide*

biodegradation **3**:1231–1235, 1239–1243, 1246, 1256, 1258–1259
compounds with polyesters **3**:1243
synthesis **3**:1231–1232, 1236, 1238–1242, 1257
Polylactide **3**:1231
  *see also Poly(lactic acid)*
Polylaurolactam **2**:699, 714, *see Nylon 12*
Polylite **2**:786
Poly(menthyl methacrylate)
  physical properties of **2**:886–887
Polymer-analogous reaction **2**:665
  conductive **3**:1227
  engineering thermoplastics **3**:1209, 1226
  heat resistant **3**:1221
  heterogeneous **1**:134, **1**:217
  homogeneous **1**:130, **1**:213
  polyesters in **3**:1221
  recycling **3**:1197, 1200, 1223–1225
  technologies **3**:1197, 1201
  thermodynamics of **3**:1203–1204
  ultrahigh molecular mass **3**:1227
Polymer composites
  fabrication of **3**:1325, 1333–1334, 1337
  laminar **3**:1325–1326, 1328–1329, 1338
  particulate **3**:1325–1326, 1328, 1337
Polymer concrete **1**:412, 414, 418–419
Polymer crystals
  degree of crystallinity **1**:475–476
  morphology **1**:471–472, 475–477, 493, 495, 511
Polymer dispersion **2**:675, 680, 685–687, 690–692
Polymerization
  of poly(acrylic acids) **2**:661
  ethylene **2**:829
  solid-phase (SPP) **2**:709
  theta condition **2**:666
Polymerization, batch
  of acrylamide and acrylic acid **2**:662
Polymerization, bulk **1**:46, **1**:162
  ABS production by **3**:1005, **3**:1011
  production of matrix copolymers **3**:1009
  production of PMMA **2**:891
  production of SAN copolymers **3**:993
Polymerization, cationic
  in production o polystyrene **3**:983
Polymerization, chain growth **1**:37, **1**:39, **1**:153, **1**:155
  condensative **1**:38, **1**:154
Polymerization, continuous **2**:663
  production of PMMA **2**:891
Polymerization, continuous bulk
  in production of polystyrene **3**:984
Polymerization, coordination
  polyethylene production **2**:833

Polymerization, emulsion **1**:46, **1**:162
  for ABS production **3**:1004, **3**:1005
  batch, VDC copolymers **3**:1182–1184, 1186,
    1188–1189
  continuous feeding **3**:1181, 1184
  polymethacrylates **2**:885–888, 892, 896, 1181,
    1182
  poly(vinylidene chloride) production by **3**:1181
  production of matrix copolymers **3**:1010
  production of poly(vinyl esters) **3**:1165, 1170,
    1172–1173
  tetrafluoroethylene-ethylene copolymers **2**:625
  of vinyl chloride **3**:1125
Polymerization, gas-phase **1**:46, **1**:162
  polyethylene production **2**:847
Polymerization, inverse emulsion
  of polyacrylamides and poly(acrylic acids) **2**:664
Polymerization, mass
  of vinyl chloride **3**:1123
Polymerization, precipitation
  acrylonitrile **4**:1513–1516, 1518–1526
  of polyacrylamides and poly(acrylic acids) **2**:665
Polymerization processes
  modeling of processes and reactors **1**:315
  with solvent **1**:338
  types of reactors **1**:317
  without solvent **1**:337
Polymerization processes **1**:265
  branched polymers **1**:303
  chain length distribution **1**:268
  classification of polymerization reactions **1**:266
  crosslinked polymers **1**:298
  kinetics **1**:267
  polydispersity index **1**:281
  weight fraction distribution **1**:269
Polymerization, radical
  acrylonitrile **4**:1513–1516, 1518–1526
  controlled **1**:43, **1**:159
  initiators **2**:832
  of MMA **2**:888–889, 892
Polymerization, solution
  acrylonitrile **4**:1513–1516, 1518–1526
  of polyacrylamides and poly(acrylic acids) **2**:662
  polyethylene production **2**:849
Polymerization, suspension **1**:46, **1**:162
  of polyacrylamides and poly(acrylic acids) **2**:664
  polyethylene production **2**:845
  poly(vinylidene chloride) production **3**:1181
  production of matrix copolymers **3**:1009
  production of PMMA **2**:891
  in production of polystyrene **3**:986
  tetrafluoroethylene-ethylene copolymers **2**:625
  vinyl chloride **3**:1114

Polymerization, tube
  production of PMMA **2**:891
Polymers
  adhesion **1**:439, 444–446, 453–455, 457–459,
    461–464, 466–467
  amorphous **1**:471–478, 480, 482, 493, 496,
    498–502, 510, 512
  copolymer **1**:471–472, 477, 480, 488, 491,
    497–498, 506, 517
  cross-linked **1**:22
  elastic aftereffects **1**:480
  electrets **1**:471, 506
  electric properties **1**:116
  fiber-reinforced **3**:1325–1328, 1337
  flammability **1**:234, 236
  flow curve **1**:517–520
  glass transition temperature **1**:474–475, 477, 479
  heat resistance **1**:458, 466
  heterogeneity of copolymer **1**:258, 260
  interfacial tension **1**:83, **1**:201
  isolation **1**:231–232, 250–251
  machining guidelines **1**:439, 442
  mechanical bonding **1**:439, 445–446, 454
  mechanical properties **1**:101
  molecular orientation **1**:496
  optical properties **1**:120
  ordered chain assemblies **1**:21
  oriented **1**:384, 392, 401–402
  photodegradation **1**:535–537, 544–545
  pyrolysis **1**:231, 234, 237–238, 258
  saponification number **1**:237–238
  rheological properties **1**:94
  saponification number **1**:237–238
  solubility **1**:231–233, 235, 238–239, 250
  synthesis **1**:37, **1**:152
  thermal properties **1**:84
  thermoforming **1**:367, 402–404
  thermooxidative degradation **1**:531, 534
  welding **1**:439–440, 445–453, 468
Polymers, biodegradable **3**:1231–1259
  agricultural application **3**:1253
  commercially available **3**:1241
  mulch films **3**:1241–1242, 1249, 1255
  registration scheme **3**:1235
Polymers, electrically conducting **3**:1261
  counterion **3**:1264, 1268–1269, 1275–1277
Polymers, high-temperature
  amorphous **3**:1282, 1285, 1287, 1289, 1293–1294,
    1297, 1302, 1306, 1308, 1315
  in automotive industry **3**:1289
  in houshold appliances **3**:1290, 1306
  in medical equipment **3**:1290
  (semi)crystalline **3**:1282

Polymer solution 2:675, 684–685
Polymethacrylamide 2:667
Polymethacrylamide potassium salt 2:667
Polymethacrylamide sodium salt 2:667
Polymethacrylates 2:885–888, 892, 896
   toxicology 2:885, 897
Poly(methacrylic acid) 2:667
Poly(methacrylic acid) sodium salt 2:667
Poly(methyl methacrylate) (PMMA)
   cast 2:885–886, 889–892, 895–896
   extruded 2:885–886, 891, 893, 895–896
   molding compounds 2:885–886, 889–895
   physical properties of 2:886–887
Poly(methyl methacrylate) [9011-14-7] (PMMA) 2:885, 887
Poly(N-methyl methacrylimide) (PMMI)
   heat-resistant molding compound 2:885, 893
Polymide 4:1771
Poly(m-phenyleneisophthalamide) (PMIA) [24938–60–1] 4:1441, 4:1543
Poly(1-naphthyl methacrylate)
   physical properties of 2:886–887
Poly(N-isopropylacrylamide) [25189-55-3] 2:668
Poly(N-methyl methacrylimide) [119499-71-7] (PMMI) 2:893
Poly(N,N-dimethylacrylamide) [26793-34-0] 2:668
Poly(octadecyl methacrylate)
   physical properties of 2:886–887
Poly(n-octyl methacrylate)
   physical properties of 2:886–887
Polyolefin fibers
   film and split yarns 4:1495, 1505, 1509
   high-modulus yarns 4:1495
Polyolefin film yarns 4:1495, 1504–1505
   high modulus 4:1509
   production 4:1495, 1497–1501, 1504–1510
Polyolefins
   antioxidants 1:527–535, 546, 568
Polyolefin split yarns 4:1504–1505
Polyoxadiazoles fibers 4:1549
Polyoxyalkylenes 2:899
   *see also Poly(alkylene oxides)*
Polyoxymethylenes 2:911–912, 914–916, 919–923
   in automotive industry 2:921
   as blend component 3:1210, 1212
   as construction materials 2:915
   copolymers 2:911–912, 914–917, 919–921, 924
   in electrical engineering 2:916
   electrical properties 2:911, 916
   fire behavior 2:911, 916
   from formaldehyde and trioxane 2:916
   with HALS stabilizers 2:920
   high-strength copolymer 2:911, 921
   homopolymers 2:911–912, 914, 916, 919, 921–922
   impact-resistance-modified 2:920–922
   low volatile organic carbon grades 2:911, 921
   in mechanical engineering 2:923
   nucleation agents 2:911, 914, 920
   optical properties of 2:916
   physical properties 2:912–914
   production from formaldehyde and trioxane 2:916
   recycling 2:911, 923
   release agents 2:911, 920
   stabilizer 2:911, 914–915, 917, 919–920, 923
   toxicology 2:911, 923
Poly(oxy-1,4-phenylenesulfonyl-1,4-phenylene) 3:1296
Polyphenylene
   as electrically conducting polymer 3:1261
   side reactions to cross-linked graphite-like lattices 3:1270
   in specialty plastics 3:1343
Poly(1,4-phenylene-1,2-ethanediyl) (PPX-N) 3:1346
Poly(phenylene oxides) 2:927
Poly(phenylene sulfide) (PPS) [9016-75-5] 3:1272, 4:1549
   films from 3:1295, 1298, 1300, 1306
   nonreinforced moldings 3:1302
   polymers, high-temperature 3:1281
   reinforced 3:1281, 1285, 1287–1289, 1295, 1302–1307, 1312–1314, 1317
   reinforced moldings 3:1281, 1302–1303
Poly(p-phenylene telluride) 3:1272
Poly(p-phenylene vinylene)
   as electrically conducting polymer 3:1261
Poly(phenyl methacrylate)
   physical properties of 2:886–887
Polyphosphonate 1:564
Poly(potassium acrylate) 2:666
Poly(p-phenylene-2,6-benzobisoxazole) (PBO) [60871-72-9] 4:1554
Poly(p-phenylene benzobisthiazole) (PBZT) [69794-31-6] 4:1554
Poly(p-phenyleneterephthalamide) (PPTA) [24938-64-5] 4:1441, 4:1543
   fiber structure 4:1553–1554
   polymerization 4:1551
Polypropylene (PP) 2:937
   as blend component 3:1210, 1212
   blends 2:965
   catalysts for 2:944
   copolymers 2:961
   crystal characteristics 2:942
   crystallization 2:941

Polypropylene (PP) (*Continued*)
  enthalpy curve **1:**499
  film and split yarns from 1495, 1505, 1509
  tapes **4:**1495, 1497, 1506–1510
  homopolymers **2:**960
  impact copolymers **2:**961
  monofilaments **4:**1495, 1505–1506, 1508–1510
  multiphase copolymers **2:**942
  polymerization processes **2:**949
  properties **4:**1495, 1497–1498, 1503–1507, 1509
  recycling **2:**976
  relaxation spectra **1:**482–483, 521
  stabilization **1:**528–529, 532–534, 536–537, 544–547, 549, 551, 553, 573, 1495, 1501, 1503
  thermal conductivity **1:**471, 489, 498–500, 522
  types **2:**940
Polypropylene adipate
  fractionation of **1:**251, 254
Polypropylene–ethylene–acrylate rubber
  thermoplastic vulcanizates **3:**1373
Polypropylene fiber process **2:**968
Polypropylene fibers **4:**1495, 1497–1498, 1503–1504, 1511
  pigmentation **4:**1495, 1502–1504
  production **4:**1495, 1497–1501, 1504–1510
  properties **4:**1495, 1497–1498, 1503–1507, 1509
  spinning **4:**1495, 1497–1501, 1504, 1508–1510
  structure **4:**1495, 1497–1498, 1505, 1508, 1510
Poly(propylene oxide) (PPO) **2:**899
  *see also Poly(propylene glycol)*
  raw material for thermoplastic elastomers **3:**1377
Poly(*n*-propyl methacrylate)
  physical properties of **2:**886–887
  as film **3:**1275–1276
  redox character of **3:**1269
Polyquinazolinediones
  polyureas as starting materials for **3:**1029
Poly(sodium acrylate) **2:**666
Polystyrene (PS) **3:**981
  enthalpy curve **1:**499
  isotactic **3:**984
  recycling **3:**991
  syndiotactic **3:**984
  thermal conductivity **1:**471, 489, 498–500, 522
Polystyrene, expandable (EPS) **3:**983
Polystyrene foams **3:**983
Polystyrene, impact-resistant (IPS) **3:**982
Polysuccinimide (PSI)
  production **2:**733–736, 739, 742
Polysuccinimide (PSI)(**2:**733–745)
Polysulfones
  as blend component **3:**1210, 1212
  polymers, high-temperature **3:**1281

Polytetrafluoroethylene (PTFE) [9002-84-0] **2:**607
  dispersions **2:**614
  modified granular **2:**615
  powder **2:**614
  production **2:**608
  properties **2:**610
Polytetrafluoroethylene fibers
  production **4:**1539–1540
  properties **4:**1539–1540
Polytetrafluoroethylene, high molar mass **2:**607
Polytetrafluoroethylene, low molar mass **2:**608
Poly(2,2,3,3-tetrafluoropropyl methacrylate)
  physical properties of **2:**886–887
Poly(tetrahydrofurfuryl methacrylate)
  physical properties of **2:**886–887
Poly(tetramethylene ether glycol)
  for synthetic fibers **4:**1489
Poly(tetramethylene oxide) (PTMO) **2:**899–902, 904–907
  raw material for thermoplastic elastomers **3:**1381
  soft segment in thermoplastic elastomers **3:**1383
Poly(tetramethylene terephthalate) **4:**1453, 1468, 1476
Polythiophene
  as electrically conducting polymer **3:**1261
Poly(2,4,6-tribromostyrene) **1:**564
Poly(tricyclodecyl methacrylate)
  physical properties of **2:**886–887
Polytriketoimidazolidines
  polyureas as starting materials for **3:**1029
Poly(3,3,5-trimethylcyclohexyl methacrylate)
  physical properties of **2:**886–887
Polytrope **3:**1369
Polyureas **3:**1029–1031, 1033–1049
Polyurethane fibers
  chain extension **4:**1491–1492
  isocyanate prepolymer **4:**1491–1492
  production **4:**1487–1488, 1490–1494
  spinning **4:**1487–1492
Polyurethane foam **1:**378
Polyurethanes (PUR) **3:**1099, 1103–1104, 1106–1108
  production **4:**1487–1488, 1490–1494
Polyvest **1:**569
Poly(vinyl acetal) films **3:**1154
Poly(vinyl acetals) **3:**1146
Poly(vinyl acetate)
  saponification **3:**1151
Poly(vinyl acetate) homopolymers **3:**1171
Poly(vinyl acetates)
  as granulating agents **3:**1116
Poly(vinyl alcohol)
  acetalization **3:**1153
  fibers from **4:**1529–1530, 1532–1533

Poly(vinyl alcohol) (PVA)
  as protective colloid in poly(vinyl ester) production
    **3**:1168, 1171, 1173
  as protective colloid in poly(vinyl ester) production
    **3**:1171
Poly(vinyl alcohol) (PVOH) **3**:1242, 1258
Poly(vinyl alcohol) fibers **4**:1529, 1533
  production **4**:1529–1531, 1533–1537
  properties **4**:1529–1534, 1536
Poly(vinyl alcohols)
  as adhesives **3**:1145
  as granulating agents **3**:1116
  in paper industry **3**:1145
  production of **3**:1142
  as protective colloid **3**:1145
  toxicology **3**:1146
Poly(vinyl alhohols) **3**:1141
Polyvinylamines **3**:1160
Poly(vinyl butyral) **3**:1147
Polyvinylcarbazole **3**:1160
Poly(vinyl chloride) (PVC) **3**:1111
  as blend component **3**:1210, 1212
  fillers **3**:1134
  grain porosity **3**:1119
  grain size **3**:1119
  heat stabilizers for **3**:1135
  industrial polymerization techniques **1**:337
  K-value **3**:1121
  lubricants **1**:527–528, 550–554, 567, **3**:1134
  photodegradation 535–537, 544–545
  slush molding **1**:373
  thermal degradation **1**:548, 571
  toxicology **3**:1139
Poly(vinyl chloride), chlorinated (CPVC) **3**:1132
Poly(vinyl chloride) fibers **4**:1529, 1532, 1536
  production **4**:1529–1531, 1533–1537
Poly(vinyl chloride), flexible
  properties **3**:1113
Poly(vinyl chloride), postchlorinated
  fibers from **4**:1529–1530, 1532
Poly(vinyl chloride), rigid (UPVC)
  properties **3**:1113
Polyvinyl compounds **3**:1141
Poly(vinyl esters) **3**:1165, 1170, 1172–1173
  produced by emulsion polymerization **3**:1172
  toxicology **3**:1165, 1173
Poly(vinyl ethers) **3**:1175–1179
  toxicology **3**:1175, 1178
Poly(vinyl fluoride) (PVF) [24981-14-4] **2**:635
  production **2**:635
  properties **2**:635
Poly(vinyl formal) **3**:1156
Poly(vinylidene chloride) (PVDC) **3**:1181

crystallinity **3**:1188–1193
extrusion grades **3**:1190, 1194
fibers from **4**:1529–1530, 1532
films **3**:1181, 1189–1195
homopolymers **3**:1181, 1189
in lacquering and coating **3**:1181, 1191
permeability **3**:1189–1193
suspension polymerization in production of **3**:1182, 1185–1187
toxicology **3**:1181, 1194–1196
Poly(vinylidene chloride) copolymers
  fibers from **4**:1529–1530, 1532
Poly(vinylidene chloride) fibers **4**:1529
  properties **4**:1529–1534, 1536
Poly(vinylidene fluoride) (PVDF) [9002-85-1] **2**:632
  production **2**:632
  properties **2**:632
Polyvinylimidazole **3**:1160
Poly(vinyl methyl ether) 1177, as plasticizer
    **3**:1177–1178
Poly(vinyl octadecyl ether)
  as polish for floors **3**:1178
Polyvinylpyrrolidone **3**:1157.
  *see also Povidone*
  as adhesives **3**:1159
  by bulk polymerization **3**:1157
  by radical polymerization **3**:1157
  as thickening agent **3**:1158
  toxicology **3**:1159
Polyviol **3**:1144
Porofor **1**:577
Poval L8 **3**:1116
Poval L9 **3**:1116
Poval L10 **3**:1116
Poval LM10HD **3**:1116
Powder immersion **1**:375
Precipitation polymerization **1**:337
Prepreg
  for polymer composite fabrication **3**:1325–1326, 1328, 1339
Prevex **2**:927
Prifrac **1**:554
Primal **2**:690–692
Primef **3**:1307
  of polymers **1**:439, 442, 447, 459, 462
Priolube **1**:554
Pristerene **1**:554
Prodox **1**:535
Propene
  polymerization **2**:948
  polypropylene production **2**:942
Propiofan **2**:690–691, 1172
Prosil **1**:569

Prosper **1**:550
Protective garments **4**:1547
Pseudomolecular mass
    of polymers **2**:669
PSI *see Polysuccinimide*
Pulse **2**:778
Pultrusion
    for polymer composite fabrication **3**:1325–1326, 1328, 1339
Pyrolysis
    of polyesters **2**:792–793, 795–796, 813

**Q**
*o*-Quinodimethane (OQDM) **3**:1350

**R**
Radel A and R **3**:1297
Radel R polyphenylsulfone **3**:1296
Ralox **1**:535
Reablend **1**:550
Reaction, chemical
    polymer-analogous **2**:665
Reaction injection molding (RIM) **2**:709, 713
    for polymer composite fabrication **3**:1325, 1333, 1337
    of polyureas **3**:1029, 1031, 1035, 1040–1042, 1046–1048
Reaction spinning
    polyurethane fibers **4**:1487–1488, 1492
Reactor Granules Technology **3**:1367
Reactor thermoplastic polyolefins **3**:1367
Realube **1**:554
Reamide **4**:1771
Reatinor **1**:550
Recart process **2**:973
Red phosphorus
    as flame retardant **1**:559, 568
Reinforced plastics **1**:488, 493, 497–498, 510, 1325, 1327, 1334, 1337, 1339–1341
    fillers for **3**:1332, 1337–1338
    polymeric matrices **3**:1325, 1331–1332
    static modulus of elasticity **1**:487–488
Reinforced reaction injection molding (RRIM)
    for polymer composite fabrication **3**:1325–1326, 1328, 1339
Reny **2**:728
Reoplast **1**:550
Reppe vinylation
    of lactams **3**:1157
Resamine **3**:1380
Resin acids **4**:1771
Resin RA **2**:692
Resins

acetone–formaldehyde **4**:1766–1767
acetophenone–formaldehyde **4**:1764–1766
aldehyde **4**:1751–1774
    ketone **4**:1751, 1755, 1759, 1762–1770
aliphatic-aromatic ketones **4**:1751–1774
aliphatic hydrocarbon **4**:1751–1774
aromatic hydrocarbon **4**:1751, 1757, 1760
$C_5$ olefin–diolefin **4**:1756, 1751–1774 cyclic
    diolefin **4**:1753
cyclohexanone **4**:1751, 1766–1768
cyclohexanone–formaldehyde **4**:1767–1768
from aliphatic ketones **4**:1751, 1766
furan **4**:1751, 1762–1763, 1751–1773
furfural and furfural-based **4**:1762
hydrocarbon **4**:1751–1761, 1764, 1770, 1772–1774
indene–coumarone **4**:1752–1753, 1757, 1759–1761
isobutyraldehyde–formaldehyde **4**:1751–1774
isocyanate-modified **4**:1766
methyl ethyl ketone–formaldehyde **4**:1751, 1766–1767
modified **4**:1751, 1755, 1757–1761, 1764–1766, 1768, 1771
petroleum **4**:1751–1753, 1756–1757, 1759, 1772–1773
piperylene **4**:1753, 1756
polyamide **4**:1751, 1760, 1769–1770, 1772, 1774
saponification number **1**:237–238
terpene **4**:1751–1752, 1754–1761, 1771–1773
toxicology 1771, 1751,
    legislation **4**:1751, 1771–1772
water reducible **4**:1609
Resins, soluble **3**:1185
Resins, synthetic **4**:1751, 1759
    *see also Resins*
    types of **4**:1763
Resin transfer molding (RTM)
    for polymer composite fabrication **3**:1325–1326, 1328, 1339
Resistat **1**:574
Revacryl **2**:690
Reversible-addition-fragmentation chain-transfer (RAFT) polymerization **1**:291, **1**:358
Reversible-deactivation radical polymerization **1**:355
Reversible-deactivation radical polymerization **1**:291
Revertex **2**:690, 692
Rhodianox **1**:535
Rhodiastab **1**:550
Rhodopas **2**:690, 692, 1172
Rhodoviol **3**:1144
Rhoplex **2**:690–691
Rilsan **2**:728
Riteflex **3**:1386
Rohagit **2**:896

Rolling and kneading
  of polymers **1**:368, 372, 379
Roskydal **2**:786
Rotational molding **2**:854
Rovel **3**:1021
Rubber modified polystyrene **3**:982
Ruggli–Ziegler dilution principle **3**:1346
Rynite **2**:810
Ryton **3**:1272, 1302–1304, 1307

**S**
Safflower oil
  typical fatty acid composition **4**:1601
Saflex **3**:1150
Salflex **3**:1369
Sandin **1**:574
Sandoflam **1**:564
Sandoflam 5087 **1**:564
Sandostab **1**:535, 572
Sanrex **3**:996
Santonox **1**:535
Santoprene **3**:1376
Santowhite **1**:535
Saponification
  in poly(vinyl alcohol) production **3**:1143
Saran dispersions **3**:1181
Saran resins **3**:1181
Sarlink 2000 **3**:1376
Sarlink 3000 **3**:1376
SCAS test
  for polyaspartates **2**:743
Schulz-Zimm distribution **1**:25
SCS **1**:572
Sealants
  SBC in **3**:1398
Seenox **1**:535
Selcion **3**:1321
Selectron **2**:786
Self dopants **3**:1265
Semibatch process
  with emulsion feed **2**:680
Sequel **3**:1369
Setalux **2**:690
Sewing yarn
  from poly(ethylene terephthalate) **4**:1453–1455,
    1459–1460, 1464, 1468, 1480, 1484
Sheet extrusion
  biodegradable polyesters **3**:1231, 1236–1237,
    1243, 1245–1251, 1257
Sheet molding compounds (SMC) **2**:785–786,
  789
  for polymer composite fabrication **3**:1325–1326,
    1328, 1339

Shinkolite **2**:889, 893–894, 896
Short spinning
  polypropylene fibers **4**:1500, 1511
Showa ACS **3**:1021
Sicolub **1**:554
Sicostab **1**:535, 550
Silac **3**:1369
Silane
  coupling agents **1**:527, 565–569
Silicone alkyds **4**:1608, 1612
Silk screen printing **1**:460
Siloxane–etherimide block copolymer (PSEI) **3**:
  1290
Sirester **2**:786
Siveras **3**:1311, 1321
Skythane **3**:1380
Slip agents
  for plastics **1**:532, 554, 568, 570–571, 576
Small-angle neutron scattering (SANS) **1**:472
Smith–Ewart assumption
  kinetics in polystyrene production **3**:1007
Sniatron **2**:786
Soarblen **1**:575
Sodium antimonates
  $Na_2Sb_2O_6 \cdot xH_2O$ as flame retardant **1**:459
Sodium peroxodisulfate
  as initiator for poly(vinyl ester) productions
    **3**:1165, 1170, 1172–1173
Solef **2**:632
Solution spinning
  polyurethane fibers **4**:1487–1488, 1491
Soybean oil
  typical fatty acid composition **4**:1601
Spandex **4**:1487–1493
  chain extenders **4**:1489–1490
  polyester **4**:1489–1490
  polyether **4**:1489–1490
  properties **4**:1487–1493
  solution **4**:1490–1492
  yarn **4**:1487–1493
Specialty fibers **4**:1541
Specialty plastics
  acidic copolymers as **3**:1343, 1353–1354,
    1357
  ethylene–methacrylic acid ionomers in
    **3**:1354–1357, 1360, 1343
  ethylene–methacrylic acid ionomers in
    **3**:1354–1357, 1360
  fluorinated ionomers in **3**:1343, 1361
  from higher aromatic hydrocarbons **3**:1344
  ionomers as **3**:1343, 1351, 1354, 1357–1358,
    1360–1364
  polyelectrolytes as **3**:1343, 1352

Specialty plastics (*Continued*)
  polyphenylenes for **3**:1343–1344, 1364
  sulfochlorinated polyethylene in **3**:1362
  sulfonated EPDM polymers in **3**:1362–1363
Spherilene process **2**:847
Spheripol process **2**:951
Spherizone multizone circulating reactor (MZCR) **2**:953
Spin dyeing
  polyacrylonitrile fibers **4**:1513–1514, 1518, 1520–1521, 1523–1524, 1526
Spinning of synthetic fibers
  melt spinning of poly(ethylene terephthalate) **4**:1453–1455, 1459–1460, 1464, 1468, 1480, 1484
Split yarns
  from polyolefines **4**:1508
Spray-up **3**:1325, 1330, 1333–1334
Stabilox **1**:550
Stabiol **1**:550
Stable-radical-mediated polymerization (SRMP) **1**:291, **1**:356
Stanclere **1**:550
Stann **1**:550, 556, 563–564
Stanyl **2**:728
Starch
  feedstock for monomer or polymer production **3**:1236
Starch compounds
  with polyesters **3**:1243
States of order
  polymers **1**:471–482, 486, 489, 492, 496–501, 504, 506–508, 510, 512, 514, 516–519
Statexan **1**:574
StatRite **1**:574
Stature **1**:574
Stavinor **1**:554
Step polymerization
  modeling of processes and reactors **1**:320
Stille reaction
  synthesis of electrically conducting polymers **3**:1262
Strandex **1**:550
Sturm test
  for polyaspartates **2**:743
  for cross-linking UP resins **2**:783
  polymerization **3**:983
Styrene–acrylonitrile copolymers (SAN) **3**:982, **3**:1009
  filled/unfilled **3**:995
Styrene–acrylonitrile–NPMI terpolymers **3**:1011
Styrene block copolymers (SBC) **3**:1394
  BR or IR mid blocks **3**:1397

  BR or IR mid blocks melt and solution processing **3**:1396
  coagulation with steam **3**:1395
  direct desolventizing **3**:1395
  properties **3**:1395, **3**:1400
  saturated rubber mid block **3**:1399
  service performance **3**:1397
Styrene copolymers **3**:981
Styrene–ethylene–butadiene–styrene copolymers
  in blends **3**:1206, 1208, 1210–1212, 1216, 1218, 1222
Styrene–maleic anhydride copolymers **3**:983, **3**:998
Styrene–maleimide copolymers **3**:999
Styrene–methyl methacrylate copolymers **3**:998
Styrene polymers
  antioxidants for **1**:531–532
Styrenic block copolymers **3**:1394
  BR or IR mid blocks **3**:1397
  coagulation with steam **3**:1395
  direct desolventizing **3**:1395
  melt processing **3**:1396
  properties **3**:1395
  saturated rubber mid block **3**:1399
  service performance **3**:1397
Styrolux **3**:1269
Styronal **2**:691
2-Sulfoethyl methacrylate
5-Sulfoisophthalic acid
  comonomer of poly(ethylene terephthalate) **4**:1453–1455, 1459–1460, 1464, 1468, 1480, 1484
Sulfur compounds, organic
  antioxidants **1**:527–535, 546, 568
Sumicon **3**:1307
Sumikasuper **3**:1311, 1321
Sumilizer **1**:535
Sumipex **2**:889, 896
Sumitomo TPE **3**:1376
Sunflower oil
  typical fatty acid composition **4**:1601
Supec **3**:1303, 1307
Surlyn **1**:572
Surlyn A **3**:1356–1357, 1360–1361
Suspension polymerization **1**:334, **1**:339
  kinetics **1**:341
Susteel **3**:1307
Swedlub **1**:554
Swedstab **1**:550, 554, 574
Swedstat **1**:574
Sylobloc **1**:554
Synesol **1**:550
Synocryl **2**:690
Synpron **1**:550

Synthacryl **2**:690
Synthalat **2**:690
Synthetic Resin 1201 **4**:1766
Synthetic Resin AFS **4**:1768
Synthetic Resin AP **4**:1766
Synthetic Resin CA **4**:1768
Synthetic Resin SK **4**:1766
Synthewax **1**:554
Szwarc process
　synthesis of poly(*p*-xylylenes) **3**:1343, 1345–1350

**T**
Tackiness
　of UP resins **2**:781–783, 785–788, 790
Taktene **1**:575
　as nucleation agent **2**:920
Tamfer **3**:1369
Tarflon **2**:778
Tariloy **2**:778
Tarnoform **2**:912
Tebestat **1**:554, 574
Technora **4**:1552, **4**:1553
Technyl **2**:728
Techster **2**:810
Tecnoflon **2**:641
Tedlar **2**:607, **2**:635
Tedur **3**:1303–1304, 1307
Tefabloc **3**:1369
Teflon **2**:608, **2**:617, **2**:620, **4**:1540
Teflon AF [37626-13-4] **2**:636
Tefzel **2**:624, 1539–1540
Tegin **1**:554
Tego **1**:554
Tego SML 20 **3**:1116
Tejinconex **4**:1544
Telcar **3**:1369
Tenite **2**:810
Tenox **1**:535
Tensylon **4**:1556, **4**:1557
　for poly(ethylene terephthalate) **4**:1453–1455,
　　1459–1460, 1464, 1468, 1480, 1484
　raw material for thermoplastic elastomers **3**:1381
　in resins **2**:782–784
Tergal **2**:810
Terluran **2**:894
Tetrabromobisphenol A [79-94-7] (TBBPA) **1**:556,
　563–564, **2**:764
Tetrabromophthalic acid
　in resins **2**:782–784
Tetrafluoroethylene (TFE) **2**:617
Tetrafluoroethylene-ethylene copolymers
　[25038-71-5] **2**:624
　production **2**:624

　properties **2**:625
　Tefzel grades **2**:628
Tetrafluoroethylene-hexafluoropropene copolymers
　**2**:617
　production **2**:617
　properties **2**:617
Tetrafluoroethylene-hexafluoropropene-vinylidene
　fluoride terpolymers **2**:638
　production **2**:638
　properties **2**:638
Tetrafluoroethylene-perfluoro(alkyl vinyl ether)
　copolymers **2**:620
　properties **2**:621
Tetrafluoroethylene-perfluoro(methyl vinyl ether)
　copolymers (FFKM) **2**:648
　production **2**:648
　properties **2**:648
Tetrafluoroethylene-propene copolymers (TFE-P)
　**2**:650
　production **2**:650
　properties **2**:650
Tetrahydrophthalic anhydride
　in resins **2**:782–784
3,3′,5,5′-Tetramethyldiphenoquinone (TMDPQ)
　**2**:928
Tetramethylene terephthalate
　hard segment in thermoplastic elastomers **3**:1382
Texicryl **2**:692
Texin **3**:1380
Textile dyeing
　poly(ethylene terephthalate) **4**:1453–1455,
　　1459–1460, 1464–1465, 1468, 1480, 1484
Thermal analysis, differential (DTA)
　of fibers **4**:1466
Therm Chek **1**:550
Thermoforming
　biodegradable polyesters **3**:1231, 1236–1237,
　　1243, 1245–1251, 1257
Thermoguard **1**:555
Thermolite **1**:550
Thermoplastic copolyester elastomers **3**:1381
　solid-phase polymerization **3**:1382
　segments **3**:1382
　uses **3**:1385, **3**:1386
Thermoplastic copolyesters **2**:796, 805
Thermoplastic elastomers **2**:651
Thermoplastic elastomers (TPE) **1**:81, **1**:198, **3**:
　1365
　hydrocarbon soft segments
　polyester soft segments **3**:1383
　short-chain ester units **3**:1382, **3**:1383
　processing **3**:1374
　uses **3**:1375

Thermoplastic foam **1**:378–379, 400
Thermoplastic natural rubbers
 properties of **3**:1374
Thermoplastic polyamide elastomers **3**:1386
 morphology **3**:1390
Thermoplastic polyesters **2**:791, 796–797, 799, 802, 804–806, 811
Thermoplastic polyetheresters **2**:806–807
Thermoplastic polyolefin blends
 mechanically prepared **3**:1367
 reactor TPO **3**:1367
Thermoplastic polyolefin elastomers (TPO) **3**:1366
 morphology **3**:1367
 soft and hard domain **3**:1368
Thermoplastic polyurethane elastomers (TPU) **3**:1376
 belt process **3**:1378
 reaction extruder process **3**:1378
 uses **3**:1380
Thermoplastics
 commodity 1211, 1198
 engineering 1198, 1199, 1206, 1207, 1209, 1213, 1216, 1221, 1222, 1224, 1226, 1227
 transitional **3**:1198
 mechanical properties **1**:471–472, 475, 478, 480, 494–496, 510
 shaving formation **1**:439–440
 pressureless **1**:367, 372–373
 under pressure **1**:367, 377, 379, 381, 383, 389, 393
Thermoplastic vulcanizates (TPV) **3**:1369
 processing **3**:1374
 types **3**:1373
Thermoran **3**:1369
Thermosets
 mechanical properties **1**:471–472, 475, 478, 480, 494–496, 510
 molds for **1**:412, 433, 435
 reinforced **1**:471, 477, 487–488, 493–494, 497–498, 510
Thermosetting plastics **2**:781
Thickening agents
Thick molding compounds (TMC)
 for polymer composite fabrication **3**:1325–1326, 1328, 1339
Thiodipropionic acid
Thioethers
 as antioxidant **3**:1015
Thixotropic alkyds **4**:1608, 1612
Timonox **1**:555
Tinopal **1**:577
Tinstab **1**:550
Titanium dioxide [13463-67-7]
 delustering polyamide fibers **4**:1444
Titanium dioxide pigment

 in plastics **1**:534, 545, 551, 555, 560, 565, 569–571
α-Tocopherol
 *see Vitamin E*
Tonen **3**:1307
Topanol **1**:535
Torelina **3**:1307
Torolithe **2**:786
Tower reactor
 for polystyrene production **3**:986
Transamidation **2**:707
 *see also Amide interchange*
Transmittance
 of plastics **1**:471–472, 474, 477–478, 481, 483, 489, 491, 493, 498–499, 503–504, 506–507, 510, 512–513, 515–516, 518
Trefsin **3**:1376
Tregalon **3**:1307
Trevira **2**:810
Triallyl isocyanurate (TAIC) [1025-15-6] **2**:647
1,1,2-Trichloroethane
 vinylidene chloride from **3**:1181–1183, 1187, 1190–1191, 1193, 1195–1196
Tricresyl phosphate (TCP)
 as flame retardant **1**:559, 568
Tricresyl phthalate
 as plasticizer **3**:1129
Triethylaluminum (TEA)
 polypropylene production **2**:946
Triethylene glycol bis[3-(3,5-di-*tert*-butyl-4-hydroxyphenyl)-propionate]
 as antioxidant **2**:917, 919
Triethylene glycol-bis-2-ethylhexanoate [94-28-0] **3**:1153
Triisobutylaluminum (TIBA)
 polypropylene production **2**:946
Triloy **2**:778
2,2,4-Trimethyl-1,2-dihydroquinoline (TMQ)
 antioxidant **1**:527–535, 546, 568, 570
Trioxane
 polymerization **2**:911, 913, 916–919, 921–922, 925
Trioxane block copolymers
 as nucleation agents **2**:920
Triphenyl phosphate
 as flame retardant **1**:559, 568
Trirex **2**:778
Tris(2-chloroethyl) phosphate **1**:564
Tris(2-chloroisopropyl) phosphate **1**:564
1,3,5-Tris(2,3-dibromopropoxy)-2,4,6-triazine **1**:536
Tris(2,4-di-*tert*-butyl)phosphite
 antioxidant **1**:527–535, 546, 568, 570

Tris(nonylphenyl)phosphite
  antioxidant **1:**527–535, 546, 568, 570
Trogamid **2:**728
Trosifol **3:**1150
Trusurf **2:**786
Tubular reactor
  polyethylene production **2:**844
  for polystyrene production **3:**986
Tung distribution **1:**26
Tyndall effect **1:**247
Tyril **3:**996

**U**
Ubbelohde viscometer **1:**249
Ube **2:**711–714, 728
Ubesta **3:**1394
Ucarsil PC **1:**569
Ucecryl **2:**690, 692
Ucefix **2:**692
Udel **3:**1296–1298
Ultem **3:**1283, 1285–1288, 1290
Ultracentrifugation **1:**231, 243, 245, 249, 254–255
Ultradur **2:**810
Ultrafine **1:**555
Ultraform **2:**912, 924
Ultra-high-molecular-mass polyethylene
  (UHMWPE) **2:**826
Ultramarine pigments
  in plastics **1:**534, 545, 551, 555, 560, 565,
    569–571
Ultramid **2:**728
Ultranox **1:**535
Ultranyl **2:**927
Ultraphor **1:**577
Ultrason **3:**1269
Ultrason E Polyethersulfone **3:**1296
Ultrasonic welding **1:**447, 451–452
Ultrason S Polysulfone **3:**1296
Ultraviolet absorber
  in plastics **1:**534, 545, 551, 555, 560, 565,
    569–571
Unibrite **3:**1021
Unicell **1:**577
Unifoam **1:**577
Unipol process **2:**847, **2:**952
Unirez **4:**1771
Unislip **1:**554
Unitika **2:**728
Uniwax **1:**554
Upper critical solution temperature (UCST) **2:**666
Uralam **2:**786
Uralkyds **4:**1608
Uramul **2:**690

Urethane alkyds **4:**1608
Uvitex **1:**577

**V**
Valox **2:**810
Vanox **1:**535
Vanstray **1:**550
Vapor pressure osmometry **1:**245
Varstat **1:**574
Vectra **2:**803, 808–812, 1311–1319, 1321
Vectra LCP **3:**1311, 1314, 1316, 1321
Veo Va 9 **3:**1165–1166, 1171–1172
  as monomer for poly(vinyl esters) **3:**1165, 1170,
    1172–1173
Veo Va 10 **3:**1165–1166, 1168, 1170–1174
  as monomer for poly(vinyl esters) **3:**1165, 1170,
    1172–1173
Versamag **1:**555
Versamid **4:**1771
Versicon **3:**1227
Vestamelt **2:**810
Vestamid **2:**728, **3:**1394
Vestodur **2:**810
Vestolen EM **3:**1369
Vestolit **1:**575
Vestopal **2:**786
Vestoprene **3:**1369
Vestoran **2:**927
Vestowax **1:**554
Viapal **2:**786
Vibrathane **2:**902
Victrex **3:**1291, 1294, 1296–1298
Victrex HTA **3:**1296–1297
Victrex PES **3:**1296–1298
Vinacryl **2:**690
Vinamul **2:**690, 1172
Vinavil **3:**1172
Vinlub **1:**554
Vinnapas **2:**690, 692, 1172
Vinnol **1:**575
Vinstab **1:**550, 577
Vinuran **3:**1022
Vinyfor **1:**577
  as ethylene copolymer **2:**828, 1165–1174
  as monomer for poly(vinyl esters) **3:**1165, 1170,
    1172–1173
  produced by acetoxylation **3:**1165
Vinyl acetate copolymers **3:**1165, 1168, 1171–1172
*N*-Vinylcaprolactam
  polymerization **3:**1159
Vinyl chloride
  polymerization, protective colloids **3:**1116
  suspension polymerization of **3:**1114

Vinyl chloride monomer (VCM) **3**:1112
  initiators for polymerization of **3**:1117
  toxicology **3**:1136
Vinylec **3**:1156
Vinyl ester resins **2**:781
  reinforcement with glass fibers **1**:417–418, 433, 436
Vinylether copolymers **2**:652
Vinyl ethyl ether **3**:1175, 1177
  continuous bulk polymerization of **3**:1175–1177
Vinyl 2-ethyl-hexanoate 1165,1166
  as monomer for poly(vinyl esters) **3**:1165, 1170, 1172–1173
Vinylidene chloride (1,1-dichloroethylene) **3**:1181, 1194, 1196, 1882, 1995
  preparation **3**:1191
  toxicology **3**:1181, 1194–1196
Vinylidenefluoride
  elastomers based on **2**:641
Vinyl isobutyl ether **3**:1175–1178
  continuous solution polymerization of **3**:1175
Vinyl laurate
  as monomer for poly(vinyl esters) **3**:1165, 1170–1173
Vinyl methyl ether **3**:1175
  batch bulk polymerization of **3**:1176
*seealso Methyl vinyl ether*
Vinyl octadecyl ether **3**:1177
Vinylon **4**:1533, 1536–1537
*N*-Vinylpiperidone
  polymerization **3**:1159
Vinyl pivalate 1165,1166
  as monomer for poly(vinyl esters) **3**:1165, 1170–1173
  as monomer for poly(vinyl esters) **3**:1165, 1170, 1172–1173
Vinyl propionate copolymers **3**:1172
2-Vinylpyridine **3**:1161
4-Vinylpyridine **3**:1161
Vipolit **3**:1172
Viscoelasticity **4**:1497–1498
  of polymers **1**:471–472, 475, 478, 480, 489, 500, 517–518
  of polymers **1**:471–472, 475, 478, 480, 489, 500, 517–518
Vistaflex **3**:1369
Vistalon **1**:575
Vistanex LM-MS **2**:760
Vitamin E **1**:530
Vitax **3**:1021
Vitel **2**:810

Viton **1**:554
Viton A [114–″116] **2**:641
Viton B **2**:641
VK tube **2**:713–714
Voltalef **2**:629
Voncoat **2**:691
VT alkyds **4**:1608
Vulkanox **1**:535
Vulkollan **2**:796
Vyflex **3**:1369

## W

Wachs BASF **1**:554
Walpol **2**:690
Waste disposal
  biodegradable polymers **3**:1231–1237, 1239, 1241, 1243–1244, 1247–1249, 1251–1259
Wastewater (sewage) 1165, 1172, 1173 poly(vinyl esters) in **3**:1165, 1170, 1172–1173
Welding
  biodegradable polymers **3**:1231–1237, 1239, 1241, 1243–1244, 1247–1249, 1251–1259
  friction welding **1**:439, 447, 451
  heated-tool welding **1**:439, 447, 449
  hot-gas welding **1**:439, 450
  induction welding **1**:439, 452
  of polymers **1**:439, 442, 447, 459, 462
  radiation welding **1**:439, 447, 450–451
Weston **1**:535
Wet spinning
  polyacrylonitrile fibers **4**:1513–1514, 1517–1518, 1521, 1523–1524, 1526
  polyurethane fibers **4**:1487–1488, 1492
  poly(vinyl alcohol) fiber **4**:1529, 1533–1534
  poly(vinyl chloride) fiber **4**:1529–1530, 1532, 1536
Wheatstone bridge **3**:1264
Whitefluor **1**:577
Whitex **1**:577
Winlite **3**:1150
Witten process **4**:1454
Wonderlite **2**:778
Wonderloy **2**:778
WPP TPO **3**:1369
Wytox **1**:535

## X

X-Fiper **4**:1544
X-ray diffraction
  studying fiber structure by **4**:1480
Xydar **2**:803, 808–812, 1310–1311, 1321

## Z

Zahn-Wellens test
  for polyaspartates **2**:743
ZB-223 **1**:555
Zenite **3**:1311, 1321
Zeonex **3**:1369
Zeonor TPO **3**:1369
Zerogen **1**:555
Ziegler catalysts
  polyethylene production **2**:836, 1182
  in poly(vinylidene chloride) production **3**:1181
Ziegler–Natta catalysts
  for poly(1-butene) production **2**:747–748
  in poly(vinyl ether) production **3**:1175–1179

  in thermoplastic polyolefin production **3**:1367
Zimm diagram **1**:248
Zinc
  zinc burning **1**:549
Zinc borates
  4 $ZnO \cdot 6\ B_2O_3 \cdot 7\ H_2O$ [*12536-65-1*] as flame retardant **1**:559, 568
Zinc pigments
  in plastics **1**:534, 545, 551, 555, 560, 565, 569–571
Zipper monofilaments **4**:1476
Zn-Stearate **1**:529
Zonyl **1**:554
Zytel **2**:728